Problems and Solutions in Rotational Mechanics

Online at: https://doi.org/10.1088/978-0-7503-6472-0

Problems and Solutions in Rotational Mechanics

Pradeep Kumar Sharma MSc, PhD, CEng(I) MIET, MInstP, SMIEEE
Consultant physicist and researcher, (India)

IOP Publishing, Bristol, UK

© IOP Publishing Ltd 2024. All rights, including for text and data mining (TDM), artificial intelligence (AI) training, and similar technologies, are reserved.

This book is available under the terms of the IOP-Standard Books License

No part of this publication may be reproduced, stored in a retrieval system, subjected to any form of TDM or used for the training of any AI systems or similar technologies, or transmitted in any form or by any means, electronic, mechanical, photocopying, recording or otherwise, without the prior permission of the publisher, or as expressly permitted by law or under terms agreed with the appropriate rights organization. Certain types of copying may be permitted in accordance with the terms of licences issued by the Copyright Licensing Agency, the Copyright Clearance Centre and other reproduction rights organizations.

Permission to make use of IOP Publishing content other than as set out above may be sought at permissions@ioppublishing.org.

Pradeep Kumar Sharma has asserted his right to be identified as the author of this work in accordance with sections 77 and 78 of the Copyright, Designs and Patents Act 1988.

ISBN   978-0-7503-6472-0 (ebook)
ISBN   978-0-7503-6468-3 (print)
ISBN   978-0-7503-6469-0 (myPrint)
ISBN   978-0-7503-6471-3 (mobi)

DOI   10.1088/978-0-7503-6472-0

Version: 20241101

IOP ebooks

British Library Cataloguing-in-Publication Data: A catalogue record for this book is available from the British Library.

Published by IOP Publishing, wholly owned by The Institute of Physics, London

IOP Publishing, No.2 The Distillery, Glassfields, Avon Street, Bristol, BS2 0GR, UK

US Office: IOP Publishing, Inc., 190 North Independence Mall West, Suite 601, Philadelphia, PA 19106, USA

This book is dedicated to
All the teachers, physicists and lovers of physics
And
My Revered Guru of Metaphysics (Vedic Philosophy)
Professor Shankha Purohit, MA(Odia), MA(English), MA(Sanskrit).

Contents

Foreword	xi
Preface	xii
Acknowledgements	xvi
Author biography	xx

1	**Kinematics of rotation**	**1-1**
1.1	Definition of a rigid body	1-1
1.2	Translation of a rigid body	1-2
1.3	Angular motion or rotation of a rigid body	1-3
1.4	Angular velocity of a rigid body	1-4
1.5	Calculation of the angular velocity	1-4
	1.5.1 Method 1: ground frame method	1-4
	1.5.2 Method 2: relative motion method	1-5
1.6	Angular acceleration and its calculation	1-6
	1.6.1 The ground frame method	1-6
	1.6.2 Relative motion method	1-7
	1.6.3 Gyroscopic motion	1-9
1.7	Fixed axis rotation	1-9
	1.7.1 Centroidal	1-10
	1.7.2 Non-centroidal	1-10
1.8	Combined motion	1-11
1.9	Finding the velocity and acceleration of a point of a rigid body under combined motion	1-12
1.10	The kinematics of rolling	1-16
1.11	General kinematical misconceptions	1-21
1.12	Kinematical equations for uniform angular acceleration	1-22
1.13	The instantaneous axis of rotation (IAR)	1-24
	1.13.1 Calculation of the position of the IAR	1-24
1.14	Rotation about a point and gyroscopic motion	1-31
	Problems	1-33
2	**Torque and angular momentum for a point mass**	**2-1**
2.1	Torque	2-1
	2.1.1 Definition	2-1

vii

		2.1.2 Factors governing the torque	2-2
		2.1.3 The mathematical expression of torque	2-2
		2.1.4 Net torque when many forces act at a point	2-4
		2.1.5 Torque about an axis	2-6
	2.2	Angular momentum	2-9
		2.2.1 Nomenclature	2-9
		2.2.2 The magnitude of $\vec{L}$	2-10
		2.2.3 The direction of angular momentum	2-11
		2.2.4 The concept of turning	2-12
		2.2.5 The relation between angular momentum L and angular velocity ω	2-13
		2.2.6 Angular momentum about an axis	2-15
	2.3	Relation between torque and angular momentum	2-17
	2.4	Conservation of the angular momentum of a particle	2-17
		Problems	2-18

3 The statics and dynamics of rotation — 3-1

3.1	Torque acting on a group/system of particles	3-1
3.2	Torque about the center of mass	3-3
3.3	The angular momentum of a system of particles	3-6
3.4	Angular momentum of a two-particle system about the center of mass	3-8
3.5	The relation between the relative and absolute values of angular momentum relative to two coinciding reference frames	3-10
3.6	The relation between torque and angular momentum	3-12
3.7	Newton's laws for a system of particles	3-14
3.8	Conservation of the angular momentum of a system of particles	3-16
3.9	The angular momentum of a rigid body about the center of mass	3-19
3.10	The moment of inertia and its calculation	3-20
	3.10.1 Discrete particle system	3-20
	3.10.2 The moment of inertia due to continuous mass distribution	3-23
	3.10.3 Parallel axis theorem	3-30
	3.10.4 Perpendicular axis theorem	3-33
3.11	Newton's laws of motion of rigid bodies	3-35
	3.11.1 Law of rotation	3-35
	3.11.2 Law of translation	3-36
3.12	Equilibrium of rigid bodies	3-39
	3.12.1 Translation of rigid bodies	3-39
	3.12.2 Rotational equilibrium	3-42

3.13	Fixed axis rotation	3-43
	3.13.1 Newton's law of rotation	3-43
	3.13.2 Newton's law of translation	3-45
	3.13.3 Kinematics	3-46
3.14	Dynamics of the combined motion of rigid bodies	3-47
	3.14.1 Planner motion: Newton's law of rotation (torque equation)	3-47
	3.14.2 Newton's law of translation (force equation)	3-47
3.15	Gyroscopic motion	3-48
	Problems	3-51

4 The energy of rigid bodies — 4-1

4.1	The gravitational potential energy of rigid bodies	4-1
4.2	Kinetic energy of rigid of a body	4-2
	4.2.1 Discrete mass distribution	4-2
	4.2.2 Continuous mass distribution	4-4
	4.2.3 Rotational and translational kinetic energy	4-6
4.3	The kinetic energy of rolling bodies on a fixed surface	4-9
4.4	Work done by a force on a rigid body	4-12
4.5	The work–energy theorem for rigid bodies	4-14
	4.5.1 Net work done on a rigid body	4-14
	4.5.2 Work–kinetic energy theorem	4-18
4.6	Conservation of energy of rigid bodies	4-20
4.7	Power delivered by a force acting on a rigid body	4-21
	Problems	4-23

5 Impulse and momentum of rigid bodies — 5-1

5.1	The linear momentum of a rigid body	5-1
5.2	Angular momentum of a rigid body	5-1
	5.2.1 Spin angular momentum	5-1
	5.2.2 Orbital angular momentum	5-2
	5.2.3 Total angular momentum	5-2
5.3	The impulse–momentum equation	5-3
	5.3.1 The linear impulse-momentum equation	5-3
	5.3.2 The angular impulse–momentum equation	5-4
5.4	Conservation of angular momentum	5-7

5.5 The collision of rigid bodies   5-10
    5.5.1 The collision of a particle with a rigid body   5-10
    5.5.2 Collison of a rigid body with another rigid body   5-15
    Problems   5-15

Foreword

It is my pleasure to write a foreword for this book authored by Dr Pradeep Kumar Sharma. The author of this book knew me while he was a lecturer of Physics at Brilliant Tutorials (BT), Chennai, and I was a professor of Physics at the Indian Institute of Technology (IIT), Madras. I also know the author for his series of five books—*GRB Understanding Physics*. The author has gained vast experience and expertise in training the best students for various competitive examinations such as IIT-JEE, Physics Olympiads, etc. In 2020, the author personally visited my house in Chennai and requested me to edit this book. Although I could not edit, I spent some time in reviewing the matter and gave some suggestions to improve the overall quality and content of this book.

This book is well-written and well-balanced with theories and problems. All concepts are covered in-depth. The explanations have been presented in detail in a simple and lucid style. The most appropriate examples are given by the author based on his vast experience of teaching physics in reputed institutes across the country such as BT, FIIT-JEE, etc. This book has an impressive layout and excellent quality of production. In each chapter, the author has included systematic theories with the best examples and scholarly set solved problems. This will assist both teachers and students in building and strengthening their concepts of physics.

This book will be immensely useful to all college students preparing for entrance examinations such as IIT-JEE and Physics Olympiads, etc. Furthermore, this book could be useful for the physics GRE and PhD qualifying examinations for top universities across the world. Teachers can also use this book to enhance their subject knowledge so as to impart a better physics education. Each topic is dealt with scrupulously so as to enhance the student's understanding of the subject to a great extent. I thank the author for this splendid work as the result of his hard work and determination. I wish him all the best for his forthcoming books.

Dr Jagabandhu Majhi
Professor of Physics (Retired)
Indian Institute of Technology, Madras

Preface

Overview of the series

There are a lot of books on the market; each has its own merits and limitations. It is not possible to get everything into a single book. However, a potential student expects everything in a book, such as a variety of adequate problems with lucid theories and solved examples. Most books are bulky with too many unnecessary theories and problems. Currently, the market is flooded with numerous textbooks with unscholarly repeated questions, many containing only exercises without any depth or sequence; additionally, thousands of common repeated problems are freely available on the Internet, and many institutions have their own study materials, mostly compiled by unscrupulous cut-and-paste methods over many years. Students are confused by the scattered variety of sources for problems, as well as their provenance and the validity of solutions. Thus, not only do the helpless students but also their teachers need some concise and precise books containing exercises where the primary concepts are covered systematically. This series of books is a sincere effort to minimize all the above limitations of existing books and resources and maximize the potential of our books in terms of content, quality, rigor, and depth of theories, questions, and problems.

Currently, undergraduates prepare for entrance examinations to get in to leading institutes such as Oxford, Cambridge, MIT, Princeton, etc. Millions of students all over the world participate in Physics Olympiads, both national and international, and other international physics examinations such as Physics GRE etc. In India, millions of students take national level examinations for the prestigious Indian Institutes of Technology (IITs) through a tough Joint Entrance Examination called IIT-JEE with a 0.5% success rate. There are no books available on the market that are balanced in terms of the problems and theories covered, and the quality and quantity of the problems, for such students.

In this series, I have made a sincere attempt to provide balanced books with theories, examples, and exercises, and with a systematic approach to concept building. By the virtue of my experience and expertise, suggestions from and association with my colleagues and hundreds of gifted students, this series of books is the outcome of my sincere effort to transform physics-phobia into a physics-loving attitude for students.

Readership

This series of books is a mixture of important, creative, and brain-storming problems arranged systematically (by sub-topic), so that even an average student can easily refer to the books. Detailed solutions are provided for each problem. Each book of this series has different types of problems with a gradual increase in the standard and difficulty level. Each problem is multi-conceptual, with a variety of small questions. These books could be ideal for an aspirant hoping to master the concepts of physics. In this sense, it would be highly suitable for students (undergraduates) preparing for national and international Physics Olympiads as it contains 50 years of national and international Olympiad problems in modified and general forms. For students preparing for PhD examinations for top universities and physics GRE examinations, this book could also be useful. In India, students preparing for UPSC examinations and semester examinations, and physics majors in BSc or BEng (of any branch of engineering), these books could be useful to an extent, as handbooks of physics problems.

The entire series will ultimately be useful for millions of Indian students preparing for JEE (mains and advanced). Any IIT-JEE student could be well-versed with the concepts of physics with a strong analytical skill in problem-solving, even after completing only 60%–70% of this book. Eventually, he or she could achieve JEE-Advanced (which students can take after passing JEE-Mains) with good ranks.

To educate their students, teachers need to be well-versed in physics and deliver in a stipulated time frame. That is possible only when we have suitable books with scholarly set questions arranged systematically. Thus, keeping in mind all the merits of my book series, I am hopeful that this series could become popular reference book for thousands of potential JEE educators in India. I received unprecedented support and recognition from educators in India for my earlier published book *GRB Understanding Physics*, because it contained original problems in systematic way. These books are out of print but still available from many e-commerce sites. However, in this series I have resolved some of the limitations of my previous publication, and it is balanced in many respects and hence unmatched by other available books.

The series consists of six individual books, providing almost an entire physics course. Each will contain nearly 400 problems with examples and exercises. This series presents an unprecedented effort to provide the maximum variety of problems arranged systematically according to the common syllabus of reputed undergraduate physics books such as *Halliday and Resnick's Principles of Physics*, *Physics for Scientists and Engineers* by R A Serway and R J Beichner, *University Physics* (known as 'Sears & Zemansky') by H Young and R Freedman, *Physics* by H C Ohanian, etc. The combination of my experience and expertise over three decade of teaching and writing, along with the contents of hundreds of books I have followed, will form the basis for this series of books.

The common features of the series

I have had a long standing desire to prepare a set of problem books/booklets by dividing physics into these smaller segments. This will provide almost all varieties of problems for students in a particular small section of physics, and the students need not purchase a complete physics book, only those for a specific area of physics.

I have handpicked many problems from various sources. Each problem has been observed being asked in different ways in various examinations all over the world. I analyzed each problem's most original version and then put it in my series. Furthermore, I prefer to put all possible questions linked with the original problem in one place, either as separate problems or as sub-questions within the original problem.

Each volume features a mixture of small, tricky, conceptual, introductory, complex, and lengthy brain teasing problems. Together these problems span 70%–80% of the first year of any undergraduate course of physics or engineering in US, UK, and Indian universities. All the questions would ultimately be useful for national and international Physics Olympiad examinations. This book covers 70% of the physics GRE and 50% of the PhD qualifying examinations for leading US universities.

Each book is made to be student-friendly, balanced with theory and problems. The quality and quantity of the content is balanced with systematic placement of the concepts, immediate examples under each concept (the heading/sub-heading of each chapter), scholarly set problems under each section, with a detailed solution immediately after each problem. The main request of most students is to access a large number of quality problems in a book arranged in a logical sequence. Thus, these books could act as an 'all-in-one' resource for aspirants of various entrance examinations fulfilling the above conditions.

All the problems can be solved using simple calculus, algebra, trigonometry, and geometry. Therefore, in the global market these books could be regarded as an unprecedented venture to present the entire physics curriculum with adequate examples and problems in a student-friendly way.

About this book

Rotational mechanics is a rank-deciding area in examinations for aspiring students. Due to its complexity average students have a fear of this area. Thus I have tried to present the subject matter in an interesting way without the use of higher mathematics such as Langrangian, Hamiltonian, etc. I present the examples and problems in a proper sequence following the theories. Rotational mechanics is the most challenging area in mechanics because it needs many concepts to solve a single problem. The correct solution of a single problem in this area will greatly improve the ranking of a student. Thus this portion is a rank-deciding one. Thus each aspiring student seeks an appropriate book of rotational mechanics that could fulfill all the criteria, but to my knowledge there is no such single book. Thus I hope this book could be of unique value for those students.

This book has five chapters:
1. Kinematics of rotation
2. Torque and angular momentum for a point mass
3. The statics and dynamics of rigid bodies
4. The energy of rigid bodies
5. Impulse and momentum of rigid bodies.

The first chapter discusses the angular quantities, such as angular displacement, angular velocity, and angular acceleration, the kinematics of rolling, and the instantaneous axis of rotation. The second chapter establishes two key concepts, namely torque and angular momentum, for the motion of a particle. The third chapter extends the application of torque and angular momentum for a system of particles and then for rigid bodies. The fourth chapter deals with the energy aspect of a rigid body, such as translational and rotational kinetic energy. The application of the work–energy theorem and energy conservation are described in this chapter. The last chapter deals with the impulse–momentum equations applied for a rigid body. The principle of conservation of angular momentum is explained with many interesting examples including the collision of rigid bodies.

How to use this book

It is my sincere suggestion to students to first complete the given theory with the examples to obtain proper understanding. Then, they should try to solve the problems in their own way. Sometimes students obtain the answers better following their own methods, so students should not be biased by my approach. If a student falters after repeated attempts, then they can refer to the given solutions. Due to the creativity and modification of many of the questions, I cannot proclaim that the book is error-free, in the first edition. However, I have tried to make it error-less. To make the book error-free, I request the readers to critically analyze my work and point out any mistake in the book which will be rectified as soon as possible. I will be grateful to all my readers for their comments and suggestions for the overall improvement of this book.

Acknowledgements

It goes without saying that a teacher is worth millions of books. An ideal teacher is built by their association with students. I acquired my experience and expertise through association with gifted students at premier institutes. Thus, first of all, I would like to express my gratitude to the directors of all the institutes at which I worked. Most illustrious are Brilliant Tutorials Private Limited (BT), Madras (Chennai), FIIT-JEE Ltd, New Delhi, and the Narayana Group of Educational Institutions, Hyderabad.

I am indebted to the late Professor P C Samal for introducing me to the world of education at his institute in 1988. Within a couple of years my students, such as Er Premananda Acharya (now a senior computer scientist and physics educator) in Bangalore, were instrumental in enhancing my problem-solving skills and setting up my own institute comprising outstanding students at that time. The clarity of the concepts and the style of approach towards presentations and problem-solving has been inherited from some of my venerable professors, such as Professor P K Dhir and Professor Sarat Mahapatra for electromagnetism, Professor Jatin Dash for general physics, and Professor Kulamani Samal for modern physics, who taught me to present modern physics with a historical evidence. All the attributes of my gifted teachers (many of whom have left their mortal bodies) and students (who are now global leaders in their own fields) are reflected in all the volumes of my problem books to be published by IOP Publishing, in addition to the current volume. Thus I express my sincere thanks to all my revered teachers (gurus) and past students.

I am grateful to Professor T Surya Kumar for sharing his problem-solving skills while I was actively involved in the Doubt Letter Cell (department engaged in clearing physics doubts in the form of letters sent by students across the country) in BT from 1004 to 1999. He discovered a talent for writing in me and inspired me to write in addition to teaching. He edited a collection of creative problems in mechanics, framed by me. I am also grateful to the principal Professor M S Subramanian in BT, Madras, who motivated me to handle the classes and doubt letters most effectively. This helped me to dispel the most crucial doubts in mechanics and was ultimately helpful in preparing the manuscript of this book after two and half decades. My personal contact with the late Professor G N Subramanian of BT, who was well-versed in both physics and mathematics, helped me greatly in understanding some intricacies of mechanics. He was my senior colleague at BT and Narayana Group, Hyderabad.

I extend my sincere thanks to Professor Bhanumati at BT for encouraging me to further my professional career at FIIT-JEE Limited, New Delhi. I thank my Head of Department Er Srikant Kumar and other seniors Er P K Mishra and Er Nagrath, etc, for imparting the best possible teacher training for my physics knowledge enhancement program at FIIT-JEE between 1999 and 2001. It has had a great positive impact on this current series of problem books.

At FIIE-JEE, I am grateful to my colleagues Er B M Sharma (author of the Cengage publication), Er D C Pandey (author of the Arihant publication),

Mr M Husain (now heading the physics department of a leading chain of schools), Er Sunil Deshpande (author and owner of his institute), Mr M Pallai, and Mr S Rout (now owner of Aryabhatta Institute, Berhampur, Odisha) for their kind cooperation in completing the highest selling study material for tough competitive examinations, popularly known as Rankers Study Material (RSM), in 2001. After three years, I started writing five volumes of a book with the title *GRB Understanding Physics*. I am grateful to the late Mr Prakash Chand Bathla for offering me the opportunity to write these books under his nationally leading publication house (G R Bathla and Sons). Based upon my experience as a national level author and educator, I was able to attain my present status of working under an international publishing house (IOP Publishing). I am grateful to the senior scientist and educator Mr S M Pathak for his promise to edit the books to make them error-free in their future editions.

I would like to thank all my previous top students who edited my books while I headed the Department of Physics at the Narayana Group of educational institutions. Furthermore, I express my deepest gratitude to the principals and deans of Nayayana Group, Hyderabad, in particular Mr Krishna Reddy and Mr Ramalinga Reddy, with whom I worked a major portion of my professional career, for giving me operational freedom, status, stability, and respect. There, I was able to complete some theories and examples of the current series in rudimentary form. In Hyderabad, leading educators such as Mr Aditya Sachan, Er L N Prusty, Er Sekhar Somnath, and Mr Monoj Pandey were constant motivators of my writing skills. Another leading physics educator Mr S K Singh of Chaitanya Educational Group was a good critic of my work and spent some time editing my manuscripts. I am also grateful to Professor Kundal Rao of Narayana IIT and Professor Srinivasa Chary of Sri-Mega for their suggestions and inspiration for my published works.

Thus, I have been directly and indirectly inspired by many people. Primarily, I would like to express my profound gratitude to my wife Usha for supporting me and bearing with me during the pandemic in 2020 when I started conceptualizing this book. As this project is comprised of six books covering the entire field of physics, it has been an arduous task, thus it required full-time effort to complete it in time. It was my long standing dream to deliver these books for a global audience as an output of my vast experience and expertise gained in the afore-mentioned ways. I had to honestly decline lucrative offers from reputed institutions in order to complete this project as soon as possible in best possible way. Although it depleted my financial position and my wife was not doing well in the pandemic, I preferred to occupy the side of propagating knowledge rather than money-making, because *knowledge is far more valuable than money*. The authors of great books such as *Problems in Physics* by I E Irodov and *Concepts of Physics* by H C Verma are brilliant examples of advocates of *knowledge over the vanities of the world*. Therefore, I offer my humble obeisance to these professors who inspired me to write books to impart knowledge to millions of students.

I remain obliged to the commissioning editor of IOP Publishing Mr John Navas for his insightful comments, suggestions, and expertise, which have enhanced the rigor and depth of this work. He took special interest in getting my book proposal reviewed by the appropriate experts who could sense something interestingly and useful for students and provided their positive feedback. Unfortunately, after signing the contract, he left IOP Publishing. Then Ms Phoebe Hooper was put in charge of my book project and Mr David McDade replaced Mr John Navas. These two new people took some time to streamline the work process due to the abrupt change in management. I thank both of them for their best efforts in handling the publication work of my book.

I express my gratitude to my ex-publisher Mr Monoj Bathla who recommended me to accept the offer from IOP Publishing after realizing the suitability of the publisher for my work. As I taught thousands of students in last two decades, it is not possible to mention the name of each student who reviewed my original manuscript again and again, and I am grateful to each one. Furthermore, I am grateful to the leading IIT-JEE faculty members of India and peers who have provided valuable feedback, stimulating discussions, and intellectual support during the preparation of this book. Some of them are worth noting: Er Anurag Mishra (the author of *Physics*) and Er Ashish Arora (the author of *Physics Galaxy*), both from Allen Career Institute, Kota, who are admirers of my previous books. The personal motivation and expectations of these top-notch physics educators were instrumental in making this problem book series remarkable.

Additionally, I would like to acknowledge the professional and financial support provided by IOP Publishing, which not only alleviated my financial hardship to an extent but also enabled me to deliver this project in time.

Four typists prepared my manuscripts at different places and times. I sincerely thank Mr Bismay Parida (Readers Institute, Balasore, Odisha State) for his continuous effort in typing the manuscripts in time. Furthermore, Mr Sai, Mr Venkatesh and Mr Bharat (from Hyderabad, Telangana State) typed a considerable portion of the manuscript of the problems sections of the entire series.

Finally, I am deeply grateful to my family and friends for their unwavering love, encouragement, and understanding throughout this challenging yet fulfilling publishing journey.

Recapitulating, the completion of this series of problem books will be the culmination of an arduous yet rewarding journey, and I owe my deepest gratitude to the afore-mentioned individuals and organizations who have supported me all along my career. Finally, I again express my profound appreciation to all my directors, principals, deans, seniors, colleagues, students, and teachers all over the country for their unwavering guidance, encouragement, and scholarly mentorship throughout the entire preparation process of the series. Their expertise, patience, and constructive feedback have been invaluable in shaping the direction and quality of this work in my physics education research in the mode of active teaching and active learning (ATAL), which was coined by Mr S K Singh to whom I am deeply indebted.

As this project of six books has been a huge task, I had to take leave as an extramural researcher from Sofia University, Bulgaria. I am grateful to the university for granting leave of one year. Furthermore, I express my sincere gratitude for the thoughtful suggestion by my professor Dr Alexander Gungov to support my book publication with ongoing research at the university.

I have many friends and professors in the UK who encouraged me to write these books. As an alumnus of the Universities of Oxford and Strathclyde I have deep respect for all my professors, in particular Professor Peter Dobson (Oxford University) who taught me how to do things to perfection and I am applying this idea in the present publication. I am deeply indebted to Dr Benjamin Hourahine and Professor Yu Chen of the University of Strathclyde for imparting a standard knowledge of nano-science so that I could include this fast growing field in my problem book series.

One of my notable friends is Mr Rajinder Sehra, director of S&RJ Ltd and Foot Print Media Production near Glasgow. His constant encouragement in writing this series is also praiseworthy. I will remain grateful to him for his moral and physical support during my stay in Glasgow and constant encouragement for my book writing. The admiration from my professors at Strathclyde University and Strath— student union representative James Higgins, two times Strathclyde University Student Union president Mr Matt Crilly, Gary Paterson (Ex-Vice president of Starth Union) and Gerry McDonnell (Ex-President of Mature Student Association and Ex-Vice president, Strath Union, Strathclyde University toward my five volumes of physics books kept in the library of the Mature Student Association of the University of Strathclyde has also been a great source of my motivation for this current series. I express my deepest gratitude to them.

Last but not least, I am wholeheartedly grateful to Mr Himanshu Sekhar Dey, one of the directors of Readers Institute, Balasore, for taking the greatest portion of the burden of the institute and freeing me to devote myself entirely to this project.

This book would not have been possible without the collective contributions and support of all those mentioned above. Their guidance and encouragement have been instrumental in the completion of this significant milestone in my publishing career. Taking this as a blessing of the Almighty, I pray for the attainment of knowledge that would be an ultimate solution to all problems of human beings and other living entities.

Pradeep Kumar Sharma
19 June 2024

Author biography

Pradeep Kumar Sharma

He is a well-known physics educator in India possessing more than three decades of experience in physics education and research in training the aspirants of the joint entrance examination conducted by prestigious Indian Institutes of Technology, popularly known as IIT-JEE. He has trained thousands of students for undergraduate entrance examinations such as IIT-JEE etc. Many of his students also qualified in national and international physics Olympiads. His vast experience as a potential teacher, team leader and head of the department in some premier institutes like Brilliant tutorials, (Chennai), FIIT-JEE Ltd (New Delhi), Narayana Group (Andhra and Telangana) etc, made him extend his service as a consultant physicist to mentor both students and teachers of reputed groups in India.

He has authored bestselling study materials, five books known as GRB *Understanding Physics* for the entrance examinations. He has been associating as a research scholar of physics education, nanoscience, metaphysics and management in some Indian and foreign universities such as Oxford University, Strathclyde University, Sofia University and Indian Institute of Technology, Patna. Furthermore, he is continuing his research while affiliated with various national and international organizations such as IEEE (USA), IET (UK), IE(I), IOP(UK) etc. He has published dozens of papers in national and international journals like IEEE-Scopus journals and journals published by Institute of Physics (UK). He is currently busy in completing the problems and solutions of a series of six books which will be ready to publish very shortly.

At present, he is actively involved with a team of top-notch educators, to design a new method of interactive education called Active Teaching and *Active Learning (ATAL)* that will make things easy for an average student to learn physics.

IOP Publishing

Problems and Solutions in Rotational Mechanics

Pradeep Kumar Sharma

Chapter 1

Kinematics of rotation

1.1 Definition of a rigid body

A body is said to be ideally rigid when its deformation (elongation, compression, and twisting) under the application of forces and torques is zero. However, nothing is perfectly rigid. If the deformation of the body is negligible compared to its size, you can call it a practically rigid body. So the shape and size of a rigid body remain practically constant. For this to happen, the distance r_{AB} between any two points of a rigid body remains practically constant. This means that,

$$\frac{\mathrm{d} r_{AB}}{\mathrm{d} t} = 0.$$

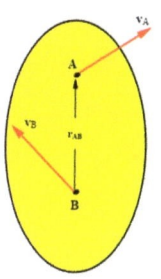

Since the distance between A and B = r_{AB} = constant, dr/dt = 0

- The component of relative velocity v_{AB} between these two points along the line AB will be zero.
- Then, A seems to move perpendicular to AB relative to B.
- Similarly, B seems to move perpendicular to AB relative to A.

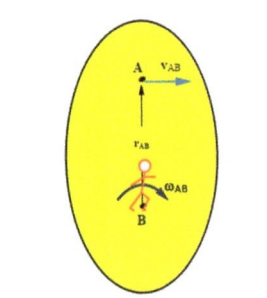

The observer at B sees A moving perpendicular to AB

Example 1 A rod AB is kept against a vertical wall making an angle of 30° with vertical as shown in the figure. If the end A of the rod moves with a velocity of $v_A = 5$ m s^{-1} downwards, find the linear velocity of its end B.

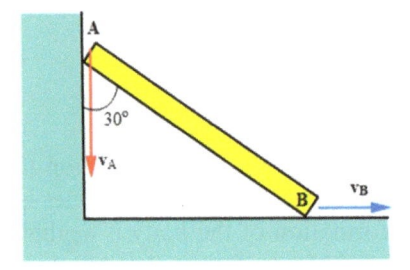

Solution

Since the rod is rigid, the component of v_{AB} along the rod is zero. This means we can equate the velocities of the end points A and B along the length of the rod.

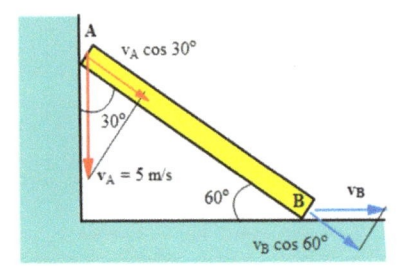

So, $5 \cos 30° = v_B \cos 60°$. This gives $v_B = 5\sqrt{3}$ ms^{-1}.

1.2 Translation of a rigid body

Let us consider a moving rectangular plate and draw a straight line AB on it. Take a fixed straight line CD outside the plate. Let AB make an angle θ with the reference line CD. If the straight line AB does not change its angle of orientation θ relative to the fixed reference straight line CD when the plate moves, the plate is said to be translating without any rotation. This is known as *pure translation* or simply *translation*.

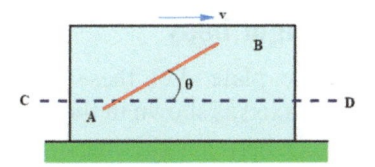

The straight line AB does not change the angle of orientation θ in translation

In pure translation, all points of the body:
- undergo equal displacements.
- have the same velocity.
- have the same acceleration.

This means that the relative motion (that is, relative displacement, relative velocity, and relative acceleration) between any two points of the rigid body is zero in pure translation.

There are two types of translation, namely rectilinear translation and curvilinear translation:
- In rectilinear translation any point of the rigid body moves in a straight line.
- In curvilinear translation any point of the rigid body moves in a curve.

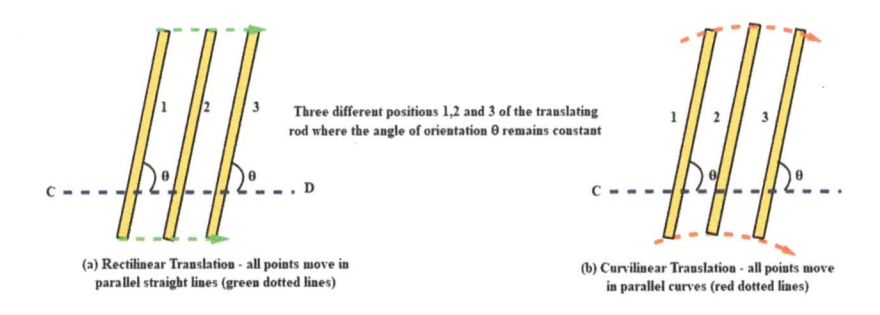

Three different positions 1,2 and 3 of the translating rod where the angle of orientation θ remains constant

(a) Rectilinear Translation - all points move in parallel straight lines (green dotted lines)

(b) Curvilinear Translation - all points move in parallel curves (red dotted lines)

1.3 Angular motion or rotation of a rigid body

For the sake of simplicity, let us consider a thin plate (a planar rigid body). Draw a line AB on the plate. We can see in the following figure that the line changes its angle of its orientation from θ_1 to θ_2 relative to a fixed reference line CD. Then the body is said to be rotating (or undergoing an angular motion).

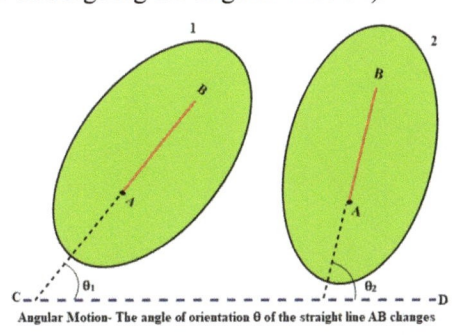

Angular Motion- The angle of orientation θ of the straight line AB changes

1.4 Angular velocity of a rigid body

Draw two lines 1 and 2 on the plate. Let these lines make angles θ_1 and θ_2, respectively, with a horizontal (x-axis) as shown in the figure. Since the lines are fixed with the rigid body, the angle β between the lines does not change. So $\frac{d\beta}{dt} = 0$. Putting $\beta = (\theta_1 - \theta_2)$ and differentiating both sides we have $\frac{d\theta_1}{dt} - \frac{d\theta_2}{dt} = 0$. This gives $\frac{d\theta_1}{dt} = \frac{d\theta_2}{dt}$. This states that, at a given instant, all lines drawn inside the plate change their angle of orientation with respect to a given reference line at the same rate. This is known as the angular velocity of the rigid body.

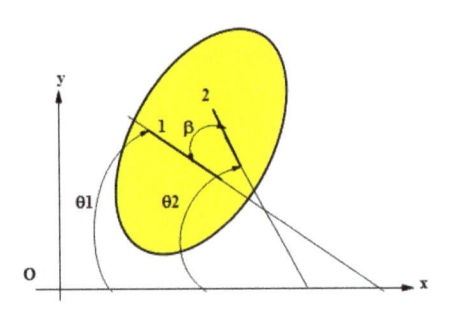

1.5 Calculation of the angular velocity

1.5.1 Method 1: ground frame method

Draw a line on the rigid body. Let its angle of inclination be θ with the x- or y-axis. Express either the x- or y-coordinate of any point of the rigid body in terms of θ. Then differentiating both sides find $\omega = \frac{d\theta}{dt}$ when $\frac{dx}{dt}$ or $\frac{dy}{dt}$ is given. Hence the angular velocity of a rigid body is equal to the rate of change of the angle made by a straight line drawn inside the rigid body with any fixed reference line. The formula of angular velocity is $\omega = \frac{d\theta}{dt}$. If θ increases with time, the sense of ω is in the increasing order of θ and vice versa.

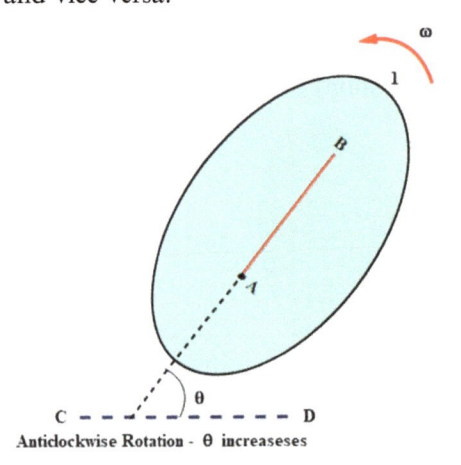

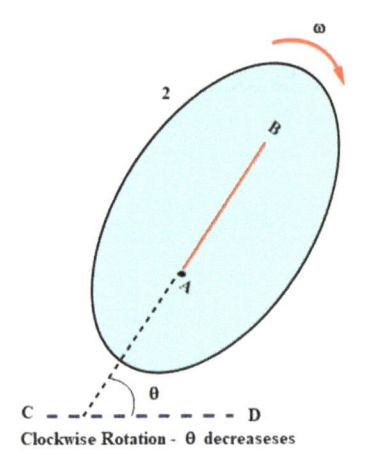

Anticlockwise Rotation - θ increaseses

Clockwise Rotation - θ decreaseses

1.5.2 Method 2: relative motion method

In this method first of all we consider two points (A and B, say) of the rigid body. Since the distance between these points is r, the component of relative velocity between A and B along the line AB is equal to zero, as discussed in section 1.1. Then A and B seem to move relative to each other perpendicular to the line AB in opposite directions. This means

$$\vec{v}_{AB} = -\vec{v}_{BA}.$$

However, A and B seem to turn relative to each other in the same sense (in a clockwise manner, as shown in the figure) with the same magnitude of angular velocity. In other words, their relative angular velocities are exactly equal to each other (in both magnitude and direction). So we can write

$$\vec{\omega}_{AB} = \vec{\omega}_{BA} = \vec{\omega}_{rel} = \vec{\omega},$$

where $\omega = \frac{v_{rel}}{r}$ is known as the angular velocity of the rigid body and $v_{rel} = v_{AB} = v_{BA}$.

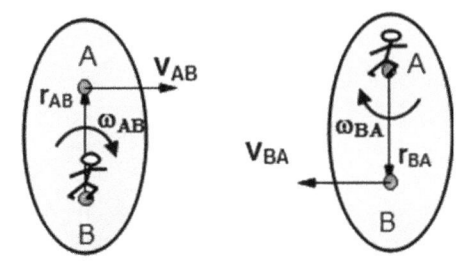

Hence in this method:
- The angular velocity of a rigid body is equal to the relative angular velocity between any two points of the rigid body.
- The relative velocities between any two points of a rotating rigid body are negative vectors (equal in magnitude and oppositely directed).
- The relative angular velocities between any two points of a rotating rigid body are equal vectors (equal in both magnitude and direction).
- At a given instant of time, the relative angular velocities between all possible pairs of points of a rotating rigid body are equal, which is defined as the angular velocity of the rigid body, denoted as ω. Hence, the angular velocity of a rigid body does not depend upon any reference point or reference frame we choose.
- The vector relation between $\vec{v}_{AB}$, $\vec{r}_{AB}$, and $\vec{\omega}_{AB}$ can be given as $\vec{\omega}_{AB} \times \vec{r}_{AB} = \vec{v}_{AB}$.

Example 2 A rod AB of length l is leaning against the wall making an angle θ with the wall. If the top end A of the rod moves with a downward velocity u, find the angular velocity of the rod.

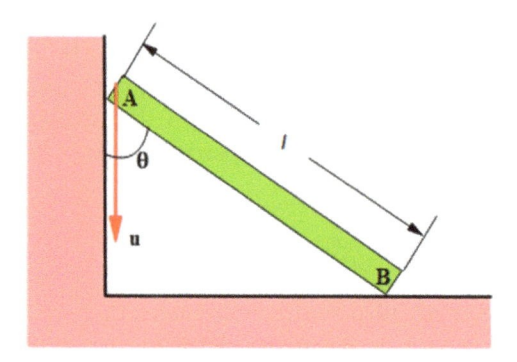

Problems and Solutions in Rotational Mechanics

Solution
Method 1 (ground frame method)
The position of the point A of the rod is given as $y = l \cos \theta$. Differentiating both sides with time, we have $\frac{dy}{dt} = -l \sin \theta \frac{d\theta}{dt}$. Put $\frac{dy}{dt} = -u$ (as y decreases) and $\frac{d\theta}{dt} = \omega$ (because θ increases) as the bottom moves towards the right. Then we have $\omega = \frac{u}{l \sin \theta}$. The positive value of ω means physically an anti-clockwise sense of rotation of the rod.

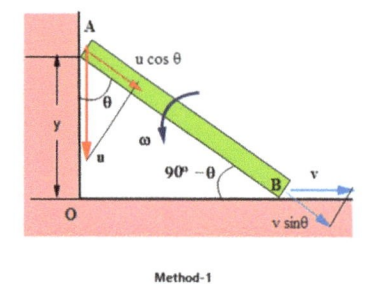

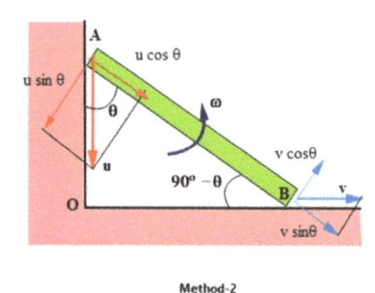

Method-1   Method-2

Method 2 (relative motion method)
Let the velocity of the bottom end of the rod be v. By equating the components of velocities of the ends of the rod (along the rod), we have $v \sin \theta = u \cos \theta$. So $v = u \cot \theta$. The relative velocity between the ends is $v_{rel} = u \sin \theta + v \cos \theta$, where $v = u \cot \theta$. Then, simplifying the terms, we can find the same result $\omega = v_{rel}/l = \frac{u}{l \sin \theta}$.

1.6 Angular acceleration and its calculation

1.6.1 The ground frame method

We define the linear acceleration as the rate of change in linear velocity given by the formula $v = dx/dt$. Similarly, we define the angular acceleration of a rigid body as the rate of change of angular velocity of the rigid body given as

$$\alpha = \frac{d\omega}{dt} = \omega \frac{d\omega}{d\theta}.$$

Since $\omega = \frac{d\theta}{dt}$ and $\alpha = \frac{d\omega}{dt}$ we can write

$$\alpha = \frac{d\omega}{dt} = \frac{d^2\theta}{dt^2}.$$

This expression of angular acceleration is given based on the ground frame method. However, in the relative motion method we can also determine the angular acceleration, as follows.

1.6.2 Relative motion method

If you stand at the point B and look at any point A on the rigid body (thin plate) you can see that A is moving in a circular path with a velocity v_{AB}. As the distance between A and B does not change, the radius of the circular path is $r_{AB} = r$, say. Let the tangential and radial or normal acceleration of A observed by (or relative to B) be $\vec{a}_t$ and $\vec{a}_n$, respectively. By applying the kinematics of circular motion of A relative to B, we will obtain the following equations:

$$a_n = r\omega^2, \quad a_t = \frac{dv_{AB}}{dt}.$$

We define angular acceleration as the rate of change in angular velocity. In the relative motion method, the angular velocity of the rigid body is given:

$$\omega_{BA} = \omega_{rel} = \omega = v_{BA}/r.$$

So the angular acceleration of the rigid body is given as

$$\alpha = \frac{d\omega}{dt} = \frac{d\left(\frac{v_{BA}}{r}\right)}{dt} = \frac{d(v_{BA})}{rdt}.$$

Putting $\frac{d(v_{BA})}{rdt} = a_t$, we have

$$\alpha = \frac{a_t}{r}.$$

We can use the following vector equations of the tangential and normal accelerations of a point relative to any other point of the rigid body borrowed from the kinematics of circular motion:

$$\vec{a}_t = \vec{\alpha} \times \vec{r} \text{ and } \vec{a}_n = \vec{\omega} \times \vec{v}_{BA}.$$

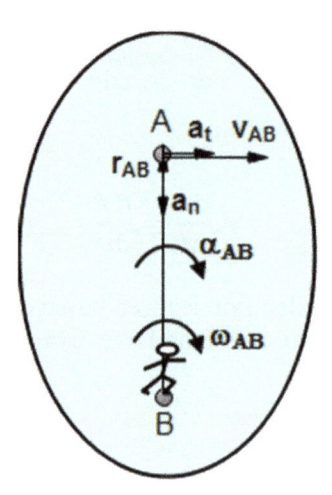

So the net acceleration of A relative to B can be given by adding the above two components:

$$\vec{a}_{AB} = \vec{a}_t + \vec{a}_n.$$

Example 3 In example 2, if the velocity u of the end A of the rod remains constant, find (a) the angular acceleration of the rod and (b) the linear acceleration of the end B of the rod.

Solution

(a) The angular acceleration is given as

$$\alpha = \frac{d\omega}{dt}, \quad \text{where } \omega = \frac{u}{l \sin \theta}.$$

Then we have

$$\alpha = u \frac{d(\text{cosec } \theta)}{l \, dt} = -(u/l) \, \text{cosec } \theta \cot \theta \left(\frac{d\theta}{dt} \right).$$

Putting the value of $\frac{d\theta}{dt} = \omega = \frac{u}{l \sin \theta}$ and simplifying the factors, we have

$$\alpha = -(u/l)^2 \, \text{cosec}^2 \theta \cot \theta.$$

Since the answer is a positive quantity, the direction (sense) of α is the same as that of the angle θ which increases in an anti-clockwise sense.

(b) From example 2 the velocity of B is given as

$$v = u \cot \theta.$$

So the acceleration of B is

$$a' = \frac{dv}{dt} = \frac{d(u \cot \theta)}{dt}.$$

So $a' = -u \operatorname{cosec}^2\theta (d\theta/dt)$.
Putting the value of $\frac{d\theta}{dt} = \omega = \frac{u}{l\cos\theta}$ and simplifying the factors, we have

$$a' = -\frac{u^2}{l} \sec\theta \operatorname{cosec}^2\theta.$$

Negative value signifies the point B is retarding.

1.6.3 Gyroscopic motion

In planar motion, the directions of the angular velocity and angular acceleration are axial, which is perpendicular to the plane of the rigid body. However, in three-dimensional rotational motion (gyroscopic motion), the direction of angular velocity also changes. Thus another component of angular acceleration comes into play due to the change in the direction of the angular velocity, which can be given as

$$\alpha = \omega\omega',$$

where ω is the angular velocity of the rigid body (cylinder) and ω' is the angular velocity of the axis of rotation of the rigid body (cylinder). You can refer to Problem 28 for its derivation.

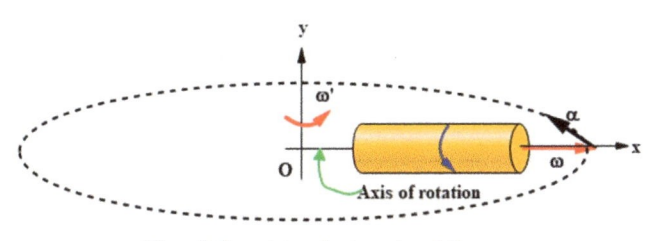

The cylinder rotates about x-axis and the x-axis rotates about y-axis

1.7 Fixed axis rotation

This is the rotation of a rigid body around a fixed axis. The meaning of fixed axis is that the axis does not move. A rotating ceiling fan, the rotation of the hands of a wall clock, and the rotating blades of a mixer are few examples of the fixed axis rotation.

In fixed axis rotation each point moves in a circular path about the axis of rotation. If we take a point P from the axis O, so that PO $= r$, the velocity of P is given as $v = r\omega$. The tangential and radial acceleration of P can be given as $a_t = dv/dt$ and $a_r = v^2/r = r\omega^2$, respectively, by applying the kinematics of circular motion.

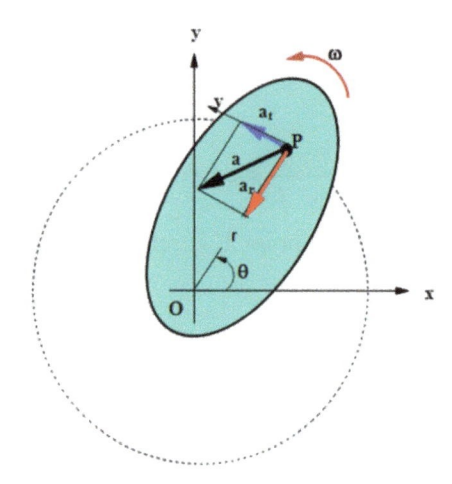

Fixed axis rotation is a special case of rotation where we can directly apply the kinematics of circular motion for each point of the body. There are two types of fixed axis rotation, as follows.

1.7.1 Centroidal

Let the uniform disc be pivoted with a fixed point at its center. As its axis of rotation is fixed and it passes through the center of mass of the rigid body, this fixed axis rotation is centroidal. Hence, in the centroidal fixed axis rotation the center of mass does not move. This type of rotation is called *pure rotation*.

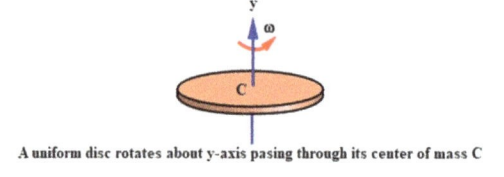

A uniform disc rotates about y-axis pasing through its center of mass C

1.7.2 Non-centroidal

If the uniform disc is pivoted at its end P, this means the axis of rotation does not pass through the center of mass C of the rod. Thus this fixed axis rotation is non-centroidal.

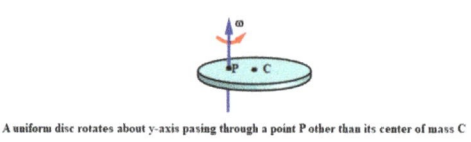

A uniform disc rotates about y-axis pasing through a point P other than its center of mass C

The velocity of any point of the rigid body in fixed axis rotation varies linearly with radial distance which is given as $v = r\omega$.

Example 4 In the given figure a disc rotates with a constant angular velocity about a vertical axis passing through the point P, the maximum speeds and accelerations of the points C and Q of the disc are u, v and a, A. respectively. Find the value of (a) v/u and (b) a/A. Assume that in each case the angular acceleration is zero and the angular velocities are equal.

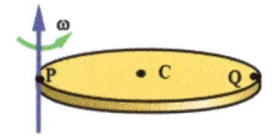

Solution

(a) Since the angular acceleration is zero, all points of the discs execute uniform circular motion. So the speeds of C and Q are, respectively, given as $u = R\omega$ and $v = 2R\omega$. Then $v/u = 2$.

(b) For uniform circular motion, the total acceleration is centripetal. Then, the ratio of centripetal accelerations of C and Q are given as

$$a/A = (R\omega^2)/(2R\omega^2) = 1/2.$$

1.8 Combined motion

It is a combination of pure translation and pure rotation. Both phenomena occur simultaneously. In pure rotation, the center of mass remains at rest and the body spins about the center of mass axis. In pure translation, the body does not rotate and hence all points move with the same velocity **v**. If we combine both phenomena (pure translation and pure rotation), we can see that the center of mass has a certain velocity **v** and the body spins with an angular velocity ω, as shown in the figure.

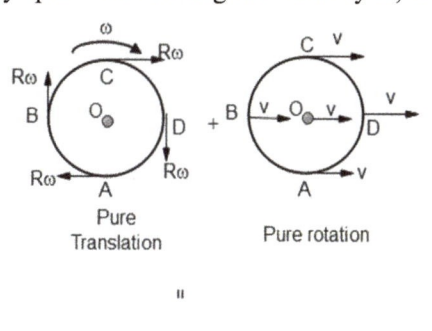

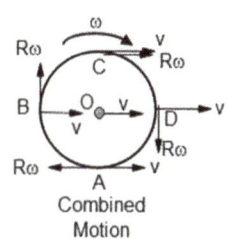

Pure Translation   Pure rotation

Combined Motion

Examples of combined motion are the rotating blades of a flying helicopter, a drone, rolling bodies, and the orbiting of the spinning earth, etc.

1.9 Finding the velocity and acceleration of a point of a rigid body under combined motion

Let a lamina (plate) undergo a combined motion with the angular velocity and angular acceleration ω and α, respectively. Let us chose two points A and B on the plate. If the velocity and acceleration of A are given as v_A and a_A, respectively, how do we find the velocity and acceleration of the other point B? Let us see.

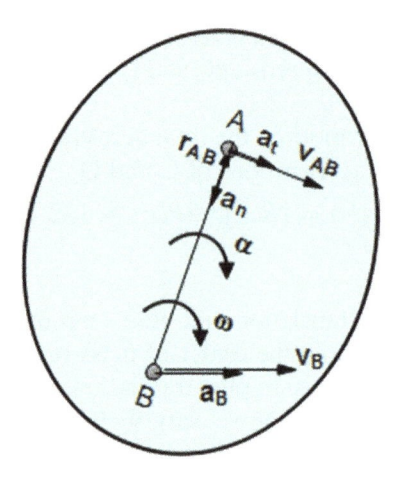

Velocity of A: The velocity of A is given as

$$\vec{v}_A = \vec{v}_{AB} + \vec{v}_B,$$

where $\vec{v}_{AB} = \vec{\omega} \times \vec{r}_{AB}$.

The magnitude of $\vec{v}_{AB}$ is given as

$$v_{AB} = r_{AB}\omega.$$

Acceleration of A: The acceleration of A relative to ground is given as

$$\vec{a}_A = \vec{a}_{AB} + \vec{a}_B,$$

where $\vec{a}_{AB}$ is equal to the vector sum of its tangential and radial components, given as

$$\vec{a}_{AB} = \vec{a}_t + \vec{a}_r.$$

Since the point A moves in a circle relative to B, the radial and tangential accelerations of B relative to A are, respectively, given as follows:

$$\vec{a}_t = \vec{\alpha} \times \vec{r}_{AB},$$

$$\vec{a}_r = \vec{\omega} \times \vec{v}_{AB}.$$

Their magnitudes are given as

$$a_t = r_{AB}\alpha \text{ and } a_r = r\omega^2.$$

Example 5 A uniform disc of radius R lying in the x–y plane spins with a constant angular velocity ω in anticlockwise sense. The center C of the disc moves with a velocity v. Find the velocities of the points A, B, D, and E of the disc.

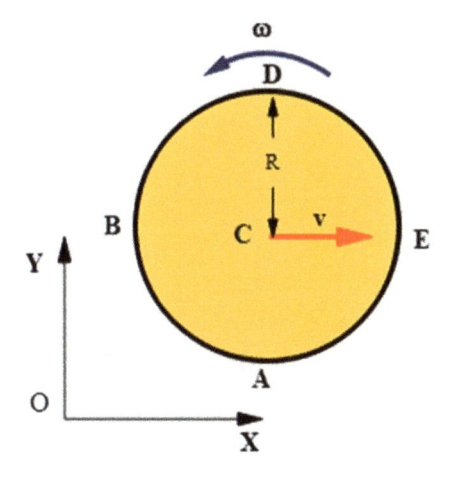

Solution
The velocity of A is given as

$$\vec{v}_A = \vec{v}_{AC} + \vec{v}_C$$
$$= +R\omega\hat{i} + v\hat{i} = (v + R\omega)\hat{i}.$$

The velocity of B is given as

$$\vec{v}_B = \vec{v}_{BC} + \vec{v}_C.$$
$$= -R\omega\hat{j} + v\hat{i}$$
$$= v\hat{i} - R\omega\hat{j}.$$

The velocity of D is given as

$$\vec{v}_D = \vec{v}_{DC} + \vec{v}_C.$$
$$= -R\omega\hat{i} + v\hat{i} = (v - R\omega)\hat{i}.$$

The velocity of E is given as

$$\vec{v}_E = \vec{v}_{EC} + \vec{v}_C.$$
$$= R\omega\hat{j} + v\hat{i}$$
$$= v\hat{i} + R\omega\hat{j}.$$

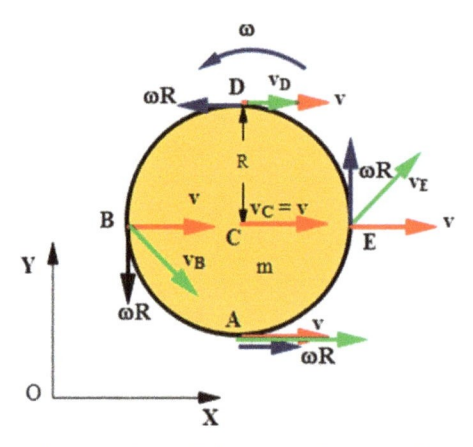

Example 6 A uniform disc of radius R lying in the x–y plane spins with a constant angular velocity $\vec{\omega} = \omega\hat{k}$. The center of the disc moves with an acceleration $\vec{a}_C = a\hat{i}$. Find the accelerations of the points A, B, D, and E of the disc.

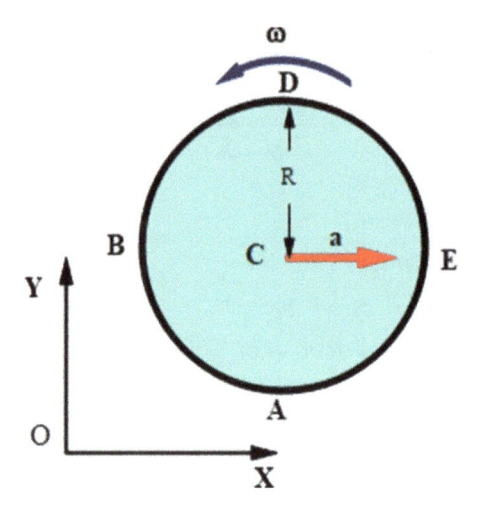

Solution
The acceleration of A is given as
$$\vec{a}_A = \vec{a}_{AC} + \vec{a}_C$$
$$= R\omega^2 \hat{j} + a\hat{i}$$
$$= a\hat{i} + R\omega^2 \hat{j}.$$

The acceleration of B is given as
$$\vec{a}_B = \vec{a}_{BC} + \vec{a}_C$$
$$= R\omega^2 \hat{i} + a\hat{i}$$
$$= (a + R\omega^2)\hat{i}.$$

1-14

The acceleration of D is given as

$$\vec{a}_D = \vec{a}_{DC} + \vec{a}_C$$
$$= R\omega^2 \hat{j} + a\hat{i}$$
$$= a\hat{i} - R\omega^2 \hat{j}.$$

The acceleration of E is given as

$$\vec{a}_E = \vec{a}_{EC} + \vec{a}_C$$
$$= R\omega^2 \hat{i} + a\hat{i}$$
$$= (a - R\omega^2)\hat{i}.$$

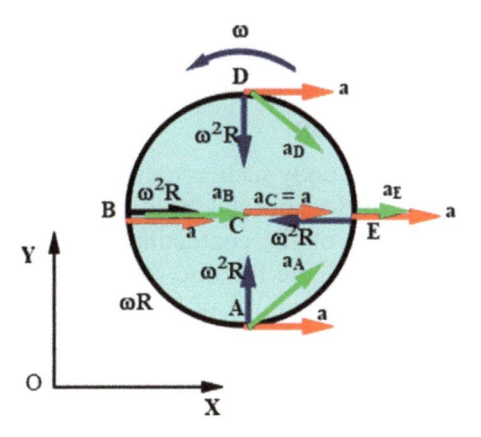

Example 7 In the previous example if we give an angular acceleration $\vec{\alpha} = \alpha \hat{k}$ to the disc: (a) draw each component of acceleration at A, B, D and E, and (b) find the acceleration at the given points.

Solution

(a) The additional acceleration a_t comes from the angular acceleration $\vec{\alpha}$ whose magnitude is given as $a_t = r\alpha$ and the direction is tangential to the disc, as shown in the figure.

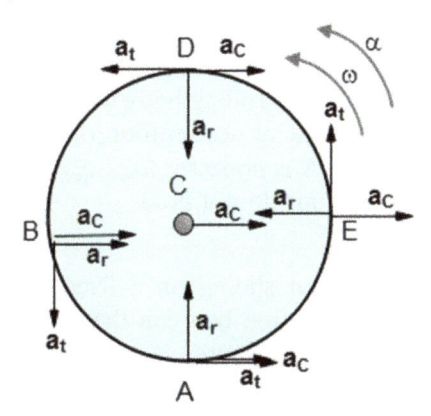

(b) The accelerations at the given points are given as follows:
$$\vec{a}_A = (a_t + a_C)\hat{i} + a_r\hat{j},$$
$$\vec{a}_B = (a_C + a_r)\hat{i} - a_t\hat{j},$$
$$\vec{a}_D = (a_C - a_t)\hat{i} - a_r\hat{j},$$
$$\vec{a}_E = (a_C - a_r)\hat{i} + a_t\hat{j},$$

where $a_t = R\alpha$ and
$$a_r = R\omega^2 \text{ and } a_C = a.$$

1.10 The kinematics of rolling

This is the combined motion of a body such as a disc, sphere, etc, on a surface such that the lowest point of the body does not slide relative to the surface in contact. This means that the relative velocity of the contacting points between the body and surface is zero. In the following figure the disc rolls on a horizontal surface. This needs the condition that the lowest point A of the disc does not slide with the fixed horizontal surface. Since the horizontal surface is stationary, the lowest point of the disc must remain at rest instantaneously. So $v_A = 0$. If the center of mass of the disc moves with a velocity v_C and the disc spins with an angular velocity ω, referring to the example 5, the velocity of A is given as

$$\vec{v}_A = \vec{v}_{AC} + \vec{v}_C.$$
$$= -R\omega\hat{i} + v\hat{i}$$
$$= (v - R\omega)\hat{i}.$$

Putting $v_A = 0$ in the above expression, we have $v_C = R\omega$. Differentiating both sides with time, we have $dv_C/dt = Rd\omega/dt$. This means $a_C = R\alpha$.

To summarize, for a rolling body on a fixed surface, if the velocity and acceleration of the center of mass of the body are v_C and a_C, respectively, the angular velocity ω and angular acceleration α of the body can be related with v_c and $\vec{a}_c$ as $v_C = R\omega$ and $a_C = R\alpha$. In rolling on a fixed surface, the velocity and tangential acceleration of the lowest point of the rolling body will be zero. This means that the tangential or horizontal component of acceleration of the lowest point A is zero. However, the net acceleration of A is nonzero: $(a_A)_{\text{tangential}} = 0$, but $\vec{a}_A = R\omega^2\hat{j}$. If the surface moves, these expressions do not hold.

Example 8 Let a disc roll without sliding on a fixed horizontal surface with a constant velocity. Establish the relation between the linear velocity v and angular velocity ω of the disc in the following figures.

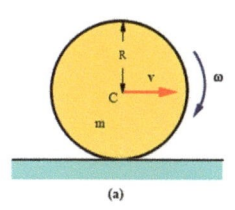

Solution

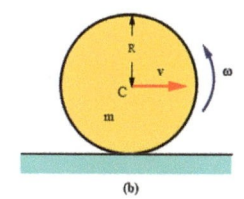

(a) The velocity of the lowest point A is
$$\vec{v}_A = \vec{v}_{AC} + \vec{v}_C.$$
$$= -R\omega\hat{i} + v\hat{i}$$
$$= (v - R\omega)\hat{i}.$$

For rolling, putting $v_A = 0$, we have
$$(v - R\omega)\hat{i} = 0.$$

Then we have
$$v = R\omega.$$

(b) The velocity of the lowest point A is
$$\vec{v}_A = \vec{v}_{AC} + \vec{v}_C$$
$$= R\omega\hat{i} + v\hat{i}$$
$$= (v + R\omega)\hat{i}.$$

For rolling, putting $v_A = 0$, we have
$$(v + R\omega)\hat{i} = 0.$$

Then we have
$$v = -R\omega.$$

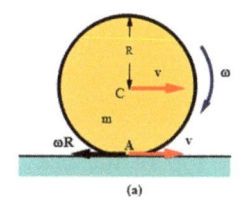

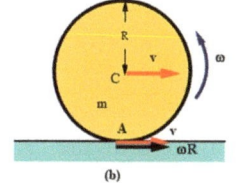

Example 9 Prove that rolling is a mixture of pure translation and pure rotation.

Solution

In pure translation, all points of the disc move with same velocity v towards the right. In pure rotation the center of mass is fixed and the disc must spin clockwise with an angular velocity

$$\vec{\omega} = -\omega\hat{k} = -v/R\hat{k} \text{ (clockwise)}.$$

Then the velocity of the lowest point A is given as

$$\vec{v}_A = \vec{v}_{AC} + \vec{v}_C$$
$$= -R\omega\hat{i} + v\hat{i}$$
$$= (v - R\omega)\hat{i}.$$

Putting $\omega = v/R$, we have $v_A = 0$ which is the required condition for rolling of the disc as discussed earlier.

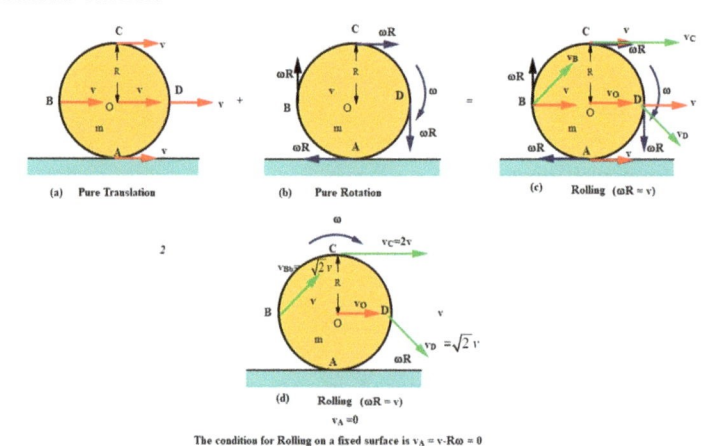

(a) Pure Translation (b) Pure Rotation (c) Rolling ($\omega R = v$)

(d) Rolling ($\omega R = v$)
$v_A = 0$
The condition for Rolling on a fixed surface is $v_A = v - R\omega = 0$

We cannot see these two effects (pure rotation and pure translation) separately with the naked eye. What we can see is the combination of these two effects. When we combine pure translation and pure rotation, the velocity distribution in the rolling disc will be completely different from that in both pure rotation and pure translation. By combining the velocities using the formula given in section 1.9, we can find the velocity of any point of the rolling body. We can see that the velocity at A is zero. This means that the lowest point of the rolling body always remains at rest instantaneously. Hence, the lowest point of a rolling body on a fixed surface is the instantaneous center (IC) of rotation. This is also called the instantaneous axis of rotation (IAR) as discussed in the next section.

For the sake of simplicity, the velocity distribution (profile) on the vertical diameter of the rolling disc is split into pure translation and pure rotation. You can see that velocity is uniform in pure translation. In pure rotation velocity changes linearly from $-v$ at the bottom, zero at the center and $+v$ at the top of the disc. However, the addition of these two effects gives rise to pure rolling in which velocity increases from zero at the bottom to $+2v$ at the top of the disc, as shown in the figure.

- The body is said to be rolling on a surface when the relative velocity of the contacting points between the body and the surface is zero.
- If the surface is fixed/stationary, for rolling, the lowest point of the body does not move instantaneously.
- The lowest point of a body rolling on a fixed surface is the IAR (please see the next section).

Example 10 Referring to the last figure, a disc of radius R rolls on the horizontal surface along the $+x$-axis. The center O of the disc moves with a constant velocity $\vec{v} = v\hat{i}$. Find the (a) velocities and (b) accelerations of the points A, B, C, and D of the disc.

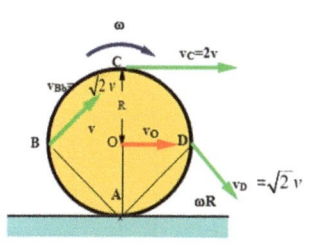

Each point move perpendicular to the line joining it to the lowest point A

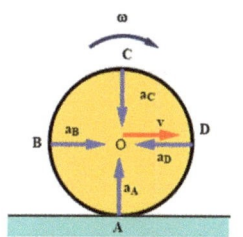

Each point accelerate towards the center O of the disc

Solution

(a) The velocity of A is given as

$$\vec{v}_A = \vec{v}_{AO} + \vec{v}_O$$
$$= -R\omega\hat{i} + v\hat{i}$$
$$= (v - R\omega)\hat{i} = 0.$$

So we have

$$R\omega = v. \tag{1.1}$$

The velocity of B is given as

$$\vec{v}_B = \vec{v}_{BO} + \vec{v}_O$$
$$= R\omega\hat{j} + v\hat{i}. \tag{1.2}$$

Putting $R\omega = v$ from equation (1.1) in equation (1.2) we have

$$\vec{v}_B = v\hat{i} + v\hat{j}$$
$$= v(\hat{i} + \hat{j}).$$

The velocity of C is given as

$$\vec{v}_C = \vec{v}_{CO} + \vec{v}_O$$
$$= R\omega\hat{i} + v\hat{i}$$
$$= (v + v)\hat{i} = 2v\hat{i}.$$

The velocity of D is given as

$$\vec{v}_D = \vec{v}_{DO} + \vec{v}_O$$
$$= -R\omega\hat{j} + v\hat{i}$$
$$= -v\hat{j} + v\hat{i} = v(+\hat{i} - \hat{j}).$$

(b) Differentiating both sides of equation (1.1) we have
$$a_t = R\frac{d\omega}{dt} = \frac{dv}{dt} = 0,$$
then we can write
$$a_t = a_o = 0. \tag{1.3}$$
So, the acceleration of A is
$$\vec{a}_A = (-a_t + a_o)\hat{i} + a_r \hat{j}. \tag{1.4}$$
Using equations (1.3) and (1.4)
$$\vec{a}_A = a_r \hat{j} = (v^2/R)\hat{j}.$$
The acceleration of B, C and D can be given as (please refer to example 7)
$$\vec{a}_B = (a_o + a_r)\hat{i} + a_t \hat{j}$$
$$\vec{a}_C = (a_o + a_t)\hat{i} - a_r \hat{j}$$
$$\vec{a}_D = (a_o - a_r)\hat{i} - a_t \hat{j},$$
where $a_t = R\alpha = 0$ and
$$a_r = v^2/R \text{ and } a_o = 0.$$
Then we have
$$\vec{a}_B = (v^2/R)\hat{i}$$
$$\vec{a}_C = -(v^2/R)\hat{j}$$
$$\vec{a}_D = -(v^2/R)\hat{i}.$$

Example 11 A uniform disc of radius R rolls on a horizontal surface so that its center C moves with a constant velocity v. Find (a) the velocity and (b) the acceleration of an arbitrary point A on the periphery of the disc.

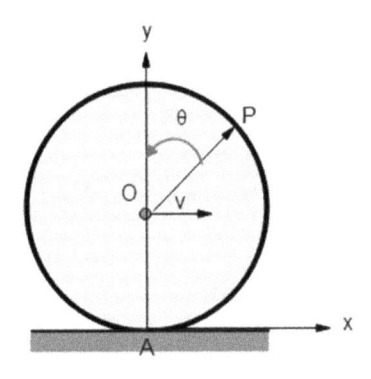

Solution

(a) Since point P moves perpendicular to the line AP with an angular velocity ω, the velocity of the point A is given as $v_P = (AP)\omega$, where $\omega = \frac{v}{R}$ as per the condition of rolling and $AP = 2R\cos\varnothing$, where $\varnothing = \theta/2$. Then $v_P = 2v\cos(\theta/2)$.

(b) The acceleration of point P is given as $\vec{a}_P = \vec{a}_{PO} + \vec{a}_O$, where $\vec{a}_O = 0$ because $\vec{v}_O =$ constant. Then

$$\vec{a}_P = \vec{a}_{PO}.$$

Since $\vec{a}_{PO}$ does not have a tangential component $a_t = R\alpha = a = 0$, it must be radially inward. So $a_P = R\omega^2 = v^2/R$ (because $\omega = \frac{v}{R}$ as per the condition of rolling).

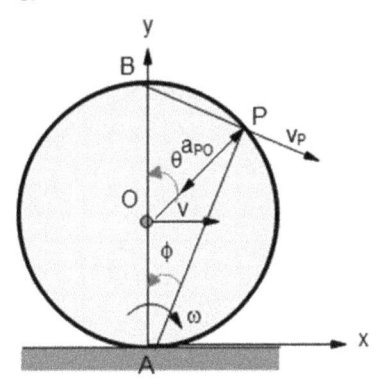

1.11 General kinematical misconceptions

By now you might have understood that an angular motion or *rotation* is *not* a *circular motion*. In fixed axis rotation, it is true that each particle in the body moves in a circular path relative to the ground. But, generally in combined motion, points of a rigid body move in arbitrary curves. So the rotation of a rigid body should not be misinterpreted as a circular motion. Circular motion is used for a particle or point, whereas rotation is used for a rigid body (or a straight line drawn inside the rigid body).

There is another misconception between fixed axis rotation and pure rotation. Both are conceptually different. Non-centroidal fixed axis rotation is a combination of translation and rotation because the center of mass moves. However, in centroidal fixed axis rotation there is no translation as the center of mass remains fixed because the body is pivoted at the center of mass.

The third misconception is that 'for a rigid body to rotate it must be pivoted at a fixed point'. If you throw a ball by giving it a spin, this a combined unconstrained motion. The body is not pivoted anywhere but is still rotating with a moving center of mass. You may wonder how this is possible? Then, you may ask 'who causes the rotation?' which will be discussed in chapter 3. So let us accept the fact (without knowing its cause) that a rigid body can rotate independently, possessing angular velocity and angular acceleration without being pivoted at a fixed support.

The fourth important fact which is often misinterpreted is that angular quantities (displacement, velocity, and acceleration) are the internal properties (or phenomena) of the body but not the properties of any point of the rigid body relative to outside points.

1.12 Kinematical equations for uniform angular acceleration

Then, for the sake of broader understanding, let us consider a spinning disc undergoing a general or combined motion. Let, at time $t = 0$, the velocity of its center of mass C be v_o and the angular velocity of the body be ω_o. After a time t, let the velocity of the center of mass be v and the angular velocity of the disc be ω, as shown in the figure. Then, the general kinematical equations for rotation and translation of the disc are given as follows.

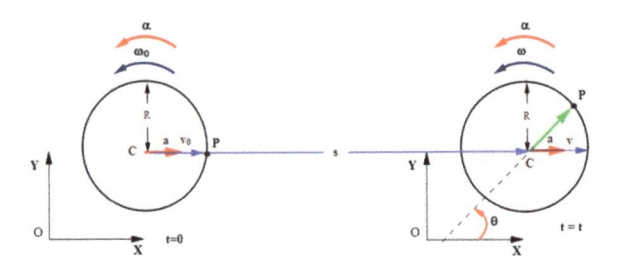

For rotation we have

$$\omega = d\theta/dt.$$

By cross-multiplying, the angular displacement of the disc during an elementary time dt is

$$d\theta = \omega dt.$$

Integrating both sides, the net angular displacement is

$$\theta = \int_0^t \omega dt \quad (1.5)$$

The angular acceleration of the disc is given as

$$\alpha = \frac{d\omega}{dt} \quad \text{and} \quad \alpha = \omega\frac{d\omega}{d\theta}.$$

If we assume that the direction of angular acceleration is opposite to that of the angular velocity, we just impose a negative sign on the right-hand side of the equations. For a small time interval dt the angular velocity is given as

$$d\omega = \alpha dt.$$

It is given that at time $t = 0$, $\omega = \omega_o$, and at time $t = t$, $\omega = \omega$. So, integrating both sides, we have

$$\int_{\omega_o}^{\omega} d\omega = \int_0^t \alpha dt.$$

If you assume a constant angular acceleration of the disc, just take α out of the integral to obtain

$$\int_{\omega_o}^{\omega} d\omega = \alpha \int_0^t dt.$$

After evaluating the integral, we have

$$\omega = \omega_o + \alpha t. \tag{1.6}$$

Putting ω from equation (1.6) in equation (1.5) we have

$$\theta = \int_0^t (\omega_o + \alpha t) dt.$$

Evaluating the integral, finally we have

$$\theta = \omega_o t + \frac{1}{2}\alpha t^2.$$

From the general equation $\alpha = \omega \frac{d\omega}{d\theta}$ it can be written in the differential form as

$$d\theta = \omega d\omega.$$

After undergoing an angular displacement θ the angular velocity of the disc changes from ω_o to ω. After integrating both sides of the last expression, we have

$$\int_0^\theta d\theta = \int_{\omega_o}^{\omega} \omega d\omega.$$

Evaluating the integral we have

$$\omega^2 - \omega_0^2 = 2\alpha\theta. \tag{1.7}$$

Example 12 A ceiling fan rotates with a frequency of 300 revolutions per minute (RPMs). If it is switched off at $t = 0$, it slows down with a constant angular retardation. If it stops after a minute, find (a) the angular acceleration, (b) the average angular speed, and (c) the total (i) angular distance covered and (ii) number of rotations until the fan stops.
Solution

(a) The angular acceleration is

$$\alpha = (\omega - \omega_o)/t.$$

Put $\omega = 0$ as it stops after $t = 60$ s and $\omega_o = 2\pi f = 2(22/7)(300/60) = (220/7)$ rad s^{-1} in the above equation to obtain

$$\alpha = -11/21 \text{ rad s}^{-2}.$$

(b) The formula for the average speed is
$$\omega_{av} = (\omega + \omega_o)/2,$$
where $\omega = 0$ and $\omega_o = (220/7)$ rad s^{-1}. This yields
$$\omega_{av} = (110/7) \text{ rad s}^{-1}.$$

(c)
(i) The total angular distance covered by the fan is given as
$$\theta = \omega_{av} t = \left(\frac{110}{7}\right)(60) = 6600/7 \text{ rad}.$$

(ii) The total number of rotation of the fan is
$$N = \frac{\theta}{2\pi} = \frac{\frac{6600}{7}}{2 \times \left(\frac{22}{7}\right)} = 150.$$

1.13 The instantaneous axis of rotation (IAR)

Inside a rotating rigid body, there exists a point either inside the physical body (or imaginary extended body) that does not move at a given instant. This is called the instantaneous axis (center) of rotation.

1.13.1 Calculation of the position of the IAR

(i) v_P and ω are given:

Let a planer rigid body perform a combined motion with an angular velocity ω. If the velocity of a point Q of the rigid body is given as v_Q, how do we find the location of IAR? Let us see.

Let us assume that O is the IAR of the rigid body located at a distance x from the P whose velocity is given as v. As per the definition the velocity of the IAR must be zero at the given instant. So, applying the addition of velocities, the velocity of the IAR at O is
$$\vec{v}_O = \vec{v}_{OP} + \vec{v}_P,$$
where $\vec{v}_{OP} = -x\omega \hat{i}$ and $\vec{v}_P = v\hat{i}$ (given). Then, we have
$$x = v/\omega.$$

Otherwise: We can think of the point O as instantaneously (but not permanently) at rest. So each point of the body (the point P, say) turns around the point O with the same angular velocity $\omega = v/x$, because v_P is perpendicular to OP. So $x = \frac{v}{\omega}$.

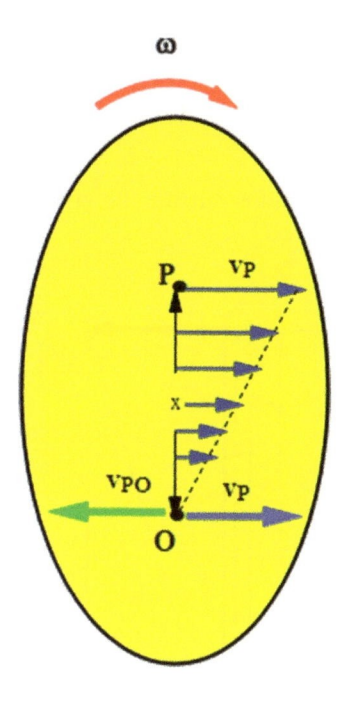

Example 14 The top end A of a rod of length is given a velocity v and simultaneously the rod is given an anti-clockwise spin $\omega = \frac{v}{2l}$, as shown in the figure. Locate the IAR.

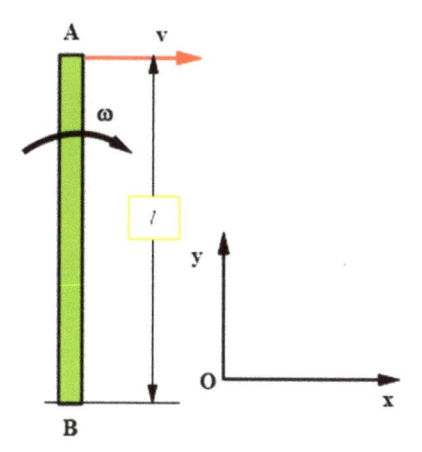

Solution

Let the IAR (point P) be located at a distance r from the end A of the rod. As the point P does not move at the given instant, $\vec{v}_P = \vec{v}_{PA} + \vec{v}_A = 0$, where $\vec{v}_{PA} = -r\omega\hat{i}$ and $\vec{v}_A = v\hat{i}$.

This yields

$$r = \frac{v}{\omega} = \frac{v}{\left(\frac{v}{2L}\right)} = 2l.$$

In this case the IC remains outside the given rod but we can imagine it to be inside the extended rod.

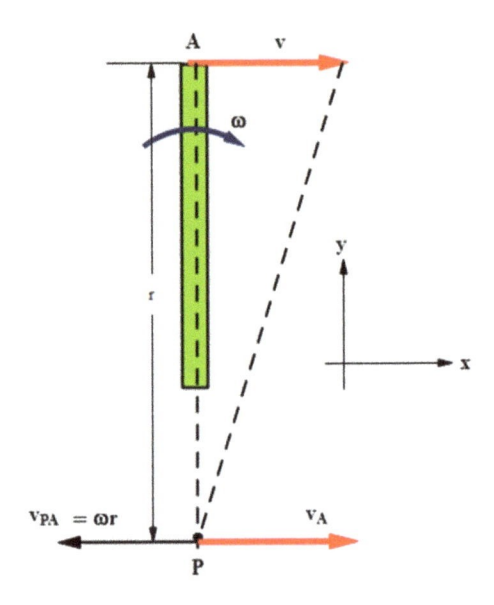

(ii) v_P and v_Q are given:

When velocities of any two points are given, let us find the IC. We have three following cases in this category.

Case 1. v_P and v_Q are arbitrary:

Let the point O be the IC taken inside the body for the sake of simplicity. As O is instantaneously stationary, both the given points P and Q must move perpendicular to OP and OQ, respectively. So, the angular velocity of the body can be given as

$$\omega = (v_P/\text{OP}) = v_Q/\text{OQ}.$$

This tells us that if we drop the perpendiculars from P and Q with their line of motion (or velocities v_P and v_Q) their point of intersection O would be our instantaneous axis or center.

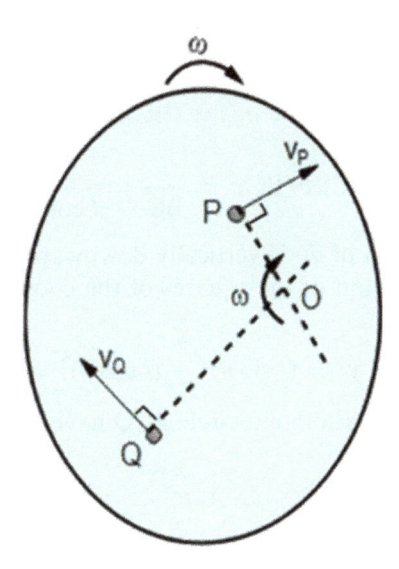

Problems and Solutions in Rotational Mechanics

Example 15 A rod AB of length l is leaning against the wall making an angle θ with the wall. If the bottom end of the rod moves with a constant velocity of v towards the right, find (a) the location of the IC, (b) the angular velocity of the rod, (c) the acceleration of the IC, and (d) the locus of the path followed by the IC.

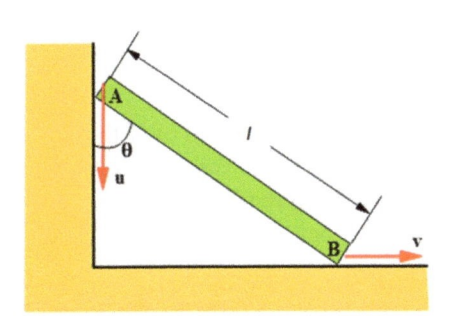

Solution

(a) When we drop the perpendiculars to u and v at the ends A and B of the rod, they intersect at the point C. Then C is the IC. The coordinates of C are given as $x = l \sin\theta$ and $y = l \cos\theta$.

(b) Since the IC of the rod is C and it is at rest instantaneously, the angular velocity of the rotating rod is $\omega = \frac{u}{AC}(=\frac{v}{BC}) = \frac{v}{l\cos\theta}$ anticlockwise.

(c) The acceleration of C is given as
$$\vec{a}_C = \vec{a}_{CB} + \vec{a}_B.$$

Since the point B moves with a constant velocity, putting $\vec{a}_B = 0$, we have
$$a_C = a_{CB} = (BC)\omega^2$$
$$= (BC)(v/BC)^2 = \frac{v^2}{BC} = \frac{v^2}{l\cos\theta}.$$

The direction of $\vec{a}_C$ is vertically downwards.

(d) Taking the sum of the squares of the coordinates, the locus of C is given as
$$x^2 + y^2 = (l\sin\theta)^2 + (l\cos\theta)^2 = l^2.$$

This is an equation of a circle PCQ having its center at the origin O.

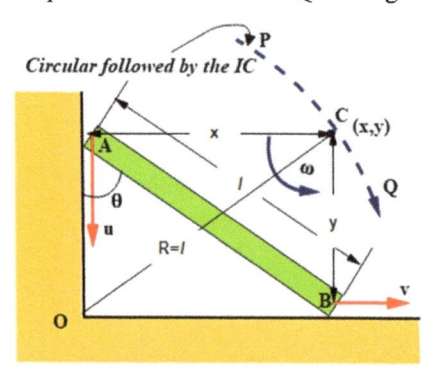

Case 2. v_P and v_Q are parallel:

Let us assume that two points P and Q of the plate move in same direction with velocities v_P and v_Q, respectively. Following the last procedure if we drop the perpendiculars from the line of motion of the given points P and Q, they meet at infinity. Does this mean that the IC is located at infinity? Obviously not. So we cannot adopt the previous method in which we considered the point of intersection of the perpendiculars as the IC.

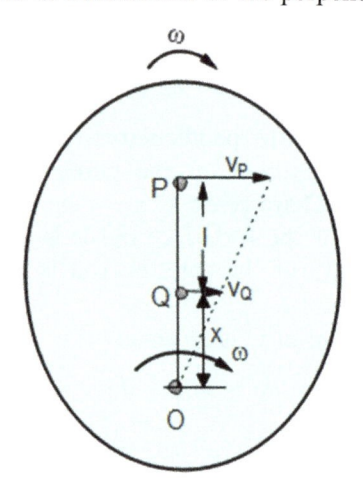

Then we need to choose the IC point O taken inside or outside the body. As O is instantaneously stationary, both the given points P and Q must move perpendicular to OP and OQ, respectively. So the angular velocity of the body can be given as

$$\omega = v_P/OP = v_Q/OQ,$$

or

$$OP/OQ = v_P/v_Q.$$

This tells us that the dotted line joining the tips of the velocities will intersect the extended line PQ at O, which is the IC. This means the location of the IC can also be calculated geometrically using two similar triangles, as shown in the figure.

Case 3. v_P and v_Q are anti-parallel:

In this case also, if we drop the perpendiculars from the velocities at P and Q, they meet at infinity. So we have to adopt the last procedure of connecting the tips of the velocity vectors by a straight line (dotted) that intersects the line PQ at the point O which is our IC. The angular velocity of the body can be given as

$$\omega = v_P/OP = v_Q/OQ,$$

or

$$OP/OQ = v_P/v_Q.$$

So also in this case we can find the location of the IC by using two similar triangles, as shown in the figure.

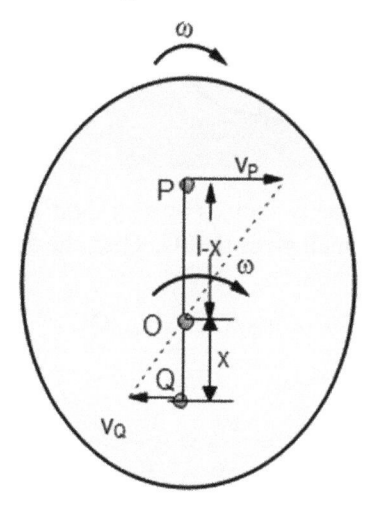

Example 16 A cylinder of radius R rolls without sliding on a plank P which moves with a velocity $3v_o$ on a horizontal ground. If the center O of the cylinder moves with

a velocity v_o as shown in the figure, find (a) the position of the IAR from the center of the cylinder and (b) the angular velocity of the cylinder.

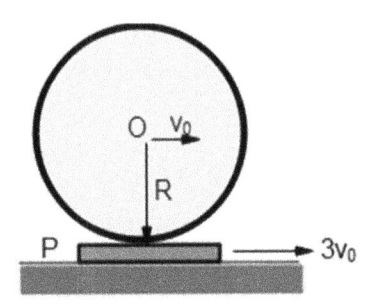

Solution

(a) As the body rolls on the plank P, the lowest point Q of the cylinder will move with a velocity equal to the velocity of the plank. Thus the velocities of the points P and O of the cylinder are given as $v_Q = 3v_o$ and $v_O = v_o$ in the $+x$-direction.

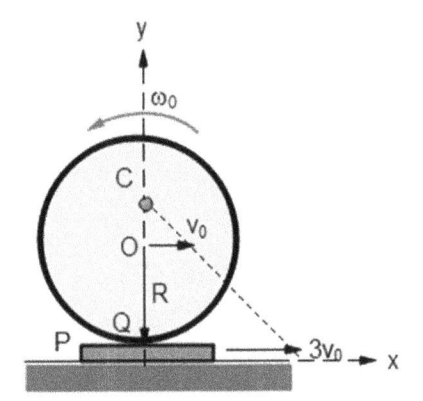

Connecting the tips of the velocity vectors by a straight line (dotted) which intersects the line OQ at the point C which gives the IC. Then the angular velocity of the body can be given as

$$\omega = v_O/OC = v_Q/QC,$$

or

$$\frac{QC}{OC} = \frac{v_Q}{v_o} = \frac{3v_o}{v_o} = 3,$$

or

$$\frac{QO + OC}{OC} = 3, \quad \text{where } QO = R.$$

This tells us that the IC is located at a height $OC = R/2$ as shown in the figure.

(b) Putting $v_O = v_o$ and $OC = R/2$ in the expression $\omega = (v_O/OC)$, we have $\omega = 2v_O/R$ in an anti-clockwise sense.

Example 17 The IC of a rolling body on a fixed surface:

If we photograph a running bicycle wheel from the ground frame using an ultrafast hi-tech camera, how does bicycle wheel appear?

Solution

The condition of rolling of a body on a fixed surface is that the lowest point of the body must be instantaneously stationary. So any point of the body must move perpendicular to the line joining the point to the lowest point of the body, as shown in the figure. If the distance of any point Q, say, of the body from its lowest point P is r, the velocity of Q is given as $v = r\omega$. Since the angular velocity ω is the same for all points relative to the lowest point P, the speed of the point Q is directly proportional to its distance r from A. In other words, the farther points from P move faster than the nearer points. In other words, the points farther from P look more blurred compared to the points nearer to the lowest point P, as shown in the figure. Hence, we can see the lowest point P as a clear dot in the photograph.

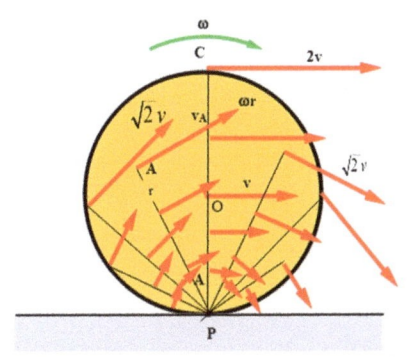

Each point move perpendicular to the line joining it to the lowest point A; so, the lowest point P is the IC and appears a distinct points. Farther points from P appear more blurred due to their greater speeds than the nearer points to P.

1.14 Rotation about a point and gyroscopic motion

There are many instances such as a spinning top, pedestal, or table fan in which the body rotates about a rotating axis. In these types of motion, one point of the body is fixed. For the spinning top the lowest point does not move. This sort of motion is called unconstrained motion about a point. We will discuss this motion as a gyroscope in the dynamics of a rigid body.

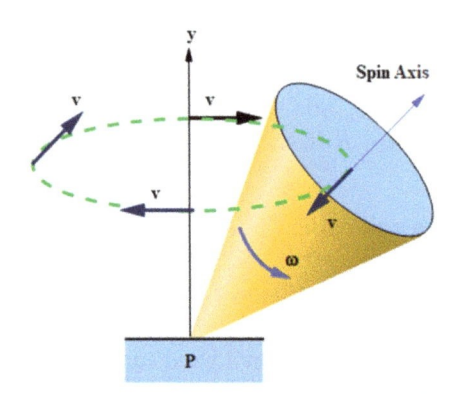

Rotation about the point P has both spinning and precession

Example 18 The bob of a conical pendulum with a bob of mass m has an angular velocity ω_1 and ω_2 relative to the center O of the circle and the point of suspension P, respectively. Find the value of ω_1/ω_2.

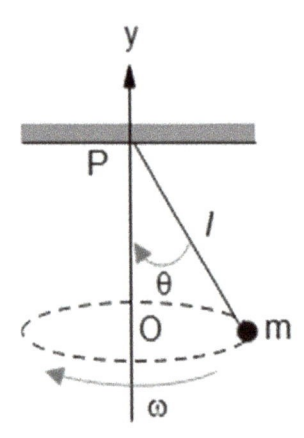

Solution

During the time δt let the bob swing at an angle $\delta\varnothing$ and $\delta\beta$ relative to O and P, respectively. This means the string rotates at an angle $\delta\varnothing$ about the point P and an angle $\delta\beta$ about the y-axis. So the angular velocity of the string is equal to the angular velocity of the bob relative to the point of suspension P. Then we can write

$$\frac{\delta\beta}{\delta t} = \omega_1 \text{ and } \frac{\delta\varnothing}{\delta t} = \omega_2,$$

or

$$\frac{\omega_1}{\omega_2} = \frac{\delta\beta}{\delta\varnothing}. \tag{1.8}$$

Let the bob undergo a small displacement δs during a time δt, which can be given as

$$\delta s = l\delta\varnothing = R\delta\beta,$$

where R is the radius of the circle described by the bob. Then we have

$$\frac{\delta\beta}{\delta\varnothing} = \frac{l}{R} = \mathrm{cosec}\theta. \qquad (1.9)$$

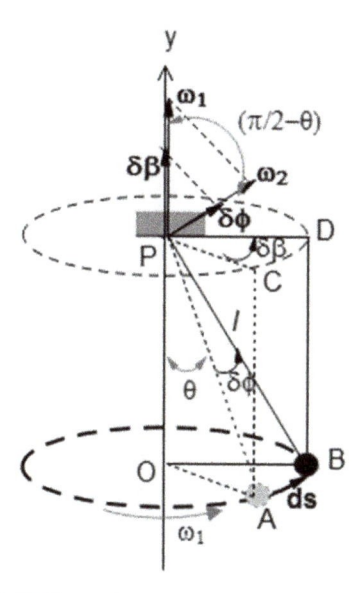

Using equations (1.8) and (1.9) we have

$$\frac{\omega_1}{\omega_2} = \mathrm{cosec}\theta.$$

Problems

Calculation of the velocity and accelerations of rigid bodies

Problem 1 A rod AB of length $l = 2$ m is leaning against a vertical wall. If the end B moves with a velocity of $v = 4$ m s^{-1} towards the right, find (a) the angular velocity of the rod, and (b) the angular acceleration of the rod assuming $a =$ the acceleration of the point B $= 8.5$ m s^{-2} to the right.

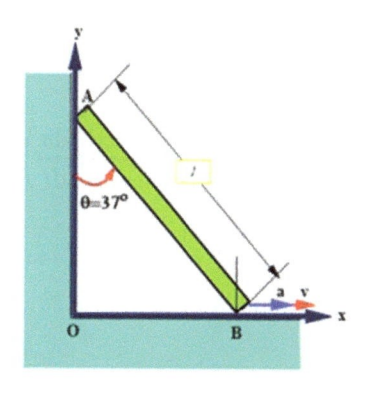

Solution

(a) The position of **B** is given by
$$x = l \sin \theta$$
$$\Rightarrow \frac{dx}{dt} = l \cos \theta \frac{d\theta}{dt}.$$

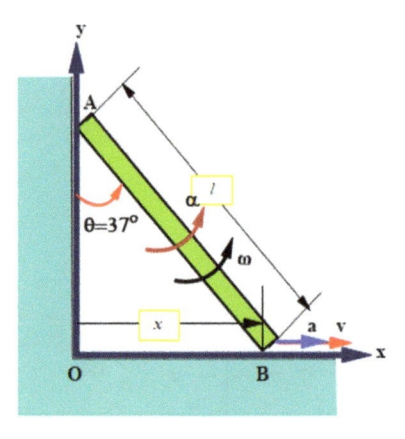

$$\Rightarrow v = \omega l \cos \theta$$
$$\Rightarrow \omega = \frac{v}{l \cos \theta}. \tag{1.10}$$

Putting the numerical values, we have
$$\Rightarrow \omega = \frac{v}{l \cos \theta} = \frac{4}{2 \cos 37°}$$
$$= \frac{4}{2 \times 4/5} = 2.5 \text{ rad s}^{-1}.$$

(b) Differentiating ω with respect to time, we have
$$\alpha = \frac{d\omega}{dt} = \frac{1}{l}\frac{d}{dt}(v\sec\theta)$$
$$= \frac{v}{l}\left(\sec\theta \tan\theta \frac{d\theta}{dt}\right) + \frac{\sec\theta}{l}\left(\frac{dv}{dt}\right)$$
$$= \frac{v}{l}\sec\theta \tan\theta \frac{d\theta}{dt} + \frac{a\sec\theta}{l}$$
$$= \frac{\sec\theta}{l}(v\omega \tan\theta + a). \tag{1.11}$$

Using the last equations (1.10) and (1.11),
$$\alpha = \frac{\sec\theta}{l}\left(v\left(\frac{v\sec\theta}{l}\right)\tan\theta + a\right)$$

$$\Rightarrow \alpha = \frac{\sec\theta}{l}\left(\frac{v^2\sec\theta\tan\theta}{l} + a\right).$$

Evaluating the factors, we have
$$\Rightarrow \alpha = \frac{\sec 37°}{2}\left(\frac{(4)^2\sec 37°\tan 37°}{2} + 8.5\right)$$

$$\Rightarrow \alpha = 10 \text{ rad s}^{-2}.$$

Problem 2 In the previous problem, if the end B is moved with a constant velocity of 4 m s^{-1} to the right, find (a) the velocity and (b) the acceleration of A. You can use all required data of the last problem.

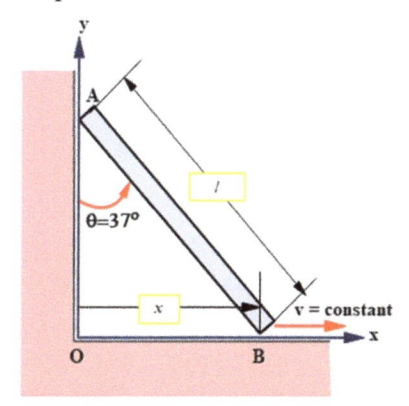

Solution

(a) Referring to the last problem, we have
$$\omega = \frac{v}{l\cos\theta}. \tag{1.12}$$

The position of A is given by
$$y = l\cos\theta$$

$$\Rightarrow \frac{dy}{dt} = -l\sin\theta\frac{d\theta}{dt}.$$

$$\Rightarrow v' = -\omega l\sin\theta. \tag{1.13}$$

Using equations (1.12) and (1.13)
$$v' = -(v/l\cos\theta)l\sin\theta = -v\tan\theta. \tag{1.14}$$

Putting in the numerical values, we have
$$v' = -v\tan\theta = -(4)\tan 37°$$
$$= -(4)(3/4) = -3 \text{ m s}^{-1}.$$

The negative sign signifies that the end A moves down.

(b) Differentiating v' with respect to time, we have
$$a_A = \frac{dv'}{dt} = \frac{d}{dt}(v\tan\theta)$$
$$= v\left(\sec^2\theta \frac{d\theta}{dt}\right) = v\omega\sec^2\theta$$
$$= v\left(\frac{v\sec\theta}{l}\right)\sec^2\theta = \frac{v^2\sec^3\theta}{l}.$$

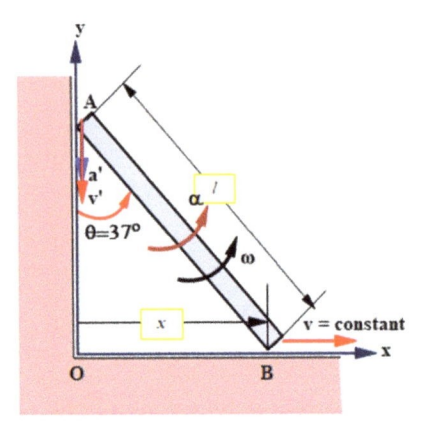

Evaluating, we have
$$a_A = a'\frac{(4)^2(5/4)^3}{2} = (125/8) \text{ m s}^{-2}.$$

N.B. we obtained the velocity of A as
$$v' = -v\tan\theta.$$

As the end B moves towards the right, the angle θ increases; so $\tan\theta$ will increase. This means that the point A speeds up in the downward direction and hence its acceleration will point vertically down.

Problem 3 Referring to the last diagram, let us assume that the rod is massless and the particles A and B have masses m and $2m$, respectively. Find the velocity of the center of mass of the rod–particles system for $\theta = 45°$. Put $v = 3$ m s^{-1}.

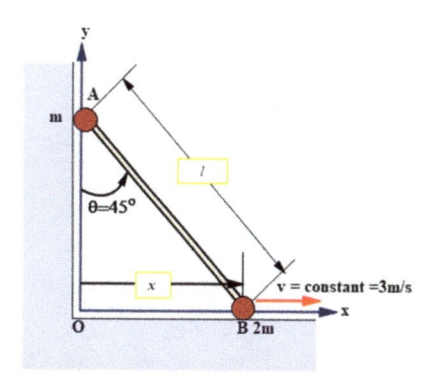

Solution

In the last problem, we have obtained the velocities of the points A and B as follows:

$$v' = -v \tan \theta (\downarrow)$$

$$v_B = +v (\rightarrow).$$

Then the velocity of the center of mass of the system is

$$\vec{v}_C = \frac{m_B v \hat{i} + m_A v'(-\hat{j})}{m_A + m_B}$$

$$\Rightarrow \vec{v}_C = \frac{2m v \hat{i} + m v'(-\hat{j})}{2m + m}$$

$$\Rightarrow \vec{v}_C = \frac{2}{3} v \hat{i} - \frac{v'}{3} \hat{j}. \tag{1.15}$$

Alternatively, by equating the velocities at the ends of the rod along its length, we have

$$v' \cos \theta = v \sin \theta$$

$$\Rightarrow v' = \frac{v \sin \theta}{\cos \theta} = v \tan \theta. \tag{1.16}$$

Using equations (1.15) and (1.16) we have

$$\vec{v}_C = \frac{v}{3} [2\hat{i} - \tan \theta \hat{j}], \text{ where}$$

$v = 3$ m s^{-1} and $\tan \theta = 1$; we have

$$\vec{v}_C = (2\hat{i} - \hat{j}) \text{ m s}^{-1}.$$

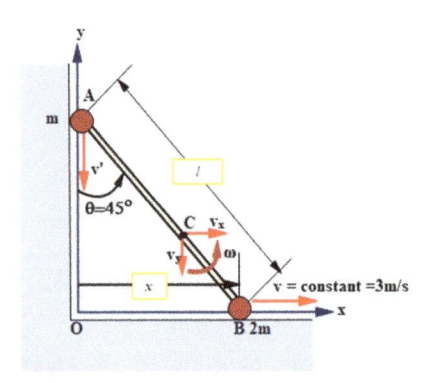

Problem 4 A bar AB of length l has angular velocity $\vec{\omega} = \omega\hat{k}$ and angular acceleration $\vec{\alpha} = \alpha\hat{k}$. Find (a) the velocity, and (b) the acceleration of the mid-point of the rod.

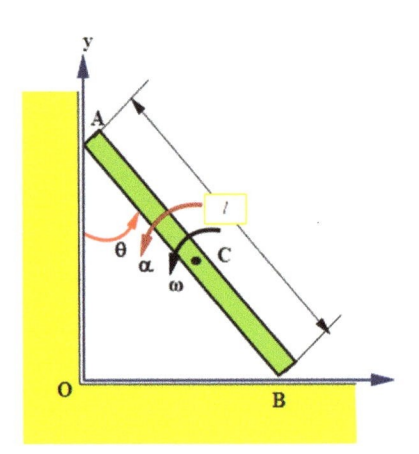

Solution

(a) Let the coordinates of the mid-point of the rod be x and y, respectively, given as

$$x = (l/2)\sin\theta \qquad (1.17)$$

$$y = (l/2)\cos\theta. \qquad (1.18)$$

Differentiating both sides with time, we have

$$v_x = \frac{\omega l}{2}\cos\theta. \qquad (1.19)$$

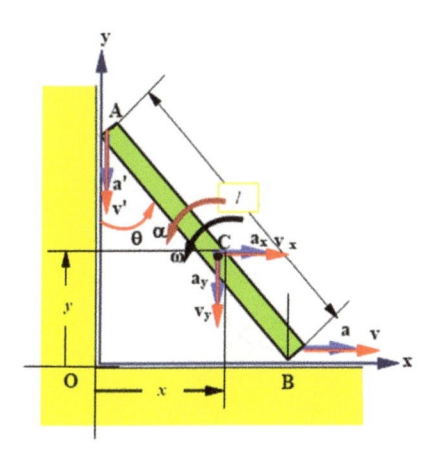

$$v_y = \frac{\omega l}{2} \sin \theta. \tag{1.20}$$

Then the center of mass has the velocity

$$\begin{aligned}\vec{v} &= \vec{v}_x + \vec{v}_y \\ &= \frac{\omega l}{2} \cos\theta \hat{i} - \frac{\omega l}{2} \sin\theta \hat{j} \\ &= \frac{\omega l}{2}(\cos\theta \hat{i} - \sin\theta \hat{j}).\end{aligned} \tag{1.21}$$

(b) Differentiating the velocities in equations (1.19) and (1.20) with time, we have

$$a_x = \frac{l}{2}(\alpha \cos\theta - \omega^2 \sin\theta) \tag{1.22}$$

$$a_y = +\frac{l}{2}(\omega^2 \cos\theta + \alpha \sin\theta). \tag{1.23}$$

The acceleration of the center of mass is

$$\begin{aligned}\vec{a}_C &= \vec{a}_x + \vec{a}_y = a_x \hat{i} - a_y \hat{j} \\ &= \frac{l}{2}(\alpha\cos\theta - \omega^2\sin\theta)\hat{i} - \frac{l}{2}(\omega^2\cos\theta + \alpha\sin\theta)\hat{j} \\ &= \frac{l}{2}[(\alpha\cos\theta - \omega^2\sin\theta)\hat{i} - (\alpha\sin\theta + \omega^2\cos\theta)\hat{j}].\end{aligned}$$

Problem 5 A slender bar AB of length $l = 0.5$ m is moving in a vertical plane as its lowest point B is pulled with a constant (a) velocity $v = 0.5$ m s^{-1} and

(b) acceleration $a = 1$ m s^{-2} towards the right. Find the angular acceleration of the rod as a function of time. Assume that the rod was initially in a near vertical position and after a time t it makes an angle θ with the horizontal.

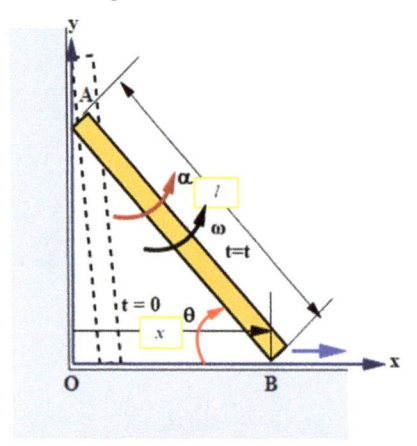

Solution

(a) For the end B moving with $v =$ constant,

$$x = vt$$

$$\Rightarrow l \cos \theta = vt$$

$$\Rightarrow -l \sin \theta \frac{d\theta}{dt} = v$$

$$\Rightarrow \frac{d\theta}{dt} = -\frac{v}{l} \operatorname{cosec} \theta.$$

Then

$$\frac{d^2\theta}{dt^2} = -\frac{v}{l} \frac{d}{dt}(\operatorname{cosec} \theta)$$

$$= -\frac{v}{l}(-\operatorname{cosec} \theta \cot \theta)\frac{d\theta}{dt}$$

$$= +\frac{v}{l} \operatorname{cosec} \theta \cot \theta \left(-\frac{v}{l} \operatorname{cosec} \theta\right)$$

$$= -\frac{v^2}{l^2}\operatorname{cosec}^2\theta \cot \theta.$$

Since θ decreases $\alpha = -\frac{d^2\theta}{dt^2}$

$$\Rightarrow \alpha = \frac{v^2}{l^2}\operatorname{cosec}^2\theta \cot \theta$$

$$= \frac{v^2}{l^2}\frac{\cos\theta}{\sin^3\theta}, \text{ where}$$

$$\cos\theta = \frac{vt}{l} \text{ and } \sin\theta = \sqrt{1 - \frac{v^2t^2}{l^2}}$$

$$\Rightarrow \alpha = \frac{v^2}{l^2}\frac{vt}{l\frac{(l^2-v^2t^2)^{\frac{3}{2}}}{l^3}}$$

$$\Rightarrow \vec{\alpha} = \frac{v^3 t}{\left(l^2 - v^2t^2\right)^{\frac{3}{2}}}\hat{k}$$

$$\Rightarrow \vec{\alpha} = \frac{(1/2)^3 t}{\{(1/2)^2 - (1/2)^2 t^2\}^{\frac{3}{2}}}\hat{k}\left(\because v = 1/2 \text{ and } l = \frac{1}{2}\right) \Rightarrow \vec{\alpha} = \frac{t}{(1-t^2)^{\frac{3}{2}}}\hat{k}.$$

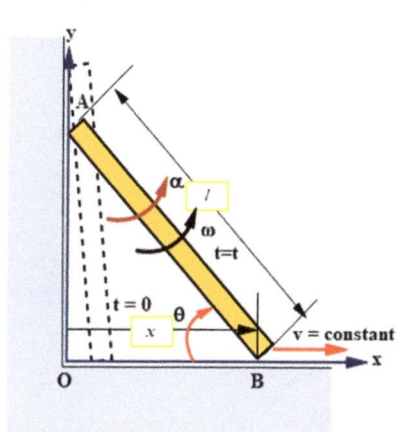

(b) For constant acceleration, $x = \frac{1}{2}at^2$,

$$\Rightarrow l\cos\theta = \frac{1}{2}at^2$$

$$\Rightarrow -l\sin\theta\frac{d\theta}{dt} = at$$

$$\Rightarrow \frac{d\theta}{dt} = -\frac{at}{l}\operatorname{cosec}\theta.$$

1-41

Since $\cos\theta = \dfrac{at^2}{2l}$, $\sin\theta = \dfrac{\sqrt{4l^2 - a^2t^4}}{2l}$,

then $\operatorname{cosec}\theta = \dfrac{2l}{\sqrt{4l^2 - a^2t^4}}$

$$\Rightarrow \frac{d\theta}{dt} = -\frac{at}{l} \cdot \frac{2l}{\sqrt{4l^2 - a^2t^4}}$$

$$= -\frac{2at}{\sqrt{4l^2 - a^2t^4}}$$

$$= \frac{2t}{\sqrt{1 - t^4}} \quad \left(\because a = 1 \text{ and } l = \frac{1}{2}\right)$$

$$\Rightarrow \frac{d^2\theta}{dt^2} = -\frac{d}{dt}\left(\frac{2t}{\sqrt{1 - t^4}}\right)$$

$$= -2 \cdot \frac{\sqrt{1 - t^4} - t/2\left(\dfrac{-4t^3}{\sqrt{1 - t^4}}\right)}{1 - t^4}$$

$$= -2\frac{1 - t^4 + 4t^4}{(1 - t^4)^{\frac{3}{2}}}$$

$$= -2\frac{(1 + t^4)}{(1 - t^4)^{\frac{3}{2}}}.$$

Since $\alpha = -\dfrac{d^2\theta}{dt^2}$ (because θ decreases with time t) we have

$$\alpha = \frac{2(1 + t^4)}{(1 - t^4)^{\frac{3}{2}}}.$$

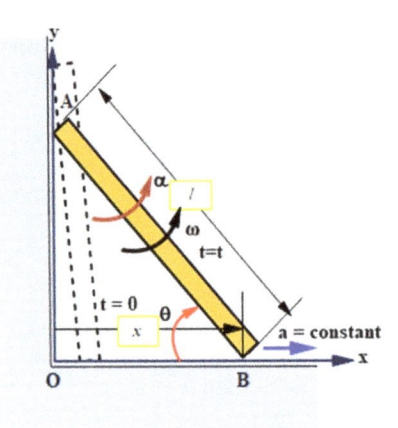

1-42

Problem 6 A rod of length l is inclined at an angle θ, as shown in the figure. Find the locus of (a) the mid-point and (b) the point at a distance b from the top end.

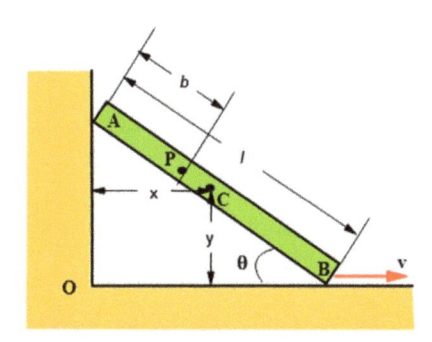

Solution

(a) If the point C is the mid-point of the rod, let the coordinates of C be x, y,

$$\text{then } x = \frac{l}{2} \cos \theta$$

$$\text{and } y = \frac{l}{2} \sin \theta.$$

$$\text{Then } \cos^2\theta + \sin^2\theta = \left(\frac{2x}{l}\right)^2 + \left(\frac{2y}{l}\right)^2$$

$$\Rightarrow x^2 + y^2 = \frac{l^2}{4}.$$

The locus is a circle.

(b) If P is any other point except the mid-point,

$$\text{then } AP = mAB \text{ and } BP = nAB,$$

where m and n are the fractions ($m + n = 1$).
The coordinates are given as

$$x = AP\cos\theta = mAB\cos\theta = ml \cos \theta$$

$$y = BP\sin\theta = nAB\sin\theta = nl \sin \theta.$$

Then $\left(\frac{x}{m}\right)^2 + \left(\frac{y}{n}\right)^2 = l^2(\cos^2 \theta + \sin^2 \theta)$

$$\Rightarrow \frac{x^2}{m^2} + \frac{y^2}{n^2} = l^2.$$

It is an ellipse, where $m = \frac{b}{l}$ and $n = \frac{l-b}{l}$

$$\Rightarrow \frac{x^2}{\left(\frac{b}{l}\right)^2} + \frac{y^2}{\left(1 - \frac{b}{l}\right)^2} = l^2$$

$$\Rightarrow \frac{x^2}{b^2} + \frac{y^2}{(l-b)^2} = 1.$$

Problem 7 The bottom of the rod AB of length l is pulled with a constant velocity v while its other end slides over the edge of a fixed cubical wedge W of height h. Find (a) the angular velocity of the rod as the function of θ and (b) the angular velocity of the rod when it leaves the edge of the wedge by assuming $h = \frac{l}{\sqrt{3}}$.

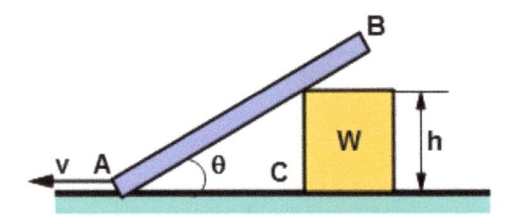

Solution

(a) Let AC $= x$. Since $\frac{h}{x} = \tan\theta$, we have

$$x = h \cos\theta$$

$$\Rightarrow \frac{dx}{dt} = h\frac{d}{dt}(\cot\theta)$$

$$\Rightarrow v = -h\cot\theta\csc^2\theta \frac{d\theta}{dt}$$

$$\Rightarrow \frac{d\theta}{dt} = -\frac{v}{h\cot\theta\csc^2\theta} = -\frac{v\sin^3\theta}{h\cos\theta} \text{ (negative sign signifies a clockwise rotation).} \quad (1.24)$$

(b) When the rod remains tangential to the wedge for the last time,

$$\tan\theta = \frac{h}{l} = \frac{1}{\sqrt{3}} \quad (\because h = l/\sqrt{3})$$

$\Rightarrow \theta = 30°$, then we have

$$\sin\theta = \frac{1}{2} \text{ and } \cos\theta = \frac{\sqrt{3}}{2}.$$

Putting these values in equation (1.24), we have

$$\Rightarrow \frac{d\theta}{dt} = \frac{v\sin^3 30°}{h\cos 30°}$$

$$= \frac{v\left(\frac{1}{2}\right)^3}{h\left(\frac{\sqrt{3}}{2}\right)} = \frac{v}{8h} \times \frac{2}{\sqrt{3}}$$

$$\vec{\omega} = \frac{v}{4\sqrt{3}R}(-\hat{k}).$$

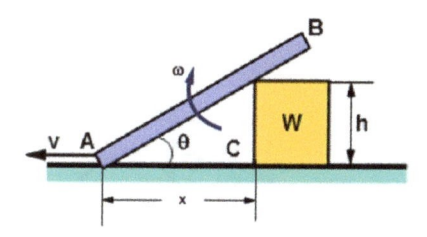

Problem 8 The bottom of the rod of length l is pulled with a constant velocity v as its other end slides over a semi-cylindrical wedge of radius R. The wedge is moved with a velocity of $2v$. Find the angular (a) velocity and (b) acceleration of the rod. $\theta = 37°$.

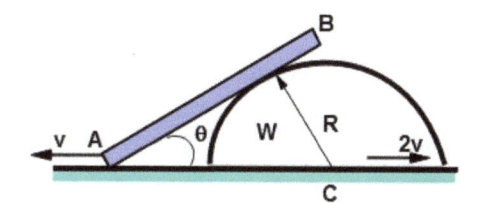

Solution

(a) Let $AC = x$. Since $\frac{R}{x} = \sin\theta$, we have

$$x = R\csc\theta$$

$$\Rightarrow \frac{dx}{dt} = R\frac{d}{dt}(\csc\theta)$$

$$\Rightarrow v_{rel} = -R\csc\theta \cot\theta \frac{d\theta}{dt}$$

$$\Rightarrow \frac{d\theta}{dt} = -\frac{v_{rel}}{R\cot\theta \csc\theta} = -\frac{v_{rel}\sin^2\theta}{R\cos\theta}.$$

1-45

Putting the $v_{rel} = v_A + v_C = v + 2v = 3v$ and $\theta = 37°$, we have

$$\Rightarrow \frac{d\theta}{dt} = -\frac{3v\sin^2(37°)}{R\cos(37°)} = (-27v/20)R \text{ (negative sign signifies a clockwise rotation).}$$

(b) The angular acceleration of the rod is

$$= \Rightarrow \frac{d^2\theta}{dt^2} = -\frac{d}{dt}\left(\frac{3v\sin\theta\tan\theta}{R}\right)$$
$$= (3v/R)\{(3/5)(5/4)2 + (3/4)(4/5)\}(- 29v/20R)$$
$$= (9963/1600)(v/R)^2;$$

positive sign signifies an anticlockwise angular acceleration.

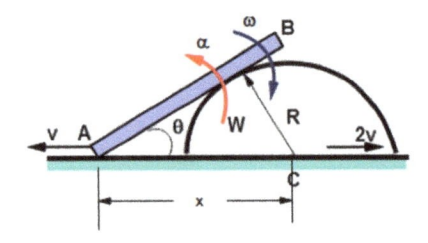

Problem 9 A point source P of light moves horizontally at a height H from ground level. A step MOQ of height h casts its shadow MQ in the light emanating from the point source. (a) Find the angular velocity of the ray POQ that sweeps the edge O of the step, as a function of θ.
  (i) if the point Q moves with a constant velocity $v_Q = u$;
  (ii) if the point P moves with a constant velocity v; (b) find the velocity of the point Q if the point moves with a velocity v at the given instant.

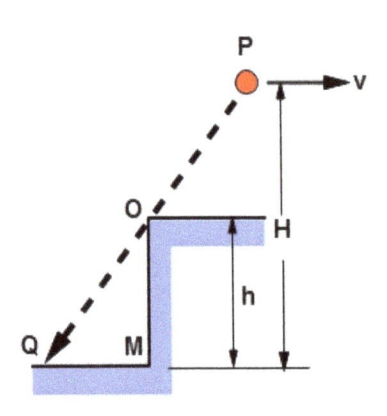

1-46

Solution

(a)

(i) Let MQ $= x$. Since $\frac{x}{h} = \tan\theta$, we have

$$x = h\tan\theta$$

$$\Rightarrow \frac{dx}{dt} = h\frac{d}{dt}(\tan\theta)$$

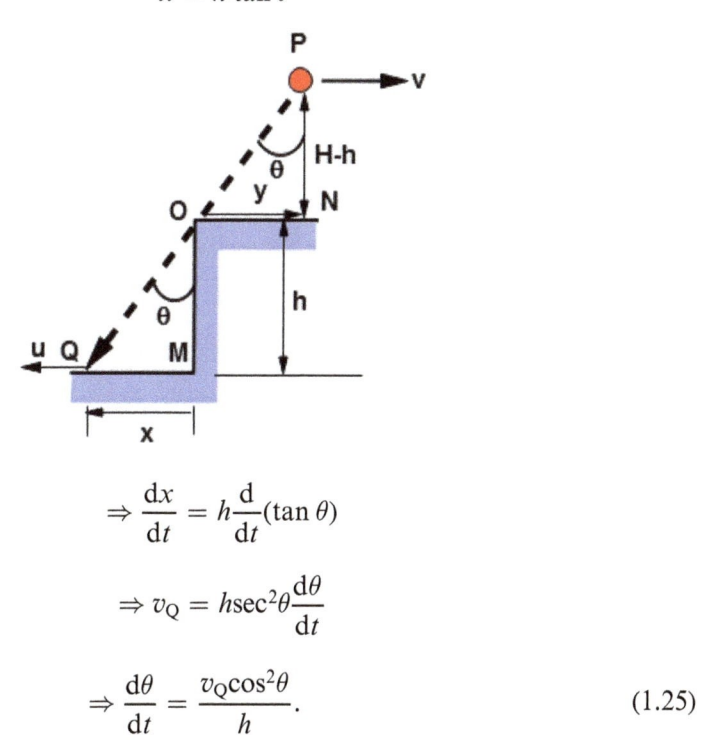

$$\Rightarrow v_Q = h\sec^2\theta\frac{d\theta}{dt}$$

$$\Rightarrow \frac{d\theta}{dt} = \frac{v_Q\cos^2\theta}{h}. \tag{1.25}$$

(ii) Let ON $= y$. Since $\frac{y}{H-h} = \tan\theta$ we have

$$y = (H-h)\tan\theta$$

$$\Rightarrow \frac{dy}{dt} = (H-h)\frac{d}{dt}(\tan\theta)$$

$$\Rightarrow v_P = v = (H-h)\sec^2\theta\frac{d\theta}{dt}$$

$$\Rightarrow \frac{d\theta}{dt} = \frac{v\cos^2\theta}{H-h}. \tag{1.26}$$

(b) Using equations (1.25) and (1.26) we have

$$\Rightarrow \frac{u}{h} = \frac{v}{H-h}. \tag{1.27}$$

Equation (1.27) can be obtained by using the properties of a triangle. Between two similar triangles PQM and OQN, we have

$$\Rightarrow \frac{ON}{MQ}\left(=\frac{y}{x}\right) = \frac{H-h}{h}$$

$$\Rightarrow \frac{dy/dt}{dx/dt} = \frac{H-h}{h}$$

$$\Rightarrow \frac{v}{u} = \frac{H-h}{h}$$

$$\Rightarrow u = \frac{h}{H-h} v.$$

Problem 10 In a slider crank arrangement, a small ball P is constrained to move along a rotating tube OP and a fixed circle of radius R. The tube is pivoted smoothly at O. The angle β made by OP with the horizontal (x-axis) changes with time as $\beta = kt^3$. Find the acceleration of the ball (a) as the function of time and (b) at $t = 1$ s. Put the radius of the circle as $R = 0.5$ m and $k = 1/3$.

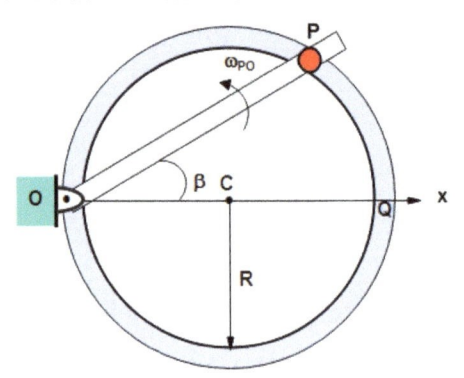

Solution

(a) It is given that

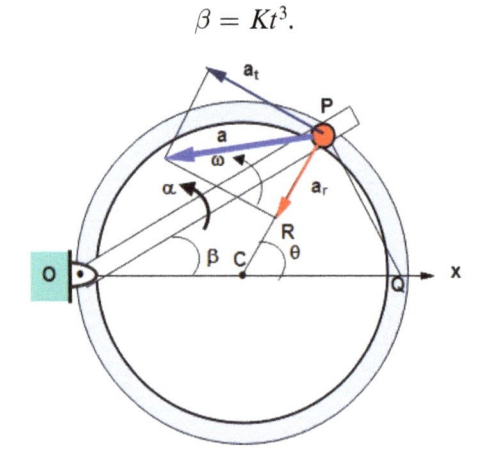

$$\beta = Kt^3.$$

By geometry $\theta = 2\beta$ and $\beta = kt^3$ (given), so we have
$$\theta = 2kt^3$$
$$\frac{d\theta}{dt} = w_1 = 6kt^2 \tag{1.28}$$

$$\Rightarrow \frac{d^2\theta}{dt^2} = 12kt$$
$$\Rightarrow \alpha_1 = 12kt. \tag{1.29}$$

Then the total acceleration of the ball is the sum of tangential and radial (centripetal) acceleration, given as

$$\Rightarrow a = \sqrt{(R\alpha_1)^2 + (Rw_1^2)^2}$$
$$\Rightarrow a = R\sqrt{\alpha_1^2 + w_1^4}. \tag{1.30}$$

From equations (1.28), (1.29), and (1.30), we have

$$\Rightarrow a = R\sqrt{(12kt)^2 + (6kt^2)^4}$$
$$\Rightarrow a = 12R\,kt\sqrt{1 + 9k^2t^6}. \tag{1.31}$$

Putting $R = 1/2$, $k = 1/3$, and $t = 1$, we have

$$\Rightarrow a = 12(1/2)(1/3)(1)\sqrt{1 + 9(1/3)^2(1)^6}$$
$$\Rightarrow a = 2\sqrt{2}\,\text{m s}^{-2}.$$

Problem 11 A laser torch is located at the origin O at a distance d from the vertical wall. At time $t = 0$, the laser beam falls perpendicular to a vertical wall situated at a distance d from the origin. (a) If the torch rotates at a constant clockwise angular velocity w, find (i) the velocity and (ii) the acceleration of the laser point P on the vertical wall. (b) If the velocity of the laser point P on the vertical wall moves with a constant upward velocity v, find the angular velocity of the torch.

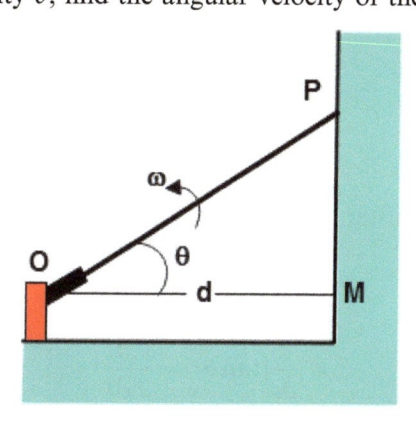

Solution

(a)

(i) If $\omega =$ constant, $\theta = \omega t$,

then $y = d \tan \theta$,

where

$$\theta = \omega t$$

$$\Rightarrow y = d \tan \omega t$$

$$\Rightarrow v_y = \frac{dy}{dt} = \omega d \sec^2 \omega t.$$

(ii) Differentiating velocity with time, we have

$$a_y = \frac{dv_y}{dt} = \omega d \frac{d}{dt}(\sec^2 \omega t)$$

$$= \omega \left\{ 2\sec\omega t, \frac{d}{dt}(\sec\omega t) \right\}$$

$$= 2d\omega^2 \sec^2 \omega t \tan \omega t.$$

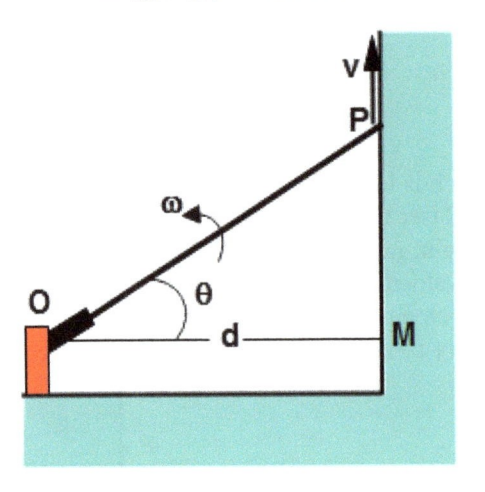

(b)

(i) If v is a constant, $PM = y = vt$

$$\Rightarrow \tan\theta = \frac{y}{d} = \frac{vt}{d}$$

$$\Rightarrow \frac{d}{dt}(\tan\theta) = \frac{v}{d}\frac{dt}{dt}$$

$$\Rightarrow \sec^2\frac{d\theta}{dt} = \frac{v}{d}$$

$$\Rightarrow \omega = \frac{v\cos^2\theta}{d}.$$

(ii) The angular acceleration is

$$\alpha = \frac{\omega d\omega}{d\theta} = \frac{1}{2}d(\omega^2)$$

$$= \frac{v^2}{2d^2}\frac{d}{d\theta}(\cos^4\theta)$$

$$= -\frac{2v^2}{d^2}\sin\theta\cos^3\theta.$$

Problem 12 Referring to a crank slider mechanism, P is constrained to move along the straight inclined groove of the cranks OP and QP with angular velocities ω_1 and ω_2, as shown in the figure. Write the relevant kinematical equations of P.

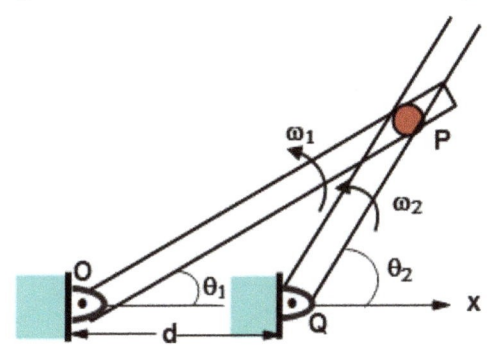

Solution

Let the components of the velocity of P along the x- and y-axes be v_x and v_y, respectively.

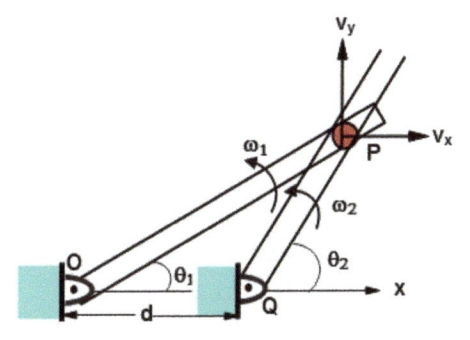

Resolving v_x and v_y along (parallel to) OP and QP, we have

$$\frac{dl_1}{dt} = v_x \cos\theta_1 + v_y \sin\theta_1 \tag{1.32}$$

$$\frac{dl_2}{dt} = v_x \cos\theta_2 + v_y \sin\theta_2. \tag{1.33}$$

Resolving v_x and v_y perpendicular (transverse) to OP and QP, we have

$$l_1\frac{d\theta_1}{dt} = l_1\omega_1 = v_x \sin\theta_1 - v_y \cos\theta_1 \tag{1.34}$$

$$l_2\frac{d\theta_2}{dt} = l_2\omega_2 = v_x \sin\theta_2 - v_y \cos\theta_2. \tag{1.35}$$

We have six unknown quantities given as ω_1, ω_2, v_x, v_y, $\frac{dl_1}{dt}$, and $\frac{dl_2}{dt}$.

If any two unknown quantities are given, we can solve for the other four unknown quantities using the last four equations.

Pure rotation: a pulley–belt system

Problem 13 Three discs of radii a, b, and c are pivoted at fixed points so that they can rotate with angular velocities of same direction. The vector sum of their angular velocities is given as $\vec{\omega} = \omega\,\vec{k}$. Find (a) the speed of any point of the belt connecting these three discs and (b) angular velocities of the discs.

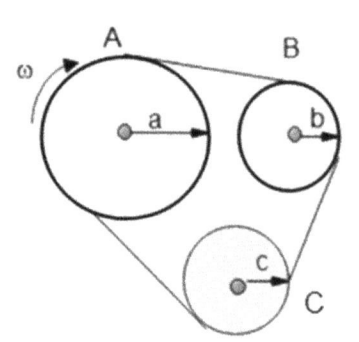

Solution

(a) Let the speed of any point on the string be v. Since the belt does not slip, the angular velocity of each will be directly proportional to the reciprocal of its radius. So we can write the angular velocities of A, B, and C, respectively, as follows:

$$\omega_1 = \frac{v}{a}, \quad \omega_2 = \frac{v}{b}, \quad \omega_3 = \frac{v}{c}.$$

Since $\omega = \omega_1 + \omega_2 + \omega_3$

$$\omega = v\left(\frac{1}{a} + \frac{1}{b} + \frac{1}{c}\right)$$

$$\Rightarrow v = \frac{abc\,\omega}{ab + bc + ac}.$$

(b) Then the angular velocity of A is

$$\omega_1 = \frac{v}{a} = \frac{bc\,\omega}{ab + bc + ac}.$$

Likewise, the angular velocities of B and C are given as follows:

$$\omega_2 = \frac{v}{b} = \frac{ac\,\omega}{ab + bc + ac}$$

$$\omega_3 = \frac{v}{c} = \frac{ab\,\omega}{ab + bc + ac}.$$

Problem 14 Two spinning discs A and B of radii a and b are connected by a belt. If the belt does not slip on the discs and the maximum magnitude of the relative acceleration between two points on the perimeters of the discs is a_o, find (a) speed of the belt, (b) the angular velocities ω_1 and ω_2 of the discs.

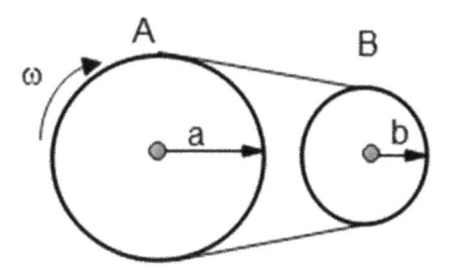

Solution

(a) Let the speed of any point on the string be v. Since the belt does not slip, the angular velocity of each will be directly proportional to the reciprocal of its radius. So we can write the angular velocities of A and B, respectively, as follows:

$$\omega_1 = \frac{v}{a} \text{ and } \omega_2 = \frac{v}{b}.$$

The relative acceleration between any two points of the discs will be maximum when the perpendicular distance between the points is minimum and maximum. Hence the chosen points can be P and Q. The relative acceleration between P and Q is the sum of their centripetal accelerations because they point opposite to each other:

$$a_{rel} = a_P + a_Q = u^2/a + v^2/b.$$

Putting $u = v$ as the belt does not slip with the discs, we have

$$a_{rel} = v^2/a + v^2/b = v^2(1/a + 1/b).$$

Putting $a_{rel} = a_o$, we have

$$\Rightarrow v^2(1/a + 1/b) = a_o$$

$$\Rightarrow v = \sqrt{a_o ab/(a+b)}.$$

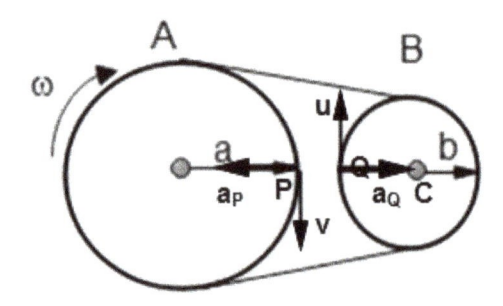

(b) Then the angular velocity of A is

$$\omega_1 = v/a = \sqrt{a_o b/a(a+b)}.$$

The angular velocity of B is

$$\omega_2 = v/b = \sqrt{a_o a/b(a+b)}.$$

Problem 15 A rod of length l is pivoted at the point O as shown in the figure. The angle θ made by the rod with the horizontal changes with time as $\theta = \theta_o \sin bt$. Find (a) the acceleration of a point P of the rod as a function of time, (b) the acceleration of point P at the mean and extreme positions, and (c) the maximum angular frequency of revolution of the point P. Assume that at time $t = 0$, $\theta = 0$.

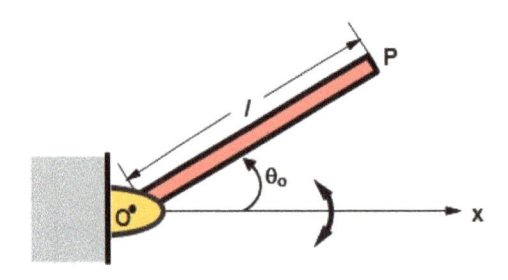

Solution

(a) The point P oscillates obeying the relation
$$\theta = \theta_o \cos bt.$$

The angular velocity of P is
$$\omega_{\text{rev}} = \frac{d\theta}{dt} = -b\theta_o \sin bt.$$

The angular acceleration is
$$\alpha = \frac{d\omega_{\text{rev}}}{dt} = -b^2\theta_o \cos bt.$$

The point P moves in a circle of radius l with variable angular velocity and angular acceleration.

Then the radial acceleration of P is
$$a_r = l\omega_{\text{rev}}^2 = -l\{b\theta_o \sin bt\}^2.$$

Then the tangential acceleration of P is
$$a_t = l\alpha = -lb^2\theta_o \cos bt.$$

Then the net acceleration of P at any point of its circular path is given as
$$a_{\text{total}} = \sqrt{a_t^2 + a_r^2}$$
$$= \sqrt{\{lb^2\theta_o \cos bt\}^2 + \{l(b\theta_o \sin bt)^2\}^2}$$
$$= lb^2\theta_o\sqrt{\cos^2 bt + l^2b^2\theta_o^2\sin^4 bt}.$$

(b) At the mean position, putting $\cos bt = 1$, we have
$$a_{\text{total}} = lb^2\theta_o.$$

At the extreme position, putting $\cos bt = 0$, we have
$$a_{\text{total}} = lb^3\theta_o^2.$$

(c) The maximum angular frequency of revolution is
$$\{\omega_{\text{rev}}\}_{\max} = \{-b\theta_o \sin bt\big|_{\max} = b\theta_o.$$

1-55

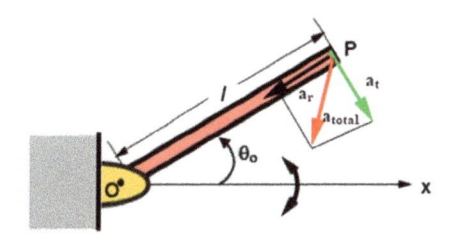

Problem 16 A disc of radius R is rotating about an axis passing through its center of mass perpendicular to its plane. The axis is fixed and the disc rotates with a constant angular acceleration α. Find the acceleration of a point of the perimeter of the disc as the function of time t.

Solution

If the disc rotates with a constant acceleration α, its angular distance is

$$\theta = \alpha t^2.$$

The angular velocity is

$$\frac{d\theta}{dt} = \alpha t.$$

The angular acceleration is

$$\frac{d^2\theta}{dt^2} = \alpha \text{ (given)}.$$

Then the tangential and radial acceleration of the point P can be respectively given as $a_t = R\alpha$ and $a_r = R\omega^2$. Then the total acceleration of P is $\vec{a_P} = \vec{a_t} + \vec{a_r}$.

$$\Rightarrow a_P = \sqrt{a_t^2 + a_r^2}; \text{ where } a_t = R\alpha \text{ and } a_r = R\omega^2$$

$$= R(\alpha t)^2 = R\alpha^2 t^2, \text{ then we have}$$

$$a_P = \sqrt{(R\alpha)^2 + (R\alpha^2 t^2)^2}$$

$$= R\alpha\sqrt{1 + \left(\frac{\alpha t^2}{R}\right)^2}.$$

Problem 17 Three fixed pulleys A, B, and C are connected by two belts as shown in the figure. If the belts do not slip on the pulleys and the angular velocity of the pulley A is ω, find the angular velocities of the pulleys B and C and the speed of the connecting belts.

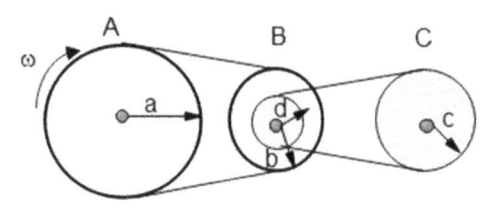

Solution

The angular velocity of the disc A is ω (given). Since it is connected with the pulley B by the belt whose speed at any point is

$$v_1 = a\omega,$$

the angular velocity of the disc B is

$$\omega_1 = v_1/b. \tag{1.36}$$

The stepped pulley B is connected with the pulley C by another belt whose speed is equal to v_2, given as

$$v_2 = d\omega_1. \tag{1.37}$$

Then the angular velocity of the pulley C is

$$\omega_2 = v_2/c. \tag{1.38}$$

Using the last three equations, we have

$$\omega_1 = a\omega/b$$

$$v_2 = ad\omega/b$$

$$\omega_2 = ad\omega/bc.$$

Example 18 A disc spinning about a fixed axis with an initial angular speed ω_o experiences an angular acceleration given as $\alpha = -k\sqrt{\omega}$, where k is a positive constant. Find the total number of rotations of the disc until its final speed is equal to the initial speed.

Solution

Put the given $\alpha = -k\sqrt{\omega}$ in the equation $\alpha = \omega \frac{d\omega}{d\theta}$ to have

$$\omega \frac{d\omega}{d\theta} = -k\sqrt{\omega}.$$

After separating the variables, we have

$$\sqrt{\omega}\, d\omega = -k\, d\theta.$$

At $\theta = 0$, $\omega = \omega_o$, and at $\theta = \varnothing$, $\omega = 0$, as the disc will stop momentarily before reversing its sense of rotation.

Then, integrating both sides,

$$\int_{\omega_o}^{0} \sqrt{\omega}\, d\omega = -k \int_{0}^{\varnothing} d\theta.$$

Evaluating the integral and simplifying the factors, we have

$$\varnothing = 2(\omega_o)^{\frac{3}{2}}/3k.$$

As the disc reverses its angular velocity from zero to $+\omega_o$, it must cover the same angular distance $\varnothing$. So the total angular distance is

$$\theta_{\text{total}} = 2\varnothing = 4(\omega_o)^{\frac{3}{2}}/3k.$$

Then the total number of rotations of the disc will be given as

$$N = \theta_{\text{total}}/2\pi = 2(\omega_o)^{\frac{3}{2}}/3\pi k.$$

Note: In this example the net angular displacement is zero but the net angular distance is nonzero. Hence, the average speed will be nonzero even though the average velocity will be zero.

Combined motion

Problem 19 In a slider crank system, at the given instant, angular velocities ω_1 and ω_2 of the rods PQ and RQ, respectively if at the given instant the velocity of the block R which is constrained to move along the x-axis (the horizontal surface) is given as v (pointing towards the right). Assume that the rods PQ and QR make acute angles θ_1 and θ_2, respectively, with the horizontal. Assume that $PQ = l_1$ and $RQ = l_2$.

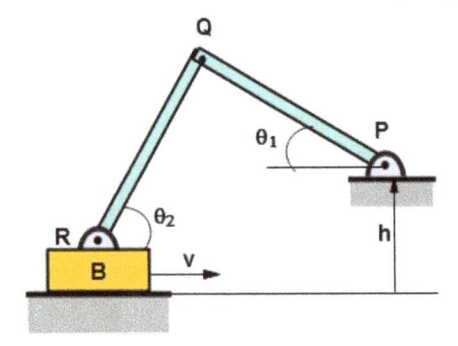

Solution

Let the horizontal positions of Q relative to P and R relative to Q be x_1 and x_2, respectively. Then the horizontal position of R relative to P is

$$x = x_1 + x$$

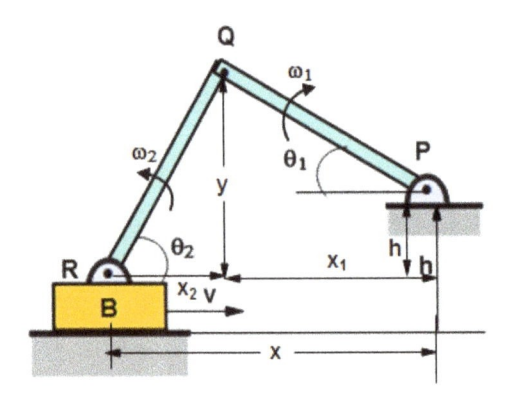

$$\Rightarrow x = l_1 \cos \theta_1 + l_2 \cos \theta_2.$$

Differentiating both sides with time,

$$\frac{dx}{dt} = -\left(l_1 \sin \theta_1 \frac{d\theta_1}{dt} + l_2 \sin \theta_2 \frac{d\theta_2}{dt} \right)$$

$$\Rightarrow v = l_1\omega_1 \sin\theta_1 + l_2\omega_2 \sin\theta_2. \tag{1.39}$$

The vertical positions

$$y = l_1 \sin\theta_1 + h = l_2 \sin\theta_2$$

$$\Rightarrow l_1 \cos\theta_1 \frac{d\theta_1}{dt} = l_2 \cos\theta_2 \frac{d\theta_2}{dt}$$

$$\Rightarrow l_1\omega_1 \cos\theta_1 = l_2\omega_2 \cos\theta_2. \tag{1.40}$$

Using equations (1.39) and (1.40)

$$v\left(=\frac{dx}{dt}\right) = l_1\omega_1 \sin\theta_1 + l_2 \sin\theta_2\left(\frac{l_1\omega_1 \cos\theta_1}{l_2 \cos\theta_2}\right)$$

$$\Rightarrow v = l_1\omega_1\left(\frac{\sin\theta_1 \cos\theta_2 + \cos\theta_1 \sin\theta_2}{\cos\theta_2}\right)$$

$$\Rightarrow \omega_1 = \frac{v \cos\theta_2}{l_1 \sin(\theta_1 + \theta_2)}, \quad \omega_2 = \frac{v \cos\theta_1}{l_2 \sin(\theta_1 + \theta_2)}.$$

Problem 20 Two rods are hinged smoothly to form a composite L-shaped rod. The upper rod is pivoted at O. The angular velocities and angular accelerations of the rods are given. Find the velocity of Q.

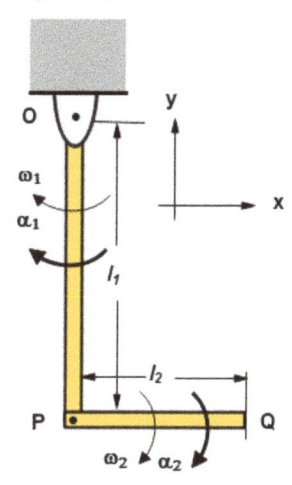

Solution

(a) The velocity of Q is

$$\vec{v}_Q = \vec{v}_{QP} + \vec{v}_P$$

$$\vec{v}_Q = \vec{v}_{QP} + \vec{v}_P$$
$$= l_2\omega_2(-\hat{j}) + l_1\omega_1(-\hat{i})$$
$$= -\{l_2\omega_2\hat{j} + l_1\omega_1\hat{i}\}.$$

(b) The acceleration of P relative to O is
$$(\vec{a}_{PO}) = (\vec{a}_{PO})_t + (\vec{a}_{PO})_r = -l_1\alpha_1\hat{i} + l_1\omega_1^2\hat{j}$$

The acceleration of Q relative to P is
$$(\vec{a}_{QP}) = (\vec{a}_{QP})_r + (\vec{a}_{QP})_t = -l_2\omega_2^2\hat{i} - l_2\alpha_2\hat{j}$$

The acceleration of Q relative to O is
$$(\vec{a}_{QO}) = (\vec{a}_{QP}) + (\vec{a}_{PO})$$

Using last three equations, we have

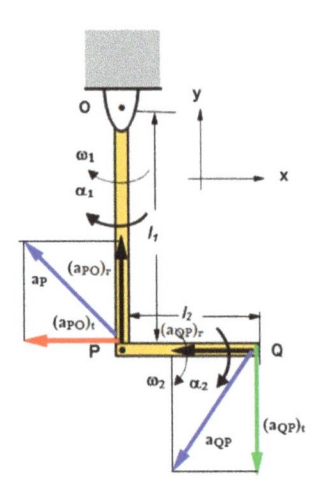

$$(\vec{a}_{QO}) = -l_2\omega_2^2\hat{i} - l_2\alpha_2\hat{j} - l_1\alpha_1\hat{i} + l_1\omega_1^2\hat{j}$$
$$= -(l_1\alpha_1 + l_2\omega_2^2)\hat{i} + (l_1\omega_1^2 - l_2\alpha_2)\hat{j}$$

Problem 21 In the following guided slider crank system the rod PQ has an angular velocity of 3 rad s^{-1} as shown in the figure. Find the angular velocity of the rod QR. Put PQ/QR = a/b = 2/3 and the angle made by QR with horizontal is 60°.

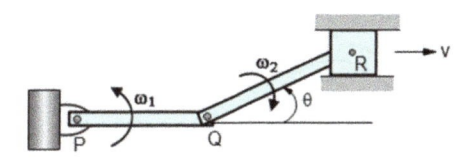

Solution

The velocity of Q is $a\omega_1$ and the velocity of R is v. Then, resolving the velocity along the rod QR and equating them, we have

$$\omega_1 a \sin\theta = v \cos\theta. \qquad (1.41)$$

Resolving the velocities at Q and R perpendicular to the rod QR,

$$v_1 + v_2 = b\omega_2$$

$$\Rightarrow \omega_1 a \cos\theta + v \sin\theta = b\omega_2. \qquad (1.42)$$

From equations (1.41) and (1.42),

$$\omega_1 a \cos\theta + \left(\omega_1 a \frac{\sin\theta}{\cos\theta}\right)\sin\theta = b\omega_2$$

$$\Rightarrow \omega_1 \frac{a(\cos^2\theta + \sin^2\theta)}{\cos\theta} = b\omega_2$$

$$\Rightarrow \omega_2 = \frac{\omega_1 a}{b \cos\theta} = 3(2/3)(2) = 4 \text{ rad s}^{-1}.$$

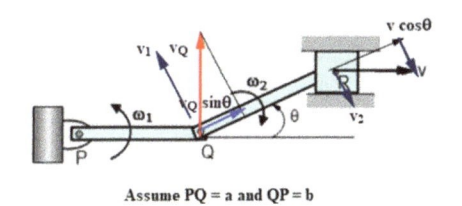

Assume PQ = a and QP = b

Kinematics of rolling

Problem 22 A disc of radius R rolls without sliding on a fixed horizontal surface. The velocity of the center O of the disc is v. Let us consider a point P at the lowest point of the disc. At $t = 0$, let the point P coincide with the origin O. Find (a) the locus equation of the point P and (b) the velocity of the point P, (c) the total distance covered during one complete rotation of the disc, and (d) the radius of curvature of the path traced by the point P, at the given angular position, after a time t.

1-61

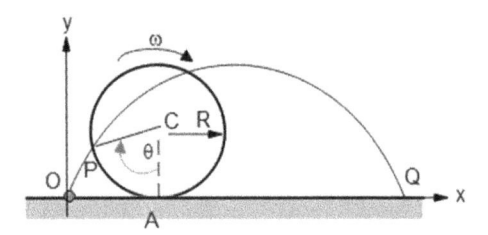

Solution

(a) Let us mark the lowest point P by a black marker. At time $t = 0$, the point P was lying at the origin. At time t, the center C of the disc moves through a distance vt. At the same time the point P revolves by an angle of θ relative to C during time t. As the disc rolls without sliding, the length of the arc AP is equal to OA. Then we can write $R\theta = vt$ and $\theta = \omega t$.

The coordinates of P can be given as

$$x = vt - R\sin\theta$$

$$y = R(1 - \cos\theta).$$

Putting $v = R\omega$ and $\theta = \omega t$, we have

$$x = R\omega t - R\sin\omega t = R(\theta - \sin\theta)$$

$$y = R(1 - \cos\omega t) = R(1 - \cos\theta).$$

So the locus equation of P is a cycloid.

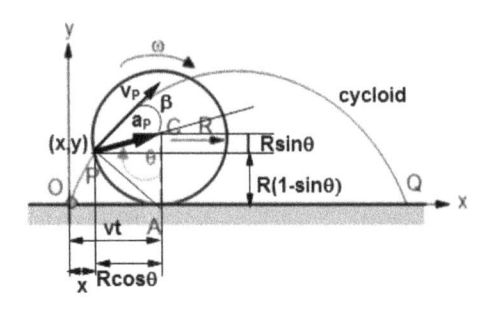

(b) The horizontal velocity of P is

$$v_x = \frac{dx}{dt} = R\left(\frac{d\theta}{dt} - \cos\theta\frac{d\theta}{dt}\right) = R\omega(1 - \cos\theta)$$

$$\Rightarrow v_x = v(1 - \cos\theta)$$

1-62

$$v_y = \frac{dy}{dt} = R\sin\theta\frac{d\theta}{dt} = \omega R\sin\theta = v\sin\theta \Rightarrow v_P = \sqrt{v_x^2 + v_y^2} = v\sqrt{(1-\cos\theta)^2 + \sin^2\theta}$$

$$= 2v\sin\frac{\theta}{2} = 2v\sin\frac{vt}{R}.$$

(c) The distance covered is

$$D = \int v_P \, dt$$
$$= \int_0^{T=\frac{2\pi R}{v}} 2v\sin\frac{vt}{2R} dt$$
$$= 8R.$$

(d) The radius of curvature is given as

$$r = \frac{v_P^2}{a_\perp}.$$

The velocity of P at time t is

$$\vec{v}_P = v(1-\cos\theta)\hat{i} + v\sin\theta\hat{j}.$$

We can show that the velocity of P is perpendicular to OP.
The acceleration of P at time t is

$$\vec{a}_P = \frac{v^2}{R}(\sin\theta\hat{i} + \cos\theta\hat{j}).$$

The acceleration is radially inward.
Then we can write

$$\vec{a}_P \cdot \vec{v}_P = a_P v_P \cos\beta$$

$$\frac{v^3}{R}(1-\cos\theta)\sin\theta + \frac{v^3}{R}\sin\theta\cos\theta$$

$$= \frac{v^2}{R}\left\{2v\sin\frac{\theta}{2}\right\}\cos\beta$$

$$\Rightarrow \frac{v^3}{R}\sin\theta = \frac{2v^3}{R}\sin\frac{\theta}{2}\cos\beta$$

$$\Rightarrow \cos\beta = \frac{\sin\theta}{2\sin\frac{\theta}{2}} = \cos\frac{\theta}{2}$$

$$\Rightarrow \beta = \frac{\theta}{2}$$ (we can also prove it geometrically in one step referring to the last figure).

Then $a_\perp = a_P \sin\beta = \left(\frac{v^2}{R}\right)\sin\frac{\theta}{2}.$

1-63

Hence the radius of curvature at P is $r = \dfrac{v_P^2}{a_\perp} = \dfrac{\left(2v \sin \frac{\theta}{2}\right)^2}{\frac{v^2}{R} \sin \frac{\theta}{2}}$

$$\Rightarrow r = 4R \sin \frac{\theta}{2}.$$

Problem 23 A disc of radius R rolls on a horizontal surface. The center of the disc moves with a constant acceleration $\vec{a} = a\hat{i}$. Find (a) the velocities and (b) the accelerations of the points A, B, C, and D of the disc as the function of time.

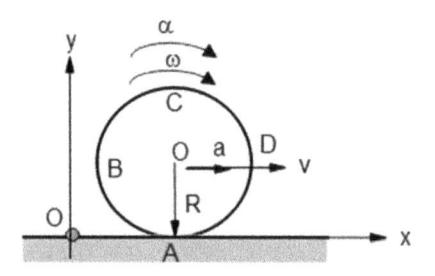

Solution

(a) For pure rolling of a body on a fixed surface $v = R\omega$ and $v = at$.
The velocity of A is given as
$$\vec{v}_A = \vec{v}_{AO} + \vec{v}_O$$
$$= -R\omega\,\hat{i} + v\hat{i}$$
$$= (-R\omega + v)\,\hat{i} = 0.$$

The velocity of B is given as
$$\vec{v}_B = \vec{v}_{BO} + \vec{v}_O$$
$$= R\omega\,\hat{j} + v\hat{i}$$
$$= v\hat{j} + v\hat{i} = v\left(\hat{i} + \hat{j}\right) = at\left(\hat{i} + \hat{j}\right)$$

The velocity of D is given as
$$\vec{v}_C = \vec{v}_{CO} + \vec{v}_O$$
$$= R\omega\hat{i} + v\hat{i} = v\hat{i} + v\hat{i} = 2v\hat{i} = 2at\hat{i}$$

The velocity of D is given as
$$\vec{v}_D = \vec{v}_{DO} + \vec{v}_O = -R\omega\,\hat{j} + v\hat{i} = -v\hat{j} + v\hat{i} = v(\hat{i} - \hat{j}) = at(\hat{i} - \hat{j})$$

(b) For pure rolling on a fixed surface
$$a_O = a = R\alpha.$$
For uniform accelerated motion of the center of mass,
$$v = at.$$
The acceleration of the points A, B, C and D are given as following:
Put $R\omega^2 = v^2/R$, where $v = at$ and $R\alpha = a$.

$$\vec{a}_A = \vec{a}_{AO} + \vec{a}_O = -R\alpha\hat{i} + R\omega^2\hat{j} + a\hat{i} = (a - R\alpha)\hat{i} + R\omega^2\hat{j}$$
$$= +R\omega^2\hat{j} \text{ because } a = R\alpha$$
So, $\vec{a}_A = \left(\frac{v^2}{R}\right)\hat{j} = \left(a^2t^2/R\right)\hat{j}$

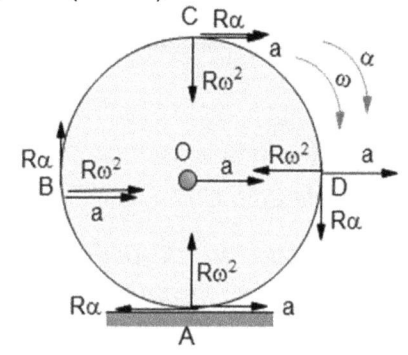

$$\vec{a}_B = \vec{a}_{BO} + \vec{a}_O = (R\omega^2 + a)\hat{i} - R\alpha\hat{j} = (R\omega^2 + a)\hat{i} - a\hat{j}$$
because $a = R\alpha$
$$= \left\{\left(\frac{v^2}{R}\right) + a\right\}\hat{i} - a\hat{j} = \left\{\frac{a^2t^2}{R} + a\right\}\hat{i} - a\hat{j}$$

$$\vec{a}_C = \vec{a}_{CO} + \vec{a}_O = (R\alpha + a)\hat{i} - R\omega^2\hat{j} = 2a\hat{i} - R\omega^2\hat{j}$$
(because $a = R\alpha$) $= 2a\hat{i} - R\omega^2\hat{j}$
$$= \left(2a\hat{i} - \frac{v^2}{R}\hat{j}\right) = \left(2a\hat{i} - \frac{a^2t^2}{R}\hat{j}\right)$$

$$\vec{a}_D = \vec{a}_{DO} + \vec{a}_O = (a - R\omega^2)\hat{i} - R\alpha\hat{j} = (R - R\omega^2 + a)\hat{i} - a\hat{j}$$
because $a = R\alpha$
$$= \left(a - \frac{v^2}{R}\right)\hat{i} - a\hat{j} = \left(a - \frac{a^2t^2}{R}\right)\hat{i} - a\hat{j}$$

Problem 24 A wheel of radius R is rolling on a horizontal surface with a constant velocity v in the rain. The water particles leave the wheel and fly as freely falling bodies under gravity. Find the maximum possible height attained by the water particles.

Solution

Let the water particle P leave the wheel at an angular position θ as shown in the figure. As discussed earlier, the vertical velocity of the particle P is
$$\vec{v}_y = v \sin\theta \hat{j}.$$

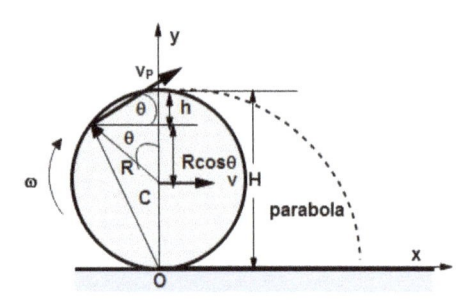

The horizontal velocity of P is
$$\vec{v}_x = (v \cos\theta + v)\hat{i}.$$

The height attained by the particle from C is
$$H = R\cos\theta + h = R\cos\theta + \frac{v_y^2}{2g}$$
$$= R\cos\theta + \frac{v^2 \sin^2\theta}{2g}. \tag{1.43}$$

For M to be maximum $\frac{dH}{d\theta} = 0$
$$\Rightarrow 0 = -R\sin\theta + \left\{\frac{v^2}{2g}\right\} 2\sin\theta\cos\theta$$
$$\Rightarrow \cos\theta = \frac{gR}{v^2}.$$

Then, putting $\cos\theta = \frac{gR}{v^2}$ and $\sin\theta = \sqrt{1 - \left(\frac{gR}{v^2}\right)^2}$ in equation (1.43), we have
$$H = \frac{v^2 \sin^2\theta}{2g} + R\cos\theta$$
$$= \frac{v^2}{2g}\left\{1 - \left(\frac{g^2 R^2}{v^4}\right)\right\} + R\left\{\frac{gR}{v^2}\right\}$$
$$= \frac{v^2(v^4 - g^2 R^2)}{2g \, v^4} + \frac{gR^2}{v^2}.$$

$$H_{max} = \frac{v^4 - g^2 R^2}{2gv^2} + \frac{gR^2}{v^2}.$$

1-66

$$\Rightarrow H_{\max} = \frac{v^2}{2g} + \frac{gR^2}{2v^2}.$$

Problem 25 A disc of radius R is rolling on a plank while the plank moves with a velocity $-2v$ and acceleration a. If the point B of the disc moves with a velocity u, find (a) the angular velocity of the disc and (b) the acceleration of B. Put $R = 2r$, $R = 0.5$ m, $a = 2.5$ m s^{-2}, and $v = 0.5$ m s^{-1} and $\alpha = 2$ rad s^{-2}.

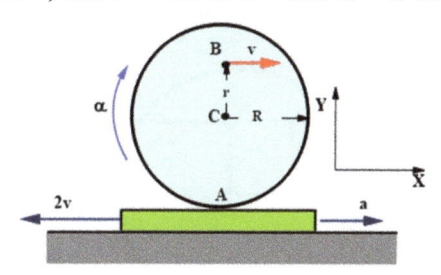

Solution

(a) As the disc rolls on the plank, the velocity of B is

$$\vec{\omega} = -\frac{2v + u}{R + r}\hat{k} = -\frac{2v + v}{R + R/2}\hat{k}$$

$$= -\frac{2v}{R}\hat{k} = -\frac{2(0.5)}{0.5}\hat{k} = -2\hat{k}$$

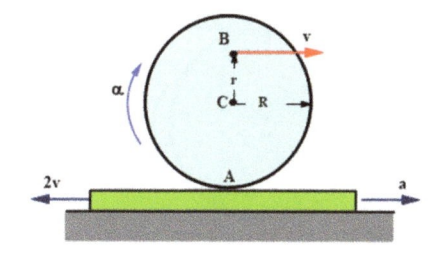

(b) The vertical (radial) acceleration of B is

$$\vec{a}_y = -r\omega^2\hat{j}$$

The horizontal acceleration of B is

$$\vec{a}_x = \{(R + r)\alpha + a\}\hat{i}$$

Then, the total acceleration of B is

$$\vec{a}_B = \vec{a}_x + \vec{a}_y$$
$$= \{(R + r)\alpha + a\}\hat{i} - r\omega^2\hat{j}$$

1-67

$$= \{(R + R/2)\alpha + a\}\hat{i} - (R/2)\omega^2\hat{j}$$

$$= \frac{1}{2}\{(3R\alpha + 2a)\hat{i} - R\omega^2\hat{j}\}$$

$$= \frac{1}{2}\left[\{3(1/2)(2) + 2(2.5)\}\hat{i} - (1/2)(2)^2\hat{j}\right]$$

$$= 4\hat{i} - \hat{j} \text{ m s}^{-2}$$

Problem 26 Three uniform cylinders of radii 0.6R, R, and 1.5R are loaded on to each other by three planks, as shown in the figure. If the planks 1, 2, and 3 move with velocities $+2v$, $-v$, and $+3v$, respectively, and the cylinders roll without sliding. Find the angular velocities of the cylinders and the velocities of the center of mass of the cylinders.

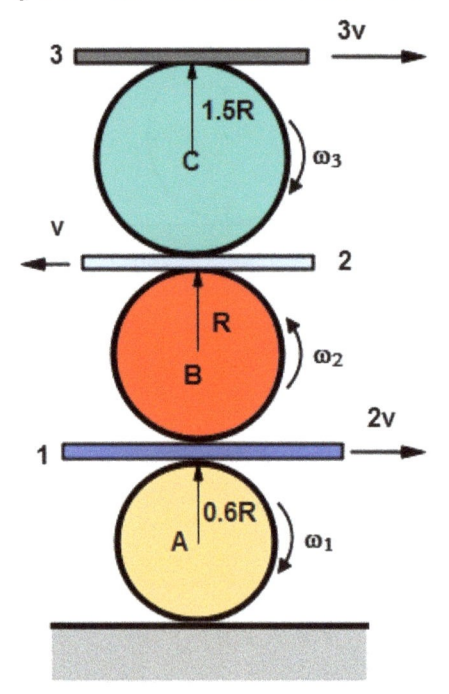

Solution

As the cylinders roll without sliding, the velocities of the top and bottom of the lowest cylinder are $2v$ and 0, respectively. So the angular velocity of the lowest cylinder is

$$\vec{\omega}_1 = \frac{2v + 0}{2 \times \frac{3}{5}R} = \frac{-5}{3}v\hat{k}.$$

The velocity of A is

$$\vec{v}_A = \frac{2v + 0}{2}\hat{i} = v\hat{i}.$$

The velocity of the top and bottom of the middle cylinder are $-v$ and $2v$, respectively. So the angular velocity of the cylinder is

$$\vec{\omega}_2 = \frac{v + 2v}{2R}\hat{k} = \frac{3v}{2R}\hat{k}.$$

The velocity of B is

$$v_B = \frac{-v + 2v}{2}\hat{i} = \frac{v}{2}\hat{i}.$$

The velocity of the top and bottom of the topmost cylinder are $3v$ and $-v$, respectively. So the angular velocity of the cylinder is

$$\vec{\omega}_3 = -\frac{3v + v}{2 \times \frac{3}{2}R}\hat{k} = -\frac{4v}{3R}\hat{k}.$$

The velocity of C is

$$v_C = \frac{3v - v}{2}\hat{i} = v\hat{i}.$$

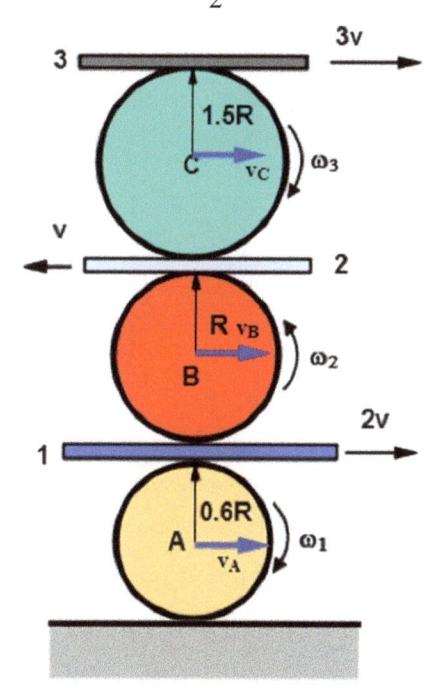

Instantaneous axis of rotation

Problem 27 A rod of length l is leaning against a horizontal surface and an inclined plane. The rod is undergoing a combined motion in a vertical plane. When the rod makes angles θ and β with the planes, as shown in the figure, the lowest point of the rod moves towards the right with a velocity v. Find the angular velocity of the rod.

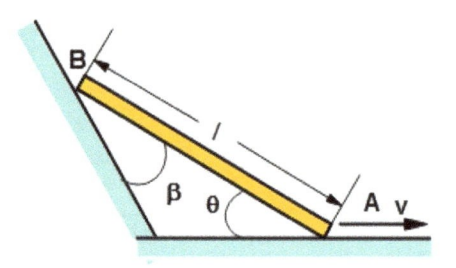

Solution
Method 1
If the end A of the rod AB moves with a velocity v, let the other end B of the rod move with a velocity v'. Resolving the velocities along the rod and equating them, we have

$$v \cos \theta = v' \cos \beta$$

$$\Rightarrow v' = v \cos \theta / \cos \beta. \tag{1.44}$$

Since the components of v and v' perpendicular to the rod are oppositely directed, the relative velocity between the end A and B is

$$v_{\text{rel}} = v \sin \theta + v' \sin \beta.$$

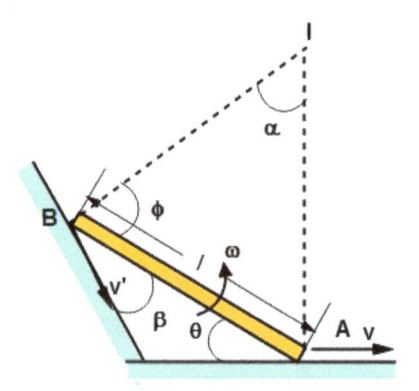

Then the angular velocity of the rod is

$$\omega = v_{\text{rel}}/l = (v \sin \theta + v' \sin \beta)/l. \tag{1.45}$$

Using equations (1.44) and (1.45), we have
$$\omega = \{v\sin\theta + (v\cos\theta/\cos\beta)\sin\beta\}/l$$
$$= v(\sin\theta\cos\beta + \cos\theta\sin\beta)/l\cos\beta$$
$$= \frac{v\sin(\theta+\beta)}{l\cos\beta}.$$

Method 2

We drop two perpendiculars to the velocities v and v' at A and B, respectively. They intersect at I, which is known as the IAR. In the triangle ABI using the triangle property (sine rule), we have
$$\Rightarrow AI/\sin\phi = AB/\sin\alpha$$
$$\Rightarrow AI = AB\sin\phi/\sin\alpha.$$
Putting $AB = l$, $\alpha = \theta + \beta$, and $\varphi = 90° - \beta$, we have
$$AI = l\sin(90° - \beta)/\sin(\theta + \beta)$$
$$\Rightarrow AI = l\cos\beta/\sin(\theta + \beta).$$
Since the velocity of instantaneous axis of rotation I is zero, the angular velocity of the rod can be written as
$$\omega = v/AI = v/\{l\cos\beta)/\sin(\theta+\beta)\}$$
$$\omega = \frac{v\sin(\theta+\beta)}{l\cos\beta}.$$

Problem 28 A sphere is spinning about the horizontal axis OQ with an angular velocity $\vec{\omega}_1$. The axis OQ rotates about the vertical axis AB with an angular velocity $\vec{\omega}_1$. Find the (a) angular velocity of a point P, say, on the vertical diameter of the sphere about the point O (b) angular acceleration of the sphere.

Solution

(a) The angular velocity $\vec{\omega}_P$ of a point P on the vertical diameter of the sphere about the point O is equal to the angular velocity $\vec{\omega}_1$ of a point on the vertical diameter of the sphere about the axis OQ plus the angular velocity $\vec{\omega}_2$ of OQ about the vertical axis AB. So, we can write
$$\vec{\omega}_P = \vec{\omega}_1 + \vec{\omega}_2$$

Since $\vec{\omega}_1$ and $\vec{\omega}_2$ are mutually perpendicular, the magnitude of $\vec{\omega}_P$ can be given as
$$\omega_P = \sqrt{\omega_1^2 + \omega_2^2}$$

The angle of orientation of $\vec{\omega}_P$ with upward vertical or $\vec{\omega}_2$ is given as
$$\phi = \tan^{-1}\left(\frac{\omega_1}{\omega_2}\right)$$

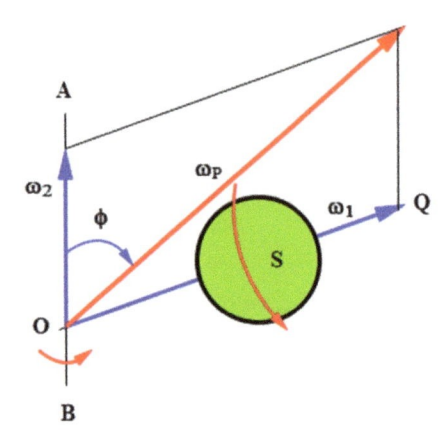

(b) Let the sphere change its spin (angular velocity from $\vec{\omega}_1(t)$ to from $\vec{\omega}_1(t + dt)$) during a time dt. As the sphere rotates about the axis OQ, its angular velocity is radially outward. The tip of the angular velocity vector $\vec{\omega}_1$ rotates with an angular velocity $\vec{\omega}_2$ which is equal to the angular velocity of the axis OQ about the fixed vertical axis AB. We can see from the vector diagram that the change in angular velocity $\vec{\omega}_1$, that is, $d\vec{\omega}_1$ is tangential because the angular velocity $\vec{\omega}_1$ is directed radially outward.

So, the angular accelertaion

$$\vec{\alpha} = \frac{d\vec{\omega}_1}{dt} = \frac{d\omega_1}{dt}\hat{\varnothing}$$

tangential to the circular dotted path traced by the sphere as shown in the figure. Then by putting

$$d\omega_1 = \omega_1 d\theta$$

we have the angular acceleration

$$\vec{\alpha} = \frac{d\vec{\omega}_1}{dt} = \frac{d\omega_1}{dt}\hat{\varnothing} = \frac{\omega_1 d\theta}{dt}\hat{\varnothing},$$

where $\frac{d\theta}{dt} = \omega_2$. Then, we have the final expression

$$\vec{\alpha} = \omega_1 \omega_2 \hat{\varnothing}$$

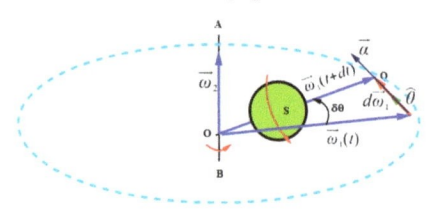

This signifies that the angular acceleration due to the change in the direction of angular velocity points tangentially to the path (dotted blue circle) traced by the tip of the spin angular velocity vector $\vec{\omega}_1$.

IOP Publishing

Problems and Solutions in Rotational Mechanics

Pradeep Kumar Sharma

Chapter 2

Torque and angular momentum for a point mass

2.1 Torque

2.1.1 Definition

- Torque is a Greek word.
- It literally means a 'twist'.
- It is a twisting/turning/rotational effect of a force.
- It is also called the moment of force.
- It is an axial (pseudo)vector.

Illustration

The most familiar example to understand the concept of torque is the rotation or turning of a stationary rod (or door) about a hinge by applying a force $\vec{F}$, as shown in the following figure.

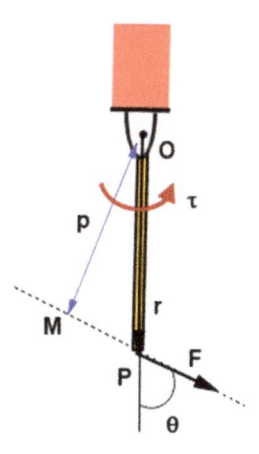

A force $\vec{F}$ acts on the hinged rod to cause a torque.

We can see that the force $\vec{F}$ produces a twist/turning/rotation of the rod which is called a *torque* of the force F on the rod about the hinge O. The greater the torque is, the more quickly the rod picks up its angular velocity. If the angular velocity of the rod does not change, the torque acting on the rod is zero.

To produce a torque on the rod, first we have to apply a force F on it. The component of the force along the rod, that is $F_\parallel = F \cos \theta$, is balanced partially by the hinge force N acting on the rod. The component of the force ($F_n = F_\perp = F \sin \theta$) normal to the rod generates shearing forces at each point of the rod which eventually produces a twisting/turning/rotation of the rod. Furthermore, the line of action of the force must not pass through the hinge. Otherwise the rod will not rotate even if you apply a huge force on the rod. In this sense, by merely applying a force we cannot cause rotation of the rod. In addition to a nonzero force, the following extra conditions must be met.

The force F must:
- Have a component $F_n(=F \sin \theta)$
  normal to the rod.
- Act at a nonzero distance r from the hinge.

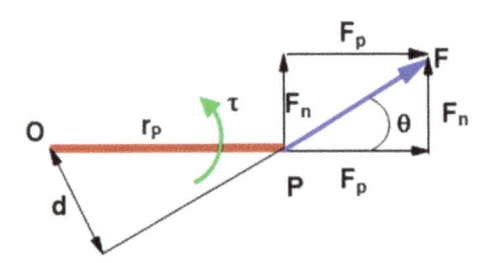

A nonzero torque is possible when r_P ($= r$) and $F_\perp$ or F_n are nonzero.

2.1.2 Factors governing the torque

Considering the above two effects, the torque of a force is governed by the following three factors:
- How much force is acting? The magnitude of the force is F.
- In which direction the force is acting? The direction of the force is given by $\sin \theta$.
- Where is the force is acting? The position vector of the point P of application of the force F has a magnitude r.

2.1.3 The mathematical expression of torque

2.1.3.1 The magnitude of torque

By combining the above three factors, for the sake of simplicity, we can say that the torque is directly proportional to both r and $F \sin \theta$. Fixing the constant of proportionality as unity (one), the magnitude of torque is given as

$$\tau = rF \sin \theta.$$

2.1.3.2 The direction of torque

When we apply a force F on the rod it will tend to rotate in a clockwise manner. However, if we reverse the direction of the force, the rod will tend to rotate in an anti-clockwise manner. Thus torque has a directional nature. According to the right-hand thumb rule, the torque $\vec{\tau}$ points in the direction of $\vec{r} \times \vec{F}$, as shown in the figure. Then the vector expression of torque (which gives both magnitude and direction) is given as

$$\vec{\tau} = \vec{r} \times \vec{F}$$

as shown in the figure.

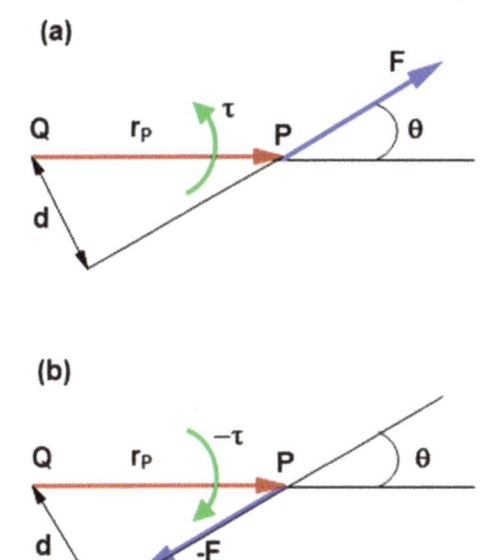

Reversing the force reverses the torque

The force $\vec{F}$ produces a clockwise (↷) or inward torque and force $-\vec{F}$ produces an anti-clockwise (↶) or outward torque. Thus reversing the direction of force reverses the direction of torque.

The moment/lever arm

The magnitude of the torque can also be given as

$$\tau = (r \sin \theta) F = dF,$$

where $d(=r \sin \theta)$ is the perpendicular distance from the pivot (axis of rotation) to the line of action, called the *moment arm*. Then the vector expression of torque of the force $\vec{F}$ acting at a point P relative to a point O is given as

$$\vec{\tau}_O = \vec{r}_{OP} \times \vec{F} = \vec{r}_P \times \vec{F}.$$

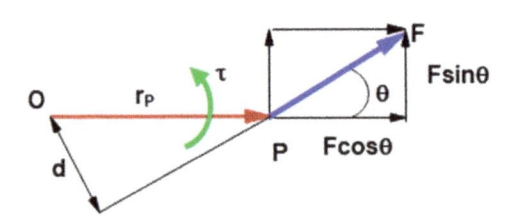

Example 1 The minimum torque required to open a small left-handed nut is equal to 20 N-m. You are given a wrench of arm length $L = 60$ cm. How much force F_t should you apply on the arm of the wrench at a distance $d = 50$ cm from the center of the nut in order to rotate it?

Solution

To produce a maximum turning effect (torque), we have to apply the force perpendicular to the rod in the plane of the face of the nut to produce a clockwise torque. At a distance d if we apply a force F, the maximum torque produced to oppose the frictional torque is

$$\tau_F = dF. \tag{2.1}$$

This is equal to the frictional torque, that is the minimum torque to rotate the nut. So we can write

$$\tau_F = 20. \tag{2.2}$$

Using the two equations, we have

$$F = \frac{\tau_F}{d} = \frac{20}{\frac{1}{2}} = 40 \text{ N}.$$

However, the minimum possible force is equal to $F = 20/0.6 = 100/3$ N because the force will be minimum when it is applied at the end of the arm of the wrench perpendicular to the arm.

2.1.4 Net torque when many forces act at a point

If many forces act at a point or particle P, each force will produce its own torque. Summing up the torques produced by all forces, we can obtain the net torque acting on the rod relative to its pivot/hinge. So the net torque can be given as

$$\vec{\tau}_{net} = \sum \vec{\tau}_i,$$

where $\vec{\tau}_i = \vec{r}_i \times \vec{F}_i$ = torque of the ith force $\vec{F}_i$:

$$\vec{\tau}_{net} = \sum \vec{r}_i \times \vec{F}_i.$$

Since all forces act at the same place, the position of the point of application of each force is same. Then

$$\vec{\tau}_{net} = \vec{r} \times \sum \vec{F}_i.$$

Putting $\sum \vec{F_i} = \vec{F}_{net}$, finally, we have

$$\vec{\tau}_{net} = \vec{r} \times \vec{F}_{net}.$$

This tells us that:
- If many forces act on a point/particle, the net torque is equal to the torque of the net force, relative to a given reference point.

However, the above relation may not hold good for a system of particles, which will be clarified in the next chapter.

Example 2 Find the torque on the bob P of mass m of a simple pendulum of string length l about the point of suspension O.

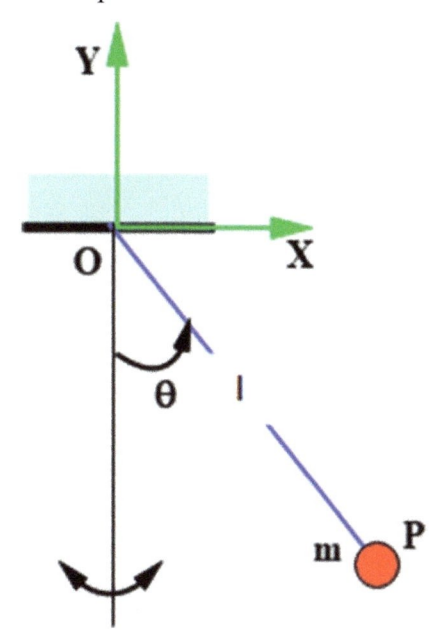

Solution
The forces acting on the pendulum bob are tension T and weight mg. Then, the net torque acting on the bob is equal to the sum of torque due to tension plus the torque due to gravity:

$$\vec{\tau}_{net} = \vec{\tau}_T + \vec{\tau}_g.$$

As the tension passes through the point of suspension P, it cannot produce a torque about P. Putting $\vec{\tau}_T = 0$ and $\vec{\tau}_g = -mgl \sin\theta \hat{k}$ in the above equation, we have

$$\vec{\tau}_{net} = -mgl \sin\theta \hat{k}.$$

Otherwise

The net force acting on the bob is

$$F_{net} = mg \sin \theta, \quad (2.3)$$

which is perpendicular to the string.

Hence, the net torque is

$$\vec{\tau}_{net} = lF_{net}\hat{k}. \quad (2.4)$$

Using equations (2.3) and (2.4), we have

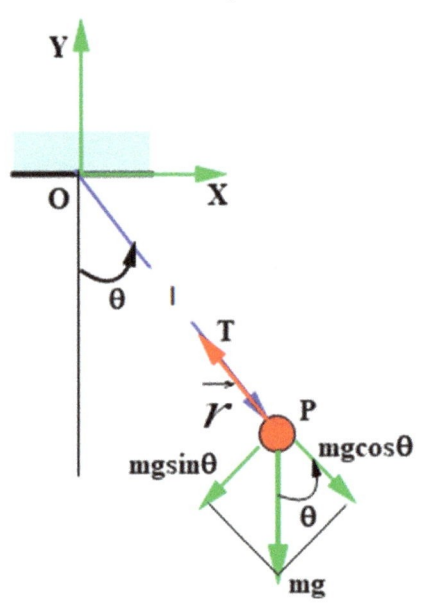

$$\vec{\tau}_{net} = -l(mg \sin \theta)\hat{k}.$$

2.1.5 Torque about an axis

Let a force F act at a point P and we want to find its torque τ_F of the force F relative to a point Q. First of all, find $\vec{r}_{PQ} = \vec{r}_P - \vec{r}_Q$ by putting the coordinates of P and Q, respectively. Let us assume that

$$\vec{r}_{PQ} = \vec{r} = x\hat{i} + y\hat{j} + z\hat{k} \quad (2.5)$$

$$\vec{F} = F_x\hat{i} + F_y\hat{j} + F_z\hat{k}. \quad (2.6)$$

Then the torque is

$$\tau_F = \vec{r}_{PQ} \times \vec{F}. \quad (2.7)$$

Using the last three equations,

$$= (x\hat{i} + y\hat{j} + z\hat{k}) \times (F_x\hat{i} + F_y\hat{j} + F_z\hat{k})$$

$$= (yF_z - zF_x)\hat{i} + (zF_x - xF_z)\hat{j} + (xF_y - yF_y)\hat{k}.$$

Putting $(yF_z - zF_y)\hat{i} = \vec{\tau}_x$, $(zF_x - xF_z)\hat{j} = \vec{\tau}_y$, and $(xF_y - yF_y)\hat{k} = \vec{\tau}_z$, where $\vec{\tau}_x$, $\vec{\tau}_y$, and $\vec{\tau}_z$ are the components of the torque $\vec{\tau}_F$ relative to the origin O about the x-, y-, and z-axes, respectively. However, using the determinant, we have a handy formula of torque given as

$$\vec{\tau}_F = \begin{vmatrix} \hat{i} & \hat{j} & \hat{k} \\ x & y & z \\ F_x & F_y & F_z \end{vmatrix}.$$

The torque of a force F relative to a reference point P:
- Depends on the position of the reference point relative to the point of application of the force.
- Can have the x-, y-, and z-components.
- Will be zero if each of its component will vanish.
- Will be zero if $\vec{r}$ and $\vec{F}$ are parallel or anti-parallel.

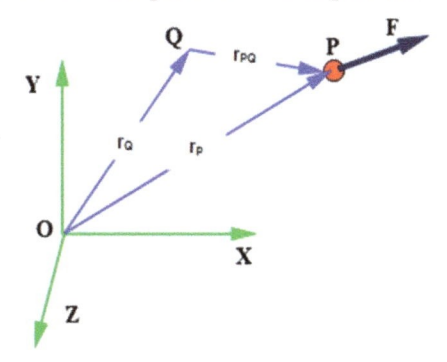

Example 4 A force $\vec{F} = 4\hat{i} + 3\hat{j} + \hat{k}$ N acts at a point P (3, 2, 1) m. (a) What is the torque of the force relative to a point Q (1, −2, 4)? (b) Find the components of the torque about the axes.

Solution
(a) The torque of the force $\vec{F}$ relative to a point Q is given as

$$\vec{\tau}_F = \vec{r}_{PQ} \times \vec{F}, \qquad (2.8)$$

where $\vec{r}_{PQ} = \vec{r}_P - \vec{r}_Q$

$$= 3\hat{i} + 2\hat{j} + \hat{k} - (\hat{i} - 2\hat{j} + 4\hat{k})$$

$$= 2\hat{i} + 4\hat{j} - 3\hat{k}. \qquad (2.9)$$

Using the above determinant formula of torque,

$$\vec{\tau}_Q = \begin{vmatrix} \hat{i} & \hat{j} & \hat{k} \\ 2 & 4 & -3 \\ 4 & 3 & 1 \end{vmatrix}.$$

Expanding the determinant and simplifying the factors, we have

$$\vec{\tau}_Q = 13\hat{i} - 14\hat{j} - 10\hat{k}.$$

So the components of torque about the x-, y-, and z-axes are $13\hat{i}$, $-14\hat{j}$, and $-10\hat{k}$, respectively.

Example 5 Couple. What is meant by a couple? Show that a couple does not depend upon the reference point. What is the couple of an action–reaction pair? If two forces **F** and −**F** act along parallel lines with the shortest distance d between them, find the couple.

Solution

Let two equal and opposite forces $\vec{F}$ and $-\vec{F}$ act on two points or particles P and Q, respectively. Then the net torque about any arbitrary fixed point O is given as a net torque

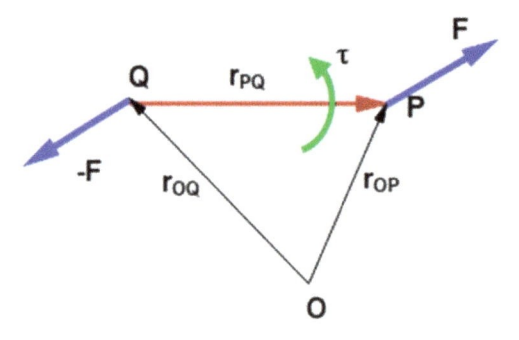

$$\vec{\tau}_{net} = \vec{\tau}_1 + \vec{\tau}_2 = \vec{r}_{OP} \times \vec{F} + \vec{r}_{OQ} \times (-\vec{F}).$$

The net torque is given as

$$= (\vec{r}_{OP} - \vec{r}_{OQ}) \times \vec{F} = \vec{r}_{PQ} \times \vec{F}.$$

This signifies that the net torque does not depend on the reference point O.

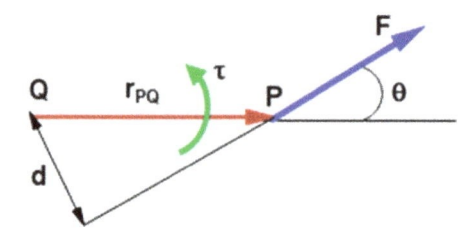

$$\text{Putting } \vec{r}_{PQ} \times \vec{F} = r_{PQ} \times F \sin\theta \hat{k} = (r_{PQ}\sin\theta)F\hat{k}$$

$$= -Fd\hat{k}, \text{ finally, wehave}$$

$$\vec{\tau}_{net} = \vec{\tau} \text{ (or } \vec{C}) = -Fd\hat{k},$$

where $d =$ the perpendicular distance from Q to the line of action of the force at P it is the shortest distance between the lines of action of the forces F and $-F$ acting at these two points. The net torque due to two equal and opposite forces is called a *couple*.

This tells us that, a couple is:
- The net torque of two forces having equal magnitude and opposite directions is called a couple.
- A couple is independent of any reference point. It depends upon the shortest distance d between the lines of action of the forces.
- The net torque of an action–reaction pair is zero.

Example 6 A force F acts at the end of a rod of length L. Another force $-F$ acts on the rod at the other end of the rod. If the rod makes an angle of $\theta = 30°$ with the force F, find the couple experienced by the rod.

Solution
The couple $C = Fd$, where d is the distance between the lines of action of the forces $= L\sin 30° = \frac{L}{2}$. Then, the couple experienced by the rod is $C = \frac{FL}{2}$.

It is not difficult to understand how a torque acting on the rod causes its rotation. But, if we say that a torque causes rotation of a point mass, it makes no sense because we cannot use the word *rotation* for a particle. What then is the physical meaning of 'torque acting on a particle'? What does a torque do when it acts on a particle? Let us answer these questions by defining a new term (concept) known as *angular momentum*.

2.2 Angular momentum

2.2.1 Nomenclature

It will be meaningless for us to define and interpret the term angular momentum if we do not know exactly from where this term emerges. Let us begin with the following reasoning:

'A force changes linear momentum. A torque is a rotational or angular analogue of force. Therefore, we should expect that the torque would change something called *angular momentum*.'

Let us now try to develop a mathematical expression for angular momentum. The torque of the net force F_{net} acting on a particle of mass m relative to a fixed origin is

$$\vec{\tau}_{net} = \vec{r} \times \vec{F}_{net}.$$

The net force $\vec{F}_{net}$ changes the linear momentum of the particle at a rate of $d\vec{P}/dt$. Then the above expression can be written as

$$\vec{\tau}_{net} = \vec{r} \times \frac{d\vec{P}}{dt} = \frac{d(\vec{r} \times \vec{P})}{dt}.$$

We can notice that the torque has the capacity of changing something called the angular momentum of the particle defined by the term $\vec{r} \times \vec{P}$ and denoted by the symbol $\vec{L}$.

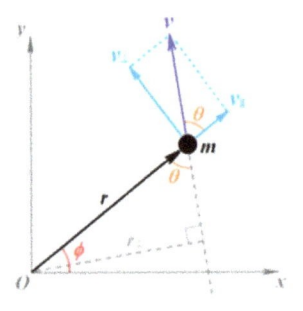

The particles have an anti-clockwise angular momentum given by $\vec{r} \times \vec{p}$.

The angular momentum of a particle of mass m, velocity $\vec{v}$, and linear momentum $\vec{p}$ is given as

$$\vec{L} = \vec{r} \times \vec{p} = \vec{r} \times m\vec{v}.$$

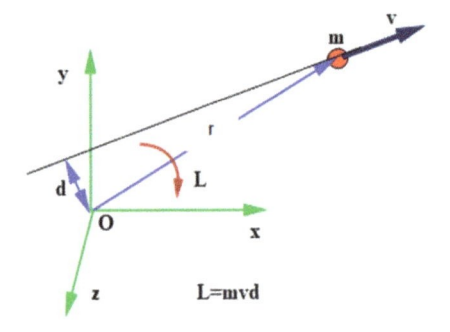

2.2.2 The magnitude of $\vec{L}$

The magnitude of angular momentum is given as

$$L = mvr \sin\theta,$$

where $r\sin\theta = d =$ the shortest or perpendicular distance from the line of motion of the particle from the point of reference. Then we have

$$L = mvd.$$

In another way, we can put $v \sin\theta = v_n = r\omega$ to obtain

$$\vec{L} = mr^2 \vec{\omega},$$

Problems and Solutions in Rotational Mechanics

where ω = the magnitude of angular velocity of the particle relative to the point of reference. This will be discussed in section 2.2.5 in detail.

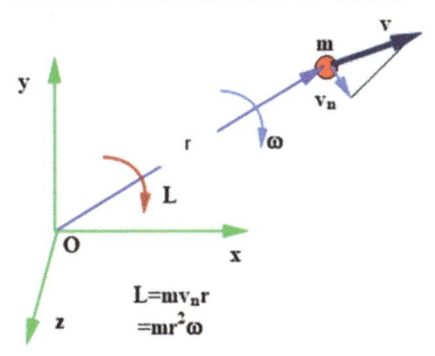

$L = mv_n r$
$= mr^2\omega$

2.2.3 The direction of angular momentum

The angular momentum points in the direction of $\vec{r} \times \vec{v}$ following the right-hand thumb or screw rule.

The angular momentum:
- Is given by the formula $\vec{L} = \vec{r} \times \vec{P}$.
- Is known as a moment of momentum.
- Is an axial/pseudo vector.
- Depends on the position and velocity of the particle relative to the reference point.
- Defines the state angular motion/turning of a particle or system of particles and the rotation of rigid bodies.

Example 7 A bird of mass 2 kg is flying with a velocity $\vec{u} = 12\hat{i}$ m s^{-1} at position P (20, 30, 50) m. A little girl is located at the origin. If she moves with a velocity $\vec{v} = 5\hat{k}$ m s^{-1} on the ground, find the angular momentum of the bird relative to the girl at the given position. Assume that the ground is in the x–z plane.

Solution
For the sake of simplicity let the girl be at the origin. Then the position of the bird relative to the girl is given as

$$\vec{r}_{bg} = 20\hat{i} + 30\hat{j} + 50\hat{k}.$$

The velocity of the bird relative to the girl is given as

$$\vec{v}_{bg} = \vec{v}_b - \vec{v}_g = (12\hat{i} - 5\hat{k}) \text{ m s}^{-1}.$$

The angular momentum of the bird relative to the girl is given as $\vec{L} = \vec{r}_{bg} \times m\vec{v}_{bg}$

$$= (20\hat{i} + 30\hat{j} + 50\hat{k}) \times 2(12\hat{i} - 5\hat{k})$$

$$= 20(-15\hat{i} + 70\hat{j} - 36\hat{k}) \text{ kg m s}^{-1}.$$

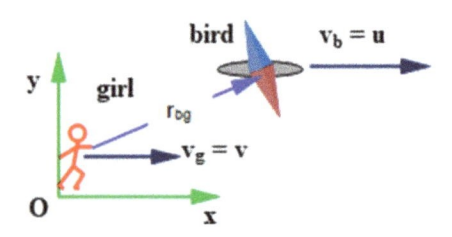

2.2.4 The concept of turning

What is turning? Is it something related to circular motion or motion in a curve? What does *turning* signify physically? How is the term *turning* connected with angular momentum? Let us try to answer these basic questions.

As said earlier, the word *rotation* is used for a straight line, system of particles, and (rigid and non-rigid) bodies. The rotation of a point is meaningless. Then, instead of rotation we can use the term *turning* for (or of) a point mass. However, the turning of a point/particle bears a meaning. The turning of a point P about a reference point O is defined as an angular motion of P about that point. As you know, linear motion is characterized by a linear velocity of P relative to O. Similarly, the angular motion of P is characterized by the angular velocity of P relative to O. Then it is wrong to state that 'a particle turns only when it moves in a circle or in a curve'. The *turning* of a point has nothing to do with the types of motion or trajectory of the point. Even though a point moves in a straight line it can also turn around some chosen points. Turning and circular motion are totally different concepts. Turning should not be misinterpreted as circular motion *or* motion in a curve. Then, what exactly does *turning* mean? Let us discuss.

Illustration: Turning around a point

Let us assume that a particle P of mass m has a velocity $\vec{v}$ at a position vector $\vec{r}$ as shown in the figure. Let the line of motion ($\vec{v}$) make an angle θ with the r-axis. Then components of velocity parallel and perpendicular to $\vec{r}$ are given as v_r and v_n (the transverse or normal component of $\vec{v}$). The particle approaches O or recedes (going away) from O with a velocity $v_r = dr/dt$. On the other hand, the particle P 'turns around the point O' with a linear velocity

$$v_n = v_\emptyset = v \sin\theta.$$

Then the particle turns with a nonzero angular velocity given as

$$\omega_{PO} = \frac{v_\emptyset}{r} = \frac{d\emptyset}{dt}.$$

Since the turning of the point P is given by an angular velocity, turning is also termed as the angular motion of the point/particle P about/around/relative to the reference point O. But you should not use the words *revolution* or *rotation* of O about P which convey wrong ideas. Rotation is used for a straight line and revolution is used for a point moving in a closed loop. Hence, the concept of 'turning' or 'angular motion' of a point has nothing to do with the type/nature of motion (circular motion, parabolic motion, etc) of the point/particle.

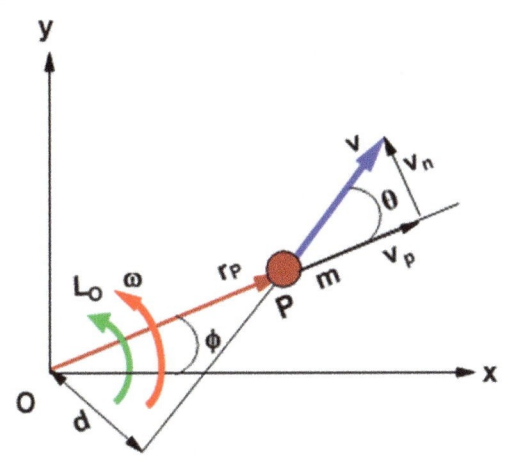

If ($v_\perp \neq 0$), the point/particle P turns around a point O. Hence, angular velocity of the point P relative to O is nonzero ($\omega_{PO} \neq 0$).

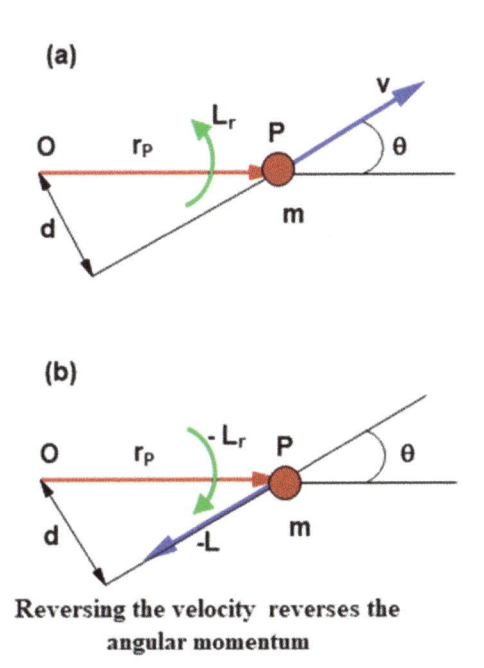

Reversing the velocity reverses the angular momentum

2.2.5 The relation between angular momentum L and angular velocity ω

We will now show you how this angular velocity is related to angular momentum. The formula for the angular momentum of a particle is given as

$$L = rv_\emptyset, \text{ where } v_\emptyset = r\omega_{PO}.$$

2-13

If we write $\omega_{PO} = \omega$ for the sake of simplicity, we have

$$L = mr^2\omega,$$

where mr^2 is defined as the second moment of inertia or moment of inertia (MI or MOI) denoted by the symbol I. We will talk about the MI in the next chapter in detail.

The vector expression of the above formula can be written as

$$\vec{L} = I\vec{\omega},$$

where I = the MI of the particle P about O and ω = the angular velocity of the particle P about the reference point (the origin, say). This is analogous to the linear momentum equation

$$\vec{P} = m\vec{v}.$$

Comparing the last two equations, we can state that linear momentum defines the state of linear motion, whereas the angular momentum defines the state of angular motion (or turning) of a point mass about (or around) a reference point.

Example 8 Turning of a particle moving in a straight line. A particle of mass m moves with a constant velocity v parallel to the x-axis as shown in the figure. (a) At what rate is the particle going away from the origin? (b) With what angular velocity is the particle turning around the origin? (c) What is the angular momentum of the particle relative to the origin?

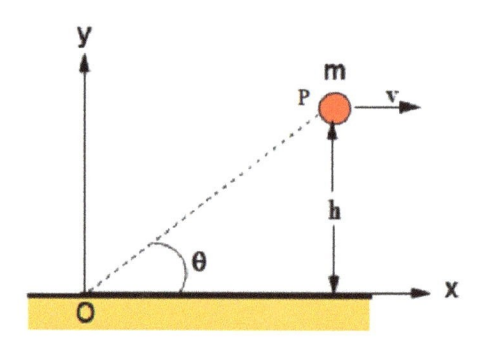

Solution
(a) The particle goes away from the origin O at a rate of $v_r = \frac{dr}{dt} = v \sin\theta$.
(b) The particle turns around the origin O with velocity

$$v_\theta = v \cos\theta.$$

So the angular velocity of the particle relative to O is

$$\omega_{PO} = \frac{d\theta}{dt} = v\cos\theta/r,$$

where $r = d/\cos\theta$.

2-14

Then we have
$$\omega_{PO} = \frac{d\theta}{dt} = \frac{v}{d}\cos^2\theta.$$

(c) The angular momentum of the particle relative to Q is
$$\vec{L}_O = -mvd\hat{k}.$$

N.B. This tells us that:
- The angular momentum of a particle moving with a constant velocity is constant with respect to a given reference point.
- In straight line motion a particle can also turn around a point having an angular velocity.
- If a particle has angular velocity relative to a point, it is said to be turning about that point.
- The angular momentum relates to angular velocity by the formula $L = I\omega$.
- Angular momentum (but not angular velocity alone) defines the state of angular motion of a point mass.

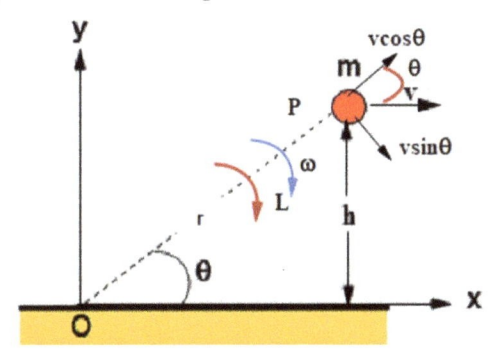

2.2.6 Angular momentum about an axis

Let a point mass P of mass m move with a velocity $\vec{v} = v_x\hat{i} + v_y\hat{j} + v_z\hat{k}$. We want to find the angular momentum $\vec{L}_{PQ}$ of the particle relative to a point Q. Let the position of P relative to Q be
$$\vec{r}_{PQ} = x\hat{i} + y\hat{j} + z\hat{k}.$$
Furthermore, let the velocity of the particle P relative to Q be
$$\vec{v}_{PQ} = \vec{v} = v_x\hat{i} + v_y\hat{j} + v_z\hat{k}.$$
Then, putting these values in the formula
$$\vec{L} = \vec{r} \times \vec{p} = \vec{r} \times m\vec{v}$$

we have
$$\vec{L}_{PO} = \vec{L} = m(x\hat{i} + y\hat{j} + z\hat{k})$$
$$x(v_x\hat{i} + v_y\hat{j} + v_z\hat{k})$$
$$= m(yv_z - zv_y)\hat{i} + m(zv_x - xv_z)\hat{j}$$
$$+ m(xv_y - yv_x)\hat{k}.$$

Putting $m(yv_z - zv_y)\hat{i} = \vec{L}_x$, $m(zv_x - xv_z)\hat{j} = \vec{L}_y$, and $m(xv_y - yv_x)\hat{k} = \vec{L}_z$, where L_x, L_y, and L_z are the components of the angular momentum L relative to the origin about the x-, y-, and z-axes, respectively. However, using the determinant, we have a handy formula of angular momentum given as

$$\vec{L}_{PO} = m \begin{vmatrix} \hat{i} & \hat{j} & \hat{k} \\ x & y & z \\ v_x & v_y & v_z \end{vmatrix}.$$

The angular momentum of a particle P relative to a reference point O:
- Depends on the position and velocity.
- Can have x-, y-, and z-components.
- Will be zero if the position and velocity vectors are parallel or anti-parallel.
- Will be zero if each of its components is zero.
- Will be a constant if v remains constant.

Example 9 A particle of mass $m = 2$ kg and position P $(1, -2, 1)$ moves with a velocity $\vec{v}_P = (4\hat{i} + 3\hat{j} + 2\hat{k})$ ms^{-1}. (a) What is the angular momentum of the particle P relative to a point Q $(1, -2, 4)$ which moves with a velocity $\vec{v}_Q = (2\hat{i} - \hat{j} + \hat{k})$ ms^{-1}. (b) Find the components of the angular momentum about the axes.

Solution

(a) The angular momentum of the particle P relative to a point Q is given as
$$\vec{L}_Q = \vec{r}_{PQ} \times m\vec{v}_{PQ},$$
where $\vec{r}_{PQ} = \vec{r}_P - \vec{r}_Q$
$$= (\hat{i} - 2\hat{j} + \hat{k}) - (\hat{i} - 2\hat{j} + 4\hat{k}) = -3\hat{k}.$$

Furthermore, velocity of the force P relative to a point Q is given as
$$\vec{v}_{PQ} = \vec{v}_P - \vec{v}_Q = (4\hat{i} + 3\hat{j} + 2\hat{k}) - (2\hat{i} - \hat{j} + \hat{k}) = 2\hat{i} + 4\hat{j} + \hat{k}$$

Using the above determinant formula of torque,

$$\vec{L}_{PQ} = 2\begin{vmatrix} \hat{i} & \hat{j} & \hat{k} \\ 0 & 0 & -3 \\ 2 & 4 & 1 \end{vmatrix}$$

Expanding the determinant and simplifying the factors, we have $\vec{L}_{PQ} = 12(2\hat{i} - \hat{j} + 0\hat{k})$ N m^{-1}.

(b) So, the components of angular momentum about x, y and z-axes are 24, -12 and 0, respectively in N m^{-1}.

2.3 Relation between torque and angular momentum

With the help of the foregoing examples and illustrations we understand that when a particle experiences a torque, its state of angular motion which is characterized by its angular momentum, will change. The relation between torque and angular momentum is given as

$$\vec{\tau}_O = \frac{d\vec{L}_O}{dt}.$$

This states that:
- Torque changes the angular momentum.
- Torque about any reference point/axis is numerically equal to the time rate of change of angular momentum of the particle about same point/axis of reference.

2.4 Conservation of the angular momentum of a particle

If we put $\vec{\tau}_O = 0$ in the expression $\vec{\tau}_O = \frac{d\vec{L}_O}{dt}$, we have $\frac{d\vec{L}_O}{dt} = 0$. This gives us a condition,

$$\vec{L}_O = \vec{L} = I\vec{\omega} = \text{constant}.$$

This states that:
- If the torque experienced by a particle (or a non-rigid system of particles) is zero ($\vec{\tau}_O = 0$) about any reference point or axis, angular momentum about the corresponding point or axis remains conserved.

Example 12 Variable MI of a particle. Justify the statement, 'since $\vec{\tau}_O = I\frac{d\vec{\omega}}{dt} = I\vec{\alpha}$, and $\vec{\alpha} = \frac{d\vec{\omega}}{dt} \neq 0$ about the origin, the net torque will be nonzero about the origin O'.

Solution

Since the MI varies with time, the rate of change of the MI is $\frac{dI}{dt} \neq 0$. Taking the derivative of MI, we have to include another term $\vec{\omega}\frac{dI}{dt}$. Hence, the correct formula of torque is given as

$$\vec{\tau}_O = I\frac{d\vec{\omega}}{dt} + \vec{\omega}\frac{dI}{dt} = I\vec{\alpha} + \vec{\omega}\frac{dI}{dt} \neq I\vec{\alpha}.$$

The statement $\vec{\tau}_O = I\vec{\alpha}$ is wrong:
- If the MI is applied for a single particle or non-rigid system we cannot write $\vec{\tau}_O = I\vec{\alpha}$.
- However, the general formula $\vec{\tau}_O = \frac{d\vec{L}}{dt}$ always holds good, where $\vec{L} = I\vec{\omega}$.

Example 13 Before Newton, Kepler gave us the kinematical formula that 'the areal velocity of planet remains constant'. This is known as Kepler's third law of planetary motion. Does this law satisfy Newtonian angular momentum conservation? Explain.

Solution

Kepler's third law states that the areal velocity of a planet remains constant. So the formula $\frac{dA}{dt} = C$, where the area swept by the planet during a small time dt is $dA = \frac{1}{2}r^2 d\theta$. Then, $\frac{dA}{dt} = \frac{1}{2}r^2\frac{d\theta}{dt} = C$. Putting $\frac{d\theta}{dt} = \omega$ and multiplying both sides by $2m$, we obtain $mr^2\omega = I\omega =$ constant. This is what is known as conservation of angular momentum of the planet about the center of mass of the Sun–planet system. So Kepler's third law obeys the Newtonian mechanics of angular momentum conservation.

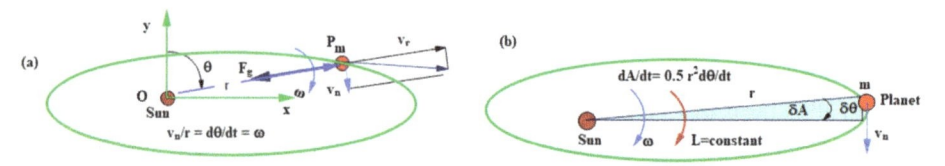

Since the gravity force F_g passes through the sun, its torque is zero about the sun; so, angular momentum of the planet is conserved

Example 15 A particle revolves in a circle with a constant speed. Can we conserve the angular momentum of the particle about (i) the center and (ii) a point on the perimeter of the circle?

Solution

In uniform circular motion, the net force acting on the particle is centripetal which is given as $F_{CP} = m\frac{v^2}{R}$. Since this force passes through the center of the circle, it cannot produce any torque about the center. So the angular momentum of the particle about the center of the circle remains constant which can be given as $L = mvR$. However, about any other point, the centripetal force does not always pass. Therefore, angular momentum of the particle about any other point does not remain constant.

Problems

Torque and angular momentum for a point mass

Problem 1 Let us twist the right-handed nut of average radius R using a wrench of arm length L. If we apply a force of magnitude F on the arm of the wrench using our hands as shown in the figure, find (a) the maximum possible torque that can be produced by the force and (b) the position(s) where the force should be applied to open the nut. Assume τ_o = the minimum torque required to open the nut.

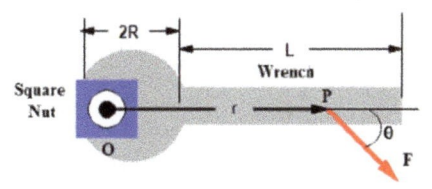

Solution
(a) The maximum value of the moment arm is $L + R$ and R, respectively. To produce a maximum turning effect, at any point of application of the force, F must always act (i) perpendicular to the arm and (ii) at the end of the arm of the wrench. Hence, the maximum value of the torque is equal to $F(L + R)$. The turning effect or torque will be maximum when the force acts at the end point Q of the wrench.
(b) Let the force act at a distance x from the center of the nut. Then the torque produced is given as

$$\tau_F = x F \sin \theta.$$

In order to just rotate the nut the applied torque needs to be equal to τ_o. Then, putting $\tau_F = \tau_o$, we have $x \geq \tau_o / F \sin \theta$.

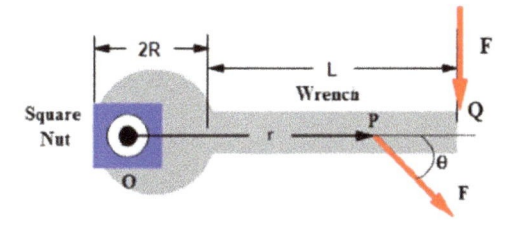

Problem 2 A pendulum bob of mass m is released from rest when a light string of length l is just taut and horizontal. Find (a) the angular momentum of the bob, (b) the torque acting on the bob, and (c) verify that

$$\vec{\tau} = \frac{d\vec{L}}{dt}.$$

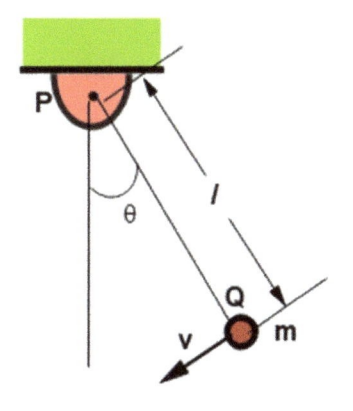

2-19

Solution

(a) For a simple pendulum the point of suspension and the center of the circle traced by the bob are the same. Since the velocity is perpendicular to the length of the string, the angular momentum about P is

$$\vec{L} = -mvl\hat{k}. \quad (2.10)$$

This is a vector of constant direction until the bob reaches the left-hand side extreme point. Then its direction is reversed.

(b) The torque about the P is given as

$$\vec{\tau} = -mgl\cos\theta\hat{k}. \quad (2.11)$$

This is a vector of constant direction until the bob reaches its lowest position. Then its direction is reversed.

(c) The magnitude of angular momentum varies with a rate of

$$dL/dt = ml\,dv/dt.$$

So we can write

$$d\vec{L}/dt = -m\left(\frac{dv}{dt}\right)l\hat{k}. \quad (2.12)$$

From the free body diagram, applying Newton's second law, we have

$$a_t = \frac{dv}{dt} = g\cos\theta. \quad (2.13)$$

Using equations (2.12) and (2.13) we have

$$\frac{d\vec{L}}{dt} = -m(g\cos\theta)l\hat{k} = -mgl\cos\theta\hat{k}. \quad (2.14)$$

Using equations (2.11) and (2.14) we can verify that

$$\vec{\tau} = \frac{d\vec{L}}{dt}.$$

Alternative method for the net torque

After falling through a vertical distance h, the velocity of the bob is given by conserving energy as

$$\frac{1}{2}mv^2 = mgh = mgl\cos\theta.$$

Differentiating with θ we have

$$\frac{v\,dv}{d\theta} = gl\sin\theta.$$

Then the tangential acceleration is

$$\frac{vdv}{ld\theta} = g \sin\theta = a_t.$$

The net torque acting on the bob is

$$\tau = ma_t l = mgl \sin\theta.$$

The net radial force can not produce a torque as it passes through the point of suspension P. So, the net tangential force is responsible for producing the torque.

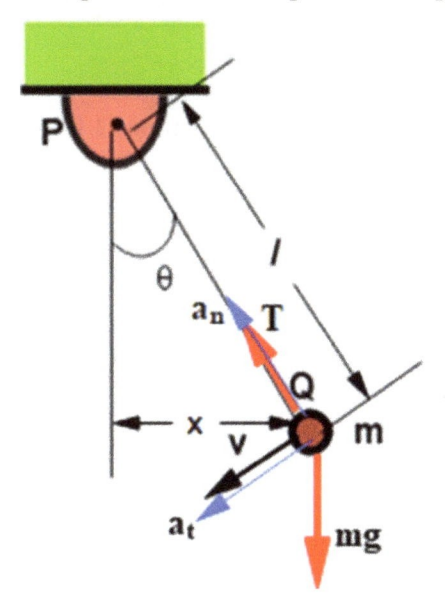

Problem 3 A stone is thrown from the origin O with a velocity u at an angle θ with the horizontal. Relative to the origin, find the angular momentum of the stone in air as the function of time t and horizontal distance x. Neglect air resistance and buoyant forces on the stone.

Solution

The torque acting on the stone due to gravity about O is

$$\tau_O = mgx. \tag{2.15}$$

Since the horizontal velocity u_x of a projectile remains constant, we can write

$$x = ut \cos\theta. \tag{2.16}$$

Using equations (2.15) and (2.16)

$$\tau_O = mgut \cos\theta. \tag{2.17}$$

We know that

$$\tau_O = \frac{dL_O}{dt}.$$

Integrating both sides, we have

$$L_O = \int_0^t \tau_O dt. \tag{2.18}$$

Using equations (2.17) and (2.18) we have

$$L_O = \int_0^t mgut \cos\theta.$$

After evaluating integration, we have

$$L_O = \frac{1}{2}mgut^2 \cos\theta. \tag{2.19}$$

Putting $t = \frac{x}{u\cos\theta}$ from equation (2.16) in equation (2.19) we have

$$L_O = \frac{mg}{2u\cos\theta}x^2.$$

The direction of angular momentum is inward or clockwise.

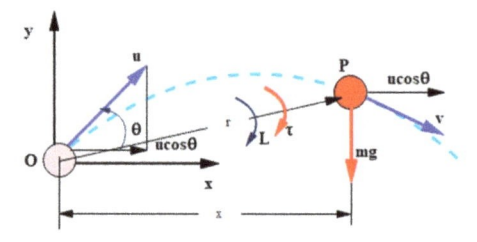

Problem 4 A stone of mass m is projected with a velocity u at an angle $\theta = 30°$ with the horizontal from a point O situated at a height h from the ground. Find the angular momentum of the stone just before hitting the ground relative to the point of projection O. Put $m = 1$ kg, $u = 10$ m s^{-1}, $h = 10$ m and take $g = 10$ m s^{-2}.

Solution

Let $t =$ the time of flight of the stone. Referring to the previous problem, the angular momentum of the stone can be given as

$$L_O = \frac{1}{2}mgut^2 \cos\theta. \tag{2.20}$$

Let us find the time of flight by using the kinematical equation for free fall as

$$-h = ut\sin\theta - \frac{1}{2}gt^2$$

$$\Rightarrow gt^2 - 2(u\sin\theta)t - 2h = 0$$

$$\Rightarrow t = \frac{2(u\sin\theta) \pm \sqrt{4u^2\sin^2\theta + 8gh}}{2g}$$

2-22

$$\Rightarrow t = \frac{u\sin\theta + \sqrt{u^2\sin^2\theta + 2gh}}{g}$$

$$= \frac{10\sin 30° + \sqrt{10^2\sin^2 30° + 2(10)(10)}}{10} = \frac{5 + \sqrt{25 + 200}}{10} = 2\text{ s}. \qquad (2.21)$$

Using the equations (2.20) and (2.21) we have

$$L_O = \frac{1}{2}mgut^2\cos\theta$$

$$L_O = \frac{1}{2}(1)(10)(10)(2)^2\cos 30°$$

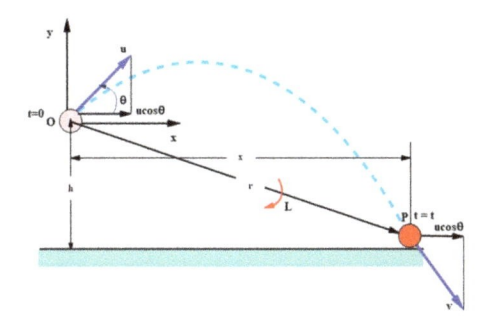

$$L_O = 100\sqrt{3} \text{ kg m}^2\text{ s}^{-1}.$$

Problem 5 A ball of mass m is projected with a velocity u at an angle θ with the horizontal from a point O on the ground. Find (a) the angle of projection θ for which the angular momentum of the ball will be maximum at the highest position relative to the point of projection, (b) the value of maximum angular momentum, and (c) the ratio of the magnitude of angular momenta at the highest position and just before hitting the horizontal ground.

Solution

(a) At the highest point, the angular momentum of the stone is

$$\vec{L} = -mv_x H\hat{k}, \qquad (2.22)$$

where the maximum height is given as

$$H = \frac{u^2\sin^2\theta}{2g}. \qquad (2.23)$$

The horizontal velocity v_x is given as

$$v_x = u\cos\theta. \qquad (2.24)$$

Using the last three equations, we have

$$\vec{L} = -mu\left(\frac{u^2 \sin^2\theta}{2g}\right)\cos\theta \hat{k}. \qquad (2.25)$$

For a maximum value of L,

$$\frac{dL}{d\theta} = \frac{d}{d\theta}\left\{-mu\left(\frac{u^2 \sin^2\theta \cos\theta}{2g}\right)\right\} = 0$$

$$\Rightarrow \frac{d}{d\theta}(\sin^2\theta \cos\theta) = 0$$

$$\Rightarrow (-\sin^3\theta + 2\cos^2\theta \sin\theta) = 0$$

$$\Rightarrow (2\cos^2\theta - \sin^2\theta)\sin\theta = 0$$

$$\Rightarrow \sin\theta = 0 \text{ and } \tan^2\theta = 2$$

$$\Rightarrow \theta = \tan^{-1}\sqrt{2}.$$

(b) Since $\tan^2\theta = 2$, $\cos^2\theta = 1/3$, and $\sin^2\theta = 2/3$. Putting these values in equation (2.25), the maximum value of angular momentum is

$$\vec{L} = -mu\left(\frac{u^2(2/3)}{2g}\right)(1/\sqrt{3})\hat{k}$$

$$= -\frac{mu^3}{3\sqrt{3}g}\hat{k}.$$

(c) As obtained in the previous problem, we have

$$L_0 = \frac{mg}{2u\cos\theta}x^2.$$

Then the ratio of angular momenta at the top and ground level is

$$\frac{L_1}{L_2} = \left(\frac{x_1}{x_2}\right)^2 = 1/4.$$

Problem 6 A bob of mass m is moving with a speed v_o in a circle of radius R on a table. It is attached to an inextensible light string which passes through a hole made on the table as shown in the figure. The free end of the string is pulled down slowly until the bob moves in a circle of radius r. Find the (a) tension of the string and (b) the work done by the external agent to pull the string.

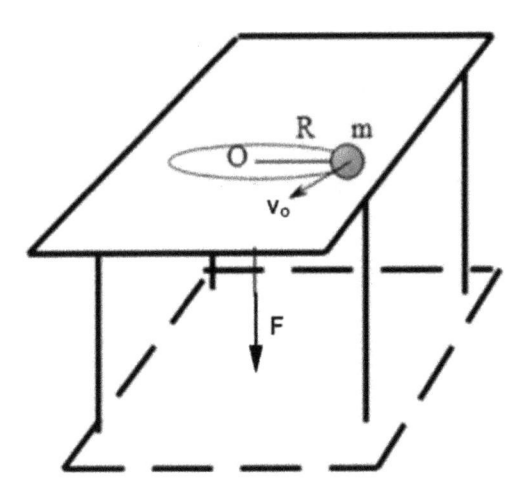

Solution

(a) Let us choose the hole as the reference point. The net force acting on the bob is the tension of the string which passes through the hole. So the net torque about the hole is zero.

Then, conserving angular momentum about the hole for any radius r, we have

$$mv_o R = mvr.$$

This gives

$$v = \frac{v_o R}{r}. \tag{2.26}$$

Then the tension in the string is

$$T = \frac{mv^2}{r}. \tag{2.27}$$

Putting the obtained value of v from equation (2.26) in equation (2.27), we have

$$T = \frac{mv_o R^2}{r^3}. \tag{2.28}$$

(b) Using the work–energy theorem, the work done by the external agent is given as

$$W = K_f - K_i, \tag{2.29}$$

where

$$K_i = \frac{mv_o^2}{2} \text{ and } K_f = \frac{mv^2}{2},$$

then we have

$$W = \frac{mv^2}{2} - \frac{mv_0^2}{2} - \ldots \quad (2.30)$$

Putting $v = \frac{v_0 R}{r}$ from equation (2.26) in equation (2.30) and simplifying the expressions, we obtain

$$W = \frac{mv_0^2}{2}\left(\frac{R^2}{r^2} - 1\right).$$

We can see that positive work is done in pulling the point by a downward force which is equal to the tension in the string that varies with time as the string is pulled down.

Alternative method

You can use the equation (2.28) and find the work done by the tension in displacing the lowest point of the string by a distance dy. This means that $dW = Tdy$. Then integrate it with distance to get the total work done by putting $dy = -dr$ as the radial distance decreases. Put initial and final radial distance as R and r, respectively.

Problem 7 In the previous problem if the bob has an initial angular velocity ω_0 when the radius of the circular path is R, and if we pull the string slowly down, find (a) the tension of the string and (b) the work done by the external agent to pull the string.

Solution

(a) Let us choose the hole as the reference point. The net force acting on the bob is the tension of the string which passes through the hole. So the net torque about the hole is zero.

Then, conserving angular momentum about the hole for any radius r, we have

$$I\omega = I_0\omega_0. \quad (2.31)$$

Putting $I = mr^2$ and $I_0 = mR^2$ in equation (2.31) and simplifying the expression, we have

$$\omega = \frac{\omega_0 R^2}{r^2}. \quad (2.32)$$

Then the tension in the string is

$$T = mr\omega^2. \quad (2.33)$$

Putting the obtained value of ω from equation (2.31) in equation (2.32), we have

$$T = mr\left(\frac{\omega_0 R^2}{r^2}\right)^2 = m\frac{\omega_0^2 R^4}{r^3}. \quad (2.34)$$

(b) Using the work–energy theorem, the work done by the external agent is given as
$$W = K_f - K_i, \qquad (2.35)$$

where $K_i = \dfrac{mR^2\omega_o^2}{2}$ and $K_f = \dfrac{mr^2\omega^2}{2}$.

Then we have
$$W = \frac{mr^2\omega^2}{2} - \frac{mR^2\omega_o^2}{2}. \qquad (2.36)$$

Putting $\omega = \dfrac{\omega_o R^2}{r^2}$ from equation (2.32) in equation (2.36) and simplifying the expressions, we obtain
$$W = \frac{mR^2\omega_o^2}{2}\left(\frac{R^2}{r^2} - 1\right).$$

We can see that positive work is done in pulling the point by a downward force which is equal to the tension in the string that varies with time as the string is pulled down.

Alternative method

You can use the equation (2.34) and find the work done by the tension in displacing the lowest point of the string by a distance dy. This means that $dW = T\,dy$. Then integrate it with distance to get the total work done by putting $dy = -dr$ as the radial distance decreases. Put initial and final radial distance as R and r, respectively.

Problem 8 A smooth small sphere of mass m is kept on a table. It is connected to a block of mass $M(=3m)$ by an ideal string which passes through the hole made on the table top as shown in the figure. Initially the block is resting on the ground and the ball is at a distance R from the hole such that the string is just taut. The sphere is given a horizontal velocity v_o perpendicular to the string at a radial distance R so that it will go to a maximum radial distance equal to $2R$. Find (a) the tension in the string, (b) the differential equation for the radial motion of the ball, (c) the magnitude of v_o, and (d) the radial speed of the ball as the function of radial distance r.

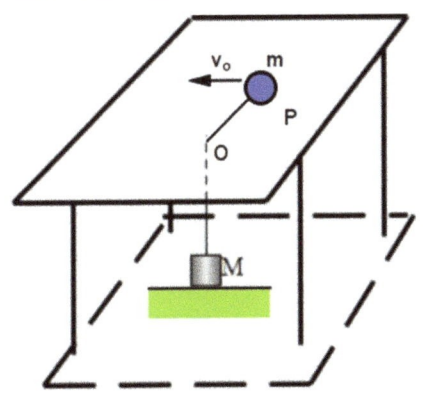

Solution
(a) Let us choose the hole as the reference point. The net force acting on the bob is the tension of the string which passes through the hole. So, the net torque about the hole is zero.

Then, conserving angular momentum about the hole for any radius r, we have
$$mv_o R = mvr$$

This gives
$$v = \frac{v_o R}{r} \tag{2.37}$$

Let the tension in the string at any radial distance r be T. At any radial distance r, by writing the force equation on m and M in radial direction, we have

$$-T = m\left(\frac{d^2 r}{dt^2} - \frac{v^2}{r}\right) \tag{2.38}$$

$$T - Mg = M\frac{d^2 r}{dt^2}, \tag{2.39}$$

Using the last two equations, we have

$$T = \frac{Mm}{M+m}\left(\frac{v^2}{r} + g\right) \tag{2.40}$$

Using equations (2.37) and (2.40), we have

$$T = \frac{Mm}{M+m}\left(\frac{R^2 v_o^2}{r^3} + g\right)$$

(b) Putting the obtained expression of tension T in equation (2.38), the differential equation along radial direction is given as

$$\frac{d^2 r}{dt^2} - \frac{m}{M+m}\frac{R^2 v_o^2}{r^3} + \frac{M}{M+m}g = 0$$

(c) Using the work–energy theorem, the sum of work done by gravity and tension is numerically equal to the change in KE of the system. Since the tension as a whole does not perform any work because it is a constraint force, then, the work done by gravity is equal to change in KE. This is given as

$$W_{gr} = K_f - K_i, \tag{2.41}$$

where $K_i = \frac{mv_o^2}{2}$ and $K_f = \frac{mv^2}{2}$.

2-28

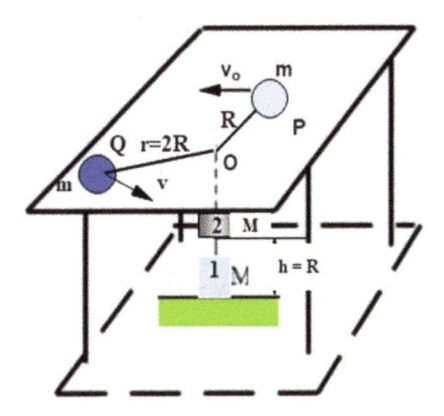

Problems and Solutions in Rotational Mechanics

Then, we have

$$W_{gr} = \frac{mv^2}{2} - \frac{mv_0^2}{2} \tag{2.42}$$

Putting $v = \frac{v_0 R}{r}$ from equation (2.37) in equation (2.42) and simplifying the expressions, we obtain

$$W_{gr} = \frac{mv_0^2}{2}\left(\frac{R^2}{r^2} - 1\right) = \frac{mv_0^2}{2}\left(\frac{R^2}{4R^2} - 1\right) = -\frac{3mv_0^2}{8} \tag{2.43}$$

Since the block M rises by a distance $h = 2R - R = R$, the work done by gravity is

$$W_{gr} = -MgR \tag{2.44}$$

Putting the obtained value of W_{gr} from equation (2.44) in equation (2.43), we have

$$-\frac{3mv_0^2}{8} = -MgR$$

$$v_0 = \sqrt{\frac{8MgR}{3m}} \tag{2.45}$$

Putting $M = 3m$, we have

$$v_0 = \sqrt{\frac{8(3m)gR}{3m}} = 2\sqrt{2gR}$$

Alternative method

From (b) let us recast the differential equation along radial direction which is given as

$$\frac{d^2 r}{dt^2} - \frac{m}{M+m}\frac{R^2 v_0^2}{r^3} + \frac{m}{M+m}g = 0$$

2-29

At the extreme position put $\frac{d^2r}{dt^2} = 0$ and $r = 2R$ in the differential equation to obtain

$$0 - \frac{m}{M+m} \frac{R^2 v_0^2}{(2)^3} + \frac{m}{M+m} g = 0$$

This gives us the same result

$$v_0 = 2\sqrt{2gR}$$

(d) Let the radial velocity of the bodies (m and M) be u. If the transverse velocity of the sphere is v, the change in kinetic energy of the system ($M + m$) is

$$\Delta K = \frac{m(u^2 + v^2)}{2} + \frac{Mu^2}{2} - \frac{mv_0^2}{2} \quad (2.46)$$

Putting $v = \frac{v_0 R}{r}$ from equation (2.37) in equation (2.46), we obtain

$$\Delta K = \frac{(M+m)u^2}{2} + \frac{mv_0^2}{2}\left(\frac{R^2}{r^2} - 1\right) \quad (2.47)$$

The work done by gravity is

$$W_{gr} = -Mg(r - R) \quad (2.48)$$

Using the work–energy theorem, we have

$$W_{gr} = \Delta K \quad (2.49)$$

Using the last three equations, we have

$$\frac{(M+m)u^2}{2} - \frac{mv_0^2}{2}\left(1 - \frac{R^2}{r^2}\right) = -Mg(r - R)$$

Then, we have

$$\frac{(M+m)u^2}{2} = \frac{mv_0^2}{2}\left(1 - \frac{R^2}{r^2}\right) - Mg(r - R) \quad (2.50)$$

So, the radial velocity is given as

$$u = \sqrt{\frac{m\left(1 - \frac{R^2}{r^2}\right)v_0^2 - 2Mg(r - R)}{M + m}}$$

N.B: Putting $u = 0$, $r = 2R$ and $M = 3m$, we can also get $v_0 = 2\sqrt{2gR}$

Alternative method to derive the differential equation
Differentiating the equation (2.50) with time, we have

$$\frac{2(M+m)udu}{2dt} = \frac{mv_0^2}{2}\left(\frac{-2R^2}{r^3}\right)\frac{dr}{dt} - Mg\frac{dr}{dt}, \text{ where } \frac{dr}{dt} = u \text{ and } \frac{du}{dt} = \frac{d}{dt}\left(\frac{dr}{dt}\right) = \frac{d^2r}{dt^2}$$

Then, we have

$$(M+m)\frac{d^2r}{dt^2} = mv_0^2\left(\frac{R^2}{r^3}\right) - Mg$$

Or,

$$(M+m)\frac{d^2r}{dt^2} - mv_0^2\left(\frac{R^2}{r^3}\right) + Mg = 0$$

Or,

$$\frac{d^2r}{dt^2} - \frac{m}{M+m}\frac{R^2v_0^2}{r^3} + \frac{m}{M+m}g = 0$$

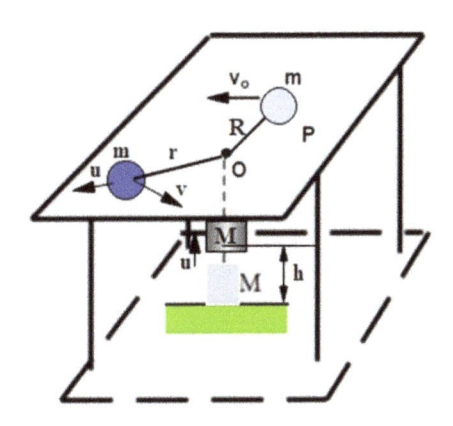

Problem 9 A small and smooth ball of mass M is kept on a table. It is connected to a block of the same mass M by an ideal string which passes through the hole made on the table top as shown in the figure. Initially the block is resting on the ground and the ball is placed at a distance $OP = R$ from the hole O such that the string is just taut. A bead of mass m collides with the ball with a horizontal velocity v_0 perpendicular to the string. If the bead sticks to the ball just after the collision,

find the maximum radial distance moved by the combined mass $M + m$. Put $v_o = \sqrt{16/3gR}$ and $M = m$.

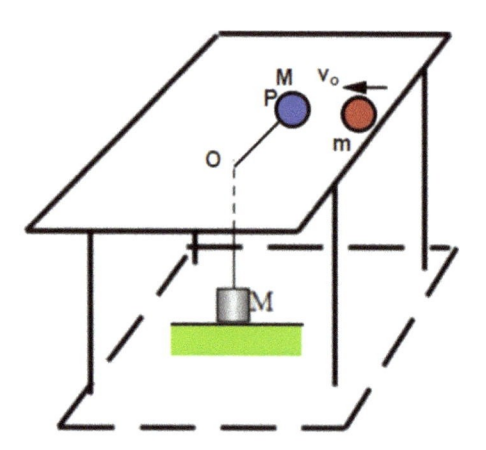

Solution
Let the velocity of the combined mass just after the collision be u and that at maximum radial distance r be v.

By conserving angular momentum about the hole between three positions, such as just before the collision, just after the collision, and at the maximum radial distance of the combined mass $(M + m)$, we have

$$mv_o R = (M + m)uR = (M + m)vr. \qquad (2.51)$$

This gives following three equations:

$$v = \frac{mv_o R}{(M + m)r} \qquad (2.52)$$

$$u = \frac{mv_o}{(M + m)} \qquad (2.53)$$

$$v = \frac{uR}{r}. \qquad (2.54)$$

Using the work–energy theorem, the sum of work done by gravity and tension is numerically equal to the change in the kinetic energy of the system. Since the tension as a whole does not perform any work because it is a constraint force, then the work done by gravity is equal to the change in the kinetic energy. This is given as

$$W_{gr} = K_f - K_i, \qquad (2.55)$$

where
$$K_i = \frac{(M+m)u^2}{2} \text{ and } K_f = \frac{(M+m)v^2}{2},$$

then we have
$$W_{gr} = \frac{(M+m)v^2}{2} - \frac{(M+m)u^2}{2}. \tag{2.56}$$

Putting $v = \frac{mv_0 R}{(M+m)r}$ from equation (2.52) in equation (2.56) and simplifying the expressions, we obtain

$$W_{gr} = \frac{M+m}{2}(v^2 - u^2). \tag{2.57}$$

Using equations (2.54) and (2.57), we have
$$W_{gr} = \frac{M+m}{2}\left\{\left(\frac{uR}{r}\right)^2 - u^2\right\}$$

$$= \frac{(M+m)u^2}{2}\left\{\left(\frac{R^2}{r^2}\right) - 1\right\}. \tag{2.58}$$

Since the block M rises up by a distance $h = r - R$, the work done by gravity is
$$W_{gr} = -Mg(r - R). \tag{2.59}$$

Using equations (2.58) and (2.59), we have
$$\frac{(M+m)u^2}{2}\left\{\left(\frac{R^2}{r^2}\right) - 1\right\} = -Mg(r - R). \tag{2.60}$$

Using equations (2.53) and (2.60), we have
$$\frac{(M+m)}{2}\left\{\frac{mv_0}{(M+m)}\right\}^2\left\{\left(\frac{R^2}{r^2}\right) - 1\right\} + Mg(r - R) = 0$$

or $\quad \dfrac{\{mv_0\}^2}{2(M+m)}\left\{1 - \left(\dfrac{R^2}{r^2}\right)\right\} = Mg(r - R)$

or $\quad \dfrac{\{mv_0\}^2}{2(M+m)}\left\{1 - \left(\dfrac{R^2}{r^2}\right)\right\} - Mg(r - R) = 0$

or $\quad (r - R)\left\{\dfrac{\{mv_0\}^2}{2(M+m)}\dfrac{R+r}{r^2} - Mg\right\} = 0$

2-33

or $\left\{\dfrac{\{mv_o\}^2}{2(M+m)}\dfrac{R+r}{r^2} - Mg\right\} = 0$

or $\dfrac{\{mv_o\}^2}{2(M+m)}\dfrac{R+r}{r^2} = Mg.$ (2.61)

Putting $M = m$ in equation (2.61), we have

$$\dfrac{\{mv_o\}^2}{2(m+m)}\dfrac{R+r}{r^2} = mg$$

or $\dfrac{R+r}{4r^2}v_o^2 = g.$ (2.62)

Putting $v_o^2 = 16gR/3$ in equation (2.62), we have

$$\dfrac{R+r}{4r^2}\left(\dfrac{16gR}{3}\right) = g$$

or $4R(R+r) = 3r^2$

or $3r^2 - 4Rr - 4R^2 = 0.$

Solving this quadratic equation, we have

$$r = 2R.$$

Problem 10 A small sphere of mass m is revolving in a circular path of radius R on a smooth table with a speed v_o. It is connected to a hanging block of mass M by an ideal string which passes through the hole made on the table top, as shown in the figure. (a) If the block is slowly pulled down by a distance $R/2$, find the work done by the external agent in pulling the block down. (b) If the block is released by withdrawing the applied force, what will be the acceleration of the block just after the release? (c) Discuss the nature of motion of the bodies and find the ratio of maximum and minimum tension in the string during the subsequent motion.

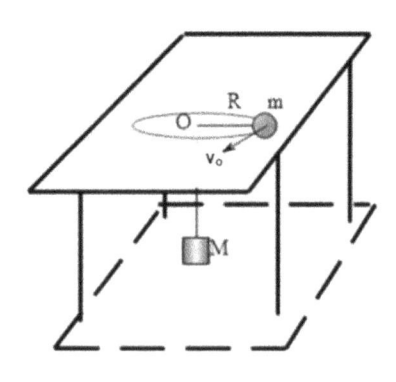

2-34

Solution
(a) Referring to problem 8, the change in kinetic energy is given as

$$\Delta K = \frac{mv_0^2}{2}\left(\frac{R^2}{r^2} - 1\right). \tag{2.63}$$

Since the block M is pulled down by a distance $h = R - r$, the change in gravitational potential energy is

$$\Delta U_{gr} = -Mg(R - r). \tag{2.64}$$

According to the work–energy theorem

$$W_{ext} = \Delta U_{gr} + \Delta K. \tag{2.65}$$

Using the last three equations, we have

$$W_{ext} = \frac{mv_0^2}{2}\left(\frac{R^2}{r^2} - 1\right) - Mg(R - r). \tag{2.66}$$

Since the particle is initially moving in a circle of radius R, the centripetal force is the initial tension T_o in the string which provides the necessary centripetal acceleration

$$a_{cp} = v_0^2/R. \tag{2.67}$$

Applying Newton's Second Law, we have

$$T_o = ma_{cp} = mv_0^2/R. \tag{2.68}$$

Since the mass M is in equilibrium initially, its acceleration and velocity are both zero. So the initial tension balances gravity (the weight of the block M). Then we can write

$$T_o = Mg. \tag{2.69}$$

From equations (2.68) and (2.69) we have

$$v_0 = \sqrt{\frac{MgR}{m}}. \tag{2.70}$$

Putting the value of v_0 from equation (2.70) in equation (2.66), we have

$$W_{ext} = \frac{m\left(\frac{MgR}{m}\right)}{2}\left(\frac{R^2}{r^2} - 1\right) - Mg(R - r)$$

$$= \frac{MgR}{2}\left(\frac{R^2}{r^2} - 1\right) - Mg(R - r)$$

$$W_{ext} = Mg(R - r)\left(\frac{R(R + r)}{2r^2} - 1\right). \tag{2.71}$$

Putting $r = R/2$, we have

$$W_{ext} = Mg\left(R - \frac{R}{2}\right)\left(\frac{R(R + R/2)}{2(\frac{R}{2})^2} - 1\right)$$

$$W_{ext} = MgR. \tag{2.72}$$

The net work done by the external agent is positive as it pulls the string down against its tension.

(b) As derived in problem 6, the tension of the string is

$$T = \frac{mR^2 v_o^2}{r^3}. \tag{2.73}$$

At the lowest position of the block M, putting $r = R/2$, we have

$$T = \frac{8mv_o^2}{R}. \tag{2.74}$$

So the net upward force acting on the block is

$$T - Mg = \frac{8mv_o^2}{R} - Mg = Ma. \tag{2.75}$$

Putting the value of v_o from equation (2.70) in equation (2.75), we have

$$\frac{8m\left(\frac{MgR}{m}\right)}{R} - Mg = Ma.$$

This gives us the upward acceleration of the block as

$$a = 7g \text{ (up)}.$$

(c) If the block is released by withdrawing the applied force at its lowest position, it will oscillate up and down by a one way displacement of $R/2$ and as a result the particle m will move in a spiral path cyclically having maximum and minimum radial distances of R and $R/2$, respectively.

Problem 11 Conical pendulum. A simple pendulum has a bob of mass m and a string PQ of length l. The bob revolves about the vertical axis passing through the point of suspension P with a constant speed v. Find (a) the torque acting on the bob, (b) the angular momentum, (c) the angular velocity of the bob about the point of suspension P, the center C of the circular path traced by the bob, and the axis of revolution PC of the bob, and (d) the ratio of torque acting on the bob, the angular momentum, and the angular velocity of the bob relative to P and C (or the axis PC).

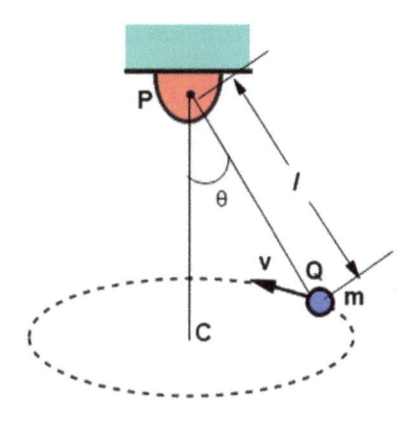

Solution

(a) Let the speed of the bob be v and the radius of the circular path be R.

Then the torques acting on the particle relative to P, C, and about the axis PQ are equal, which can be given as

$$\vec{\tau}_P = \vec{\tau}_C = \vec{\tau}_y = mgR\hat{\omega},$$

where $R = l\sin\theta$.

The magnitude of the torque remains constant, but its direction keeps on changing as the torque can be imagined as a tangentially moving vector, as shown in the following figure.

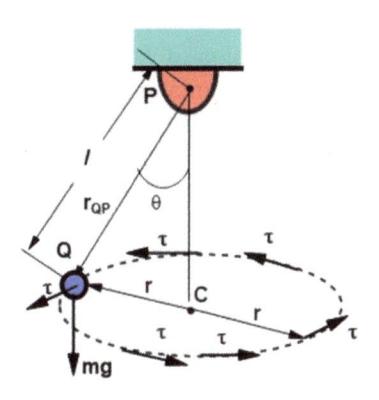

(b) The angular momentum of the bob relative to the point of suspension P is

$$L_P = mvl.$$

2-37

This vector rotates in space forming a cone, shown as a dotted red circle in the following figure.

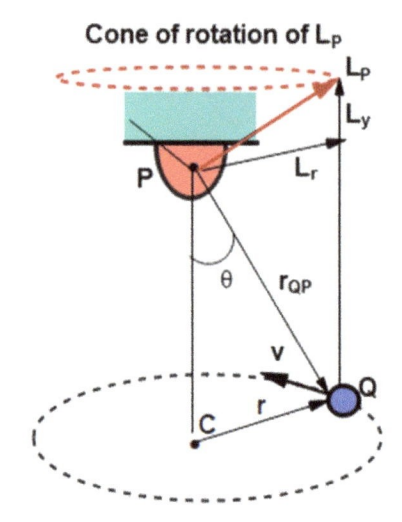

Cone of rotation of L_p

However, the component of L_P in the radial direction, that is $\vec{L}_r$, keeps on rotating about the vertical axis where as its vertical component $\vec{L}_y$ remains constant, which is given as

$$L_y = mvR = mvl \sin \theta.$$

(c) The angular velocity of the bob relative to P is given as

$$\omega_P = v/l.$$

This vector points in the direction of L_P that rotates in space about the axis of rotation forming a cone with an angular velocity equal to that of the bob relative to the center C of the circle.

However, the angular velocity of the bob relative to C and the vertical axis is

$$\omega_y = \frac{v}{R} = \frac{v}{l \sin \theta}.$$

(d) The net torque about the axis and any point on the axis such as the center C etc, are equal. So the ratio of their magnitudes is one.

The relation between L_P and L_y can be given as

$$\frac{L_y}{L_P} = \frac{mvR}{mvl} = \frac{R}{l} = \sin \theta.$$

2-38

The relation between the angular velocities is given as

$$\frac{\omega_y}{\omega_P} = \frac{\frac{v}{R}}{\frac{v}{l}} = \frac{l}{R} = \csc\theta.$$

Problem 12 A conical pendulum has a bob of mass m and a string of length l suspended from a fixed point P. The angle made by the string with the vertical is θ and the bob executes a horizontal circle around the center O. (a) Find the angular momentum of the bob relative to (i) the point of suspension P and (ii) the center of the circle O traced by the bob. (b) Verify that the torque experienced by the bob about P is equal to the rate of change of angular momentum about P.

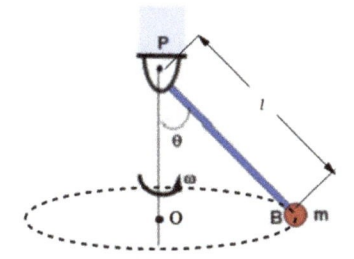

Solution

(a)   (i) Let us first calculate the speed of the bob for a given angle of inclination θ of the string with vertical. As the bob moves in a horizontal circle, the vertical acceleration is zero. So the net vertical force is

$$F_y = T\cos\theta - mg = 0$$

$$T\cos\theta = mg. \tag{2.76}$$

The centripetal force acting on the circulating bob is

$$T\sin\theta = \frac{mv^2}{r}. \tag{2.77}$$

Using equations (2.76) and (2.77), we have

$$mg\sin\theta = \frac{mv^2}{r}\cos\theta. \tag{2.78}$$

Putting $r = l\sin\theta$ in the equation (2.78), we have

$$mg\sin\theta = \frac{mv^2}{l\sin\theta}\cos\theta$$

$$\Rightarrow v = (\sqrt{gl\sec\theta})\sin\theta. \tag{2.79}$$

The angular momentum of the bob about the point of suspension P is

$$L_P = mvl. \tag{2.80}$$

This is *not* a constant vector because its direction changes in space as it rotates about the y-axis, forming a cone of semi-vertical angle $90°- \theta$ as shown in the following figure.

Using equations (2.79) and (2.80), we have

$$L_P = m\{(\sqrt{gl\sec\theta})\sin\theta\}l$$

$$= ml(\sqrt{gl/\cos\theta})\sin\theta. \tag{2.81}$$

(ii) The component of angular momentum of the bob about the y (vertical) axis is

$$L_y = L_P \sin\theta = ml(\sqrt{gl/\cos\theta})\sin^2\theta. \tag{2.82}$$

This is a constant vector as its direction is vertically upward and its magnitude is a constant quantity.

(b) The component of angular momentum of the bob about the r-axis (radial axis) is

$$L_r = L_P \cos\theta = mvl\cos\theta = mvr.$$

As it is always pointing radially outward, we can write it vectorially as

$$\vec{L_r} = mvl\cos\theta\hat{r}.$$

This is a not a constant vector as its direction is always pointed upward and its magnitude is a constant quantity. This component of L_P changes at a rate

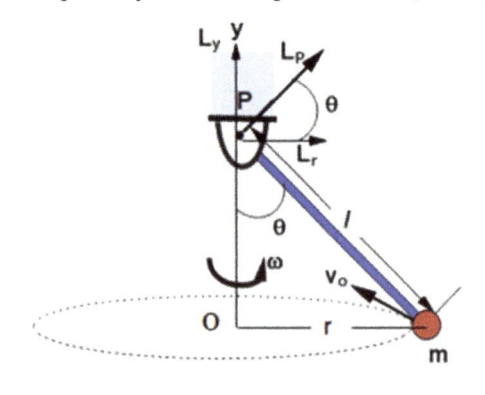

$$\frac{d\vec{L_r}}{dt} = mvl\cos\theta\frac{d\hat{r}}{dt}.$$

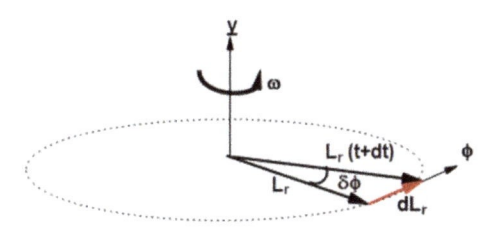

Putting $d\vec{L}_r = L_r \delta\phi \hat{\phi}$ in the last expression, we have

$$\frac{d\vec{L}_r}{dt} = \frac{(L_r \delta\phi \hat{\phi})}{dt} = L_r \frac{(\delta \hat{\phi})}{\delta t}.$$

Putting $\frac{(\delta\phi)}{\delta t} = \omega$ we have

$$\frac{d\vec{L}_r}{dt} = L_r \omega \hat{\phi}.$$

Putting $L_r = L_P \cos\theta = mvl \cos\theta$,

$$\frac{d\vec{L}_r}{dt} = L_P \cos\theta \omega \hat{\phi} = mvl \cos\theta \omega \hat{\phi}. \qquad (2.83)$$

Then the torque about P is numerically equal to the rate of change of momentum of the bob about P, given as

$$\vec{\tau}_P = \frac{d\vec{L}_P}{dt} = \frac{d\vec{L}_r}{dt} + \frac{d\vec{L}_y}{dt}$$

$$\Rightarrow \vec{\tau}_P = \frac{d\vec{L}_r}{dt} \quad (\because \vec{L}_y = \text{constant}). \qquad (2.84)$$

Using equations (2.83) and (2.94) we have

$$\Rightarrow \vec{\tau}_P = mvl \cos\theta \omega \hat{\phi}. \qquad (2.85)$$

We know that the angular velocity of the bob about the y-axis is

$$\omega = v/r. \qquad (2.86)$$

Using equations (2.85) and (2.86) we have

$$\Rightarrow \vec{\tau}_P = mvl \cos\theta \{v/r\} \hat{\phi}$$

$$\Rightarrow \vec{\tau}_P = \frac{mv^2 l \cos\theta}{r} \hat{\phi}. \qquad (2.87)$$

As the bob moves in a horizontal circle, the vertical acceleration is zero. So the net vertical force is
$$F_y = T\cos\theta - mg = 0$$
$$T\cos\theta = mg. \tag{2.88}$$

The centripetal force acting on the circulating bob is
$$T\sin\theta = \frac{mv^2}{r}. \tag{2.89}$$

Using equations (2.88) and (2.89) we have
$$mg\sin\theta = \frac{mv^2}{r}\cos\theta. \tag{2.90}$$

Using equations (2.87) and (2.90) we have
$$\Rightarrow \vec{\tau}_P = mgl\sin\theta\,\hat{\phi}. \tag{2.91}$$

By using the basic formula of torque, the net acting on the bob about P is
$$\Rightarrow \vec{\tau}_P = \vec{r}_{QP} \times \vec{F}_{net} = \vec{r}_{QP} \times (m\vec{g} + \vec{T}).$$

Since the tension passes through the point of suspension P, its torque about P is zero. Then the torque about P is equal to the gravitational torque, given as
$$\Rightarrow \vec{\tau}_P = \vec{r}_{QP} \times m\vec{g}$$
$$\Rightarrow \vec{\tau}_P = mgl\sin\theta\,\hat{\phi}.$$

We can see that the torque vector also rotates along the perimeter of a horizontal circle about the y-axis, whereas its magnitude remains constant. Thus, the relation $\vec{\tau}_P = \frac{d\vec{L}_P}{dt}$ is verified.

Problem 13 A bob of mass m of a conical pendulum revolves with an angular speed $\omega = \sqrt{2g/l}$ about the vertical (y-axis) on an ideal string of length l. What is (a) the angular momentum of the bob about its point of suspension P and (b) the net torque acting on the bob about (i) the axis of revolution of the bob and the center of the circular path traced by the bob and (ii) the point of suspension and any other point on the vertical axis? Put $m = 2$ kg and $l = 5$ m.

Solution
(a) Referring to the previous problem, the force equation on the bob can be given as following:
$$T\cos\theta = mg \tag{2.92}$$

$$T \sin \theta = \frac{mv^2}{r} = mr\omega^2. \tag{2.93}$$

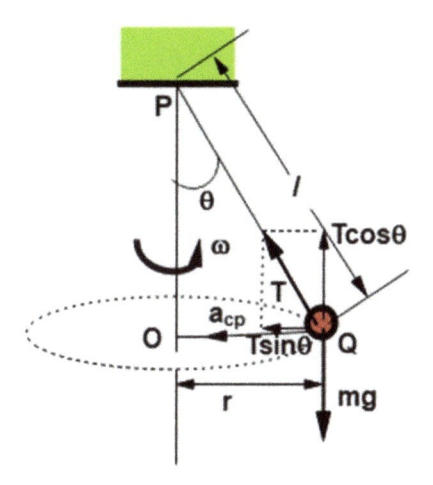

Putting $r = l\sin\theta$ in equation (2.93) we have

$$T \sin \theta = m(l \sin \theta)\omega^2$$

$$\Rightarrow T = ml\omega^2. \tag{2.94}$$

Using equations (2.92) and (2.94) we have

$$ml\omega^2 \cos \theta = mg$$

$$\Rightarrow \cos \theta = \frac{g}{l\omega^2}. \tag{2.95}$$

It is given that

$$\Rightarrow \omega = \sqrt{2g/l}. \tag{2.96}$$

Using equations (2.95) and (2.96)

$$\Rightarrow \cos \theta = \frac{g}{l(2g/l)} = 1/2$$

$$\Rightarrow \theta = 60°.$$

Then the net torque acting on the bob about P is mgr, where $r =$ radius of the circle traced by the bob as calculated earlier. However, the net torque experienced by the bob about the center of the circle O is zero as the net force, that is the centripetal force $F_{cp} = T\sin\theta = ma_{cp} = mv^2/r = mr\omega^2$, passes through the point O as shown in the figure. So the angular momentum about O (or about the y-axis) remains conserved, which is given as

$$L_O = mvr = m(r\omega)r = mr^2\omega$$
$$= m(l\sin\theta)^2\,\omega$$
$$= (ml^2\sin^2\theta)\omega$$
$$= (2)(5)^2(3/4)(\sqrt{(2\times 10)/5})$$
$$= 75\,\text{kgms}^{-1}\ (\text{directed up}).$$

The angular momentum about the point P is given as
$$L_O = mvl = m(r\omega)l = m\{(l\sin\theta)(\omega)\}l$$
$$= ml^2\omega\sin\theta = (2)(5)^2\left(\sqrt{2\times 10/5}\right)\left(\sqrt{3}/2\right)$$
$$= 50\sqrt{3}\,\text{kgms}^{-1}\ (\text{rotates about } y\text{-axis forming a cone}).$$

(b)
 (i) The net torque about O (or y-axis) is zero because the net force ma passes through this point.
 (ii) However, the net torque about any other point on the y-axis will not be zero. The net torque about P is given as

$$\vec{\tau}_P = mgl\sin\theta\hat{\phi} = (2)(10)(5)(0.866)\hat{\phi} = 86.6\hat{\phi}\,\text{Nm (nearly)}.$$

The torque vector moves along the perimeter of a horizontal circle and rotates about the y-axis as shown in the figure of the previous problem.

Problem 14 In a conical pendulum of string of length l, the bob of mass m has a velocity $v = \sqrt{3gl/2}$. Find (a) the angle made by the string with the vertical and (b) the torque acting on the bob about the point of suspension and the center of the circular path traced by the bob.

Solution
(a) From the previous problem we have
$$\cos\theta = \frac{g}{l\omega^2} \qquad (2.97)$$

$$\Rightarrow \cos\theta = \frac{g}{l(v/r)^2} = \frac{gr^2}{lv^2}. \qquad (2.98)$$

Putting $r = l \sin \theta$ in equation (2.98) we have

$$\Rightarrow \cos \theta = \frac{g(l \sin \theta)^2}{lv^2}$$

$$\Rightarrow \cos \theta = \frac{gl}{v^2}(1 - \cos^2 \theta)$$

$$\Rightarrow \cos^2 \theta + \frac{v^2}{gl} \cos \theta - 1 = 0$$

$$\Rightarrow \cos \theta = \frac{1}{2}\left\{ -\frac{v^2}{gl} \pm \sqrt{\left(\frac{v^2}{gl}\right)^2 + 4} \right\}.$$

Disregarding the negative value of $\cos\theta$ because it is always less than one for an acute angle, we accept the other value given as

$$\cos \theta = \frac{1}{2}\left\{ \sqrt{\left(\frac{v^2}{gl}\right)^2 + 4} - \frac{v^2}{gl} \right\} \qquad (2.99)$$

$$\frac{v^2}{gl} = 3/2 \text{ (given)}. \qquad (2.100)$$

Using equations (2.99) and (2.100) we have

$$\cos \theta = \frac{1}{2}\left\{ \sqrt{\left(\frac{3}{2}\right)^2 + 4} - \frac{3}{2} \right\} = \frac{1}{2}$$

$$\Rightarrow \theta = 60°.$$

An alternative method for finding θ

The formula for

$$\tan \theta = \frac{a}{g} = \frac{v^2/r}{g} = \frac{v^2/l \sin \theta}{g}$$

$$\Rightarrow \frac{\sin^2 \theta}{\cos \theta} = = \frac{v^2}{gl}$$

$$\Rightarrow \cos \theta = \frac{gl}{v^2}(1 - \cos^2 \theta)$$

$$\Rightarrow \cos^2 \theta + \frac{v^2}{gl} \cos \theta - 1 = 0.$$

Putting $\frac{v^2}{gl} = 3/2$ in the last equation and solving this equation, we can obtain $\theta = 60°$.

(b) About P, the torque due to tension is zero as it passes through this point. The gravitational torque about P is

$$\vec{\tau}_{gr} = -mgl\sin\theta\hat{k} = -mgl\sin 60°\hat{k} = -\frac{\sqrt{3}\,mgl}{2}\hat{k}.$$

Then the net torque about P is equal to that of gravity.
About O, the torque due to tension is

$$\Rightarrow \vec{\tau}_T = r(T\cos\theta)\hat{k}. \tag{2.101}$$

Referring to the free body diagram of problem 13, the net vertical force acting on the bob is

$$F_y = T\cos\theta - mg = ma_y. \tag{2.102}$$

Since the bob is revolving in a horizontal plane, its vertical acceleration is zero. Putting $a_y = 0$ in equation (2.101) we have

$$T\cos\theta - mg = 0$$

$$\Rightarrow T = mg\sec\theta. \tag{2.102}$$

Using equations (2.101) and (2.102) we have

$$\Rightarrow \vec{\tau}_T = r(mg\sec\theta \times \cos\theta)\hat{k} = mgr\hat{k}. \tag{2.103}$$

Putting $r = l\sin\theta$ in equation (2.103) we have

$$\Rightarrow \vec{\tau}_T = mgl\sin\theta\hat{k}. \tag{2.104}$$

Putting this value of angle in equation (2.104) we have

$$\Rightarrow \vec{\tau}_T = mgl\sin 60°\hat{k} = \frac{\sqrt{3}}{2}mgl\hat{k}.$$

The gravitational torque about O is

$$\Rightarrow \vec{\tau}_{gr} = -mgr\hat{k}$$

$$= -mgl\sin 60°\hat{k} = -\frac{\sqrt{3}}{2}mgl\hat{k}.$$

So the net torque about O is

$$\vec{\tau}_O = \vec{\tau}_{gr} + \vec{\tau}_T = 0.$$

An alternative method for finding the torque about the point of suspension P and the center O of the circle traced by the bob

Since the acceleration of the bob is directed towards the center O of the circle, the net force ma passes through O. So the net torque about O is zero. However, about P the net torque is due to gravity only because the tension cannot produce any torque about P as it passes through that point.

The horizontal component of tension, that is $T\sin\theta$, produces a horizontal acceleration of the bob relative to the ground. So the net force acting on the bob is

$$\Rightarrow F_{horizontal} = T\sin\theta = ma. \tag{2.105}$$

Since the net torque is equal to the torque of the net force, the torque acting on the bob about P (relative to the ground) is

$$\Rightarrow \vec{\tau_P} = h(T\sin\theta)\hat{k} = -mah\hat{k}. \tag{2.106}$$

Using equations (2.105) and (2.106), we have the net torque acting on the bob about P:

$$\vec{\tau_P} = -(l\cos\theta)(mg\sec\theta)(\sin\theta)\hat{k}$$

$$= -mgl\sin\theta\hat{k} = -mgl\sin 60°\hat{k} = -\frac{\sqrt{3}mgl}{2}\hat{k}.$$

Problem 15 A boy of mass m starts moving from rest from the position shown in the figure. He moves with a horizontal velocity which varies with time as $\vec{v} = (u - at)\hat{i}$, where a is a positive constant. Find the net torque experienced by the particle about the origin O.

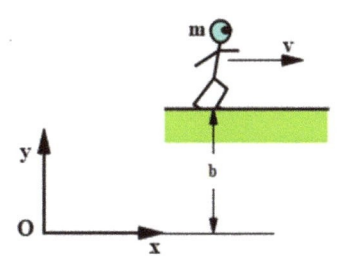

Solution
The angular momentum of the boy about O is

$$\vec{L_O} = -mvr\sin\theta\hat{k}. \tag{2.107}$$

Putting $v = u - at$ in equation (2.107) we have

$$\vec{L_O} = -m(u - at)r\sin\theta\hat{k}.$$

The rate of change of the angular momentum is

$$\frac{d\vec{L_O}}{dt} = \frac{d}{dt}\{-m(u - at)r\sin\theta\hat{k}\}$$

$$\frac{d\vec{L_O}}{dt} = +mar\sin\theta\hat{k}. \tag{2.108}$$

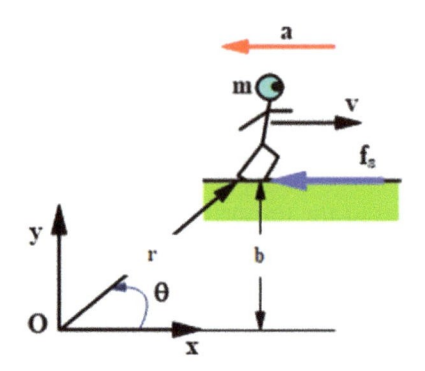

The general dynamical relation between torque and angular momentum about any fixed point O is

$$\vec{\tau_O} = \frac{d\vec{L_O}}{dt}. \tag{2.109}$$

Using equations (2.108) and (2.109) we have

$$\vec{\tau_O} = + mar \sin\theta \hat{k}$$

$$\vec{\tau_O} = + mab\hat{k} (\because r \sin\theta = b).$$

Here the acceleration actually points to the left because the boy is decelerating. So the net force, that is, the static friction is acting to the left. Hence the net torque or the torque of the net force will be counter-clockwise.

Problem 16 A boy B moves with a constant horizontal velocity v. When the position vector BO makes an angle θ with the horizontal, find (a) the angular momentum, (b) the angular velocity and (c) the areal velocity of the particle about the origin.

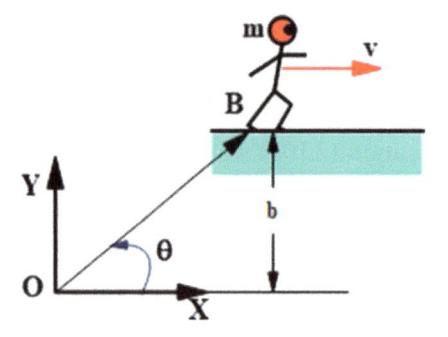

Solution
(a) The angular momentum of the boy about O is

$$\vec{L_O} = -mvr \sin\theta \hat{k}. \tag{2.110}$$

Putting $r \sin \theta = b$ in equation (2.110) we have
$$\vec{L}_O = -mvb\hat{k}.$$
This is a constant vector.
(b) The x-displacement of the boy is
$$x = r \cos \theta = vt$$
$$\Rightarrow \frac{dx}{dt} = \frac{d}{dt}(r \cos \theta) = v.$$
Putting $r = b/\sin \theta$ in the last equation, we have
$$\Rightarrow \frac{d}{dt}\{(b/\sin \theta)\cos \theta)\} = v$$
$$\Rightarrow \frac{d}{dt}(b \cot \theta) = v$$
$$\Rightarrow b \csc^2 \theta \frac{d\theta}{dt} = v$$
$$\Rightarrow -\frac{d\theta}{dt} = \omega = \frac{v}{b \csc^2 \theta} = \frac{v \sin^2 \theta}{b}.$$

The angular velocity of the boy relative to the origin is in the clockwise sense or $-$ve z-direction. So we can state that the angular velocity of the boy relative to the origin is equal to the rate of change of the angle θ, which can be given as
$$\vec{\omega} = -\frac{v \sin^2 \theta}{b}\hat{k}.$$

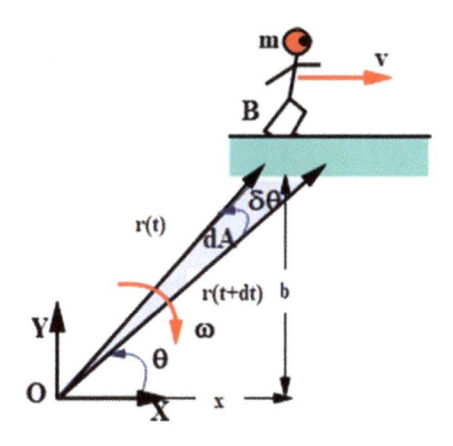

(c) The areal velocity can be given as the rate of area swept by the position vector of the boy relative to the origin, which can be given as

$$\frac{dA}{dt} = \frac{(r^2 d\theta/2)}{dt} = \frac{r^2 d\theta}{2dt} = \frac{r^2 \omega}{2}.$$

Putting $\omega = \frac{v \sin^2 \theta}{b}$ in the last equation we have

$$\frac{dA}{dt} = \frac{r^2 \omega}{2} = \frac{r^2}{2}\left(\frac{v \sin^2 \theta}{b}\right).$$

Putting $r = b/\sin \theta$ in the last equation we have

$$\frac{dA}{dt} = \frac{(b \csc \theta)^2}{2}\left(\frac{v \sin^2 \theta}{b}\right)$$

$$\Rightarrow \frac{dA}{dt} = \frac{vb}{2}.$$

Problem 17 At time $t = 0$, the boy moves from rest with a constant acceleration α from rest from the bottom of the incline. Find (a) the angular momentum and (b) the torque experienced by the boy (c) the torque due to the friction as a function of time about the origin O.

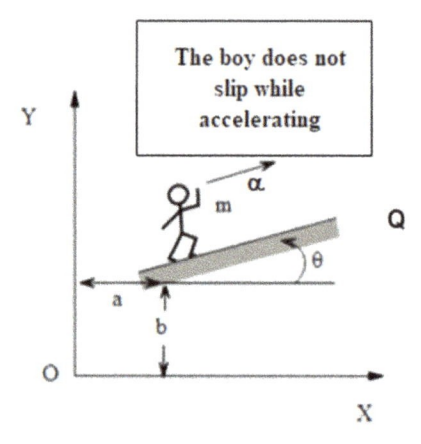

Solution

(a) The angular momentum of the boy about O is

$$\vec{L_O} = -mvp\hat{k}, \qquad (2.111)$$

where the perpendicular distance p is given as

$$p = OM = OP\cos\theta = c \cos \theta (\because OP = c). \qquad (2.112)$$

The value of c can be calculated by using the equation of a straight line PQ, given as
$$y = mx + c. \tag{2.113}$$

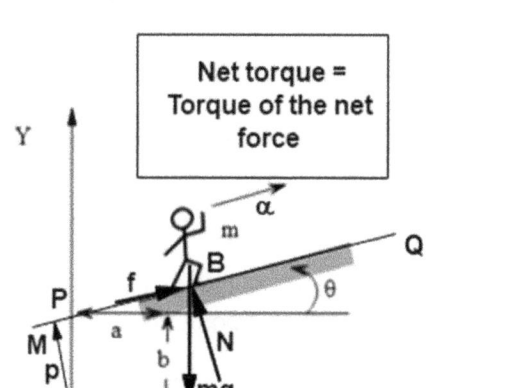

Putting $m = \tan\theta$, $x = a$, and $y = b$, which satisfies the straight line equation, we have
$$b = (\tan\theta)a + c$$
$$\Rightarrow c = b - a\tan\theta. \tag{2.114}$$

Using equations (2.112) and (2.114) we have
$$p = c\cos\theta = (b - a\tan\theta)\cos\theta$$
$$\Rightarrow p = b\cos\theta - a\sin\theta. \tag{2.115}$$

Then the angular momentum of the boy at time t, relative to O, is
$$\vec{L_O} = -mvp\hat{k} \tag{2.116}$$

$$\vec{L_O} = -mv(b\cos\theta - a\sin\theta)\hat{k}. \tag{2.117}$$

Putting $v = at$ we have
$$\vec{L_O} = -ma(b\cos\theta - a\sin\theta)t\hat{k}. \tag{2.118}$$

(b) The rate of change of the angular momentum is
$$\frac{d\vec{L_O}}{dt} = \frac{d}{dt}\{-ma(b\cos\theta - a\sin\theta)t\hat{k}\}$$

$$\frac{d\vec{L_O}}{dt} = -ma(b\cos\theta - a\sin\theta)\hat{k}. \tag{2.119}$$

The general dynamical relation between torque and angular momentum is

$$\vec{\tau_P} = \frac{\overrightarrow{dL_P}}{dt}. \tag{2.120}$$

Using equations (2.119) and (2.120) we have

$$\vec{\tau_P} = -m\alpha(b\cos\theta - a\sin\theta)\hat{k}$$

$$\vec{\tau_P} = -m\alpha(a\sin\theta - b\cos\theta)\hat{k}. \tag{2.121}$$

This physically signifies that the net torque is equal to the torque of the net force when many forces act on a point mass. This torque remains constant in both magnitude and direction.

(c) Resolving the forces along the slant, the static friction that prevents relative sliding of the boy and helps him to accelerate up can be given as

$$f - mg\sin\theta = m\alpha$$

$$f = m(g\sin\theta + \alpha). \tag{2.122}$$

Then the torque due to the friction is

$$\vec{\tau_f} = -fp\hat{k}. \tag{2.123}$$

Using equations (2.122) and (2.123)

$$\vec{\tau_f} = -m(g\sin\theta + \alpha)p\hat{k}. \tag{2.124}$$

From equation (2.115) putting

$$p = b\cos\theta - a\sin\theta,$$

in equation (2.124) we have

$$\vec{\tau_f} = -m(g\sin\theta + \alpha)(b\cos\theta - a\sin\theta)\hat{k}.$$

This torque remains constant in both magnitude and direction.

Problem 18 A particle of mass m is connected to a fixed disc of radius R by a light inextensible string of length l_o. If the particle is given a velocity v perpendicular to the string: (a) discuss the possibility of conservation of momentum and energy and (b) find (i) the tension of the string and (iii) the frictional torque acting on the disc, as a function of time.

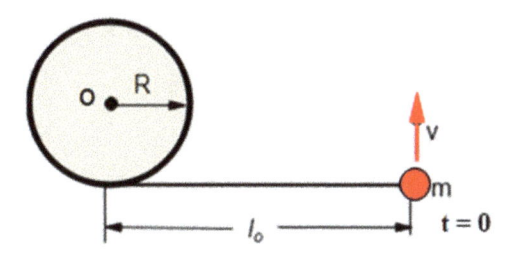

Solution

(a) Since the disc is stationary and the tension in the string tries to rotate the disc, the frictional torque acting on the disc about its center of mass cancels the torque due to the tension. So the frictional torque is nonzero; then, we cannot conserve the angular momentum of the disc–particle system about the center of the disc. Symbolically, if $\tau_f \neq 0$, $\vec{L} \neq C$.

As the tension is always perpendicular to the length of the string, it does not perform work. So, the change in kinetic energy of the particle is zero. In other words, the kinetic energy (speed) of the particle remains conserved.

(b)

(i) As the particle revolves in a curved path, the tension in the string is nonzero. The torque of this tension about O is

$$\tau_O = TR = \frac{dL_O}{dt}.$$

This torque changes the angular momentum L_O of the particle about the point O. Since the torque is equal to the rate of change of the angular momentum, we can write

$$\Rightarrow \frac{dL_O}{dt} = \frac{mv^2}{l} R,$$

where $L_O = mvl$ (anti-clockwise)

$$\Rightarrow \frac{d}{dt}(mvl) = \frac{mv^2}{l} R$$

$$\Rightarrow mv \frac{dl}{dt} = \frac{mv^2}{l} R.$$

As the string is wrapped around the disc, its length will decrease with time. So we have

$$\int_{l_0}^{l} l \, dl = vR \int dt$$

$$\Rightarrow l_0^2 - l^2 = 2vRt.$$

Then the length of the string after a time t is

$$l = \sqrt{l_0^2 - 2vRt}.$$

Then the tension in the string is

$$T = \frac{mv^2}{l} = \frac{mv^2}{\sqrt{l^2 - 2vRt}}.$$

(ii) Then the frictional torque acting on the disc is equal in magnitude and opposite in direction to the torque of the tension. So the frictional torque is

$$\tau_O = TR = \frac{dL_O}{dt}.$$

Putting the value of tension T, we have

$$\Rightarrow \frac{dL_O}{dt} = \frac{mRv^2}{\sqrt{l^2 - 2vRt}}.$$

This means that the angular momentum is not conserved about any point such as the center of the disc and instantaneous point of contact of the unwrapped string with the disc.

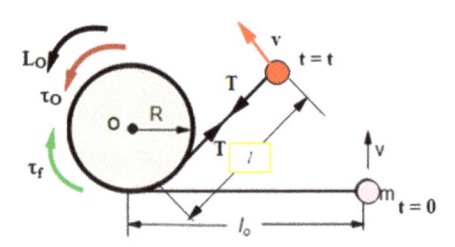

Problem 19 A particle of mass m is connected to a fixed disc of radius R by a light inextensible string of length l_o. If the particle is given a velocity v perpendicular to the string, find (a) the length of the unwrapped string verses time, (b) the distance covered by the bead when the length of the string is l, and (c) the total distance covered by the bead.

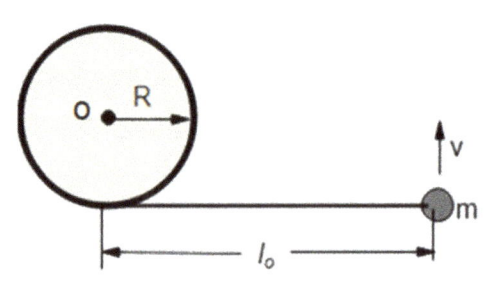

Solution

(a) Let the string have a length l at any time t. Let an elementary length MN = dl become wrapped during an elementary time dt subtending an angle $d\varphi$ at the center of the disc. During this time dt the length of the distance traversed by the particle is $ds = PQ = v dt$. By comparing the similar triangles OMN and QPM, we can write

$$d\varphi = \frac{dl}{R} = \frac{v dt}{l}.$$

Please note that the negative sign is given as the length l of the string decreases with time.

Then, integrating both sides, we have $-\int_{l_0}^{l} l\, dl = vR \int_0^t dt$

$$\Rightarrow l = \sqrt{l_0^2 - 2vRt}.$$

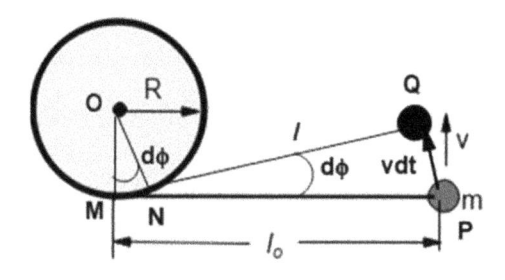

(b) Squaring both sides we have

$$l^2 = l_0^2 - 2vRt.$$

The time after which the length of the string will be equal to l is given as

$$\Rightarrow t = (l_0^2 - l^2)/2vR.$$

Then the distance covered during the time t is

$$\Rightarrow D = vt = (l_0^2 - l^2)/2R.$$

(c) When the entire string is wrapped, putting $l = 0$ in the last expression, the total time is

$$T = \frac{l_0^2}{2vR}.$$

Then the total distance covered during the time T is

$$D = vT = l_0^2/2R.$$

Problem 20 A particle of mass m is connected to a fixed disc of radius R, by a long light inextensible string of length l_o. The string is wrapped over the disc and it does not slip on the disc. Let the particle be given a velocity v_o perpendicular to the string at time $t = 0$ at the given position as shown in the figure. Find (a) the tension (b) the rate of change of angular momentum of the particle relative to the point C as the function of time. The motion of the particle takes place on a smooth horizontal surface.

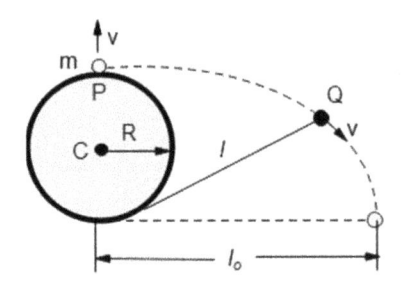

Solution

(a) The angular momentum of the particle about the center O of the disc is

$$L_O = mvl$$

The external frictional torque acting on the disc is

$$\tau_f = \frac{dL_O}{dt} = \frac{d(mvl)}{dt} = mv\frac{dl}{dt} \quad (\because v = \text{constant}) \quad (2.125)$$

Since $\tau_f = TR = \left(\frac{mv^2}{l}\right)R$, we have

$$\frac{mv^2}{l}R = mv\frac{dl}{dt}$$

$$\Rightarrow Rv\,dt = l\,dl$$

Integrating both sides, we have

$$Rv\int_0^t dt\,dt = \int_0^l l\,dl$$

This gives us

$$t = \frac{l^2}{2vR}$$

2-56

Then the tension is given as

$$T = \frac{mv^2}{l} = \frac{mv^2}{\sqrt{2vRt}} = \frac{mv^{3/2}}{\sqrt{2Rt}}$$

(b) Then the rate of change of angular momentum about the center C is equal to the frictional torque whose magnitude is equal and to the frictional torque given as

$$\tau_f = TR = \frac{mv^{3/2}}{\sqrt{2Rt}} R = \frac{mv^{3/2} R^{1/2}}{\sqrt{2t}}$$

Tension decreases from infinite value to a finite value $\frac{mv^2}{l_0}$ and the torque also decreases from an infinite value to a finite value $\frac{mv^2 R}{l_0}$.

Problem 21 A small smooth ball of mass m is projected horizontally from a point A inside a fixed hemispherical bowl with a velocity v_o. As a result it goes to point B before it comes down. Find v_o.

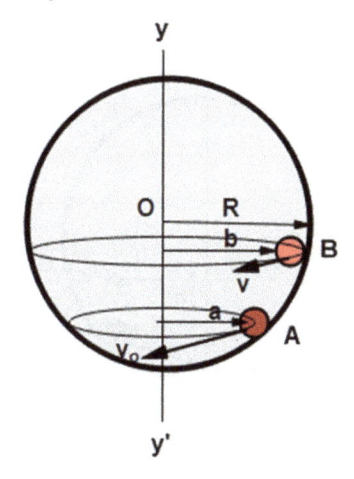

Solution
The velocities of the ball at the top point B is horizontal because all its vertical velocity will be reduced to zero. Since the normal reaction and gravity cannot produce any torque about the vertical axis, the angular momentum of the ball about the vertical axis remains conserved. If v is the velocity at A, and the radii of the circular path of the ball at A and B are a and b, respectively, conserving angular momentum about the vertical axis, we have

$$mv_o a = mvb$$

$$\Rightarrow v_o a = vb. \qquad (2.126)$$

Conserving the energy of the ball between A and B, we have
$$\Delta K + \Delta U = 0$$
$$\Rightarrow -\left(\frac{1}{2}mv_0^2 - \frac{1}{2}mv^2\right) + (+mgh) = 0$$
$$\Rightarrow v^2 = v_0^2 - 2gh. \tag{2.127}$$

Using equations (2.126) and (2.127)
$$\Rightarrow (av_0/b)^2 = v_0^2 - 2gh$$
$$\Rightarrow v_0^2 - (av_0/b)^2 = 2gh$$
$$\Rightarrow v_0^2\left\{1 - \left(\frac{a}{b}\right)^2\right\} = 2gh$$
$$\Rightarrow v_0^2 = 2gh/\left\{1 - \left(\frac{a}{b}\right)^2\right\}. \tag{2.128}$$

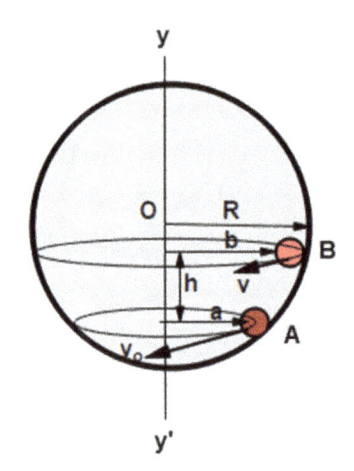

The vertical distance shifted by the ball is given as
$$\Rightarrow h = \sqrt{(R^2 - a^2)} - \sqrt{(R^2 - b^2)}. \tag{2.129}$$

Using equations (2.128) and (2.129), we have
$$\Rightarrow v_0^2 = 2g\left\{\sqrt{(R^2 - a^2)} - \sqrt{(R^2 - b^2)}\right\}/\left\{1 - \left(\frac{a}{b}\right)^2\right\} = 2gb^2\frac{\sqrt{(R^2 - a^2)} - \sqrt{(R^2 - b^2)}}{(b^2 - a^2)}$$

$$v_0 = \sqrt{2gb^2\frac{\sqrt{(R^2 - a^2)} - \sqrt{(R^2 - b^2)}}{(b^2 - a^2)}}.$$

2-58

Problem 22 A small smooth ball of mass m is projected horizontally at a height $h = R/5$ inside a fixed hemispherical bowl with a velocity v_o. As a result it goes to the height $H = 2R/5$. Find (a) v_o and (b) the velocity v of the block at the top of the cone.

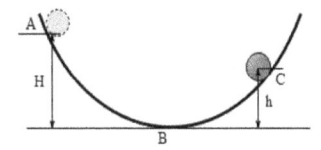

Solution
(a) From the last problem we have

$$\Rightarrow v_o^2 = 2gh/\left\{1 - \left(\frac{a}{b}\right)^2\right\}.$$

Put $h = 4R/5 - 2R/5 = 2R/5$ and

$$\left(\frac{a}{b}\right)^2 = \frac{a^2}{b^2}$$

$$= \frac{R^2 - (R-h)^2}{R^2 - (R-H)^2}$$

$$= \frac{R^2 - \{R - (R/5)\}^2}{R^2 - \{R - (2R/5)\}^2}$$

$$= \frac{R^2 - (4R/5)^2}{R^2 - (3R/5)^2}$$

$$= \frac{9}{16}.$$

Then we have

$$\Rightarrow v_o^2 = 2g(2R/5)/\{1 - 9/16\}$$

$$\Rightarrow v_o^2 = 2g(2R/5)/(7/16)$$

$$\Rightarrow v_o^2 = 64gR/35$$

$$\Rightarrow v_o = 8\sqrt{gR/35}.$$

(b) The velocity at the top is

$$v = av_o/b = \frac{a}{b}\left(8\sqrt{gR/35}\right)$$

$$v = \frac{3}{4}\left(8\sqrt{gR/35}\right)$$

$$v = 6\sqrt{gR/35}.$$

2-59

Problem 23 A pendulum bob of mass m is hanging by an inextensible light string of length l from a ceiling. The bob is shifted horizontally so that the string makes an angle θ with the vertical. Now the bob is projected horizontally with a velocity v_o along the tangent drawn to the dotted circle. As a result, the bob goes up to the level of the ceiling. Find the (a) magnitude of v_o (b) speed of the bob when it touches the ceiling.

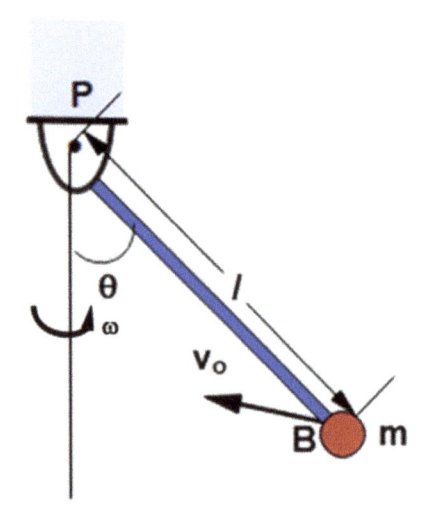

Solution
(a) The velocity of the bob at its top point A is horizontal because all its vertical velocity will be reduced to zero. The normal reaction and gravity cannot produce any torque about the vertical axis, so the angular momentum of the ball about the vertical axis remains conserved. If v and R are the velocity at A, and the radii of the circular path of the ball at A and B are a and b, respectively, conserving angular momentum about the vertical axis passing through point P, we have

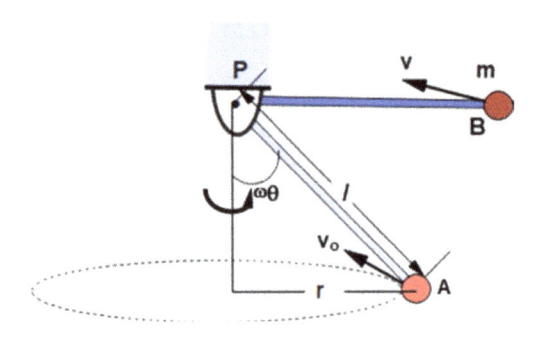

$$mv_o = mvl$$

$$\Rightarrow v_o r = vl.$$

Putting $r = l \sin \theta$ in the last equation, we have
$$\Rightarrow v_o l \sin \theta = vl$$
$$\Rightarrow v = v_o \sin \theta. \tag{2.130}$$

Conserving the energy of the ball between A and B, we have
$$\Delta K + \Delta U = 0$$
$$\Rightarrow \left(\frac{1}{2}mv^2 - \frac{1}{2}mv_o^2\right) + (mgh) = 0$$
$$\Rightarrow v^2 = v_o^2 - 2gh. \tag{2.131}$$

Putting $r = l \cos \theta$ in the last equation, we have
$$\Rightarrow v^2 = v_o^2 - 2gl \cos \theta. \tag{2.132}$$

Using equations (2.130) and (2.132)
$$\Rightarrow (v_o \sin \theta)^2 = v_o^2 - 2gl \cos \theta$$
$$\Rightarrow v_o^2 - (v_o \sin \theta)^2 = 2gl \cos \theta$$
$$\Rightarrow v_o^2(1 - \sin^2 \theta) = 2gl \cos \theta$$
$$\Rightarrow v_o = \sqrt{2gl/\cos \theta}.$$

(b) From equation (2.130), the velocity at the top is
$$v = v_o \sin \theta = (\sqrt{2gl/\cos \theta})\sin \theta.$$

Problem 24 A small ball is projected from a point on the internal surface of a smooth hemispherical bowl with horizontal velocity v_1 so that the bead will go to the top of the bowl. If the ball is projected with a velocity v_2 from the same initial point it just reaches the top of the bowl moving in a vertical plane. If the bead is released from rest from the top of the bowl it reaches the bottom of the bowl with a speed v_3. Relate these three velocities.

Solution
Let the initial distance of the ball from the vertical axis be r and the velocity of the ball at the top of the bowl be v. Then, conserving the angular momentum, we have
$$mv_1 r = mvR$$
$$\Rightarrow v_1 r = vR. \tag{2.133}$$

Putting $r = R \cos \theta$ in the last equation we have
$$\Rightarrow v_1 R \cos \theta = vR$$

2-61

$$\Rightarrow v = v_1 \cos \theta. \quad (2.134)$$

Conserving the energy of the ball between these two positions we have

$$\Delta K + \Delta U = 0$$

$$\Rightarrow \left(\frac{1}{2}mv^2 - \frac{1}{2}mv_1^2\right) + (mgh) = 0$$

$$\Rightarrow v^2 = v_1^2 - 2gh. \quad (2.135)$$

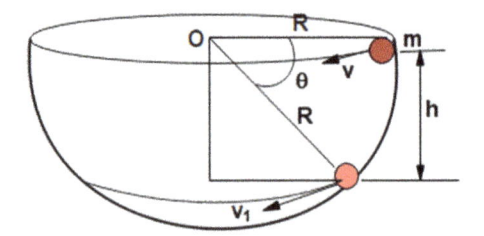

Putting $h = R \sin \theta$ in equation (2.135) we have

$$\Rightarrow v^2 = v_o^2 - 2gR \sin \theta. \quad (2.136)$$

Using equations (2.133) and (2.136)

$$\Rightarrow (v_o \cos \theta)^2 = v_o^2 - 2gR \sin \theta$$

$$\Rightarrow v_o^2 - (v_o \cos \theta)^2 = 2gR \sin \theta$$

$$\Rightarrow v_o^2(1 - \cos^2 \theta) = 2gR \sin \theta$$

$$\Rightarrow v_1^2 = 2gR/\sin \theta. \quad (2.137)$$

If the ball is projected in a vertical plane with a velocity v_2, it just reaches the top of the bowl with zero velocity. So, conserving energy, we have

$$\Delta K + \Delta U = 0$$

$$\Rightarrow \left(0 - \frac{1}{2}mv_2^2\right) + (mgh) = 0$$

$$\Rightarrow v_2^2 = 2gR \sin \theta \; (\because h = R \sin \theta). \quad (2.138)$$

If the ball is released from rest from the top of the bowl, in moves a vertical plane and reaches the bottom with a velocity v_3. Conserving energy, we have

$$\Delta K + \Delta U = 0$$

$$\Rightarrow \left(\frac{1}{2}mv_3^2\right) + (-mgR) = 0$$

$$\Rightarrow v_3^2 = 2gR. \tag{2.139}$$

Using last three equations we have

$$v_1 v_2 = v_3^2.$$

Problem 25 A small ball of mass m is projected horizontally from a height h inside a cone with a velocity v_o. As a result, it goes to the top of the cone acquires a velocity v at the top position. Find (a) v_o and (b) v in terms of R, r and semi-vertical angle of the cone.

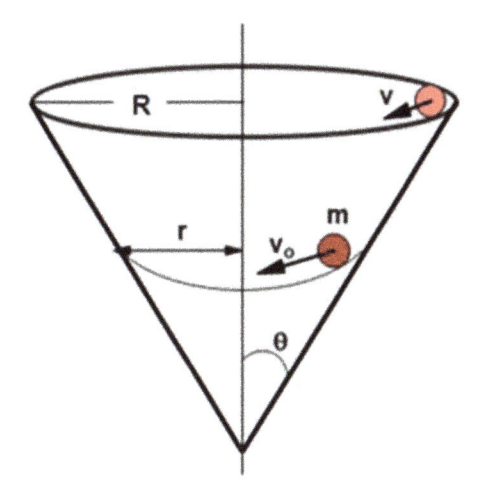

Solution
(a) The normal reaction and gravity cannot produce any torque about the vertical axis. So the angular momentum of the ball about the vertical axis remains conserved. Conserving angular momentum between the lowest and highest positions of the ball, we have

$$mv_o r = mvR$$

$$\Rightarrow v_o r = vR. \tag{2.140}$$

Conserving the energy of the ball between these two positions, we have

$$\Delta K + \Delta U = 0$$

$$\Rightarrow \left(\frac{1}{2}mv^2 - \frac{1}{2}mv_o^2\right) + (mgh) = 0$$

$$\Rightarrow v^2 = v_o^2 - 2gh. \tag{2.141}$$

Putting $h = (R - r)\cot\theta$ in the equation (2.141) we have

$$\Rightarrow v^2 = v_o^2 - 2g(R - r)\cot\theta. \tag{2.142}$$

Using equations (2.140) and (2.142)

$$\Rightarrow \frac{v_0^2 r^2}{R^2} = v_0^2 - 2g(R-r)\cot\theta$$

$$\Rightarrow v_0^2 - \frac{v_0^2 r^2}{R^2} = 2g(R-r)\cot\theta$$

$$\Rightarrow v_0^2\left(1 - \frac{r^2}{R^2}\right) = 2g(R-r)\cot\theta$$

$$\Rightarrow v_0^2 = \frac{2gR^2 \cot\theta}{R+r}$$

$$\Rightarrow v_0 = \sqrt{\frac{2gR^2 \cot\theta}{R+r}}.$$

(b) From equation (2.140) the velocity at the top is

$$v = \left(\sqrt{\frac{2gR^2 \cot\theta}{R+r}}\right)(r/R)$$

$$\Rightarrow v = \left(\sqrt{\frac{2gr^2 \cot\theta}{R+r}}\right).$$

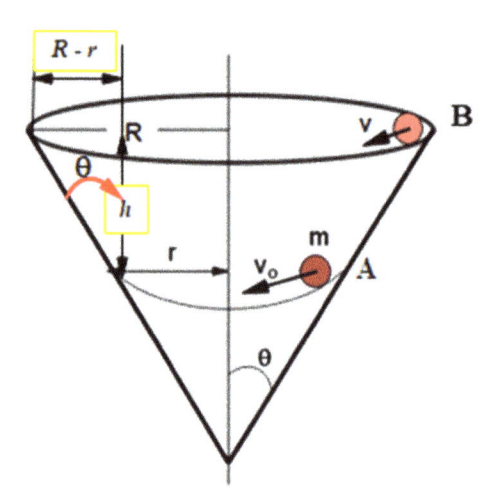

Problem 26 A bead of mass m is released from rest from $r = l/2$ in a light smooth tube. If the tube is given an initial angular velocity ω_0 and released, find the velocity of the bead relative to ground at the time of escaping from the tube.

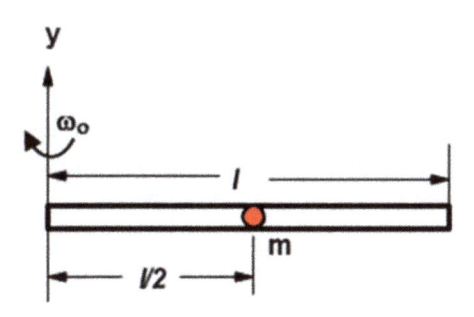

Solution

In this case there is no external torque acting on the tube–bead system. So the energy of the system remains constant. Since the tube is light, we can conserve the kinetic energy and angular momentum of the bead.

Conserving the kinetic energy of the bead at the time of escaping of the tube we have

$$\frac{m}{2}\left(\frac{l}{2}\omega_o\right)^2 = \frac{1}{2}mv^2,$$

where $v =$ the speed of the bead at the time of escaping the tube:

$$v = l\omega_o/2 \tag{2.143}$$

Conservation of angular momentum between A and B yields, $\omega_o\left(\frac{ml^2}{4}\right) = \omega(ml^2)$

$$\Rightarrow \omega = \frac{\omega_o}{4}. \tag{2.144}$$

If the velocity of the bead relative to the tube is v_r, it can be given as

$$v^2 = v_r^2 + (l\omega)^2. \tag{2.145}$$

Using equations (2.143), (2.144), and (2.145) we have

$$(l\omega_o/2)^2 = v_r^2 + (l\omega_o/4)^2$$

$$\Rightarrow v_r^2 = 3(l\omega_o)^2/16$$

$$v_r = \sqrt{3}(l\omega_o)/4.$$

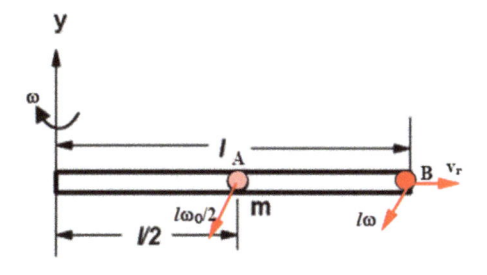

Problem 27 A small ball of mass m is released from rest from $r_o = b$ in a light smooth tube. If the the tube is given an initial angular velocity ω_o and released, at any radial distance r, find (a) the velocity of the ball relative to the tube, (b) the horizontal contact force acting on the ball by the tube, (c) the tangential acceleration of the ball, and (d) the radius of curvature of the path followed by the ball (e) the differential equation of the ball relative to the tube.

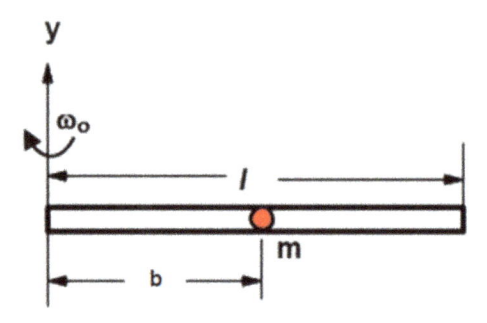

Solution
(a) In this case there is no external torque acting on the tube–ball system. So the energy of the system remains constant. Since the tube is light and smooth, we can conserve the kinetic energy and angular momentum of the ball.

Conserving the kinetic energy of the ball at the time of escaping from the tube we have

$$\frac{m}{2}(b\omega_o)^2 = \frac{1}{2}mv^2,$$

where $v =$ the speed of the bead at the radial distance r,

$$v = b\omega_o. \tag{2.146}$$

Conservation of angular momentum yields $\omega_o(mb^2) = \omega(mr^2)$

$$\omega = \omega_o(b^2/r^2). \tag{2.147}$$

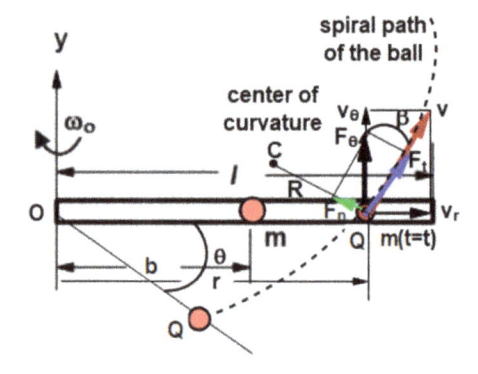

If the velocity of the ball relative to the tube is v_r, it can be given as

$$v^2 = v_r^2 + (r\omega)^2. \tag{2.148}$$

2-66

Using equations (2.146), (2.147), and (2.148) we have
$$(b\omega_o)^2 = v_r^2 + (b^2\omega_o/r^2)^2(r^2)$$
$$\Rightarrow v_r^2 = (b\omega_o)^2 - (b^2\omega_o/r^2)^2(r^2)$$
$$\Rightarrow v_r^2 = (b\omega_o)^2 - (b^4\omega_o^2)/r^2$$
$$\Rightarrow v_r = (b\omega_o)\sqrt{\left(1 - \frac{b^2}{r^2}\right)}. \tag{2.149}$$

(b) The horizontal contact force (reaction force) of the tube on the ball is
$$F_\theta = 2m\omega v_r. \tag{2.150}$$
Using equations (2.149) and (2.150) we have
$$F_\theta = 2m\omega(b\omega_o)\sqrt{\left(1 - \frac{b^2}{r^2}\right)}. \tag{2.151}$$

Using equations (2.147) and (2.151) we have
$$F_\theta = 2m\{\omega_o(b^2/r^2)\}\left\{(b\omega_o)\sqrt{\left(1 - \frac{b^2}{r^2}\right)}\right\}$$
$$F_\theta = \frac{2m\omega_o^2 b^3 \sqrt{r^2 - b^2}}{r^3}. \tag{2.152}$$

(c) The tangential force is
$$F_t = F_\theta \cos\beta = ma_t$$
$$a_t = F_t/m = F_\theta \cos\beta/m$$
$$= \frac{2\omega_o^2 b^3 \sqrt{r^2 - b^2}}{r^3} \cos\beta$$
$$= \frac{2\omega_o^2 b^3 \sqrt{r^2 - b^2}}{r^3}\left(\frac{v_\theta}{v}\right)$$
$$= \frac{2\omega_o^2 b^3 \sqrt{r^2 - b^2}}{r^3}\left(\frac{r\omega}{b\omega_o}\right). \tag{2.153}$$

Putting $\omega/\omega_o = (b^2/r^2)$ from equation (2.147) in equation (2.153) we have
$$a_t = \frac{2\omega_o^2 b^3 \sqrt{r^2 - b^2}}{r^3}\left(\frac{rb^2}{br^2}\right)$$

$$\Rightarrow a_t = \frac{2w_o^2 b^4 \sqrt{r^2 - b^2}}{r^4}.$$

(d) The normal or centripetal force is
If R is the radius of curvature, the centripetal force is equal to mv^2/R. So we can write

$$F_{cp} = F_\theta \sin \beta = mv^2/R$$

$$\Rightarrow \frac{2mw_o^2 b^3 \sqrt{r^2 - b^2}}{r^3} \left(\frac{v_r}{v}\right) = \frac{mv^2}{R}$$

$$\Rightarrow \frac{2w_o^2 b^3 v_r \sqrt{r^2 - b^2}}{r^3} = \frac{v^3}{R}$$

$$\Rightarrow R = \frac{v^3 r^3}{2w_o^2 b^3 v_r \sqrt{r^2 - b^2}}. \tag{2.154}$$

Putting $v = bw_o$ from equation (2.146) and $v/v_r = r/\sqrt{r^2 - b^2}$ from equation (2.149) in equation (2.154), we have
Using equations (2.146) and (2.153) we have

$$\Rightarrow R = \frac{(bw_o)^2 r^3}{2w_o^2 b^3 \sqrt{r^2 - b^2}} \frac{v}{v_r}$$

$$= \frac{r^3}{2b\sqrt{r^2 - b^2} \sqrt{r^2 - b^2}} = \frac{r}{\sqrt{r^2 - b^2}} = \frac{r^4}{2b(r^2 - b^2)}.$$

(e) The differential equation for the ball is given as

$$\frac{d^2 r}{dt^2} - rw^2 = 0, \tag{2.155}$$

where the angular velocity of the tube is given by equation (2.147) as

$$w = \frac{b^2 w_o}{r^2}.$$

In equation (2.155) we have

$$\frac{d^2 r}{dt^2} = r \left(\frac{b^2 w_o}{r^2}\right)^2$$

$$= \frac{d^2 r}{dt^2} = \frac{b^4 w_o^2}{r^3}.$$

Problem 28 In the previous problem, find (a) the time taken by the ball to reach the end of the tube, (b) velocity of the ball relative to the tube as the function of time and (c) the position, velocity, and acceleration of the ball with time.

Solution
(a) Recasting equation (2.149) of the previous problem we have

$$v_r = (b\omega_o)\sqrt{\left(1 - \frac{b^2}{r^2}\right)}.$$

Putting $v_r = dr/dt$ we have

$$\frac{dr}{dt} = \left(\frac{b\omega_o}{r}\right)\sqrt{(r^2 - b^2)}.$$

Separating the variables we have

$$b\omega_o dt = \frac{rdr}{\sqrt{(r^2 - b^2)}}.$$

Integrating both sides,

$$\int_0^t b\omega_o dt = \int_b^r \frac{rdr}{\sqrt{(r^2 - b^2)}}$$

$$\Rightarrow b\omega_o t = \frac{1}{2}\int_b^r \frac{2rdr}{\sqrt{(r^2 - b^2)}}.$$

Evaluating the integral we have

$$b\omega_o t = \sqrt{(r^2 - b^2)} \qquad (2.156)$$

So putting $r = l$ in equation (2.156), the time after which the ball reaches the end of the tube is given as

$$\Rightarrow t = \frac{\sqrt{l^2 - b^2}}{b\omega_o}.$$

(b) The radial position as the function of time is

$$r = \sqrt{b^2 + (b\omega_o t)^2}. \qquad (2.157)$$

Differentiating equation (2.157) with time, the velocity of the ball relative to tube is

$$v_r = \frac{dr}{dt} = \frac{d}{dt}\left(\sqrt{b^2 + b^2\omega_o^2 t^2}\right)$$

$$\Rightarrow v_r = \frac{b\omega_o^2 t}{\sqrt{1 + \omega_o^2 t^2}}. \qquad (2.158)$$

(c) The expression for radial acceleration is given as

$$a_r = \left(\frac{d^2r}{dt^2} - \omega^2 r\right) = F_r/m. \qquad (2.159)$$

Since the tube is smooth, the force along the tube, that is the radial force, is zero. So the radial acceleration is zero. Putting $F_r = 0$ or $a_r = 0$ in equation (2.159), we have

$$\frac{d^2r}{dt^2} = \omega^2 r,$$

as obtained as equation (2.155), where the angular velocity of the rod is given by equation (2.147) of the previous problem as

$$\omega = = \frac{b^2 \omega_0}{r^2}.$$

Since there is no net force acting on the ball along the tube, the net force acting on the ball is equal to the transverse force which is given by equation (2.152) of the previous problem, as

$$F_{net} = F_\theta = \frac{2m\omega_0^2 b^3 \sqrt{r^2 - b^2}}{r^3}.$$

So the acceleration of the ball is

$$a = F_{net}/m = \frac{2\omega_0^2 b^3 \sqrt{r^2 - b^2}}{r^3}. \qquad (2.160)$$

Putting $r = f(t)$ from equation (2.157) in equation (2.160) we can find the acceleration as the function of time.

Problem 29 A small iron ball of mass m is connected to one end of a light spring of natural length l and stiffness k. The other end of the spring is fixed in a smooth horizontal plane so that it can rotate freely about its free end. Let the particle be given a horizontal velocity v_0 perpendicular to the spring. (a) If v_0 is small, find the maximum elongation x of the spring. (b) If the maximum length of the spring is equal to $l/2$, find the velocity v_0.

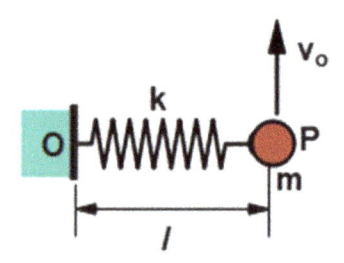

Solution
(a) At the maximum elongation x of the spring, the radial velocity of the ball is zero. Then the velocity of the ball will be perpendicular to the spring. Conserving the angular momentum of the ball about point O, we have

$$mv_0 l = mv(l + x)$$

$$\Rightarrow v = \frac{v_0 l}{(l + x)}. \tag{2.161}$$

Conserving the energy between the two extreme positions,

$$\frac{1}{2}mv_0^2 = \frac{1}{2}mv^2 + \frac{1}{2}kx^2$$

$$\Rightarrow kx^2 = m(v_0^2 - v^2). \tag{2.162}$$

Putting v from equation (2.161) in equation (2.162),

$$kx^2 = m\left\{v_0^2 - \left(\frac{v_0 l}{l+x}\right)^2\right\}$$

$$\Rightarrow kx^2 = mv_0^2\left\{1 - \left(\frac{l}{l+x}\right)^2\right\}$$

$$\Rightarrow kx^2 = mv_0^2\left\{1 - \left(1 + \frac{x}{l}\right)^{-2}\right\}. \tag{2.163}$$

We know that when $X \ll 1$

$$(1 + X)^n \cong 1 + nX, \tag{2.164}$$

where $X = x/l$ and $n = -2$.

Using this binomial expansion, we have

$$\Rightarrow kx^2 \cong mv_0^2\left\{1 - \left(1 - 2\frac{x}{l}\right)\right\}$$

$$\Rightarrow kx^2 \cong mv_0^2\left(2\frac{x}{l}\right)$$

2-71

$$\Rightarrow x \simeq \frac{2mv_o^2}{kl}.$$

(b) Putting $x = l/2$ in equation (2.163), we have

$$k(l/2)^2 = mv_o^2 \left\{ 1 - \left(1 + \frac{l/2}{l}\right)^{-2} \right\}$$

$$\Rightarrow \frac{kl^2}{4} = \frac{5}{9} mv_o^2$$

$$\Rightarrow v_o = \frac{3}{2}\sqrt{\frac{k}{5m}} l.$$

Problem 30 In the previous problem, if the ball is given an initial velocity $v_o = \sqrt{\frac{4k}{3m}} l$, find the (a) maximum length of the spring (b) radius of curvature of the path of the ball at the time of maximum elongation of the spring (c) velocity of the ball relative to the ground (d) velocity of the ball along the spring at the extension of the spring $x = l/2$.

Solution

(a) From the previous problem we have

$$\Rightarrow kx^2 = mv_o^2 \left\{ 1 - \left(1 + \frac{x}{l}\right)^{-2} \right\}. \qquad (2.165)$$

Putting $v_o = \sqrt{\frac{4k}{3m}} l$ in equation (2.165) we have

$$kx^2 = m \left(\sqrt{\frac{4k}{3m}} l\right)^2 \left\{ 1 - \left(1 + \frac{x}{l}\right)^{-2} \right\}$$

$$\Rightarrow kx^2 = m \frac{4k}{3m} l^2 \left\{ 1 - \left(\frac{l}{l+x}\right)^2 \right\}$$

$$\Rightarrow x^2 = \frac{4}{3} l^2 \left\{ 1 - \left(\frac{l}{l+x}\right)^2 \right\}$$

$$\Rightarrow 3x^3 + 6lx^2 - l^2 x - 8l^3 = 0$$

$$\Rightarrow (x - l)(3x^2 + 9lx + 8l^2) = 0$$

$$\Rightarrow x = l, \text{ because the equation}$$

$(3x^2 + 9lx + 8l^2) = 0$ gives imaginary roots.

2-72

So the maximum length of the spring is
$$l_{max} = l + l = 2l.$$

(b) From the previous problem, at the maximum length of the spring the velocity of the ball is given as
$$\Rightarrow v = \frac{v_o l}{(l+x)} = \frac{v_o l}{(l+l)} = \frac{v_o}{2}. \tag{2.166}$$

The force acting on the ball is
$$F = kx = kl. \tag{2.167}$$

The spring force is the centripetal force. If the radius of curvature is r at the point R as shown in the figure of the last problem, then we can write
$$F_{cp} = kl = mv_R^2/r. \tag{2.168}$$

Putting $v_R = v = v_o/2$ from equation (2.166) in equation (2.168) we have
$$kl = \frac{mv_o^2}{4r}$$

$$\Rightarrow r = \frac{mv_o^2}{4kl}. \tag{2.169}$$

Putting $v_o = \sqrt{\frac{4k}{3m}}\, l$ in equation (2.169) we have
$$\Rightarrow r = \frac{m(4kl^2/3m)}{4kl} = l/3.$$

(c) At $x = l/2$, let the radial and transverse velocities of the ball be v_r and v', respectively, as shown in the last figure.

Conserving the angular momentum about O, we have
$$mv_o l = mv'(l + l/2)$$

$$\Rightarrow v' = \frac{2v_o}{3}. \tag{2.170}$$

Conserving energy between two positions P and Q we have
$$\frac{1}{2}mv_o^2 = \frac{1}{2}mv^2 + \frac{1}{2}kx^2$$

$$\Rightarrow kx^2 = m(v_o^2 - v^2). \tag{2.171}$$

Putting $x = l/2$ in equation (2.171) we have
$$\Rightarrow kl^2/4 = m(v_0^2 - v^2). \tag{2.172}$$

Putting $v_0 = \sqrt{\frac{4k}{3m}} l$ (given) in equation (2.172) we have
$$kl^2/4 = m\left\{\left(\sqrt{\frac{4k}{3m}} l\right)^2 - v^2\right\}$$

$$\Rightarrow \frac{kl^2}{4} = \frac{4k}{3} l^2 - mv^2$$

$$\Rightarrow mv^2 = \frac{4k}{3} l^2 - \frac{kl^2}{4}$$

$$\Rightarrow v = \sqrt{\frac{13k}{12m}} l. \tag{2.173}$$

(d) The kinematical relation between the radial, transverse, and total velocity at Q is given as
$$\Rightarrow v^2 = v_r^2 + v'^2. \tag{2.174}$$

Putting $v' = \frac{2v_0}{3}$ from equation (2.170) in equation (2.174) we have
$$\Rightarrow v^2 = v_r^2 + \left(\frac{2v_0}{3}\right)^2$$

$$\Rightarrow v^2 = v_r^2 + \frac{4v_0^2}{9}$$

$$\Rightarrow v^2 = v_r^2 + \frac{4v_0^2}{9}. \tag{2.175}$$

Putting $v = \sqrt{\frac{13k}{12m}} l$ from equation (2.173) and $v_0 = \sqrt{\frac{4k}{3m}} l$ (given) in equation (2.175) we have

$$\Rightarrow \left(\sqrt{\frac{13k}{12m}} l\right)^2 = v_r^2 + \frac{4}{9}\left(\sqrt{\frac{4k}{3m}} l\right)^2$$

$$\Rightarrow v_r^2 = \frac{k}{m} l^2 \left(\frac{13}{12} - \frac{16}{27}\right)$$

$$\Rightarrow v_r = \sqrt{\frac{53k}{108m}} l.$$

2-74

Problems and Solutions in Rotational Mechanics

Problem 31 A smooth block of mass m is connected with one end of a light spring of natural length $r_0 = l$ and stiffness k. The other end of the spring is fixed with the other end of the horizontal tube. Let the tube be given an angular velocity w_0. If the maximum elongation of the spring will be equal to $l/3$, find (a) w_0 and the angular velocity of the tube at the time of maximum elongation of the spring w, (b) the work done by the spring on the block, and (c) the work done by the tube on the block (d) differential equation of the block as the function of radial distance r from the axis of rotation AB. Put $l = 1$ m, $k = 81$ N m^{-1}, and $m = 1/7$ m.

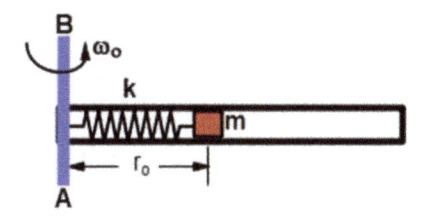

Solution
(a) At the maximum elongation of the spring, the radial velocity of the block is zero. Then the velocity of the block will be perpendicular to the spring. Conserving the angular momentum of the block between the positions A and B about

$$ml^2 w_0 = m(l + l/3)^2 w$$

$$\Rightarrow w = 9w_0/16. \tag{2.176}$$

Conserving the energy between the two extreme positions,

$$K_o = K + U_{sp}$$

$$\Rightarrow \frac{1}{2}ml^2 w_0^2 = \frac{1}{2}m(l + l/3)^2 w^2 + \frac{1}{2}k(l/3)^2$$

$$\Rightarrow w_0^2 = \frac{16}{9}w^2 + \frac{k}{9m}. \tag{2.177}$$

Putting $w = 9w_0/16$ from equation (2.176) in equation (2.177) we have

$$w_0^2 = \frac{16}{9}(9w_0/16)^2 + \frac{k}{9m}$$

2-75

$$\Rightarrow \omega_o^2 = \frac{9}{16}\omega_o^2 + \frac{k}{9m}$$

$$\Rightarrow \omega_o^2 - \frac{9}{16}\omega_o^2 = \frac{k}{9m}$$

$$\Rightarrow \frac{7}{16}\omega_o^2 = \frac{k}{9m}$$

$$\Rightarrow \omega_o = \sqrt{\frac{16k}{63m}} = \frac{4}{3}\sqrt{\frac{k}{7m}}$$

$$\Rightarrow \omega_o = \frac{4}{3}\sqrt{\frac{81}{7(1/7)}} = 12 \text{ rad s}^{-1}.$$

Then from equation (2.176)

$$\Rightarrow \omega = 9\omega_o/16 = 9 \times 12/16 = 27/4 \text{ rad s}^{-1}.$$

(b) The work done by the spring is
$$W_{sp} = K - K_o = -\Delta U_{sp}$$

$$= -\frac{1}{2}k(l/3)^2 = -\frac{1}{2}(81)(1/3)^2 = -4.5\text{J}.$$

(c) The work done by the rod on the block is
$$W_{rod} = -W_{sp} = 4.5 \text{ J}$$

because the net work done is zero.

(d) The net force acting on the block along the tube is given as

$$F_r = ma_r = -kx$$

$$\Rightarrow m\left(\frac{d^2r}{dt^2} - r\omega^2\right) = -kx, \qquad (2.178)$$

where the angular velocity of the rod is given by equation (2.177) as

$$\omega = = \frac{l^2\omega_o}{r^2} \text{ and } x = r - l$$

$$\Rightarrow \frac{d^2r}{dt^2} - r\left(\frac{l^2\omega_o}{r^2}\right)^2 = -\frac{k}{m}(r - l)$$

$$\Rightarrow \frac{d^2r}{dt^2} - \frac{l^4\omega_0^2}{r^3} = -\frac{k}{m}(r-l)$$

$$\Rightarrow \frac{d^2r}{dt^2} - \frac{l^4\omega_0^2}{r^3} + \frac{kr}{m} = \frac{kl}{m}.$$

Putting in the given values of the parameters we have

$$\frac{d^2r}{dt^2} - \frac{(1)^4(12)^2}{r^3} + \frac{81r}{1/7} = \frac{81(1)}{1/7}$$

$$\frac{d^2r}{dt^2} - \frac{144}{r^3} + 567r = 567.$$

Problem 32 Two blocks of mass M and m are connected to the ends of a light spring of stiffness k and natural length l. The spring–particle system is placed in a smooth horizontal tube. The axis of rotation of the tube coincides with the center of mass of the spring–particle system. Let the tube be given an angular velocity ω_o. What is the maximum elongation x of the spring if $(x \ll l)$ as the spring is very stiff?

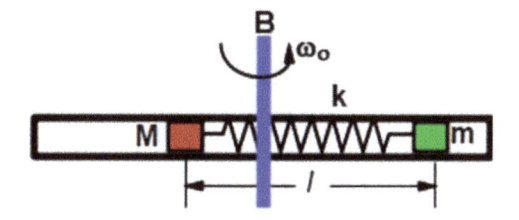

Solution
Since there is no external force acting on the system $(M+m)$, its center of mass does not move. So, the kinetic energy of translation is zero. Then the kinetic energy of the system is equal to the kinetic energy of rotation, given as

$$K_i = \frac{Mml^2}{2(M+m)}\omega_o^2.$$

At the maximum elongation x of the spring, the kinetic energy of the system is

$$K_f = \frac{Mm(l+x)^2}{2(M+m)}\omega^2.$$

Then, conserving the energy of the system we have

$$\Delta K_{\text{system}} + \Delta U_{\text{spring}} = 0$$

$$\Rightarrow \frac{Mm}{2(M+m)}(l+x)^2\omega^2 - \frac{Mm}{2(M+m)}l^2\omega_o^2 + \frac{1}{2}kx^2 = 0$$

$$\Rightarrow \frac{Mml^2}{(M+m)}\{l^2\omega_0^2 - (l+x)^2\omega^2\} = kx^2. \tag{2.179}$$

At the maximum elongation of the spring, the radial velocity of the block is zero. Then the velocity of the block will be perpendicular to the spring. Conserving the angular momentum of the block about

$$\frac{Mm}{M+m}l^2\omega_0 = \frac{Mm}{M+m}(l+x)^2\omega$$

$$\Rightarrow \omega = \frac{\omega_0 l^2}{(l+x)^2}. \tag{2.180}$$

Using equations (2.179) and (2.180) we have

$$\Rightarrow \frac{Mm}{(M+m)}\left[l^2\omega_0^2 - \left\{\frac{\omega_0 l^2}{(l+x)^2}\right\}^2 (l+x)^2\right] = kx^2$$

$$\Rightarrow \frac{Mml^2\omega_0^2}{(M+m)}\left[1 - \frac{l^2}{(l+x)^2}\right] = kx^2$$

$$\Rightarrow \frac{Mm\omega_0^2 l^2}{(M+m)}\left\{1 - \left(1 + \frac{x}{l}\right)^{-2}\right\} = kx^2. \tag{2.181}$$

We know that when $X \ll 1$

$$(1+X)^n \cong 1 + nX, \tag{2.182}$$

where $X = x/l$ and $n = -2$.
Using this binomial expansion we have

$$\frac{Mm\omega_0^2 l^2}{(M+m)}\left\{1 - \left(1 - 2\frac{x}{l}\right)\right\} \cong kx^2$$

$$\Rightarrow \frac{2Mm\omega_0^2 lx}{(M+m)} \cong kx^2$$

$$\Rightarrow x \cong \frac{2Mm\omega_0^2 l}{(M+m)k}.$$

Problem 33 In the previous problem, if $M = 2m$ and $\omega_0 = \sqrt{\frac{2k}{m}}$, find the maximum elongation of the spring.

Solution
From the previous problem we have

$$\frac{Mm\omega_o^2 l^2}{(M+m)}\left\{1 - \left(1 + \frac{x}{l}\right)^{-2}\right\} = kx^2.$$

Putting $M = 2m$ we have

$$kx^2 = \left(\frac{(2m)(m)}{2m+m}\right)m\omega_o^2 l^2\left\{1 - \left(1 + \frac{x}{l}\right)^{-2}\right\}$$

$$\Rightarrow kx^2 = \frac{2}{3}m\omega_o^2 l^2\left\{1 - \left(1 + \frac{x}{l}\right)^{-2}\right\}. \qquad (2.183)$$

Putting $\omega_o = \sqrt{\frac{2k}{m}}$ in equation (2.183) we have

$$kx^2 = \frac{2}{3}ml^2\left(\sqrt{\frac{2k}{m}}\right)^2\left\{1 - \left(1 + \frac{x}{l}\right)^{-2}\right\}$$

$$\Rightarrow x^2 = \frac{4}{3}l^2\left\{1 - \left(\frac{l}{l+x}\right)^2\right\}$$

$$\Rightarrow 3x^2(l+x)^2 = 4l^2(2l+x)x$$

$$\Rightarrow 3x^3 + 6lx^2 - l^2x - 8l^3 = 0$$

$$\Rightarrow (x - l)(3x^2 + 9lx + 8l^2) = 0$$

$$\Rightarrow x = l,$$

because the equation $(3x^2 + 9lx + 8l^2) = 0$ gives imaginary roots.
So the maximum elongation of the spring is

$$x_{\max} = l.$$

Problem 34 Two particles of mass m_1 and m_2 are connected to the ends of a light spring of stiffness k and natural length l. The spring–particle system is placed in a smooth horizontal plane. Let the particles be given velocities u_1 and u_2, respectively, in arbitrary directions. As a result, the particles move in the horizontal plane. Discuss the nature of the motion of the particles with the possibility of all conservation principles.

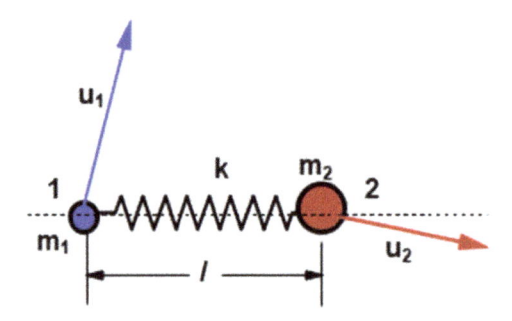

Solution

Linear momentum conservation

The net force acting on the spring–particle system is zero. So its net momentum is conserved. In other words, the center of mass C will move with a velocity given as

$$\vec{v}_C = \frac{m_1 \vec{u}_1 + m_2 \vec{u}_2}{m_1 + m_2}. \tag{2.184}$$

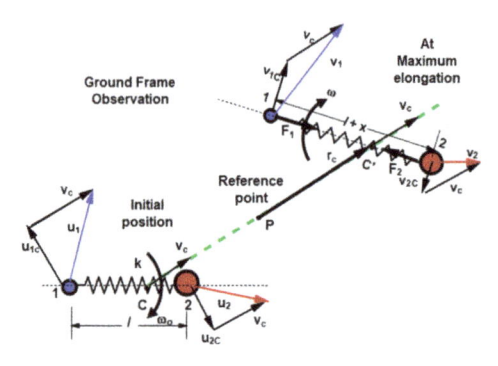

Let the center of mass move from the initial position C to a position C′ in a straight line CC′ with the constant velocity $\vec{v}_C$. At the position C′ of the center of mass the spring has maximum elongation. This means that the relative velocity between the balls along the spring will be zero at this position. As a result, the velocities of the balls relative to the new position C′ of the center of mass will be perpendicular to the spring. However, at each point of time the velocity of each particle is always equal to the velocity of that particle relative to the center of mass plus the velocity of the center of mass. This is given as

$$\vec{u}_1 = \vec{u}_{1C} + \vec{u}_C = \vec{u}_{1C} + \vec{v}_C \ (\because \vec{u}_C = \vec{v}_C)$$

$$\vec{u}_2 = \vec{u}_{2C} + \vec{u}_C = \vec{u}_{2C} + \vec{v}_C \ (\because \vec{u}_C = \vec{v}_C)$$

$$\vec{v}_1 = \vec{v}_{1C} + \vec{v}_C, \ \vec{v}_2 = \vec{v}_{2C} + \vec{v}_C.$$

Since the velocity of the center of mass relative to itself is zero, we can write

$$\vec{v}_{CC} = \frac{m_1 \vec{u}_{1C} + m_2 \vec{u}_{2C}}{m_1 + m_2} = 0$$

$$m_1 \vec{u}_{1C} = -m_2 \vec{u}_{2C}. \tag{2.185}$$

Similarly, at C' we have

$$m_1 \vec{v}_{1C} = -m_2 \vec{v}_{2C}. \tag{2.186}$$

This means that the net momentum of the system relative to the center of mass is zero. In other words, the particles seem to move opposite to each other from the center of mass frame as shown in the figure.

Angular momentum conservation

Since the spring forces $\vec{F}_1$ and $\vec{F}_2$ are an action–reaction pair, the net torque acting on the system is zero about any inertial point. We know that the angular momentum of a system of a particle about any point P is equal to the angular momentum of the system about the center of mass $\vec{L'}_C$ plus the angular momentum $\vec{L}_C$ of the center of mass about the chosen reference point. Then we can write

$$\vec{L}_P = \vec{L}_C + \vec{L'}_C. \tag{2.187}$$

If the point of reference P is taken on the line CC' of motion of the center of mass

$$\vec{L}_C = \vec{r}_C \times (m_1 + m_2)\vec{v}_C = 0. \tag{2.188}$$

Using equations (2.187) and (2.188) the angular momentum of the system about P can be given as

$$\vec{L}_P = \vec{L'}_C = \text{constant}. \tag{2.189}$$

The angular momentum of a two-particle system relative to its center of mass C-frame is given as

$$\vec{L'}_C = \frac{m_1 m_2}{m_1 + m_2} \vec{r}_{12} \times \vec{v}_{12}$$

$$= \frac{m_1 m_2}{m_1 + m_2} \vec{r}_{12} \times \{(\vec{v}_{12})_p + (\vec{v}_{12})_n\}.$$

where $(v_{12})_p$ and $(v_{12})_n$ are the components of the relative velocity between the particles 1 and 2 parallel and perpendicular to the spring in the horizontal plane. Since $\vec{r}_{12} \times (\vec{v}_{12})_p = 0$ we can write

$$\vec{L'}_C = \frac{m_1 m_2}{m_1 + m_2} \vec{r}_{12} \times (\vec{v}_{12})_n$$

$$L'_C = \frac{m_1 m_2}{m_1 + m_2} r_{12}(v_{12})_n. \tag{2.190}$$

Conserving the angular momentum between the initial and final (maximum elongated spring) positions relative to the center of mass, we have

$$\frac{m_1 m_2}{m_1 + m_2} r_0 (u_{12})_n = \frac{m_1 m_2}{m_1 + m_2} r(v_{12})_n$$

$$\Rightarrow r_0 (u_{12})_n = r(v_{12})_n$$

$$\Rightarrow l(u_{12})_n = (l + x)(v_{12})_n. \tag{2.191}$$

Conservation of energy

As there are no non-conservative forces acting on the system, the net force is zero, and the spring forces are conservative forces, we can conserve the mechanical energy of the system.

We know that the kinetic energy of a system of a particle about any point P is equal to the kinetic energy of the system about the center of mass K'_C plus the kinetic energy K_C of the center of mass about the chosen reference point. Then we can write

$$K_P = K_C + K'_C. \tag{2.192}$$

Since the center of mass moves with constant velocity v_C, the kinetic energy K_C of the center of mass about the chosen reference point remains constant. So the kinetic energy of the system relative to the C-frame changes with time. Since the total mechanical energy remains conserved, the sum of the kinetic energy of the system relative to its center of mass frame and the spring potential energy of the system remains constant:

The kinetic energy of the system about any fixed point P is $\Delta K_P = \Delta K_C + \Delta K'_C$

$$\Rightarrow \Delta K'_C + \Delta U_{spring} = 0 \; (\because \Delta K_C = 0). \tag{2.193}$$

Now the change in the kinetic energy of the system in the C-frame is

$$\Delta K'_C = (K'_C)_f - (K'_C)_i. \tag{2.194}$$

The final kinetic energy in the C-frame is

$$(K'_C)_f = \frac{m_1 m_2}{2(m_1 + m_2)} v_{12}^2. \tag{2.195}$$

The initial kinetic energy in the C-frame is

$$(K'_C)_i = \frac{m_1 m_2}{2(m_1 + m_2)} u_{12}^2. \tag{2.196}$$

Using equations (2.194), (2.195), and (2.196) we have

$$\Delta K_P = \Delta K'_C = \frac{m_1 m_2}{2(m_1 + m_2)} (u_{12}^2 - v_{12}^2). \tag{2.197}$$

The change in spring potential energy is

$$\Delta U_{spring} = \frac{1}{2}kx^2. \qquad (2.198)$$

Using equations (2.193), (2.197), and (2.198) we have

$$\frac{m_1 m_2}{2(m_1 + m_2)}(u_{12}^2 - v_{12}^2) = \frac{1}{2}kx^2$$

$$\Rightarrow \frac{m_1 m_2}{(m_1 + m_2)}(u_{12}^2 - v_{12}^2) = kx^2. \qquad (2.199)$$

We know that

$$(u_{12})^2 = (u_{12})_p^2 + (u_{12})_n^2 \qquad (2.200)$$

$$(v_{12})^2 = (v_{12})_p^2 + (v_{12})_n^2. \qquad (2.201)$$

At the maximum elongation of the spring, put $(v_{12})_p = 0$, so $v_{12} = (v_{12})_n$. Then we have

$$\{(u_{12})_p^2 + (u_{12})_n^2 - v_{12}^2\} = \frac{k}{\mu}x^2, \qquad (2.202)$$

where $\mu = \frac{m_1 m_2}{(m_1 + m_2)}$ = the reduced mass of the two-particle system.

Using equations (2.192) and (2.202) we have

$$\left\{(u_{12})_p^2 + (u_{12})_n^2 - \left(\frac{l}{l+x}\right)^2 (u_{12})_n^2\right\} = \frac{k}{\mu}x^2$$

$$\Rightarrow (u_{12})_p^2 + (u_{12})_n^2\left\{1 - \left(\frac{l}{l+x}\right)^2\right\} = \frac{k}{\mu}x^2. \qquad (2.203)$$

The spring–particle system undergoes three types of motion, namely rotation, translation, and vibration. The center of mass of the system moves with a constant velocity v_C, given by equation (2.184). The spring with particles rotates about the center of mass with variable angular velocity given as

$$\omega = \frac{(v_{12})_n}{r},$$

where r = the length of the spring at any instant and $(v_{12})_n$ = the component of relative velocity between the particles normal to the spring. Furthermore, the spring–particle system oscillates about the center of mass C, with a maximum and minimum length of the spring l and $l + x$, respectively. The value of x can be calculated by solving equations (2.191) and (2.202). The underlying principles working in this case are the conservation of energy and momentum (both linear and angular).

Problem 35 Two particles of mass 1 and 2 having masses $3m$ and $6m$, respectively, are connected to the ends of a light spring of stiffness k and natural length $l = 1$ m. The spring–particle system is placed in a smooth horizontal plane. Let the particles be given with velocities $u_1 = 5$ ms^{-1} and $u_2 = 5\sqrt{2}$ m s^{-1}, respectively, as shown in the figure. Furthermore, the velocities make $\theta_1 = 37°$ and $\theta_2 = 45°$. If the elongation of the spring is $x = 0.25$ m, find the stiffness of the spring.

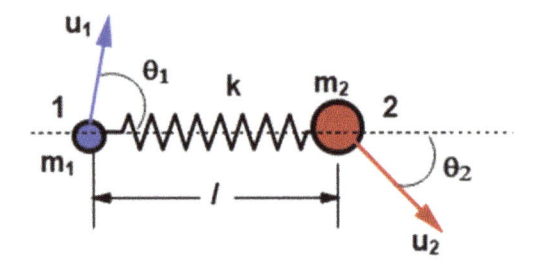

Solution
The component of initial relative velocity between the particles along (parallel to) the spring is

$$|\vec{u_{12}}|_p = |(\vec{u_1})_p - (\vec{u_2})_p|$$

$$= |u_1 \cos\theta_1 - u_2 \cos\theta_2|$$

$$= |(5)\cos 37° - 5\sqrt{2} \cos 45°|$$

$$= |(5)(4/5) - 5\sqrt{2}(1/\sqrt{2})| = 1$$

$$\Rightarrow |\vec{u_{12}}|_p = 1 \text{ ms}^{-1}. \tag{2.204}$$

This physically signifies that the spring will stretch. The component of initial relative velocity between the particles normal to the spring is

$$|\vec{u_{12}}|_n = |(\vec{u_1})_n - (\vec{u_2})_n|$$

$$= |u_1 \sin\theta_1 - (-u_2 \sin\theta_2)|$$

$$= |(5)\sin 37° + 5\sqrt{2} \sin 45°|$$

$$= |(5)(3/5) + 5\sqrt{2}(1/\sqrt{2})|$$

$$\Rightarrow |\vec{u_{12}}|_n = 8 \text{ms}^{-1}. \tag{2.205}$$

From the last problem we have

$$(u_{12})_p^2 + (u_{12})_n^2 \left\{ 1 - \left(\frac{l}{l+x}\right)^2 \right\} = \frac{k}{\mu}x^2, \tag{2.206}$$

where $\mu = \dfrac{m_1 m_2}{(m_1 + m_2)} = \dfrac{(3)(6)}{(3+6)} = 2$ kg.

By putting $l = 1$ m and $x = 1/4$ m (given), and the values of the components of the relative velocities between the particles from equations (2.204) and (2.205) in equation (2.206), we have

$$(1)^2 + (8)^2 \left\{ 1 - \left(\frac{l}{l + 0.25l} \right)^2 \right\} = \frac{k}{2}(0.25)^2.$$

This gives $k = 769.28$ N m^{-1}.

Problem 36 Two particles of mass $m_1 = 1$ kg and $m_2 = 2$ kg are connected to the ends of a light spring of stiffness k and natural length $l = 1$ m. The spring–particle system is placed in a smooth horizontal plane. Let the particles be given with velocities $u_1 = 3$ ms^{-1} and $u_2 = 6$ ms^{-1}, respectively, as shown in the figure. Find the maximum deformation of the spring and describe the motion of the spring-particle system. Assume that $k = 1000$ N m^{-1}.

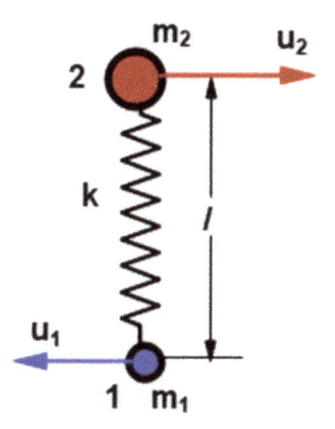

Solution
In this problem, both particles are given velocities perpendicular to the spring. So the component of initial relative velocity along the spring is zero. Then, putting $(u_{12})_p = 0$ and $(u_{12})_n = u_{12}$ in the equation

$$(u_{12})_p^2 + (u_{12})_n^2 \left\{ 1 - \left(\frac{l}{l + x} \right)^2 \right\} = \frac{k}{\mu} x^2, \qquad (2.207)$$

we have

$$u_{12}^2 \left\{ 1 - \left(\frac{l}{l + x} \right)^2 \right\} = \frac{k}{\mu} x^2. \qquad (2.208)$$

Since the term kl^2 (= 1000) is much greater than the term $\mu u_{12}^2 = (2/3)(3 + 6)^2 = 54$, the stiffness of the spring is so large that the elongation

2-85

of the spring will be nearly 10% of the length of the spring. So, as discussed earlier, after binomial expansion

$$1 - \left(\frac{l}{l+x}\right)^2 = 1 - \left(1 + \frac{x}{l}\right)^{-2} \cong \frac{2x}{l}. \tag{2.209}$$

From equations (2.207) and (2.208) we have

$$u_{12}^2 \frac{2x}{l} \cong \frac{k}{\mu} x^2$$

$$\Rightarrow x \cong \frac{2\mu u_{12}^2}{kl} = \frac{2(2/3)(3+6)^2}{(1000)(1)} = 0.108 \text{ m}.$$

The center of mass of the system will move towards the right with a constant velocity

$$\vec{v}_C = \frac{m_1 \vec{u_1} + m_2 \vec{u_2}}{m_1 + m_2}$$

$$= \frac{(1)(-3) + (2)(+6)}{1+2} = 3\hat{i} \text{ ms}^{-1}.$$

Furthermore, the particles will revolve about the common center of mass with variable angular velocity and simultaneously the length of the spring varies between $l = 1$ m to $1 + 0.108 = 1.108$ m.

Problem 37 Two particles of mass $m_1 = 3$ kg and $m_2 = 6$ kg are connected to the ends of a light spring of stiffness k and natural length $l = 1$ m. The spring–particle system is placed in a smooth horizontal plane. Let the particles be given with velocities $u_1 = 4$ ms^{-1} and $u_2 = 6$ ms^{-1}, respectively, as shown in the figure. Find the maximum deformation of the spring and describe the motion of spring–particle system. Assume that $k = 1800$ N m^{-1}.

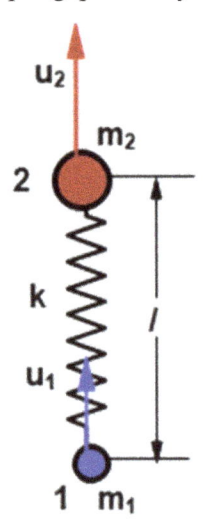

2-86

Solution

In this problem both particles are given velocities along the length of the spring. So the component of initial relative velocity normal to the spring is zero. Then, putting $(u_{12})_n = 0$ and $(u_{12})_p = u_{12}$ in the equation

$$(u_{12})_p^2 + (u_{12})_n^2 \left\{ 1 - \left(\frac{l}{l+x}\right)^2 \right\} = \frac{k}{\mu}x^2, \qquad (2.210)$$

we obtain the familiar expression

$$u_{12}^2 = \frac{k}{\mu}x^2$$

$$\Rightarrow x = \sqrt{\frac{\mu}{k}} u_{12} = \sqrt{\frac{2}{1800}}(6-4) = \frac{1}{15} \text{m}.$$

The center of mass of the system will move up with a constant velocity

$$\vec{v}_C = \frac{m_1 \vec{u_1} + m_2 \vec{u_2}}{m_1 + m_2}$$

$$= \frac{(1)(4) + (2)(6)}{1+2} = \frac{16}{3}\hat{j} \text{ ms}^{-1}.$$

In this case the particles will not revolve about the common center of mass because the initial angular momentum about the C-frame is zero, but the length of the spring varies between $l = 1$ m to $l + l/15 = 16l/15 = 16/15$ m. In other words, the particles will oscillate about the moving center of mass.

Problem 38 Two particles 1 and 2 each of mass m are connected to the ends of a light spring of stiffness k and natural length l. Another particle 3 of mass m collides with particle 1 with a velocity v_0 and sticks to it, as shown in the figure. Find the (a) maximum deformation of the spring (b) fraction of energy lost in the collision (c) kinetic energy relative to the center of mass frame (d) angular momentum relative to the center of mass frame (e) initial angular velocity of the spring assuming that the maximum elongation of the spring is much less than the length of the spring.

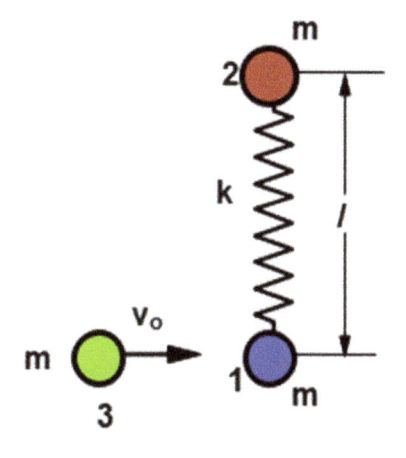

Solution

(a) Let the particle 1 attain a velocity u_1 just after particle 3 collides with it. Conserving the linear momentum of the system of two particles (1 and 3) we have

$$u_1 = \frac{mv_o}{m+m} = \frac{v_o}{2}. \qquad (2.211)$$

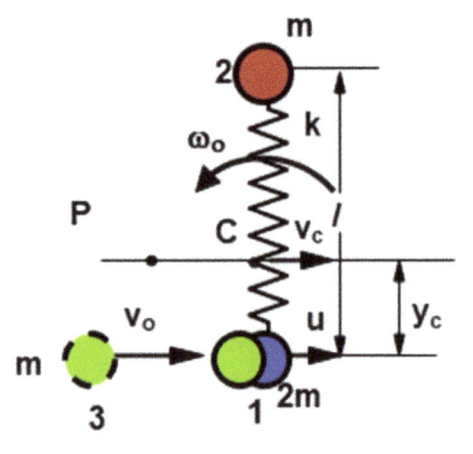

However, the center f mass of the system of the three particles will move with a velocity

$$\vec{v_C} = \frac{m_1 \vec{u_1}}{m_1 + m_2 + m_3}$$

$$= \frac{mv_o}{m+m+m}\hat{i} = \frac{v_o}{3}\hat{i}. \qquad (2.212)$$

Now the problem is reduced to a two-particle problem, where the first particle (a combination of particle 1 and 2) has mass m_1 ($= 2m$) and velocity $u_1 = u$ perpendicular to the spring and the second particle (particle 2) of mass $m_2 = m$ which is at rest initially ($u_2 = 0$). So the component of initial relative velocity along the spring is zero. Then, putting $(u_{12})_p = 0$ and $(u_{12})_n = u_{12} = u$ in the equation

$$(u_{12})_p^2 + (u_{12})_n^2\left\{1 - \left(\frac{l}{l+x}\right)^2\right\} = \frac{k}{\mu}x^2, \qquad (2.213)$$

we have

$$u_{12}^2\left\{1 - \left(\frac{l}{l+x}\right)^2\right\} = \frac{k}{\mu}x^2. \qquad (2.214)$$

2-88

As the stiffness of the spring is so large that the elongation of the spring will be much less than the length of the spring, as mentioned in the last problem, after binomial expansion

$$1 - \left(\frac{l}{l+x}\right)^2 = 1 - \left(1 + \frac{x}{l}\right)^{-2} \cong \frac{2x}{l}. \qquad (2.215)$$

From equations (2.214) and (2.215) we have

$$u_{12}^2 \frac{2x}{l} \cong \frac{k}{\mu} x^2$$

$$\Rightarrow x \cong \frac{2\mu u_{12}^2}{kl} = \frac{2\mu u_1^2}{kl}. \qquad (2.216)$$

From equations (2.211) and (2.216) we have

$$x \cong \frac{2\mu \left(\frac{v_0}{2}\right)^2}{kl} = \frac{\mu v_0^2}{2kl} = \frac{(2m/3)v_0^2}{2kl}$$

$$\Rightarrow x = \frac{mv_0^2}{3kl}.$$

(b) The energy lost in the collision is

$$\Delta K = -\frac{\mu u_{rel}^2}{2} = -\frac{(m/2)v_0^2}{2} = -\frac{mv_0^2}{4}.$$

Then the fraction of energy lost in the collision is

$$+\frac{\Delta K}{K} = = \frac{mv_0^2/4}{mv_0^2/2} = 1/2.$$

(c) The internal energy of the system is the kinetic energy relative to the C-frame, which is given as

$$K' = \frac{\mu u_{rel}^2}{2}.$$

The reduced mass of the system (2m and m) is

$$\mu = \frac{(m)(2m)}{2m + m} = 2m/3.$$

The initial relative velocity between the particles of mass 2m and m is equal to $v_0/2$. Then we have

$$K' = \frac{(2m/3)(v_0/2)^2}{2} = \frac{mv_0^2}{12}.$$

(d) The angular momentum of the system relative to the C-frame is
$$L'_C = -\frac{m_1 m_2}{m_1 + m_2} r_{12}(v_{12})_n$$
and
$$= \frac{2m}{3} l(v_o/2)\hat{k} = \frac{mv_o l}{3}\hat{k}.$$

Alternative method

If we chose a point P on the line of motion of the center of mass, $\vec{L}_C = \vec{r}_C \times M\vec{v}_C = 0$, so
$$\vec{L'}_C = mv_o y_C \hat{k}, \qquad (2.217)$$
putting $y_C = \frac{ml}{3m} = \frac{l}{3}$ in equation (2.217) we have
$$\vec{L'}_C = \frac{mv_o l}{3}\hat{k} = I_C \vec{\omega} = \text{constant}.$$

(e) Since I_C changes, $\vec{\omega}$ will change. The initial angular velocity is given as
$$\frac{mv_o l}{3}\hat{k} = I_C \vec{\omega} = \mu l^2 \vec{\omega} = \frac{(2m)(m)}{2m+m}\vec{\omega}$$
$$\vec{\omega} = \frac{v_o}{2l}\hat{k}.$$

Alternatively, we can also find $\vec{\omega}$ by using its basic formula
$$\vec{\omega}_o = (\vec{\omega}_{12})_o = \frac{(u_{12})_n}{l}\hat{k} = \frac{v_o/2}{l}\hat{k} = \frac{v_o}{2l}\hat{k}.$$

Problem 39 Two particles 1 and 2 of masses m_1 and m_2 are connected by a string of length l which is just slack. The particles are given velocities v_1 and v_2, as shown in the figure. Find (a) the internal energy just before collision (b) the internal energy just after collision (c) the change in kinetic energy in the collision (d) the fraction of energy lost in the collision.

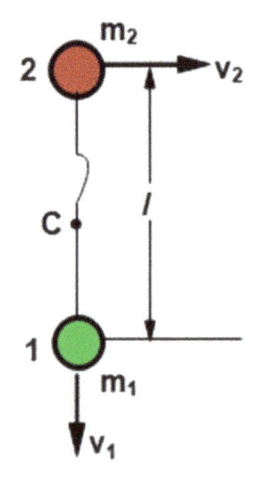

Solution
(a) The internal energy just before the collision is equal to the kinetic energy in the C-frame, given as

$$= \frac{\mu \, |\vec{u_1} - \vec{u_2}|^2}{2}$$

$$= \frac{\mu \, |-v_1 \hat{j} - v_2 \hat{i}|^2}{2}$$

$$= \frac{\mu}{2}(v_1^2 + v_2^2)$$

$$= \frac{m_1 m_2 (v_1^2 + v_2^2)}{2(m_1 + m_2)}. \qquad (2.218)$$

(b) Just after the inelastic collision, the relative velocity along the string will be zero. So the component of internal energy along the string will be zero. However, the component of kinetic energy relative to the C-frame, that is the internal energy perpendicular to the string, will remain conserved. Then the internal energy just after the collision is equal to the kinetic energy in the C-frame perpendicular to the string, given as

$$K'_f = \frac{\mu \, |(\vec{u}_{12})_n|^2}{2}$$

$$= \frac{\mu v_2^2}{2} = \frac{m_1 m_2 v_2^2}{2(m_1 + m_2)}. \qquad (2.219)$$

(c) The change in energy in the collision is

$$\Delta K = \Delta K_C + \Delta K' = \Delta K'$$

$$\Delta K = (K')_f - (K')_i. \qquad (2.220)$$

Using the last three equations, we have

$$\Delta K = \frac{m_1 m_2 v_2^2}{2(m_1 + m_2)} - \frac{m_1 m_2 (v_1^2 + v_2^2)}{2(m_1 + m_2)}$$

$$\Rightarrow \Delta K = -\frac{m_1 m_2 v_1^2}{2(m_1 + m_2)}. \qquad (2.221)$$

Otherwise:
We can use the direct expression for the change in energy in a collision, given as

$$\Delta K = -\frac{\mu(u_{\text{rel}}^2)_p}{2}(1 - e^2)$$

$$= -\frac{m_1 m_2 v_1^2}{2(m_1 + m_2)} \quad (\because e = 0).$$

(d) Then the fraction of energy lost in the collision is

$$\eta = -\frac{\Delta K}{K_i}. \tag{2.222}$$

Using equations (2.218), (2.221), and (2.222) we have

$$\eta = -\frac{-\frac{m_1 m_2 v_1^2}{2(m_1+m_2)}}{\frac{m_1 m_2 (v_1^2 + v_2^2)}{2(m_1+m_2)}} = \frac{v_1^2}{v_1^2 + v_2^2}.$$

Problem 40 A boy walks with a constant velocity v on a light horizontal rod of length L. The rod rotates with a constant angular velocity ω about the vertical z-axis, as shown in the figure. If the boy starts from the middle of the rod, find (a) the torque experienced by the boy, (b) the magnitude of the force that provides the torque, the power delivered by the reaction force acting on the boy due to the rod, (d) the rate of change of kinetic energy of the boy-rod system (e) the rate of energy loss and (f) the total energy loss until the boy reaches the free end of the rod.

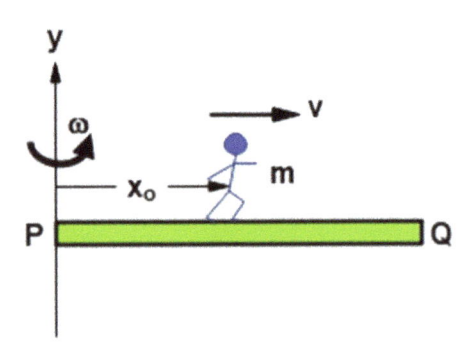

Solution
(a) Referring to the free-body diagram, the forces acting on the boy are the downward gravity mg, the upward normal reaction, and a transverse force F_t and a radial force F_r. These radial and transverse contact forces are the components of static friction of the rod acting on the boy.

As the boy moves outward, due to the change (in both magnitude and direction) in the transverse velocity of the boy, a transverse acceleration is generated which is given as

$$a_\theta = 2v\omega. \tag{2.223}$$

So a transverse force that produces the transverse acceleration is given as
$$F_t(= F_\theta) = ma_\theta = 2mv\omega. \tag{2.224}$$
This force is responsible for producing a torque on the boy about the y-axis of rotation, given as
$$\tau = xF_t = 2mvx\omega. \tag{2.225}$$
At any instant, the distance of the boy from the axis of rotation is
$$x = x_o + vt. \tag{2.226}$$
Using all the last four equations, we have
$$\tau = 2mv(x_o + vt)\omega.$$

Alternative method
The MI of the boy is
$$I = mx^2 = m(x_o + vt)^2. \tag{2.227}$$
The angular momentum of the boy is
$$L = I\omega = m(x_o + vt)^2\omega. \tag{2.228}$$
The torque acting on the boy is
$$\tau = \frac{dL}{dt} = \frac{d}{dt}\{m(x_o + vt)^2\omega\}$$
$$= 2mv(x_o + vt)\omega. \tag{2.229}$$

(b) Let the transverse force acting on the boy that produces a torque about the axis of rotation be F. Then, at any radial distance x, the torque produced by this force is
$$\tau = xF = (x_o + vt)F. \tag{2.230}$$
Using equations (2.226) and (2.227) we have
$$(x_o + vt)F = 2mv(x_o + vt)\omega$$
$$\Rightarrow F = 2mv\omega. \tag{2.231}$$

(c) The power delivered by this force is
$$P = Fv_t = F(x\omega) \ (\because v_t = x\omega). \tag{2.232}$$
Using equations (2.231) and (2.232)
$$P = (2mv\omega)(x\omega) = 2mv\omega^2 x$$
$$= (2mv\omega)(x\omega) = 2mv\omega^2(x_o + vt).$$

2-93

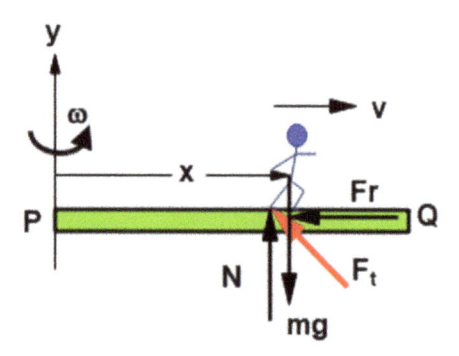

(d) The rate of change in kinetic energy is

$$\frac{dK}{dt} = \frac{d}{dt}\left(\frac{I\omega^2}{2}\right) = \frac{\omega^2}{2}\frac{dI}{dt}$$

$$= \frac{\omega^2}{2}\frac{d}{dt}(mx^2)$$

$$= m\omega^2 x \frac{dx}{dt} = m\omega^2 v\, x.$$

(e) The rate of energy loss is

$$\frac{dE}{dt} = P - \frac{dK}{dt}$$

$$= 2m\omega^2 vx - m\omega^2 vx$$

$$= m\omega^2 vx = m\omega^2 v(x_o + vt).$$

(f) The total energy loss until the boy reaches the other end of the rod is

$$\Delta E = \int_0^t \frac{dE}{dt}dt = \int_0^t m\omega^2 v(x_o + vt)dt = m\omega^2 v\left(x_o t + \frac{vt^2}{2}\right)$$

Putting $t = (l - x_o)/v$, we have

$$\Delta E = m\omega^2 v\left[x_o \frac{l-x_o}{v} + \frac{v}{2}\left(\frac{l-x_o}{v}\right)^2\right]$$

$$= m\omega^2(l - x_o)\left\{x_o + \left(\frac{l-x_o}{2}\right)\right\}$$

$$= m\omega^2(l - x_o)\left(\frac{l+x_o}{2}\right) = \tfrac{1}{2}m\omega^2(l^2 - x_o^2) \quad \text{Ans.}$$

Problem 41 A small ball of mass m collides with the smooth wall at O with a coefficient of restitution $e = 0.5$ with a velocity v, as shown in the figure. Find the change in angular momentum of the ball about the points B and C located on the vertical wall.

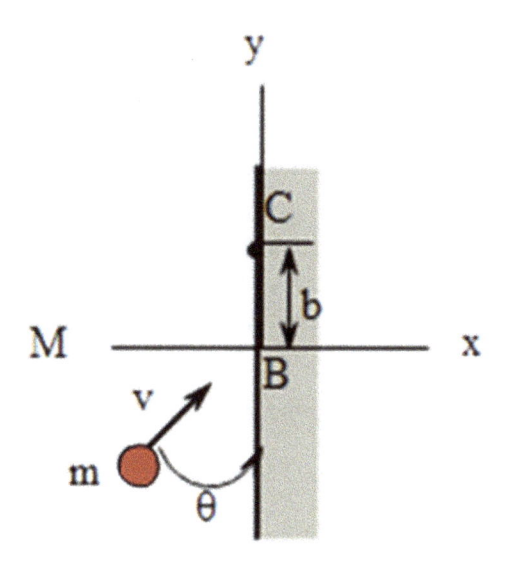

Solution

If we take the reference point B, the velocity vectors just before and after the collision pass through the point of collision B, so the angular momentum just before and after the collision will be zero about B. Then the change in angular momentum about B is zero.

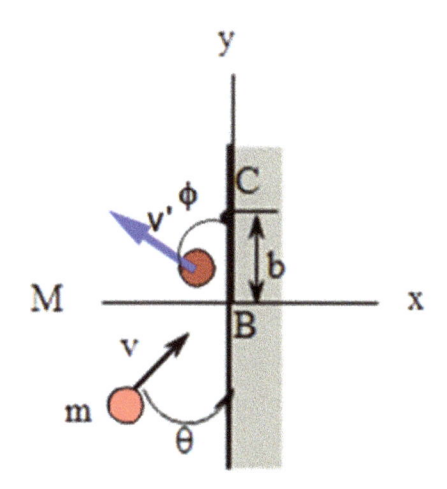

2-95

However, the change in angular momentum about C will not be zero because the point C is not zero for the following reason.

As the wall is smooth, the impact does not take place vertically. So the change in linear momentum along the vertical line BC is zero. In other words, the change in angular momentum about C due to the vertical components of the momenta of the ball just before and after the impact will be zero:

$$\Delta \vec{P}_x = (-mv' \sin \varnothing \hat{k}) - (+mv \sin \theta \hat{k}) = 0.$$

Then we have

$$v' \cos \varnothing = v \cos \theta. \qquad (2.233)$$

Due to the normal reaction offered by the wall to the ball, which is directed towards the left, the horizontal plane is responsible for changing the linear momentum of the ball from $+mv \sin \theta \hat{i}$ to $-mv' \sin \varnothing \hat{i}$. Now, the angular momentum about C changes from $+mvb \sin \theta \hat{k}$ to $-emvb \sin \theta \hat{k}$. Then the change in angular momentum about C is

$$\Delta \vec{L}_C = (-mv'b \sin \varnothing \hat{k}) - (+mvb \sin \theta \hat{k})$$

$$= -mb(v' \sin \varnothing + v \sin \theta)\hat{k}. \qquad (2.234)$$

Applying Newton's impact formula, we have

$$(v' \sin \varnothing = ev \sin \theta). \qquad (2.235)$$

Using equations (2.234) and (2.235) we have

$$\Delta \vec{L}_C = -mvb \sin \theta (1 + e)\hat{k}.$$

IOP Publishing

Problems and Solutions in Rotational Mechanics

Pradeep Kumar Sharma

Chapter 3

The statics and dynamics of rotation

3.1 Torque acting on a group/system of particles

Let us chose a system of n particles in which each particle experiences internal and external forces. If the external force F_i acts on the ith particle, the net torque acting on the system is equal to the vector sum of the torques of all forces acting on all particles of the system, relative to a fixed reference point O.

$$\vec{\tau}_{net} = \sum \vec{\tau}_i = \vec{\tau}_{int} + \vec{\tau}_{ext},$$

where $\vec{\tau}_{int}$ and $\vec{\tau}_{ext}$ are the net torque due to internal and external forces. Since the internal forces acting between any two particles are action–reaction pairs (equal in magnitude, oppositely directed, and collinear), the couple of each action–reaction pair ($\vec{f}$ and $-\vec{f}$) will be equal to zero. This means that the net torque due to all internal forces is zero:

$$\vec{\tau}_{int} = 0.$$

Using the last two equations we can write

$$\vec{\tau}_{net} = \sum \vec{\tau}_i = \vec{\tau}_{ext}.$$

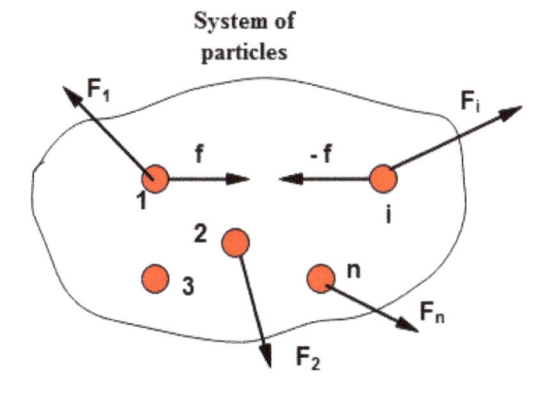

doi:10.1088/978-0-7503-6472-0ch3

This states that:
- The net torque acting on the system of particles is equal to the resultant of torques produced by the external forces acting on the system.
- The net torque due to internal forces is zero.

Example 1 Freely falling spring–mass system. Two particles of masses M and m are interconnected by an ideal spring as shown in the figure. The spring–particle system is released from rest from a certain height while the spring is initially compressed. What is the net torque due to (a) the spring forces and (b) the external forces when the coordinates of the center of mass of the system are x and y?

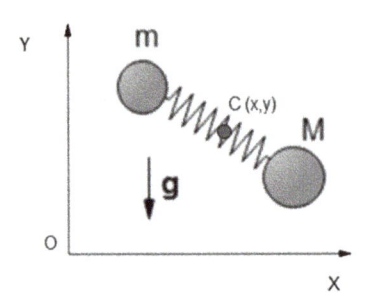

Solution
(a) The spring forces are internal forces to the spring–particle system. Therefore, the net torque due to internal forces is zero. Then the net torque is due to the external forces, that is, gravity. This is given as

$$\vec{\tau}_{net} = (\vec{r}_1 \times m_1\vec{g} + \vec{r}_2 \times m_2\vec{g})$$
$$= (m_1\vec{r}_1 + m_2\vec{r}_2) \times \vec{g},$$

where $m_1\vec{r}_1 + m_2\vec{r}_2 = (m_1 + m_2)\vec{r}_c$.
Then $\vec{\tau}_{net} = (m_1 + m_2)\vec{r}_c \times \vec{g}$

$$\vec{\tau}_{net} = (m_1 + m_1)(x\hat{i} + y\hat{j}) \times (-g\hat{j})$$
$$= -(m_1 + m_1)gx(\hat{i} \times \hat{j})$$
$$= -(m_1 + m_1)g \cdot \hat{k}.$$

3-2

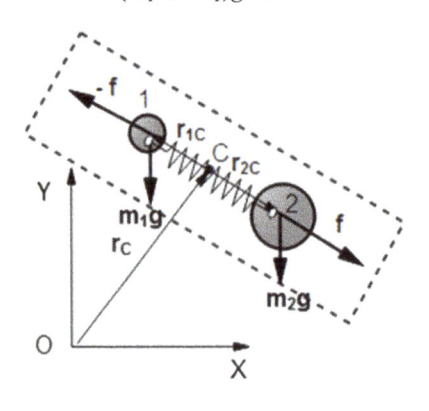

3.2 Torque about the center of mass

Let a system of particles experience a net torque $\vec{\tau}_{net}$ relative to any reference point P. Let the net torque due to all external forces taken about the center of mass C of the system be $\vec{\tau}_C$. Furthermore, let the torque of the net force $\vec{F}_{net}$ (imagined to be acting at the center of mass) taken about the given reference point P be $\vec{r}_C \times \vec{F}_{net}$. We can show that

$$\vec{\tau}_{net} = \vec{\tau}_C + \vec{r}_C \times \vec{F}_{net}.$$

Proof: Let $\vec{F}_i$ be the external force acting on the ith particle/point having a position vector $\vec{r}_i$. The net torque acting on the system about P is given as

$$\vec{\tau}_{net} = \sum \vec{\tau}_i = \sum \vec{r}_i \times \vec{F}_i.$$

Putting $\vec{r}_i = \vec{r}_{iC} + \vec{r}_C$, we have

$$\vec{\tau}_{net} = \sum \vec{\tau}_i = \sum (\vec{r}_{iC} + \vec{r}_C) \times \vec{F}_i$$
$$= \sum (\vec{r}_{iC} \times \vec{F}_i + \vec{r}_C \times \vec{F}_i)$$
$$= \sum \vec{r}_{iC} \times \vec{F}_i + \vec{r}_C \times \sum \vec{F}_i.$$

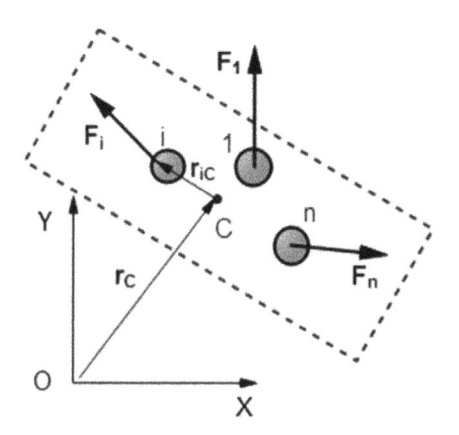

The first term is the net torque $\vec{\tau}_C$ due to all external forces taken relative to the center of mass of the system and the second term $\sum \vec{F}_i = \vec{F}_{net}$ = the net force acting on the system because the net internal force is zero. Then we have

$$\vec{\tau}_{net} = \vec{\tau}_C + \vec{r}_C \times \vec{F}_{net}.$$

This tells us that the net torque on a system of particles relative to any reference point P is equal to 'the torque $\vec{\tau}_C$ (= the sum of the torques due to all external forces taken relative to the center of mass)' plus (+) 'the torque of the net external force $\vec{F}_{net}$' (imagined to be acting at its center of mass) taken relative to the given reference point P.

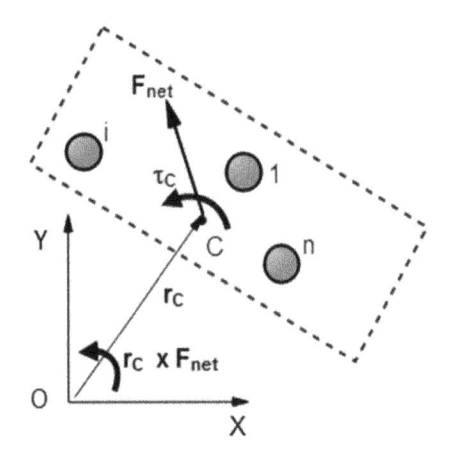

Example 2 Two particles of mass $m_1 = 1$ kg and $m_2 = 2$ kg are connected by a light spring which is initially compressed. If we release the spring–particle system, find the torque experienced by the spring–particle system relative to (a) the center of mass and (b) the origin, when their positions are $P(4, -1, 0)$ and $Q(1, -2, 3)$, respectively.

Solution

(a) The spring forces acting on the particles are internal forces (an action–reaction pair). We know that the net torque due to internal forces is zero. Then the net torque about the center of mass due to gravity is given as

$$\vec{\tau}_C = (\vec{r}_{1C} \times m_1 \vec{g} + \vec{r}_{2C} \times m_2 \vec{g})$$
$$= (m_1 \vec{r}_{1C} + m_2 \vec{r}_{2C}) \times \vec{g},$$

where $m_1 \vec{r}_{1C} + m_2 \vec{r}_{2C} = (m_1 + m_2) \vec{r}_{CC}$.

Since $\vec{r}_{CC} = 0$, $\vec{\tau}_C = 0$.

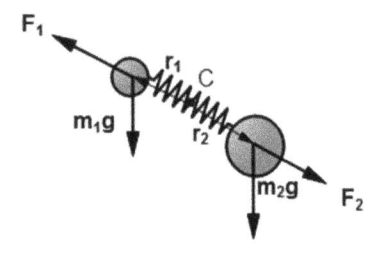

(b) Putting $\vec{\tau}_C = 0$ in the expression in

$$\vec{\tau}_{net} = \vec{\tau}_C + \vec{r}_C \times \vec{F}_{net},$$

we have

$$\vec{\tau}_{net} = \vec{r}_C \times \vec{F}_{net},$$

where we have

$$\vec{r}_C = (m_1\vec{r}_1 + m_2\vec{r}_2)/(m_1 + m_2)$$
$$= \frac{(1)(4\hat{i} - \hat{j}) + (2)(\hat{i} - 2\hat{j} + 3\hat{k})}{(1 + 2)}$$
$$= \left(2\hat{i} - \frac{5}{3}\hat{j} + 2\hat{k}\right)$$

$$\vec{F}_{net} = (m_1 + m_2)\vec{g}$$
$$= (1 + 2)(-10\hat{j}) = -30\hat{j} \text{ N}.$$

After substitution and simplification we have $\vec{\tau}_O = -60(\hat{i} - \hat{k})$ N-m.
- Torque due to gravity about the center of mass of any system of particles is zero. However, it will be non-zero when the size of the system of particles is comparable with the radius of the Earth as gravity becomes non-uniform.

Example 3 Couple. Two forces $\vec{F}_1$ and $\vec{F}_2$ are acting on particles of mass m_1 and m_2, respectively. The locations of the particles are given by the position vectors $\vec{r}_1$ and $\vec{r}_2$, respectively. Find the net torque of the forces relative to the center of mass of the system of two particles. Assume (a) $\vec{F}_1 = \vec{F}_2$ and (b) $\vec{F}_1 = -\vec{F}_2$.

Solution
(a) The net torque of the forces relative to the center of mass of the system is given as

$$\vec{\tau}_C = \vec{r}_{1C} \times \vec{F} + \vec{r}_{2C} \times \vec{F}$$
$$= (\vec{r}_{1C} + \vec{r}_{2C}) \times \vec{F}.$$

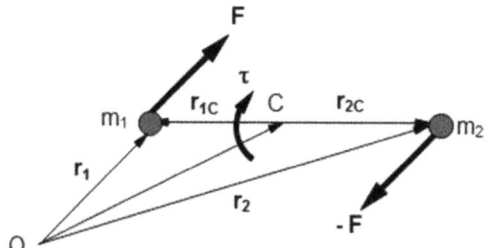

Putting $\vec{r}_{1C} = \frac{m_2 \vec{r}_{12}}{m_1 + m_2}$ and $\vec{r}_{2C} = -\frac{m_1 \vec{r}_{12}}{m_1 + m_2}$ in the last expression and simplifying the factors, we have

$$\vec{\tau}_C = \frac{(m_2 - m_1)\vec{r}_{12} \times \vec{F}}{m_1 + m_2}.$$

If $m_1 = m_2$ we have $\vec{\tau}_C = 0$.

(b) The net torque of the forces relative to the center of mass of the system is given as

$$\vec{\tau}_C = \vec{r}_{1C} \times \vec{F} + \vec{r}_{2C} \times (-\vec{F})$$
$$= (\vec{r}_{1C} - \vec{r}_{2C}) \times \vec{F}.$$

Putting $\vec{r}_{1C} - \vec{r}_{2C} = \vec{r}_{12}$ in the last expression we have

$$\vec{\tau}_C = \vec{r}_{12} \times \vec{F}.$$

Otherwise: The net torque of the forces relative to the point O is given as

$$\vec{\tau}_C = \vec{r}_1 \times \vec{F} + \vec{r}_2 \times (-\vec{F})$$
$$= (\vec{r}_1 - \vec{r}_2) \times \vec{F}$$
$$= \vec{r}_{12} \times \vec{F}.$$

The torque due to two equal and opposite forces is called a couple and does not depend upon the reference point.

3.3 The angular momentum of a system of particles

Let v_i be the velocity of the ith particle/point of mass m_i and position vector r_i. The momentum of the ith particle about O is given as

$$\vec{L}_i = \vec{r}_i \times m_i \vec{v}_i.$$

Summing up the angular momenta of all particles, the net angular momentum of the system of particles about O is given as

$$\vec{L}_O = \sum \vec{L}_i = \sum \vec{r}_i \times m_i \vec{v}_i,$$

where $\vec{r}_i = \vec{r}_{iC} + \vec{r}_C$ and $\vec{v}_i = \vec{v}_{iC} + \vec{v}_C$.

Then we have

$$\vec{L}_i = \sum (\vec{r}_{iC} + \vec{r}_C) \times m_i (\vec{v}_{iC} + \vec{v}_C).$$

After vector multiplication we have the following four terms:

$$\vec{L}_i = \sum \vec{r}_{iC} \times m_i \vec{v}_{iC} + \sum \vec{r}_{iC} \times m_i \vec{v}_C$$
$$+ \sum \vec{r}_C \times m_i \vec{v}_{iC} + \sum \vec{r}_C \times m_i \vec{v}_C.$$

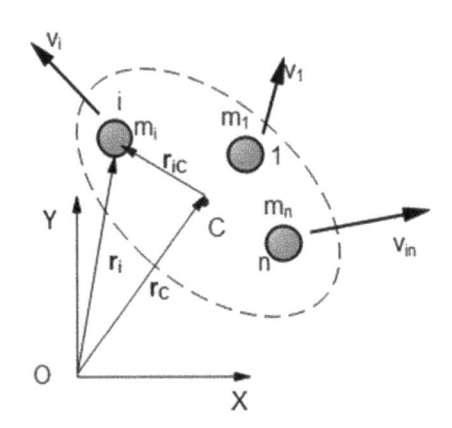

The first term is the net angular momentum measured relative to the center of mass of the system. Let it be denoted by $\vec{L}_C$. In the second term since $\vec{v}_C$ is a constant quantity for all particles, the term $\vec{x}\vec{v}_C$ can be taken out of the sum to obtain the term $\sum m_i \vec{r}_{iC} \times \vec{v}_C$. Since $\sum m_i \vec{r}_{iC} = 0$, the second term will vanish. In the third term since $\vec{r}_C$ is a constant quantity for all particles, the term $(\vec{r}_{iC} x)$ can be taken out of the sum which gives us $\vec{r}_C \times (\sum m_i \vec{v}_{iC})$. Since $\sum m_i \vec{v}_{iC} = 0$, the third term will also vanish. In the fourth term, since $\vec{r}_C$ and $\vec{v}_C$ are constant quantities for all particles, the term $\vec{r}_C \times \vec{v}_C$ can be taken out of the sum to obtain the term

$$\left(\sum m_i\right)\left(\vec{r}_C \times \vec{v}_C\right),$$

where $(\sum m_i) = M =$ the total mass of the system. Then the final expression of $\vec{L}_P$ can be given as

$$\vec{L}_O = \vec{L}_C + \vec{r}_C \times M\vec{v}_C.$$

The first term is known as the spin/internal/intrinsic angular momentum $\vec{L}_S$ and the second term is called orbital angular momentum $\vec{L}_O$ (the angular momentum of the center of mass relative to the given reference point).

This tells us that the angular momentum $\vec{L}_O$ of a system of particles relative to any reference point O is equal to the sum of 'the angular momentum $\vec{L}_C$ of the system measured relative to the center of mass' and the 'angular momentum of the total mass' (imagined to be centered at its center of mass) taken relative to the given reference point O.

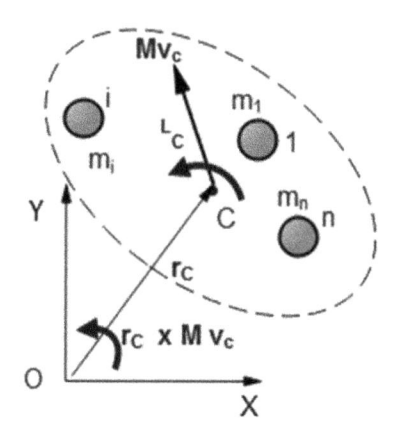

Example 4 It is interesting to watch a whirlwind (rapidly whirling dust particles and air) that moves towards the east with a velocity $v = 20$ m s^{-1}. Let us assume that you are sitting on the ground and a whirlwind passes over you at time $t = 0$. The angular momentum of the rotating stuff about its centroidal axis is $\vec{L}_C = 4000\hat{j}$ kg ms^{-2}. The center of mass of the whirlwind is located at a height of 5 m from the ground level. If the mass of the spinning stuff is $m = 30$ kg, find the magnitude of angular momentum of the whirlwind at $t = 2.5$ s relative to you.

Solution

The whirlpool is a system of particles of dust and air etc. It is given that the spin angular momentum of the whirlpool is $\vec{L}_C = 4000\hat{j}$ kg m s^{-2}.

The value of angular momentum of the centre of mass of the whirlpool relative to the origin O (you) is given as

$$m\vec{r}_C \times \vec{v}_C = \vec{L}'_C(let) = -mrv\sin\theta\hat{k} = -m(r\sin\theta)v\hat{k}$$
$$= -mhv\hat{k} = -(30)(5)(20)\hat{k} = -3000\hat{k} \text{ kg m s}^{-2}.$$

Then, net angular momentum relative to you is the vector sum of these two angular momenta, given as

$$\vec{L}_O = \vec{L}_C + m\vec{r}_C \times \vec{v}_C = 4000\hat{j} - 3000\hat{k} \text{ kg m s}^{-2}$$

Since these two components of total angular momentum are perpendicular to each other, its magnitude is $L_O = 5000$ kg m s^{-2} and it is pointed at an angle of $\phi = \tan^{-1}(3/4)$ with vertical

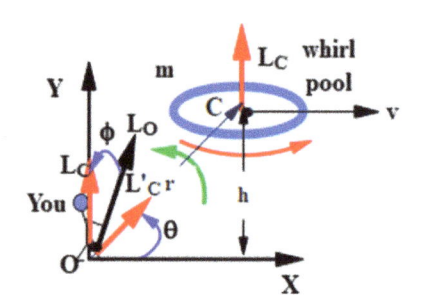

3.4 Angular momentum of a two-particle system about the center of mass

Two particles of mass m_1 and m_2 have positions $\vec{r}_1$ and $\vec{r}_2$ and move with velocities of $\vec{v}_1$ and $\vec{v}_2$, respectively. Find the angular momentum of the system about its center of mass.

Solution: The angular momentum of the system of particles relative to the center of mass is given as

$$\vec{L}_C = \sum \vec{r}_{iC} \times m_i \vec{v}_{iC}$$
$$= \vec{r}_{1C} \times m_1 \vec{v}_{1C} + \vec{r}_{2C} \times m_2 \vec{v}_{2C}.$$

Putting the expression

$$m_2 \vec{v}_{2C} = -m_1 \vec{v}_{1C}$$

in the last expression, we have

then $\vec{L}_C = \vec{r}_{1C} \times m_1\vec{v}_{1C} - \vec{r}_{2C} \times m_1\vec{v}_{1C}$
$$= (\vec{r}_{1C} - \vec{r}_{2C}) \times m_1\vec{v}_{1C} \qquad (3.1)$$
$$= \vec{r}_{12} \times m_1\vec{v}_{1C}.$$

We know that
$$\vec{v}_{1C} = \vec{v}_1 - \vec{v}_C$$
$$= \vec{v}_1 - \frac{m_1\vec{v}_1 + m_2\vec{v}_2}{m_1 + m_2}$$
$$= \frac{m_1 m_2(\vec{v}_1 - \vec{v}_2)}{m_1 + m_2}$$

$$\vec{v}_{1C} = \frac{m_1 m_2}{m_1 + m_2} \vec{v}_{12}. \qquad (3.2)$$

Using equations (3.1) and (3.2) we have
$$\vec{L}_C = \frac{m_1 m_2}{m_1 + m_2}(\vec{r}_{12} \times \vec{v}_{12}),$$

where
$$\frac{m_1 m_2}{m_1 + m_2} = \mu \text{ (reduced mass)}.$$

Then we have a handy formula,
$$\vec{L}_C = \mu(\vec{r}_{12} \times \vec{v}_{12}),$$

The angular momentum of the two-particle system about the center of mass does not depend upon any reference frame. Therefore, it is called 'internal or intrinsic' angular momentum.

Example 5 Two particles A and B each of mass m are connected with the ends of a light inextensible string of length l. The string–particle system is placed in a smooth horizontal plane. Another particle of mass m collides with a velocity v_0 perpendicular to the spring, as shown in the figure. Find (a) the angular momentum and (b) the kinetic energy of the system (A + B) relative to its center of mass frame.

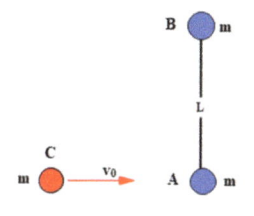

Solution

Just after the elastic collision the velocity of A will be equal to $v_0\hat{i}$ and that of B is zero. So we have

$$\vec{v}_{AB} = \vec{v}_A - \vec{v}_B = v_0\hat{i}.$$

The angular momentum about the center of mass of the two-particle system is

$$\vec{L}_C = \mu(\vec{r}_{AB} \times \vec{v}_{AB}),$$

where $\vec{r}_{AB} = -L\hat{j}$, $\vec{v}_{AB} = v_0\hat{i}$ and

$$\mu = \frac{m \cdot m}{m+m} = m/2.$$

Then we have

$$\vec{L}_C = \mu(\vec{r}_{AB} \times \vec{v}_{AB}),$$
$$= (m/2)\left(-L\hat{j} \times v_0\hat{i}\right)$$
$$= \frac{mLv_0}{2}\hat{k}.$$

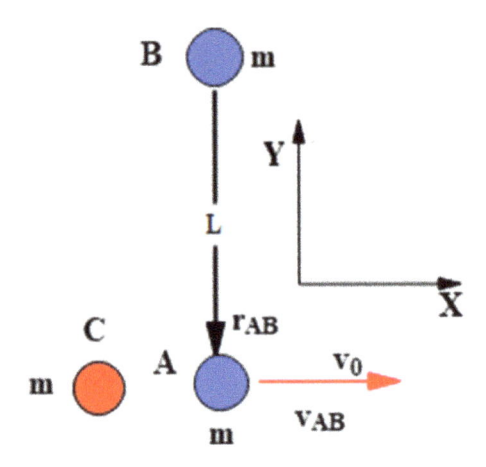

3.5 The relation between the relative and absolute values of angular momentum relative to two coinciding reference frames

Let two reference frames (persons A and B, say) be located at the origin O. At time $t = t_1$, let the persons measure the angular momentum of a bird of mass m and position $\vec{r}$ as $\vec{L}_1$ and $\vec{L}_2$, respectively. Let the velocity of the bird measured in the reference frames A and B be $\vec{v}_1$ and $\vec{v}_2$, respectively. Let the velocity of the bird be $\vec{v}_b$ relative to ground frame. Then the angular momenta of the bird relative to A and B are given as follows:

$$\vec{L}_1 = \vec{r} \times m(\vec{v}_b - \vec{v}_1)$$

$$\vec{L}_2 = \vec{r} \times m(\vec{v}_b - \vec{v}_2).$$

Then $\vec{L}_1 - \vec{L}_2 = \vec{r} \times m(\vec{v}_2 - \vec{v}_1)$.
This tells us that:
- If two observers located at the same place move with different velocities, they measure different angular momenta of the same object.

Example 6 If a stationary observer A measures the angular momentum of an object as $\vec{L}_1$, it is called the absolute value of $\vec{L}_1 = \vec{L}_{abs}$. If the observer B located at the same place as A, moves with a velocity $\vec{v}$, the value of $\vec{L}_2$ is called relative angular momentum. So we can write $\vec{L}_2 = \vec{L}_{rel}$. Establish a relation between $\vec{L}_{abs}$ and $\vec{L}_{rel}$.

Solution
Putting $\vec{L}_1 = \vec{L}_{abs}$, $\vec{L}_2 = \vec{L}_{rel}$, and $\vec{v}_2 - \vec{v}_1 = \vec{v}$ in the previously derived expression,
$$\vec{L}_1 - \vec{L}_2 = \vec{r} \times m(\vec{v}_2 - \vec{v}_1),$$
we obtain the relation between absolute and relative angular momentum from the same location of reference frames (observers), which can be given as
$$\vec{L}_{rel} = \vec{L}_{abs} + \vec{r} \times m\vec{v}.$$

Example 7 Three insects each of mass m are located at the vertices of an equilateral triangular frame of side l. The insects move along the sides in a counter-clockwise sense with the same speed v, as shown in the figure. The frame is lying in the x–y-plane. Find the net angular momentum L_C of the insects about the center of mass of the system of three insects.

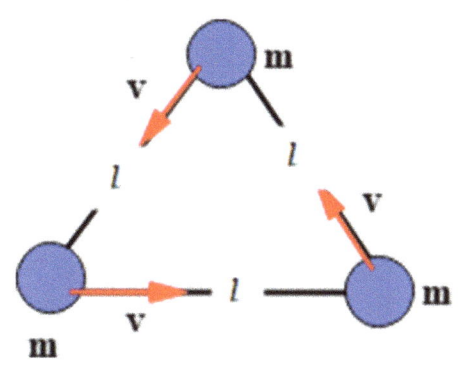

Solution
As the insects have the same mass the center of mass of the system of three insects coincides with the geometrical center C of the equilateral triangle. So the perpendicular distance from C on to the line of motion of each insect (the sides of the triangle) is equal to $p = \frac{l}{2\sqrt{3}}$. Then the angular momentum due to each insect about C is equal to mvp (anticlockwise). Hence, the net angular momentum due to three insects is $L = mvp$ (anticlockwise). Putting the value of p, $L = \frac{\sqrt{3}mvl}{2}$ (anticlockwise).

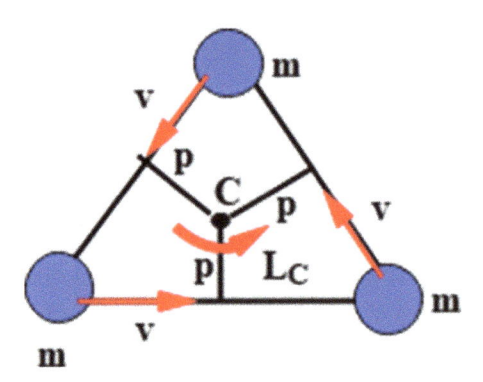

- If L_C is non-zero, the system of particles is said to be rotating (spinning) about the center of mass.
- L_C is called spin or internal angular momentum.

3.6 The relation between torque and angular momentum

The angular momentum of the system of particles is given as

$$\vec{L}_O = \sum \vec{L}_i = \sum \vec{r}_i \times m_i \vec{v}_i,$$

where $\vec{L}_i$ is the angular momentum of the ith particle about P.

Taking the time derivative of both sides,

$$\frac{d\vec{L}_P}{dt} = \frac{d}{dt}\left(\sum \vec{r}_i \times m_i \vec{v}_i\right)$$

$$\sum \frac{d}{dt}(\vec{r}_i \times \vec{p}_i) = \sum \vec{r}_i \times \frac{d\vec{p}_i}{dt} + \sum \frac{d\vec{r}_i}{dt} \times \vec{p}_i,$$

where $\frac{d\vec{p}_i}{dt} = \vec{F}_i$ and $\frac{d\vec{r}_i}{dt} = \vec{v}_i$.

Then we can write

$$\frac{d\vec{L}_P}{dt} = \sum \vec{r}_i \times \vec{F}_i + \sum \vec{v}_i \times \vec{p}_i$$

Since $\vec{v}_i \times \vec{p}_i = 0$ (because $\vec{v}_i$ and $\vec{p}_i$ are parallel), now we have

$$\frac{d\vec{L}_P}{dt} = \sum \vec{r}_i \times \vec{F}_i,$$

where $\sum \vec{r}_i \times \vec{F}_i$ = the sum of the torques of all forces (internal and external) acting on all particles of the system. Since the net torque produced by internal forces is zero, we can write

$$\frac{d\vec{L}_P}{dt} = \vec{\tau}_{\text{ext}},$$

where
$\vec{\tau}_{ext}$ = the net torque produced by the external forces. This tells us that:
- The net torque, that is, the sum of torque produced by all external forces acting on the system of particles is equal to the rate of change of the angular momentum of the system of particles relative to any reference frame.

Example 8 A firecracker is thrown from the top of a tall tower with a velocity v at an angle of θ with the horizontal. It explodes at its highest position into several fragments. Find (a) the torque experienced by the exploded firecracker about (i) the center of mass and (ii) the point of projection, after a time t from the projection. Assume that the cracker moves in the vertical x–y-plane and air friction and buoyancy are negligible.

Solution
(a)
(i) The net torque of the exploded cracker about the point of projection O can be given as
$$\vec{\tau}_O = \vec{\tau}_C + \vec{r}_C \times \vec{F}_{net}.$$
We have learnt that the gravitational torque about the center of mass of a system of particle is zero:
$$\vec{\tau}_C = 0.$$

(ii) Then we can write
$$\vec{\tau}_O = \vec{r}_C \times \vec{F}_{net}.$$
Since $\vec{F}_{net} = M\vec{g}$ we can write
$$\vec{\tau}_O = \vec{r}_C \times M\vec{g}.$$
As firecracker explodes, the center of mass of the firecracker follows the original path. Then by putting $\vec{r}_C = x\hat{i} + y\hat{j}$ and $\vec{g} = g\hat{j}$, we have
$$\vec{\tau}_O = -Mgx\hat{k},$$
where $x = vt\cos\theta$.
Then
$$\vec{\tau}_O = -Mgvt\cos\theta\hat{k}.$$

3.7 Newton's laws for a system of particles

We need to derive the Newton's law of translation and rotation. For this purpose, let us recast the following relations: the net torque experienced by a system of particle relative to any reference point O is equal to the rate of change of angular momentum of the system about O

$$\text{so } \vec{\tau}_{net} = \vec{\tau}_{ext} = \frac{d\vec{L}_P}{dt}.$$

Putting $\vec{L}_O = \vec{L}_C + \vec{r}_C \times M\vec{v}_C$, we have that the net torque about O is

$$\vec{\tau}_{net} = \frac{d\vec{L}_O}{dt} = \frac{d(\vec{L}_C + \vec{r}_C \times M\vec{v}_C)}{dt}$$

$$= \frac{d\vec{L}_C}{dt} + \frac{d(\vec{r}_C \times M\vec{v}_C)}{dt}$$

$$= \frac{d\vec{L}_C}{dt} + \vec{r}_C \times \frac{d(M\vec{v}_C)}{dt} + M\vec{v}_C \times \frac{d\vec{r}_C}{dt}.$$

Since $\vec{v}_C \times \frac{d\vec{r}_C}{dt} = \vec{v}_C \times \vec{v}_C = 0$, rearranging the terms we have,

$$\vec{\tau}_{net} = \frac{d\vec{L}_C}{dt} + \vec{r}_C \times \frac{d(M\vec{v}_C)}{dt},$$

where

$$\vec{\tau}_{net} = \vec{\tau}_C + \vec{r}_C \times \vec{F}_{net}.$$

Then we have

$$\vec{\tau}_C + \vec{r}_C \times \vec{F}_{net} = \frac{d\vec{L}_C}{dt} + \vec{r}_C \times \frac{d(M\vec{v}_C)}{dt}.$$

Comparing both sides, we obtain the following two equations:

$$\vec{\tau}_C = \frac{d\vec{L}_C}{dt}$$

$$\vec{F}_{net} = \frac{d(M\vec{v}_C)}{dt} = M\frac{d(\vec{v}_C)}{dt} = M\vec{a}_C.$$

The first one is known as Newton's law of rotation, which states that:
- The net torque about the center of mass is numerically equal to the rate of change of angular momentum $\vec{L}_C$ about the center of mass.

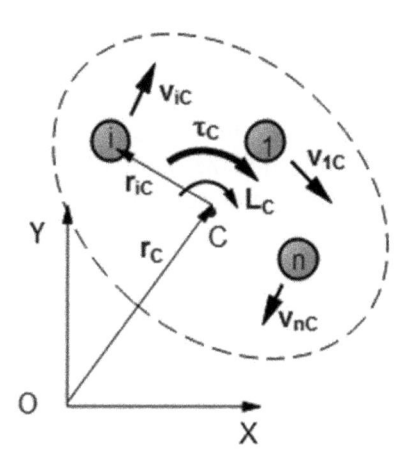

The second one is called Newton's law of translation for a system of particles, which states that:
- The net force acting on the system is numerically equal to the rate of change of linear momentum of the center of mass, that is, the product of total mass M of the system and acceleration $\vec{a}_C$ of the center of mass of the system.

Then we can conclude that:
- Translation of a system of particles is characterized by the linear momentum of the center of mass.
- $\vec{F}_{net} = M\vec{a}_C$ is Newton's law of translation of a system of particles.
- Torque about the center of mass of the system causes the rotation of the system of particles.
- The rotation of a system of particles is characterized by its angular momentum about the center of mass, that is, spin angular momentum $L_s = L_C$.
- $\vec{\tau}_C = \frac{d\vec{L}_C}{dt}$ is the law of rotation of the system of particles.

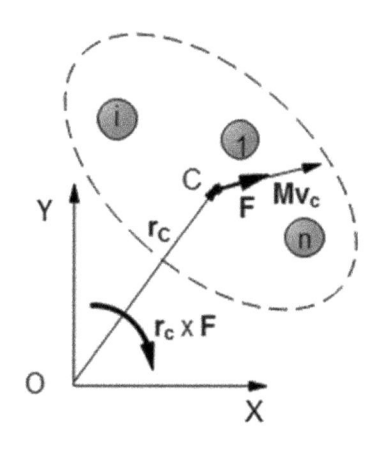

3-15

Example 9 Four identical particles each of mass m are lying at the vertices of a square of length l. Find the spin angular momentum L_C and τ_C of the system of four particles A, B, C, and D, if (a) A heads towards B, B towards C, and so on, at a constant speed v, (b) they head towards the center of mass following a straight line, and (c) move along the sides of the same square from A to B, B to C, and so on.

Solution

As each particle has equal mass, the center of mass of the system coincides with the geometrical center of the square.

(a) Since each particle heads towards the other with equal speed, they will stay at the vertices of a square of continuously reducing length. After a time t, the new length of the square is given as $b = l - vt$, where $b =$ the shortest distance between the line of motion and the center of mass. The angular momentum about the center of mass due to each particle at the center of mass is equal to $mbv/2$. Then the total angular momentum due to the four particles is $L = 4 \times mbv/2 = 2mbv$.

In an anticlockwise sense, the angular momentum is

$$\vec{L} = 2mv(l-vt)\hat{k}.$$

Then the torque acting on the system of particles is

$$\vec{\tau}_C = \frac{d\vec{L}_C}{dt} = \frac{d}{dt}\{2mv(l-vt)\hat{k}.\} = 2mv^2\hat{k}.$$

(b) As each particle moves diagonally, it does not contribute spin angular momentum. So $L_C = 0$ and hence $\tau_C = 0$.

(c) As each particle moves with a constant speed along the side of the square, the value of $b = l/2$, which remains constant. Hence the angular momentum of each about the center of mass is equal to $mlv/2$. Then the total angular momentum due to four particles is $L = 4 \times mlv/2 = 2mlv$. For the anticlockwise sense, $\vec{L} = 2mlv\hat{k}$, and hence $\tau_C = 0$.

3.8 Conservation of the angular momentum of a system of particles

Let us recast the formula

$$\vec{\tau}_{ext} = \frac{d\vec{L}_P}{dt}.$$

If no net force acts on a system, we put $\vec{\tau}_{ext} = 0$ in the previous expression to obtain

$$\frac{d\vec{L}_P}{dt} = 0.$$

Then the angular momentum $\vec{L}_P$ must remain constant. In other words:

- If the net torque acting on the system is zero relative to a given reference frame, the angular momentum of the system of particles relative to the same reference frame (point/axis) remains conserved.

Example 10 In example 8, (a) can you conserve the angular momentum of the firecracker relative to (i) the center of mass of the firecracker and (ii) the point of projection? (b) If the firecracker explodes at the highest position, find the angular momentum of the firecracker just after the explosion.

Solution
(a)
  (i) Yes. The internal forces act on the particles during the collision. Since the torque due to internal forces is zero about any inertial reference point, the angular momentum of the firecracker remains conserved about the center of mass of the firecracker. So we can say that $L_C = 0$.
  (ii) No. Referring to example 8, by integrating the expression of tourque with time, the angular momentum of the cracker about the origin is given as
  $$L_O = -\frac{1}{2}mgvt^2 \cos\theta.$$
  This means that there is not a constant due to the gravitational torque experienced by the firecracker about the point of projection O.

(b) Putting $t = v\sin\theta/g$ in the above expression and simplifying the factors, we have
$$\vec{L}_O = \frac{mv^3\sin^2\theta \cos\theta}{2g}\hat{k}.$$

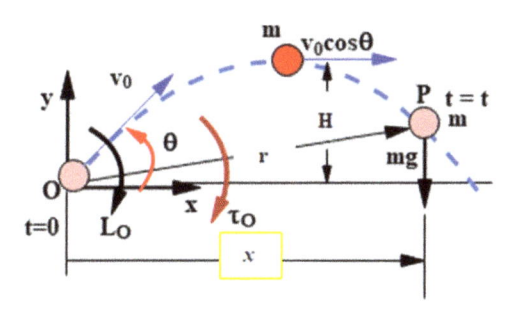

Otherwise: $\vec{L}_O = -mvH\cos\theta\hat{k}$, where,
$H = \frac{u^2\sin^2\theta}{2g}$ which gives us the same result.

Example 11 Two particles each of mass m are connected with the ends of a light spring of stiffness k and natural length l. The spring–particle system is placed in a smooth, horizontal, light, rigid tube. The tube is free to rotate in a horizontal plane about a vertical axis passing through the mid-point of the spring. Let the tube be given an angular velocity ω_0 and released. As the spring extends due to the centrifugal effects, find the maximum elongation x of the spring if $(x \ll l)$ as the spring is very stiff.

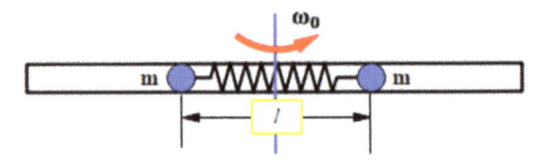

Solution
As there is no external torque involved, the angular momentum of the system is conserved: $L_f = L_i$.

Putting $L_f = \frac{m(l+x)^2 \omega}{2}$ and $L_i = \frac{ml^2 \omega_0}{2}$, we have

$$l^2 \omega = (l+x)^2 \omega_0. \tag{3.3}$$

As the spring elongates by a distance x, the spring potential energy increases by $\frac{kx^2}{2}$ and the kinetic energy of the system decreases from $\frac{ml^2\omega_0^2}{4}$ to $\frac{m\omega^2(l+x)^2}{4}$. Then, conserving the total energy of the system, we have

$$\frac{ml^2\omega_0^2}{4} = \frac{m\omega^2(l+x)^2}{4} + \frac{kx^2}{2}$$

$$\frac{ml^2\omega_0^2}{4}\left(1 - \frac{\omega^2(l+x)^2}{\omega_0^2}\right) = \frac{kx^2}{2}. \tag{3.4}$$

Putting the value of ω from equation (3.3) in equation (3.4) and simplifying the factors, we have

$$\frac{ml^2\omega_0^2}{4}\left\{1 - \left(\frac{l}{l+x}\right)^2\right\} = \frac{kx^2}{2}. \tag{3.5}$$

Since $x \ll l$, $\left(\frac{l}{l+x}\right)^2 = \left(1 + \frac{x}{l}\right)^{-2} \cong 1 - 2x/l$, putting this value in equation (3.5) and simplifying the factors, we have

$$x \cong \frac{ml\omega_0^2}{k}.$$

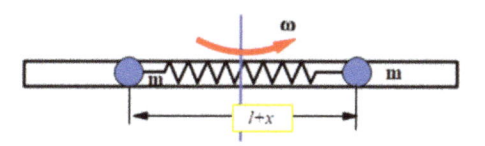

3.9 The angular momentum of a rigid body about the center of mass

For the sake of simplicity let us take a plate which is free to rotate about its centroidal axis C perpendicular to the plane of the plate. The angular momentum of an element of mass dm located at a radial distance r from the axis of rotation of the plate is $dL = (dm)(v)(r)$. Since each element moves in a circle about the center of mass axis with an angular velocity $\vec{\omega}$, putting $v = r\omega$, we have

$$d\vec{L}_C = (dm)(r\vec{\omega})(r) = (r^2 dm)\vec{\omega}.$$

By summing $d\vec{L}_C$, the net angular momentum of the disc about the center of mass axis is

$$\vec{L}_C = \int d\vec{L}_C = \int r^2 dm \vec{\omega}.$$

Since each element has same $\vec{\omega}$, take it out of the integral to obtain

$$\vec{L}_C = \vec{\omega} \int r^2 dm,$$

where the term $\int r^2 dm$ is termed as the second moment of mass or inertia, or popularly known as the moment of inertia, denoted by the letter I. Then we can write

$$\vec{L}_C = I_C \vec{\omega},$$

where I_C = the moment of inertia of the body about the centroidal axis.

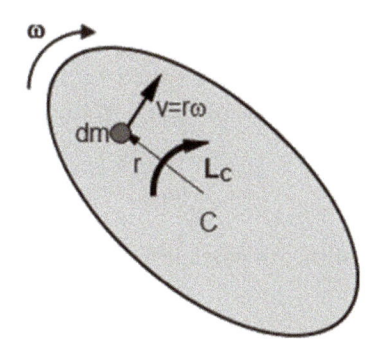

Example 12 A disc of mass $m = 2$ kg and radius $R = 0.4$ m is rotating with a spin angular momentum of 10π kg m^{-2} s^{-2}. Find the moment of inertia of the disc about its center of mass if the disc spins at a frequency of three rotations per second.

Solution
The spin angular momentum is
$$L_C = I_C \omega, \quad \text{where } \omega = 2\pi f.$$
Then the moment of inertia is
$$I_C = \frac{L_C}{2\pi f} = \frac{10\pi}{2\pi f} = \frac{5}{f},$$
where
$$f = 3, \quad \text{so } I_C = \frac{5}{3} \text{kg m}^2.$$

Students may be tempted to use the traditional formula for the moment of inertia of a disc, that is, $I = mR^2/2$, but be careful! It is the formula for a uniform distribution of mass. Here the disc might be a non-uniform one!

3.10 The moment of inertia and its calculation
3.10.1 Discrete particle system

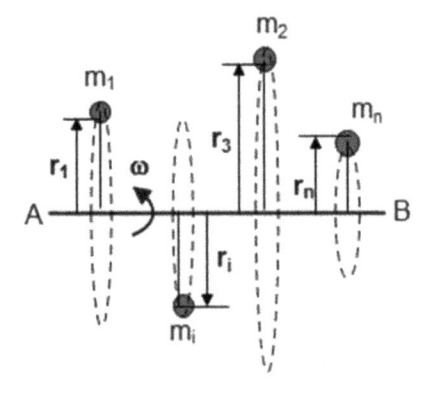

Let the particles 1, 2, 3...n be rigidly connected with the axis of rotation AB. Each particle moves in a circle. The moment of inertia of the system of particles about the axis of rotation is given as
$$I = \sum_{1}^{n} m_i r_i^2,$$
where m_i and r_i are the mass of the ith particle and radius of the particle, respectively.

Example 13 What is the moment of inertia of (a) a simple pendulum and (b) conical pendulum, if the mass of the bob is m and length of the string is l? Assume that the string makes an angle θ in a conical pendulum.

Solution

(a) In a simple pendulum the axis of the circular motion is the horizontal (z-axis) passing through the point O. So the radius of the circle is equal to the length of the string. Putting $r = l$, we have $I = mr^2 = ml^2$.

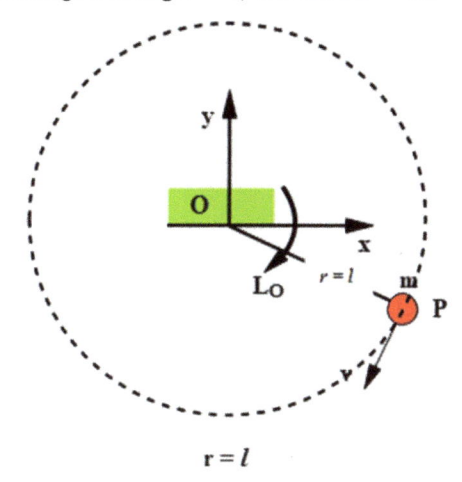

(b) In the conical pendulum the bob revolves in a horizontal circle in the x–y-plane and the axis of the circular motion is vertical (the z-axis). So the radius of the circle $r = l\sin\theta$. Then $I = mr^2 = ml^2\sin^2\theta$.

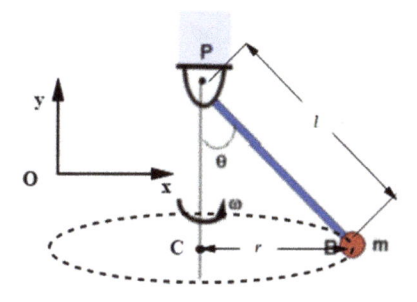

Example 14 Three particles each of mass m are placed at the vertices of a light rigid frame with the shape of an equilateral triangle ABC of side l. The plane of the frame is in the x–y-plane. Find the moment of inertia of the system of three particles about an axis passing through (a) the center of mass O and perpendicular to the plane of the frame, (b) the perpendicular bisector of the side AB, and (c) any side (AB, say) of the triangle.

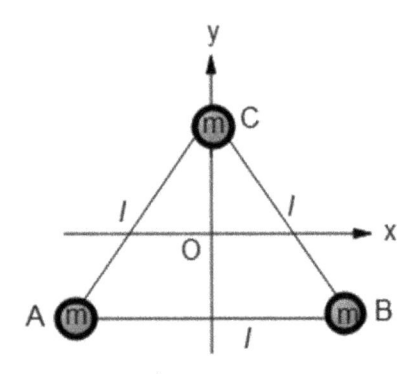

Solution

(a) In this case, each particle revolves in a circle about the z-axis passing through the point O. The perpendicular distance of each particle from the axis of rotation of the system is

$OA = OB = OC = r$. So, $I = 3mr^2$, where $r = \dfrac{l}{\sqrt{3}}$. Then $I = ml^2$.

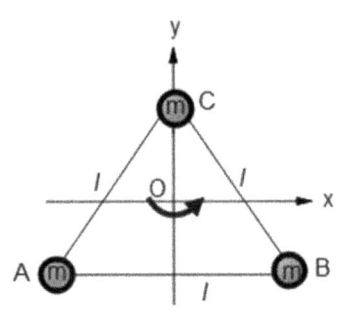

(b) The axis passes through the particle C for which $r = 0$. For the other two particles A and B, $r = \dfrac{l}{2}$. Then $I = 2mr^2 = ml^2/2$.

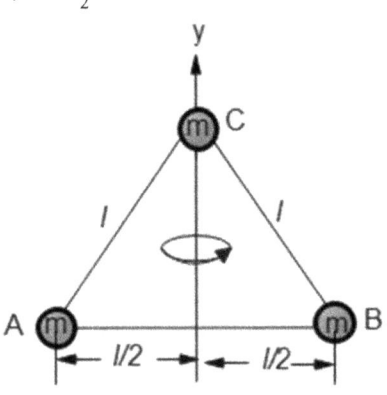

3-22

(c) The axis passes through two particles for which $r = 0$ for two particles. Then by putting $r = l\sin 60° = \frac{\sqrt{3}l}{2}$ for the third particle, we have $I = mr^2 = \frac{3ml^2}{4}$.

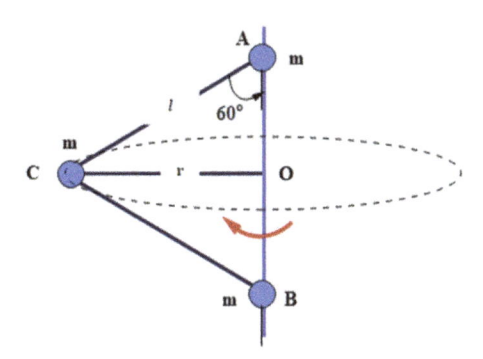

3.10.2 The moment of inertia due to continuous mass distribution

Linear mass: The basic formula for the moment of inertia is $I = \int r^2 dm$, where $dm = \lambda dr$, where λ is the linear mass density ($dm/dl = \lambda$). This gives

$$I = \int \lambda r^2 dr.$$

Example 15 Rod. Find the moment of inertia of a rod of mass m and length L about the vertical axis AB, if it is (a) uniform and (b) non-uniform having a linear mass density which varies with x as $\lambda = \lambda_0(1 + \frac{x}{L})$.

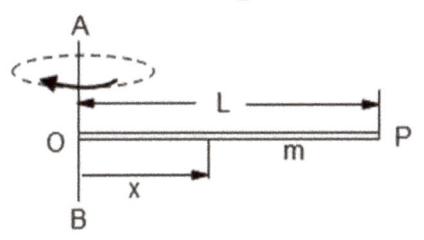

Solution
(a) Let us use the last expression $I = \int \lambda r^2 dr$. Since the rod is uniform, λ is a constant throughout the length of the rod. So we can take λ out of the integral to obtain

$$I = \lambda \int_0^L r^2 dr = \lambda L^3/3,$$

where $\lambda = \frac{m}{L}$. Then $I = mL^2/3$.

(b) Putting $\lambda = \lambda_0(1 + \frac{x}{L})$ and $r = x$ in the expression $I = \int \lambda r^2 dr$, we have

$$I = \int_0^l x^2 dx \lambda_0 \left(1 + \frac{x}{L}\right) = \frac{7\lambda_0 l^3}{12}. \tag{3.6}$$

The mass of the rod is

$$m = \int_0^l \lambda dx = \int_0^l \lambda_0 \left(1 + \frac{x}{L}\right) dx = 3\lambda_0 l/2. \tag{3.7}$$

Substituting λ_0 in equation (3.6) from equation (3.7), we obtain $I = \frac{7ml^2}{18}$.

Example 16 Ring. Find the moment of inertia of the ring about the vertical (z-axis) assuming that the plane of the ring lies in the x–y-plane.

Solution
The general formula for the moment of inertia is $I = \int r^2 dm$. Since each element dm of the ring is equidistant from the axis, $r = R$. So taking R out of the integral, we have

$$I = \int R^2 dm = R^2 \int dm = mR^2.$$

This expression does not depend upon the uniformity of the ring.

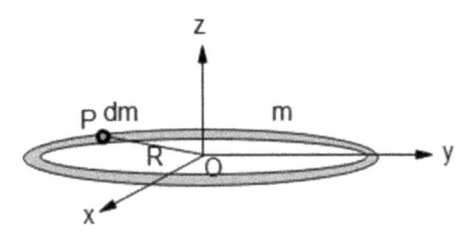

Areal mass distribution: In areal or surface mass distribution we take an element of mass $dm = \sigma dA$, where dA is the area of the element. Then we can use the general expression

$$I = \int r^2 dm.$$

Putting $dm = \int \sigma dA$, we have

$$I = \int \sigma r^2 dA.$$

Example 16 Disc. Find the moment of inertia of a disc of mass m and radius R about an axis passing through its center of mass and perpendicular to the plane of the disc, if the disc is (a) uniform, (b) annular having inner and outer radii r and R, respectively, (c) non-uniform, whose areal mass density varies with radial distance as $\sigma = \frac{kr}{R}$.

Solution

(a) Let us take a thin ring of radius r and thickness dr. Since the area of the ring is $dA = 2\pi r dr$, putting the value of dA in the above equation, we have

$$I = \int \sigma r^2 dA = \int \sigma r^2 dA = \sigma r^2 2\pi r dr.$$

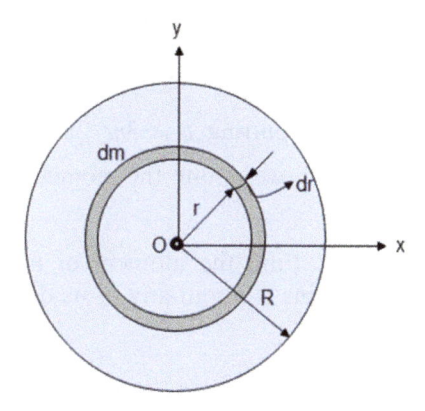

Since the disc is a combination of thin rings of radii ranging from 0 to R, we can write

$$I = \int_0^R \sigma r^2 2\pi r dr.$$

As the disc is uniform, σ is a constant quantity. So we can take σ out of the integral to obtain

$$I = 2\pi\sigma \int_0^R r^3 dr = \pi\sigma R^4/2.$$

Putting $\sigma = m/\pi R^2$, we have

$$I = mR^2/2.$$

(b) For the annular ring, we need to put the lower limit as $r = a$ and the upper limit $r = b$ in the previous equation to obtain

$$I = 2\pi\sigma \int_r^R r^3 dr = \pi\sigma(R^4 - r^4)/2,$$

where $\sigma = \dfrac{m}{\pi(R^2 - r^2)}$.

then $I = \dfrac{m(R^2 + r^2)}{2}$.

(c) If $\sigma = kr/R$, we cannot take σ it out of the integral. Then the expression can be given as

$$I = 2\pi \int_0^R \sigma r^3 dr = 2\pi \int_0^R \left(\frac{kr}{R}\right) r^3 dr.$$

Evaluating the integral and simplifying the factors, we have

3-25

$$I = \frac{2\pi k}{R} \int_0^R r^4 dr = \frac{2\pi k R^4}{5}. \tag{3.8}$$

The mass of the disc is given as

$$m = \int_0^R \sigma dA = \int_0^R \left(\frac{kr}{R}\right)(2\pi r dr)$$

$$m = 2\pi k R^2/3. \tag{3.9}$$

We can eliminate the term k by putting $k = 3m/2\pi R^2$ from equation (3.9) in equation (3.8). After simplification, we obtain the moment of inertia $I = \frac{3mR^2}{5}$.

Example 17 Thin spherical shell. Find the moment of inertia of a uniform thin spherical shell of mass m and radius R about any of its diametrical axis.

Solution
Let us take a thin ring of area $dA = 2\pi rt$, where the radius of the ring is $r = R\sin\theta$ and its thickness is $t = Rd\theta$. Hence $dA = (2\pi R\sin\theta)(Rd\theta) = 2\pi R^2 \sin\theta d\theta$.

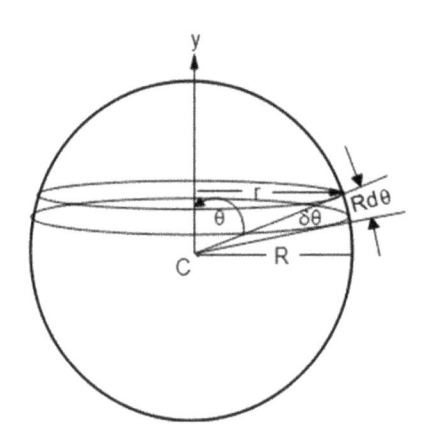

Putting the values of dA in the general expression $I = \int \sigma r^2 dA$, we have

$$I = \int \sigma (R^2 \sin^2\theta)(2\pi R \sin\theta)(Rd\theta).$$

So

$$I = 2\pi R^4 \int \sigma \sin^3\theta d\theta. \tag{3.10}$$

For uniform mass putting $\sigma = m/4\pi R^2$ and taking the constant terms out of the integral, we have

$$I = \frac{mR^2}{2} \int \sin^3\theta d\theta. \tag{3.11}$$

Since the thin sphere is the combination of thin rings of radii ranging from 0 to R for each hemisphere, when θ ranges from 0 to π for one hemisphere, putting zero (0) as the lower limit and π as the upper limit of the integration, we have

$$I = \frac{mR^2}{2} \int_0^{\pi} \sin^3\theta d\theta.$$

After evaluating the integral to be equal to 4/3, we have $I = \frac{2mR^2}{3}$.

Example 18 Hollow cone. Find the moment of inertia of a uniform thin hollow cone of mass m and radius R of the base about its symmetrical axis.

Solution

The hollow cone is a combination of thin coaxial rings of radii ranging from 0 to R. Hence, the moment of inertia of the hollow cone is equal to the sum (or integration) of the moment of inertia of the thin rings. So we have

$$I = \int I_{ring}, \qquad (3.12)$$

where I_{ring} is the moment of inertia of the thin rings of radius r and thickness dx. Since $I_{ring} = m_r r^2$, where m_r is the mass of the thin ring, given as

$$m_r = \sigma(2\pi r dx).$$

Then we have

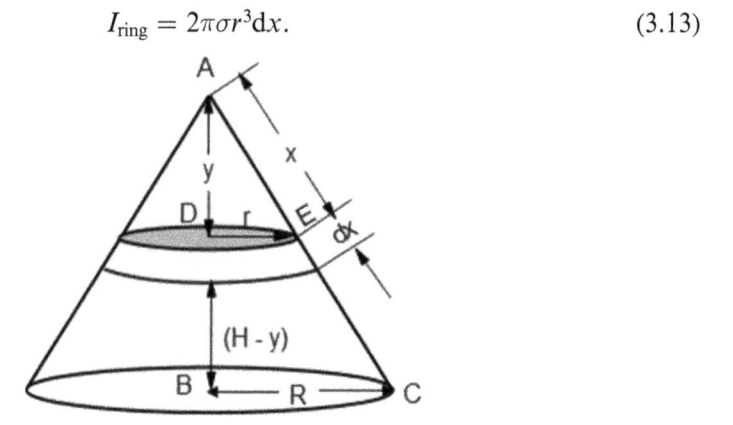

$$I_{ring} = 2\pi\sigma r^3 dx. \qquad (3.13)$$

Putting I_{ring} from equation (3.13) in equation (3.12) we have

$$I = 2\pi\sigma \int r^3 dx,$$

where $AE/AC = x/L = r/R$.

$$\text{So, } I = \frac{2\pi\sigma L}{R} \int r^3 dr.$$

Since the hollow cone is a combination of thin rings of radii ranging from 0 to R, setting the limits of the integral, we have

$$I = \frac{2\pi\sigma L}{R} \int_0^R r^3 dr = \frac{\pi\sigma L R^3}{2}. \qquad (3.14)$$

Putting $\sigma = \frac{m}{A} = \frac{m}{\pi RL}$ in equation (3.14) we have $I = I_{cone} = \frac{mR^2}{2}$.

Volume mass distribution: The general expression of the moment of inertia is $I = \int r^2 dm$. If the mass is distributed throughout the volume we can write $dm = \rho dV$. Then $I = \int r^2 \rho dV$. However, the direct use of this formula may not be so straightforward. Therefore, let us follow a tricky approach to find the moment of inertia of some uniform familiar structures in the following examples.

Example 19 Solid cone. Find the moment of inertia of a uniform solid cone of mass m and radius of the base R about its symmetrical axis.

Solution

The cone is a combination of thin coaxial discs piled one over the other whose radii vary from 0 to R. Hence, the moment of inertia of the cone is equal to the sum or integration of the moment of inertia of the thin discs. So

$$I_{cone} = \int I_{disc}, \qquad (3.15)$$

where I_{disc} is the moment of inertia of the thin disc of radius r and thickness dy. Since $I_{disc} = \frac{m_d r^2}{2}$, where $m_d = \rho(\pi r^2 dy)$. Then we have

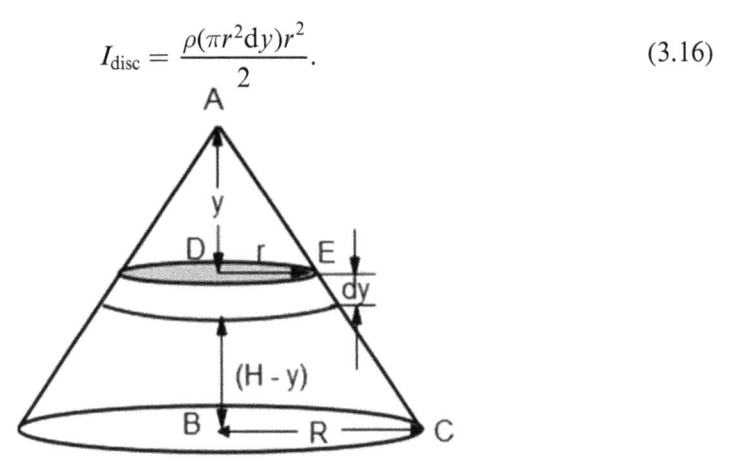

$$I_{disc} = \frac{\rho(\pi r^2 dy)r^2}{2}. \qquad (3.16)$$

Putting I_{disc} from equation (3.16) in equation (3.15) we have

$$I = \frac{\pi \rho}{2} \int r^4 dy,$$

where $y/H = r/R$ (by comparing the similar triangles ADE and ABC). Then we have

$$I = \frac{\pi \rho H}{2R} \int r^4 dr.$$

Since the cone is a combination of thin discs of radii ranging from 0 to R, setting the limits of the integral, we have

$$I = \frac{\pi \rho H}{2R} \int_0^R r^4 dr = \frac{\pi \rho H R^4}{10}. \qquad (3.17)$$

Putting $\rho = \frac{m}{V} = \frac{m}{\pi R^2 H / 3} = \frac{3m}{\pi R^2 H}$ in equation (3.17), we have $I = \frac{3mR^2}{10}$.

Example 20 Solid sphere. Find the moment of inertia of a uniform solid sphere of mass m and radius of the base R about its diametrical axis.

Solution

A solid sphere is combination of thin concentric spherical shells piled one over the other of radii ranging from 0 to R. Hence, the moment of inertia of the sphere is equal to the sum or integration of the moment of inertia of the thin spherical shells which can be given as

$$I = \int I_{\text{shell}}, \tag{3.18}$$

where I_{shell} is the moment of inertia of the thin spherical shell of radius r and thickness dr.

We know that

$$I_{\text{shell}} = \frac{2(m_{\text{shell}})r^2}{3},$$

where $m_{\text{shell}} = \rho(4\pi r^2 dr)$.

Then we have

$$I_{\text{shell}} = \frac{2\rho(4\pi r^2 dr)r^2}{3} = \frac{8\pi \rho r^4 dr}{3}. \tag{3.19}$$

Putting I_{shell} from equation (3.19) in equation (3.18) we have

$$I = \frac{8\pi \rho}{3} \int r^4 dr.$$

As the radius of the spherical shells varies from 0 to R we have

$$I = \frac{8\pi \rho}{3} \int_0^R r^4 dr = \frac{8\pi \rho R^5}{15}. \tag{3.20}$$

Putting $\rho = \frac{m}{V} = \frac{m}{4\pi R^3 / 3} = \frac{3m}{4\pi R^3}$ in equation (3.20), finally we obtain $I = \frac{2mR^2}{5}$.

3-29

Otherwise: Let us take a thin disc (a horizontal slice of the sphere) of thickness dy at a distance y from the center. Its mass is given as

$$m_{disc} = dm = \rho dV = \rho \pi r^2 dy. \tag{3.21}$$

The moment of inertia of the disc about the y-axis is

$$I_{disc} = \frac{m_{disc} r^2}{2}. \tag{3.22}$$

Putting the value of m_{disc} from equation (3.21) in equation (3.22) we have

$$I_{disc} = \frac{\rho \pi r^4 dy}{2}. \tag{3.23}$$

The moment of inertia of the sphere is the sum of the moments of inertia of all the thin discs. So, integrating I_{disc} from $y = -R$ to $y = R$, we have

$$I_{sphere} = \int_{-R}^{R} I_{disc} = 2 \int_{0}^{R} \frac{\rho \pi r^4 dy}{2}.$$

For the uniform sphere take ρ out of the integral to obtain

$$I_{sphere} = \rho \pi \int_{0}^{R} r^4 dy.$$

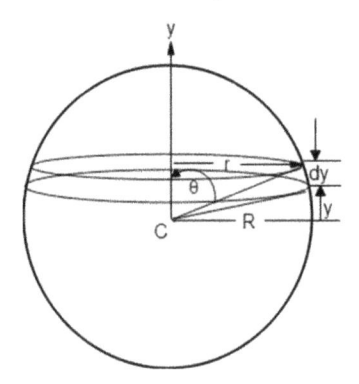

Putting $r^4 = (R^2 - y^2)^2$ and evaluating the integral to be 8/15, in the last equation, we have

$$I_{sphere} = (\pi)(\rho)\left(\frac{8R^5}{15}\right)$$

Putting $\rho = \frac{3m}{4\pi R^3}$ and evaluating the integral we can obtain the previous expression for the moment of inertia of the sphere, which is equal to $\frac{2mR^5}{5}$.

3.10.3 Parallel axis theorem

Until now we have derived the moment of inertia (I_C) of some familiar uniform rigid bodies about their centroidal axis. If we want to find the moment of inertia of a body about an axis which is parallel to the centroidal axis, we need to derive a general expression known as the parallel axis theorem. It states that:

- If the moment of inertia of a rigid body about its centroidal axis is given as I_C, its moment of inertia I about an axis parallel to the centroidal axis is given as

$$I = I_C + mr^2,$$

where $r =$ the distance between two (centroidal axis and given axis) parallel axes and $m =$ the mass of the body.

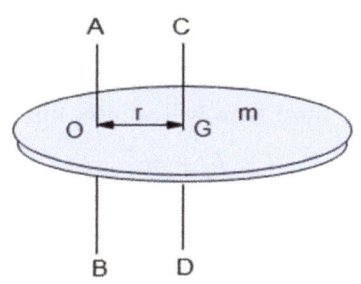

Proof

For the sake of simplicity let us take a thin plate which can rotate about an axis perpendicular to the plane of the plate. This called the *axis of rotation*. Let us consider an element of mass dm at a point P on the plate at a position $OP = r'$ from the axis of rotation. If the position of the element from the centroidal axis is r_c and the position of the centroidal axis from the axis of rotation is r, we can write

$$OP = |\vec{r} + \vec{R}|. \tag{3.24}$$

Then the moment of inertia of the plate about the axis of rotation is

$$I = \int OP^2 dm. \tag{3.25}$$

Putting OP from equation (3.24) in equation (3.25) we have

$$I = \int |\vec{r} + \vec{R}|^2 dm$$
$$= \int (R^2 + r^2 + 2\vec{R}\cdot\vec{r}) dm$$
$$= \int R^2 dm + \int r^2 dm + \int 2\vec{R}\cdot\vec{r}\, dm.$$

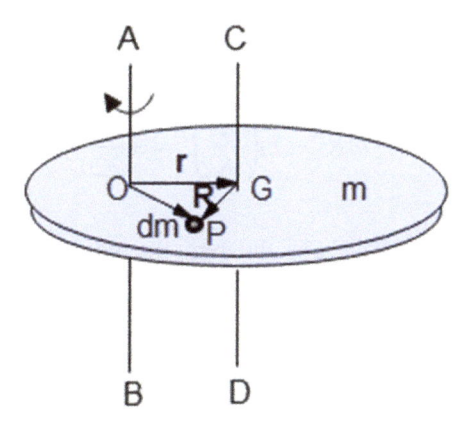

The first term is $\int R^2 dm$ = the moment of inertia of the body about the center of mass axis = I_C. The second term r is a fixed distance between two axes. So pulling it out of the integral, the second term can be written as $\int r^2 dm = r^2 \int dm = mr^2$, where m = the mass of the body. The third term can be written as

$$\int 2\vec{R}\cdot\vec{r}\, dm = 2\vec{r}\cdot\int \vec{R}\, dm = 0,$$

where $\int \vec{R}\, dm$ = the sum of the first moment of mass relative to the center of mass which equals to zero. As the third term will vanish, finally, we have

$$I = I_C + mr^2.$$

Example 21 Find the moment of inertia of a solid hemisphere of mass m and radius R about the axis (a) DCE passing through the center of mass C, (b) MPQ and (c) AB.

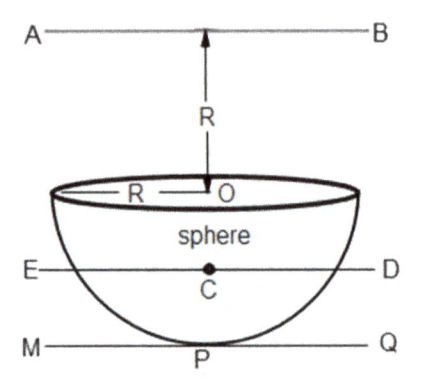

Solution
(a) The moment of inertia of the hemisphere about the vertical axis passing through its flat face is $I_O = \frac{2mR^2}{5}$. According to the parallel axis theorem

$$I_C = I_O - mr^2,$$

where $r = OC = 3R/8$. Then we have

$$I_C = \frac{2}{5}mR^2 - m\left(\frac{3R}{8}\right)^2 = \frac{83mR^2}{320}.$$

(b) Again applying the parallel axis theorem, the moment of inertia about MPQ is

$$I_P = I_C + m(PC)^2, \quad \text{where } PC = 5R/8.$$

$$\text{Then } I_P = I_C + m\left(\frac{5R}{8}\right)^2. \tag{3.26}$$

Recasting the previous equation we have
$$I_O = I_C + m(OC)^2, \text{ where } OC = 3R/8.$$

Then $I_O = I_C + m\left(\dfrac{3R}{8}\right)^2.$ (3.27)

Solving equations (3.26) and (3.27) we have
$$I_Q = I_O + m\left\{\left(\dfrac{5R}{8}\right)^2 - \left(\dfrac{3R}{8}\right)^2\right\}. \tag{3.28}$$

Putting $I_O = \dfrac{2}{5}mR^2$ in equation (3.28) and simplifying the factors, we have
$I = \dfrac{13}{20}mR^2.$

(c) Applying parallel axis theorem between the axes AB and ECD, we have
$$I_{AB} = I_C + mr^2,$$

where $r = R + OC = R + 3R/8 = 11R/8$ and $I_C = \dfrac{80mR^2}{320}$.
Then, finally, we have

$$I_{AB} = \dfrac{83mR^2}{320} + m\left(\dfrac{11R}{8}\right)^2 = \dfrac{43mR^2}{20}.$$

3.10.4 Perpendicular axis theorem

Let us consider a lamina (a plate-like object) in the x–y-plane. Consider any point O of the lamina. Let the moment of inertia of the plate about an axis perpendicular to its plane, that is, the z-axis is I_z. If the moments of inertia of the lamina about the x- and y-axes are I_x and I_y, respectively, we can state that:

'The sum of the moment of inertia about the x- and y-axis is numerically equal to the moment of inertia of the body passing through the z-axis.'

Symbolically, $I_z = I_x + I_y$.
This is known as perpendicular axis theorem.

Proof: Consider an element of mass dm of the plate at a distance r from the z-axis. The moment of inertia of the plate about the z-axis is

$$I_z = \int r^2 dm,$$

where $r^2 = x^2 + y^2$.
Then we have

$$I_z = \int (x^2 + y^2)dm = \int x^2 dm + \int y^2 dm,$$

where $\int x^2 dm = I_x =$ and $\int y^2 dm = I_y$.

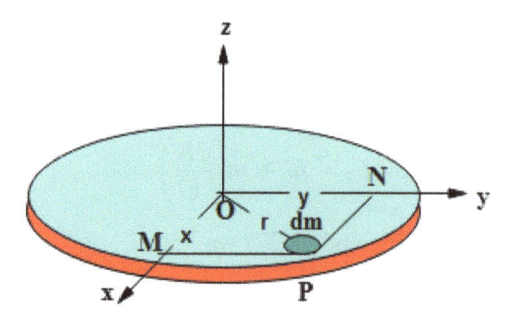

Then we have $I_z = I_x + I_y$.

Example 22 Find the moment of inertia of the half disc of mass m and radius R about the center of mass axis (ACB) parallel to the x-axis, as shown in the figure.

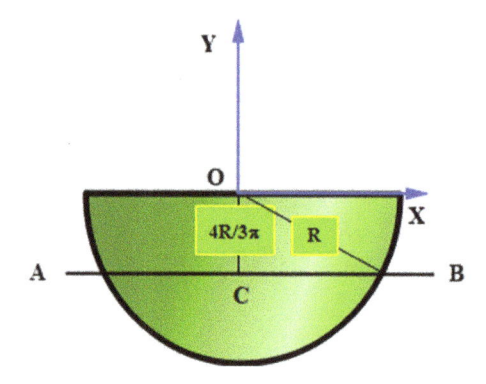

Solution
Let the desired moment of inertia of the half disc about the axis ACB be I. Using the parallel axis theorem between two parallel axes ACB and OX, we have

$$I_x = I + m(OC^2).$$

Since $CO = \frac{4R}{3\pi}$.

By putting the value of OC we have

$$I = I_x - m\left(\frac{4R}{3\pi}\right)^2. \qquad (3.29)$$

Using perpendicular axis theorem we have

$$I_z = I_x + I_y.$$

Since $I_x = I_y$ and $I_z = \frac{mR^2}{2}$, we have $I_x = \frac{mR^2}{4}$.

Putting this value in equation (3.29) we have

$$I = I_x - m\left(\frac{4R}{3\pi}\right)^2 = \frac{mR^2}{4}\left(1 - \frac{64}{9\pi^2}\right).$$

3.11 Newton's laws of motion of rigid bodies
3.11.1 Law of rotation
Let us recast the expression
$$\vec{\tau}_C = \frac{d\vec{L}_C}{dt},$$
where $\vec{L}_C = I_C \vec{\omega}$.

Then we have
$$\vec{\tau}_C = \frac{d(I_C \vec{\omega})}{dt}.$$

Taking I_C out of the derivative we have
$$\vec{\tau}_C = I_C \frac{d\vec{\omega}}{dt}.$$

We know that
$$\frac{d\vec{\omega}}{dt} = \vec{\alpha},$$
which is the angular acceleration of the body.

Then we can write
$$\vec{\tau}_C = I_C \vec{\alpha}.$$

This is known as Newton's law for rotation of a rigid body, which states that:
- The net (external) torque taken about the center of mass of a rigid body is numerically equal to the product of the moment of inertia and angular acceleration of the body, about the center of mass axis.
- However,
$$\vec{\tau}_C = I_C \vec{\alpha}$$
is invalid when I_C varies in a non-rigid system (variable moment of inertia).

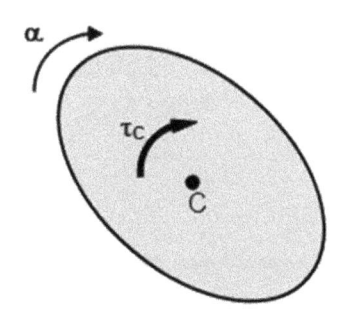

3.11.2 Law of translation

Let us recast the expression

$$\vec{F}_{net} = \frac{d\vec{P}_C}{dt}.$$

By putting $\vec{P}_C = M\vec{v}_C$ and taking M out of the derivative, we have

$$\vec{F}_{net} = M\frac{d\vec{v}_C}{dt},$$

where

$$\frac{d\vec{v}_C}{dt} = \vec{a}_C,$$

which is the acceleration of the center of mass of the body. Then we can write

$$\vec{F} = M\vec{a}_C.$$

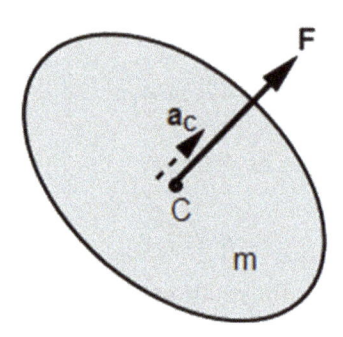

This is known as Newton's law for translation of a rigid body, which states that:

'The net (external) force acting on a rigid body is numerically equal to the product of its mass and acceleration of the center of mass of the rigid body.'

$F = Ma_C$ is invalid for a variable mass system because here M must be a constant. However, in a variable mass system, we can use this equation by adding the impact or reaction force in the left hand side of the equation.

Example 23 The uniform smooth rod of mass m and length $l = 5$ m is kept on a horizontal floor. Two forces F_1 and F_2 are acting on the rod, as shown in the figure. Find (a) the angular acceleration of the rod and (b) the linear acceleration of the center of mass of the rod. Put $F_1 = 3$ N and $F_2 = 10$ N and $m = 12$ kg.

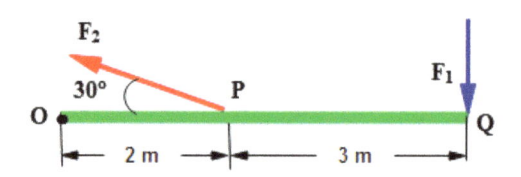

Solution

(a) The net torque about the center of mass C is equal to the sum of the torques of the forces F_1 and F_2 about the center of mass which can be given as

$$\vec{\tau}_C = F_1(2.5)(-\hat{k}) + F_2 \sin 30°(0.5)(-\hat{k}).$$

Putting the values of $F_1 = 3$ N and $F_2 = 10$ N, we have

$$\vec{\tau}_C = -10\hat{k} \text{ N m}.$$

The moment of inertia of the rod about its center of mass is
$I_C = \frac{ml^2}{12} = \frac{(12)(5^2)}{12} = 25 \text{ kg m}^2.$
Then, putting the values of $\vec{\tau}_C$ and I_C in the equation

$$\vec{\tau}_C = I_C \vec{\alpha},$$

we have

$$-10\hat{k} = (25)\vec{\alpha}.$$

This gives $\vec{\alpha} = 0.4\hat{k}$ rad s^{-2}.

(b) The net force acting on the rod is

$$\vec{F} = -F_2 \cos 30°\hat{i} + F_2 \sin 30°\hat{j} - F_1\hat{j}$$

$$= -10\left(\frac{\sqrt{3}}{2}\right)\hat{i} + 10(0.5)\hat{j} - 3\hat{j}$$

$$= -5\sqrt{3}\hat{i} + 2\hat{j} \text{ N}.$$

Putting this value in the equation, $\vec{F} = M\vec{a}_C$, and simplifying the factors, we have

$$a_C = \frac{|-5\sqrt{3}i + 2j|}{12} = (\sqrt{79}/12) \text{ m s}^{-2}.$$

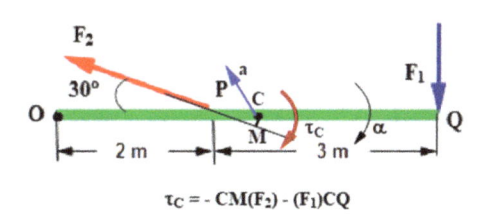

$\tau_C = -CM(F_2) - (F_1)CQ$

Example 24 Two particles A and B of mass m and M are connected rigidly to the ends of a light rigid rod of length L. The rod–particle system is kept on a smooth horizontal plane. A force F acts at A on the particle of mass m perpendicular to the rod, as shown in the figure. Find (a) the angular acceleration of the rod (b) acceleration of the center of mass of the system (M + rod + m).

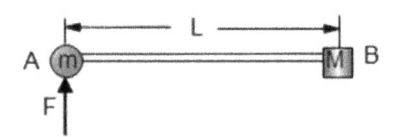

Solution
(a) Referring to the FBD, let us write the torque and force equations.
Torque equation: The torque of the force F about the center of mass of the system is

$$\vec{\tau}_C = -Fx\hat{k},$$

where $x =$ the distance of the center of mass from the particle A $= ML/(M + m)$. Then we have

$$\vec{\tau}_C = -F\{ML/(M + m)\}\hat{k}. \qquad (3.30)$$

This torque is responsible for producing an angular acceleration $\vec{\alpha}$ of the rod, given as

$$\vec{\tau}_C = I_C \vec{\alpha},$$

where I_C is given as

$$I_C = \frac{MmL^2}{M + m}.$$

Then we have

$$\vec{\tau}_C = \frac{MmL^2}{M + m}\vec{\alpha}. \qquad (3.31)$$

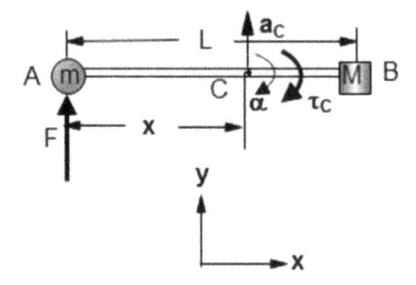

Putting the value of $\vec{\tau}_C$ from equation (3.30) in equation (3.31) and simplifying the factors,

$$\vec{\alpha} = -\left(\frac{F}{mL}\right)\hat{k}.$$

3-38

(b) *Force equation*:
$$\vec{F}_{net} = (M_{system})\vec{a}_C,$$
where $\vec{F}_{net} = Fj$ and $M_{system} = M + m$.

This yields $\vec{a}_C = \frac{F\hat{j}}{M+m}$.

3.12 Equilibrium of rigid bodies

There are two types of equilibrium of rigid bodies, namely *translational* and *rotational*.

3.12.1 Translation of rigid bodies

A rigid body undergoes translation if all its particles have same velocity at a given instant. This means, all particles have same acceleration, which is also equal to the acceleration of the center of mass of the rigid body. The equation of translation of the body is given as

$$\vec{F} = m\vec{a}_C,$$

where m = the mass of the body and $\vec{a}_C$ = the acceleration of the center of mass of the body.

For a pure translation, rotation does not occur. In other words,

$$\vec{\omega} = \frac{d\vec{\theta}}{dt} = 0.$$

Then $\theta = C$ (constant), which signifies that the angle of orientation of any line drawn inside the body does not change with time. If $\vec{\omega} = 0$, obviously the rate of change of angular velocity, that is, angular acceleration, $\vec{\alpha} = 0$. For this to happen, the net torque acting on the body about the center of mass must be zero:

$$\vec{\tau}_{net} = I\vec{\alpha} = 0.$$

- A rigid body is said to have pure translation, only when $\vec{\omega} = 0$.

Example 25 Freely falling rod. If a meter stick of mass m is released from rest, describe its equilibrium, disregarding the air friction and buoyancy etc.

Solution

The net force acting on the body is gravity. As the weight of the body acts vertically down at its center of mass, the net torque about the center of mass is zero:

$$\vec{\tau}_{net} = I_C\vec{\alpha} = 0.$$

Since $\vec{\alpha} = 0$ and the initial angular velocity $\omega_0 = 0$ (because the rod is released from rest), the rod performs a translation without rotation. So the rod is in rotational equilibrium but not in translational equilibrium because its center of mass moves with an acceleration $a_C = g$.

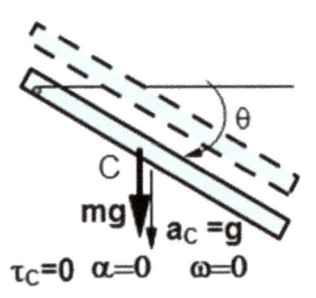

Rotation and rotational equilibrium are different concepts. Rotation means a non-zero angular velocity and rotational equilibrium requires a constant (including zero) angular velocity. A pure translation needs a zero angular velocity (no rotation). So, a pure translation possesses a rotational equilibrium but the converse may not hold true.

Example 26 Toppling. A box of mass m and height H is placed on two wooden supports at A and B, as shown in the figure. If a horizontal force F acts on the box at a height h, find (a) the normal reactions acting at A and B, (b) the net friction acting on the box, and (c) the condition for toppling of the box.

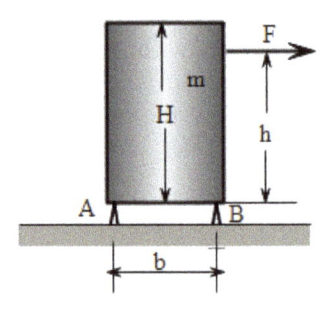

Solution
(a) Let the normal reaction forces acting at A and B be N_A and N_B, respectively. Furthermore, let the static frictions acting at A and B be f_A and f_B, respectively. Since the box does not change its orientation, $\tau_C = 0$. Since gravity W ($= mg$) acts at the center of mass, it cannot produce torque about the center of mass of the box.

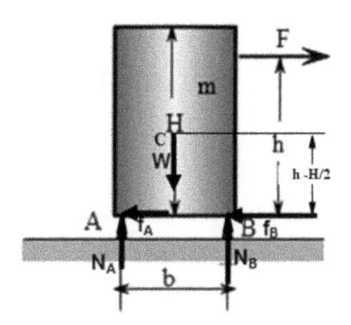

Then, taking the torques of N_A, N_B, f_A, f_B and F, about the center of mass C, and adding them, we have

$$\vec{\tau}_C = \frac{N_B b}{2}(+\hat{k}) + F\left(h - \frac{H}{2}\right)(-\hat{k})$$
$$+ \frac{N_A b}{2}(-\hat{k}) + (f_A + f_B)H/2(-\hat{k}) = 0.$$

Puting $f_A + f_B = f$ in the last expression it gives

$$(N_A - N_B)b + fH = (F2h - H). \tag{3.32}$$

As the box does not move, $F_C = 0$. The net horizontal and vertical will be zero. Then

$$F_y = N_A + N_B - mg = 0 \tag{3.33}$$

$$F_x = F - f = 0. \tag{3.34}$$

Solving these three equations for three unknown quantities such as f, N_A and N_B, we have

$$N_A = \frac{mg}{2} - \frac{F(H-h)}{b}$$

and

$$N_B = \frac{mg}{2} + \frac{F(H-h)}{b}.$$

(b) As $a_C = 0$, $F_{\text{net}} = 0$. Then we have $f(=f_A + f_B) = F$.
(c) The box will topple when the box loses contact at A. Putting $N_A = 0$ we have $=\frac{mgb}{(H-h)}$:
 • The box is in both translational and rotational equilibrium.

3-41

3.12.2 Rotational equilibrium

If the net torque acting on the rigid body is zero about the center of mass or any fixed point, we can write

$$\vec{\tau}_{net} = I\vec{\alpha} = 0.$$

So the angular acceleration of the body will be zero:

$$\vec{\alpha} = 0.$$

In other words, the angular velocity of the body will be a constant quantity (zero or non-zero). This is termed as rotational equilibrium:
- A purely translating body must be in rotational equilibrium.
- But a body in rotational equilibrium may or may not have pure translation.

Example 27 Accelerating rod. The uniform rod of mass $m = 2$ kg and length $l = 1$ m is placed on a smooth horizontal floor. Two anti-parallel forces $F_1 = 6$ N and F_2 are acting on the rod, as shown in the figure, so that the rod executes a translational motion without rotation with an acceleration 3 m s^{-2}. Find (a) the force F_2 and (b) x.

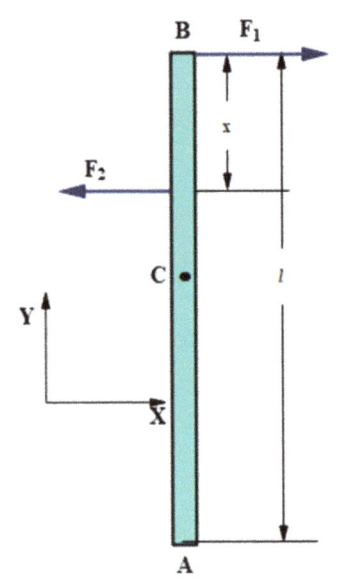

Solution
(a) As the body translates without rotation,

$$\vec{\tau}_{net} = I_C \vec{\alpha} = 0.$$

Taking the torques of F_1 and F_2 about the center of mass C and adding them, we have

$$\tau_C = \frac{F_1 l}{2}(-\hat{k}) + F_2(l/2 - x)(+\hat{k}) = 0.$$

This gives
$$x = (F_2 - F_1)l/2F_2. \qquad (3.35)$$

This tells us that F_2 must be greater than F_1. So the net force points to left and in consequence, the centre of mass C of the rod will accelerate to left. Let the acceleration of C be a C. As the rod moves with an acceleration $\vec{a}_C = -3\hat{i}$ m s^{-2}, the net horizontal force is
$$F_2 - F_1 = ma_C. \qquad (3.36)$$

By using equations (3.35) and (3.36), we have $F_2 = F_1 + ma_C = 6 + (2)(3) = 12$ N.

(b) Putting the values of $F_2 - F_1 = ma_C = (2)(3) = 6$ N and $F_2 = 12$ N from equation (3.36) in equation (3.35), we have $x = \frac{(6)(1)}{(2 \times 12)} = 0.25$ m.

- The rod is in rotational equilibrium, translation without rotation but not in translational equilibrium as its center of mass accelerates.

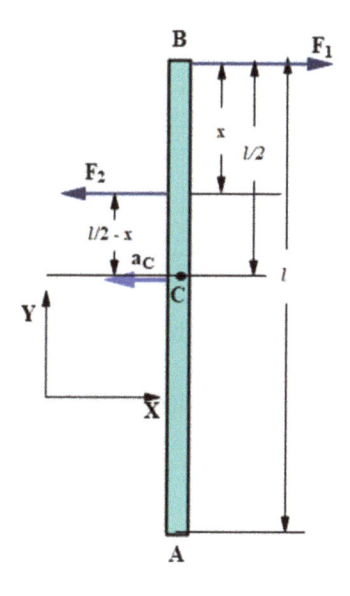

3.13 Fixed axis rotation

3.13.1 Newton's law of rotation

In fixed axis rotation, the axis of rotation is kept fixed. If the axis passes through the center of mass, we can call it the centroidal axis, otherwise the axis will be non-centroidal. If P is the fixed axis of rotation, each element of the rigid body moves in a circular path about the axis of rotation P. If dm is the mass of an element situated at a distance r from the axis, the angular momentum of the element is
$$dL_P = (dm)vr,$$

where $v = r\omega$. Writing vectorially,
$$d\vec{L}_P = dm r^2 \vec{\omega}.$$
For the fixed axis rotation, all elements move in circles with the same angular velocity $\vec{\omega}$. Hence, integrating $d\vec{L}_P$, the total angular momentum $\vec{L}_P$ of the rigid body about the fixed axis can be given as
$$\vec{L}_P = \int d\vec{L}_P = \int r^2 (dm) \vec{\omega} = \vec{\omega} \int r^2 (dm),$$
where $\int r^2 (dm)$ is the moment of inertia of the body about the fixed axis, denoted as I_P.

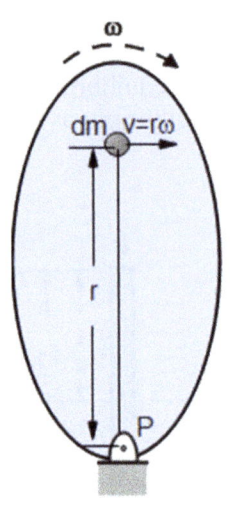

Then we can write $\vec{L}_P = I_P \vec{\omega}$. Hence, the rate of change of L_P with time is
$$\frac{d\vec{L}_P}{dt} = d(I\vec{\omega})/dt.$$
Since the moment of inertia (I) of the rigid body about an axis for fixed axis rotation does not vary with time, we can take it out of the derivative to obtain
$$\frac{d\vec{L}_P}{dt} = I_P \left(\frac{d\vec{\omega}}{dt} \right),$$
where $\frac{d\vec{\omega}}{dt} = \vec{\alpha}$ = the angular acceleration of the rigid body. Putting $\frac{d\vec{L}_P}{dt} = \vec{\tau}_P$, we have
$$\vec{\tau}_P = I_P \vec{\alpha},$$
where $I_P = I_C + md^2$ (according to the parallel axis theorem). This states that:
- The rate of change of angular momentum of a rigid body is equal to the net torque acting on the body taken relative to the given axis of rotation, is numerically equal to the product of the moment of inertia I of the rigid body about that axis and angular acceleration of the rigid body.

3-44

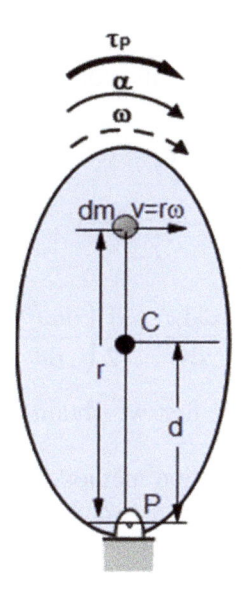

3.13.2 Newton's law of translation

Consider all external forces such as applied force, gravity, and reactions at the pivot. Gravity $m\vec{g}$ acts at the center of mass. Let R be the reaction force acting on the rigid body at its pivot denoted by the letters O, P etc. Let the tangential and radial accelerations of the center of mass of the rigid body be a_t and a_n, respectively. Then, resolving all forces along the tangential and radial directions, respectively, we can write

$$F_t = ma_t$$

$$F_n = ma_n.$$

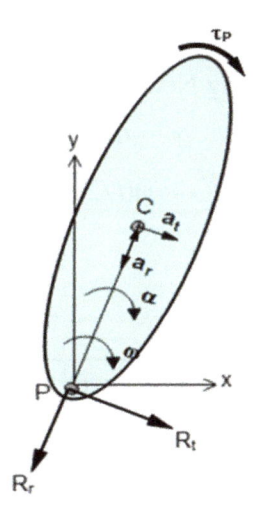

3.13.3 Kinematics

As the center of mass moves in a circle, its radial and tangential accelerations can be given as

$$a_t = r\alpha$$

$$a_n = r\omega^2,$$

where α and ω can be related as

$$\alpha = \frac{\omega d\omega}{d\theta} = \frac{1}{2}\left\{\frac{d(\omega^2)}{d\theta}\right\}.$$

- In this dynamical approach, first we obtain $\vec{\alpha}$. Then using the last equation, integrating α we obtain $\vec{\omega}$.
- However, in energy conservation method, first we obtain $\vec{\omega}$. Then, using the last equation and differentiating $\vec{\omega}$, we obtain $\vec{\alpha}$. (Please refer to the energy of rigid bodies in the previous chapter.)

Example 28 A uniform rod of mass $m = 2$ kg and length $l = 1$ m is freely pivoted at P on a smooth horizontal floor. An external horizontal force $F = 10$ N at the free end of the rod, as shown in the figure. As a result, the rod rotates in the horizontal plane. Find (a) the angular acceleration of the rod and (b) the reaction at the pivot at the given instant.

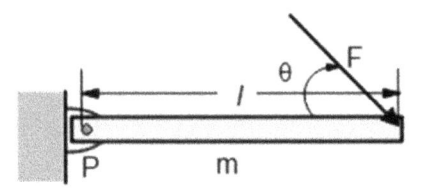

Solution
(a) *Torque equation:* Recasting Newton's equation of rotation

$$\tau_P = I_P \alpha,$$

where $I_P = I_C + md^2 = m\frac{L^2}{12} + m(l/2)^2 = \frac{ml^2}{3}$ and $\tau_P = LF \sin\theta$.
This gives $\alpha = \frac{3F \sin\theta}{mL}$.

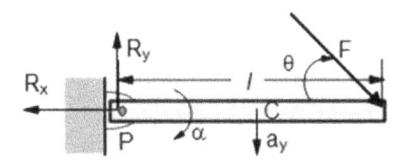

(b) *Force equation:* The horizontal and vertical components of reaction can be given:

3-46

$$-R_x + F\cos\theta = ma_x$$

$$R_y - F\sin\theta = -ma_y.$$

Kinematic: Putting $a_x = 0$ and $a_x = (l/2)\alpha$ in the force equations, we have $R_x = F\cos\theta$

$$R_y = F\sin\theta - m(l/2)\alpha,$$

where $\alpha = \frac{3F\sin\theta}{ml}$.

This gives $R_y = (1/2)F\sin\theta$. Then the net reaction force acting at the pivot is

$$R = \sqrt{R_x^2 + R_y^2} = \left(\sqrt{4 - 3\sin^2\theta}\right)F/2.$$

3.14 Dynamics of the combined motion of rigid bodies

3.14.1 Planner motion: Newton's law of rotation (torque equation)

For the sake of simplicity let us take a plate in the x–y-plane (a horizontal surface). Assume that many horizontal forces act at different points of the plate. As we learned, each force produces either a clockwise or anticlockwise torque about the center of mass of the body. The forces passing through the center of mass cannot produce any torque about the center of mass axis. Summing up all individual torques, the net torque $\vec{\tau}_C$ about the center of mass axis (z-axis) of the plate is given as

$$\sum_{i=1}^{i=N} \vec{r}_{iC} \times \vec{F}_i = \vec{\tau}_C.$$

As the net torque $\vec{\tau}_C$ produces an angular acceleration $\vec{\alpha}$ (along the z-axis or perpendicular to the page), we can write

$$\vec{\tau}_C = I_C \vec{\alpha}.$$

3.14.2 Newton's law of translation (force equation)

Summing up all forces, the net force $\vec{F}_{net} = \sum_{i=1}^{i=N} \vec{F}_i = \vec{F}_{ext}$.

The x- and y-components (F_x and F_y) of the net force $\vec{F}_{net}$ produces the acceleration components a_x and a_y of the center of mass of the rigid body along the respective (x, y) axes. Then we can write:

$$F_x = ma_x \text{ and } F_y = ma_y.$$

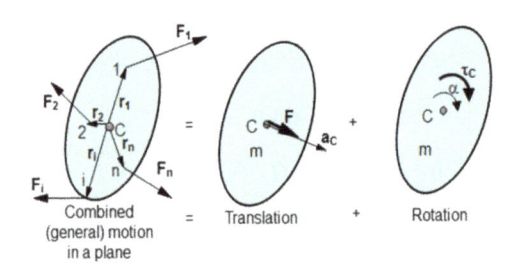

Problems and Solutions in Rotational Mechanics

- The net effect of all the forces acting on the rigid body can produce (a) translation and (b) rotation. Since both occur simultaneously, the combination of translation and rotation can be termed as combined motion as discussed in the first chapter (the kinematics of rigid bodies).
- In addition to one torque equation in *the z*-direction and two force equations in the *x* and *y* or (r—radial and t—tangential) axes, we can write the other force equations such as laws of friction, etc.
- Then, all kinematical equations can be written for accelerations so that the numbers of equations will be equal to the numbers of unknown quantities such as reaction force, a_r, and a_t, etc.
- Finally, we can solve these equations to evaluate the unknown quantities.

3.15 Gyroscopic motion

Let us suspend a bicycle wheel from one end of its axle (axis of rotation) by a light string. The other end of the axle is kept horizontal by holding it in one hand. Let us spin the wheel using the other hand and then withdraw the support of the hand from the end of the axle. We can notice that the free end of the axle rotates about the vertical string with a constant frequency called precession frequency. The change in the angle of orientation of the angular velocity (or the spin angular momentum or the axis of spinning in the horizontal plane) is called precession or gyroscopic motion. This type of device in which the rotating part keeps on changing the plane or axis of rotation is called a gyroscope.

Calculation of precession time period: The gyroscope (precession of the spinning wheel) seems to defy gravity. It is because the torque created by the spinning wheel counteracts the torque due to gravity.

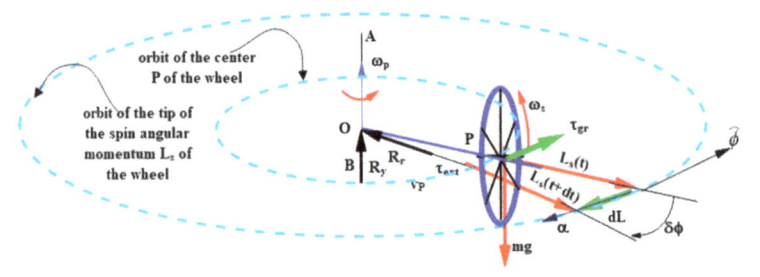

The upward reaction force at the pivot produces a torque about P or weight will produce same torque about O so as to counterbalance the graviy and rotate the spin angular momentum L_s with an angular frequency ω_p in the direction as shown in the figure.If we reverse the direction of spin of the wheel, the directionof precession will also be reversed so as to keep the directin of the mpressed torque same.

Let us assume that during a time dt the spin angular momentum $\vec{L}_s = I\vec{\omega}_s$ of the bicycle wheel changes by d$\vec{L}_s$ by rotating through the angle d$\varnothing$ along the $\varnothing$-axis. Then the rate of change of $\vec{L}_s$ with time is given as

$$\frac{dL_s}{dt} = \frac{(L_s d\varnothing)}{dt}\hat{\varnothing} = I\omega_s\left(\frac{d\varnothing}{dt}\right)\hat{\varnothing}.$$

Put $\frac{d\varnothing}{dt} = \omega_p = 2\pi/T_p$ in the last equation, where ω_p and T_p are the angular frequency and time period of precession, respectively. Then we can write

$$\frac{dL_s}{dt} = I\omega_s\omega_p\hat{\varnothing}. \tag{3.37}$$

Since the spin angular momentum L_s changes in $\hat{\varnothing}$-direction, there must be a torque acting on the body (wheel) in that direction. This torque is due to gravity about O or due to the upward reaction R_y acting at the pivot O about the center of mass P of the wheel. The weight of the wheel creates a torque about the point of suspension P of the wheel which tends to rotate the axle in the $\hat{\varnothing}$-direction. This external or gravitational torque is given as

$$\vec{\tau}_g = mgr\hat{\varnothing}, \tag{3.38}$$

where $r = $ PO = the distance of the center of the wheel from the end of its axle.
Newton's law of rotation can be given as

$$\vec{\tau} = \frac{d\vec{L}}{dt}. \tag{3.39}$$

Putting $\frac{d\vec{L}}{dt}$ from equation (3.37) and $\vec{\tau}$ from equation (3.38) in equation (3.39), we have

$$I\omega\omega_p = mgr.$$

This gives us the precession frequency as

$$\omega_p = mgr/I\omega.$$

Then the precession time period is given as

$$T_p = \frac{2\pi I\omega}{mgr}.$$

Using the last expression $T_p = \frac{2\pi I\omega}{mgr}$, where $\omega = 2\pi/T_s$ (= the time period of spin),

$$T_p T_s = \frac{4\pi^2 I}{mgr}.$$

It should be remembered that:
- The torque axis is tangential and spin axis is in the radial direction: $\vec{\omega}_s = \omega_s \hat{r}$.
- The torque axis lies in the $\varnothing$-direction: $\vec{\tau}_g = mgr\hat{\varnothing}$.
- The precession axis points in the z (upward)-direction: $\vec{\omega}_p = \omega_p \hat{z}$.
- These three axes are mutually perpendicular, as shown in the figure.
- r, $\varnothing$, and z are the axes in a spherical coordinate system.

Example 29 In a classroom demonstration, a bicycle wheel with moment of inertia $I_1 = 0.4 \text{ kg m}^2$ is spinning with an upward angular velocity of magnitude $\omega_1 = 15 \text{ rad s}^{-1}$, as shown in the figure. Suppose that a boy holds the wheel while standing on a turntable. Initially the boy is stationary with the turntable. The moment of inertia of the boy along with that of the turntable is equal to $I_2 = 4 \text{ kg m}^2$. If the boy flips the axle (axis of rotation) of the wheel by 180° so that its axis points in the downward direction, with what angular velocity will the boy rotate with the turntable about the vertical axis? Disregard the frictional effect between the axle and the turntable and assume that there is enough friction between the boy and turntable so that the boy does not slide on the turntable.

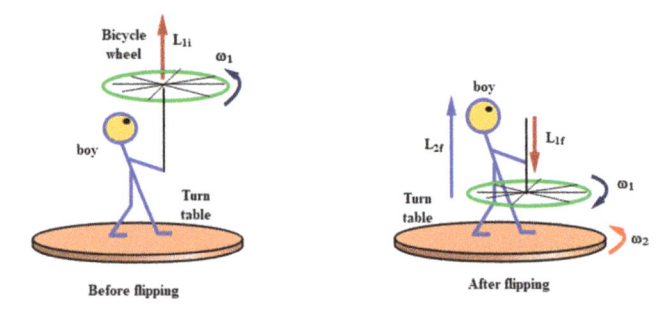

Solution
As the wheel is spinning up,

$$\vec{L}_{1i} = I_1 \omega_1 \hat{j}.$$

Since the boy is stationary along with the turntable, $\vec{L}_{2i} = 0$. Then the initial angular momentum of the system (turntable + wheel) is given as

$$\vec{L}_i = \vec{L}_{1i} + \vec{L}_{2i} = I_1 \omega_1 \hat{j}. \tag{3.40}$$

After flipping of the axle down, the angular velocity of the wheel changes its direction from upward to downward. So the angular momentum of the wheel is now given as

$$\vec{L}_{1f} = -I_1 \omega_1 \hat{j}.$$

Since there is no external torque involved in the system, the total angular momentum of the system remains constant. To conserve the angular momentum, let the combination (boy + turntable) rotate with an angular velocity $\vec{\omega}_2$. So the angular momentum of this combination is

$$\vec{L}_{2f} = I_2 \vec{\omega}_2.$$

Then, after the flipping of the axle, the angular momentum of the axle is given as

$$\vec{L}_{1f} = -I_1 \omega_1 \hat{j}.$$

Then the angular momentum of the system after flipping is

$$\vec{L}_f = \vec{L}_{1f} + \vec{L}_{2f}$$
$$= -I_1 \omega_1 \hat{j} + I_2 \vec{\omega}_2. \tag{3.41}$$

3-50

In order to conserve the angular momentum, equating the values of $\vec{L}_i$ and $\vec{L}_f$, we have

$$-I_1\omega_1\hat{j} + I_2\vec{\omega}_2 = I_1\omega_1\hat{j},$$

or

$$\vec{\omega}_2 = 2(I_1/I_2)\omega_1\hat{j}$$
$$= 2\left(\frac{0.4}{4}\right)(15)\hat{j} = 3\hat{j} \text{ rad s}^{-1}.$$

Problems

Moment of inertia

Problem 1

A circular portion of radius R is removed from a uniform equilateral triangular lamina. If the mass of the triangular lamina is m, find the moment of inertia of the remaining portion about PQ.

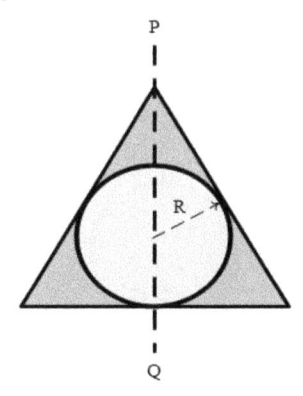

Solution

The moment of inertia of the remaining point is

$$I = I_{\text{triangle}} - I_{\text{circle}}. \tag{3.42}$$

The moment of inertia of the triangle about PQ is given as

$$I_{\text{triangle}} = \sigma h b^3/12$$
$$= \sigma(\sqrt{3}l/2)(l^3)/12 = \sqrt{3}\sigma l^4/24.$$

Similarly $I_{\text{circle}} = \frac{1}{4}m_{\text{circle}} R^2$, where $R = \frac{l}{2\sqrt{3}}$

and $m_{\text{circle}} = \sigma \pi R^2$

$$\Rightarrow I_{\text{circle}} = \frac{1}{4}(\sigma\pi R^4) = \frac{\sigma\pi}{4}\left(\frac{l}{2\sqrt{3}}\right)^4 = \frac{\sigma\pi l^4}{36 \times 16}.$$

The mass of the given triangular lamina is

$$m = \sigma A = \sigma\left(\frac{l}{2}\right)\left(\frac{\sqrt{3}\,l}{2}\right) = \frac{\sqrt{3}}{4}\sigma l^2$$

$$I = I_{triangle} - I_{circle} = \frac{\sigma l^4}{24}\left(\sqrt{3} - \frac{\pi}{24}\right).$$

Putting $\sigma l^2 = \frac{4m}{\sqrt{3}}$ we have

$$I = \frac{l^2}{24}\left(\sqrt{3} - \frac{\pi}{24}\right)\left(\frac{4m}{\sqrt{3}}\right)$$

$$\Longrightarrow I = \frac{ml^2}{6}\left(1 - \frac{\pi}{24\sqrt{3}}\right).$$

Problem 2 A spherical portion is removed from a solid cone of mass m, as shown in the figure. If the apex angle of the cone is $\pi/3$ radians, find the moment of inertia of the remaining portion about the axis PQ.

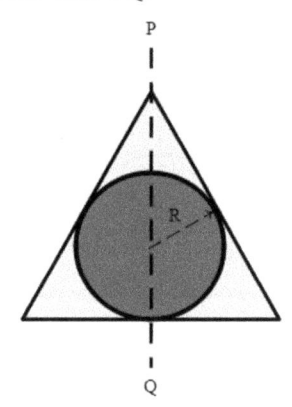

Solution
Let the mass of the sphere be m. So the ratio of the mass of the sphere and cone is

$$\frac{m_1}{m} = \frac{\frac{4}{3}\pi r^3}{\frac{\pi R^2 h}{3}} = \frac{4r^3}{R^2 h},$$

where

$$h = \sqrt{3}\,R \text{ and } r = \frac{h}{3} = \frac{\sqrt{3}\,R}{3} = \frac{R}{\sqrt{3}}$$

$$\Rightarrow \frac{m_1}{m} = \frac{4\left(\frac{R}{\sqrt{3}}\right)^3}{R^2(\sqrt{3}\,R)} = 4/9$$

3-52

$$\Rightarrow m_1 = \frac{4m}{9}. \qquad (3.43)$$

The moment of inertia of the cone about PQ is

$$I_{\text{conc}} = \frac{3mR^2}{10}.$$

The moment of inertia of the solid sphere about PQ is

$$I_{\text{sph}} = \frac{2}{5}m_1 r^2.$$

Then the moment of inertia of the remaining portion about PQ is

$$I = I_{\text{cone}} - I_{\text{sph}}$$

$$\Rightarrow I = \frac{3mR^2}{10} - \frac{2}{5}m_1 r^2$$

$$= \frac{3mR^2}{10} - \frac{2}{5}\left(\frac{4m}{9}\right)\left(\frac{R}{\sqrt{3}}\right)^4$$

$$= \frac{mR^2}{5}\left\{(3/2) - \left(8/27\right)\right\}$$

$$= \frac{mR^2}{5}(81 - 16)/54 = 13mR^2/54.$$

Problem 3 A solid sphere and a hollow sphere each of mass m (at the bottom), and a disc of mass M (at the top) are welded to form a composite body. Find the moment of inertia of the composite body about AB. Put $R/r = 1.5$ and $M/m = 46/45$.

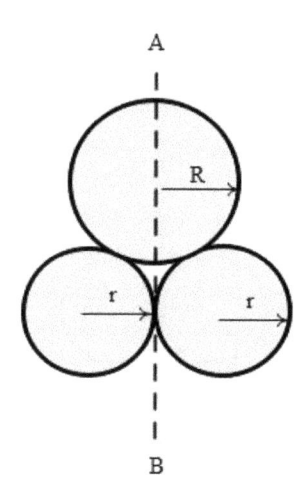

Solution
The moment of inertia of the combination is
$$I = I_{disc} + I_{shell} + I_{solid\ sphere}$$
$I_{disc} = \frac{MR^2}{4}$. By the parallel axis theorem, $I_{shell} = \frac{2}{3}mr^2 + mr^2 = \frac{5}{3}mr^2$
$$I_{solid\ sphere} = \frac{2}{5}mr^2 + mr^2 = \frac{7}{5}mr^2.$$
Then
$$I = \frac{MR^2}{4} + \frac{5}{3}mr^2 + \frac{7}{5}mr^2$$
$$= \frac{MR^2}{4} + mr^2\left(\frac{25+21}{15}\right) = \frac{MR^2}{4} + \frac{46mr^2}{15}.$$
Putting $R = \frac{3}{2}r$ we have
$$I = \frac{MR^2}{4} + \frac{46}{15}m\left(\frac{4}{9}R^2\right)$$
$$= R^2\left(\frac{M}{4} + \frac{46 \times 4m}{15 \times 9}\right).$$
Putting $m = \frac{45}{46}M$ we have
$$I = R^2\left(\frac{M}{4} + \frac{46 \times 4}{15 \times 9} \times \frac{45M}{46}\right)$$
$$\Rightarrow I = MR^2\left(\frac{1}{4} + \frac{4}{3}\right) = \frac{19MR^2}{12}.$$

Problem 4 Four bodies 1, 2, 3, and 4 are welded to form a composite body. Body 1 is a thin spherical shell completely filled with a non-viscous liquid, body 2 is a solid sphere, body 3 is a ring, and body 4 is a disc. Each body has a mass m and all solid bodies have radius R. Find the moment of inertia of the composite object about (a) AB and (b) CD.

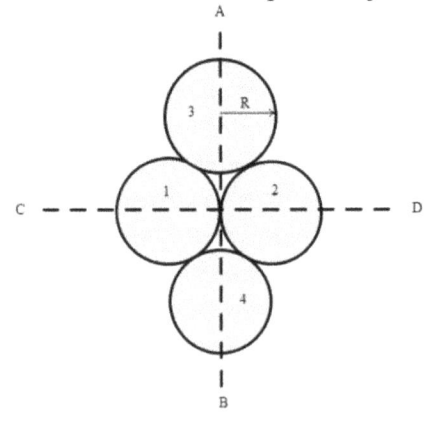

Solution

(a) Since the axis of rotation is AB, the perpendicular distance of the bodies 1 and 2 from the axis is equal to $r = R$. Then, by parallel axis theorem, we have

$$I_{shell} = \frac{2}{3}mR^2 + mr^2 = \frac{5}{3}mR^2 (\because r = R).$$

Because the non-viscous liquid behaves as a particle as it cannot rotate due to the absence of friction, its moment of inertia is

$$I_{liquid} = mR^2.$$

Then, adding the above two moments of inertia, we have

$$I_1 = I_{liquid} + I_{liquid} = 5mR^2/3 + mR^2 = 8mR^2/3.$$

Again, by applying the parallel axis theorem,

$$I_2 = \frac{2}{5}mR^2 + mR^2 = \frac{7}{5}mR^2.$$

As the axis AB passes through the center of mass of bodies 3 and 4, their moments of inertia can be written directly as follows:

$$I_3 = \frac{mR^2}{2}$$

$$I_4 = \frac{mR^2}{4}.$$

Then

$$I = I_1 + I_2 + I_3 + I_4$$
$$= \frac{8}{3}mR^2 + \frac{7}{5}mR^2 + \frac{mR^2}{2} + \frac{mR^2}{4}$$
$$= mR^2 \left(\frac{8}{3} + \frac{7}{5} + \frac{1}{2} + \frac{1}{4} \right) = \frac{289}{60}mR^2.$$

(b) As the axis AB passes through the center of mass of bodies 1 and 2, their moments of inertia can be written directly as follows:

$$I_1 = \frac{2}{3}mR^2$$

$$I_2 = \frac{2}{5}mR^2.$$

Please note that the non-viscous liquid behaves as a point mass and it passes through the axis of rotation CD, so its moment of inertia will be zero.

Since the axis of rotation is AB, the perpendicular distance of bodies 1 and 2 from the axis is equal to $r = \sqrt{3}\,R$. Then, by the parallel axis theorem, we have

$$I_3 = \frac{mR^2}{2} + mr^2 = \frac{mR^2}{2} + m(\sqrt{3}\,R)^2 = \frac{7}{2}mR^2$$

$$I_4 = \frac{mR^2}{4} + mr^2$$
$$= \frac{mR^2}{4} + m(\sqrt{3}\,R)^2$$
$$= \frac{13}{4}mR^2$$

$$\Rightarrow I = I_1 + I_2 + I_3 + I_4$$
$$= \frac{2}{3}mR^2 + \frac{2}{5}mR^2 + \frac{7}{2}mR^2 + \frac{13}{4}mR^2$$
$$= mR^2\left(\frac{7}{2} + \frac{2}{3} + \frac{2}{5} + \frac{13}{4}\right)$$
$$= mR^2\left(\frac{210 + 40 + 24}{60}\right)$$
$$= \frac{469}{60}mR^2.$$

Problem 5 A triangular lamina ABC of length L is placed on a table. If we remove the mass of the triangular portion DEF from the given lamina, find the moment of inertia of the remaining portion about a perpendicular axis passing through its center of mass. Assume that the mass of the remaining portion is M.

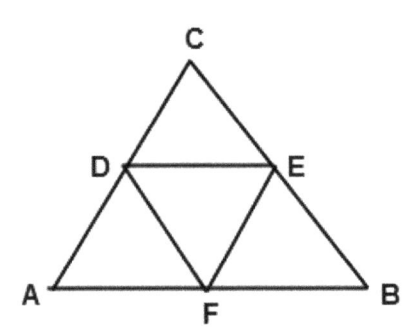

Solution
The mass of each triangle is equal. Then the mass of the central triangle is $m_1 = \frac{m}{4}$, where $m =$ the mass of the given triangular lamina. Let the moment of inertia of the original triangle be I and the moment of inertia of the middle triangle be I_1. Then

$$\frac{I_1}{I} = \left(\frac{m_1}{m}\right)\left(\frac{l_1}{l}\right)^2.$$

Since the mass is directly proportional to the area, we have

$$\frac{I_1}{I} = \left(\frac{l_1}{l}\right)^4, \text{ where } \frac{l_1}{l} = \frac{1}{2}$$

$$\Rightarrow \frac{I_1}{I} = \left(\frac{1}{2}\right)^4 = \frac{1}{16}$$

$$\Rightarrow I_1 = \frac{I}{16}.$$

Then the moment of inertia of the remaining portion is

$$I' = I - I_1 = I - \frac{I}{16} = \frac{15}{16}I,$$

where I = the moment of inertia of the total triangular plate.
This is given as $I = \frac{ml^2}{12}$

$$\Rightarrow I' = \frac{15}{16}I = \frac{15}{16} \times \frac{ml^2}{12} = \frac{5ml^2}{64}.$$

The mass of the remaining portion is given as

$$M = \frac{3}{4}m$$

$$\Rightarrow m = \frac{4M}{3}$$

then $I = \frac{5ml^2}{64} = \frac{5}{64}\left(\frac{4M}{3}\right)l^2$

$$= \frac{5}{48}Ml^2.$$

Problem 6 Four identical circular portions are removed from a thin square plate of mass m and length L. Find the moment of inertia of the remaining portion about the lines AB and CD.

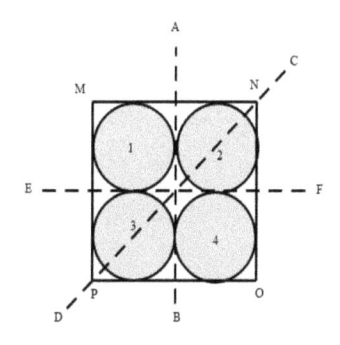

Solution

The moment of inertia of each cut portion disc about AB is

$$I_1 = \frac{m_1 R^2}{4} + m_1 R^2 = \frac{5m_1 R^2}{4},$$

where m_1 = the mass of each disc.

Then the total moment of inertia of the cut portion is

$$I_2 = 4I_1 = 5m_1 R^2.$$

If m = the mass of the square plate (given), its moment of inertia about AB is

$$I_3 = \frac{ml^2}{12}.$$

Then the moment of inertia of the remaining portions is

$$I = I_3 - I_2$$
$$= \frac{ml^2}{12} - 5m_1 R^2,$$

where

$$R = \frac{l}{4} \text{ and } \frac{m_1}{m} = \frac{\pi R^2}{l^2}$$

$$= \frac{\pi\left(\frac{l^2}{16}\right)}{l^2} = \frac{\pi}{16}$$

$$\Rightarrow I = \frac{ml^2}{12} - 5\left(\frac{\pi m}{16}\right)\left(\frac{l}{4}\right)^2$$

$$= \frac{ml^2}{4}\left(\frac{1}{3} - \frac{5\pi}{64}\right).$$

By the perpendicular axis theorem

$$I_{AB} + I_{EF} = I_z.$$

By symmetry $I_{AB} = I_{EF}$

$$\Rightarrow I_Z = 2I_{AB} = 2I_{EF}. \quad (3.44)$$

By perpendicular axis theorem,

$$I_{CD} + I_{GH} = I_Z.$$

By symmetry, $I_{CD} = I_{GH}$

$$\Rightarrow I_Z = 2I_{CD} = 2I_{GH}. \quad (3.45)$$

Using equations (3.44) and (3.45)

$$I_{CD} = I_{AB} = I = \frac{ml^2}{4}\left(\frac{1}{3} - \frac{5\pi}{64}\right).$$

Problem 7 A spherical portion of radius $R = L/2$ is removed from a cube of mass m and length L. Find the moment of inertia of the remaining portion of the cube about the axes passing through (a) the center of mass and perpendicular to two opposite faces, and (b) the edge.

Solution

(a) The mass of the cube is

$$m = \rho l^3.$$

The mass of the removed sphere is

$$m_1 = \rho \frac{4}{3}\pi R^3$$

$$\Rightarrow \frac{m_1}{m} = \frac{4\pi}{3}\left(\frac{R}{l}\right)^3, \quad \text{where } 2R = l$$

$$\Rightarrow \frac{m_1}{m} = \frac{4\pi}{3}\left(\frac{l^3}{8l^3}\right) = \frac{\pi}{6}$$

$$\Rightarrow m_1 = \frac{\pi m}{6}. \tag{3.46}$$

The moment of inertia of the cube is

$$I_1 = \frac{ml^2}{6}.$$

The moment of inertia of the sphere is

$$I_2 = \frac{2}{5}m_1 R^2 = \frac{2}{5}\left(\frac{\pi m}{6}\right)\left(\frac{l}{2}\right)^2 = \frac{\pi m l^2}{60}.$$

Then the moment of inertia of the remaining portion is

$$I = I_1 - I_2$$
$$= \frac{ml^2}{6} - \frac{\pi ml^2}{60}$$
$$= (10 - \pi)\frac{ml^2}{60}.$$

(b) The moment of inertia of the remaining portion about the axis 2 (the side) is

$I_2 = I_C + m_1 r^2$, where $r = \dfrac{l}{\sqrt{2}}$ and $\Rightarrow I_2 = I_C + \dfrac{m' l^2}{2}$, where $m' = m - m_1$

$$= I + m\left(1 - \dfrac{\pi}{6}\right)\dfrac{l^2}{2}$$

$$= (10 - \pi)\dfrac{ml^2}{60} + m\left(1 - \dfrac{\pi}{6}\right)l^2$$

$$= \dfrac{ml^2}{6}\left(7 - \dfrac{11}{10}\pi\right).$$

Problem 8 A circular cut of radius $r = R/3$ is made in a disc of mass m and radius R. Find the moment of inertia about the y-axis.

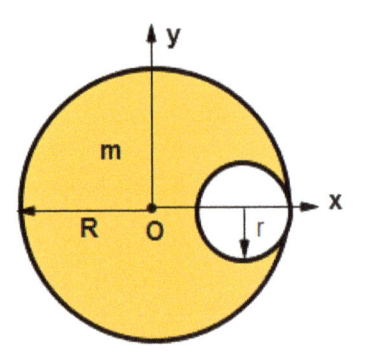

Solution
The ratio of the mass m_1 (removed portion) and mass m of the given disc is

$$\dfrac{m_1}{m} = \dfrac{\pi r^2}{\pi R^2} = \dfrac{r^2}{R^2}. \qquad (3.47)$$

The moment of inertia of m_1 about O is

$$I_1 = \dfrac{m_1 r^2}{2} + m_1(R - r)^2$$

$$\Rightarrow I_1 = m_1\left\{\dfrac{r^2}{2} + (R - r)^2\right\}. \qquad (3.48)$$

3-60

Using equations (3.47) and (3.48) we have

$$I_1 = \frac{r^2}{R^2}m\left\{\frac{r^2}{2} + (R-r)^2\right\}$$
$$= \frac{mr^4}{R^2}\left\{\frac{1}{2} + \left(\frac{R}{r} - 1\right)^2\right\}. \tag{3.49}$$

The moment of inertia of m about O is

$$I_2 = \frac{mR^2}{2}.$$

Then the moment of inertia of the remaining portion about O is

$$I = I_2 - I_1$$
$$= \frac{mR^2}{2} - \frac{mr^4}{R^2}\left\{\frac{1}{2} + \left(\frac{R}{r} - 1\right)^2\right\}$$
$$= mR^2\left[\frac{1}{2} - \frac{r^4}{R^4}\left\{\frac{1}{2} + \left(\frac{R}{r} - 1\right)^2\right\}\right].$$

Putting

$$\frac{R}{r} = \frac{R}{R/3} = 3$$

$$= mR^2\left[\frac{1}{2} - \left(\frac{1}{3}\right)^4\left\{\frac{1}{2} + (3-1)^2\right\}\right]$$
$$= mR^2\left[\frac{1}{2} - \frac{1}{81}\left\{\frac{1}{2} + 4\right\}\right]$$
$$= mR^2\left[\frac{1}{2} - \frac{1}{81}\left(\frac{9}{2}\right)\right]$$
$$= \frac{8}{18}mR^2 = \frac{4}{9}mR^2.$$

Problem 9 (a) A circular portion of radius $R/2$ is cut from a disc of mass m and radius R. Find the moment of inertia about the z-axis passing through the center of mass of the remaining portion. (b) Do the problem if the mass of the remaining portion is given as M.

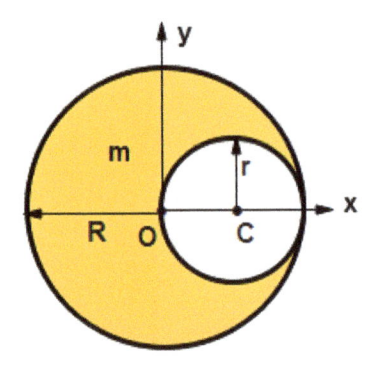

Solution

(a) Referring to the last problem,

$$I_0 = mR^2\left[\frac{1}{2} - \frac{r^4}{R^4}\left\{\frac{1}{2} + \left(\frac{R}{r} - 1\right)^2\right\}\right].$$

Putting $\frac{R}{r} = 2$ we have

$$I_0 = mR^2\left[\frac{1}{2} - \left(\frac{1}{2}\right)^4\left\{\frac{1}{2} + (2-1)^2\right\}\right]$$

$$= mR^2\left[\frac{1}{2} - \frac{1}{16}\left\{\frac{1}{2} + 1\right\}\right]$$

$$= mR^2\left\{\frac{1}{2} - \frac{3}{32}\right\} = \frac{13mR^2}{32}.$$

By the parallel axis theorem,

$$I_C + mr^2 = I_0,$$

where

$$r = -\frac{m'\frac{R}{2}}{m - m'} = -\frac{R}{2}\left(\frac{1}{\frac{m}{m'} - 1}\right).$$

By putting $\frac{m}{m'} = \left(\frac{R}{r}\right)^2 = 4$ we have $r = \frac{-R}{6}$.

Then

$$I_C = I_0 - mr^2$$

$$= \frac{13mR^2}{32} - \left(\frac{3m}{4}\right)\left(\frac{R}{6}\right)^2$$

$$= \frac{39 - 2}{96}mR^2$$

$$= \frac{37}{96}mR^2.$$

3-62

(b) As the mass of the remaining portion is $M = \frac{3}{4}$ (the mass of the bigger disc) putting $m = 4M/3$, we have

$$I_C = \frac{37}{96}MR^2$$
$$= \frac{37}{96} \times \frac{4}{3}MR^2$$
$$= \frac{37}{72}MR^2.$$

Problem 10 A spherical cavity of radius $R/2$ is made in a sphere of radius R. Find the moment of inertia about the axis passing through the center of mass of the remaining portion. Assume that M = mass of the remaining portion.

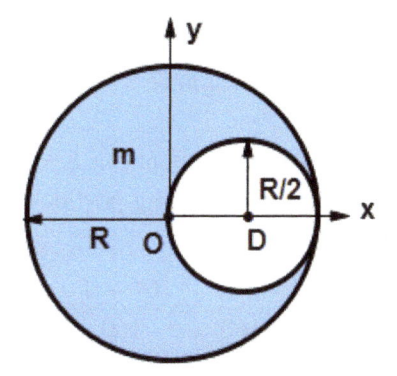

Solution
Let m_1 be the mass of the removed portion and m be the mass of the given sphere before removal of the smaller sphere. Then the position of the center of mass of the remaining portion relative to O is

$$x = \frac{(-m_1)\frac{R}{2}}{m - m_1} = -\frac{R}{2\left(\frac{m}{m_1} - 1\right)},$$

where $\frac{m}{m_1} = \left(\frac{R}{r}\right)^3 = \left(\frac{R}{\frac{R}{2}}\right)^2 = 8$

$$\Rightarrow x = -\frac{R}{2(8-1)} = -\frac{R}{14}.$$

The negative sign signifies that the center of mass C of the remaining portion is located on the left-hand side of the origin O.
Then, by the parallel axis theorem,

$$I_O = I_C + m'(CO)^2,$$

where

$$m' = m - m_1 = m - \frac{m}{8} = \frac{7m}{8}$$

and

$$CO = x = -\frac{R}{14}$$

$$\Rightarrow I_O = I_C + \left(\frac{7m}{8}\right)\left(\frac{R}{14}\right)^2$$

$$\Rightarrow I_C = I_O - \frac{7mR^2}{8 \times 14 \times 14}. \tag{3.50}$$

The moment of inertia of the original sphere m about O is

$$I_1 = \frac{2}{5}mR^2.$$

The moment of inertia of the removed sphere about O is $I_2 = \frac{2}{5}m_1 r^2 = \frac{2}{5}\left(\frac{m}{8}\right)\left(\frac{R}{2}\right)^2 = \frac{7mR^2}{160}$.

Then the moment of inertia of the remaining portion about O is

$$I_O = I_1 - I_2$$
$$= \frac{2}{5}mR^2 - \frac{7mR^2}{160} = \frac{64 - 7}{160}mR^2 \tag{3.51}$$
$$= \frac{57}{160}mR^2.$$

Using equations (3.50) and (3.51)

$$I_C = \frac{57}{160}mR^2 - \frac{7mR^2}{8 \times 14 \times 14}$$
$$= \frac{mR^2}{32}\left(\frac{57}{5} - \frac{7}{49}\right) = \frac{mR^2}{32}\left(\frac{57}{5} - \frac{1}{7}\right)$$
$$= \frac{mR^2}{32}\left(\frac{399 - 5}{35}\right) = \frac{394mR^2}{32 \times 35} = \frac{197}{560}mR^2.$$

Putting $7m/8 = M$ we have

$$I_C = \frac{197}{560}(8M/7)R^2 = \frac{197MR^2}{490}.$$

Problem 11 A truncated cone of mass m, radii R_1 and R_2, and height h is placed on a horizontal surface, as shown in the figure. Find the moment of inertia of the cone about its vertical centroidal axis.

Solution

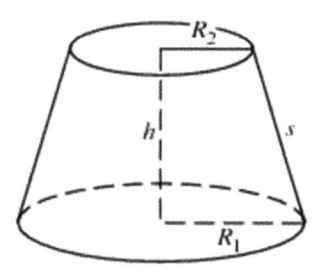

The moment of inertia of the bigger cone is

$$I_1 = \frac{3 m_1 R_1^2}{10}.$$

The moment of inertia of the smaller cone is

$$I_2 = \frac{3}{10} m_2 R_2^2.$$

Then moment of inertia of the remaining portion is

$$I = I_1 - I_2 = \frac{3}{10}(m_1 R_1^2 - m_2 R_2^2).$$

We know that

$$\frac{m_1}{m_2} = \frac{R_1^2 h_1}{R_2^2 h_2} = \left(\frac{R_1}{R_2}\right)^3 \quad (\because h \propto R)$$

$$\frac{m_1 - m_2}{m_2} = \frac{R_1^3 - R_2^3}{R_2^3}.$$

Putting $m_1 - m_2 = m$ we have

$$\Rightarrow m = \frac{R_1^3 - R_2^3}{R_2^3} m_2.$$

Then

$$I = \frac{3}{10} m_2 R_2^2 \left\{ \left(\frac{m_1}{m_2}\right)\left(\frac{R_1}{R_2}\right)^2 - 1 \right\}$$

$$= \frac{3}{10}\left(\frac{R_2^3 m}{R_1^3 - R_2^3}\right) R_2^2 \left\{ \left(\frac{R_1}{R_2}\right)^3 \left(\frac{R_1}{R_2}\right)^2 - 1 \right\}$$

$$= \frac{3}{10}\left\{\frac{m R_2^5}{R_1^3 - R_2^3}\right\}\left(\frac{R_1^5 - R_2^5}{R_2^5}\right)$$

$$= \frac{3 m (R_1^5 - R_2^5)}{10 (R_1^3 - R_2^3)}.$$

3-65

Problem 12 Find the moment of inertia of the truncated sphere (shaded portion) of radius R and thickness h about the vertical diametrical axis of the sphere.

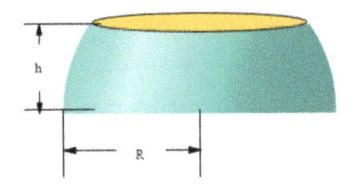

Solution
The moment of inertia of the thin disc about AB is

$$dI = \frac{dmr^2}{2},$$

where $dm = \rho(\pi r^2 dy)$,

$$r^2 = R^2 - y^2$$

$$\Rightarrow dI = \frac{\rho\pi}{2}(R^2 - y^2)^2 dy.$$

Then the total moment of inertia is

$$I = \int dI = \frac{\rho\pi}{2}\int_0^h (R^2 - y^2)^2 dy$$

$$= \frac{\rho\pi}{2}\left[R^4 y + \frac{y^5}{5} - \frac{2R^2 y^3}{3}\right]_0^{\frac{R}{2}}$$

$$= \rho\frac{\pi}{2}\left[R^4 \frac{R}{2} + \frac{R^5}{32 \times 5} - \frac{2R^2 R^3}{24}\right]$$

$$= \frac{\rho\pi}{2}\left[\frac{R^5}{2} + \frac{R^5}{160} - \frac{R^5}{12}\right]$$

$$= \frac{\rho\pi R^5}{2 \times 2}\left(1 + \frac{1}{80} - \frac{1}{6}\right)$$

$$= \frac{\pi\rho R^5}{4}\left(\frac{1}{80} + \frac{5}{6}\right)$$

$$= \frac{\pi\rho R^5}{8}\left(\frac{5}{3} + \frac{1}{40}\right)$$

$$= \frac{\rho\pi R^5}{8}\left(\frac{203}{120}\right).$$

(3.52)

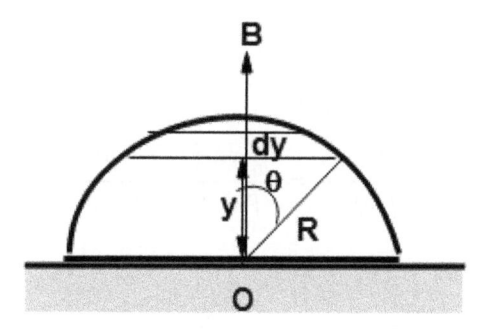

The mass of the elementary thin disc is

$$dm = \rho \pi r^2 dy = \rho \pi (R^2 - y^2) dy.$$

Then the total mass of the segment is

$$m = \int dm = \pi \rho \int_0^{R/2} (R^2 - y^2) dy$$

$$= \pi \rho \left(R^2 y - \frac{y^3}{3} \right) \Big|_0^{R/2} = \pi \rho \left(\frac{R^3}{2} - \frac{R^3}{24} \right)$$

$$= \frac{\pi \rho R^3}{2} \left(1 - \frac{1}{12} \right) = \frac{11 \pi \rho R^3}{24}$$

$$\Rightarrow \pi \rho R^3 = \frac{24}{11} m. \tag{3.53}$$

Using equations (3.52) and (3.53)

$$I = \left(\frac{24}{11} m \right) \left(\frac{203}{960} \right) R^2 = \frac{203 m R^2}{440}.$$

Problem 13 The density of a rod AB of mass m and length L increases linearly from ρ at A to $\eta \rho$ at B. Find the moment of inertia of the rod about an axis perpendicular to the rod passing through the center of mass of the rod.
Solution
The center of mass of the rod is

$$x_C = CP = \frac{\int x\,dm}{\int dm}$$

$$= \frac{\int x(\lambda dx)}{\int \lambda dx} = \frac{\int_0^l x\lambda_0\left(1 + \frac{x}{l}\right)dx}{\int_0^l \lambda_0\left(1 + \frac{x}{l}\right)dx}$$

$$= \frac{\left(\frac{x^2}{2} + \frac{x^3}{3l}\right)\Big|_0^l}{\left(x + \frac{x^2}{2l}\right)\Big|_0^l} = \frac{l^2\left(\frac{1}{2} + \frac{1}{3}\right)}{l\left(1 + \frac{1}{2}\right)}$$ (3.54)

$$= \frac{5 \times 2}{6 \times 3}l = \frac{5}{9}l.$$

The moment of inertia of the rod about P is

$$I_P = \int_0^l x^2\,dm = \int_0^l x^2\lambda_0\left(1 + \frac{x}{l}\right)dx$$

$$= \lambda_0\left(\frac{x^3}{3} + \frac{x^4}{4l}\right)\Big|_0^l = \lambda_0 l^3\left(\frac{1}{3} + \frac{1}{4}\right)$$ (3.55)

$$= \frac{7}{12}\lambda_0 l^3.$$

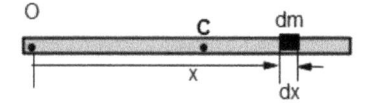

The mass of the rod is

$$m = \int x\,dx = \int \lambda_0\left(1 + \frac{x}{l}\right)dx$$

$$= \lambda_0\left(l + \frac{x^2}{2l}\right)\Big|_0^l = \frac{3\lambda_0 l}{2}$$

$$\Rightarrow \lambda_0 l = \frac{2m}{3}.$$ (3.56)

Using equation (3.55) and (3.56) we have

$$I_P = \frac{7}{12} \times \left(\frac{2m}{3}\right)l^2 = \frac{7}{18}ml^2.$$ (3.57)

According to the parallel axis theorem,

$$I_P = I_C + m(CP)^2$$

$$\Rightarrow I_C = I_P - m(CP)^2$$
$$= \frac{7}{18}ml^2 - m\left(\frac{5}{9}l\right)^2$$
$$= ml^2\left(\frac{7}{18} - \frac{25}{81}\right)$$
$$= \frac{13}{162}ml^2.$$

Problem 14 A uniform cone of mass m and height h is placed on a uniform hemisphere of radius R so that the center of mass of the system of two bodies lies at the center of curvature of the hemisphere. If the bodies are (a) hollow or (b) solid, find the moment of inertia of the system about an axis perpendicular to AB and passing through the center of mass of the system. Put $h = R$.

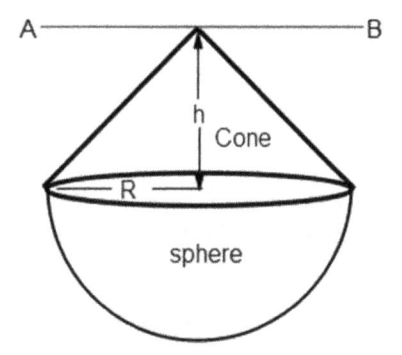

Solution
(a) Let m_1 and m_2 be the mass, and y_1 and y_2 be the center of mass of the cone and hemisphere, respectively. Since the center of mass of the system is at O, fixing the origin at O we can write
$$m_1 y_1 = m_2 y_2.$$
Putting $m_1 = m$, $y_1 = H/3$, and $y_1 = R/2$ we have
$$\Rightarrow m \cdot \frac{h}{3} = m_2 \frac{R}{2}$$
$$\Rightarrow m\frac{R}{3} = m_2 \frac{R}{2} (\because h = R).$$
$$\Rightarrow m_2 = \frac{2m}{3}.$$

Then the moment of inertia of the system about the vertical axis passing through the center of mass of the system is equal to the sum of the moment of inertia of the hollow cone and hollow hemisphere:

3-69

$$I = \tfrac{1}{2}m_1 R^2 + \tfrac{2}{3}m_2 R^2$$
$$= \tfrac{1}{2}mR^2 + \tfrac{2}{3}\left(\tfrac{2}{3}m\right)R^2$$
$$= \left(\tfrac{4}{9} + \tfrac{1}{2}\right)mR^2 = \tfrac{17}{18}mR^2.$$

A — ─────────── B

h
D ↕ Cone
─R─────O
• C
sphere

(b) Let m_1 and m_2 be the mass, and y_1 and y_2 be the center of mass of the cone and hemisphere, respectively. Since the center of mass of the system is at O, fixing the origin at O, we can write

$$m_1 y_1 = m_2 y_2.$$

Putting $m_1 = m$, $y_1 = H/4$, and $y_1 = 3R/8$ we have

$$\left(m\frac{H}{4}\right) = m_2 \frac{3}{8}R$$

$$m_2 = \frac{2}{3}m.$$

Then

$$I = \tfrac{3}{10}mR^2 + \tfrac{2}{5}m_2 R^2$$
$$= \tfrac{3}{10}mR^2 + \tfrac{2}{5} \times \tfrac{2}{3}mR^2$$
$$= \left(\tfrac{3}{10} + \tfrac{4}{15}\right)mR^2 = \tfrac{85}{150}mR^2 = \tfrac{17}{30}mR^2.$$

Problem 15 A hemisphere of mass m and radius of curvature R is kept on a horizontal surface, as shown in the figure. Find the moment of inertia of the hemisphere about the axis of rotation (a) ED, (b) PQ, and (c) AB.

3-70

Solution

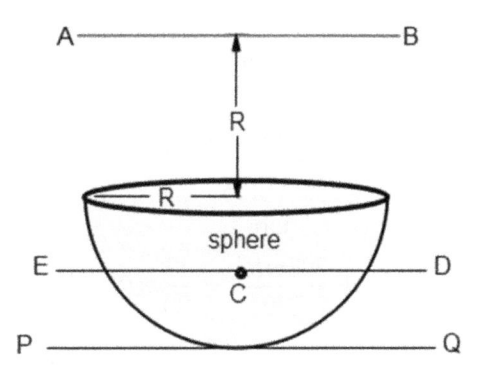

(a) The moment of inertia of the hemisphere about xx' is $I_{xx'} = \frac{2}{5}mR^2$.

The center of mass C of the solid hemisphere is at a distance $OC = 3R/8$ from the center of curvature O.

The moment of inertia of the hemisphere about ED is given by using the parallel axis theorem:

$$I_{ED} = I_{x'x} - m\left(\frac{3}{8}R\right)^2$$

$$\Rightarrow I_{ED} = \frac{2}{5}mR^2 - \frac{9}{64}mR^2$$

$$= mR^2\left(\frac{2}{5} - \frac{9}{64}\right)$$

$$\Rightarrow I_{ED} = \frac{128 - 45}{320}mR^2$$

$$\Rightarrow I_{ED} = \frac{83}{320}mR^2.$$

(b) By using the parallel axis theorem, the moment of inertia about PQ is

$$I_{PQ} = I_{ED} + m\left(\frac{5R}{8}\right)^2$$

$$= \frac{83}{320}mR^2 + \frac{25}{64}mR^2$$

$$= \frac{83 + 125}{320}mR^2$$

$$= \frac{208mR^2}{320} = \frac{13}{20}mR^2.$$

(c) The moment of inertia of the hemisphere about AB is given by using the parallel axis theorem:

$$I_{AB} = I_{CD} + m\left(R + \frac{3}{8}R\right)^2$$

$$= \frac{83mR^2}{320} + \frac{121mR^2}{64}$$

$$= \frac{83 + 605}{320}mR^2$$

$$= \frac{688}{320}mR^2 = \frac{43}{20}mR^2.$$

The equilibrium of rigid bodies and the concept of topping

Problem 16 A ball of mass M and a disc of mass m are hanging from the horizontal rod AB of mass m' and length L. The rod is smoothly pivoted at P. The distance between the points of suspension of the disc and the pivot P is b. If the system is released, the rod will remain horizontal while the disc rolls without sliding on the vertical string wrapped over it. (a) Find the value of b. (b) What is the value of b if the rod is light?

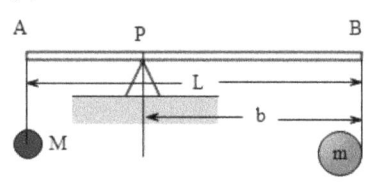

Solution
(a) The net torque about P due to M, the rod, and m is zero:

$$\vec{\tau_0} = \vec{\tau_M} + \vec{\tau_{rod}} + \vec{\tau_m} = 0. \qquad (3.58)$$

The torque due to M is

$$\vec{\tau_M} = Mg(L - b)\hat{k}. \qquad (3.59)$$

The torque due to the rod is

3-72

$$\vec{T}_{\text{rod}} = -m'g\left(b - \frac{L}{2}\right)\hat{k}. \qquad (3.60)$$

The torque due to the tension T in the right-hand side string is

$$\vec{T}_m = -Tb\hat{k}. \qquad (3.61)$$

For the cylinder m:
Force equation:

$$\vec{F}_y = (+T - mg)\hat{j} = m(-a_C)\hat{j}. \qquad (3.62)$$

Torque equation:

$$TR\hat{k} = \frac{mR^2}{2}\alpha\hat{k}$$

$$\Rightarrow T = m\frac{R\alpha}{2}. \qquad (3.63)$$

For rolling: $a_C = R\alpha$. $\qquad (3.64)$

Using equations (3.62), (3.63), and (3.64)

$$T = \frac{mg}{3}. \qquad (3.65)$$

Using equations (3.58), (3.59), (3.60), and (3.61)

$$Mg(L - b) = m'g\left(b - \frac{L}{2}\right) + Tb$$

$$\Rightarrow b = \frac{\left(M + \frac{m'}{2}\right)gl}{(M + m')g + T}. \qquad (3.66)$$

Using equation (3.65) and (3.66)

$$b = \frac{\left(M + \frac{m'}{2}\right)L}{M + m' + \frac{m}{3}}.$$

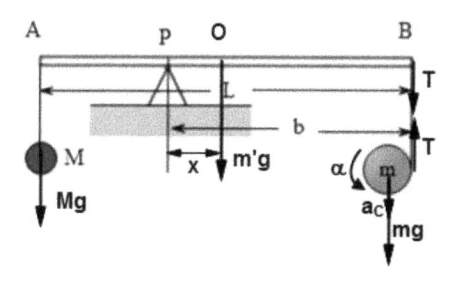

3-73

(b) If the rod is light, $m' \simeq 0$,

$$\Rightarrow b = \frac{3ML}{3M+m}.$$

Problem 17 A vertical L-shaped rod of mass m is smoothly pivoted at O and the end Q is tied to a horizontal string. The rod remains at rest in the vertical plane. Find (a) the tension in the string, (b) the reaction forces at the pivot, and (c) the angle made by the reaction force with the horizontal. Put $b = l/2$ and a mass of PS $= m/3$.

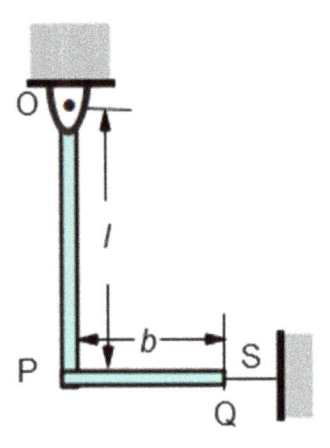

Solution

(a) *Torque equation*: Referring to FBD, if $T =$ the tension in the string, for equilibrium, the torque about O is zero. Since $m_1 g$ passes through the pivot O, it cannot produce a torque about O. Then

$$\Rightarrow -m_2 g \frac{l}{4}\hat{k} + Tl\hat{k} = 0$$

$$\Rightarrow T = \frac{m_2 g}{4}, \quad \text{where } m_2 = m/3 \text{ so, } T = mg/12.$$

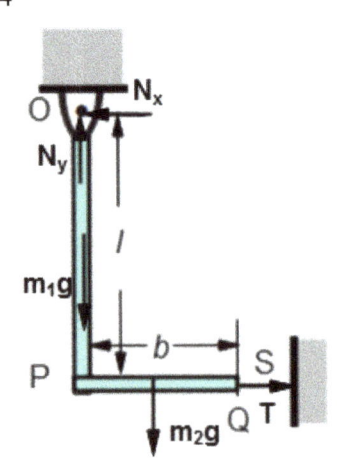

(b) *Force equation*:
Referring to the free-body diagram (FBD), along the horizontal
$$F_x = -N_x + T = 0$$
$$\Rightarrow N_x = T = \frac{mg}{12}.$$

Along the vertical,
$$N_y - mg = 0 \Rightarrow N_y = mg.$$

Then the reaction is
$$R = \sqrt{N_x^2 + N_y^2}$$
$$= mg\frac{\sqrt{145}}{12}.$$

(c) $\tan \phi = \frac{N_x}{N_y} = \frac{\frac{mg}{12}}{mg}$

$$\Rightarrow \phi = \tan^{-1}\left(\frac{1}{12}\right) \text{ with vertical.}$$

Problem 18 Four light rods each of length L are smoothly hinged to form a rhombus. A block of mass m hangs from the bottom of the rhombus. Two equal and opposite forces of magnitude F act so as to equilibrate the system in the vertical plane. Find the magnitude of F.

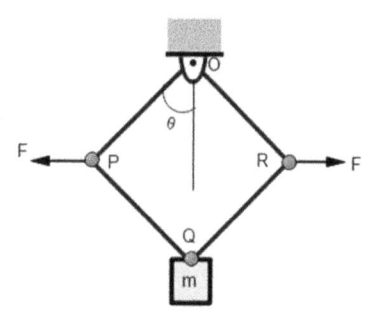

Solution
Let N_x and N_y be the horizontal and vertical reactions, respectively, at O acting on the rod OP. Referring to FBD (free-body diagram), since $F_y = 0$, on the entire system (four rods and the block/particle):
$$F_y = 2N_y - mg = 0. \tag{3.67}$$

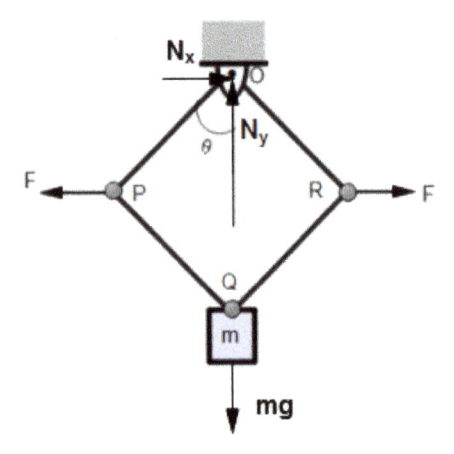

The net horizontal force on the rod OP is zero. Then we can write the force equation on the rod OP, given as

$$F/2 - N_x = 0. \tag{3.68}$$

The net torque acting on the rod OP about O is zero. Then we have

$$\tau_P = N_x L \cos\theta - N_y L \sin\theta = 0$$

$$\Rightarrow \frac{N_x}{N_y} = \tan\theta. \tag{3.69}$$

Using the last three equations,

$$\Rightarrow (F/2) = (m/2)\tan\theta$$

$$\Rightarrow F = mg\tan\theta.$$

Problem 19 A uniform cylindrical wooden log of mass m and radius R is kept in front of a step. It is pulled by a horizontal force F. (a) If the log will remain stationary, find the normal reaction offered by the horizontal surface to the log. (b) What minimum force F is needed to surmount the step?

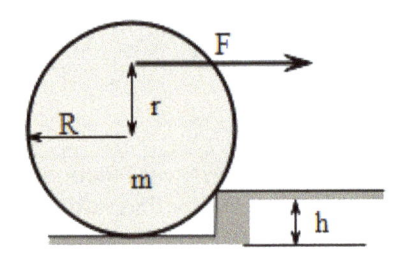

Solution

(a) If $N =$ the normal reaction offered by the horizontal surface to the cylinder, it must be positive as the cylinder is stationary. Taking the net torque about the edge P of the step, we have

$$\Rightarrow -F(r + R - h)\hat{k} - (N - mg)x\hat{k} = 0$$

$$\Rightarrow N = mg - F\frac{(R + r - h)}{x},$$

where $x = \sqrt{R^2 - (R - h)^2} = \sqrt{(2R - h)h}$

$$\Rightarrow N = mg - \frac{F(R + r - h)}{\sqrt{(2R - h)h}}.$$

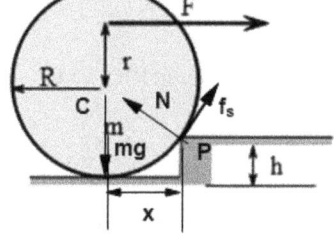

(b) To surmount the step, the net torque about the edge P of the step must be equal to or greater than zero and the normal reaction offered by the horizontal surface to the log must be zero because it must lose contact with the horizontal surface. Since $N = 0$ we have

$$mg - \frac{F(R + r - h)}{\sqrt{(2R - h)h}} = 0$$

$$\Rightarrow F = \frac{mg\sqrt{(2R - h)h}}{R + r - h}.$$

Problem 20 A rod of mass m and length L is in equilibrium while it leans against the smooth edge of the step of height h. Find (a) the reaction offered by the step to the rod and (b) the coefficient of friction between the rod and horizontal surface.

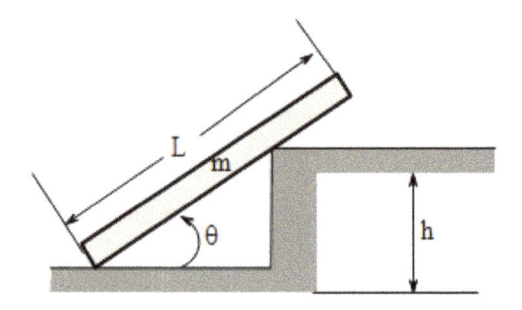

Solution
(a) For the equilibrium of the rod, the torque about the edge Q of the step is zero. So put

$$\tau_P = 0,$$

or

$$N(PQ) - mg(PR) = 0$$

$$\Rightarrow N = mg\frac{PR}{PQ} = mg\frac{\left(\frac{L}{2}\cos\theta\right)}{h\operatorname{cosec}\theta}$$

$$= \frac{mgL\sin\theta\cos\theta}{2h}.$$

Then

$$f_s = N\sin\theta = \frac{mgL\sin^2\theta\cos\theta}{h}$$

and

$$N_y + N\cos\theta = mg$$

$$\Rightarrow N_y = mg - N\cos\theta$$

$$= mg - \frac{mgL\sin\theta\cos^2\theta}{2h}$$

$$= mg\left(1 - \frac{L\sin\theta\cos^2\theta}{2h}\right).$$

3-78

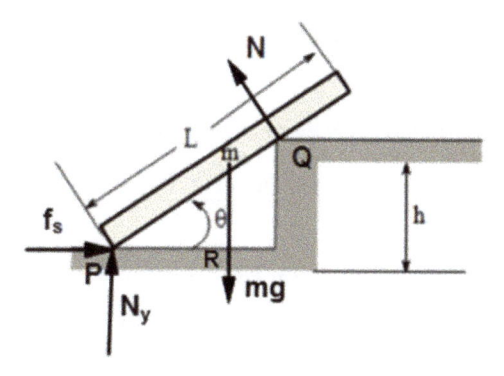

(b) *The law of static friction:* $f_s \leq \mu_s N$

$\Rightarrow \mu_s \geq \frac{f_s}{N}$. Putting the obtained values of f_s and N in this expression, we have

$$\mu_s \geq \frac{mgL\sin^2\theta \cos\theta/2h}{mg(1 - L\sin\theta\cos^2\theta/2h)}$$

$$\Rightarrow \mu_s \geq \frac{\sin^2\theta \cos\theta}{2h - L\sin\theta\cos^2\theta}.$$

Problem 21 A cylinder of mass m is resting on a step of height h which is fixed with an inclined plane of angle of elevation θ. Find the normal reaction offered by the inclined plane on the cylinder. No slipping is allowed between the cylinder and step.

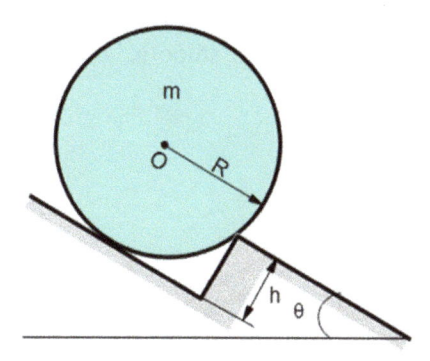

Solution
If the cylinder does not slip on the step, the friction must be static, as shown in the FBD. For equilibrium of the cylinder,
The torque about $P = 0$

$$\vec{T_P} = -(mg\sin\theta)y\hat{k} + (mg\cos\theta - N)R\hat{k} = 0,$$

where $y = R - h$:

$$\Rightarrow (mg\sin\theta)(R-h) = (mg\cos\theta - N)x$$

$$\Rightarrow N = mg\cos\theta - \frac{mg\sin\theta(R-h)}{x}, \text{ where}$$

$$x = \sqrt{h(2R-h)},$$

$$\Rightarrow N = mg\cos\theta - \frac{mg\sin\theta(R-h)}{\sqrt{(2R-h)h}}$$

$$\Rightarrow N = mg\cos\theta\left\{1 - \frac{\tan\theta(R-h)}{\sqrt{(2R-h)h}}\right\}.$$

Problem 22 A uniform rod of mass m and length l leans against the vertical wall making a maximum angle θ. The coefficient of friction between the rod and the horizontal surface is μ_1 and that between the rod and vertical plane is μ_2. If the rod does not slip, find (a) the normal reactions and frictional forces offered by the horizontal and vertical surfaces on the rod and (b) the value/s of θ if (i) both surfaces are rough, (ii) the horizontal surface is smooth, and (iii) the vertical surface is smooth.

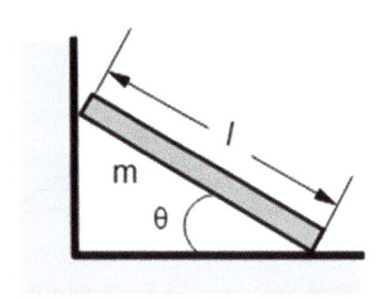

Solution

(a) As the rod is at rest, the torque about Q is zero, $\tau_Q = 0$

$$\Rightarrow -mg\frac{l}{2}\cos\theta \hat{k} - f_1 l \sin\theta \hat{k} + N_1 l \cos\theta \hat{k} = 0$$

$$\Rightarrow f_1 = \left(N_1 - \frac{mg}{2}\right)\frac{\cos\theta}{\sin\theta}. \tag{3.70}$$

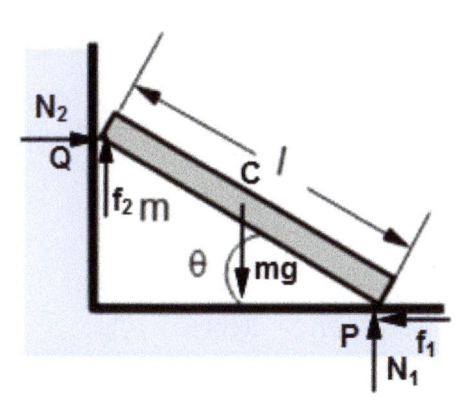

The net horizontal force $F_x = 0$

$$\Rightarrow N_2 - f_1 = 0$$

$$f_1 = N_2. \tag{3.71}$$

The net horizontal force $F_y = 0$

$$f_2 + N_1 - mg = 0$$

$$N_1 + f_2 = mg. \tag{3.72}$$

At the verge of sliding,

$$f_1 = \mu_1 N_1 \tag{3.73}$$

$$f_2 = \mu_2 N_2. \tag{3.74}$$

Now we have five equations and five unknown quantities, namely N_1, N_2, f_1, f_2, and θ. Let us solve them.

From equations (3.70) and (3.73)

$$f_1 = \mu_1 N_1$$

$$\Rightarrow \left(N_1 - \frac{mg}{2}\right)\frac{\cos\theta}{\sin\theta} = \mu_1 N_1$$

3-81

$$\Rightarrow N_1(\cot\theta - \mu_1) = \frac{mg}{2}\cot\theta$$

$$\Rightarrow N_1 = \frac{mg\cot\theta}{2(\cot\theta - \mu_1)}.$$

Then $f_2 = mg - N_1$ (equation (3.72))

$$= mg - \frac{mg\cot\theta}{2(\cot\theta - \mu_1)}$$

$$= \frac{mg(\cot\theta - 2\mu_1)}{2(\cot\theta - \mu_1)}.$$

From equation (3.74),

$$N_2 = \frac{f_2}{\mu_2} = \frac{(\cot\theta - 2\mu_1)}{2(\cot\theta - \mu_1)\mu_2}.$$

Using equation (3.71)

$$f_1 = N_2 = \frac{(\cos\theta - 2\mu_1)mg}{2(\cos\theta - \mu_1)\mu_2}.$$

Again $f_1 = \mu_1 N_1 = \mu_1 \left\{ \frac{mg\cot\theta}{2(\cot\theta - \mu_1)} \right\}$.

(b)

(i) Using the last two expressions,

$$\frac{(\cot\theta - 2\mu_1)mg}{2(\cot\theta - \mu_1)\mu_2} = \frac{\mu_1 mg \cot\theta}{2(\cot\theta - \mu_1)}$$

$$\Rightarrow \cos\theta - 2\mu_1 = \mu_1\mu_2\cos\theta$$

$$\Rightarrow \cos\theta(1 - \mu_1\mu_2) = 2\mu_1$$

$$\Rightarrow \theta = \cot^{-1}\left(\frac{2\mu_1}{1 - \mu_1\mu_2}\right).$$

(ii) If $\mu_1 = 0$ and $\mu_2 = \mu$,

$$\theta = \cot^{-1} 0 = \frac{\pi}{2}.$$

(iii) If $\mu_2 = 0$ and $\mu_1 = \mu$

$$\theta = \cot^{-1}(2\mu_1) = \cot^{-1}(2\mu).$$

Problem 23 A horizontal force F acts on a box of mass m, breadth b, and height H. If the box does not slide, find (a) the position of application of the normal reaction and (b) the coefficient of friction for (i) toppling before sliding, (ii) simultaneous sliding and toppling, and (iii) sliding before toppling.

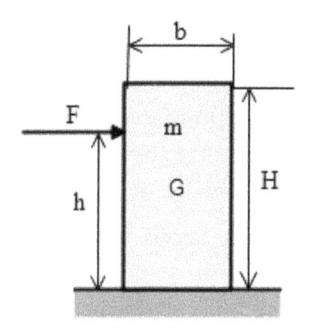

Solution

(a) The net force along the horizontal and vertical directions will be zero, since the net force $\vec{F} = 0$. We have $f_s = F$ and $N = mg$.
The net torque about the center of mass G is zero:

$$\vec{\tau_G} = \vec{\tau_F} + \vec{\tau_N} + \vec{\tau}_{gr} + \vec{\tau_f} = 0$$

$$\Rightarrow -F\left(h - \frac{H}{2}\right)\hat{k} + Nx\hat{k} + f_s\frac{H}{2}(-\hat{k}) = 0.$$

3-83

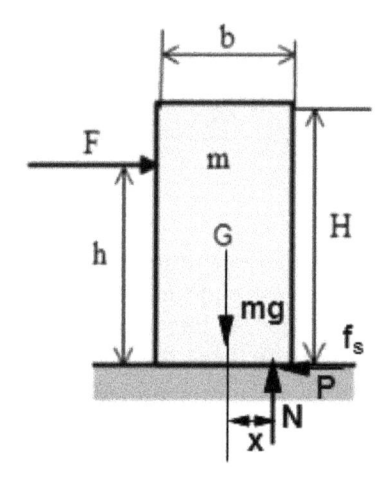

Since $f_s = F$ and $N = mg$ we have

$$F\left\{-\frac{H}{2} - h + \frac{H}{2}\right\} + mgx = 0$$

$$\Rightarrow x = \frac{Fh}{mg}.$$

(b) F is maximum when x is maximum. Putting $x_{max} = \frac{b}{2}$ for the beginning of toppling, we have

$$F_{min} = mg\frac{b}{2h}.$$

Put $F'_{min} = \mu_s mg$ for the beginning of sliding.

(i) If $F_{min} > F'_{min}$ the body topples before the body slides and when $\mu_s > \frac{b}{2h}$ topples then slides or before sliding.

(ii) When $\mu_s = \frac{b}{2h}$ simultaneous topping and sliding take place at any value of F.

(iii) When $\mu_s < \frac{b}{2h}$, the body slides before toppling.

Problem 24 A box of mass m with a square base of side b and height H is placed on a trolley car of mass M. A horizontal force F is acting on the plank. If the ground is smooth, find (a) the value(s) of force F in order to topple the plank, (b) the position of application of the normal reaction on the box, and (c) the coefficient of friction between the plank and box so that the box topples before it slides.

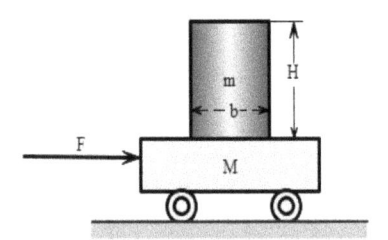

Solution

(a) Let the plank move with an acceleration A and the box does not slide on the plank. Then we can take the pseudo-force mA acting towards left on the box at its center of mass. For the toppling about the edge Q, the normal reaction will pass through the edge Q. Then the torque of the pseudo-force must be equal to or greater than the gravitational torque taken from the plank, about the edge Q. Then we can write

$$mAy \geqslant mgx$$

$$\Rightarrow mA\frac{H}{2} \geqslant mg\frac{b}{2}$$

$$\Rightarrow A \geqslant \frac{b}{H}g, \text{ where } A = \frac{F}{M+m}$$

$$\Rightarrow F \geqslant (M+m)g\frac{b}{H}.$$

(b) The net torque about the edge Q of the box taken relative to the plank is zero:

$$\vec{\tau} = mA\left(\frac{H}{2}\right)\hat{k} - Nx\hat{k} - mg\frac{b}{2}\hat{k} = 0$$

$$\Rightarrow x = \frac{\left(mg\frac{b}{2} - mA\frac{H}{2}\right)}{N}.$$

Putting $N = mg$ and $A = F/(M+m)$ in the last expression we have

$$x = \frac{mgb}{2(mg)}\left[1 - \frac{\left(\frac{F}{M+m}\right)H}{gb}\right]$$

$$\Rightarrow x = \frac{b}{2}\left(1 - \frac{FH}{(M+m)gb}\right).$$

Problem 25 A box of mass m, breadth b, and height H is placed on two sharp edges of wooden supports at A and B fixed on an inclined plane. Find (a) the normal reaction acting at A and B, and (b) the condition for toppling of the box.

3-85

Solution

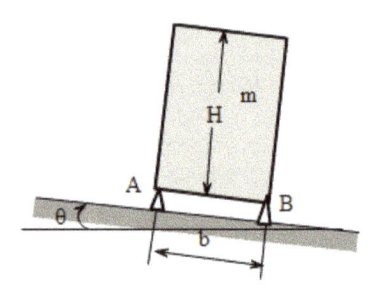

(a) The net force normal to the inclined plane is zero. So we have

$$N_1 + N_2 = mg \cos \theta. \tag{3.75}$$

The net force parallel to the inclined plane is zero. So we have

$$f_s = mg \sin \theta. \tag{3.76}$$

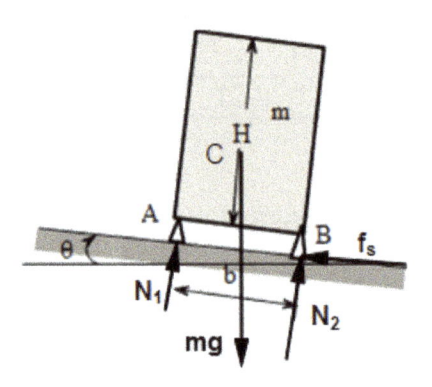

The net torque about the center of mass of the box is zero:

$$\vec{\tau}_C = -f_s h \hat{k} - N_1 \frac{b}{2} \hat{k} + N_2 \frac{b}{2} \hat{k} = 0$$

$$\Rightarrow f_s \frac{H}{2} + N_1 \frac{b}{2} = N_2 \frac{b}{2}$$

$$\Rightarrow N_2 = N_1 + f_s \frac{H}{b} \tag{3.77}$$

From equations (3.76) and (3.77)

$$N_2 = N_1 + \frac{mgH \sin \theta}{b}. \tag{3.78}$$

From equations (3.75) and (3.78)

$$mg \cos \theta - N_1 = N_1 + mg \sin \theta \frac{H}{b}$$

$$\Rightarrow N_1 = \frac{mg}{2}\left(\cos\theta - \frac{H}{b}\sin\theta\right)$$

$$N_2 = \frac{mg}{2}\left(\cos\theta + \frac{H\sin\theta}{b}\right).$$

(b) For toppling, put $N_1 = 0$ in the expression

$$\Rightarrow N_1 = \frac{mg}{2}\left(\cos\theta - \frac{H}{b}\sin\theta\right)$$

to obtain

$$\tan\theta \geqslant \frac{b}{H}.$$

Problem 26 A box of mass m, breadth b, and height H is placed on an inclined plane, as shown in the figure. Find (a) the position of the normal reaction acting on the box, (b) the condition for toppling of the box, and (c) the condition for (i) first sliding then toppling, (ii) simultaneous toppling and sliding, and (iii) first toppling then sliding.

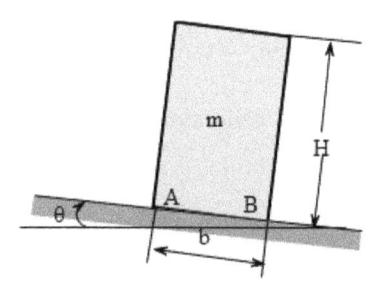

Solution
(a) Since the net force normal to the inclined plane is zero,

$$N = mg\cos\theta. \tag{3.79}$$

Since the net force normal to the inclined plane is zero,

$$f_s = mg\sin\theta. \tag{3.80}$$

The net torque about the center of mass of the box is zero,

$$\vec{\tau_C} = -f_s(H/2)\hat{k} + Nx\hat{k} = 0$$

$$x = \frac{Hf_s}{2N}. \tag{3.81}$$

3-87

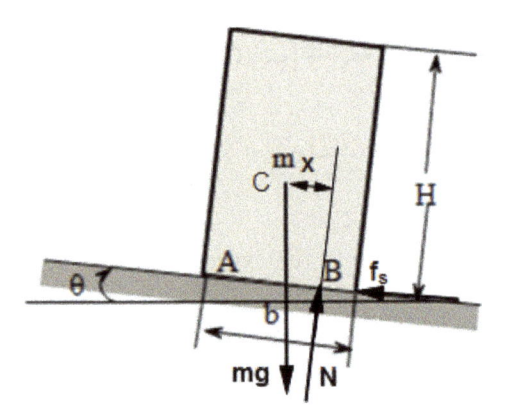

Using the last three equations,

$$x = \frac{Hf_s}{2N} = \frac{Hmg\sin\theta}{2mg\cos\theta}$$

$$\Rightarrow x = \frac{H}{2}\tan\theta.$$

(b) To just topple, put $x = \frac{b}{2}$,

$$\text{then } \frac{H}{2}\tan\theta = \frac{b}{2}.$$

So, for toppling, we have a general condition (provided sufficient friction is present to avoid sliding)

$$\tan\theta \geqslant \frac{b}{H}$$

$$\text{so } \theta \geqslant \tan^{-1}\left(\frac{b}{H}\right).$$

(c) Let $\theta_{min} = \tan^{-1}(b/H) = \theta_1$ for toppling; for sliding we have

$$\theta_{min} = \tan^{-1}\mu_s = \phi = \theta_2.$$

(i) If $\theta_2 < \theta_1$ or $\mu_s < \frac{b}{H}$ the body slides before toppling.
(ii) If $\theta_1 = \theta_2$ the body starts toppling and sliding simultaneously:

$$\phi = \frac{b}{H} = \tan^{-1}(\mu_s).$$

(iii) If $\theta_1 > \theta_2$ or $\mu_s > \frac{b}{H}$ the body topples then slides.

Problem 27 A car of mass m is negotiating a planar curve of radius of curvature R, with a constant speed v. The perpendicular distance of the center of mass of the car is h from the road and the distance between the front tyres is equal to b. Find (a) the normal reaction forces acting on the front and rear wheels and (b) the maximum speed without toppling.

Solution
Referring to the FBD (free-body diagram) drawn relative to the vehicle, we have the normal reactions N_A, N_B, frictions f_A, f_B, weight mg, and pseudo (centrifugal) force $mR\omega^2$. Let us write the force and torque equations as follows.
(a) *Force equations:* $F_x = 0$

$$f_A + f_B = mR\omega^2 \tag{3.82}$$

$$F_y = 0$$

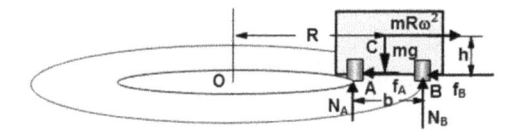

$$N_A + N_B - mg = 0. \tag{3.83}$$

Torque equation: $\tau_B = 0$

$$N_A(b)(-\mathbf{k}) + mg\left(\frac{b}{2}\right)(\mathbf{k}) + mR\omega^2(h)(-\mathbf{k}) = 0 \tag{3.84}$$

$$\text{or } N_A = \frac{mg}{2} - mR\omega^2 h/b.$$

Putting the value of N_A in equation (3.83),

$$N_B = \frac{mg}{2} + mR\omega^2 h/b, \text{ where the angular speed } \omega = v/R.$$

(b)
(i) For no toppling $N_A \geqslant 0$

$$\text{or } \frac{mg}{2} - mR\omega^2 h/b \geqslant 0$$

$$\text{or } \omega \leqslant \sqrt{\frac{gb}{2hR}}.$$

Then the optimum speed is the maximum speed without overturning or toppling is given as

$$v = \omega R = \sqrt{\frac{gbR}{2h}}.$$

3-89

Problem 28 A box of mass m, thickness $L = b$, and height H is placed on a wedge of mass M. A horizontal force F acts on the wedge so that the box does not slide on the wedge. Find the value(s) of the force so as to topple about the rear or left hand side edge of the box. Assume that the ground is smooth.

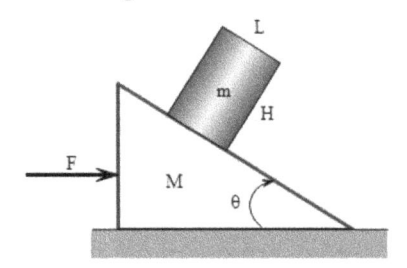

Solution
The applied force will accelerate the wedge–block system with a common acceleration A, say. Then we can take the pseudo-force mA acting towards the left on the box at its center of mass. For the toppling about the edge P, the normal reaction will pass through the edge P. Then the torque of the pseudo-force must be equal to or greater than the gravitational torque taken from the plank, about the edge P. Then we can write

$$\tau_P \geqslant 0. \qquad (3.85)$$

The gravitational torque and torque due to the pseudo-force about P are given as follows:

$$\vec{\tau}_{gr} = -(mg\cos\theta)(L/2) + (mg\sin\theta)h)\hat{k}$$

$$\vec{\tau}_{pseudo} = (mA\cos\theta)h - (mA\sin\theta)(L/2))\hat{k}.$$

Then the net torque about P is

$$\vec{\tau}_P = \vec{\tau}_{gr} + \vec{\tau}_{pseudo}$$

$$\vec{\tau}_P = -(mg\cos\theta)(L/2) + (mg\sin\theta)h)\hat{k} +$$

$$(mA\cos\theta)h - (mA\sin\theta)(L/2))\hat{k}. \qquad (3.86)$$

Using equations (3.85) and (3.86) we have

$$-(mg\cos\theta + mA\sin\theta)\frac{b}{2} + m(A\cos\theta - g\sin\theta)h \geqslant 0.$$

Rearranging the terms we have

$$m(A\cos\theta - g\sin\theta)h \geqslant m(\cos\theta + A\sin\theta)L/2.$$

Putting $h = H/2$ in the last expression, $A[H\cos\theta - L\sin\theta] \geqslant g(L\cos\theta + H\sin\theta)$

$$\Rightarrow A(H\cos\theta - L\sin\theta) \geqslant g(H\sin\theta + L\cos\theta) \Rightarrow A \geqslant \frac{g(H\sin\theta + L\cos\theta)}{(H\cos\theta - L\sin\theta)}$$

$$\Rightarrow A \geqslant g \frac{\tan\theta + \frac{b}{H}}{1 - \frac{b}{H}\tan\theta}$$

$$\Rightarrow A \geqslant g\tan(\theta + \phi), \tag{3.87}$$

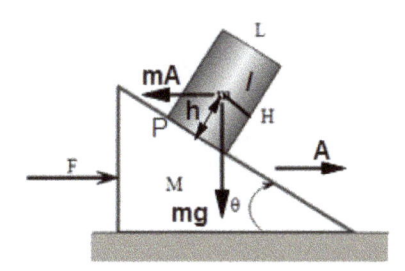

where $\tan\phi = \frac{b}{H}$.

Since $F = (M + m)A$, because both bodies move with a common acceleration A without any relative sliding, we have

$$A = \frac{F}{M+m}. \tag{3.88}$$

Using equations (3.87) and (3.88) we have

$$F \leqslant (M+m)g\tan(\theta + \phi),$$

where $\phi = \tan^{-1}\frac{L}{H}$.

Problem 29 A uniform smooth rod of mass M fitted with two smooth wheels each of mass m is kept at rest against fixed inclined planes. Find (a) the reactions offered by the right- and left-hand sides of the inclined plane on the rod, (b) the ratio of these reaction forces, and (c) the angle of inclination θ of the rod with the horizontal.

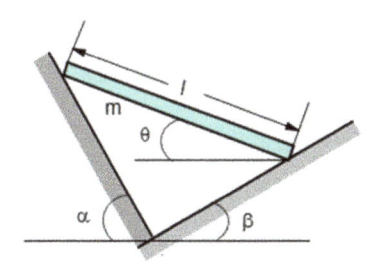

3-91

Solution
Method 1

(a) Let N_1 and N_2 be the normal reactions offered by the inclined planes on the rod.

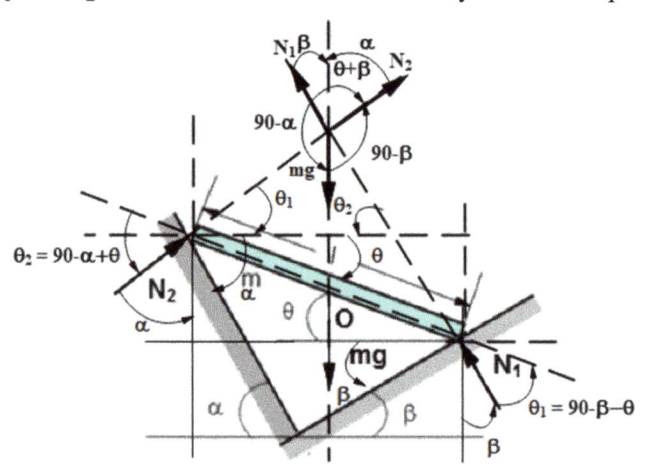

For equilibrium of the rod,
$$F_x = N_2 \sin \alpha - N_1 \sin \beta = 0$$
$$N_2 \sin \alpha = N_1 \sin \beta. \tag{3.89}$$

Furthermore, $F_y = 0$
$$N_2 \cos \alpha + N_1 \cos \beta - mg = 0. \tag{3.90}$$

Solving equations (3.89) and (3.90), we have
$$N_1 = \frac{mg\sin\alpha}{\sin(\alpha + \beta)} \text{ and } N_2 = \frac{mg\sin\beta}{\sin(\alpha + \beta)}$$

(b) Then we have
$$\Rightarrow \frac{N_1}{N_2} = \frac{\sin \alpha}{\sin \beta}.$$

Method 2
According to Lami's theorem,
$$\frac{N_1}{\sin A} = \frac{mg}{\sin B} = \frac{N_2}{\sin C}, \text{ where}$$
$$C = 180° - \alpha, \quad A = 180° - \beta, \text{ and } B = \alpha + \beta.$$

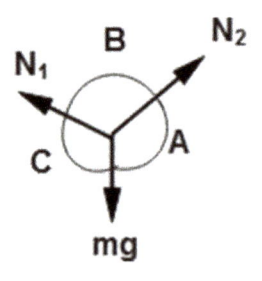

Then we can write

$$\frac{N_1}{\sin(180-\alpha)} = \frac{N_2}{\sin(180-\beta)} = \frac{mg}{\sin(\alpha+\beta)}$$

$$\Rightarrow N_1 = \frac{mg \sin \alpha}{\sin(\alpha+\beta)} \text{ and } N_2 = \frac{mg \sin \beta}{\sin(\alpha+\beta)}$$

$$\Rightarrow \frac{N_1}{N_2} = \frac{\sin \alpha}{\sin \beta}. \qquad (3.91)$$

(c) For rotational equilibrium, the net torque about the center of mass C of the rod will be zero.

$$\vec{\tau}_P = 0$$

$$\Rightarrow (N_1 \sin \theta_1)\frac{l}{2} = N_2 \sin \theta_2 \frac{l}{2}$$

$$\Rightarrow N_1 \sin \theta_1 = N_2 \sin \theta_2 \Rightarrow N_1 \sin(90-\beta-\theta) = N_2 \sin(90-\alpha+\theta)$$

$$\Rightarrow \frac{N_1}{N_2} = \frac{\cos(\alpha-\theta)}{\cos(\beta+\theta)}. \qquad (3.92)$$

By using equations (3.91) and (3.92), we have, $\theta = \tan^{-1}\left(\frac{\cot\beta - \cot\alpha}{2}\right)$.

Problem 30 A uniform rod AB of mass m and length L (= $a + b + c$) is smoothly pivoted at P. The end B is connected with the ceiling by a thread. Find (a) the tension in the string and (b) the reaction at the pivot. Put $M = 2m$ and $a = c = b/2$.

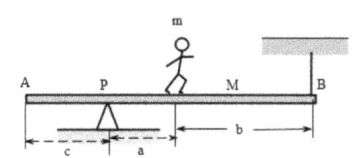

Solution
(a) Taking the torque about P and equating it to zero, we have

$$T(b+a) = mga + Mg\left(\frac{a+b+c}{2} - c\right)$$

$$\Rightarrow T = \frac{1}{b+c}\left\{mga + Mg\left(\frac{a+b-c}{2}\right)\right\}$$

3-93

$$\Rightarrow T = \frac{\frac{Mb}{2} + ma}{b + c} g$$

$$= \frac{2m(b/2) + m(b/2)}{b + (b/2)} = mg.$$

(b) The net vertical force is zero. So

$$N + T = (M + m)g = 3mg.$$

Then $N = (3m - m)g = 2mg.$

Problem 31 A bar of mass $M = 2$ kg and length l is rigidly fitted to a rigid vertical wall at P. The other end of the bar is connected with a small disc of mass m. A light inextensible string which passes over the disc is connected to a hanging body of mass m. Find (a) the torque offered by the wall on the rod and (b) (i) the horizontal and (ii) vertical reaction force at the joint P. Put AP/AC = 0.5.

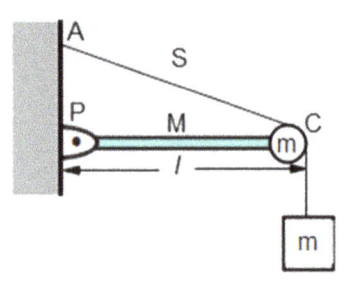

Solution
(a) Let $\tau=$ the torque offered by the wall to equilibrate the rod. Taking the torque about P and equating it to zero, we have

$$\vec{\tau_p} = 0$$

$$\Rightarrow + Mg\frac{l}{2}(-\hat{k}) + (T \sin\theta)l(-\hat{k}) + (mg + mg)l(-\hat{k}) + \vec{\tau} = 0$$

3-94

$$\Rightarrow \vec{\tau} = \left(2mgl + Tl\sin\theta + Mg\frac{l}{2}\right)\hat{k},$$

where $T = mg$ and $\sin\theta = 0.5$,

$$\text{then } \vec{\tau} = \left(\frac{5}{2}mgl + Mg\frac{l}{2}\right)\hat{k} = \frac{(M+5m)gl}{2}\hat{k}.$$

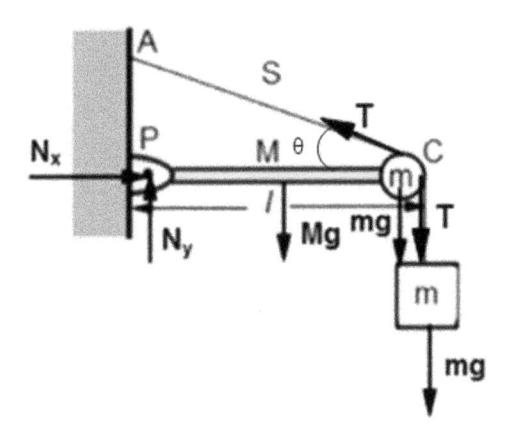

(b)

(i) The net horizontal force is zero on the rod; $F_x = 0$. So we have

$$N_x - T\cos\theta = 0$$

$$\Rightarrow N_x = T\cos\theta = mg\cos\theta (\because T = mg)$$

$$\Rightarrow N_x = m\cos 30° = \frac{\sqrt{3}}{2}mg.$$

(ii) The net vertical force on the rod is zero:

$$F_y = T\sin\theta + N_y - 2mg - Mg = 0$$

$$\Rightarrow N_y = Mg + 2mg - T\sin\theta, \text{ where } T = mg \text{ and } \sin\theta = \frac{1}{2}.$$

$$\Rightarrow N_y = Mg + 2mg - mg \times \frac{1}{2} = \frac{(2M+3m)g}{2}.$$

Problem 32 A uniform rectangular wooden block of mass m and height H is resting on a horizontal surface. A bullet of mass m collides with the block with a horizontal velocity v_0. If the block stops after sliding a distance L, find (a) the coefficient of kinetic friction, (b) the position of the line of action of the normal reaction from the center of mass of the box, and (c) the torque due to the normal reaction about the center of mass of the block.

Solution

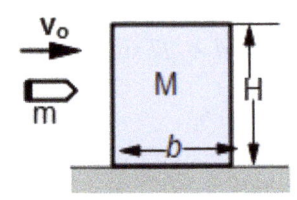

(a) The conservation of linear momentum yields

$$mv_0 = (M+m)v$$

$$\Rightarrow v = \frac{Mv_0}{M+m}. \qquad (3.93)$$

The coefficient of kinetic friction is

$$\mu_k = \frac{v^2}{2gL}. \qquad (3.94)$$

Using equations (3.93) and (3.94)

$$\mu_k = \left(\frac{mv_0}{M+m}\right)^2 \frac{1}{2gL} = \frac{m^2 v_0^2}{2(M+m)^2 gL}.$$

(b) For sliding without toppling, the net torque about the center of mass will be zero:

$$\vec{\tau}_C = \vec{\tau}_N + \vec{\tau}_{gr} + \vec{\tau}_f = 0$$

$$\Rightarrow Nx\hat{k} + (f_k)\frac{H}{2}(-\hat{k}) = 0$$

$$\Rightarrow Nx\hat{k} + \mu_k(M+m)g\frac{H}{2}(-\hat{k}) = 0$$

$$\Rightarrow x = \mu_k \frac{(M+m)gH}{2N}, \text{ where } N = (M+m)g$$

$$\Rightarrow x = \frac{\mu_k H}{2} = \left\{\frac{m^2 v_0^2}{2(M+m)^2 gL}\right\}(H/2)$$

$$\Rightarrow x = \frac{m^2 v_0^2 H}{4(M+m)^2 gL}.$$

3-96

(c) The torque due to the normal reaction about the center of mass is

$$\vec{\tau}_N = -\vec{\tau}_f = \mu_k(M+m)g\frac{H}{2}\hat{k}.$$

Putting the obtained value of μ_k

$$\vec{\tau}_N = (M+m)g\left(\frac{m^2v_0^2}{2(M+m)^2gL}\right)\left(\frac{H}{2}\right)\hat{k}$$

$$= \frac{m^2v_0^2H}{4(M+m)L}\hat{k}.$$

Problem 33 A uniform rod of mass m and length L is smoothly pivoted at the point P of a wedge of mass M. The wedge is kept on a smooth horizontal surface. When a horizontal force F acts on the wedge, find the reactions at (a) the bottom and (b) the pivot P of the rod, (c) the minimum force F if the rod loses contact with the ground, and (d) the force F so that the vertical reaction force at the pivot will be zero for the angle of inclination of the rod with vertical is equal to $37°$, $M = 3$ kg and $m = 1$ kg.

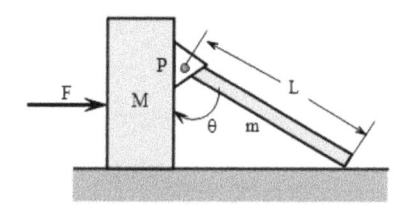

Solution
(a) The torque about the pivot P on the rod is zero due to gravity mg and the normal reaction N, and the pseudo-force ma acts on the center of mass of the rod towards the left as the wedge–rod system $(M+m)$ accelerates towards the right with an acceleration $a = F/(M+m)$; $\vec{\tau}_P = 0$:

$$\vec{\tau}_P = \vec{\tau}_N + \vec{\tau}_{gr} + \vec{\tau}_{pseudo} = 0$$

$$\Rightarrow NL\sin\theta\hat{k} - mg\frac{L}{2}\sin\theta\hat{k} - ma\frac{L}{2}\cos\theta\hat{k} = 0$$

$$\Rightarrow N = \frac{m\frac{L}{2}(a\cos\theta + g\sin\theta)}{L\sin\theta}$$

$$\Rightarrow N = \frac{m(a\cot\theta + g)}{2}, \text{ where } a = \frac{F}{M+m}$$

$$\Rightarrow N = \frac{mg}{2}\left[\frac{F\cot\theta}{(M+m)g} + 1\right].$$

3-97

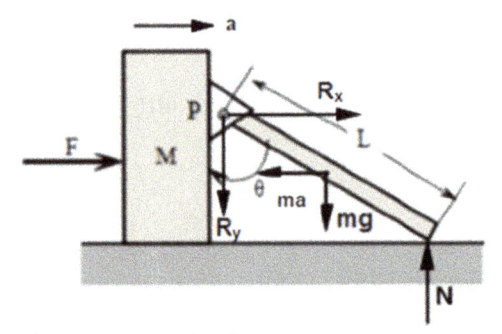

(b) The horizontal reaction R_x is given as
$$R_x = ma = m\{F/(M+m)\} = mF/(M+m).$$
The vertical reaction R_y is given as
$$R_y = N - mg$$
$$= \frac{mg}{2}\left[1 + \frac{F\cot\theta}{(M+m)g}\right] - mg$$
$$= \frac{mg}{2}\left[\frac{F\cot\theta}{(M+m)g} - 1\right].$$

(c) For losing contact, put $N = 0$ in last expression to obtain
$$\Rightarrow N = \frac{mg}{2}\left[\frac{F\cot\theta}{(M+m)g} + 1\right] = 0$$
$$F = -(M+m)g\tan\theta.$$
The −ve sign means that the force must act towards the left.

(d) By putting $R_y = 0$, we have $F = (M+m)\tan 37° = 3(3+1)/4 = 3N$

Problem 34 A small ball of mass m is welded at the bottom end of the rod of mass m and length l which is smoothly pivoted with a block of mass M at point P. The block can slide along the smooth horizontal rod by the action of a horizontal force F acting on it. If the rod makes a constant angle θ with the vertical, find the magnitude of the force F.

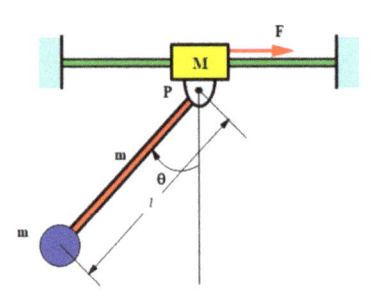

3-98

Solution

The system $(M + m)$ accelerates towards the right with an acceleration A, which can be given as

$$A = \frac{F}{m + m + M} = \frac{F}{(M + 2m)}. \tag{3.95}$$

The net torque about the pivot P on the rod is zero due to gravity mg, the reaction forces at P, and the left-ward pseudo-force mA acting on the center of mass C of the rod and the ball.

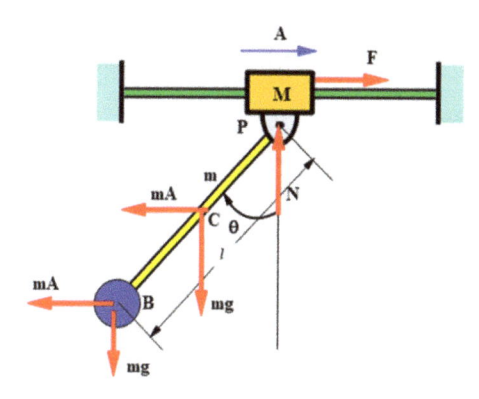

Taking the moments of all the forces acting on the rod about the pivot P measured from the reference frame of the block, we have

$$\vec{\tau}_P = 0.$$

$$\Rightarrow \vec{\tau}_P = \vec{\tau}_{Reaction} + \vec{\tau}_{gr} + \vec{\tau}_{pseudo} = 0.$$

Since the total reaction force at the pivot acts at P, its torque about P is zero. Therefore we have the torque due to gravity and pseudo-forces:

$$\Rightarrow \vec{\tau}_P = \vec{\tau}_{gr} + \vec{\tau}_{pseudo} = 0$$

$$\Rightarrow mg\frac{l}{2}\sin\theta\hat{k} + mgl\sin\theta\hat{k} - mA\frac{l}{2}\cos\theta\hat{k} - mAl\cos\theta\hat{k} = 0 \Rightarrow 3mg\frac{l}{2}\sin\theta = 3mA\frac{l}{2}\cos\theta$$

$$\Rightarrow A = g\tan\theta. \tag{3.96}$$

Using equations (3.95) and (3.96)

$$\frac{F}{(M + 2m)} = g\tan\theta$$

$$\Rightarrow F = (M + 2m)\tan\theta.$$

Fixed axis rotation

Problem 35 A uniform disc of mass m and radius R is rotating with an angular velocity ω_0 about a vertical axis passing through its center of mass. The plane of the disc is horizontal. The disc is gently placed on a horizontal surface. The coefficient of friction between the disc and surface is μ. Find (a) the frictional torque experienced by the disc, (b) the angular acceleration of the disc, and (c) the time after which the disc stops.

Solution
(a) Let us take a thin ring of the disc whose radius is r and thickness is dr. Then the frictional torque on each element dm of the ring is given as
$$d\tau_r = r\,df, \quad \text{where } df = \mu\,dmg.$$

Then we have
$$d\tau_r = r\mu\,dmg. \tag{3.97}$$

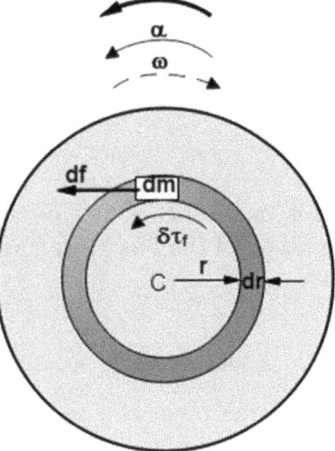

Since the $d\tau_r$ has the same vertical direction on each element of the ring, the net frictional torque τ_r acting on the ring will be equal to the sum (integration) of all elementary torques $d\tau_r$:
$$\tau_r = \int d\tau_r. \tag{3.98}$$

Using equations (3.97) and (3.98)
$$\tau_r = \int r\mu\,dmg = r\mu g \int dm = r\mu m_r g,$$

where the mass of the ring is
$$m_r = m/(\pi R^2)(2\pi r\,dr) = 2mr\,dr/R^2.$$

Then we have

$$T_r = \frac{2\mu mgr^2 dr}{R^2}. \tag{3.99}$$

Since the disc is a combination of concentric thin rings, the torque acting on the disc is given as the integration of torque acting on all the thin rings:

$$T_d = \int T_r. \tag{3.100}$$

Using equations (3.99) and (3.100) we have

$$T_d = \int_0^R \frac{2\mu mgr^2 dr}{R^2} = \frac{2\mu mgR}{3}.$$

(b) The frictional torque will provide an angular deceleration α given as

$$\alpha = \frac{T}{I} = \frac{\frac{2\mu mgR}{3}}{mR^2/2} = \frac{4\mu g}{3R}.$$

(c) The torque decreases the angular speed of the disc from ω_0 to 0 during a time $t = \frac{\omega_0}{\alpha}$. Putting the value of α we have $t = \frac{3R\omega_0}{4\mu g}$.

Problem 36 A thin hollow sphere of mass m_1 and radius R is completely filled with water of mass m_2 is given an initial angular velocity ω_0 near a wall as shown in the figure. The coefficient of friction between all the contacting surfaces is μ. Find (a) the normal reactions offered by the surfaces on the sphere, (b) the angular acceleration of the sphere, and (c) the time taken by the sphere to stop rotation.

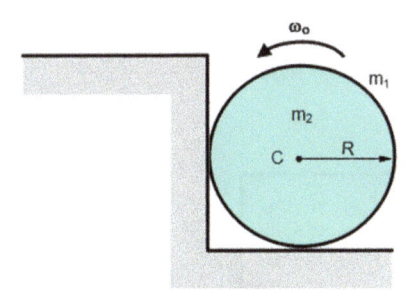

Solution
(a) *Torque equation:*
Since the water is non-viscous, it does not rotate. So we can consider the rotation of the spherical shell only. The frictional torque about C is

$$T_C = (f_1 + f_2)R = I_C \alpha$$

3-101

$$\Rightarrow (f_1 + f_2)R = \left(\frac{2}{3}m_2 R^2\right)\alpha$$

$$\Rightarrow \mu(N_1 + N_2)R = \frac{2}{3}m_2 R^2 \alpha$$

$$(\because f_1 = \mu N_1 \text{ and } f_2 = \mu N_2)$$

$$\Rightarrow N_1 + N_2 = \frac{2}{3\mu}m_2 R\alpha. \tag{3.101}$$

Force equation:

$$F_x = 0$$

$$\Rightarrow N_2 = f_1 = \mu N_1 \tag{3.102}$$

$$F_y = 0$$

$$N_1 + f_2 = (m_1 + m_2)g,$$

where $f_2 = \mu N_2$

$$\Rightarrow N_1 + \mu N_2 = (m_1 + m_2)g. \tag{3.103}$$

By using equations (3.102) and (3.103),

$$N_1 + \mu(\mu N_1) = (m_1 + m_2)g$$

$$\Rightarrow N_1 = \frac{(m_1 + m_2)g}{1 + \mu^2}$$

and

$$N_2 = \mu N_1 = \frac{(m_1 + m_2)\mu g}{1 + \mu^2}.$$

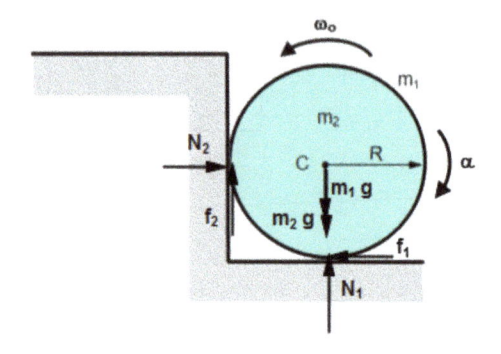

(b) Putting the values of N_1 and N_2 in equation (3.101) we have

$$\frac{(m_1 + m_2)g}{1 + \mu^2} + \frac{(m_1 + m_2)\mu g}{1 + \mu^2} = \frac{2}{3\mu}m_2 R\alpha$$

$$\Rightarrow \frac{(m_1 + m_2)g}{(1 + \mu^2)}(1 + \mu) = \frac{2}{3\mu}m_2 R\alpha$$

$$\Rightarrow \alpha = \frac{3(m_1 + m_2)\mu(1 + \mu)g}{2m_2(1 + \mu^2)R}.$$

(c) The time after which the body will stop rotation is

$$t = \frac{\omega_0}{\alpha} = \frac{2m_2(1 + \mu^2)R\omega_0}{3(m_1 + m_2)\mu(1 + \mu)g}.$$

Example 37 A uniform rod PQ of mass $m = 2$ kg and length L is smoothly pivoted at O. It is released from rest when it is horizontal. As a result, the rod rotates in the vertical plane. Find (a) the angular (i) acceleration and (ii) velocity of the rod, and (b) the reaction at the pivot when the rod swings through an angle θ.

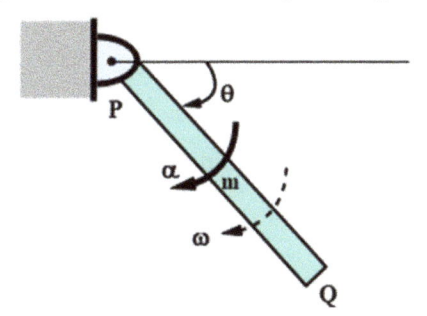

Solution
(a)
(i) *Torque equation*:
Recasting Newton's equation of rotation:

$$\tau_O = I_O \alpha, \text{ where } I_O = I_C + md^2 = m\frac{L^2}{12} + m\,(L/2)^2 = \frac{mL^2}{3}.$$

At any angular position θ, taking the torque of gravity about the pivot O, we have

$$\tau_O = mg\left(\frac{L}{2}\right)\cos\theta = \frac{mL^2}{3}\alpha.$$

This gives $\alpha = \frac{3g\cos\theta}{2L}$.

(ii) Using the kinematical equation $\alpha = \frac{3g\sin\theta}{2L} = \frac{d(\omega^2)}{2d\theta}$ we have

$$\int_0^\omega d\omega^2 = \int_0^\theta 3\left(\frac{g}{L}\right)\cos\theta d\theta.$$

3-103

Evaluating the integral we have
$$\omega = \sqrt{3(g/L)\sin\theta}.$$

(b) *Force equation*:
Let the horizontal and vertical components of the reaction be
$$R_r - mg\cos\theta = ma_r$$
$$mg\sin\theta - R_t = ma_t.$$

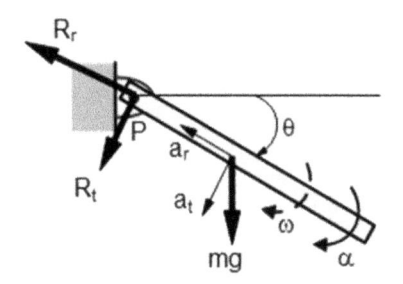

Kinematics:
Putting
$$a_r = (L/2)\omega^2$$
and $a_x = (L/2)\alpha$ in the force equations.
We have

$$R_r = m\left(g\sin\theta + \frac{L\omega^2}{2}\right)$$

and

$$R_t = m\left(g\cos\theta - \frac{L\alpha}{2}\right),$$

where

$$\alpha = \frac{3g\cos\theta}{2L} \quad \text{and} \quad \omega = \sqrt{3(g/L)\sin\theta}.$$

This gives $R_t = \frac{mg\cos\theta}{4}$ and

$$R_r = \frac{5mg\sin\theta}{2}.$$

Then $R = \sqrt{R_r^2 + R_t^2} = \left\{\frac{(\sqrt{1+99\sin^2\theta})}{4}\right\}mg.$

3-104

Problem 38 A uniform rod AB of mass M and length l is smoothly pivoted at C at a distance of $a = l/3$ from the end A. The rod is kept horizontal by a vertical thread connected at a distance of $c = l/5$ from the end B. (a) Find the tension in the string and the reaction at the pivot. (b) If the string is cut, find the acceleration of the center of the rod and (c) the reaction at the pivot just after cutting the string (assume $a + b + c = l$).

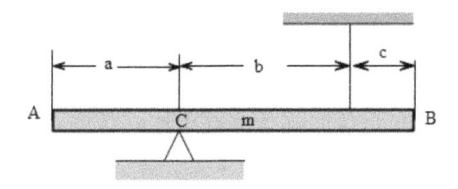

Solution
(a) The rod is in equilibrium under the action of the forces of tension T, the normal reaction N, and gravity mg.

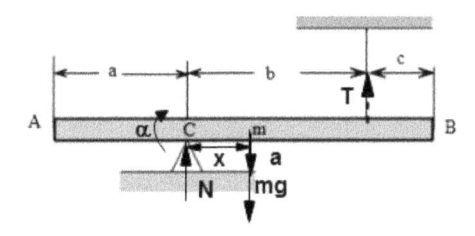

Torque equation: $\tau_C = 0$

$$\Rightarrow T b \hat{k} - mgx\hat{k} = 0$$

$$\Rightarrow T = mg\frac{x}{b}, \text{ where } x = (b + c) - \frac{a+b+c}{2} = \frac{b+c-a}{2}$$

$$\Rightarrow T = mg\left(1 + \frac{c}{b} - \frac{a}{b}\right)$$
$$= mg(b + c - a)/b = mg(l - 2a)/b$$
$$= 5\,mg/7.$$

The net vertical force is
$$F_y = N + T - mg = 0$$

$$\Rightarrow N = mg - T = mg\left(\frac{a-c}{b}\right) = 2mg/7.$$

If the string is cut, the net torque about C is

$$\tau_C = I_C \alpha$$

$$\Rightarrow mgx = \left(\frac{ml^2}{12} + mx^2\right)\alpha$$

$$\Rightarrow \alpha = \frac{12gx}{l^2 + 12x^2},$$

where

$$x = \left(\frac{b+c-a}{2}\right).$$

Force equation:
The net vertical force is

$$F_y = mg - N = ma, \text{ where } a = x\alpha$$

$$\Rightarrow N = m(g - x\alpha) = m\left\{g - \frac{12gx^2}{l^2 + 12x^2}\right\}$$

$$= \frac{mgl^2}{l^2 + 12x^2} = \frac{mg}{1 + 12(x/l)^2}$$

$$\Rightarrow N = \frac{mg}{3\left\{\frac{b+c-a}{a+b+c}\right\}^2 + 1}$$

$$= \frac{mg}{3\left\{\frac{l-2a}{l}\right\}^2 + 1} = 3mg/4.$$

Problem 39 A disc of mass m and radius R can rotate in the vertical plane about a smooth pivot P. A block of mass m is hanging by an inextensible light string which is wrapped over the disc so that it does not slip. If the block is released from the given position, find (a) the tension in the string, (b) the angular acceleration of the disc, (c) the acceleration of the block and the center of mass of the disc at the given position, (d) the reaction at the pivot, and (e) the acceleration of the center of mass of $M + m$.

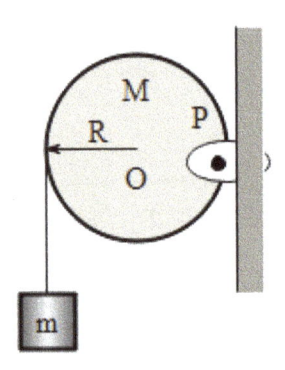

Solution
(a) The torque about P is
$$T_P = mg(2R) + MgR = I_P\alpha$$
$$\Rightarrow (2m + M)gR = \left\{\frac{3MR^2}{2} + m(2R)^2\right\}\alpha$$
$$\Rightarrow \alpha = \frac{2(2m + M)g}{(3M + 8m)R}.$$

The acceleration of m is
$$a_m = 2R\alpha = \frac{4(2m + M)g}{3M + 8m} > g.$$

Since $a_m > g$, the tension in the string is
$$T = m(g - a_m) < 0.$$
This means that the string will get slack ($T = 0$).

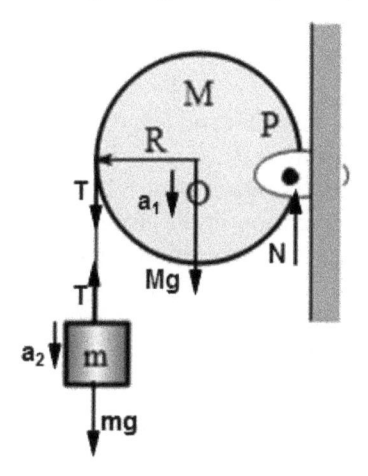

(b) Then the torque about P is
$$T_P = MgR$$
$$\Rightarrow \frac{3MR^2}{2}\alpha = MgR$$
$$\alpha = \frac{2g}{3R}.$$

3-107

(c) Then the acceleration of the block is equal to free fall acceleration $g = 10 \text{ m s}^{-2}$ beacause the tension in the strng is zero at the given position; the acceleration of the center of mass of the disc is

$$a_M = a_1 = R\alpha = \frac{2}{3}g.$$

(d) The reaction of the pivot is equal to the normal or vertical reaction N because the horizontal reaction is zero because the horizontal acceleration of the center of mass of the disc is zero. So $N = M(g - a_m) = M(g - \frac{2}{3}g)$

$$\Rightarrow N = \frac{Mg}{3}.$$

(e) Then the center of mass of the system $(M + m)$ has an acceleration

$$\vec{a}_C = \frac{m\vec{a}_m + M\vec{a}_M}{M + m} = \frac{mg + M\left(\frac{2}{3}g\right)}{M + m}(-\hat{j})$$

$$\Rightarrow a_C = \frac{2M + 3m}{3(M + m)}g \text{ (points vertically down)}.$$

Problem 40 A square plate of mass $2m$ and length l is cut into two equal parts. Let us take one part and pivot it smoothly at P, as shown in the figure, so that it is free to rotate in the vertical plane. Initially the triangular plate is held stationary by a horizontal string tied at a height b. (a) Find the tension in the string. (b) If the string is cut find (i) the angular acceleration of the plate and (ii) reaction at the pivot just after cutting the string.

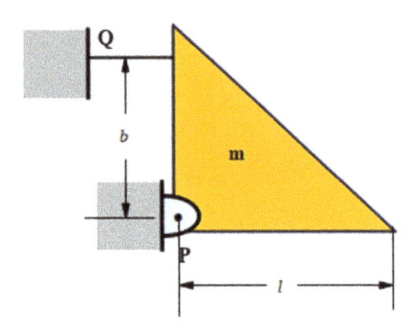

Solution
(a) The moment of inertia of the plate about O is

$$I_O = m\left(\frac{l^2 + l^2}{12}\right) = \frac{ml^2}{6}$$

By parallel axis theorem, we have $I_P = I_C + m(CP)^2$ and $I_O = I_C + m(CO)^2$
Then,

$$\begin{aligned}
I_P &= I_0 + m(CP^2 - CO^2) \\
&= \tfrac{ml^2}{6} + m(CP + CO)(CP - CO) \\
&= \tfrac{ml^2}{6} + m(OP)(2CP - OP), \text{ where } OP = \tfrac{l}{\sqrt{2}} \\
&= \tfrac{ml^2}{6} + m\tfrac{l}{\sqrt{2}}(2CP - OP), \text{ where } CP = \tfrac{\sqrt{2}l}{3} \\
&= \tfrac{ml^2}{6} + m\tfrac{l}{\sqrt{2}}\left(2\tfrac{\sqrt{2}l}{3} - \tfrac{l}{\sqrt{2}}\right) \\
&= \tfrac{ml^2}{6} + \tfrac{ml^2}{6} = \tfrac{ml^2}{3}
\end{aligned}$$

Let T = tension in the string. For equilibrium of the plate, the torque about P is zero. So we can write

$$\vec{\tau}_P = Tb\hat{k} - mg\tfrac{l}{3}\hat{k} = 0$$

$$\Rightarrow T = \frac{mgl}{3b}$$

(b)

(i) The torque about P is

$$\tau_P = I_P \alpha$$

$$mg\frac{l}{3} = I_P \alpha. \tag{3.104}$$

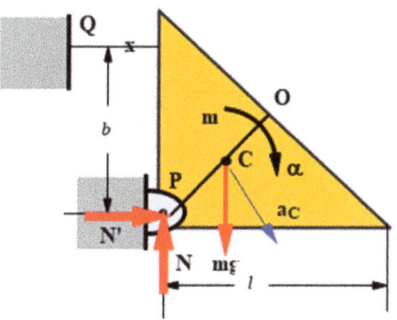

After breaking the string

Putting $I_P = \tfrac{ml^2}{3}$ we have

$$\alpha = \frac{\tfrac{mgl}{3}}{\tfrac{ml^2}{3}} = \frac{g}{l}.$$

(ii) The horizontal reaction is

$$N' = ma_x = m\frac{a_C}{\sqrt{2}} = \frac{m(r\alpha)}{\sqrt{2}}.$$

3-109

Putting $r = \left(\frac{2}{3}\right)\left(\frac{l}{\sqrt{2}}\right)$ and $\alpha = \frac{g}{l}$

$$N' = m\left\{\left(\frac{2}{3}\right)\left(\frac{l}{\sqrt{2}}\right)\right\}\left\{\frac{g}{l}\right\}/\sqrt{2} = \frac{mg}{3}.$$

The vertical reaction is given as

$$mg - N = ma_y$$

$$\Rightarrow N = m(g - a_y) = m\left(g - \frac{r\alpha}{\sqrt{2}}\right)$$

$$= m\left[g - \frac{\sqrt{2}l}{3} \cdot \frac{g}{l} \times \frac{1}{\sqrt{2}}\right] = \frac{2mg}{3}.$$

Then the total reaction torque

$$R = \sqrt{N^2 + N'^2} = \frac{mg}{3}\sqrt{1+4} = \sqrt{5}\frac{mg}{3}.$$

Problem 41 A rod PQ of mass M and length l is smoothly hinged at the fixed point O. It is placed leaning against a smooth cubical wedge of mass M at an angle θ with the vertical. (a) If the rod does not move due to friction after releasing the system from rest, find the maximum value of θ. (b) If the wedge is smooth it will move towards right after being released. Just after the release of the system $(M + m)$, find (i) the angular acceleration of the rod, (ii) the normal reaction between the wedge and the rod, and (iii) the acceleration of the wedge.

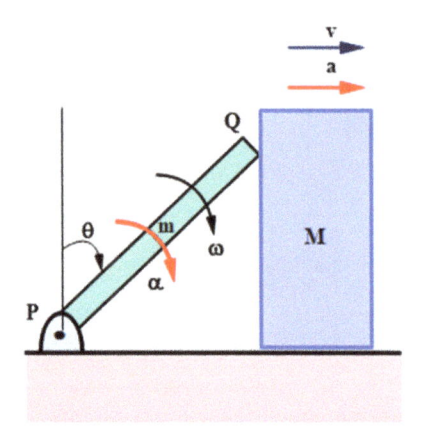

Solution
(a) Let N = the normal reaction between M and m. In this case the body M does not move and static friction is present between the wedge and ground. So the net torque about the pivot P is zero:

$$T_P = Nl\cos\theta - mg\frac{l}{2}\sin\theta = 0.$$

3-110

Then we have
$$N = \frac{1}{2}mg \tan \theta. \qquad (3.105)$$

Equating the horizontal forces,
$$N = f_s. \qquad (3.106)$$

Law of static friction:
$$f_s \leq \mu_s N'. \qquad (3.107)$$

Equating vertical forces
$$N_y = mg \qquad (3.108)$$

$$\Rightarrow \frac{1}{2}mg \tan \theta \leq \mu_s Mg$$

$$\Rightarrow \theta \leq \tan^{-1}\left(2\mu_s \frac{M}{m} \theta\right)$$

$$\Rightarrow \theta_{max} = \tan^{-1}\left(2\mu_s \frac{M}{m} \theta\right).$$

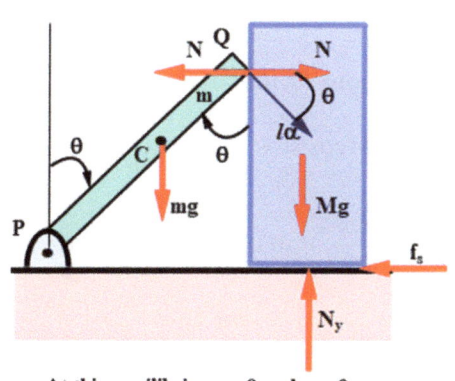

At this equilibrium $v=0$ and $\omega =0$

(b)

(i) If the friction is absent, the net torque about the pivot is
$$\tau_P = I\alpha$$
$$= mg\frac{l}{2} \sin \theta - Nl \cos \theta = =ml^2\alpha/3$$
$$\Rightarrow \left(-N \cos \theta + \frac{mg}{2} \sin \theta\right) = \frac{ml}{3}\alpha. \qquad (3.109)$$

3-111

Kinematics:
$$l\alpha \cos\theta = a. \tag{3.110}$$

The net horizontal force on M is
$$N = Ma. \tag{3.111}$$

By using equations (3.110) and (3.111)
$$N = Ml\alpha \cos\theta. \tag{3.112}$$

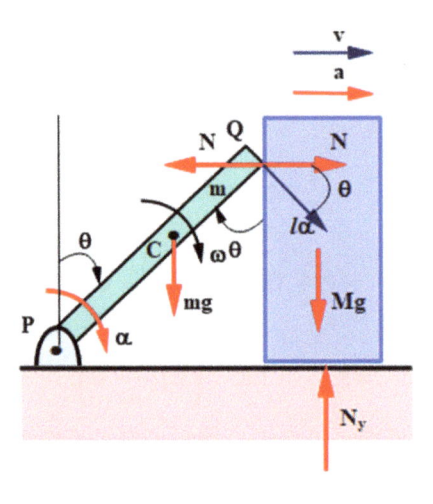

Using equations (3.109) and (3.112)
$$\left(-Ml\alpha\cos^2\theta + \frac{mg}{2}\sin\theta\right) = \frac{ml\alpha}{3}$$

$$\Rightarrow l\alpha\left(\frac{m}{3} + M\cos^2\theta\right) = \frac{mg}{2}\sin\theta$$

$$\Rightarrow \alpha = \frac{mg\sin\theta}{2\left(M\cos^2\theta + \frac{m}{3}\right)l}.$$

(ii) Then the value of N is $N = Ml\alpha \cos\theta = \dfrac{Mmg\sin\theta\cos\theta}{2\left(M\cos^2\theta + \frac{m}{3}\right)}$.

(iii) The acceleration of the wedge is $a = l\alpha\cos\theta = \dfrac{N}{M} = \dfrac{mg\sin\theta\cos\theta}{2\left(M\cos^2\theta + \frac{m}{3}\right)}$.

Problem 42 A hollow and a solid sphere of the same radius R, masses M and bM, respectively, are joined to form a compound sphere which is pivoted at the point O of a light rod of length L. The rod is fitted rigidly with the vertical wall at P. A particle of mass

m is attached to the compound cylinder at Q as shown in the figure. (a) If this system does not rotate after releasing it from the given position, find the value of b. (b) If $M = 2m$ and $b = 4$, just after releasing the system find (i) the angular acceleration of the system, (ii) the reactions at the point O, and (iii) the moment offered by the wall to the rod.

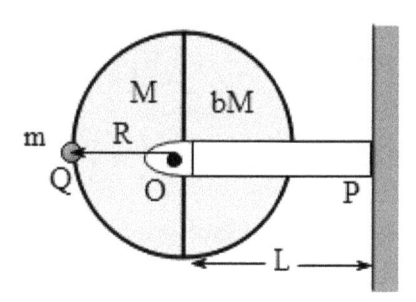

Solution

(a) The net torque about O is given as

$$\vec{\tau_0} = I_0 \vec{\alpha}$$

$$\Rightarrow mgR\hat{k} + Mg\frac{R}{2}\hat{k} - bMg\left(\frac{3R}{8}\right) = I_0 \vec{\alpha}. \tag{3.113}$$

The moment of inertia about O is

$$I_0 = mR^2 + \frac{2}{3}MR^2 + \frac{2}{5}(bM)R^2. \tag{3.114}$$

Using equations (3.113) and (3.114)

$$\vec{\alpha} = \frac{\left\{\left(\frac{M}{2} - \frac{3bM}{8}\right) + m\right\}g}{\left(\frac{2bM}{5} + \frac{2}{3}M + m\right)R}\hat{k}.$$

If $\alpha = 0$, $\frac{M}{2} - \frac{3bM}{8} + m = 0$

$$\Rightarrow \frac{M + 2m}{2} = \frac{3b}{8}M$$

$$\Rightarrow b = \frac{4(M + 2m)}{3M}.$$

(b)

(i) If $M = 2m$ and $b = 4$ we have

$$\vec{\alpha} = \frac{\left\{\frac{2m}{2} - \frac{3 \times 4 \times 2m}{8} + m\right\}g}{\left(\frac{2 \times 4 \times 2m}{5} + \frac{2}{3} \times 2m + m\right)R}\hat{k}$$

$$= \frac{-mg}{m(16/5 + 4/3 + 1)R}\hat{k} = -\frac{15g\hat{k}}{83R}.$$

3-113

(ii) The net force acting on the entire system is

$$\vec{F} = m_1\vec{a_1} + m_2\vec{a_2} + m_3\vec{a_3} = (m_{total})\vec{a_C}$$

$$\Rightarrow N - (M + m + bM)g$$

$$= mR\alpha + M\frac{R\alpha}{2} + bM\left(-\frac{3R\alpha}{8}\right)$$

$$\Rightarrow N = \{M(b+1) + m\}g + R\alpha\left(m + \frac{M}{2} - \frac{3bM}{8}\right).$$

Putting $M = 2m$, $b = 4$, and $R\alpha = \frac{15g}{83}$ we have

$$N = \{2m(4+1) + m\}g + \frac{15g}{83}\left(m + \frac{2m}{2} - \frac{3 \times 4 \times 2m}{8}\right) = 11mg + \left(-\frac{15mg}{83}\right) = mg\left(11 - \frac{15}{83}\right) = \frac{898mg}{83}.$$

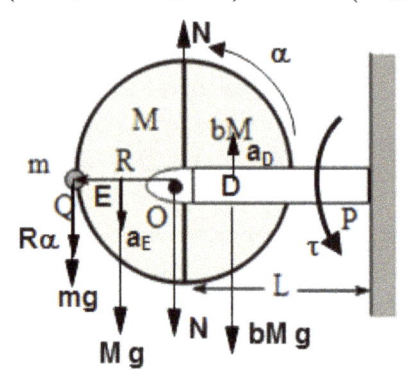

(iii) The moment offered by the wall to the rod is given as

$$\vec{\tau} = +Nl\hat{k} = +mgl\left(\frac{898}{83}\right)\hat{k}.$$

Problem 43 A vertical rod of mass m and length l is pivoted smoothly at P. If the rod is given a gentle tap it rotates in the vertical plane. Find (a) ω, (b) α, (c) the horizontal reaction force, and (d) the normal reaction acting on the rod by the pivot as a function of angle θ made by the rod with the vertical.

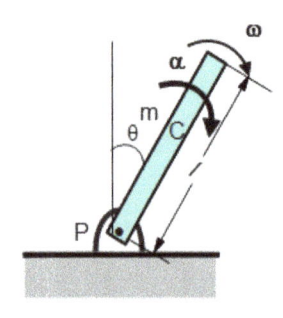

Solution

(a) As the rod rotates about the fixed point P, the net torque about P is

$$\vec{\tau_P} = I_P \vec{\alpha}$$

$$\Rightarrow mg\frac{l}{2}\sin\theta = \frac{ml^2}{3}\alpha$$

$$\Rightarrow \alpha = \frac{3g\sin\theta}{l}.$$

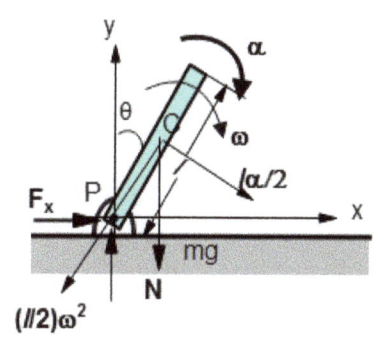

(b) Conserving energy,

$$\frac{1}{2}\frac{ml^2}{3}\omega^2 = mg\frac{l}{2}(1-\cos\theta)$$

$$\Rightarrow \omega = \sqrt{\frac{3g}{l}(1-\cos\theta)}.$$

We can also find angular velocity by integrating the angular acceleration with the angular displacement.

(c) The horizontal reaction force acting on the rod is

$$F_x = ma_x = m\left(\frac{l}{2}\alpha\cos\theta - \omega^2\frac{l}{2}\sin\theta\right).$$

where a_x is the horizontal acceleration of C.
Putting α and ω, we have

$$F_x = \frac{3mg}{4}(3\cos\theta - 2)\sin\theta.$$

(d) The net vertical force is $mg - N = ma_y = (\frac{\omega^2 l}{2}\cos\theta + \frac{l\alpha}{2}\sin\theta)$
where a_y is the vertical acceleration of C.

$$\Rightarrow N = m\left[g - \frac{l}{2}(\omega^2\cos\theta + \alpha\sin\theta)\right]$$

$$= \left[g - \frac{l}{2}\left\{\frac{3g}{l}(1-\cos\theta)\cos\theta + \frac{3g\sin\theta}{2l}\sin\theta\right\}\right]$$

$$= mg\left[1 - \frac{3\{(1-\cos\theta)(2\cos\theta + 1 + \cos\theta)\}}{2}\right]$$

$$= mg\left[1 - \frac{3}{4}(1-\cos\theta)(1+3\cos\theta)\right]$$

$$= \frac{mg}{4}[4 - 3(1 + 3\cos\theta - \cos\theta - 3\cos^2\theta)]$$

$$= \frac{mg}{4}[1 - 6\cos\theta + 9\cos^2\theta]$$

$$\Rightarrow N = \frac{mg(1 - 3\cos\theta)^2}{4}.$$

Problem 44 A uniform rod of mass m and length L is placed vertically at the edge of a table. If the rod is released by giving a gentle tap so that it rotates in the vertical plane about the edge P, find (a) the static friction and (b) the normal reaction N acting on the rod by the rough edge as a function of the angle θ made by the rod with the horizontal. (c) At what value of value of θ does the rod begin to slip if the coefficient of static friction is equal to (i) 0.75 and (ii) very large?

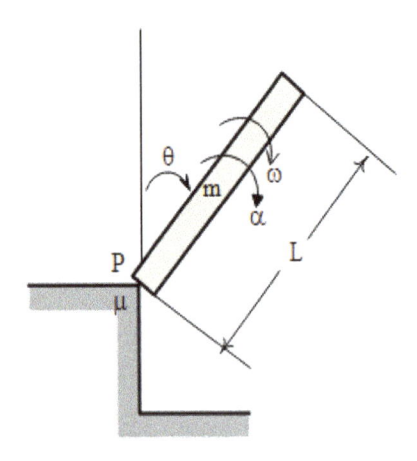

Solution

(a) By conserving energy, we have

$$mg\frac{l}{2}(1-\cos\theta) = \frac{ml^2}{6}\omega^2$$

$$\Rightarrow \omega = \sqrt{\frac{3g(1-\cos\theta)}{l}}.$$

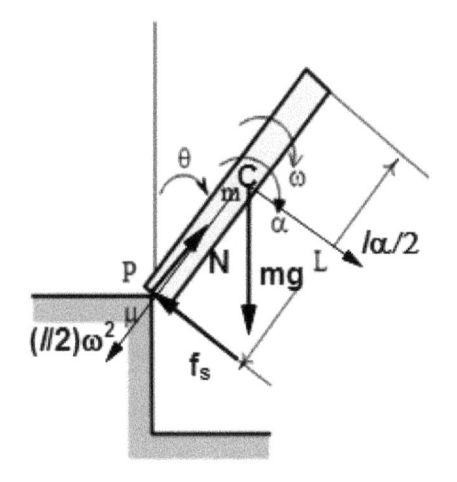

The torque about P is

$$\tau_P = I_P\alpha$$

$$mg\frac{l}{2}\sin\theta = \frac{ml^2}{3}\alpha$$

$$\Rightarrow \alpha = \frac{3g\sin\theta}{2l}.$$

The tangential acceleration of C is

$$\Rightarrow a_t = \frac{l}{2}\alpha = \frac{3g\sin\theta}{4}.$$

The net force acting on the rod in the tangential direction is

$$mg\sin\theta - f_s = ma_t$$

$$\Rightarrow f_s = m(g\sin\theta - a_t)$$

$$= m\left(g\sin\theta - \frac{3g\sin\theta}{4}\right)$$

3-117

$$\Rightarrow f_s = \frac{1}{4} mg \sin \theta.$$

(b) The net force acting on the rod in the radial direction is

$$mg \cos \theta - N = a_r = m\frac{l}{2}\omega^2$$

$$\Rightarrow N = mg \cos \theta - \frac{l}{2}\omega^2$$

$$= mg \cos \theta - \frac{l}{2}\frac{3g(1 - \cos \theta)}{l}$$

$$\Rightarrow N = \frac{mg}{2}(5 \cos \theta - 3).$$

(c)

(i) The coefficient of static friction is

$$\mu_s \geqslant \frac{f_s}{N}.$$

Putting the values of f_s and N in the last expression, we have

$$\mu_s \geqslant \frac{\frac{1}{4} mg \sin \theta}{\frac{mg}{2}(5 \cos \theta - 3)}$$

$$\Rightarrow \mu_s \geqslant \frac{\sin \theta}{2(5 \cos \theta - 3)}.$$

Putting $\mu_s = \frac{3}{10}$ we have

$$\frac{3}{10} = \frac{\sin \theta}{2(5 \cos \theta - 3)}$$

$$\Rightarrow 15 \cos \theta - 9 = 5 \sin \theta.$$

By solving this equation $\sin \theta = \frac{3}{5} \Rightarrow \theta = 36°$.

(ii) Putting μ_s is very large we have

$$(5 \cos \theta - 3) = 0. \text{ So } \theta = \cos^{-1}\frac{3}{5} = 53°.$$

Problem 45 A pulley of mass m, radius R, and radius of gyration K is pivoted smoothly at its center. A block of mass m and a ladder of mass $M = 2m$ and length l are connected with a light inextensible string that passes over the pulley. If the string does not slip over the pulley, find (a) the acceleration of the blocks, (b) time taken by the block to cross the ladder. Assume that the ladder is heavier than the block.

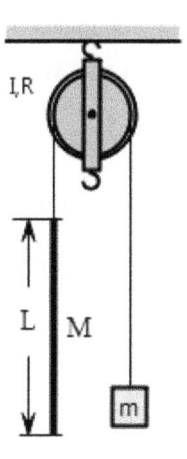

Solution
(a) The torque about the point P is

$$\tau_P = I_P \alpha$$

$$\Rightarrow (M-m)gR = [(M+m)R^2 + I]$$

$$\Rightarrow R\alpha = \frac{(M-m)g}{(M+m) + \frac{I}{R^2}}$$

$$\Rightarrow a = \frac{(M-m)}{M+m+\frac{I}{R^2}} = \frac{M-m}{M+m\left(1+\frac{k^2}{R^2}\right)} \quad (\because R\alpha = a)$$

$$= \frac{2m-m}{2m+m\left(1+\frac{1}{2}\right)} = \frac{mg}{2m+\frac{3m}{2}} = \frac{2g}{7}.$$

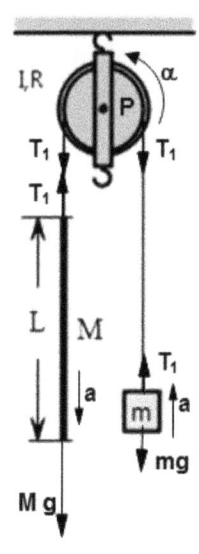

3-119

(b) The relative acceleration between the M and m blocks is
$$a_{rel} = 2a = \frac{4g}{7}.$$

Then the relative distance covered during a time t is
$$s_{rel} = \frac{1}{2}a_{rel}t^2$$

$$\Rightarrow t = \sqrt{\frac{2l}{a_{rel}}} = \sqrt{\frac{2l}{4g/7}} = \sqrt{\frac{7l}{2g}}.$$

Combined motion

Problem 46 A rolling body of mass m, radius R, and radius of gyration K is released from rest on an inclined plane of angle of inclination θ. If the coefficient of static and kinetic friction between the body and inclined plane are μ_s and μ_k, respectively. If $\mu_s < \frac{\tan\theta}{(1+\frac{R^2}{k^2})}$, find (a) the acceleration of the center of mass and (b) the angular acceleration of the sphere. Assume the rolling body is a uniform solid sphere.

Solution

(a) *Law of friction:* Referring to Problem 66, if $\mu_s < \frac{\tan\theta}{(1+\frac{R^2}{k^2})}$, the body will not roll without sliding. Then kinetic friction comes into play, which is given as

$$f_k = \mu_s N. \tag{3.115}$$

Force equation: The net force along the slant is

$$F_x = mg\sin\theta - f_k = ma_C. \tag{3.116}$$

Since the center of mass does not move along the y-axis, $a_y = 0$. Then the net vertical force $F_y = N - mg = ma_y = 0$. This gives

$$N = mg\cos\theta. \tag{3.117}$$

Put the value of N from equation (3.117) in equation (3.115) to obtain

$$f_k = \mu_k mg\cos\theta \tag{3.118}$$

Putting the value of f_k from equation (3.118) in equation (3.116), we have

$$a_C = g(\sin\theta - \mu_k\cos\theta).$$

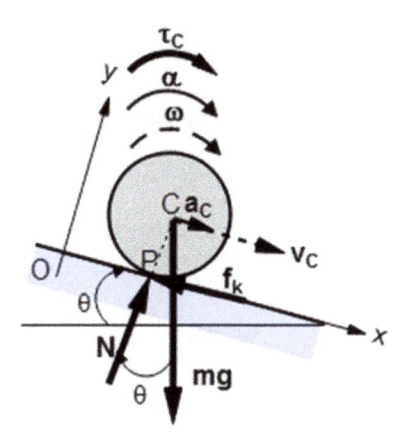

(b) *Torque equation:* Then the net torque is produced by the kinetic friction f_k, which can be given as

$$\tau_C(=f_k R k) = I_C \alpha,$$

where

$$I_C = 2mR^2/5.$$

Then we have

$$f_k R = \left(\frac{2mR^2}{5}\right)\alpha$$

or

$$\alpha = \frac{5f_k}{2mR}. \qquad (3.119)$$

Putting the value of f_k from equation (3.118) in equation (3.119) we have
$$\alpha = \frac{5\mu_k g \cos\theta}{2R}.$$

Problem 47 A ball of mass m, radius R, and radius of gyration K is given an initial velocity v_0 towards the right and a counter-clockwise angular velocity ω_0. After what time will the body roll without sliding?

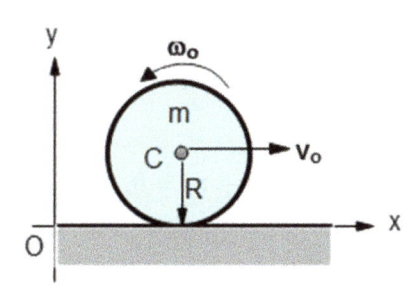

3-121

Solution

Force equations: The lowest point of the ball moves forward with a velocity $v_P = (R\omega_0 + v_0)$. So kinetic friction $f_k = \mu_k N = \mu_k mg$ acts backward on the body towards the left. This force produces an acceleration a_C given as $-\mu_k mg = ma_C$. This gives

$$a_C = -\mu_k g. \tag{3.120}$$

Torque equation: The friction can also produce a counter-clockwise torque $\tau_C = \mu_k mgR = I_C \alpha$, where $I_C = mK^2$. Then

$$\vec{\alpha} = -\frac{\mu_k gR}{K^2}\vec{k}. \tag{3.121}$$

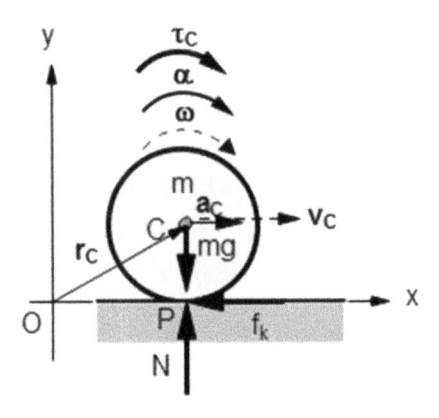

Kinematics of rolling: Let the body stop sliding at time t. If v and ω are the linear and angular velocities of the body at time t, they are given as

$$\vec{v} = (\vec{v}_0 + \vec{a}_C t) \tag{3.122}$$

$$\vec{\omega} = (\vec{\omega}_0 - \vec{\alpha} t). \tag{3.123}$$

Using equations (3.120) and (3.122)

$$v = \left(v_0 - \mu_k gt\right)\vec{i}. \tag{3.124}$$

Using equations (3.120) and (3.122)

$$\omega = -\left(\omega_0 - \frac{\mu_k gRt}{K^2}\right). \tag{3.125}$$

As the point P stops sliding

$$v_P = (-R\omega + v) = 0. \tag{3.126}$$

Putting the values of v and ω from equations (3.124) and (3.125), respectively, in equation (3.126) and simplifying the factors, we have

$$t = \frac{(R\omega_0 + v_0)}{\left(1 + \frac{R^2}{k^2}\right)\mu_K g}.$$

Problem 48 A uniform thin hollow sphere of mass m, radius r, and radius of gyration k is placed gently on a horizontal conveyor belt which moves with a constant velocity v_0. Find (a) the time after which the sphere begins to roll without sliding, (b) the velocity of the sphere, and (c) the distance traversed by the sphere relative to (i) the ground and (ii) the belt until the ball stops sliding on the conveyor belt.

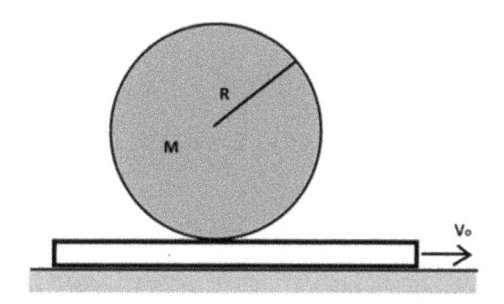

Solution

(a) *Force equations:* Let us look at the ball from the conveyor belt. The lowest point of the ball moves back with a velocity $v_P = v_0$ relative to the conveyor belt. So the kinetic friction $f_k = \mu_k N = \mu_k mg$ acts on the body towards the right. This force produces an acceleration a_C given as $\mu_k mg = ma_C$. This gives

$$a_C = \mu_k g \text{ (to right)}. \qquad (3.127)$$

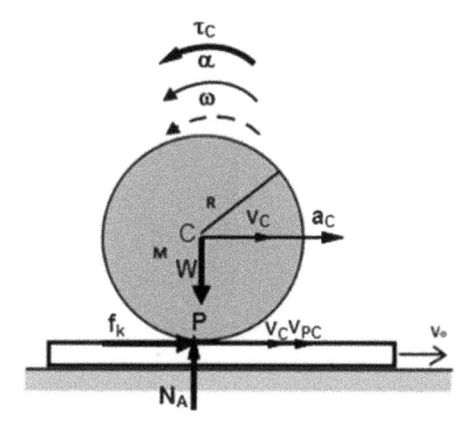

3-123

Torque equation: The friction can also produce a counter-clockwise torque $\tau_C = \mu_k mgR = I_C\alpha$, where $I_C = 2mR^2/3$. Then

$$\alpha = \frac{3\mu_k g}{2R}. \tag{3.128}$$

Kinematics of rolling: Let the body stop sliding at time t. If v and ω are the linear and angular velocity of the body at time t, respectively, they are given as

$$v = a_C t \tag{3.129}$$

$$\omega = \alpha t. \tag{3.130}$$

Using equations (3.127) and (3.129) we have

$$v = \mu_k gt. \tag{3.131}$$

Using equations (3.128) and (3.130) we have

$$\omega = \frac{5\mu_k gt}{2R}. \tag{3.132}$$

As the point P stops sliding

$$v_P = (R\omega + v) = v_0. \tag{3.133}$$

Putting the values of v and ω from equations (3.131) and (3.132), respectively, in equation (3.133) and simplifying the factors, we have

$$t = \frac{2v_0}{5\mu_k g}.$$

(b)
  (i) Putting the obtained value of the time t in equation (3.131), the final velocity of the center of mass of the ball relative to the ground is

  $$v = \frac{2v_0}{5}.$$

  (ii) Since the conveyor belt moves with a velocity $\vec{v}_{bg} = v_0 \hat{i}$, relative to the belt the sphere moves back with a velocity

  $$\vec{v}_{sb} = \vec{v}_{sg} - \vec{v}_{bg} = \frac{2v_0}{5}\hat{i} - v_0\hat{i} = -\frac{3v_0}{5}\hat{i}.$$

(c)

(i) As the center of mass of the sphere speeds up from 0 to $v = \frac{2v_0}{5}$, the distance covered by its center of mass relative to the ground is

$$S_{sg} = \frac{2v_0^2}{25\mu_k g}.$$

(ii) The distance covered by the ball relative to the belt is $s_{sb} = s_{sg} - s_{bg}$, where $s_{bg} = v_0 t = v_0(\frac{2v_0}{5\mu_k g}) = \frac{2v_0^2}{5\mu_k g}$. Then, putting the value of s_{sg}, we have $s_{sb} = -\frac{8v_0^2}{25\mu_k g}$.

Problem 49 A rod of mass m and length l is placed on a rough horizontal surface. The coefficient of friction between the rod and surface is μ_k. If the rod is given a horizontal impulse I at one of its ends normal to its length, find (a) the instantaneous axis of rotation, (b) the (i) frictional force and (ii) acceleration of the center of mass of the rod, and (c) the (i) net torque and (ii) angular acceleration of the rod at the given position.

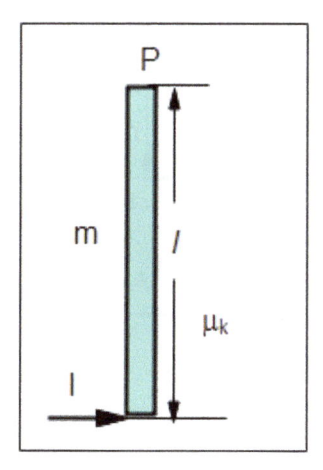

Solution
(a) The position of the instantaneous axis of rotation (point O) is given as

$$x = \frac{K^2}{y} = \frac{\frac{l^2}{12}}{\frac{l}{2}} \Rightarrow x = \frac{l}{6}$$

from the center of mass C.

(b)

(i) Let us take an element of thickness dr which experiences a frictional force

$$df = \mu_k dmg \hat{i}.$$

Then the net frictional force is

$$\vec{F}_x = \int d\vec{f} = \int \mu_k dmg \hat{i} = \int \mu_k g \left(\frac{m}{l} dr\right) \hat{i}.$$

Integrating from $r = OQ = 2l/3$ to $r = OP = l/3$,

$$\vec{F}_x = \left|\mu_k g \frac{m}{l} \int_{\frac{2l}{3}}^{\frac{l}{3}} dr\right| \hat{i}$$

$$= \mu_k g \frac{m}{l}\left(\frac{2l}{3} - \frac{l}{3}\right) \hat{i} = \frac{\mu_k mg}{3} \hat{i}.$$

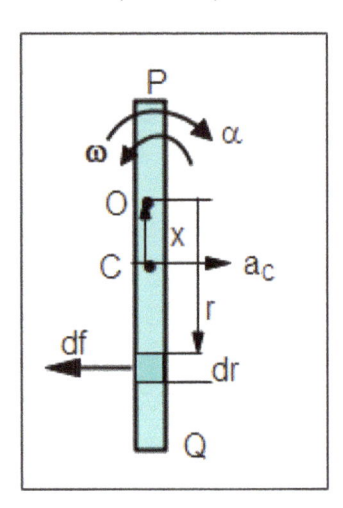

(ii) Then

$$\vec{a}_C = \vec{F}_x/m = \frac{\mu_k g \hat{i}}{3}.$$

(c)

(i) The frictional torque due to the element of mass dm about the instantaneous axis of rotation can be integrated to obtain the net frictional torque

$$(T_f)_O = \int r df = \mu_k g \frac{m}{l} \left\{ \int_0^{\frac{l}{3}} r dr + \int_0^{\frac{2l}{3}} r dr \right\}$$

$$= \mu_k g \frac{m}{l} \left(\frac{l^2}{18} + \frac{4l^2}{18}\right)$$

$$\Rightarrow (\vec{\tau_f})_O = \frac{5\mu_k mgl}{18}\hat{k}.$$

(ii) Then the angular acceleration is

$$\vec{\alpha} = (\vec{\tau_f})_O/I_O,$$

where $I_O = I_C + m(l/6)^2 = ml^2/12 + ml^2/36 = ml^2/9$

$$\Rightarrow \vec{\alpha} = -\frac{5\mu_k mgl}{18} \times \frac{9}{ml^2}\hat{k}$$

$$\Rightarrow \vec{\alpha} = -\frac{5\mu_k g}{2l}\hat{k}.$$

Problem 50 A rod of mass m and length l fitted with a small ring is placed on a rough horizontal surface. The coefficient of friction between the rod and surface is μ_k. If the rod is given a horizontal velocity v_0 normal to its length, it collides inelastically with a rigid peg P. Find (a) the time after which the rod will just stop after the collision, (b) the reaction force perpendicular to the rod given by the peg, (c) the reaction force along the rod given by the peg, and (d) the change in acceleration of the rod.

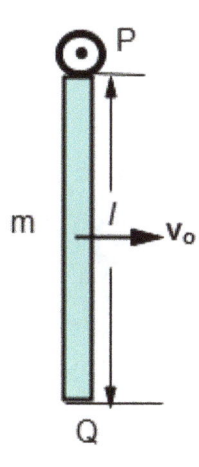

Solution
(a) Let the rod attain an angular velocity ω_0 just after the collision. Conserving the angular momentum about the peg P, just before and after the collision,

$$mv_0\frac{l}{2} = \frac{ml^2}{3}\omega$$

3-127

$$\Rightarrow \omega = \frac{3v_0}{2l}. \tag{3.133a}$$

The net frictional torque about the peg due to an element dm at a distance x from P is $d\tau = \mu_k dm xg$, where $dm = (m/l)x$. Then the net frictional torque about P is

$$\tau = \mu_k g \frac{m}{l} \int_0^l x dx$$

$$\Rightarrow \tau_f = \mu_k gm \frac{l}{2}.$$

Then the angular acceleration of the rod is

$$\vec{\alpha} = \frac{\vec{\tau_f}}{\frac{ml^2}{3}} = \frac{(\mu_k mgl/2)}{\frac{ml^2}{3}} = \frac{3\mu_k g}{2l}\hat{k}.$$

As the rod will stop after a time t, putting $\vec{\omega_f} = 0$ in the equation,

$$\vec{\omega_f} = \vec{\omega_i} + \vec{\alpha} t,$$

we have

$$0 = \omega \hat{k} - \frac{3\mu g}{2l} t \hat{k} \Rightarrow t = \frac{2\omega l}{3\mu_k g} = \frac{3v_0}{3\mu_k g} = \frac{v_0}{\mu_k g}$$

($\because 2\omega l = 3v_0$ in equation (3.133a)).

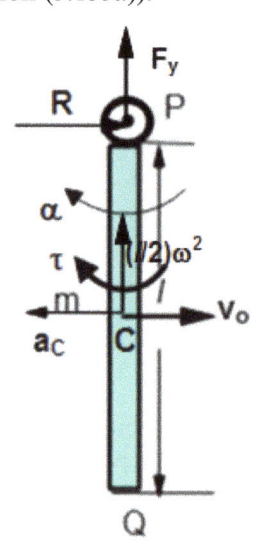

3-128

(b) If R is the reaction force normal to the rod,

$$\vec{F}_{net} = -\mu_k mg\hat{i} + \vec{R} = m\vec{a}_C$$

$$= -m\frac{l}{2}\alpha\hat{i} = -\frac{ml}{2} \times \left(\frac{3\mu_k g}{2l}\right)\hat{i}$$

$$\Rightarrow \vec{R} = \mu_k mg\hat{i} - \frac{3\mu_k mg}{4}\hat{i} = \frac{\mu_k mg}{4}\hat{i} = \vec{F}_x \text{ (say)}.$$

(c) If F_y is the reaction force along the rod,

$$F_y = ma_{C_r} = m\frac{l}{2}\omega^2 = m\frac{l}{2}\left(\frac{3v_0}{2l}\right)^2.$$

Then

$$\vec{F}_y = \frac{3mv_0^2}{8l}\hat{j}.$$

(d) The change in acceleration of the center of mass is

$$\Delta\vec{a} = \vec{a}_{C_f} - \vec{a}_{C_i}$$

$$= -\frac{l}{2}\alpha\hat{i} - (-\mu_k g\hat{i}) \quad \left(\because \alpha = \frac{3\mu_k g}{2l}\right)$$

$$= -\frac{3}{4}\mu_k g\hat{i} + \mu_k g\hat{i} = \frac{\mu_k g}{4}\hat{i}.$$

This change in acceleration is due to the pivot force F_x.
Then $\vec{F}_x = m(\Delta\vec{a}) = \frac{\mu_k mg}{4}\hat{i}.$

Problem 51 A wedge of mass m with a semi-cylindrical cut of radius of curvature R is placed on a horizontal surface. A cylinder of mass m and radius r is resting on the wedge. The coefficient of kinetic friction between all contacting surfaces is μ_k. If the bodies (wedge and cylinder) are given velocities v_1 and v_2, respectively, find the acceleration of the bodies at the given position. Assume that the cylinder does not slip relative to the wedge.

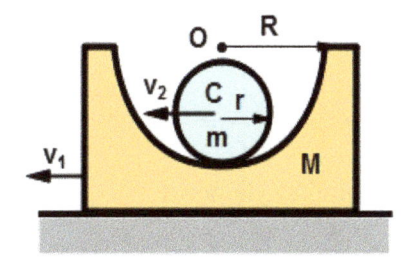

Solution

The kinetic friction acts towards the right on the wedge from the ground because the wedge moves towards the left. Let A be the acceleration of the wedge which points towards the right. Then, relative to the wedge, the horizontal pseudo-force mA acts on the cylinder towards the left. Let the static friction f_s act on the cylinder towards the right, then another static friction of the same magnitude will act on the wedge towards the left due to the cylinder. Since the friction, gravity, and normal reaction pass through the lowest point of the cylinder, their torques will be zero about the lowest point or point of contact of the cylinder. Then the torque produced by the pseudo-force mA about the point of contact P of the rolling cylinder is given as

$$mAr = I\alpha = \frac{3mr^2}{2}\alpha.$$

Then

$$\alpha = \frac{2A}{3r}. \tag{3.134}$$

The torque of the static friction f_s acting on the cylinder about its center of mass C is

$$\tau_C = I_C \alpha$$

$$\Rightarrow f_s r = \frac{mr^2}{2}\alpha$$

$$\Rightarrow f_s = mr\alpha. \tag{3.135}$$

From equations (3.134) and (3.135)

$$f_s = -mr\frac{2A}{3r} = \frac{2}{3}mA. \tag{3.136}$$

The net vertical force acting on the cylinder is

$$F_y = N - mg = ma_y,$$

3-130

where a_y = the vertical acceleration of the cylinder is

$$a_y = \frac{(v_1 - v_2)^2}{R - r}.$$

Then we have

$$N = m\left\{\frac{(v_1 - v_2)^2}{R - r} + g\right\}. \tag{3.137}$$

The net horizontal force acting on the wedge is

$$f_k - f_s = MA$$

$$\Rightarrow \mu_k N' - f_s = MA. \tag{3.138}$$

The net vertical force acting on the wedge is

$$F_y = N' - Mg - N = 0$$

$$\Rightarrow N' = Mg + N. \tag{3.139}$$

Using the equations (3.136), (3.138), and (3.139)

$$\mu_k\{Mg + N\} - \frac{2}{3}mA = MA$$

$$\Rightarrow \left(M + \frac{2}{3}m\right)A = Mg + N. \tag{3.140}$$

Using equations (3.137) and (3.140)

$$\left(M + \frac{2m}{3}\right)A = Mg + m\left(g + \frac{(v_1 - v_2)^2}{R - r}\right).$$

Then

$$A = \frac{(M + m)g + m\frac{(v_1 - v_2)^2}{R - r}}{M + (2m/3)}.$$

The acceleration of m is

$$\vec{a}_m = \vec{a}_x + \vec{a}_y$$

$$= \frac{f_s}{m}\hat{i} + \frac{|v_1 - v_2|^2}{R - r}\hat{j},$$

where

$$f_s = \frac{2}{3}mA$$

3-131

$$\Rightarrow \vec{a}_m = \frac{2}{3} A \hat{i} + \frac{|v_1 - v_2|^2}{R - r} \hat{j},$$

where

$$A = \frac{(M + m)g + \frac{m|v_1 - v_2|^2}{R - r}}{M + \frac{2m}{3}}.$$

Problem 52 Two beads, each of mass m, can slide freely along a fixed L-shaped smooth rod fixed in a horizontal plane above the ground. The beads are connected by a rigid rod of mass m and length l. If anticlockwise torque τ acts on the rod, find (a) the angular acceleration of the rod and (b) the reactions offered by the L-shaped rod on the beads at the ends (i) A (ii) B of the rod.

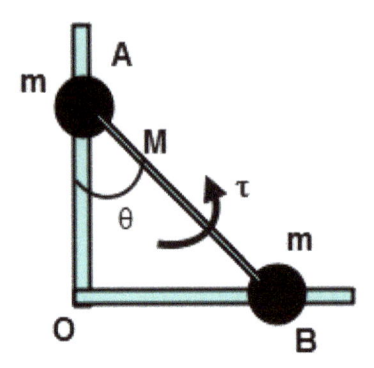

Solution
(a) Let the horizontal reactions offered by the L-shaped rod to the beads be N_1 and N_2, respectively. Furthermore, let a_x and a_y be the accelerations of the center of mass C of the rod–particle system, parallel to the OB and OA axes, respectively. If α is the angular acceleration of the rod, referring to the FBD (free-body diagram), let us write the following equations.
  The force equations of the center of mass C of the system ($M = 2m$):
  Along the x-axis
$$N_1 = (M + 2m)a_x,$$

where $(a_A)_x = a_x - \frac{l}{2}\alpha \cos\theta = 0$

$$\Rightarrow N_1 = (M + 2m)\frac{l}{2}\alpha \cos\theta. \tag{3.141}$$

Along y-axis, the net for is
$$N_2 = (M + 2m)a_y,$$

3-132

where

$$(a_B)_y = a_y - \frac{l}{2}\alpha \sin\theta = 0$$

$$\Rightarrow N_2 = (M+2m)\frac{l\alpha}{2}\sin\theta. \qquad (3.142)$$

The toque equation about C:

$$\tau_C = I_C \alpha = -N_1 \frac{l}{2}\cos\theta \hat{k} - N_2 \frac{l}{2}\sin\theta \hat{k} + \tau \hat{k}$$
$$= \left(\frac{Ml^2}{12} + \frac{ml^2}{2}\right)\alpha. \qquad (3.143)$$

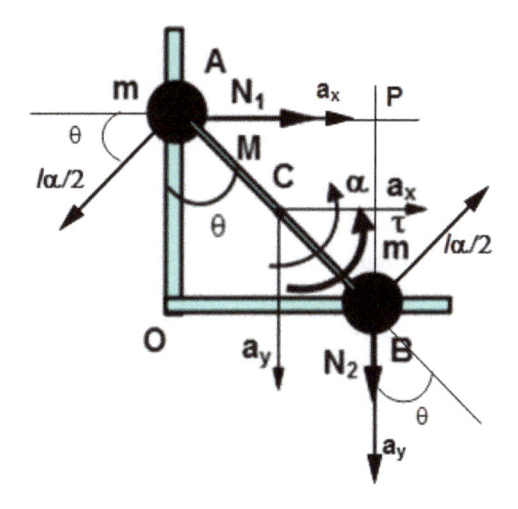

Using equations (3.141), (3.142), and (3.143),

$$\Rightarrow -(M+2m)\frac{l\alpha}{2}\frac{l}{2}\cos^2\theta - (M+2m)\frac{l\alpha}{2}\frac{l}{2}\sin^2\theta$$

$$\tau = \left(\frac{6m+M}{12}\right)l^2\alpha$$

$$\Rightarrow l^2\alpha\left(\frac{6m+M+3M+6m}{12}\right) = \tau$$

$$\Rightarrow l^2\alpha\left(m + \frac{M}{3}\right) = \tau$$

$$\Rightarrow \alpha = \frac{3\tau}{(3m+M)l^2}.$$

3-133

(b)

(i) The reaction force N_1 is

$$N_1 = (M + 2m)\frac{l\alpha}{2}\cos\theta$$

$$= (M + 2m)\frac{l}{2}\frac{3\tau\cos\theta}{(3m + M)l^2}$$

$$\Rightarrow N_1 = \frac{3(M + 2m)\tau\cos\theta}{2(3m + M)l}.$$

(ii) The reaction force N_2 is

$$N_2 = (M + 2m)\frac{l\alpha}{2}\sin\theta$$

$$= (M + 2m)\frac{l\sin\theta}{2}\frac{3\tau}{(3m + M)l^2}$$

$$= \frac{3(M + 2m)\tau\sin\theta}{2(3m + M)l}.$$

Alternative method: As the reactions N_1 and N_2 pass through the instantaneous axis of rotation P, their torques will be zero about P. Then, the net torque about P is equal to the applied torque. Then, dividing this torque by the moment of inertia I_P of the system about P, we can obtain the same value of angular acceleration. You can prove that

$$I_P = \frac{Ml^2}{3} + ml^2.$$

Then the torque equation about P is

$$\tau_P = I_P \alpha$$

$$\Rightarrow \tau = \left(\frac{Ml^2}{3} + ml^2\right)\alpha$$

$$\Rightarrow \alpha = \frac{3\tau}{(3m + M)l^2}.$$

Problem 53 A light rigid rod of length l connected to two particles of masses m_1 and m_2 is placed on a horizontal surface. The coefficient of friction between the particles and surface are μ_1 and μ_2, respectively. If the particles are given velocities v_1 and v_2, respectively, find (a) the acceleration of the center of mass of the system (rod–particles), (b) the torque about the center of mass of the system, (c) the angular acceleration of the rod, (d) the angular momentum about the center of mass, (e) the angular velocity, (f) the rotational kinetic energy, (g) the translational kinetic energy, and (h) the ratio of the rotational and total kinetic energies.

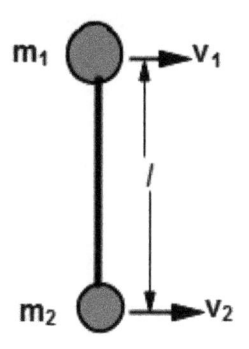

Solution
(a) As the particles move towards the right, to oppose the relative motion, the kinetic friction f_1 and f_2 acting on the particles acts towards the left.
The force equation on the system of rod–particles is

$$\vec{F} = (m_1 + m_2)\vec{a_C}$$

$$\vec{f_1} + \vec{f_2} = (-\mu_1 m_1 g \hat{i} - \mu_2 m_2 g \hat{i}) = (m_1 + m_2)\vec{a_C} \Rightarrow \vec{a_C} = -\frac{(\mu_1 m_1 + \mu_2 m_2)}{m_1 + m_2} g \hat{i}.$$

(b) Taking the torques of the frictional forces about C,

$$\vec{\tau_C} = \{f_1 x - f_2(l - x)\}\hat{k}$$

$$\vec{\tau_C} = \{(\mu_1 m_1 g)(x) - (\mu_2 m_2 g)(l - x)\}\hat{k},$$

where $x = \frac{m_2 l}{m_1 + m_2}$. Then we have

$$\vec{\tau_C} = \left\{(\mu_1 m_1 g)\frac{m_2 l}{m_1 + m_2} - (\mu_2 m_2 g)\frac{m_1 l}{m_1 + m_2}\right\}\hat{k}$$

$$\Rightarrow \vec{\tau_C} = \frac{m_1 m_2 g l}{m_1 + m_2}(\mu_1 - \mu_2)\hat{k}.$$

The torque equation about C is

$$\vec{\tau_C} = I_C \vec{\alpha}$$

$$\Rightarrow \frac{m_1 m_2 l^2}{m_1 + m_2} \vec{\alpha} = \frac{m_1 m_2 lg}{m_1 + m_2}(\mu_1 - \mu_2)\hat{k}$$

$$\Rightarrow \vec{\alpha} = \frac{g(\mu_1 - \mu_2)}{l}\hat{k}.$$

If $\mu_1 > \mu_2$, $\vec{\alpha}$ is anticlockwise.

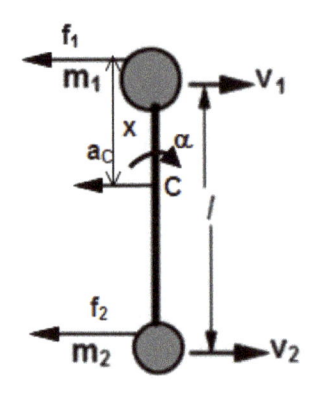

If $\mu_1 = \mu_2$, $\alpha = 0$.
If $\mu_1 < \mu_2$, $\vec{\alpha}$ is clockwise.

(c) The angular momentum of the system about C can be given as

$$\vec{L}_C = -\left\{m_1 v_1 x - m_2 v_2 (l - x)\right\}\hat{k}.$$

Putting $x = m_1 l/(m_1 + m_2)$ and simplifying the factors we have

$$\vec{L}_C = \frac{-m_1 m_2 (v_1 - v_2) l}{m_1 + m_2}\hat{k}.$$

If $v_1 > v_2$, $\vec{L}_C$ is clockwise.
If $v_1 = v_2$, $\vec{L}_C = \vec{0}$.
If $v_1 < v_2$, L_C is anticlockwise.

(d) Putting $\vec{L}_C = \frac{m_1 m_2}{m_1 + m_2} l^2 \vec{\omega}$ we have

$$\vec{\omega} = -\frac{(v_1 - v_2)}{l}\hat{k}.$$

This can also be written directly using the basic formula of relative angular velocity.
1. If $v_1 > v_2$, ω is clockwise.
2. If $v_1 = v_2$, $\omega = 0$.
3. If $v_1 < v_2$, ω is anticlockwise.

(e) The rotational kinetic energy is
$$K_{rot} = K_C = \frac{m_1 m_2}{2(m_1 + m_2)} |v_1 - v_2|^2.$$

(f) The translational kinetic energy is
$$K_{trans} = \frac{1}{2}(m_1 + m_2)v_C^2$$
$$= \frac{1}{2}(m_1 + m_2)\left(\frac{m_1 v_1 + m_2 v_2}{m_1 + m_2}\right)^2$$
$$= \frac{(m_1 v_1 + m_2 v_2)^2}{2(m_1 + m_2)}.$$

(g) Then
$$\frac{K_{rot}}{K} = \frac{m_1 m_2 (v_1 - v_2)^2}{2(m_1 + m_2)\left(\frac{m_1 v_1^2 + m_2 v_2^2}{2}\right)}$$
$$= \frac{m_1 m_2 (v_1 - v_2)^2}{(m_1 + m_2)(m_1 v_1^2 + m_2 v_2^2)}.$$

Problem 54 A light rigid rod PQ of length l connected to two particles of masses m and M at its ends, is placed on a smooth horizontal surface. If the rod is being pulled by a force F at its mid-point C, find the acceleration of (a) the center of mass of the rod–particle system, (b) the angular acceleration of the rod, and (c) the acceleration of the particles at P and Q.

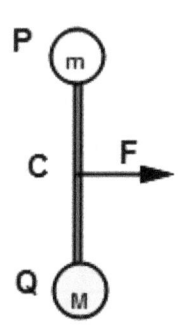

3-137

Solution

(a) The force equation on the rod–particle system is
$$\vec{F} = F\hat{i} = (M+m)\vec{a}_G$$
$$\Rightarrow \vec{a}_G = \frac{F}{M+m}\hat{i}.$$

(b) The torque equation about G is
$$\vec{\tau}_G = I_G \vec{\alpha}$$
$$\Rightarrow -F(CG)\hat{k} = \frac{Mml^2}{M+m}\vec{\alpha}$$
$$\Rightarrow -F(PG - PC)\hat{k} = \frac{Mml^2}{M+m}\vec{\alpha}$$
$$\Rightarrow -F\left(\frac{Ml}{M+m} - \frac{l}{2}\right)\hat{k} = \frac{Mml^2}{M+m}\vec{\alpha}$$
$$\Rightarrow -\left(\frac{M-m}{M+m}\right)\frac{Fl}{2}\hat{k} = \frac{Mml}{M+m}\vec{\alpha}$$
$$\Rightarrow \vec{\alpha} = -\frac{(M-m)F}{2Mml}\hat{k}.$$

(c) As we know,
$$\vec{a}_P = \vec{a}_{PG} + \vec{a}_G = \vec{a}_1 \text{ (say)}$$
$$= \{(PG)\alpha + a_G\}\hat{i}$$
$$= \left\{\frac{Ml}{M+m}\cdot\frac{(M-m)F}{2(Mml)} + \frac{F}{M+m}\right\}\hat{i}$$
$$= \frac{F}{M+m}\left[1 + \frac{M-m}{2m}\right] = \frac{F}{2m}\hat{i}.$$

Similarly, $\vec{a_Q} = \vec{a_{QG}} + \vec{a_G}$

$$\vec{a_Q} = \{-(QG)\alpha + a_G\}\hat{i} = \vec{a_2} \text{ (say)}$$

$$= \left\{-\frac{ml}{M+m} \cdot \frac{(M-m)F}{2(Mml)} + \frac{F}{M+m}\right\}\hat{i}$$

$$= \frac{F}{M+m}\left(1 - \frac{M-m}{2M}\right)$$

$$\Rightarrow \vec{a_Q} = \frac{F}{2M}\hat{i}.$$

Then $\dfrac{a_P}{a_Q} = \dfrac{\frac{F}{2m}}{\frac{F}{2M}} = \dfrac{M}{m}.$

Alternative method:

Let f_2 and f_2 be the reaction (shearing force) between the rod and particles m_1 and m_2, respectively. Forces of the same magnitude and opposite directions are also acting on the rod at its ends. The torque about the center of mass C of the rod is

$$\vec{\tau_C} = I_C \vec{\alpha}$$

$$\Rightarrow \{f_1(PC) - f_2(QC)\}\hat{k} = I_C \vec{\alpha}.$$

Since the rod is light its moment of inertia $I_C = 0$

$$\Rightarrow \{f_1(PC) - f_2(QC)\}\hat{k} = 0.$$

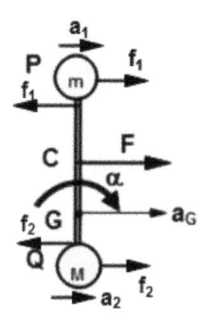

Since C is the mid-point of the rod, $PC = QC$,

$$\Rightarrow f_1 = f_2. \tag{3.144}$$

The force equation on the rod is

$$\Rightarrow F_x = F - f_1 - f_2 = m_{rod} a_C.$$

Since the rod is light, its mass $m_{rod} = 0$

$$\Rightarrow F = f_1 + f_2. \tag{3.145}$$

Using equations (3.144) and (3.145) we have
$$\Rightarrow f_1 = f_2 = F/2.$$
Then the accelerations of m and M can be given as
$$\Rightarrow a_1 = f_1/m = F/2m; \quad a_2 = f_2/M = F/2M$$
acting towards the right.

Problem 55 A uniform rigid rod PQ of mass M and length l connected to a particle of mass m at Q is placed on a smooth horizontal surface. If the rod is being pulled by a force F at Q, find (a) acceleration of the center of mass, (b) the moment of inertia of the rod–particle system about its center of mass, (c) angular acceleration of the rod, (d) the acceleration of P, and (e) acceleration of Q. Put $M = m$.

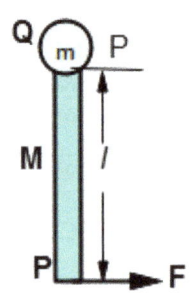

Solution
(a) Let G be the center of mass of the system $(M + m)$.
The force equation on the of rod–particle system:
$$\vec{F} = F\hat{i} = (M + m)\vec{a}_G$$
$$\Rightarrow \vec{a}_G = \frac{F}{M + m}\hat{i}.$$

(b) The moment of inertia of the system about the center of mass G is
$$I_G = I_{\text{rod}} + I_m, \quad I_{\text{rod}} = I_C + M(CG)^2$$
and $I_m = m(QG)^2$. Putting $CG = Ml/\{2(M + m)\}$, $QG = ml/\{2(M + m)\}$, and $I_C = Ml^2/12$, we have

$$I_G = m\left\{\frac{Ml}{2(M+m)}\right\}^2 + \frac{Ml^2}{12} + m\left\{\frac{ml}{(M+m)}\right\}^2$$
$$= \frac{Ml^2}{12} + \frac{Mml^2}{4(M+m)} = \frac{Ml^2}{4}\left[\frac{1}{3} + \frac{m}{M+m}\right]$$
$$= \frac{ml^2}{4}\left[\frac{1}{3} + \frac{1}{2}\right] = \frac{5ml^2}{24}(\because M = m).$$

3-140

(c) The torque about G is

$$\tau_G = I_G \alpha$$

$$\Rightarrow F(PG) = I_G \alpha$$

$$\Rightarrow F\left\{\frac{l}{2} + \frac{ml}{2(M+m)}\right\} = \frac{5ml^2}{24}\alpha$$

$$\Rightarrow F\left\{\frac{l}{2} + \frac{l}{4}\right\} = \frac{5ml^2}{24}\alpha$$

$$\Rightarrow \alpha = \frac{24F}{5ml^2}\left(\frac{3l}{4}\right) = \frac{18F}{5ml}.$$

(d) The acceleration of P is

$$\vec{a}_P = \vec{a}_{PG} + \vec{a}_G = (PG)\alpha\hat{i} + a_G\hat{i}$$

$$= \left\{\frac{l}{2} + \frac{ml}{2(M+m)}\right\}\alpha\hat{i} + a_G\hat{i}.$$

Putting $M = m$ we have

$$\vec{a}_P = \left(\frac{l}{2} + \frac{l}{4}\right)\alpha\hat{i} + a_G\hat{i} = \left(\frac{3l}{4}\alpha + a_G\right)\hat{i}$$

$$= \frac{3l}{4} \times \frac{18F}{5ml}\hat{i} + \frac{F}{2m}\hat{i} = \left(\frac{16}{5}\right)\frac{F}{m}\hat{i}$$

(e) Similarly, the acceleration of Q is given as

$$\vec{a}_Q = \vec{a}_{QG} + \vec{a}_G = -(QG)\alpha\hat{i} + a_G\hat{i}$$

$$= -\frac{Ml}{2(M+m)}\alpha\hat{i} + a_G\hat{i}.$$

3-141

Putting $M = m$ we have

$$\vec{a_Q} = \left\{-\frac{l\alpha}{4} + a_G\right\}\hat{i} = \left\{-\left(\frac{l}{4}\right)\frac{18F}{5ml} + \frac{F}{2m}\right\}\hat{i} = -\frac{2F}{5m}\hat{i}.$$

Problem 56 A rigid rod PN of mass m and length l is placed on a smooth horizontal surface. A horizontal force F acts on the rod at Q at an angle $\theta = 37°$ to the rod. If the linear mass density of the rod varies according to the relation

$$\lambda = \lambda_0\left(1 + \frac{x}{l}\right)$$

with a distance x taken from N along the rod, find (a) the position of the center of mass of the rod, (b) λ_0, (c) the moment of inertia of the rod about P, (d) the moment of inertia about the center of mass of the rod, (e) the acceleration of the center of mass of the rod, the (f) angular acceleration of the rod. Put $NQ = l/4$

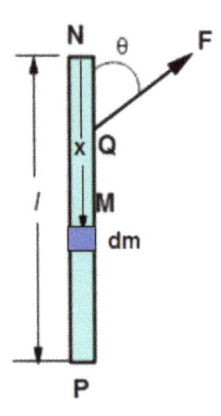

Solution
(a) The position of the center of mass of the rod is given as

$$NG = x_G = \frac{\int x\,dm}{\int dm}$$

$$= \frac{\int_0^l x\lambda_0\left(1 + \frac{x}{l}\right)dx}{\int_0^l \lambda_0\left(1 + \frac{x}{l}\right)dx} = \frac{5}{9}l.$$

(b) The mass of the rod is

$$m = \int_0^l \lambda\,dx = \int_0^l \lambda_0\left(1 + \frac{x}{l}\right)dx = \frac{3}{2}\lambda_0 l$$

$$\Rightarrow \lambda_0 = \frac{2m}{3l}.$$

3-142

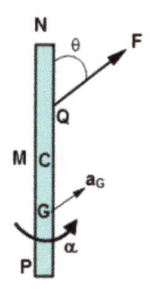

(c) The moment of inertia of the rod about P is

$$I_P = \int x^2 dm = \int_0^l x^2 \lambda_0 \left(1 + \frac{x}{l}\right) dx$$

$$= \lambda_0 \left(\frac{x^3}{3} + \frac{x^4}{4l}\right)\Big|_0^l = \frac{7\lambda_0 l^3}{12}$$

$$= \frac{7}{12}\left(\frac{2m}{3l}\right) l^3 = \frac{7ml^2}{18}.$$

(d) According to the parallel axis theorem

$$I_G = I_P - m(PG)^2.$$

Putting in the values $I_P = 7ml^2/18$ and $PG = 5l/9$, we have

$$= \frac{7ml^2}{18} - m\left(\frac{5l}{9}\right)^2$$

$$= \frac{13}{162} ml^2.$$

(e) Since we have

$$\vec{a_G} = \frac{\vec{F}}{m},$$

using component form we can write

$$\vec{a_G} = \frac{F}{m}(\sin\theta \hat{i} + \cos\theta \hat{j})$$

$$\Rightarrow \vec{a_G} = \frac{F}{m}(3\hat{i}/5 + 4\hat{j}/5) = \frac{F}{5m}(3\hat{i} + 4\hat{j}).$$

(f) The clockwise angular acceleration of the rod has the magnitude

$$\alpha = \frac{\tau_G}{I_G} = \frac{(F\sin\theta)(QG)}{\left(\frac{13}{162}ml^2\right)}$$

$$= \frac{(F\sin\theta)\left(\frac{11}{36}l\right)}{\frac{13}{162}ml^2} = \frac{99F\sin\theta}{26ml}$$

$$= \frac{99F}{26ml} \times \frac{3}{5} = \frac{297F}{130ml}.$$

Problem 57 Two blocks of mass m_1 and m_2 are connected to an inextensible light string that passes over a disc of mass M and radius R which is placed on a smooth table. Let us release the disc–block system from rest. If the string does not slip with the disc, find (a) the tensions in the string, (b) the acceleration of the center of mass of the disc, (c) the angular acceleration of the disc, and (d) the accelerations of the blocks. The masses of A, B, and the disc are 1 kg, 2 kg, and 4 kg, respectively, and $R = 0.25$ m.

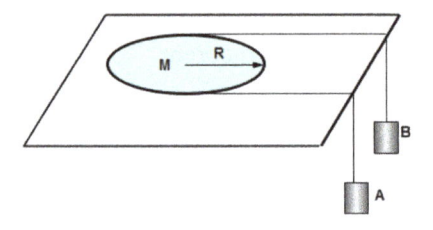

Solution
(a) Since B is heavier than A, B will accelerate down and A will accelerate up. Referring to the FBD, we can write the following equations.
The force equation on M:

$$T_1 + T_2 = Ma. \tag{3.146}$$

The torque equation on M about its center of mass:

$$(T_2 - T_1)R = \frac{MR^2}{2}\alpha. \tag{3.147}$$

The force equation on A:

$$m_1g - T_1 = m_1a_1. \tag{3.148}$$

The force equation on B:

$$m_2g - T_2 = m_2a_2. \tag{3.149}$$

3-144

Kinematics: We know that
$$a_1 = a - R\alpha \tag{3.150}$$
$$a_2 = a + R\alpha. \tag{3.151}$$
Using equations (3.150) and (3.151)
$$\Rightarrow a_1 + a_2 = 2a. \tag{3.152}$$
Putting the accelerations a_1, a_2, and a from equations (3.146), (3.148), and (3.149), respectively, in equation (3.152), we have

$$\Rightarrow g - \frac{T_1}{m_1} + g - \frac{T_2}{m_2} = 2\left(\frac{T_1 + T_2}{M}\right)$$

$$\Rightarrow 2g = T_1\left(\frac{1}{m_1} + \frac{2}{M}\right) + T_2\left(\frac{1}{m_2} + \frac{2}{M}\right)$$

$$\Rightarrow T_1\left(\frac{1}{1} + \frac{2}{4}\right) + T_2\left(\frac{1}{2} + \frac{2}{4}\right) = 2g$$

$$\Rightarrow \frac{3T_1}{2} + T_2 = 2g$$

$$\Rightarrow 3T_1 + 2T_2 = 4g. \tag{3.153}$$

From equations (3.151) and (3.152)
$$\Rightarrow a_2 - a_1 = 2R\alpha. \tag{3.154}$$
Using (3.147), (3.148), (3.149), and (3.154) we have

$$\left(g - \frac{T_2}{m_2}\right) - \left(g - \frac{T_1}{m_1}\right) = \left\{\frac{2R(T_2 - T_1)}{MR^2}\right\}2R$$

$$\Rightarrow T_1\left(\frac{1}{m_1} + \frac{4}{M}\right) = T_2\left(\frac{1}{m_2} + \frac{4}{M}\right)$$

3-145

$$\Rightarrow T_1\left(\frac{1}{1}+\frac{4}{4}\right) = T_2\left(\frac{1}{2}+\frac{4}{4}\right)$$

$$\Rightarrow 2T_1 = \frac{3}{2}T_2$$

$$\Rightarrow 4T_1 = 3T_2 \Rightarrow \frac{T_1}{T_2} = \frac{3}{4}. \qquad (3.155)$$

Using equations (3.153) and (3.155)

$$3T_1 + 2\left(\frac{4T_1}{3}\right) = 4g$$

$$\Rightarrow T_1\left(\frac{9+8}{3}\right) = 4 \times 10$$

$$\Rightarrow T_1 = \frac{120}{17} \text{ N and } T_2 = \frac{4T_1}{3} = \frac{4}{3} \times \frac{120}{17} = \frac{160}{17} \text{ N}.$$

(b) The acceleration of the disc is

$$a = \frac{T_1 + T_2}{M} = \frac{\left(\frac{120}{17}+\frac{160}{17}\right)}{4}$$

$$= \frac{280}{4 \times 17} = \frac{70}{17} \text{ m s}^{-2}.$$

(c) The angular acceleration of the disc is

$$\alpha = \frac{2(T_2 - T_1)}{mR} = \frac{2(160 - 120)}{17 \times 4 \times \frac{1}{4}}$$

$$= \frac{2 \times 40}{17} = \frac{80}{17} \text{ rad s}^{-2}.$$

(d) The acceleration of A is

$$a_1 = g - \frac{T_1}{m_1} = 10 - \frac{120}{17 \times 1}$$

$$= \frac{50}{17} \text{ m s}^{-2} \downarrow.$$

The acceleration of B is

$$a_2 = g - \frac{T_2}{m_2} = 10 - \frac{\frac{160}{17}}{2}$$

$$= \frac{170 - 80}{17} = \frac{90}{17} \text{ m s}^{-2} \downarrow.$$

Problem 58 A uniform rod PQ of mass m and length l is smoothly pivoted at P on an inclined plane of angle of inclination θ with the horizontal. The rod can be rotated by an angle $\varnothing$ and released. The coefficient of friction between the rod and surface is μ. (a) Find the maximum value of $\varnothing(=\varnothing_m)$ so that the rod remains in equilibrium after releasing it. (b) If $\varnothing > \varnothing_m$, find the angular (i) acceleration and (ii) velocity of the rod as a function of $\varnothing$.

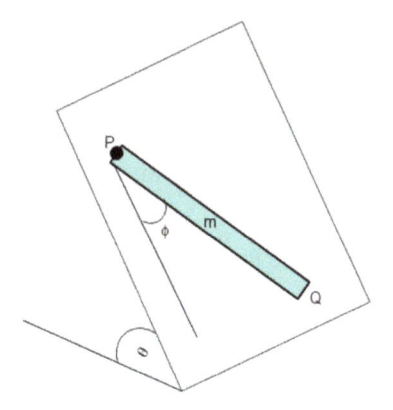

Solution
(a) Gravity acts at the center of mass of the rod. The component of gravity along the line of greatest slope is $mg\sin\theta$. The torque of gravity about P is

$$T_g = mgl \sin\theta \sin\phi/2.$$

The friction due to an element dm of the rod is df which produces a torque xdf about P in opposition to gravitational torque. Then the total frictional torque can be given as

$$T_f = \int x df = \int x \big(\mu dm g \cos\theta\big)\big(\because df = \mu dm g \cos\theta\big)$$

$$= \int_0^l x\mu\left(\frac{m}{l}dx\right)g \cos\theta \left(\because dm = \frac{m}{l}dx\right)$$

$$T_f = \mu mg \frac{l}{2} \cos\theta.$$

Then the net torque is

$$T_{net} = T_{gr} - T_f$$

$$= mg\frac{l}{2} \sin\theta \sin\phi - \mu mg\frac{l}{2} \cos\theta.$$

Let $T_{net} = 0$ at $\phi = \phi_m$

$$\Rightarrow \sin\phi_m = \mu \cot\theta$$

$$\Rightarrow \phi_m = \sin^{-1}(\mu \cot\theta).$$

3-147

(b)

(i) The angular acceleration of the rod is

$$\alpha = \frac{T_{net}}{I_P} = \frac{mg\frac{l}{2}\sin\theta\sin\phi - \mu mg\frac{l}{2}\cos\theta}{m\frac{l^2}{3}}$$

$$\Rightarrow \alpha = \frac{3g}{l}(\sin\theta\sin\phi - \mu\cos\theta).$$

(ii) Furthermore, the angular acceleration is $\alpha = \frac{\omega d\omega}{d\phi} = \frac{3g}{l}(\sin\theta\sin\phi - \mu\cos\theta)$

$$\Rightarrow \int_0^\omega \omega d\omega = \frac{3g}{l}\left\{\sin\theta\int_0^\phi \sin\phi d\phi - \mu\cos\theta\int_0^\phi d\phi\right\}$$

$$\Rightarrow \omega^2 = \frac{3g}{l}\{\sin\theta(1-\cos\phi) - \mu(\cos\theta)\phi\}$$

$$\Rightarrow \omega = \sqrt{\frac{3g}{l}(\sin\theta(1-\cos\phi) - \mu\phi\cos\theta)}.$$

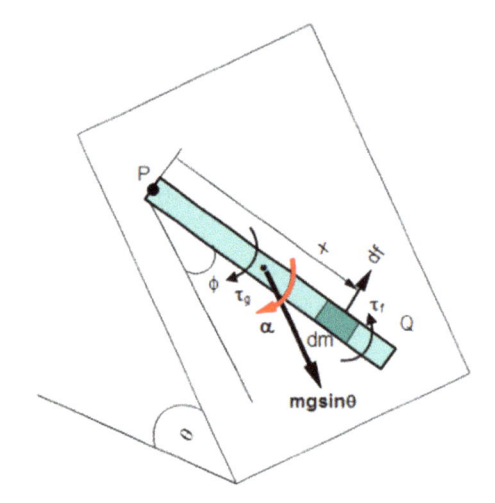

Problem 59 A uniform rod PQ of mass m and length l is placed on a horizontal plane. The coefficient of friction between the rod and surface is μ. The end P is given a velocity v and the other end Q is acted upon by a force F. Find (a) the acceleration of the center of mass of the rod, (b) the net torque about the center of mass of the rod, and (c) the acceleration of (i) P and (ii) Q.

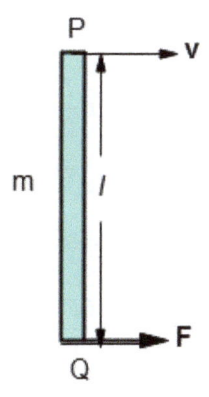

Solution
(a) The net force acting on the rod is

$$\vec{F}_{net} = -\mu_k mg\hat{i} + F\hat{i} = m\vec{a}_C$$

$$\Rightarrow \vec{a}_C = \left(\frac{F}{m} - \mu_k g\right)\hat{i}.$$

(b) The torque about the center of mass is

$$\vec{\tau}_C = (\vec{\tau}_f)_C + (\vec{\tau}_F)_C$$

On each element of the rod the friction is perpendicular to the rod along the plane. Since the frictional force is uniformly distributed along the rod, the net torque due to friction about the center of mass is zero.

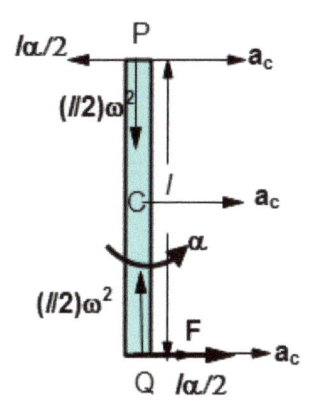

Putting $(\vec{\tau}_f)_C = 0$ and

$$(\vec{\tau}_f)_C = F\frac{l}{2}\hat{k}.$$

3-149

(c) In the last equation we have
$$\Rightarrow \vec{\tau_C} = F\frac{l}{2}\hat{k}.$$
We know that
$$\vec{\tau_C} = I_C \vec{\alpha}$$
$$\Rightarrow \frac{ml^2}{12}\vec{\alpha} = F\frac{l}{2}\hat{k}$$
$$\Rightarrow \vec{\alpha} = \frac{6F}{ml}\hat{k}.$$

(d)
(i) As we know the acceleration of Q is
$$\vec{a_P} = \vec{a_{PC}} + \vec{a_C}$$
$$\Rightarrow \vec{a_P} = \left(-\frac{l}{2}\alpha\hat{i} - \frac{l}{2}\omega^2\hat{j}\right) + \vec{a_C}$$
$$\Rightarrow \vec{a_P} = \left(a_C - \frac{l}{2}\alpha\right)\hat{i} - \frac{l}{2}\omega^2\hat{j}.$$

Putting the obtained values of the above parameters in the last equation, we have
$$= \left(\left(\frac{F}{m} - \mu_k g\right) - \frac{l}{2}\left(\frac{6F}{ml}\right)\right)\hat{i} - \frac{l}{2}\left(\frac{v^2}{l^2}\right)\hat{j}$$
$$\vec{a_P} = -\left(\frac{2F}{m} + \mu_k g\right)\hat{i} - \frac{v^2}{2l}\hat{j}.$$

(ii) Similarly, the acceleration of Q is
$$\vec{a_Q} = \vec{a_{QC}} + \vec{a_C}$$
$$\Rightarrow \vec{a_Q} = \left(\frac{l}{2}\alpha\hat{i} + \frac{l}{2}\omega^2\hat{j}\right) + a_C\hat{i}$$
$$\Rightarrow \vec{a_Q} = \left(a_C + \frac{l}{2}\alpha\right)\hat{i} + \frac{l}{2}\omega^2\hat{j}.$$

Putting the obtained values of the above parameters in the last equation, we have
$$\vec{a_Q} = \left\{\left(\frac{F}{m} - \mu_k g\right) + \frac{3F}{m}\right\}\hat{i} + \frac{l}{2}\left(\frac{v}{l}\right)^2\hat{j}$$
$$= \left(\frac{4F}{m} - \mu_k g\right)\hat{i} + \frac{v^2}{2l}\hat{j}.$$

Problem 60 A uniform rod of mass m and length l is hanging from a ceiling by two light elastic strings. Let the string connected with Q be cut. Find (a) the angular acceleration, (b) the linear acceleration of the center of mass of the plate, (c) the acceleration of A just after cutting the string, and (d) the value of b if the point A of the rod does not accelerate just after cutting the string.

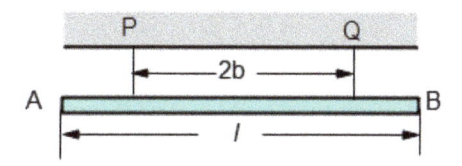

Solution
(a) Just after cutting the string, the tension in the other string PM does not change immediately after cutting the string because it is extensible. As the initial tension in each string is equal to mg/2, this tension in the string PM produces a torque about C, which is given as

$$\left(\frac{mg}{2}\right)(b) = \frac{ml^2}{12}\alpha$$

$$\Rightarrow \alpha = \frac{6gb}{l^2}.$$

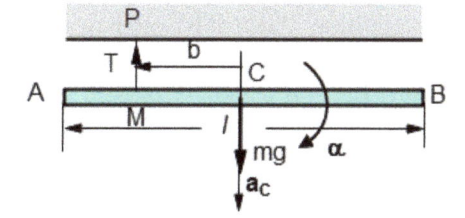

(b) *Force equation*: Let the tension in the string PM be T and the acceleration of C be a_C. Then the net downward force is

$$mg - T = ma_C. \tag{3.156}$$

Putting $T = mg/2$ we have

$$a_C = mg - (mg/2)m = g/2 \downarrow.$$

(c) The acceleration of A is

$$\vec{a_A} = \vec{a_{AC}} + \vec{a_C} = \frac{l}{2}\alpha\hat{j} - a_C\hat{j}$$

$$= \left\{(l/2)\left(\frac{6gb}{l^2}\right) - \frac{g}{2}\right\}\hat{j} = \left(\frac{3gb^2}{l^2} - \frac{g}{2}\right)\hat{j}.$$

3-151

(d) If the acceleration of A is zero just after cutting the string, putting $\vec{a_P} = 0$, we have

$$\frac{3gb^2}{l^2} = \frac{g}{2}$$

$$\Rightarrow b = \sqrt{\frac{gl^2}{6 \times g}} = \frac{l}{\sqrt{6}}.$$

Problem 61 A uniform rod of mass m and length l is hanging from a ceiling by two light inextensible strings. Let us hold the end A of the rod when the rod remains horizontal and the left-hand string is vertical. If we cut the right-hand string and release the rod from the given position, find, just after releasing the rod, (a) the tension in the other string, (b) the acceleration of the center of mass of the rod, and (c) the angular acceleration of the rod.

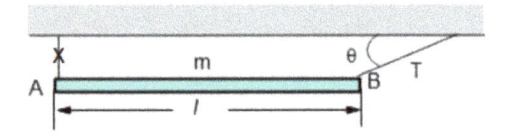

Solution
(a) Taking the torque equation about the center of mass C of the rod,

$$(T \sin \theta)\frac{l}{2} = I_C \alpha, \text{ where } I_C = m^2/12$$

$$\Rightarrow T \sin \theta = \frac{ml\alpha}{6}. \qquad (3.157)$$

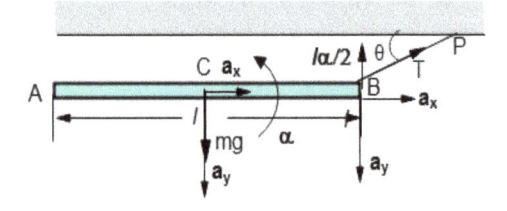

Force equations:

$$T \cos \theta = ma_x \qquad (3.158)$$

$$mg - T \sin \theta = ma_y. \qquad (3.159)$$

The accelerations of the points P and B along the string PB are equal because the string is inextensible. Since P is stationary, the acceleration of

3-152

the point B along the string must be zero. Equating, the components of the accelerations of B along the string PB, we have

$$a_y \sin \theta = a_x \cos \theta + \frac{l\alpha}{2} \sin \theta. \qquad (3.160)$$

Putting the values of a_x, a_y, and α from equations (3.157), (3.158), and (3.159) in equation (3.160), we have

$$\left(g - \frac{T}{m} \sin \theta\right) \sin \theta = \frac{T \cos^2 \theta}{m} + \frac{3T \sin^2 \theta}{m}$$

$$\Rightarrow g \sin \theta = \frac{T}{m}[\sin^2 \theta + \cos^2 \theta + 3 \sin^2 \theta]$$

$$\Rightarrow T = \frac{mg \sin \theta}{1 + 3 \sin^2 \theta}.$$

(b) Then, putting T in equation (3.158), we have

$$a_x = \frac{g \cos \theta \sin \theta}{1 + 3 \sin^2 \theta}.$$

Putting T in equation (3.159) we have

$$a_y = mg \frac{[1 + 3 \sin^2 \theta - \sin^2 \theta]}{m(1 + 3 \sin^2 \theta)}$$

$$\Rightarrow a_y = \frac{1 + 2 \sin^2 \theta}{1 + 3 \sin^2 \theta} g.$$

The acceleration of C is the vector sum of these components of accelerations, given as

$$\vec{a}_C = \frac{g}{1 + 3 \sin^2 \theta} \{\sin \theta . \cos \theta \hat{i} - (1 + 2 \sin^2 \theta) \hat{j}\}.$$

(c) Putting the value of

$$T = \frac{mg \sin \theta}{1 + 3 \sin^2 \theta}$$

in equation (3.157) we have

$$\alpha = \frac{6T \sin \theta}{ml} = \frac{6mg \sin^2 \theta}{(1 + 3 \sin^2 \theta)ml}$$

$$\Rightarrow \alpha = \frac{6g \sin^2 \theta}{(1 + 3 \sin^2 \theta)l}.$$

Problem 62 A uniform disc of mass m, radius r, and radius of gyration K is placed on a smooth horizontal surface. A horizontal force F acts at a distance r from the center C of the body. Find (a) the angular acceleration of the disc, (b) the linear acceleration of the center of mass C, (c) the velocity of the lowest point of the disc as a function of time, and (d) the value of r so that the lowest point of the disc will be at rest.

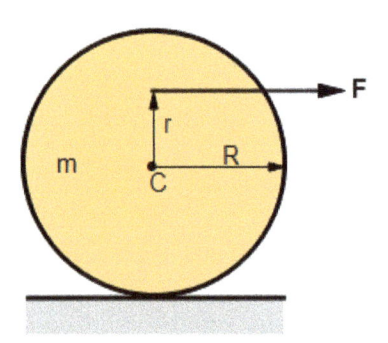

Solution
(a) *Torque equation*:
Since the weight mg and normal reaction N pass through the center of mass C, they cannot produce a net torque about C. Then the net torque about C is produced by the applied force F, which can be given as

$$\tau_C (=Fr) = I_C \alpha,$$

where

$$I_C = mR^2/2.$$

Then we have

$$\alpha = \frac{2Fr}{mR^2}.$$

(b)
Force equation:
Since the center of mass does not move in the y-axis, $a_y = 0$. Hence, the net vertical force is

$$F_y = N - mg = 0.$$

This gives $N = mg$. The net horizontal force is $F_x = Fi = ma_x$. So $a_C = \frac{Fi}{m}$ pointing towards the right.

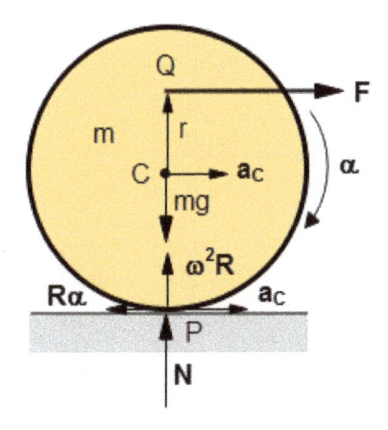

(c) The velocity of the lowest point of the disc is given as $\vec{v}_P = \vec{v}_{PC} + \vec{v}_C$, where $\vec{v}_C = \vec{a}_C t = \frac{F\hat{i}}{m}$ and $\vec{v}_{PC} = -R\omega\hat{i} = -R\alpha t\hat{i} = R\left(\frac{2Frt}{mR^2}\right)\hat{i} = -\frac{2Frt}{mR}$. Then, $\vec{v}_P = \frac{Ft\hat{i}}{m}\left(1 - \frac{2r}{R}\right)$.

(d) Putting $v_P = 0$ for rolling without sliding, we have $(1 - \frac{2r}{R}) = 0$. Then $r = \frac{R}{2}$.

Problem 63 A uniform rod of mass m and length l is smoothly pivoted with a block of mass M which is placed on a horizontal surface. When the rod is released from rest from its near vertical position, it rotates in the vertical plane. If the block does not slip on the surface, find (a) the angular acceleration, (b) the angular speed of the rod, (c) the friction acting on the block, (d) the normal reaction acting on the block by the surface, and (e) the coefficient of static friction as a function of the angle rotated by the rod.

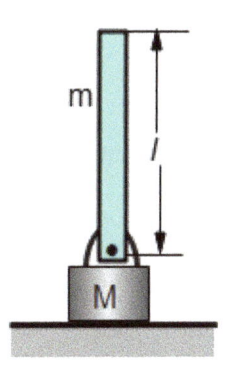

Solution
(a) As the block does not slip, static friction f_s prevails. The torque acting on the rod about its pivot P is

$$T_P = I_P\alpha \Rightarrow mg\frac{l}{2}\sin\theta = \frac{ml^2}{3}\alpha$$

$$\Rightarrow \alpha = \frac{3g\sin\theta}{l}. \tag{3.161}$$

(b) Conserving energy, we have

$$\frac{1}{2}I_P\omega^2 = mg\frac{l}{2}(1-\cos\theta)$$

$$\Rightarrow \frac{1}{2}\frac{ml^2}{3}\omega^2 = mg\frac{l}{2}(1-\cos\theta)$$

$$\Rightarrow \omega = \sqrt{\frac{3g}{l}(1-\cos\theta)}. \tag{3.162}$$

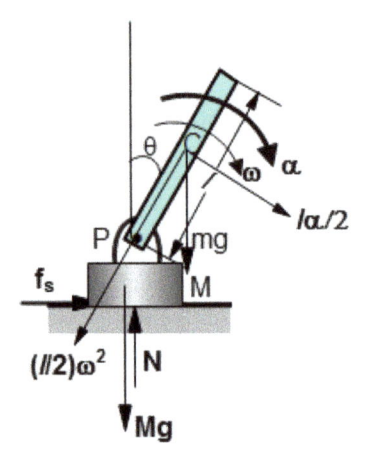

(c) The net horizontal force acting on the rod is

$$f_s = m\left(\frac{l}{2}\alpha\cos\theta - \omega^2\frac{l}{2}\sin\theta\right). \tag{3.163}$$

Using equations (3.161), (3.162), and (3.163) we have

$$f_s = \frac{3mg}{4}(3\cos\theta - 2)\sin\theta.$$

(d) The vertical force acting on the system $(M+m)$ is

$$F_y = (M+m)g - N = (M+m)a_y. \tag{3.164}$$

Kinematics: We can write

$$a_y = -\left(\frac{\omega^2 l}{2}\cos\theta + \frac{l\alpha}{2}\sin\theta\right). \tag{3.165}$$

Using equations (3.164) and (3.165) we have

$$N = (M + m)\left[g - \frac{l}{2}(\omega^2 \cos\theta + \alpha \sin\theta)\right]. \qquad (3.166)$$

Using equations (3.161), (3.162), and (3.166) we have

$$N = (M + m)\left[g - \frac{l}{2}\frac{3g}{l}(1 - \cos\theta)\cos\theta\left\{+\frac{3g \sin\theta}{2l}\sin\theta\right\}\right]$$

$$= (M + m)g\left[1 - \frac{3}{2}\frac{\{(1 - \cos\theta)(1\cos\theta + 1 + \cos\theta)\}}{2}\right]$$

$$= \frac{(M + m)g}{4}[1 - 6\cos\theta + 9\cos^2\theta]$$

$$\Rightarrow N = \frac{(M + m)g(1 - 3\cos\theta)^2}{4}.$$

(e) The law of static friction is

$$f_s \leq \mu_s N$$

$$\Rightarrow \mu_s \geq \frac{f_s}{N}.$$

Putting the obtained values of f_s and N we have

$$\Rightarrow \mu_s \geq \frac{\frac{3mg}{4}(3\cos\theta - 2)\sin\theta}{\frac{(M+m)g}{4}(1 - 3\cos\theta)^2}$$

$$\Rightarrow \mu_s \geq \left|\frac{3m(3\cos\theta - 2)\sin\theta}{(M + m)(1 - 3\cos\theta)^2}\right|.$$

Problem 64 A light rod AB is balanced on a two pulley–particle system as shown in the figure. On the right-hand side, the pulley has a moment of inertia I and radius R and on the left-hand side, the mass and radius of the pulley are M and R, respectively. There is no slipping of the string with the pulleys. Then, find the value of b/c. Put $M = 2m$. The pulleys can be assumed as uniform discs such that $I = MR^2/2$.

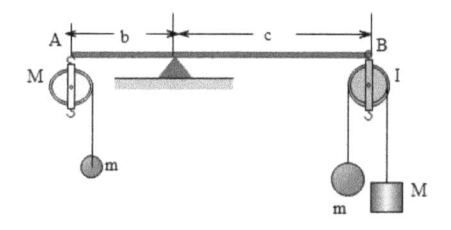

Solution

Let T_1 and T_2 be the reaction forces offered to the rod by the left-hand and right-hand pulley–particle system, respectively. Furthermore, let α_1 and α_2 be the angular accelerations of the left-hand and right-hand pulleys, respectively. Since the rod is in equilibrium, the net torque of the reactions about the pivot is zero. Then $T_1 b = T_2 c$,

$$\text{or } \frac{T_1}{T_2} = \frac{BD}{AD} = \frac{c}{b}. \tag{3.167}$$

For the left-hand pulley–particle system, the net torque of gravity about the pivot is $mgR = \left(mR^2 + \frac{MR^2}{2}\right)\alpha_1$, where

$$\alpha_1 = \frac{a}{R} \Rightarrow a_1 = R\alpha_1 = \frac{2mg}{(M+2m)}.$$

The force acting on the system $(M + m)$ is

$$(M + m) - T_1 = ma_1 = m(R\alpha_1); \text{ or } T_1 = (M+m)g - ma_1 = (M+m)g - m\left(\frac{2mg}{M+2m}\right).$$

Putting $M = 2m$ in the last expression we have

$$T_1 = \frac{5mg}{2}. \tag{3.168}$$

The net torque about the pivot of the right-hand pulley is

$$(M - m)gR = I'\alpha_2$$

$$\Rightarrow \alpha_2 = (M - m)gR/I'.$$

Putting I' = the moment of inertia about the pivot of the right-hand pulley is

$$I' = \frac{MR^2}{2} + (M + m)R^2$$

$$\Rightarrow \alpha_2 = \frac{(M - m)(gR)}{MR^2/2 + (M + m)R^2}.$$

Putting $M = 2m$ in the last expression we have

$$\alpha_2 = \frac{(2m - m)(gR)}{(2m)R^2/2 + (2m + m)R^2} = (g/4R).$$

Then the magnitude of acceleration of the particles M and m is

$$a_2 = \alpha_2 R = \frac{g}{4R} R = g/4.$$

The force equation for the right-hand pulley particle system is

$$T_2 - (M + 3m)g = M(-a_2) + m(a_2) = -(M - m)a_2.$$

3-158

Putting $M = 2m$ in the last expression we have

$$T_2 = (M + 3m)g + (M - m)a_2$$
$$= (2m + 3m)g - (2m - m)g/4 = 19mg/4. \qquad (3.169)$$

Putting $T_1 = \frac{5mg}{2}$ and $T_2 = 19mg/4$.

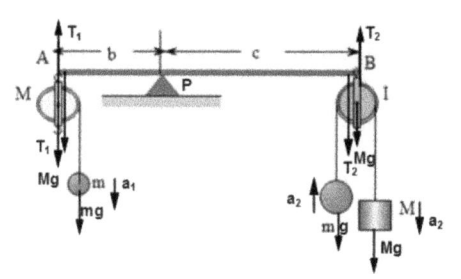

From equations (3.168) and (3.169) in equation (3.167),

$$\frac{c}{b} = \frac{T_1}{T_2} = \frac{\frac{5mg}{2}}{\frac{19}{4}mg} = \frac{10}{19}.$$

The dynamics of rolling bodies on a fixed horizontal surface

Problem 65 A disc of mass m, radius R, and radius of gyration K is pulled by a horizontal force F as shown in the figure of problem 62. The surface is rough so that the disc rolls without sliding. Find (a) the frictional force acting on the disc, (b) the acceleration of the center of mass of the disc, (c) the angular acceleration of the disc, (d) the value of the radial distance r of point of application of the force F at which the body rolls without friction, and (e) the minimum coefficient of friction required for rolling without sliding for $F = mg$, $\frac{K^2}{R^2} = 1/2$ and $r = R$.
Solution

(a) *Torque equation*: Since the disc rolls without sliding, static friction must prevail. As we do not know the exact direction of friction, let us assume that f_s acts backwards. Since the weight $W = mg$ and normal reaction N pass through the center of mass C they cannot produce a torque about C. Then the net torque about C is produced by the applied force F and static friction f_s which can be given as

$$\vec{\tau}_C \left(= -Fr\hat{k} - f_s R\,\hat{k} \right) = I_C \vec{\alpha},$$

where $I_C = mK^2$. Then we have

$$Fr + f_s R = mK^2 \alpha. \qquad (3.170)$$

3-159

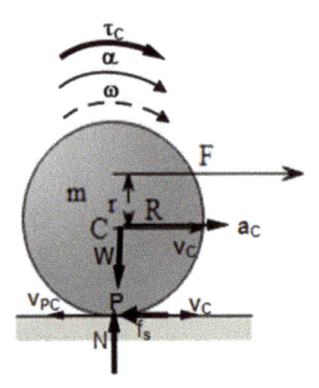

Force equation: Since the center of mass does not move along the y-axis, $a_y = 0$, the net vertical force $F_y = N - mg = ma_y = 0$. This gives

$$N = mg. \tag{3.171}$$

The net horizontal force is

$$F_x = F - f_s = ma_x = ma_C. \tag{3.172}$$

Kinematics of rolling: For pure rolling, the velocity of the lowest point of the disc is

$$\vec{v}_P = \vec{v}_{PC} + \vec{v}_C = 0, \tag{3.173}$$

where $\vec{v}_C = -v_C \hat{i}$ and $\vec{v}_{PC} = -Rw\hat{i}$. This gives

$$v_C = Rw.$$

Differentiating both sides with time,

$$\frac{dv_C}{dt} = R\frac{dw}{dt}$$

or $a_C = R\alpha$. \hfill (3.174)

Take the value of α from equation (3.170) and a_C from equation (3.172) and put these values in equation (3.173), we have

$$\frac{F - f_s}{m} = R\left(\frac{Fr + f_s R}{mK^2}\right).$$

After simplification we have

$$f_s = -F\left(1 - \frac{Rr}{K^2}\right) \Big/ \left(1 + \frac{R^2}{K^2}\right).$$

Putting the value of f_s in equation (3.171) we have

$$a_C = \frac{F}{m\left(1 + \frac{K^2}{R^2}\right)}.$$

Putting the value of a_C in equation (3.173) we have

$$\alpha = -\frac{F}{mR\left(1 + \frac{K^2}{R^2}\right)}.$$

If the static friction will be zero in pure rolling, putting $f_s = 0$, we obtain

$$1 - \frac{Rr}{K^2} = 0.$$

This gives $r = \frac{K^2}{R}$.

Law of friction: according to the law of static friction,

$$f_s \leq \mu_s N. \qquad (3.175)$$

Putting the value of $\frac{K^2}{R^2} = 1/2$ and $r = R$, we have $f_s = F/3$. Putting $N = mg$, $f_s = F/3$, and $F = mg$ in equation (3.175) we have $\mu_s \geq 1/3$:

N.B. 1. If $r < K^2/R$, f_s acts backward.
      2. If $r > K^2/R$, f_s acts backward.
      3. If $r = K^2/R$, $f_s = 0$.

Problem 66 A body of mass m, radius R, and radius of gyration K is released from rest from the given position on an inclined plane so that it rolls without sliding. Find (a) the frictional force, (b) the acceleration of the center of mass of the sphere, and (c) the value(s) of μ_s.

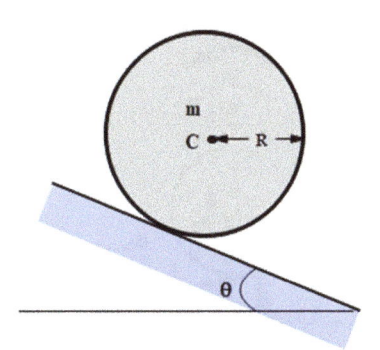

Solution
(a) *Torque equation:* Let us assume that f_s acts upwards along the inclined plane. Since the weight mg and normal reaction N pass through the center

of mass C they cannot produce a torque about C. Then the net clockwise torque is produced by the static friction f_s, which can be given as

$$\tau_C (= f_s R) = I_C \alpha,$$

where

$$I_C = mK^2.$$

Then we have

$$f_s R = mK^2 \alpha. \tag{3.176}$$

(b) *Force equation*: Set x- and y-axes parallel and perpendicular to the inclined plane, respectively. The net force along the inclined plane is

$$F_x = mg \sin\theta - f_s = ma_C. \tag{3.177}$$

Since the center of mass does not move along the y-axis, $a_y = 0$. Then the net vertical force $F_y = N - mg = ma_y = 0$. This gives

$$N = mg \cos\theta. \tag{3.178}$$

Kinematics of rolling: For pure rolling,

$$a_C = R\alpha. \tag{3.179}$$

(c) Putting the value of α from equation (3.176) and a_C from equation (3.177) in equation (3.178) we have

$$\frac{mg \sin\theta - f_s}{m} = R\left(\frac{f_s R}{mK^2}\right). \tag{3.180}$$

After simplification we have

$$f_s = mg \sin\theta \bigg/ \left(1 + \frac{R^2}{K^2}\right).$$

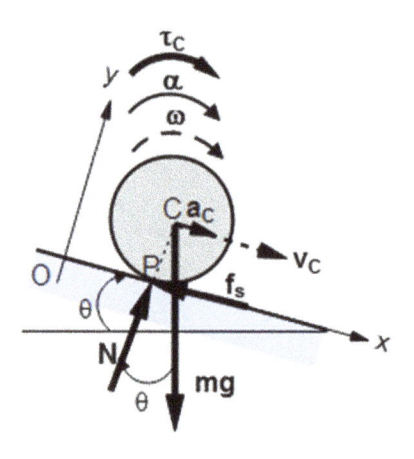

Putting the value of f_s in equation (3.177) we have

$$a_C = \frac{g \sin \theta}{m\left(1 + \frac{K^2}{R^2}\right)} \hat{i}.$$

Law of friction: according to the law of static friction,

$$f_s \leq \mu_s N. \tag{3.181}$$

Putting the value of f_s and $N = mg$ in equation (3.178) we have, $\mu_s \geq \frac{\tan \theta}{(1 + \frac{R^2}{K^2})}$.

- If $\mu_s < \frac{\tan \theta}{(1 + \frac{R^2}{K^2})}$, the body will no longer roll without sliding. So kinetic friction will come into play.
- The direction of f_s always remains upwards even if the center of mass moves up or down.

Problem 67 At time $t = 0$, let a uniform solid sphere of mass m and radius R be projected up with a speed $v_0 = 14$ m s^{-1} along the line of the greatest slope of the inclined plane. Assume that the coefficients of static and kinetic friction between the sphere and inclined plane are $\mu_s = 1/3$ and $\mu_k = 2/7$. The angle of inclination of the plane is $\theta = 37°$ and $g = 10$ m s^{-2}. (a) Will the body roll? (b) What is (i) the velocity and (ii) the acceleration of the point of contact just after projection? (c) After what time will the body (i) start rolling and (ii) attain the highest position? (d) What is the distance covered by the center of mass of the body until it attains the highest position? (e) What is (i) the velocity of the body and (ii) the time when it returns to its initial position?

Solution
(a) *Force equation:*
With reference to the last problem, the minimum coefficient of static friction for rolling without sliding is

$$\mu_s = \frac{\tan \theta}{\left(1 + \frac{R^2}{K^2}\right)} = \frac{\tan 37°}{(1 + 5)} = 1/8.$$

Since the given $\mu_s = 1/3$ is greater than the minimum μ_s, that is, 1/8, the sphere will eventually roll without sliding.

(b)
(i) As the sphere is projected up without any initial spin, the point of contact of the sphere slides up along the plane with a velocity $v_0 = 14$ m s^{-1} just after the projection.

(ii) Then the kinetic friction f_k acts on the body downwards along the slant. So the net downward force parallel to the slant is

$$F_x = -(mg \sin\theta + \mu_k mg \cos\theta) = ma_C.$$

This gives

$$a_C = -g(\sin\theta + \mu_k \cos\theta)$$
$$= -58/7 \text{ (down the slant)}(\text{m s}^{-2}). \tag{3.182}$$

Torque equation:
  The net clockwise torque produced by the kinetic friction f_k is

$$\tau_C(=f_k R) = I_C \alpha,$$

where $I_C = 2mR^2/5$. So $f_k R = \left(\dfrac{2mR^2}{5}\right)\alpha$

or $\alpha = \dfrac{5f_k}{2mR}$, where $f_k = \mu_k mg \cos\theta$.

Then we have

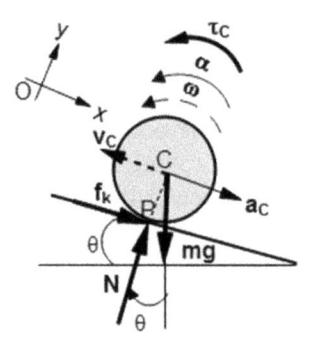

$$\alpha = \dfrac{5\mu_k g \cos\theta}{2R}. \tag{3.183}$$

The acceleration of the point of contact P is

$$\vec{a}_P = \vec{a}_{PC} + \vec{a}_C = -(R\alpha + a_C)\hat{i}. \tag{3.184}$$

Take the values of a_C from equation (3.182) and α from equation (3.183) and put these values in equation (3.184), we have

$$a_P = \dfrac{(2\sin\theta + 7\mu_k \cos\theta)}{2} g = 14 \text{ m s}^{-2} \text{ acting downwards.}$$

(c)
  (i) *Kinematics:*

The clockwise frictional torque builds up the clockwise angular velocity. On the other hand, the net downward force decreases the linear velocity. After a time t, the angular and linear velocities are adjusted so as to satisfy the relation for rolling without sliding, given as

$$v_C = R\omega,$$

where $v_C = v_0 - a_C t$ and $\omega = \alpha t$. Then, we have

$$t = \frac{v_0}{(a_C + R\alpha)}. \tag{3.185}$$

Putting $a_C = \frac{v_0}{(a_C + R\alpha)}$ and $R\alpha = \frac{5\mu_k g \cos\theta}{2}$ in equation (3.185) and simplifying the factors, we have

$$t = \frac{2v_0}{g(2\sin\theta + 7\mu_k \cos\theta)} = 1 \text{ s}.$$

This means that after one second the sphere begins to roll.

(ii) The velocity of the sphere at $t = 1$ s is given as

$$\begin{aligned} v &= v_0 - a_C t \\ &= v_0 - g(\sin\theta + \mu_k \cos\theta)t \\ &= 40/7 \text{ m s}^{-1} \text{ (up)}. \end{aligned}$$

As the sphere starts pure rolling while it is moving up with a velocity $v = 40/7$ m s^{-1}, let it stop after an additional time t_1, say. Then $0 = v - a_1 t_1$. This gives $t_1 = v/a_1$, where $a_1 =$ the downward acceleration of a rolling body on an inclined plane given as

$$a_C = \frac{g \sin\theta}{\left(1 + \frac{K^2}{R^2}\right)} = \frac{30}{7} \text{ m s}^{-2}.$$

Then we have $t = v/a_1 = (40/7)/(30/7) = 4/3$ s. So after a time $T = 1 + 4/3 = 7/3$ s, the sphere will go to its highest position.

(d) The distance traversed by the sphere until it stops relative sliding is

$$s_1 = v_0 t - a_1 t^2/2 = (14)(1) - (58/7)(1/2) = 69/7 \text{ m}.$$

The extra distance traversed by the sphere until it attains the highest position is

$$s_2 = v^2/2a_2 = (40/7)^2/2(30/7) = 80/21 \text{ m}.$$

Then the total distance covered by the body to attain its highest position is

$$s = s_1 + s_2 = 69/7 + 80/21 = (41/3) \text{ m}.$$

(e)
   (i) The speed of the body at the time of attaining its initial position is $v_1 = \sqrt{2a_2 s}$, where $a_2 = 30/7$ m s^{-2} and $s = 41/3$ m. Then $v_1 = \sqrt{820/7}$ m s^{-1}.
   (ii) The time required from the highest position to attain the initial position is $t_3 = v_1/a_2 = \sqrt{574/90}$ s. So the total time after which the body comes back to the point of projection is $T_{\text{total}} = T + t_3 = 1/3(\sqrt{574/10} + 7)$ s.

Problem 68 A uniform sphere of mass m, radius R, and radius of gyration K is placed on a horizontal surface. A horizontal torque τ acts on the sphere. (a) If the surface is smooth, find (i) the angular acceleration of the sphere and (ii) the linear acceleration of the center of the sphere. (b) If the surface is a little rough so that sliding may happen, find (i) the linear acceleration of the center of mass of the sphere, (ii) angular acceleration of the sphere. (c) If the friction is enough to prevent relative sliding, find the (i) anglular acceleration of the sphere (ii) acceleration of the center of mass of the sphere, (iii) static friction. (d) Find the minimum coefficient of static friction required for rolling of the body without sliding. (e) Discuss different cases of the given coefficient of friction between the body and surface.

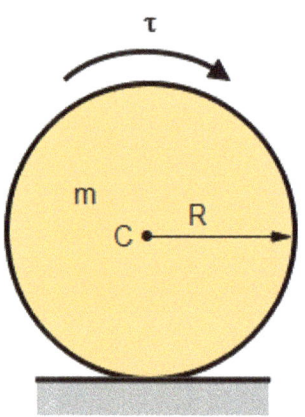

Solution
(a) (i) It the surface is smooth, there is no friction. So frictional torque is zero. Then the net torque about the center of mass is $\tau_C = \tau$

$$\Rightarrow mK^2 \vec{\alpha} = \vec{\tau} = -\tau \hat{k}$$

$$\Rightarrow \vec{\alpha} = \frac{-\tau}{mK^2} \hat{k}.$$

(ii) Since $f_s = 0$, $a_C = 0$.

3-166

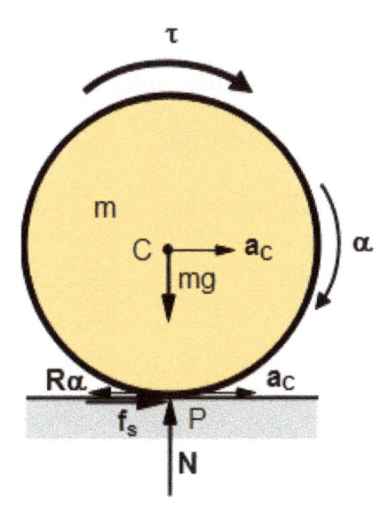

(b) (i) If the surface is a little rough so that sliding will take place, kinetic friction will prevail. The net horizontal force acting on the body is

$$f_k = \mu_k mg = ma_C$$
$$a_C = \mu_k g. \qquad (3.186)$$

(ii) The net torque acting about the center of mass C of the body is

$$\Rightarrow \vec{\tau_C} = I_C \vec{\alpha}$$
$$\Rightarrow -\tau \hat{k} = mK^2 \vec{\alpha}$$
$$\Rightarrow \vec{\alpha} = -\tau \hat{k}/mK^2.$$

Please note that the frictional torque is ignored here in comparison to the applied torque because the friction is very small. So, this is an approximated answer.

(c) (i) If the surface is sufficiently rough so that rolling will take place, then a forward static friction f_s will prevail.

Then, the net torque acting about the lowest point P of the body is

$$\vec{\tau_P} = \vec{\tau_N} + \vec{\tau_{gr}} + \vec{\tau_f} + \vec{\tau}_{applied} = I_C \vec{\alpha}.$$

Since f_s, mg, and N pass through P, their torque about P is zero. Then the net torque about P will be equal to the applied torque:

$$\Rightarrow \vec{\tau_P} = \vec{\tau}_{applied} = I_P \vec{\alpha}$$
$$\Rightarrow -\tau \hat{k} = m(R^2 + K^2)\vec{\alpha}$$

$$\Rightarrow \vec{\alpha} = -\frac{\tau}{m(K^2 + R^2)}\hat{k}. \tag{3.187}$$

(ii) Then the acceleration of the center of mass is

$$\vec{a_C} = R\alpha \hat{i}$$

$$\Rightarrow \vec{a_C} = \frac{\tau R}{m(K^2 + R^2)}\hat{i}.$$

(iii) Putting the obtained value of a_C, the frictional force is

$$f_s = ma_C = m\left\{\frac{\tau R}{m(K^2 + R^2)}\hat{i}\right\}$$

$$\Rightarrow \vec{f_s} = \frac{\tau R}{(K^2 + R^2)}\hat{i}.$$

Alternative method

The net torque acting about the center of mass C of the body is

$$\Rightarrow \vec{\tau_C} = I_C \vec{\alpha}$$

$$\vec{\tau_C} = \vec{\tau_N} + \vec{\tau_{gr}} + \vec{\tau_f} + \vec{\tau}_{applied} = I_C \vec{\alpha}.$$

Since mg and N pass through C, their torque about P is zero. Then the net torque about P will be equal to the sum of the applied torque and frictional torque:

$$\Rightarrow \vec{\tau_P} = \vec{\tau}_{applied} + \vec{\tau_f} = I_C \vec{\alpha}$$

$$\Rightarrow \vec{\tau_P} = \tau(-\hat{k}) + f_s R\hat{k} = -mK^2\alpha\hat{k}$$

$$\Rightarrow \tau - f_s R = mK^2 \alpha.$$

The net horizontal force acting on the body is

$$F_x = f_s = ma_C.$$

For rolling without sliding

$$\vec{a_P} = (a_C - R\alpha)\hat{i} = 0.$$

Using last three equations, we have the same clockwise annular acceleration, given as

$$\Rightarrow \alpha = \frac{\tau}{m(K^2 + R^2)}.$$

(d) The net horizontal force acting on the body is
$$F_y = N - mg = 0.$$
For no slipping the law of static friction is
$$f_s \leqslant \mu_s N$$
$$\Rightarrow \frac{\tau R}{(K^2 + R^2)} \leqslant \mu_s mg$$
$$\Rightarrow \mu_s \geqslant \frac{\tau R}{(K^2 + R^2)mg}.$$

(e) Let the minimum coefficient of static friction required to prevent a relative sliding of the body so that it will roll under the action of the applied torque be called the critical coefficient of friction, given as
$$\frac{\tau R}{(K^2 + R^2)mg} = \mu_C.$$
If the given coefficient of friction is less than this critical value, the sliding will happen and the kinetic friction will come into play. Symbolically,
$$\text{if } \mu < \mu_C, \ \mu = \mu_k.$$
Then, the acceleration of the center of mass of the sphere is
$$\vec{a}_C = f_k/m = \mu_k g \hat{i}.$$
The net torque about C is
$$\vec{\tau}_C = \left(\mu_k mgR - \tau\right)\hat{k} = mK^2 \vec{\alpha}$$
$$\Rightarrow \vec{\alpha} = \left(\frac{\mu_k mgR - \tau}{mK^2}\right)\hat{k}.$$

If the given coefficient of friction is greater than this critical value, the sliding will not happen and the static friction will come into play for rolling without sliding. Symbolically,
$$\text{if } \mu \geqslant \mu_C, \ \mu = \mu_s.$$
Then
$$\vec{a}_C = f_s/m = \mu_s g \hat{i}.$$
Then the angular acceleration of the body can be given as
$$\Rightarrow \vec{\alpha} = -\frac{\tau}{m(K^2 + R^2)}\hat{k}$$
as obtained in the last part of the problem.

Problem 69 A string is wrapped over a hollow cylinder of mass M and radii R and r. The free end of the string is connected to a bob of mass m. If the cylinder is released from rest, it rolls without sliding on the horizontal surface. Find (a) the linear and angular acceleration of the cylinder and bob and (b) the friction acting on the cylinder. Assume that the string does not slip on the cylinder.

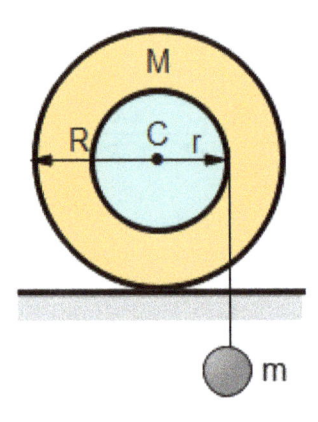

Solution
(a) The net vertical force acting on the rolling body is
$$F_y = N - T - Mg = 0$$
$$\Rightarrow T + Mg = N. \qquad (3.188)$$

The net horizontal force is
$$F_x = f_s = ma_C,$$

where the condition for rolling is
$$a_C = R\alpha$$
$$\Rightarrow f_s = MR\alpha. \qquad (3.189)$$

The net vertical force acting on the particle is $mg - T = ma_y$, where
$$a_y = r\alpha$$
$$\Rightarrow T = m(g - r\alpha). \qquad (3.190)$$

The torque about the point P is
$$\tau_P = Tr = M(R^2 + K^2)\alpha$$
$$T = \frac{M}{r}(R^2 + K^2)\alpha. \qquad (3.191)$$

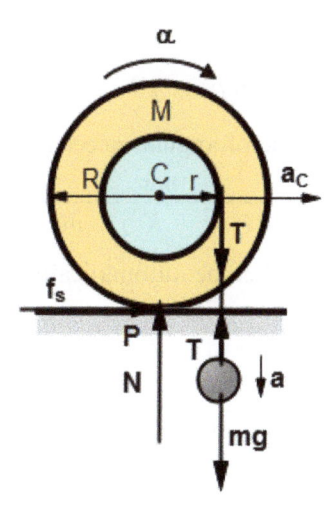

Using equations (3.190) and (3.191) we have

$$\frac{M}{r}(R^2 + K^2)\alpha = m(g - r\alpha)$$

$$\Rightarrow \alpha \left\{ mr + \frac{M(R^2 + K^2)}{r} \right\} = mg$$

$$\Rightarrow m\alpha r \left\{ 1 + \frac{M(K^2 + R^2)}{mr^2} \right\} = mg$$

$$\Rightarrow \alpha = \frac{mgr}{mr^2 + M(K^2 + R^2)}.$$

So the linear acceleration of the cylinder is

$$\Rightarrow a_C = R\alpha = \frac{mgRr}{mr^2 + M(K^2 + R^2)}.$$

(b) According to equation (3.189)

$$f_s = Ma_C = MR\alpha$$

$$\Rightarrow f_s = \frac{MmgRr}{mr^2 + M(K^2 + R^2)}.$$

Alternative method

We can also find the angular acceleration by writing the torque equation about P for the total system $(M + m)$.

The torque about the point P is

$$\tau_P(=mgr) = I_P\alpha,$$

3-171

where I_P can be given as
$$I_P = M(R^2 + K^2) + mr^2.$$
This yields the same value of angular acceleration given as
$$\alpha = \frac{mgr}{M(R^2 + K^2) + mr^2}.$$
The other parameters will come automatically as discussed above.

Problem 70 Let us form a pyramid by slowly pouring dry sand, as shown in the figure. The diameter of the base of the pyramid is D. Find the time taken by a light-weight ball (a spherical grain of sand or a small ball of stone) to roll down from the top of the pyramid to reach the ground. Assume that the rolling body has radius r and radius of gyration K.

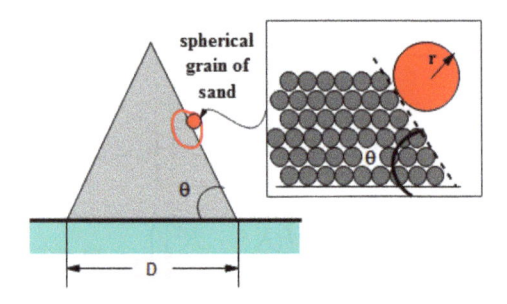

Solution
The slant length of the pyramid is
$$L = \frac{D \sec \theta}{2}. \tag{3.192}$$
As the coefficient of static friction is μ the maximum angle of the pyramid is equal to the angle of repose of the slant which is given as
$$\tan \theta = \mu. \tag{3.193}$$
For the rolling body the acceleration is
$$a = \frac{g \sin \theta}{1 + \frac{K^2}{r^2}}. \tag{3.194}$$
Then the time of rolling is
$$t = \sqrt{\frac{2L}{a}}. \tag{3.195}$$

Using the above four equations, we have

$$t = \sqrt{\frac{d(1+\mu^2)\left(1+\frac{K^2}{r^2}\right)}{\mu g}}.$$

A rolling body on a moving frame

Problem 71 A cylindrical wooden log is kept on a stationary truck, as shown in the figure. If the truck moves with a constant acceleration A, the log rolls without sliding. The coefficient of friction between the log and truck is μ_s. Find (a) the angular acceleration of the log, (b) the acceleration of the log, (c) the friction acting on the log, (d) the possible value(s) of A, and (e) the time after which the log will escape from the truck.

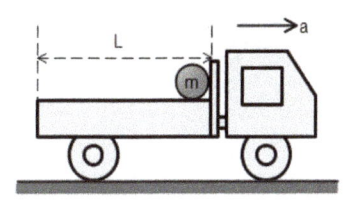

Solution

(a) Let us sit on the accelerating frame (truck) and impose a pseudo-force mA in the backward direction. This force acts at the center of mass C of the log. The net torque about the lowest point P of the rolling body is

$$\tau_P = I_P \alpha.$$

Since N, mg, and static friction f_s pass through the point P, their torque about P will be zero. Then the pseudo-force will produce a counter-clockwise torque, given as

$$mAR = m(K^2 + R^2)\alpha$$

$$\Rightarrow \alpha = \frac{AR}{K^2 + R^2}. \qquad (3.196)$$

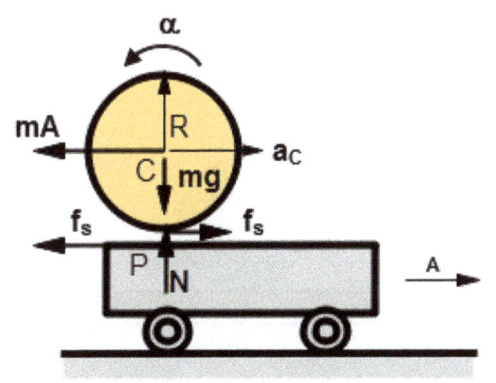

(b) For rolling, the acceleration of the point P (truck) is
$$a_P = A = R\alpha + a_C. \qquad (3.197)$$
Using equations (3.196) and (3.197) we have
$$a_C = \frac{AK^2}{K^2 + R^2}.$$

(c) The torque about the center of mass C is
$$f_s R = mK^2\alpha \qquad (3.198)$$
$$\Rightarrow f_s = m\frac{K^2}{R}\frac{AR}{K^2 + R^2}$$
$$\Rightarrow f_s = \frac{mA}{1 + \frac{R^2}{K^2}}.$$

(d) Since $f_s \leq \mu_s N$ and $N = mg$
$$\Rightarrow \frac{mA}{1 + \frac{R^2}{K^2}} \leq \mu_s mg$$
$$A \leq \left(1 + \frac{R^2}{K^2}\right)\mu_s g.$$

(e) The acceleration of the log relative to the truck is
$$a_{\text{rel}} = R\alpha = \frac{AR^2}{K^2 + R^2}.$$
Then the time it takes to escape from the truck is
$$t = \sqrt{2L/a_{\text{rel}}} = \sqrt{2L(K^2 + R^2)/AR^2}.$$

Problem 72 A uniform cylinder of mass m and radius R is kept on a plank of mass M. The plank rests on the smooth ground. The coefficient of static friction between the cylinder and plank is μ_s. If the plank is pushed by a horizontal force F, the cylinder rolls without sliding on the plank. Find (a) the acceleration of the plank and (b) the possible value(s) of the force F.

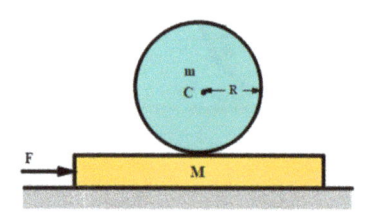

Solution

(a) Let the acceleration of the plank be A.
The net force equation on $(M+m)$ is
$$F = MA + ma_C,$$
where the condition for rolling is
$$a_C = (A - R\alpha).$$
Then we have
$$F = MA + m(A - R\alpha)$$
$$F = (M+m)A - mR\alpha. \tag{3.199}$$

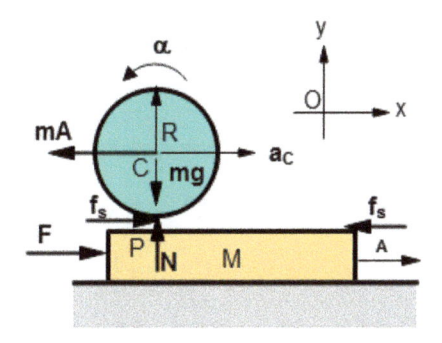

The torque about P due to the pseudo-force mA, relative to the plank is
$$\tau_P = I_P \alpha$$
$$\Rightarrow mAR = m(R^2 + K^2)\alpha$$
$$\Rightarrow \alpha = \frac{AR}{R^2 + K^2}. \tag{3.200}$$

Using equations (3.199) and (3.200)
$$F = (M+m)A - \frac{mAR^2}{R^2 + K^2}$$
$$\Rightarrow F = \left(M + \frac{m}{1 + \frac{R^2}{K^2}}\right) A$$
$$\Rightarrow A = \frac{F(K^2 + R^2)}{M(R^2 + K^2) + mK^2}. \tag{3.201}$$

3-175

(b) As obtained in the last problem,

$$A \leqslant \left(\frac{R^2}{K^2} + 1\right)\mu_s g. \qquad (3.202)$$

Putting the value of A from equation (3.201) in the inequality (3.202) we have

$$\Rightarrow A = \frac{F(K^2 + R^2)}{M(R^2 + K^2) + mK^2}$$

$$\frac{F}{M + \frac{m}{1 + \frac{R^2}{K^2}}} \leqslant \left(\frac{R^2}{K^2} + 1\right)\mu_s g$$

$$\Rightarrow F \leqslant \left\{M\left(\frac{R^2}{K^2} + 1\right) + m\right\}\mu_s g$$

$$\Rightarrow F \leqslant \mu_s M g \left\{\frac{m}{M} + \frac{R^2}{K^2} + 1\right\}.$$

For the solid uniform cylinder, putting $R^2/K^2 = 2$, we have

$$F \leqslant \mu_s M g \left\{\frac{m}{M} + 3\right\}.$$

Problem 73 A uniform solid sphere of mass m and radius R and radius of gyration K is kept on a plank of mass M. The plank is placed on smooth ground. If the horizontal force F acts on the sphere, find (a) the acceleration of the plank and (b) the maximum value of F so that the sphere can roll without sliding on the plank. Assume μ_s as the coefficient of static friction between the sphere and plank.

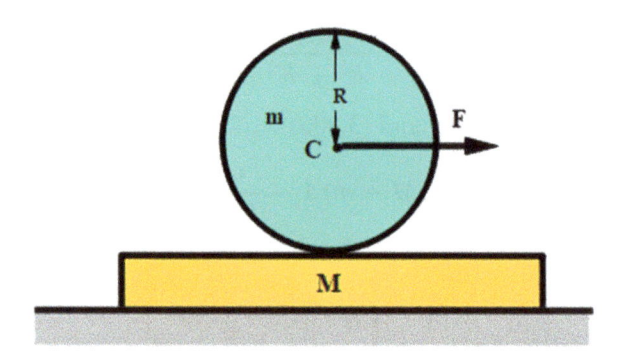

Solution
(a) Let the acceleration of the plank be A.

Force equation on $(M + m)$:

$$F = MA + ma_C,$$

where $a_C = (A + R\alpha)$ for rolling.
Then we have
$$F = MA + m(A + R\alpha)$$
$$F = (M + m)A + mR\alpha. \tag{3.203}$$

The torque about P due to the pseudo-force mA and applied force F, relative to the plank, is
$$\tau_P = I_P \alpha$$
$$\Rightarrow (F - mA)R = m(K^2 + R^2)\alpha. \tag{3.204}$$

From equations (3.203) and (3.204)
$$\Rightarrow F = (M + m)A + mR\left\{\frac{(F - mA)R}{m(K^2 + R^2)}\right\}$$

$$F\left\{1 - \frac{R^2}{K^2 + R^2}\right\} = \left\{M + \left(m - \frac{mR^2}{K^2 + R^2}\right)\right\}A$$

$$\Rightarrow \frac{FK^2}{K^2 + R^2} = \left(M + \frac{mK^2}{K^2 + R^2}\right)A$$

$$\Rightarrow F = \frac{K^2 + R^2}{K^2}\left(M + \frac{mK^2}{K^2 + R^2}\right)A$$

$$\Rightarrow F = \left\{M\left(1 + \frac{R^2}{K^2}\right) + m\right\}A$$

$$\Rightarrow A = \frac{F}{M\left(1 + \frac{R^2}{K^2}\right) + m}.$$

(b) The net force acting on the plank M is
$$f_s = MA.$$
Since $f_s \leqslant \mu_s mg$ (law of static friction) we have
$$\Rightarrow MA \leqslant \mu_s mg$$
$$\Rightarrow A \leqslant \mu_s mg/M.$$
Putting the obtained value of A,
$$\Rightarrow \frac{F}{M\left(1 + \frac{R^2}{K^2}\right) + m} \leqslant \mu_s mg/M$$

3-177

$$\Rightarrow F \leqslant \left\{ M\left(1 + \frac{R^2}{K^2}\right) + m \right\}(\mu_s mg/M)$$

$$\Rightarrow F \leqslant \mu_s mg\left(1 + \frac{R^2}{K^2} + \frac{m}{M}\right).$$

Alternative method: ground frame method

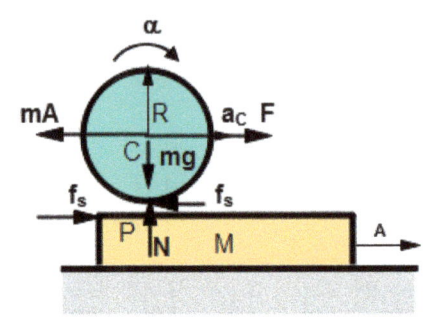

Force equation on m:
$$F - f_s = Ma_C. \tag{3.205}$$

Force equation on M:
$$f_s = MA. \tag{3.206}$$

Torque equation on m relative to C:
$$f_s R = mk^2 \alpha. \tag{3.207}$$

Condition of rolling:
$$a_C - A = R\alpha. \tag{3.208}$$

Using all the above four equations we have

$$\frac{F - f_s}{m} - \frac{f_s}{M} = R \frac{f_s R}{mK^2}$$

$$\Rightarrow f_s \left\{ \frac{1}{m} + \frac{1}{M} + \frac{R^2}{mK^2} \right\} = \frac{F}{m}$$

$$\Rightarrow f_s = \frac{F}{\left(1 + \frac{m}{M} + \frac{R^2}{K^2}\right)}.$$

Since $f_s \leqslant \mu_s mg$ (law of static friction) we have

$$\frac{F}{\left(\frac{1}{m} + \frac{1}{M} + \frac{R^2}{mk^2}\right)} \leqslant \mu_s mg$$

$$\Rightarrow F \leqslant \mu_s mg\left(1 + \frac{m}{M} + \frac{R^2}{K^2}\right).$$

Problem 74 A uniform rolling body of mass m, radius R, and radius of gyration K is pulled by a force F at an angle θ with the horizontal. (a) Find the static friction between the rolling body and plank. (b) Find the magnitude/s of the applied force F so that the body rolls without sliding relative to the plank. (c) Discuss the cases when (i) $M \gg m$ and (ii) the rolling body is replaced by a block of mass m so that it does not slide on the plank.

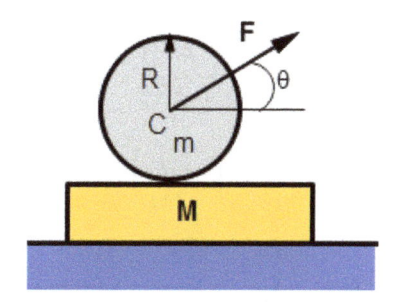

Solution
(a) The horizontal force on m is

$$F \cos \theta - f_s = ma_C. \tag{3.209}$$

The vertical force on m is

$$F_y = N + F \sin \theta - mg = 0$$

$$\Rightarrow N = mg - F \sin \theta. \tag{3.210}$$

The horizontal force on M is

$$f_s = MA. \tag{3.211}$$

The torque on m about its center of mass is

$$f_s R = mK^2 \alpha. \tag{3.212}$$

The condition of rolling is

$$a_C - A = R\alpha. \tag{3.213}$$

The law of static friction is given as

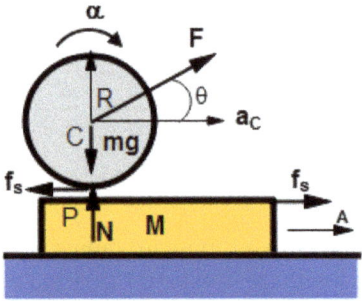

$$f_s \leqslant \mu_s N. \tag{3.214}$$

Using equations (3.209), (3.211), (3.212), and (3.213)

$$\frac{F\cos\theta - f_s}{m} - \frac{f_s}{M} = \frac{f_s R}{mK^2} R$$

$$\Rightarrow f_s\left(\frac{1}{m} + \frac{1}{M} + \frac{R^2}{mK^2}\right) = \frac{F\cos\theta}{m}$$

$$\Rightarrow f_s\left\{\frac{1}{m}\left(1 + \frac{R^2}{K^2}\right) + \frac{1}{M}\right\} = \frac{F\cos\theta}{m}$$

$$\Rightarrow f_s = \frac{F\cos\theta}{m\left\{\frac{1}{m}\left(1 + \frac{R^2}{K^2}\right) + \frac{1}{M}\right\}}$$

$$\Rightarrow f_s = \frac{F\cos\theta}{1 + \frac{m}{M} + \frac{R^2}{K^2}}. \tag{3.215}$$

(b) Using equations (3.214) and (3.215)

$$\frac{F\cos\theta}{1 + \frac{m}{M} + \frac{R^2}{K^2}} \leqslant \mu_s(mg - F\sin\theta)$$

$$\Rightarrow F\left\{\frac{\cos\theta}{\left(1 + \frac{m}{M} + \frac{R^2}{K^2}\right)} + \mu_s\sin\theta\right\} \leqslant \mu_s mg$$

$$\Rightarrow F \leqslant \frac{\mu_s mg}{\frac{\cos\theta}{1 + \frac{m}{M} + \frac{R^2}{K^2}} + \mu_s\sin\theta}.$$

3-180

(c) (i) If $M \to \infty$

$$\Rightarrow F \leqslant \frac{\mu_s mg}{\frac{\cos\theta}{1+\frac{R^2}{k^2}} + \mu_s \sin\theta}.$$

(ii) If m is a rectangular block, no rotation takes place ($K = 0$)

$$\Rightarrow F \leqslant \frac{\mu_s mg}{\frac{\cos\theta}{1+\frac{m}{M}} + \mu_s mg}.$$

Problem 75 A solid sphere of mass m is pulled by a force F so that it rolls without sliding on a horizontal surface. Find the minimum magnitude of the applied force F at the (a) given angle (b) any angle of inclination of the force with the horizontal.

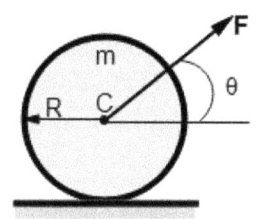

Solution
(a) The net horizontal force on m is

$$F\cos\theta - f_s = ma. \tag{3.216}$$

The vertical force on m is

$$F\sin\theta + N - mg = 0. \tag{3.217}$$

The net torque about the center of mass is

$$f_s R = mk^2 \alpha. \tag{3.218}$$

The condition of rolling is

$$a = R\alpha. \tag{3.219}$$

The law of static friction is

$$f \leqslant \mu_s N. \tag{3.220}$$

Using equations (3.216), (3.218), and (3.219) we have

$$f_s = \frac{F\cos\theta}{1+\frac{R^2}{k^2}}. \tag{3.221}$$

3-181

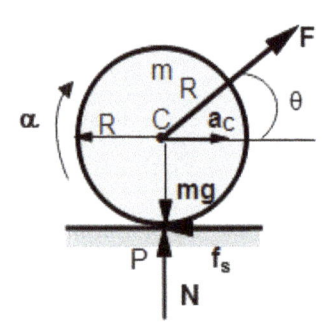

Putting f_s from equation (3.221) and N from equation (3.218) in equation (3.220) and simplifying the terms, we have

$$F \geqslant \frac{\mu mg}{\left(\mu \sin\theta + \frac{K^2 \cos\theta}{K^2 + R^2}\right)}.$$

(b) For the minimum force, the denominator must be a maximum. We can use the mathematical idea that the maximum value of the function

$$f(\theta) = a\sin\theta + b\cos\theta$$

is equal to $\sqrt{a^2 + b^2}$, where

$$\mu = a \text{ and } K^2/(K^2 + R^2) = b$$

$$\Rightarrow F_{min} = \frac{\mu mg}{\sqrt{\mu^2 + \left(\frac{K^2}{K^2 + R^2}\right)^2}}.$$

The rolling body on an inclined plane

Problem 76 A cylinder of mass m rolls down without sliding on a rough prismatic wedge of mass M which is placed on a smooth horizontal surface. Prove that the acceleration of the wedge is given as

$$A = \frac{mg\sin\theta\cos\theta}{3M + m(1 + 2\sin^2\theta)}.$$

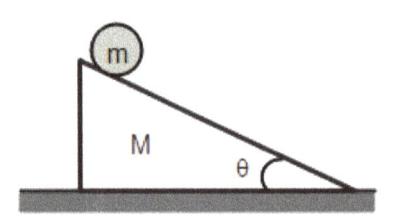

Solution

Let a_{rel} be the acceleration of m relative to M and A be the acceleration of M.
Force equation on $(M + m)$:
Since no net horizontal force acts on the system $(M + m)$, we can write

$$F_x = -MA + m(a_{rel} \cos\theta - A) = 0$$

$$\Rightarrow A = \frac{m a_{rel} \cos\theta}{M + m}. \tag{3.222}$$

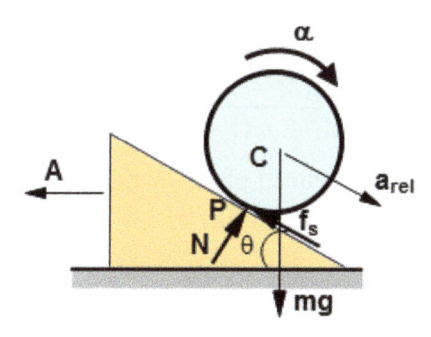

Let us impose the pseudo-force mA sitting on the accelerating wedge which must act on the cylinder at its center of mass towards the right. Now we can take the torque of all forces acting on the cylinder about P. Since the normal reaction N and friction f_s pass through P, these forces cannot produce any torque about P. So the net torque about P (due to gravity and pseudo-force mA) is

$$\tau_P = I_P \alpha$$

$$\Rightarrow (mg \sin\theta + mA \cos\theta)R = m(K^2 + R^2)\alpha$$

$$\Rightarrow (g \sin\theta + A \cos\theta)R = (K^2 + R^2)\alpha. \tag{3.223}$$

For rolling on a moving surface, the condition is

$$\Rightarrow a_{rel} = R\alpha. \tag{3.224}$$

Using equations (3.222) and (3.223) we have

$$(g \sin\theta + A \cos\theta) = \left(\frac{K^2 + R^2}{R^2}\right) a_{rel}. \tag{3.225}$$

Eliminating a_{rel} between equations (3.222) and (3.225) and putting $K^2/R^2 = 1/2$ for the cylinder and simplifying the factors, we obtain

$$A = \frac{2mg \sin\theta \cos\theta}{3M + m(1 + 2\sin^2\theta)}.$$

Problem 77 In last problem if the wedge is massive or fixed, find the ratio of the time of motion of the cylinder to travel the same distance relative to the inclined plane. Put $\theta = 45°$.

Solution
From the last problem, the acceleration of the wedge is

$$A = \frac{mg \sin \theta \cos \theta}{3M + m(1 + 2\sin^2\theta)}. \tag{3.226}$$

The acceleration of the cylinder relative to the wedge is

$$a_{rel} = \frac{(M+m)}{m \cos \theta} A. \tag{3.227}$$

Using equations (3.226) and (3.227) we have

$$a_{rel} = \frac{2(M+m)g \sin \theta}{3M + m(1 + 2\sin^2\theta)}. \tag{3.228}$$

For a distance l relative to the inclined plane (wedge), the time required is given as

$$t = \sqrt{\frac{2l}{a_{rel}}}.$$

Let the relative accelerations for the accelerating and stationary wedge be a_1 and a_2, respectively. Putting $\frac{m}{M} \simeq 0$ in the obtained expression for a massive wedge (a very large value of M), we have

$$a_2 = \left(\frac{2g}{3}\right) \sin \theta.$$

Then the ratio of times is

$$\frac{t_1}{t_2} = \sqrt{\frac{a_2}{a_1}} = \sqrt{\frac{\left(\frac{2g}{3}\right)\sin\theta \left\{\frac{M+m}{M+\frac{m}{3}(1+2\sin^2\theta)}\right\}}{\left(\frac{2g}{3}\right)\sin\theta}}$$

$$= \sqrt{\frac{M+m}{M+\frac{m}{3}(1+2\sin^2\theta)}}.$$

Putting $\theta = 45°$ we have

$$\frac{t_1}{t_2} = \sqrt{\frac{M+m}{M+\frac{2m}{3}}} = \sqrt{\frac{3(M+m)}{3M+2m}}.$$

Pulleys with connected bodies

Problem 78 An ideal string is wrapped over a stepped pulley of mass M, radius of gyration K, and radii R and r. The string does not slip over the pulley. The upper end of the string is connected to the ceiling and the lower end of the string is connected to a bob of mass m. After releasing the bob, (a) write down the equations for the tensions in the strings in terms of the angular acceleration of the pulley and (b) find the angular acceleration of the pulley.

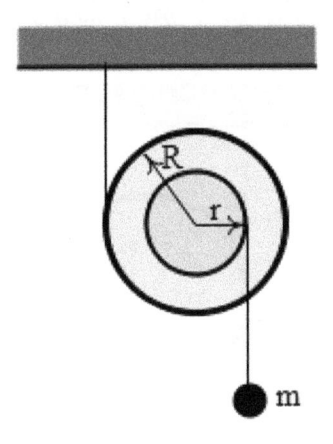

Solution

(a) The net torque acting on the center of mass of the pulley is

$$\tau_C = T_2 R + T_1 r = M K^2 \alpha. \tag{3.229}$$

The net force acting on the pulley is

$$-T_2 + T_1 + Mg = M a_1. \tag{3.230}$$

As the pulley rolls over the upper fixed string,

$$a_1 = R\alpha. \tag{3.231}$$

The net force acting on the bob is

$$mg - T_1 = m a_2. \tag{3.232}$$

As the pulley rolls over the fixed string,

$$a_2 = (R + r)\alpha. \tag{3.233}$$

Adding equations (3.230) and (3.232)

$$-T_2 + Mg + mg = M a_1 + m a_2. \tag{3.234}$$

Putting a_1 from equation (3.231) and a_2 from equation (3.233) in equation (3.234),

$$T_2 = (M + m)g - \{MR + m(R + r)\}\alpha. \tag{3.235}$$

Putting a_1 from equation (3.231) in equation (3.232),

$$T_1 = mg - (R + r)\alpha. \tag{3.236}$$

3-185

(b) Putting T_1 from equation (3.236) and T_2 from equation (3.235) in equation (3.229), and simplifying the factors, we obtain

$$\Rightarrow \alpha = \frac{[MR + m(R + r)]g}{M(K^2 + R^2) + m(R + r)^2}.$$

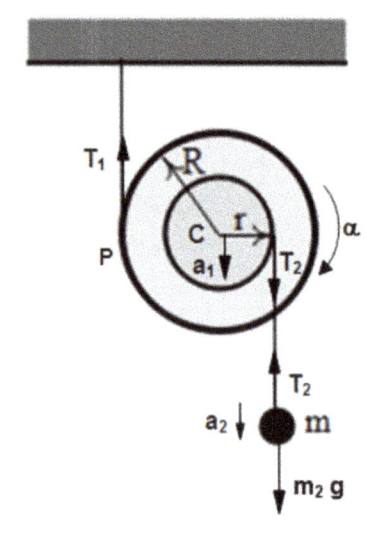

Alternative method

Taking the torques of mg and Mg about P,

$$\tau_P = MgR + mg(R + r).$$

The moment of inertia of the pulley about P is

$$I_M = M(K^2 + R^2).$$

The moment of inertia of the bob about P can be given as

$$I_m = m(R + r)^2.$$

Then the total moment of inertia about P is

$$I_P = I_M + I_m = M(K^2 + R^2) + m(R + r)^2.$$

Using the relation

$$\tau_P = I_P \alpha$$

and putting the values of torque and moment of inertia about P, we have the same expression of angular acceleration

$$\Rightarrow \alpha = \frac{[MR + m(R + r)]g}{M(K^2 + R^2) + m(R + r)^2},$$

which is the same as that obtained earlier.

Problem 79 A stepped pulley with a moment of inertia of I and radii R and r is smoothly pivoted with a fixed vertical rod. The ends of a string are connected to two particles of masses m and M, as shown in the figure. If the pulley is released from rest, find (a) the angular acceleration of the pulley and (b) the tensions in the string, assuming that the string does not slip over the pulley.

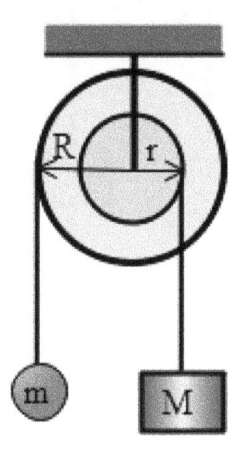

Solution
(a) Assuming the total clockwise torque about the pivot O due to mg and Mg, we have
$$\tau_P = Mgr - mgR.$$

The moment of inertia of the pulley about P is
$$I_M = M(K^2 + R^2).$$

The moment of inertia of the bodies about O can be given as
$$I_O = mR^2 + Mr^2 + I.$$

Using the relation
$$\tau_O = I_O \alpha$$
and putting the values of the torque and moment of inertia about O, we have the same expression of angular acceleration
$$\Rightarrow \alpha = \frac{[Mr - mR]g}{Mr^2 + mR^2 + I}.$$

3-187

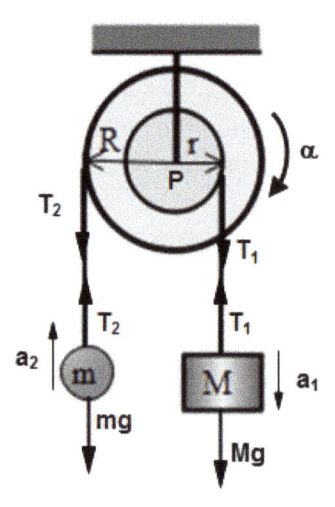

(b) The force acting on M is

$$Mg - T_1 = Ma_1 = MR\alpha$$

$$\Rightarrow T_1 = M(g - R\alpha).$$

Putting α in the last equation and simplifying the factors, we have

$$\Rightarrow T_1 = \frac{mR(R+r)+I}{Mr^2+mR^2+I}Mg.$$

The force acting on m is

$$T_2 - mg = ma_2 = mR\alpha$$

$$T_2 = m(g + R\alpha).$$

Putting α in the last equation and simplifying the factors we have

$$\Rightarrow T_2 = \frac{Mr(R+r)+I}{Mr^2+mR^2+I}mg.$$

Problem 80 A stepped pulley of mass M and radii a and b is pivoted smoothly with the ceiling by a fixed vertical rod. An inextensible string is wrapped over the pulley and a thin ring of mass m and radius c. The string does not slip over the bodies. If the pulley is released from rest from the given position, find (a) the tensions in the string and (b) angular acceleration of the bodies, (c) the acceleration of the ring.

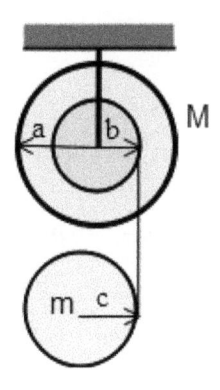

Solution

(a) Let the tension in the string be T which produces a clockwise torque about the pivot C of the pulley. It is given as

$$Tb = (Ma^2/2)\alpha_1$$

$$\Rightarrow \alpha_1 = \frac{2Tb}{Ma^2}. \tag{3.237}$$

The tension T also produces a counter-clockwise torque on the ring about its center of mass which is given as

$$Tc = (mc^2)\alpha_2$$

$$\Rightarrow \alpha_2 = T/(mc). \tag{3.238}$$

If $a_C =$ the acceleration of the center of mass of the ring, the net force acting on it is

$$mg - T = ma_C$$

$$\Rightarrow a_C = g - T/m. \tag{3.239}$$

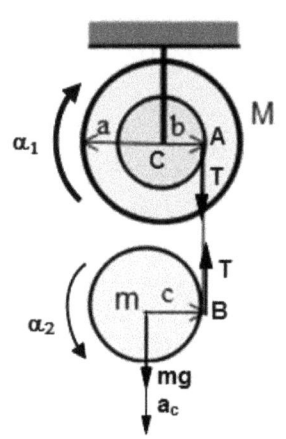

3-189

For the inextensible string,
$$b\alpha_1 = a_C - c\alpha_2. \quad (3.240)$$

Substituting the values of α_1, α_2, and a_C from equations (3.237), (3.238), and (3.239) in equation (3.240),
$$b(2Tb/Ma^2) = (g - T/m) - c(T/mc)$$
$$\Rightarrow \frac{2Tb^2}{Ma^2} = g - \frac{2T}{m}$$
$$\Rightarrow 2T\left(\frac{b^2}{Ma^2} + \frac{1}{m}\right) = g$$
$$\Rightarrow T = \frac{Mma^2 g}{2(Ma^2 + mb^2)}.$$

(b)

(i) Putting T in equation (3.237)
$$\Rightarrow \alpha_1 = \frac{2Tb}{Ma^2} = \left(\frac{2b}{Ma^2}\right)\left(\frac{Mma^2 g}{2(Ma^2 + mb^2)}\right)$$
$$\Rightarrow \alpha_1 = \frac{mbg}{Ma^2 + mb^2}.$$

(ii) Putting T in equation (3.238)
$$\Rightarrow \alpha_2 = \frac{T}{mc} = \left(\frac{1}{mc}\right)\left(\frac{Mma^2 g}{2(Ma^2 + mb^2)}\right)$$
$$\Rightarrow \alpha_2 = \frac{Ma^2 g}{2c(Ma^2 + mb^2)}.$$

(c) Putting T in equation (3.239)
$$\Rightarrow a_C = g - (1/m)\frac{Mma^2 g}{2(Ma^2 + mb^2)}$$
$$\Rightarrow a_C = \frac{(Ma^2 + 2mb^2)g}{2(Ma^2 + mb^2)}.$$

IOP Publishing

Problems and Solutions in Rotational Mechanics

Pradeep Kumar Sharma

Chapter 4

The energy of rigid bodies

4.1 The gravitational potential energy of rigid bodies

The potential energy possessed by a system of particles is given as

$$U_g = mgy_C,$$

where y_C is the vertical position of the center of mass of the body from ground level.

Proof: Let us consider an element of mass dm of the rigid body at a height y from ground level. The gravitational potential energy of the element is

$$dU = (dm)gy.$$

The gravitational potential energy of the body is

$$U = \int dU = \int (dm)gy.$$

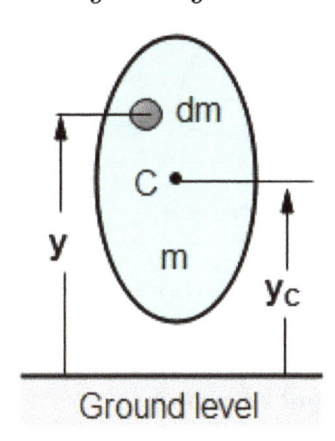

Example 1 A hemispherical wedge of mass M and radius R is kept on a horizontal surface as shown in the figure. (a) Find the gravitational potential energy of the hemisphere to ground level. (b) If we invert the hemisphere, find (i) the change in gravitational potential energy of the hemisphere and (ii) the work done by gravity.

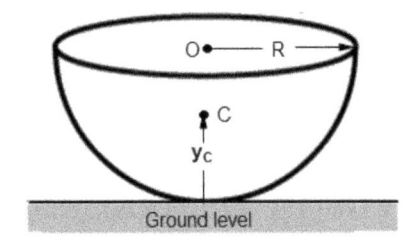

Solution
(a) The center of mass of the hemisphere is located at a depth $d = 3R/8$ from the center of curvature O. Then the height of the center of mass $= y_C = R - 3R/8 = 5R/8$. Then the gravitational potential energy of the hemisphere is

$$U_1 = mgh = mg(5R/8) = 5mgR/8.$$

(b) (i) After inverting the hemisphere the new height of the center of mass is $y'_2 = 3R/8$. Then the new gravitational potential energy of the hemisphere is $U_2 = 3mgR/8$. So the change in gravitational potential energy is

$$(U_2 - U_1) = mg(y_C - y'_C)$$
$$= mg(5R/8 - 3R/8)$$
$$= mgR/4.$$

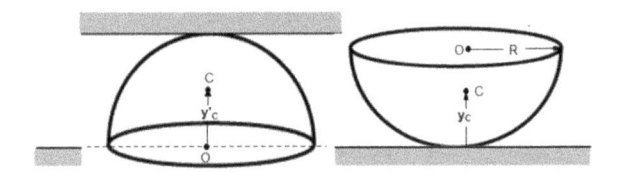

(ii) The work done by gravity is

$$W_g = -(U_2 - U_1) = -mgR/4.$$

4.2 Kinetic energy of rigid of a body
4.2.1 Discrete mass distribution
Let us take the ith particle of mass m_i for a system of discrete mass. If the system is rigid, all particles must be connected rigidly with each other by light rods so that all

particles have the same angular velocity ω about the other particles and about the point P. If the ith particle has a radius r_i, its velocity $v_i = r_i\omega$. Then the kinetic energy of m_i is

$$K_i = \frac{1}{2}m_i v_i^2 = \frac{1}{2}m_i(r_i\omega)^2 = \frac{1}{2}m_i r_i^2 \omega^2.$$

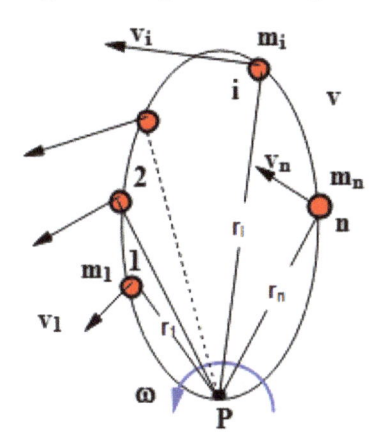

Adding the kinetic energy of all particles, the total kinetic energy of the rigid body relative to the fixed point (pivot) P is given as

$$K = \sum_1^N K_i = \sum_1^N \frac{1}{2}m_i r_i^2 \omega^2.$$

Taking $\omega^2/2$ out of the summation we have

$$K = \frac{\omega^2}{2}\sum_1^N m_i r_i^2.$$

Putting $\sum_1 m_i r_i^2 = I =$ the moment of inertia of the rigid body about the axis of rotation P, we have

$$K = \frac{I\omega^2}{2}.$$

Example 2 Two particles of mass m_1 and m_2 are connected at the ends of a light rod of length L. The rod is rotating about the vertical axis with a minimum kinetic energy E. Find the frequency of rotation of the rod.

Solution
Let the particle M be at a distance of x from the axis of rotation. Hence, the moment of inertia of the system is given as

$$I = m_1 x^2 + m_2(L - x)^2. \tag{4.1}$$

Since the kinetic energy of the rotating rod is directly proportional to the moment of inertia of the system, the kinetic energy will be minimum when the moment of inertia is minimum. For minimum I

$$\frac{dI}{dx} = 0,$$

or

$$\frac{d}{dx}\{m_1 x^2 + m_2(L-x)^2\} = 0,$$

or

$$2m_1 x - 2m_2(L-x) = 0,$$

or

$$x = \frac{m_2 L}{m_1 + m_2}.$$

Putting this value in equation (4.1) and simplifying the factors, we have

$$I_{min} = \frac{m_1 m_2}{m_1 + m_2} L^2.$$

Then the minimum kinetic energy of the rod is

$$K_{min} = \frac{I_{min}\omega^2}{2} = \frac{m_1 m_2 L^2 \omega^2}{2(m_1 + m_2)}.$$

4.2.2 Continuous mass distribution

If the rigid body is a continuous distribution of mass, we can take an element of mass dm at a radial distance r from the axis of rotation. Then the kinetic energy of the element at Q is

$$dK = \frac{1}{2}dm v^2 = \frac{1}{2}dm(r\omega)^2 = \frac{1}{2}dm r^2 \omega^2.$$

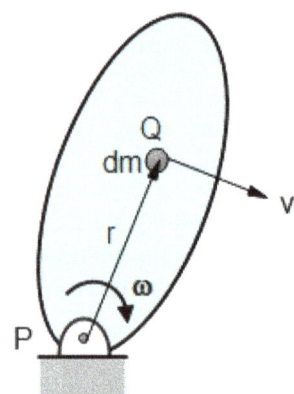

Adding the kinetic energy of all particles, the total kinetic energy of the rigid body about the axis of rotation P is given as

$$K = \int dK = \int \frac{1}{2} dm r^2 \omega^2.$$

Taking $\omega^2/2$ out of the summation we have
$$K = \frac{\omega^2}{2} \int r^2 dm.$$
Putting $\int r^2 dm = I =$ the moment of inertia of the rigid body about the axis of rotation, we have
$$K = \frac{I\omega^2}{2}.$$

Example 3 Two uniform hemispheres A and B each of mass $m/2$ and radius R are joined to form a composite sphere, as shown in the figure. One of the hemispheres is thin and the other is solid. The sphere is made to rotate with an angular velocity $\omega = \sqrt{g/R}$. Find the kinetic energy of the rotating sphere.

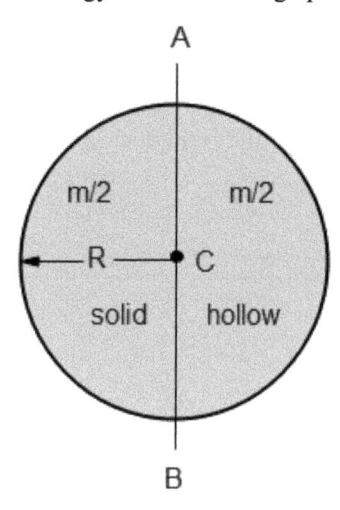

Solution
Let the moment of inertia of the thin and solid sphere about the axis AB be I_1 and I_2, respectively. Then the total moment of inertia is
$$I = I_1 + I_2. \tag{4.2}$$
Putting
$$I_1 = \frac{2(m/2)R^2}{3} = \frac{mR^2}{3}$$
and
$$I_2 = \frac{2(m/2)R^2}{5} = \frac{mR^2}{5}$$
in equation (4.2), we have
$$I = \frac{8mR^2}{15}. \tag{4.3}$$

The kinetic energy of the sphere is given as

$$K = \frac{I\omega^2}{2}. \qquad (4.4)$$

Using equations (4.2) and (4.3) and putting $\omega = \sqrt{g/R}$, we have

$$K = \frac{I\omega^2}{2} = \frac{1}{2}\left(\frac{8mR^2}{15}\right)\left(\frac{g}{R}\right) = \frac{4mgR}{15}.$$

4.2.3 Rotational and translational kinetic energy

Rotational kinetic energy: Let us recast the kinetic energy of a rotating rigid body about a fixed axis, given as

$$K = \frac{I\omega^2}{2},$$

where $I =$ the moment of inertia of the rigid body about the axis passing through the center of mass of the body.

Using parallel axis theorem put $I = I_C + mr^2$ in the last expression to obtain

$$K = \frac{(I_C + mr^2)\omega^2}{2} = \frac{I_C\omega^2}{2} + \frac{mr^2\omega^2}{2},$$

where $r\omega = v_C$.
Then

$$K = \frac{I_C\omega^2}{2} + \frac{mv_C^2}{2}.$$

The first term signifies the kinetic energy relative to the center of mass axis. This is known as the kinetic energy of *rotation*, denoted by the symbol K_r. Putting $I_C = mk^2$ we can write

$$K_r = \frac{I_C\omega^2}{2} = \frac{mk^2\omega^2}{2}.$$

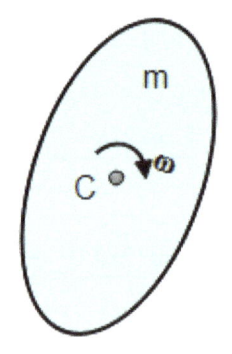

$K_{Rot} = I_C \omega^2 / 2$

Translational kinetic energy: The second term of the expression of the kinetic energy of the rotating body is called the kinetic energy of translation. This is because this term gives the kinetic energy of the center of mass. Let us replace the rigid body by a particle of mass equal to the mass of the rigid body and place it at the center of mass of the rigid body. Since it moves with a velocity equal to that of the center of mass, its kinetic energy can be given as

$$K_t = \frac{mv_C^2}{2} = \frac{mr^2\omega^2}{2}.$$

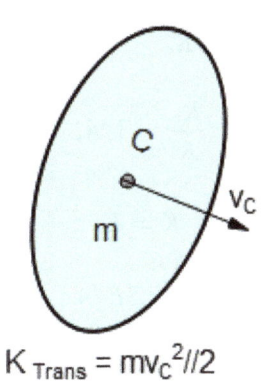

$$K_{Trans} = mv_C^2/2$$

As the center of mass is a point and there is no meaning to rotation of a particle, this kinetic energy is termed as translational kinetic energy, denoted by K_t.

The ratio of the translational, rotational, and total kinetic energy of rigid bodies
The ratio of translational and rotational kinetic energy is

$$\frac{K_t}{K_r} = \frac{\frac{mr^2\omega^2}{2}}{\frac{mk^2\omega^2}{2}} = \frac{r^2}{k^2}.$$

Then the ratio of rotational and translational kinetic energy relative to the total kinetic energy of the rigid body can be given as follows:

$$\frac{K_t}{K} = \frac{r^2}{k^2 + r^2}$$

$$\frac{K_r}{K} = \frac{k^2}{k^2 + r^2},$$

where r is distance between the fixed axis of rotation and the centroidal axis parallel to the axis of rotation:
- Rotational kinetic energy is the kinetic energy about the centroidal axis.
- Rotational kinetic energy is not equal to the kinetic energy of the rigid body in a fixed axis of rotation.
- Translational kinetic energy is the kinetic energy of the center of mass.

Example 4 A uniform horizontal rod rotates about a vertical axis passing through one of its ends. Its kinetic energy is E. Find its (a) translational and (b) rotational kinetic energy.

Solution
(a) The ratio of translational and total kinetic energy is
$$\frac{K_t}{K} = \frac{r^2}{k^2 + r^2}.$$
For a uniform rod of length L, $I_C = mk^2 = mL^2/12$. Then $k^2 = L^2/12$. The distance between the axis of rotation and center of mass $= r = L/2$. Then $r^2 = L^2/4$. Putting the values of k^2 and r^2 in the above expressions to obtain
$$\frac{K_t}{K} = 3/4$$
and
$$\frac{K_r}{K} = 1/4,$$
where
$$K = E.$$
Then, the kinetic energy of translation is
$$K_t = 3E/4$$
(b) The kinetic energy of rotation is
$$K_r = E/4.$$

Example 5 Find the kinetic energy of (a) the translation and (b) rotation of the thin hemispherical shell which is rotating about its centroidal axis with an angular speed of $\omega = \sqrt{g/2R}$, as shown in the figure.

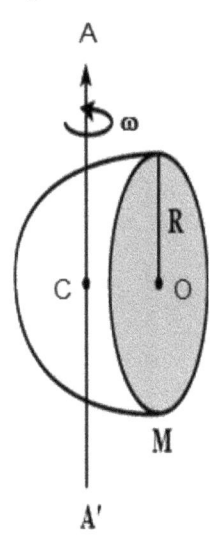

Solution

(a) The kinetic energy of translation of the rotating body is given by

$$K_t = \frac{mv_C^2}{2},$$

where v_C = the velocity of the center of mass. Since the center of mass is lying on the axis of rotation, its velocity is zero. Therefore, the translational energy is zero.

(b) Then the total kinetic energy of rotation is purely rotational, which is given as

$$K_r = \frac{I_C \omega^2}{2},$$

where $\omega = \sqrt{g/2R}$. By the parallel axis theorem

$$I_O = I_C + mr^2,$$

where $r = OC = R/2$ and $I_O = \frac{2mR^2}{3}$.

Then

$$I_C = I_O - mr^2 = \frac{2mR^2}{3} - m\left(\frac{R}{2}\right)^2 = \frac{5mR^2}{12}.$$

Then

$$K_r = \frac{1}{2}\left(\frac{5mR^2}{12}\right)\left(\frac{g}{2R}\right) = \frac{5mgR}{48}.$$

4.3 The kinetic energy of rolling bodies on a fixed surface

Moving surface

Any rolling body has translational and rotational kinetic energy. The total kinetic energy of a rolling body is given as the sum of these two kinetic energies:

$$K = \frac{I_C \omega^2}{2} + \frac{mv_C^2}{2}.$$

Fixed surface

If the body rolls on a fixed surface $v_C = R\omega$. Putting $\omega = \frac{v_C}{R}$ and $I_C = mk^2$, we have

$$K = \frac{mk\left(\frac{v_C}{R}\right)^2}{2} + \frac{mv_C^2}{2}.$$

Taking the common factor $\frac{mv_C^2}{2}$ out, finally we have

$$K = \frac{mv_C^2}{2}\left(1 + \frac{k^2}{R^2}\right).$$

Putting $\frac{mv_C^2}{2} = K_t$ we have

$$K = K_t\left(1 + \frac{k^2}{R^2}\right).$$

Then the ratio of translational and total kinetic energy is given as

$$\frac{K}{K_t} = \left(1 + \frac{k^2}{R^2}\right)$$

as obtained earlier.

Putting $v_C = R\omega$ and $\frac{mv_C^2}{2} = K_t$, in the last expression and rearranging the terms, we have $K = \frac{(I_C + mR^2)\omega^2}{2}$, where $(I_C + mR^2) = I_P =$ the moment of inertia of the rigid body about the point of contact P. Then we can write the handy formula of kinetic energy of the rolling body as

$$K = \frac{I_P \omega^2}{2}.$$

It can be shown that the ratio of rotational and total kinetic energy is given as

$$\frac{K}{K_r} = \left(1 + \frac{R^2}{k^2}\right).$$

- The above expressions are valid when the body rolls on a stationary surface.
- If the surface moves, the above formulae will no longer be applicable. We need to use the general formula.

Example 6 A uniform solid sphere rolls without sliding on a stationary surface. What fraction of the total kinetic energy the rolling sphere is translational and rotational?

Solution
Let us recast the formula

$$\frac{K}{K_t} = \left(1 + \frac{k^2}{R^2}\right),$$

where $k^2/R^2 = 2/5$ for the solid uniform sphere. Then we have

$$\frac{K}{K_t} = \left(1 + \frac{2}{5}\right) = 7/5.$$

This means that $\frac{K_t}{K} = 5/7$ and $\frac{K_r}{K} = (1 - \frac{5}{7}) = 2/7$.

Example 7 A car of mass M having four wheels each of mass m, radius R, and radius of gyration K, is moving with a kinetic energy E. Find the linear velocity of the car.

Solution
The kinetic energy of the car (excluding the wheels) is given as

$$K_{car} = \frac{m_C v_C^2}{2} = \frac{m_C v^2}{2},$$

where m_C = the mass of the car excluding the wheels and $v_C = v$.
The kinetic energy of four wheels is

$$K_{wheel} = 4 \times \frac{m v_C^2}{2}\left(1 + \frac{K^2}{R^2}\right) = 2m\left(1 + \frac{K^2}{R^2}\right)v^2.$$

Adding the last two kinetic energies, the total kinetic energy of the car is given as

$$K_{tot} = K_{car} + K_{wheel} = E \text{ (given)},$$

or

$$E = \frac{m_C v^2}{2} + 2mv^2\left(1 + \frac{K^2}{R^2}\right),$$

or

$$E = \frac{\left(m_C + 4m + 4m\frac{K^2}{R^2}\right)v^2}{2}.$$

Putting $(m_C + 4m) = M$ we have

$$v = \sqrt{\frac{2E}{\left(M + 4m\frac{K^2}{R^2}\right)}}.$$

Example 8 One end of an inextensible light string is fixed to the ceiling. It passes over a hanging movable pulley 1 and a fixed pulley 2. The mass of each pulley is m and the radius is R. An apple of mass m is hanging from the movable pulley. The free end of the string is pulled with a velocity v downwards. Find the kinetic energy of the system of two pulleys and an apple. Assume that the string does not slide over the pulleys and each pulley can be treated as a uniform disc. The strings are light and inextensible.

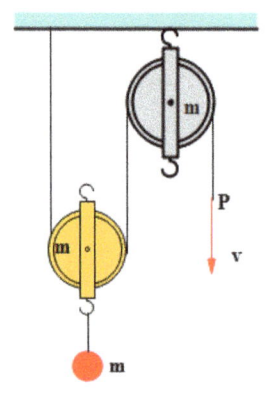

Solution

Applying the kinematics, we can find that the angular velocity of the pulleys 1 and 2 are $v/2R$ and v/R, respectively, and the velocity of the apple is $v/2$ directed upwards. Then the kinetic energy of the system is the sum of the kinetic energy of the pulleys and apple given as

$$K = K_1 + K_2 + K_{\text{apple}},$$

where

$$K_1 = \frac{1}{2}\left(\frac{3mR^2}{2}\right)\left(\frac{v}{2R}\right)^2,$$

$$K_2 = \frac{1}{2}\left(\frac{mR^2}{2}\right)\left(\frac{v}{R}\right)^2 \quad \text{and} \quad K_{\text{apple}} = \frac{1}{8}mv^2.$$

Then we have $K = \frac{9}{16}mv^2$.

4.4 Work done by a force on a rigid body

When a force acts on a rigid body it has two effects. The torque of the force about the center of mass causes rotation and the force pulls the center of mass of the rigid body causing a translation. If the point of application of the force $\vec{F}$ undergoes a displacement $d\vec{s}$, the work performed by the force $\vec{F}$ is given as

$$W_F = \int \vec{F} \cdot d\vec{s}.$$

Example 9 A body rolls down an inclined plane without sliding. (a) What is the work done by normal contact force and friction, if the center of mass of the body undergoes a displacement s? (b) What are the roles of gravity, normal reaction, and static friction in causing rotation?

Solution

(a) Let us use the expression $W_F = \int F \cdot ds$. Static friction f_s and the normal reaction N act at the point of contact P of the rolling body with the inclined plane. Their directions are shown in the FBD (free-body diagram). Since the point of contact does not slide or move, its displacement $ds = 0$. So the work done by f_s and N is zero. Students may be tempted to write $W_f = -f \cdot s$ because the friction acts opposite to the displacement of the center of mass. This is wrong because the friction does not act at the center of mass, so we cannot take the displacement of the center of mass while finding the work done by the friction.

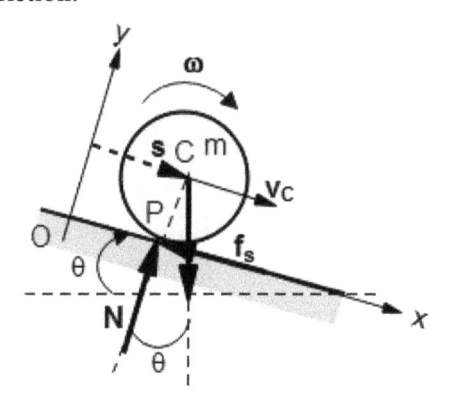

On the other hand we can consider the displacement of the center of mass to calculate the work done by gravity because gravity always acts at the center of mass of any system of particles. The work done by gravity is given as

$$W_{gr} = \int m\vec{g} \cdot d\vec{s} = m\vec{g} \cdot \int d\vec{s} = m\vec{g} \cdot \vec{s},$$

where $\vec{s}$ = the displacement of the center of mass of the body. The angle between $\vec{g}$ and $\vec{s}$ is

$$\varnothing = \pi/2 - \theta,$$

so

$$m\vec{g} \cdot \vec{s} = mg \cos \varnothing = mgs \sin \theta.$$

Then we have

$$W_{gr} = mgs \sin \theta.$$

(b) As the gravity acts at the center of mass, it cannot produce a torque about the center of mass. In this sense it cannot produce rotation. However, due to the effect of gravity the contact forces N and f_s come into play as the body tends to slide down. The normal reaction N cannot produce a torque about the center of mass as it passes through the center of mass. However, the friction produces a clockwise torque about the center of mass and causes the rotation. The friction is induced due to the normal reaction and the tendency of relative sliding. The normal reaction is caused by gravity. So in this case gravity has an indirect role in causing rotation of the body. However, in a free fall, gravity alone cannot produce a rotation.

N.B. Static friction does positive work in increasing the kinetic energy of rotation.
- Static friction does equal negative work in decreasing the kinetic energy of translation.
- The net work done by static friction on the rolling body on an inclined plane is equal to zero.

4.5 The work–energy theorem for rigid bodies

4.5.1 Net work done on a rigid body

Let a force F act on a planar rigid body at a point P. The work done by F is given as

$$W = \int \vec{F} \cdot d\vec{s}_P.$$

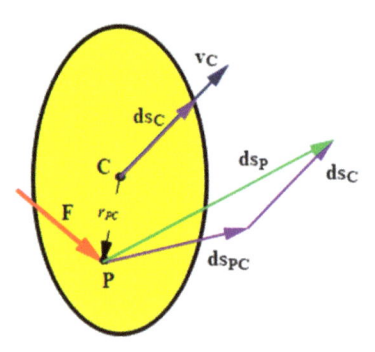

Let us write the elementary displacement $d\vec{s}_P$ as the combination of two other elementary displacements as

$$d\vec{s}_P = d\vec{s}_{PC} + d\vec{s}_C.$$

Putting the value of $d\vec{s}_P$ in the expression of work we have

$$W = \int \vec{F} \cdot (d\vec{s}_{PC} + d\vec{s}_C).$$

Expanding the terms on the right-hand side of the last expression, we have

$$W = \int \vec{F} \cdot d\vec{s}_{PC} + \int \vec{F} \cdot d\vec{s}_{C}.$$

Let us try to understand the first term $\int \vec{F} \cdot d\vec{s}_{PC} = W_1$, say. We can call it 'the elementary work done by the force $\vec{F}$ relative to the center of mass C'. As the body is rigid, the radial distance r of the point P from the center of mass C remains constant. This means that P moves in a circle of radius $r (= PC)$ relative to C. If the body rotates through a very small angle $d\theta$ during an elementary time dt, the elementary shift (displacement) of P relative to C, that is $d\vec{s}_{PC}$, has a magnitude $rd\theta$ and remains perpendicular to $\vec{r}$. So the displacement $d\vec{s}_{PC}$ is tangential or transverse, as observed from the center of mass C. Let us now resolve the force F along radial and tangential line given as

$$\vec{F} = \vec{F}_r + \vec{F}_t.$$

The radial component F_r acts along the radial line PC. The tangential or transverse component of F acts along the elementary displacement ds_{PC}. Then, the elementary work done by the force F relative to the center of mass C is given as

$$W_1 = \int \vec{F} \cdot d\vec{s}_{PC} = \int (\vec{F}_r + \vec{F}_t) \cdot d\vec{s}_{PC}.$$

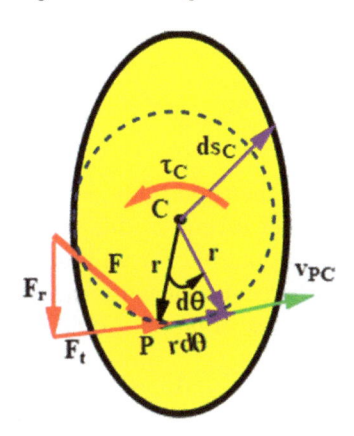

Since $\vec{F}_r$ is perpendicular to $d\vec{s}_{PC}$ we can put $\vec{F}_r \cdot d\vec{s}_{PC} = 0$. Then we have

$$W_1 = \int \vec{F}_t \cdot d\vec{s}_{PC}.$$

Since $\vec{F}_t$ is parallel to $d\vec{s}_{PC}$, we can put $\vec{F}_t \cdot d\vec{s}_{PC} = F_t \cdot ds_{PC} \cos 0 = F_t \cdot ds_{PC}$. Then we have

$$W_1 = \int F_t \cdot ds_{PC},$$

where $ds_{PC} = rd\theta$. Then we have

$$W_1 = \int F_t r d\theta.$$

As the force F_t produces a torque τ_C about the center of mass C, putting $F_t r = \tau_C$ in the last expression, we have

$$W_1 = \int \tau_C \, d\theta.$$

Since $\vec{\tau}_C$ and $d\vec{\theta}$ are vectors, the last expression can be given as

$$W_1 = \int \vec{\tau}_C \cdot d\vec{\theta}.$$

The second term $\int \vec{F} \cdot d\vec{s}_C$ is the work done by the applied force on the center of mass.

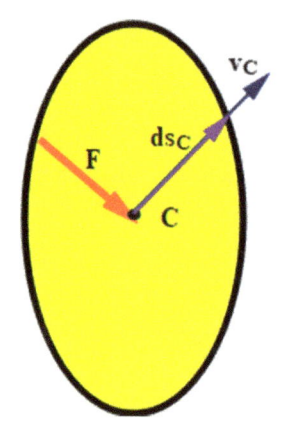

If many external forces act on the rigid body, as we learned, not only is the net internal force zero but also the net torque $\vec{\tau}_C$ due to all internal forces being zero. Therefore, the work done by the internal forces is zero in a rigid body. Then the work done and torque about the center of mass can be attributed to the external forces only. So, finally, the net work done on a rigid body can be given as

$$W_{\text{total}} = \int \vec{F}_{\text{net}} \cdot d\vec{s}_C + \int \vec{\tau}_C \cdot d\vec{\theta},$$

where the first and second terms on the right-hand side are the work done in translation and rotation of the rigid body, respectively. So, in a nutshell,

$$W_{\text{total}} = W_t + W_r,$$

where $W_t = \int \vec{F}_{\text{net}} \cdot d\vec{s}_C$,

$$W_r = \int \vec{\tau}_C \cdot d\vec{\theta}.$$

So for a rigid body:
- The total work done is due to the external forces only.
- Total work = work done in translation + work done in rotation.

Example 10 A uniform sphere of mass m, radius R, and radius of gyration k rolls on a horizontal plane when a horizontal force F acts on its center. If the center of mass of the body undergoes a displacement x, find (b) the total work done and (b) the work done in translation and rotation of the cylinder.

Solution
(a) Since mg, N, and f do not perform any work, the net work done on the body is
$$W_{total} = Fx.$$

(b) The net force acting on the center of mass of the sphere is
$$F_{net} = F - f = ma_C = ma,$$
where f = the static friction acting on the body and let us assume that the acceleration of the center of mass is $a_C = a$.

Then the work done in translation is
$$W_t = F_{net} x = max,$$
where
$$a = F / \left(1 + \frac{k^2}{R^2}\right).$$

Then
$$W_t = Fx / \left(1 + \frac{k^2}{R^2}\right).$$

The work done in rotation of the body is
$$W_r = W_{total} - W_t = Fx - F_{net} x$$
$$= (F - ma)x.$$

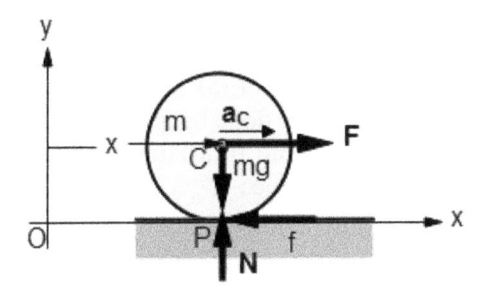

Putting the value of $a = F/(1 + \frac{k^2}{R^2})$ and simplifying the factors, we have
$W_r = Fx/(1 + \frac{R^2}{k^2}).$

4.5.2 Work–kinetic energy theorem

When many forces act on a rigid body, the sum of work done by all forces is known as net work done. Since the net work done by the internal forces is zero, the external forces can perform a net (total) work which is the sum of two works, such as the work done to translate the body and the work done in rotation of the body, as described above.

According to the work–energy theorem, the total work is numerically equal to the change in the kinetic energy of the rigid body given as

$$W_{total} = \Delta K,$$

where

$$W_{total} = W_t + W_r.$$

Then we have

$$W_t + W_r = \Delta K.$$

In the last section we learned that the kinetic energy of a rigid body is the sum of the kinetic energy of translation and the kinetic energy of rotation. Putting $K_{total} = K_t + K_r$ in the last expression, we obtain

$$W_t + W_r = \Delta K_t + \Delta K_r.$$

Now we can compare the terms on both sides and separate them to obtain following two expressions:

$$W_t = \Delta K_t$$

$$W_r = \Delta K_r.$$

This states that:
- The total work done in translation = the change in translational kinetic energy.
- The total work done in rotation = the change in rotational kinetic energy.

Example 11 Referring to example 9, find (a) the velocity of the center of mass, (b) total work done in translation, work done by gravity in rotation and translation, (i) work done by friction in rotation and translation of the rolling body when the center of mass descends a vertical distance y.

Solution
(a) The net work done is due to gravity only, given as $W_{total} = mgx \sin \theta$. Putting $x \sin \theta = y$, we have $W_{total} = mgy$.
The kinetic energy of the body increases from zero to K, so the change in kinetic energy is given as

$$\Delta K = K = \frac{mv_C^2}{2}\left(1 + \frac{k^2}{R^2}\right).$$

Using work–energy theorem

$$W_{\text{total}} = \Delta K.$$

Putting the values of W_{total} and ΔK we have

$$\frac{mv_C^2}{2}\left(1 + \frac{k^2}{R^2}\right) = mgy.$$

This gives $v_C = \sqrt{\dfrac{2gy}{\left(1 + \frac{k^2}{R^2}\right)}}$.

(b) The total work done in translation is

$$W_t = \Delta K_t,$$

where $\Delta K_t = \dfrac{mv_C^2}{2} = \dfrac{mgy}{\left(1 + \frac{k^2}{R^2}\right)}$. Then we have

$$W_t = \frac{mgy}{\left(1 + \frac{k^2}{R^2}\right)}.$$

(c)
  (i) The work done by gravity in translation is $W_{\text{gr}} = mgy$ and that of rotation is equal to zero.
  (ii) The work done by friction in translation is $W_t = -f_s s$ and that of rotation is equal to $W_r = +f_s s$, where $s = $ the displacement of the center of mass of the body. So the net work done by static friction is zero as discussed in the last example. Since

$$W_r = \Delta K_r = K_r$$

and

$$K_{\text{total}} = K_r\left(1 + \frac{R^2}{k^2}\right) = mgy,$$

finally we have

$$W_r = \frac{mgy}{\left(1 + \frac{R^2}{k^2}\right)}.$$

4.6 Conservation of energy of rigid bodies

The work–energy theorem states that net work done is equal to the change in kinetic energy of the rigid body given as

$$W = \Delta K.$$

The total work is the sum of work done by conservative and non-conservative forces. So we can write

$$W_{\text{cons}} + W_{\text{non-cons}} = \Delta K.$$

The work done by a non-conservative force can be zero when:
1. There is no non-conservative force acting on the rigid body (its magnitude is zero).
2. The force acts perpendicular to the displacement of the point of its application.
3. The point of application of the force does not move.

If the work done by the non-conservative forces is zero, then putting $W_{\text{non-cons}} = 0$, we have

$$W_{\text{cons}} = \Delta K. \tag{4.5}$$

We know that a conservative force does positive work at the expense of the potential energy of the force-field. So we can write

$$W_{\text{cons}} = -\Delta U,$$

where $-\Delta U$ = the loss of potential energy. Then we have

$$-\Delta U = \Delta K,$$

$$\text{or } \Delta U + \Delta K = 0,$$

$$\text{or } \Delta(U + K) = 0,$$

$$\text{or } \Delta E = 0.$$

This states that:
- When the work done by the non-conservative forces is zero on a rigid body, the total mechanical energy of the rigid body remains conserved.

Example 12 A uniform sphere of radius R and radius of gyration K rolls down the inclined plane without sliding so that the center of mass of the body undergoes a vertical displacement h, as shown in the figure. Find the velocity of the body as a function of h. Put $h = 7$ m.

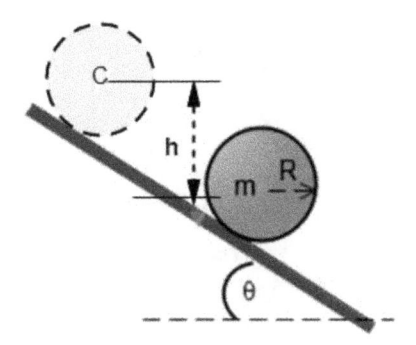

Solution

We find that the work done by the non-conservative forces such as static friction and the normal reaction is zero. Since the gravity is a conservative force, we can conserve the total energy of the rolling body. As the body rolls down, the change in gravitational potential energy is

$$\Delta U = -mgh. \tag{4.6}$$

The corresponding change in the kinetic of the body is

$$\Delta K = \frac{mv_C^2}{2}\left(1 + \frac{k^2}{R^2}\right), \text{ where } \vec{v}_C = v. \tag{4.7}$$

The energy conservation expression is

$$\Delta U + \Delta K = 0. \tag{4.8}$$

Bringing the values of ΔU from equation (4.6) and ΔK from equation (4.7), putting these values in equation (4.8) and simplifying the factors, we have

$$v = \sqrt{\frac{2gh}{1 + \frac{k^2}{R^2}}} = 10 \ m \ s^{-1}.$$

4.7 Power delivered by a force acting on a rigid body

The power delivered by a force F is given as the rate at which the force does work, which is given as $P_F = \vec{F} \cdot \vec{v}_P = F \cdot (\vec{v}_{PC} + \vec{v}_C)$, where $\vec{v}_{PC} = \vec{\omega} \times \vec{r}_{PC}$. Then we have

$$P_F = \vec{F} \cdot \vec{v}_P = \vec{F} \cdot \vec{v}_C + \vec{F} \cdot (\vec{\omega} \times \vec{r}_{PC}). \tag{4.9}$$

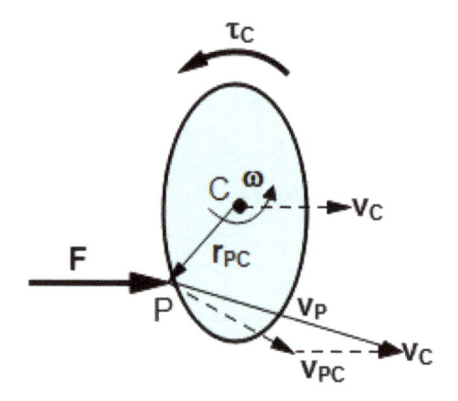

The second term of the right-hand side of equation (4.9) can be simplified as follows:

$$\vec{F} \cdot (\vec{\omega} \times \vec{r}_{PC}) = (\vec{\omega} \times \vec{r}_{PC}) \cdot \vec{F}$$
$$= (\vec{\omega} \cdot \vec{r}_{PC}) \times \vec{F}.$$

Putting $\vec{r}_{PC} \times \vec{F} = \vec{\tau}_C$, we have

$$\vec{F} \cdot (\vec{\omega} \times \vec{r}_{PC}) = \vec{\omega} \cdot \vec{\tau}_C = \vec{\tau}_C \cdot \vec{\omega}. \qquad (4.10)$$

Using equations (4.9) and (4.10) we have

$$P_F = \vec{F} \cdot \vec{v}_C + \vec{\tau}_C \cdot \vec{\omega},$$

where the first and second terms on the right-hand side are the power delivered by the force in translation and rotation of the rigid body, respectively:
- The rate of doing work, that is, the power delivered by F, is given as the sum of powers delivered for translation and rotation of the rigid body.

Example 13 A motor cycle of mass M has two wheels each of mass m, radius r, and radius of gyration K. If it starts from rest with a constant acceleration a, find the power delivered to the car at time t.

Solution

Let the frictional force act on the driving wheel of the motor cycle when the accelerator is on. If the car moves with a velocity v at an instant t, the rate of doing work is equal to the rate of change in the kinetic energy. The kinetic energy of the vehicle is

$$K = \frac{Mv_C^2}{2} + \frac{mk^2 v_C^2}{2R^2} = \frac{Mv_C^2}{2}\left(1 + \frac{mk^2}{MR^2}\right).$$

According to the work–energy theorem, the net work done is equal to the change in kinetic energy. This means that the instantaneous power delivered by the net force is numerically equal to the rate of change in kinetic energy:

$$\frac{dK}{dt} = M\left(1 + \frac{mk^2}{MR^2}\right)v_C(dv_C/dt).$$

Putting $(dv_C/dt) = a_C = a$ and $v_C = a_C t$ (if the force or acceleration remains constant), putting $a_C = a$, we have

$$P_{net} = \frac{dK}{dt} = M\left(1 + \frac{mk^2}{MR^2}\right)a^2 t.$$

Problems

Fixed axis rotation

Problem 1 A rod of mass M and length L is smoothly pivoted at point P of the ceiling as a constant horizontal force F acts on the bead which is welded to the bottom of the rod, as shown in the figure. (a) Find the angular velocity of the rod as the function of its angular displacement (the angle made by the rod with the vertical). (b) Prove that the maximum angle rotated by the rod is numerically equal to twice the equilibrium angle made by the rod with the vertical.

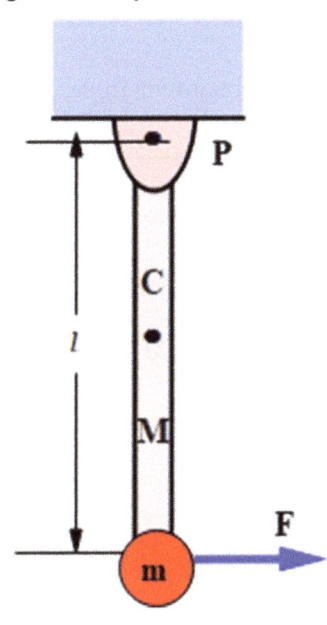

Solution
(a) The net work done by gravity and applied force in rotating the rod about the pivot is

$$W_{net} WF + Wgr = FL\sin\theta - \frac{(M + 2m)L(1 - \cos\theta)}{2}.$$

4-23

The increase in kinetic energy of the rod–particle (bead) system is

$$\Delta K = \frac{I\omega^2}{2} = \frac{(I_{rod} + I_m)\omega^2}{2}$$
$$= \frac{\{(ML^2/3) + mL^2\}\omega^2}{2}$$
$$= \frac{L^2(M + 3m)\omega^2}{6}.$$

The work–energy theorem is

$$W_{net} = \Delta K$$

$$\Rightarrow FL\sin\theta - \frac{(M + 2m)gL(1 - \cos\theta)}{2} = \frac{L^2(M + 3m)\omega^2}{6}$$

$$\Rightarrow \frac{2F\sin\theta - (M + 2m)g(1 - \cos\theta)}{1} = \frac{L(M + 3m)\omega^2}{3}$$

$$\Rightarrow \omega = \sqrt{\frac{3\{2F\sin\theta - (M + 2m)g(1 - \cos\theta)\}}{L(M + 3m)}}.$$

(b) When the rod will rotate a maximum angle, its angular velocity will be zero instantaneously:

$$\Rightarrow \omega = \sqrt{\frac{3\{2F\sin\theta - (M + 2m)g(1 - \cos\theta)\}}{L(M + 3m)}} = 0$$

$$\Rightarrow 2F\sin\theta = (M + 2m)g(1 - \cos\theta)$$

$$\Rightarrow \frac{\sin\theta}{1 - \cos\theta} = (M + 2m)g/2F$$

$$\Rightarrow \tan(\theta/2) = 2F/(M + 2m)g$$

$$\Rightarrow \theta = 2\tan^{-1}(2F/(M + 2m)g) = \theta_{max}. \quad (4.11)$$

The equilibrium position is given by putting net torque as zero. The net torque acting on the rod is

$$\Rightarrow \tau_{net} = FL\cos\theta - \frac{(M + 2m)gL\sin\theta}{2} = 0$$

$$\Rightarrow \theta = \tan^{-1}(2F/(M + 2m)g) = \theta_O. \quad (4.12)$$

From equations (4.11) and 4.122) we have

$$\theta_{max} = 2\theta_O.$$

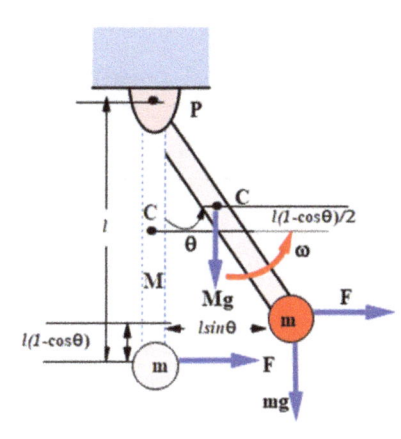

Problem 2 The stepped pulley of mass M, and radii R and r is smoothly pivoted at the point C. The ends of a string are connected with two particles of mass m and M, as shown in the figure. If the pulley is released from rest, find (a) the angular velocity of the pulley as the function of the angle rotated by it, (b) angular the acceleration of the pulley, and (c) the distance of separation between the bodies M and m as the function of time, assuming that the string does not slip over the pulley. The pulley can be considered as a uniform disc.

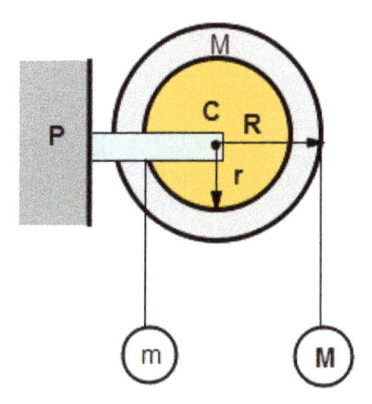

Solution
(a) Let the disc rotate clockwise with an angular velocity w. Then the bodies M and m move down (and up) with velocities wr and wR, respectively. The change in total kinetic energy of the system is

4-25

$$\Delta K = \frac{I_{\text{Pulley}}\omega^2}{2} + \frac{MR^2\omega^2}{2} + \frac{mr^2\omega^2}{2}$$

$$= \frac{(I_{\text{Pulley}} + MR^2 + mr^2)\omega^2}{2}$$

$$= \frac{\{(MR^2/2) + MR^2 + mr^2\}\omega^2}{2}$$

$$\Delta K = \frac{(3MR^2 + 2mr^2)\omega^2}{4}. \tag{4.13}$$

If the body m moves up by a distance h, the body M will come down by a distance h', say. Then the change in gravitational potential energy is

$$\Delta U = -Mgh' + mgh.$$

Putting $h' = hR/r$ we have

$$\Delta U = -MghR/r + mgh = -(MR - mr)gh/r. \tag{4.14}$$

Conservation of energy: As we know

$$\Delta U + \Delta K = 0$$

$$-(MR - mr)gh/r + \frac{(3MR^2 + 2mr^2)\omega^2}{4} = 0$$

$$\Rightarrow \omega = \sqrt{\frac{4(MR - mr)gh}{(3MR^2 + 2mr^2)r}}$$

$$= \sqrt{\frac{4(MR - mr)g\theta}{(3MR^2 + 2mr^2)}} \quad (\because h = r\theta).$$

(b) Then the angular acceleration is

$$\alpha = \frac{1}{2}\frac{d}{d\theta}(\omega^2) = \frac{1}{2}\frac{d}{d\theta}\left[\frac{4(MR - mr)gr\theta}{(3MR^2 + 2mr^2)r}\right]$$

$$= \frac{2(MR - mr)g}{3MR^2 + 2mr^2}.$$

(c) The separation between the bodies after a time t is

$$D = h + h' = \frac{1}{2}R\alpha t^2 + \frac{1}{2}r\alpha t^2$$

$$= \frac{1}{2}(R+r)\alpha t^2.$$

Putting the value of α and simplifying the factors, we have

$$D = \frac{(R+r)(MR-mr)gt^2}{3MR^2 + 2mr^2}.$$

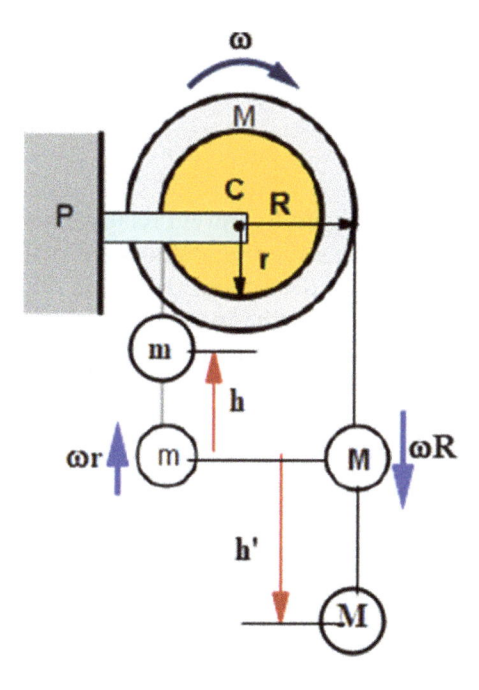

Problem 3 A uniform rod is smoothly hinged with a block of mass M at the point P, as shown in the figure. The block is free to slide along the smooth horizontal rigid bar. If the rod is released from rest when the rod is horizontal, the rod swings in the vertical plane. Find (a) (i) the frictional force acting on the block as a function of angle θ swung by the rod and (ii) the minimum and maximum static friction, (b) (i) the normal force acting on the block as a function of the angle θ swung by the rod and (ii) the minimum and maximum values of the normal reaction, (c) the coefficient of friction between the block and the horizontal rod as the function of the angle θ swung by the rod, (d) the angle at which the linear acceleration of the center of mass of the rod will be horizontal, (e) the value of friction and normal reaction at the above angle, and (f) the value of acceleration of the center of mass at the above angle.

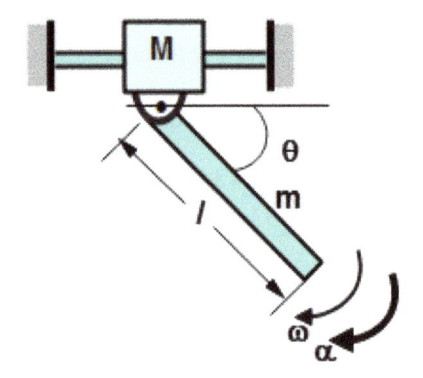

Solution

(a)

(i) Static friction is present to prevent relative sliding of the block on the horizontal rod. The static friction does not perform work because the point of application of friction, that is the block, does not undergo any displacement. Furthermore, the normal reaction does not perform work. So we can conserve the total mechanical energy of the system.

As the rod swings through an angle θ, the block M does not slip along the horizontal bar. The center of mass of the rod falls through a vertical distance $y = (l/2)\sin\theta$. Then the decrease in gravitational potential energy of the rod is equal to the increase in the kinetic energy of the rod. According to the principle of energy conservation,

$$\Delta U + \Delta K = 0$$

$$\Rightarrow \frac{1}{2}\left(\frac{ml^2}{3}\right)\omega^2 - mg\frac{l}{2}\sin\theta = 0$$

$$\Rightarrow \omega = \sqrt{\frac{3g\sin\theta}{l}}. \qquad (4.15)$$

The net torque acting on the rod about the body M is equal to the gravitational torque, which is given as

$$\tau_P = mg\frac{l}{2}\cos\theta = \frac{ml^2}{3}\alpha$$

$$\Rightarrow \alpha = \frac{3g\cos\theta}{2l}. \qquad (4.16)$$

The static friction acts on the block M is

$$f_s = ma_x = m\frac{l}{2}(\omega^2\cos\theta + \alpha\sin\theta). \qquad (4.17)$$

Using the last three equations we have

$$f_s = m\frac{l}{2}\left\{\frac{3g\sin\theta\cos\theta}{l} + \frac{3g}{2l}\cos\theta\sin\theta\right\}$$

$$\Rightarrow f_s = \frac{9}{2}mg\sin\theta\cos\theta.$$

(ii) At $\theta = 0$, $f_s = 0$ and at $\theta = 90°$, $f_s = 0$.
Then f_s is maximum at $\theta = 45°$

$$f_s|_{max} = \frac{9mg}{4}.$$

(b)
(i) The normal reaction acting on the block is given as

$$(M + m)g - N = ma_y$$

$$= m\left(\frac{l}{2}\alpha\cos\theta - \frac{l}{2}\omega^2\sin\theta\right)$$

$$= \frac{ml}{2}\left\{\left(\frac{3g}{2l}\cos\theta\right).\cos\theta - \left(\frac{3g\sin\theta}{l}\right).\sin\theta\right\}$$

$$\Rightarrow (M + m)g - N = +\frac{3mg}{4}[\cos^2\theta - 2\sin^2\theta]$$

$$\Rightarrow N = (M + m)g - \frac{3}{4}mg(\cos^2\theta - 2\sin^2\theta)$$

$$\Rightarrow N = g\left[M + \frac{m}{4}\{4 - 3\cos^2\theta + 6\sin^2\theta\}\right]$$

$$\Rightarrow N = \left[M + \frac{m}{4}\{1 + 9\sin^2\theta\}\right]g$$

(ii)

At $\theta = 0$, $N = \left\{M + \left(\frac{m}{4}\right)\right\}g$

At $\theta = 90°$, $N = (M + 2.5m)g$.

(c) The law of static friction is

$$f_s \leq \mu_s N \Rightarrow \mu_s \geq \frac{f_s}{N}$$

$$\Rightarrow \mu_s \geq \frac{9}{2}\left[\frac{mg \sin\theta \cos\theta}{\dfrac{4M+m(1+9\sin^2\theta)}{4}}g\right]$$

$$\Rightarrow \mu_s \geq \frac{18m \sin\theta \cos\theta}{4M+m(1+9\sin^2\theta)}.$$

(d) If linear acceleration of the center of mass will be horizontal, the vertical acceleration of the center of mass will be zero.
This means $a_{C_y} = 0$

$$\Rightarrow \frac{ml}{2(M+m)}(\omega^2 \sin\theta - \alpha \cos\theta) = 0$$

$$\Rightarrow \frac{3g \sin\theta}{l}\sin\theta - \frac{3g}{2l}\cos\theta \cdot \cos\theta = 0$$

$$\Rightarrow \tan^2\theta = \frac{1}{2} \Rightarrow \theta = \tan^{-1}\frac{1}{\sqrt{2}}.$$

(e) Then the friction at the above angle is $f_s = \frac{9}{2}mg \sin\theta \cos\theta = \frac{9}{2}mg\frac{1}{\sqrt{3}}\cdot\frac{\sqrt{2}}{\sqrt{3}}$

$$f_s = \frac{3mg}{\sqrt{2}}.$$

(f) The acceleration of the center of mass of the rod is
$$a_C = a_{C_x} = \frac{\frac{3mg}{\sqrt{2}}}{M+m} = \frac{3mg}{\sqrt{2}(M+m)}.$$

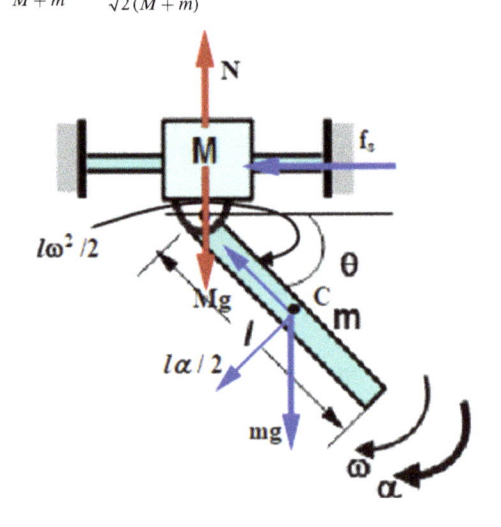

4-30

Problem 4 A uniform rod of mass m and length l is smoothly pivoted at a fixed point. It is released from rest from its horizontal position. Find (a) the radial force and (b) the tangential or shearing force, offered by the length x of the rod measured from its free end, as a function of the angle θ made by the rod with the horizontal.

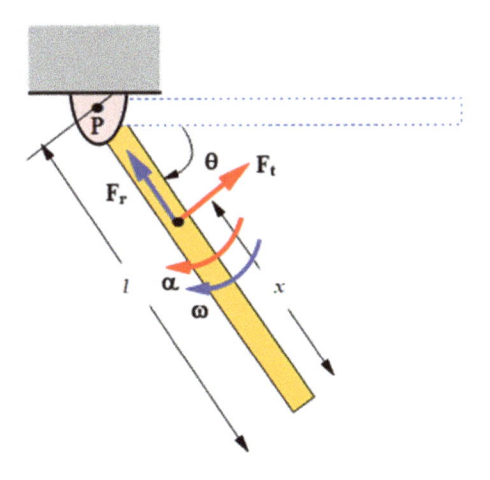

Solution
(a) The reaction force acting at the pivot P of the rod does not perform work because the point of application of the force, does not undergo any displacement. So, we can conserve the total mechanical energy of the rod.

As the rod swings through an angle θ, the center of mass of the rod falls through a vertical distance $y = (l/2)\sin\theta$. Then the decrease in gravitational potential energy of the rod is equal to the increase in the kinetic energy of the rod. According to the principle of energy conservation,

$$\Delta U + \Delta K = 0$$

$$\Rightarrow \frac{1}{2}\left(\frac{ml^2}{3}\right)\omega^2 - mg\frac{l}{2}\sin\theta = 0$$

$$\Rightarrow \omega = \sqrt{\frac{3g\sin\theta}{l}}. \tag{4.18}$$

The net torque acting on the rod about the pivot P is equal to the gravitational torque, which is given as

$$\tau_P = mg\frac{l}{2}\cos\theta = \frac{ml^2}{3}\alpha$$

$$\Rightarrow \alpha = \frac{3g \cos\theta}{2l}. \qquad (4.19)$$

Take a small segment of mass dm at a distance y from the free end of the rod. The radial force acting on the element is given by

$$dF_r = dma_r = dm((l-y)\omega^2 + g\sin\theta).$$

Then the net force acting on the segment of length x is

$$F_r = \int_{l-x}^{l} dF_r = \int_{l-x}^{l} dm(y\omega^2 + g\sin\theta)$$

$$= \int_{l-x}^{l} dmy\omega^2 + \int_{l-x}^{l} dmg\sin\theta$$

$$= \omega^2 \int_{l-x}^{l} dmy + g\sin\theta \int_{l-x}^{l} dm$$

$$= \omega^2 \int_{l-x}^{l} (mdy/l)y + g\sin\theta \int_{l-x}^{l} mdy/l$$

$$= \frac{m}{l}\omega^2 \int_{l-x}^{l} ydy + \frac{mg\sin\theta}{l} \int_{l-x}^{l} dy$$

$$= \frac{m}{2l}\omega^2\{l^2 - (l-x)^2\} + \frac{mg\sin\theta}{l}x$$

$$= \frac{m}{l}\omega^2(2l-x)x + \frac{mg\sin\theta}{l}x$$

$$F_r = \frac{mx}{2l}\{\omega^2(2l-x) + g\sin\theta\}. \qquad (4.20)$$

Using equations (4.18) and (4.20)

$$F_r = \frac{mx}{l}\left\{\frac{3g\sin\theta}{2l}(2l-x) + g\sin\theta\right\}$$

(obtained after putting the value of angular velocity from equation 4.18)

$$\Rightarrow F_r = \frac{mgx(8l-3x)\sin\theta}{2l^2}.$$

(b) The tangential or shearing force acting on the element is given by

$$dF_t = dma_t = dm\{(g\cos\theta) - y\alpha\}.$$

4-32

Then the net shearing force acting on the segment of length x is

$$F_t = \int_{l-x}^{l} dm(g\cos\theta - y\alpha)$$
$$= \int_{l-x}^{l} -(mdy/l)y\alpha + \int_{l-x}^{l} (mdy/l)g\cos\theta$$
$$= -(m\alpha/l)\int_{l-x}^{l} y\,dy + mg\cos\theta/l \int_{l-x}^{l} dy$$
$$= -(m\alpha/2l)[l^2 - (l-x)^2] + mg\cos\theta/l[l-(l-x)]$$
$$= -(m\alpha/2l)(2l-x)x + mgx\cos\theta/l$$
$$= -m(3g\cos\theta/2l)(2l-x)x/2l + mgx\cos\theta/l$$

(obtained after putting the value of angular acceleration from equation 4.19)

$$= \frac{mgx\cos\theta}{l}\left[1 - \frac{3(2l-x)}{4l}\right]$$
$$= \frac{mg\cos\theta}{4l^2}\left\{x(3x - 2l)\right\}.$$

We can see that the shearing force increases from zero to a maximum value at $x = l/3$ (because the first derivative of the function is zero for maxima), becomes zero again at $x = 2l/3$ and changes its direction from down to up (perpendicular to the rod); again increases to a maximum upward value at $x = l$.

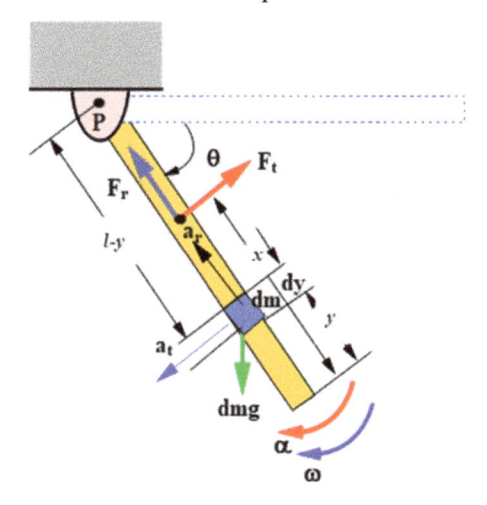

Problem 5 A block of mass m and radius r is placed at the top of a disc of mass M which is pivoted smoothly at its center O, as shown in the figure. If the disc is given a small angular speed, it continues to rotate and the block will revolve in a circular path without sliding on the disc. Find the (a) angular velocity (b) annular acceleration, (c) static friction (d) normal reaction acting on the block, (e) coefficient of static friction as the function of the angle of rotation θ of the disc, (f) angle at which the block will slip, if $M/m = 1$ and the minimum value of coefficient of friction is 3/8.

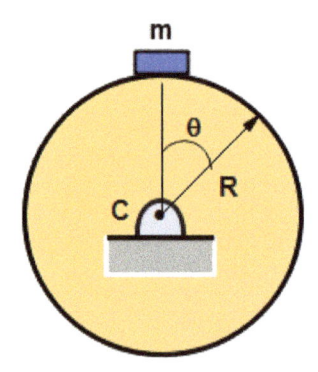

Method 1

(a) As the disc rotates through an angle θ, the block falls down through a vertical distance

$$h = l(1 - \cos\theta).$$

So the change in gravitational potential energy of the system $(M + m)$ is

$$\Delta U = -mgh = -mgl(1 - \cos\theta). \tag{4.21}$$

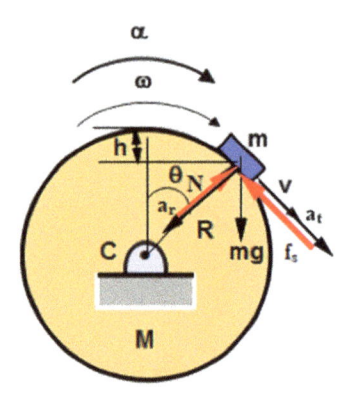

The kinetic energy of a two-particle system is

$$\Delta K = \frac{MR^2\omega^2}{4} + \frac{1}{2}mv^2, \tag{4.22}$$

where $v = R\omega$

$$\Rightarrow \Delta K = \frac{MR^2\omega^2}{4} + \frac{1}{2}mR^2\omega^2. \tag{4.23}$$

The principle of conservation of energy is

$$\Delta U + \Delta K = 0. \tag{4.24}$$

Using equations (4.21), (4.23), and (4.24) we have

$$mgR(1 - \cos\theta) = \frac{MR^2\omega^2}{4} + \frac{mR^2\omega^2}{2}$$

$$\Rightarrow \frac{R^2\omega^2}{4}[M + 2m] = mgR(1 - \cos\phi)$$

$$\Rightarrow \omega = \sqrt{\frac{4mg(1 - \cos\theta)}{(M + 2m)R}}. \tag{4.25}$$

(b) Taking the torque of gravity about the pivot,

$$mgR = \left(\frac{MR^2}{2} + mR^2\right)\alpha$$

$$\Rightarrow \alpha = \frac{2mg\sin\theta}{(M + 2m)R}.$$

(c) The torque produced by static friction on the disc about its center of mass is given as

$$f_s R = \frac{MR^2}{2}\alpha.$$

Putting the value of α we have

$$f_s = \frac{MR}{2} \times \frac{2mg\sin\theta}{(M + 2m)R}$$

$$\Rightarrow f_s = \frac{Mmg\sin\theta}{(M + 2m)}$$

(d) The normal reaction acting on the block is

$$N = m\left(-\frac{v^2}{R} + g\cos\theta\right)$$

$$= m(-R\omega^2 + g\cos\theta).$$

Putting the value of ω we have

$$N = m\left\{g\cos\theta - \frac{4mg(1-\cos\theta)}{M+2m}\right\}$$

$$= mg\left[\frac{(M+2m)\cos\theta - 4m + 4m\cos\theta}{M+2m}\right]$$

$$\Rightarrow N = mg\left[\frac{(M+6m)\cos\theta - 4m}{M+2m}\right].$$

(e) Applying the law of static friction and putting the values of f_s and N, we have

$$\mu_s \geqslant \frac{f_s}{N} = \frac{\frac{Mmg\sin\theta}{M+2m}}{mg\left[\frac{(M+6m)\cos\theta - 4m}{M+2m}\right]}$$

$$\Rightarrow \mu_s \geqslant \frac{M\sin\theta}{(M+6m)\cos\theta - 4m}.$$

(f) Putting $M = m$ and the minimum value of $\mu_s = 3/8$ in the last expression, we have

$$\Rightarrow 3/8 = \frac{\sin\theta}{7\cos\theta - 4}.$$

Solving the last equation we have

$$\theta = 36°.$$

Problem 6 A solid uniform cylinder of mass m and radius R rolls without sliding on the step with a small velocity. Find (a) the angular acceleration, (b) the frictional force, (c) the angular velocity, (d) the normal reaction, (e) the coefficient of static friction verses the angle of rotation, (f) the angle at which the body will slip and lose contact with the edge of the step, for (i) large friction (ii) finite friction having coefficient equal to 3/8 and (g) the speed of the cylinder at the time of escaping from the step assuming sufficient friction. Furthermore, assume a very small value of angular speed of the cylinder and value of height h of the step much greater than the radius R of the cylinder.

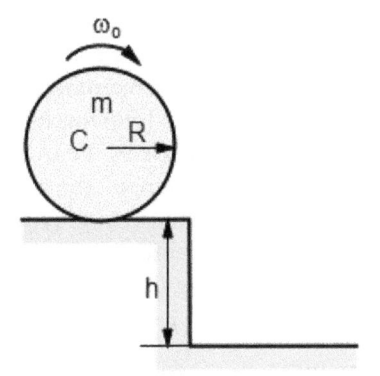

Solution

(a) At the angular position θ the cylinder the net tangential force acting on it is

$$F_t = mg \sin \theta - f_s = ma_t. \qquad (4.26)$$

The net torque acting on the cylinder about the point of contact is equal to the gravitational torque, which is given as

$$\tau_P = mgR \sin \theta = \frac{3mR^2}{2}\alpha$$

$$\Rightarrow \alpha = \frac{2g \sin \theta}{R}. \qquad (4.27)$$

(b) As the center of mass of the cylinder revolves in a circular path about the edge, the tangential acceleration is

$$a_t = R\alpha. \qquad (4.28)$$

The centripetal force is

$$F_r = mg \cos \theta - N = ma_r. \qquad (4.29)$$

The radial acceleration is

$$a_r = R\omega^2. \qquad (4.30)$$

Using the equations (4.26), (4.27), and (4.28) we have

$$f_s = m(g \sin \theta - a_t) = m(g \sin \theta - R\alpha)$$

$$\Rightarrow f_s = m\left[g \sin \theta - R\left(\frac{2g \sin \theta}{3R}\right)\right]$$

$$\Rightarrow f_s = \frac{mg \sin \theta}{3}. \qquad (4.31)$$

(c) Static friction is present to prevent relative sliding of the disc on the table top. The static friction does not perform work because the point of application of friction does not undergo any displacement. Furthermore, the normal reaction does not perform work. So we can conserve the total mechanical energy of the system.

As the rod swings through an angle θ, the cylinder does not slide on the edge of the step. The center of mass of the rod falls through a vertical distance $y = R \sin \theta$. Then the decrease in gravitational potential energy of the cylinder is equal to the increase in the kinetic energy of the disc. According to the principle of energy conservation,

$$\Delta U + \Delta K = 0$$

$$\Rightarrow \frac{1}{2}\left(\frac{3mR^2}{2}\right)\omega^2 - mgR(1 - \cos\theta) = 0$$

$$\Rightarrow \omega = \sqrt{\frac{4g(1 - \cos\theta)}{3R}}. \qquad (4.32)$$

(d) Using the equations (4.29), (4.30), and (4.32) we have

$$N = m(g \cos\theta - a_r)$$

$$\Rightarrow N = m(g \cos\theta - R\omega^2)$$

$$\Rightarrow N = m\left[g \cos\theta - R\frac{4g(1 - \cos\theta)}{3R}\right]$$

$$\Rightarrow N = mg(7 \cos\theta - 4)/3. \qquad (4.33)$$

(e) The law of static friction is

$$f_s \leq \mu_s N. \qquad (4.34)$$

Using the equations (4.31), (4.33), and (4.34) we have

$$\frac{mg \sin\theta}{3} \leq \mu_s mg(7 \cos\theta - 4)/3$$

$$\Rightarrow \sin\theta \leq \mu_s(7 \cos\theta - 4)$$

$$\Rightarrow \mu_s \geq \sin\theta/(7 \cos\theta - 4).$$

(f) (i) The minimum coefficient of static friction must vary with the angle θ so as to prevent relative sliding of the cylinder at the edge of the step. As a result, the cylinder will rotate about the edge of the step up to a maximum angle

$$\theta_m = \cos^{-1} 4/7.$$

At this angle the cylinder must lose contact with the edge. It does not matter how large the friction is. So we obtain an infinite value of the coefficient of static friction when we put the above maximum value of the angle:

$$\theta = \theta_m = \cos^{-1}4/7; \; \mu_s = \infty.$$

(ii) However, for any finite value of the coefficient of static friction, the maximum angle at which the body will slip can be given by the following equation:

$$\sin\theta = \mu_s(7\cos\theta - 4).$$

Putting $\mu_s = 3/8$ in the last equation and solving it we obtain the maximum angle that the body will rotate before it slips is

$$\theta = \theta_m = \cos^{-1}4/5 = 37°.$$

The normal reaction at this angle is

$$N = mg(7\cos\theta - 4)/3$$

$$\Rightarrow N = mg\{(7)(4/5) - 4\}/3 = 8mg/15.$$

This means that the body will begin to slip at this angle, but it will not lose contact with the table. It will rotate by an additional angle before losing contact.

(g) The velocity of the center of mass of the cylinder at the time of escaping from the step is

$$v = R\omega = \sqrt{\frac{4gR(1 - \cos\theta)}{3}}$$

$$\Rightarrow v = \sqrt{\frac{4gR(1 - 4/7)}{3}} = \sqrt{\frac{4gR}{7}}.$$

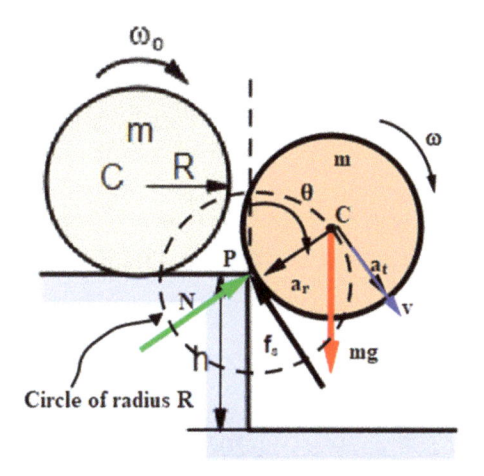

4-39

Rolling bodies

Problem 7 A uniform solid sphere of mass m and radius r and a uniform cylinder of mass M and radius R are kept on two surfaces of a step. A plank of mass m is placed on the bodies and a force F is applied on the plank. If there is no relative sliding at all contacting surfaces, find (a) the kinetic energy of the system if the speed of the plank is v and (b) the speed of the plank as a function of distance x covered by the plank. Assume that the system starts from rest from the given position put $M = 2m$.

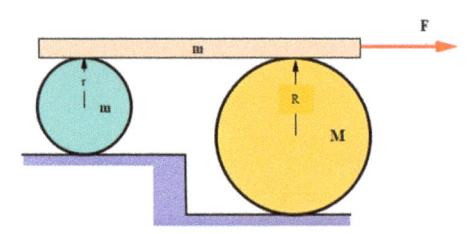

Solution

(a) If the velocity of the plank be v, the velocity of the rolling bodies will be $v/2$. The kinetic energy of the system is

$$K = K_{plank} + K_{sphere} + K_{cylinder}$$

$$= \frac{mv^2}{2} + \frac{m(v/2)^2}{2}(1 + 2/5) + \frac{M(v/2)^2}{2}(1 + 1/2)$$

$$= \frac{mv^2}{2} + \frac{7mv^2}{40} + \frac{3Mv^2}{16}$$

$$= \frac{mv^2}{2}\left(1 + \frac{7}{20} + \frac{3M}{8m}\right)$$

$$= \frac{mv^2}{2}\left(1 + \frac{7}{20} + \frac{3(2m)}{8m}\right) = \frac{21mv^2}{20}.$$

(b) The work done by gravity mg and N is zero as they are perpendicular to the displacement. The net work done by static friction is also zero. Then the work is done by the applied force F only, which is given as $W_F = Fx$. Applying the work–energy theorem we have

$$W_{total} = \Delta K.$$

Putting $W_{total} = Fx$ and $\Delta K = \frac{21mv^2}{20}$ we obtain

$$\text{and } Fx = \frac{21mv^2}{20}.$$

Then we have

$$v_C = v = \sqrt{20Fx/21m}.$$

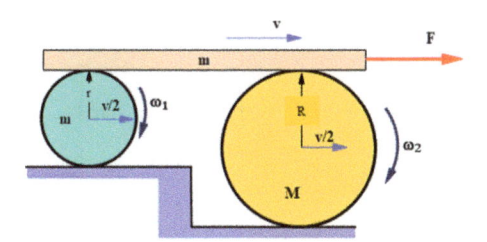

Problem 8 A uniform cylinder of mass m and radius R is kept on a smooth horizontal surface. It is wrapped with a string which does not slip over the cylinder. At time $t = 0$ the string is pulled by a horizontal force F as shown in the figure. The initial velocity of the cylinder is v_0. Applying the work–energy theorem, find (a) the linear velocity, (b) the linear acceleration, (c) the angular acceleration, and (d) the angular velocity of the cylinder as a function of displacement x of its center of mass.

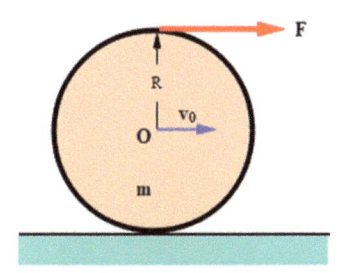

Solution
(a) The work done by gravity mg and N is zero as they are perpendicular to the displacement. Then the work is done by the applied force F only. Applying the work–energy theorem we have
$$W_t = \Delta K_t.$$
Putting $W_t = Fx$ and $\Delta K_t = \frac{mv_C^2}{2}$ we obtain
$$Fx = \frac{mv_C^2}{2} - \frac{mv_0^2}{2}.$$
Then we have
$$v_C = v = \sqrt{v_0^2 + 2Fx/m}.$$

(b) $a_C = v\frac{dv}{dx} = F/m.$

(c) The work–energy theorem for rotation is given as
$$W_r = \Delta K_r,$$
where
$$W_r = \int \tau d\theta = \int FR d\theta = FR\theta$$
and
$$\Delta K_r = \frac{I_C \omega^2}{2} - \frac{I_C \omega_0^2}{2}.$$

Putting $I_C = mR^2/2$ and simplifying the factors we have
$$FR\theta = \frac{mR^2(\omega^2 - \omega_0^2)}{4},$$
or
$$\omega = \sqrt{\omega_0^2 + 4F\theta/mR}.$$

Then the angular acceleration is $\alpha = \frac{\omega d\omega}{d\theta} = 2F/mR$.

(d) Using kinematics we can write $\frac{\omega}{v} = R\alpha/a_C$. Putting the obtained values of a_C and α, we obtain $\frac{\omega R}{v} = 2$. So finally we can write
$$\omega = \frac{2v}{R} = \left(\frac{2}{R}\right)\left(\sqrt{v_0^2 + \frac{2Fx}{m}}\right).$$

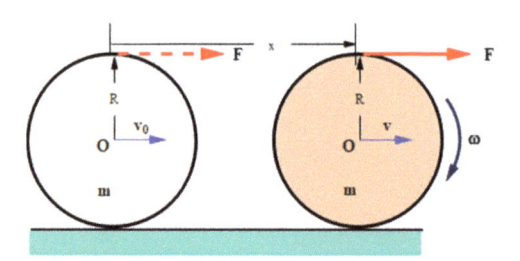

Problem 9 A sphere of mass m, radius R, and radius of gyration K rolls without sliding from a height h_A along the track AB. It climbs onto the smooth track BC up to a height h_C. Find the value of h_A/h_C.

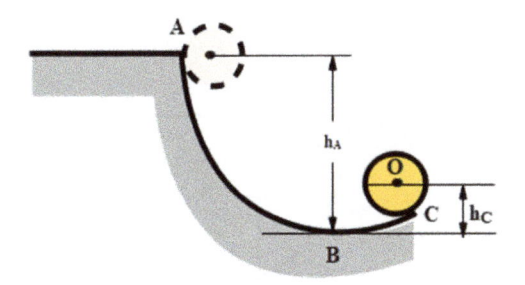

Solution

From A to B the points of action of static friction and normal reaction do not move. So these two forces do not perform work in pure rolling. From B to C there is no friction and the normal reaction is perpendicular to the displacement of the point of contact. So the friction and normal reaction do not perform work. Then we can conserve the total mechanical energy of the body from A to C via B. As the center of mass of the body falls by a vertical distance $h = h_A - h_C$, the change in potential energy of the body is

$$\Delta U = -mgh = -mg(h_A - h_C). \tag{4.35}$$

In BC the frictional torque about the center of mass is zero due to the absence of friction. So the kinetic energy of rotation K_r will not change from B to C. This means that the kinetic energy of translation decreases to zero at the highest point C, while the angular velocity of the body remains unchanged on the smooth surface.

Then the change in kinetic energy between the initial position A and final position C is

$$\Delta K = K_r. \tag{4.36}$$

Energy conservation between the initial position A and final position C can be given as

$$\Delta U + \Delta K = 0. \tag{4.37}$$

Using the last three equations we have

$$-mg(h_A - h_C) + K_r = 0. \tag{4.38}$$

Putting $\Delta U = -mgh_A$ and $\Delta K = K_r(1 + \frac{R^2}{K^2})$ in equation (4.37) we have

$$K_r = mgh_A \bigg/ \left(1 + \frac{R^2}{K^2}\right). \tag{4.39}$$

Putting K_r from equation (4.39) in equation (4.38) and simplifying the factors, finally we have

$$\frac{h_A}{h_C} = \left(1 + \frac{K^2}{R^2}\right).$$

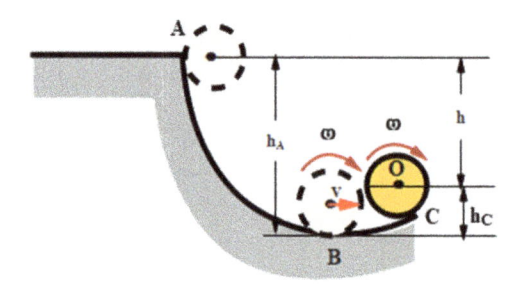

Problem 10 A carpet of mass M rolled in the form of a cylinder of radius R is kept on a rough horizontal surface as shown in the figure. If it is gently pushed towards the right, it unrolls without sliding. Find the speed of the center of mass of the unrolling carpet when its radius will become $r = R/2$. Take necessary assumptions.

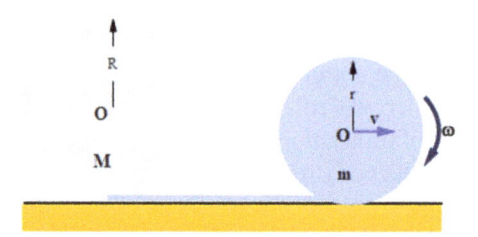

Solution

Let v and ω be the linear and angular velocities of the rolling body (carpet) when its radius is reduced to r. The center of mass of the carpet falls from a height R to r. Furthermore, the mass of the moving portion of the carpet decreases from M to m as the portion of the unrolled carpet lying on the ground does not move. So the potential energy of the moving part of the carpet decreases from MgR to mgr. Conserving the energy we have

$$\frac{1}{2}mv^2 + \frac{1}{2}mK^2\omega^2 = MgR - mgr. \tag{4.40}$$

Kinematics: We know that for rolling

$$r\omega = v. \tag{4.41}$$

Using equations (4.40) and (4.41) we have

$$\Rightarrow v = \sqrt{\frac{2g(MR - mr)}{m\left(1 + \frac{K^2}{r^2}\right)}}$$

$$\Rightarrow v = \sqrt{\frac{2gR\{(M/m) - (r/R)\}}{1 + \frac{K^2}{r^2}}}.$$

Putting $M/m = (R/r)^2$ we have

$$v = \sqrt{\frac{2gr\{(R^3/r^3) - 1\}}{1 + K^2/r^2}}.$$

Putting $r/R = 1/2$ and $k^2/R^2 = 1/2$ we have

$$v = \sqrt{\frac{2gr(8-1)}{1 + 1/2}} = \sqrt{\frac{28gr}{3}} = \sqrt{\frac{14gR}{3}}.$$

Here we have assumed that the vertical velocity of the center of mass of the carpet is very small compared to its horizontal velocity. So, we have ignored the vertical kinetic energy of the center of mass in the calculation.

Problem 11 A cylinder of mass m, radius R, and radius of gyration K placed on a prismatic wedge of mass M, is released from rest from a height h. As a result, it rolls down the wedge without sliding and the wedge recoils back on the smooth horizontal surface. When the cylinder reaches at the bottom of the wedge, find (a) the velocity of (i) the cylinder relative to the wedge and (ii) the wedge relative to the ground, (b) the acceleration of (i) the cylinder relative to the wedge and (ii) the wedge relative to the ground, (c) the normal reaction offered by the wedge to the cylinder, and (d) the values of all the above if the cylinder is uniform.

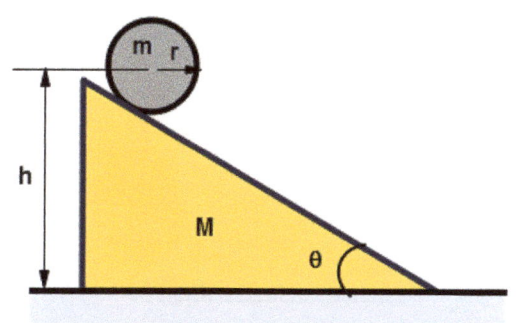

Solution

(a)

(i) Since the net horizontal force acting on the system $(M + m)$ is zero, its horizontal momentum remains conserved. Since both were at rest initially, the initial momentum is zero. Let $u =$ the velocity of m relative to M at the bottom of the wedge and v be the recoil velocity of the wedge. By conserving the horizontal momentum of the system

$$-Mv + m(u\cos\theta - v) = 0$$
$$\Rightarrow m(u\cos\theta - v) = Mv$$
$$\Rightarrow v = \frac{mu\cos\theta}{M+m}. \tag{4.42}$$

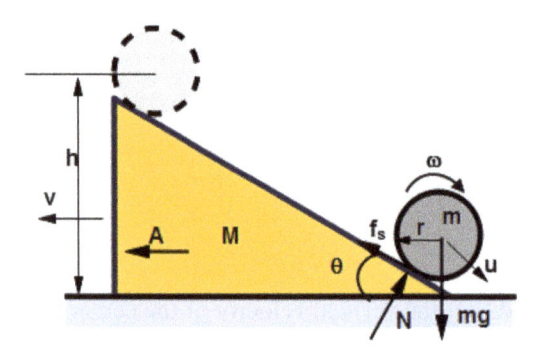

Conserving the energy we have

$$\Delta K = -\Delta U$$
$$\Rightarrow K_m + K_M = mgh,$$

where

$$K_m = \frac{m}{2}v_m^2 + \frac{mK^2}{2}\omega^2$$
$$= \frac{m}{2}v_m^2 + \frac{mk^2u^2}{2R^2}(\because \omega = u/R) \text{ and } K_M = \frac{M}{2}v^2.$$

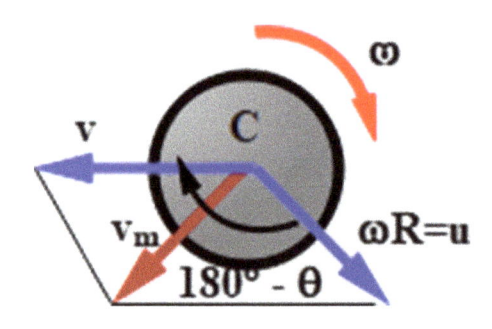

4-46

The velocity of m is given as
$$v_m^2 = u^2 + v^2 - 2uv\cos\theta$$
$$\Rightarrow K_m = \frac{1}{2}m(u^2 + v^2 - 2uv\cos\theta) + \frac{mu^2}{2}\left(\frac{K^2}{R^2}\right).$$

Then the change in kinetic energy of the system $(M + m)$ is
$$\Delta K = K_M + K_m$$
$$= \frac{mv^2}{2} + \frac{1}{2}Mv^2 + \frac{1}{2}mu^2\left(1 + \frac{k^2}{R^2}\right) - muv\cos\theta \quad (4.43)$$
$$= \frac{1}{2}(M + m)v^2 + \frac{1}{2}mu^2\left(1 + \frac{K^2}{R^2}\right) - muv\cos\theta.$$

Using equations (4.42) and (4.43)
$$\Delta K = \frac{1}{2}(M + m)\left(\frac{mv\cos\theta}{M + m}\right)^2 + \frac{1}{2}mv^2\left(1 + \frac{K^2}{R^2}\right) - mv\cos\theta\frac{mv\cos\theta}{M + m}$$
$$= \frac{1}{2}mu^2\{1 + \frac{K^2}{R^2} - \frac{m\cos^2\theta}{M + m}\}$$
$$= \frac{1}{2}\{\frac{K^2}{R^2} + \frac{M + (m - m\cos^2\theta)}{M + m}\}mu^2$$

$$\Rightarrow \Delta K = \frac{1}{2}\{\frac{K^2}{R^2} + \frac{M + m\sin^2\theta}{M + m}\}u^2.$$

As the body m comes down through a vertical distance h, the change in gravitational potential energy of the system is
$$\Delta U = -mgh.$$
Conserving the energy we have
$$\Delta K = -\Delta U$$
$$\Rightarrow \frac{1}{2}\left\{\frac{K^2}{R^2} + \frac{M + m\sin^2\theta}{M + m}\right\}mu^2 = -(-mgh)$$

$$\Rightarrow u = \sqrt{\frac{2gh}{\frac{k^2}{R^2} + \frac{M + m\sin^2\theta}{M + m}}}.$$

(ii) Putting the obtained value of u in equation (4.42) we have
$$v = \left(\frac{m\cos\theta}{M + m}\right)\sqrt{\frac{2gh}{\frac{k^2}{R^2} + \frac{M + m\sin^2\theta}{M + m}}}.$$

4-47

(b)

(i) Let $s =$ the distance traversed by m relative to M. Since $h = s \sin\theta$

$$u^2 = \frac{2gs \sin\theta}{\frac{K^2}{R^2} + \frac{M + m\sin^2\theta}{M+m}}$$

$$\Rightarrow a_{mM} = a_{rel} = u\frac{du}{ds}$$

$$\Rightarrow a_{mM} = a = \frac{g\sin\theta}{\frac{k^2}{R^2} + \frac{M+m\sin^2\theta}{M+m}}.$$

(ii) Assuming $A =$ the acceleration of the wedge M, since there is no net horizontal force acting on the system $(M + m)$, we can write

$$F_x = -MA + m(a\cos\theta - A) = 0$$

$$\Rightarrow A = \frac{ma\cos\theta}{M+m}.$$

Putting the obtained value of acceleration a of the body m,

$$A = \frac{mg\sin\theta\cos\theta}{(M+m)\left(\frac{K^2}{R^2} + \frac{M+m\sin^2\theta}{M+m}\right)}.$$

(c) The normal reaction acting on the body m is $N = m(g\cos\theta - A\sin\theta)$. Putting the value of A and simplifying the factors we have

$$N = mg\cos\theta \left\{ \frac{\frac{K^2}{R^2} + \left(\frac{M}{M+m}\right)}{\frac{K^2}{R^2} + \frac{M+m\sin^2\theta}{M+m}} \right\}.$$

(d) In the obtained expressions, for the uniform cylinder, by putting $\frac{K^2}{R^2} = \frac{1}{2}$,

$$u = \sqrt{\frac{2gh}{\frac{1}{2} + \frac{M+m\sin^2\theta}{M+m}}}$$

we have the velocity of the cylinder relative to the wedge is

$$= 2\sqrt{\frac{(M+m)gh}{3M + m(1 + 2\sin^2\theta)}}.$$

The velocity of the wedge relative to the ground is

$$v = 2m\sqrt{\frac{gh}{(M+m)\{3M + m(1 + 2\sin^2\theta)\}}}.$$

4-48

The acceleration of the wedge relative to the ground is

$$A = \frac{mg \sin\theta \cos\theta}{(M+m)\{\frac{1}{2} + \frac{M+\sin^2\theta}{M+m}\}}$$

$$\Rightarrow A = \frac{2mg \sin\theta \cos\theta}{\{3M + m(1 + 2\sin^2\theta)\}}.$$

The acceleration of m relative to M is

$$a = \frac{M+m}{m \cos\theta} A$$

$$= \frac{M+m}{m \cos\theta} \{\frac{2mg \sin\theta \cos\theta}{3M + m(1 + 2\sin^2\theta)}\}$$

$$\Rightarrow a = \frac{2(M+m)g \sin\theta}{3M + m(1 + 2\sin^2\theta)}.$$

The normal reaction between the cylinder and wedge can be obtained by applying Newton's second law normal to the slant. It is given as

$$N - mg \cos\theta + mA \sin\theta = 0$$

$$N = m(g \cos\theta - A \sin\theta)$$
$$= m\{g \cos\theta - \frac{(2mg \sin\theta \cos\theta)\sin\theta}{3M + m(1 + 2\sin^2\theta)}\}$$
$$= mg \cos\theta \{\frac{3M + m(1 + 2\sin^2\theta) - 2m \sin^2\theta}{3M + m(1 + 2\sin^2\theta)}\} \Rightarrow N = \frac{(3M+m)mg \cos\theta}{3M + m(1 + 2\sin^2\theta)}.$$

Problem 12 A disc of mass m and radius r is placed on a plank of mass M. The bodies are connected by a spring of stiffness k, as shown in the figure. The spring is kept compressed by a string connected between the smooth pivot (axle) of the disc and the vertical light rod fitted with the plank. The initial compression of the spring is x. If the string is cut, the disc will roll without sliding on the plank. Assuming a smooth ground, find the maximum (a) linear velocity of the plank and (b) angular velocity of the disc.

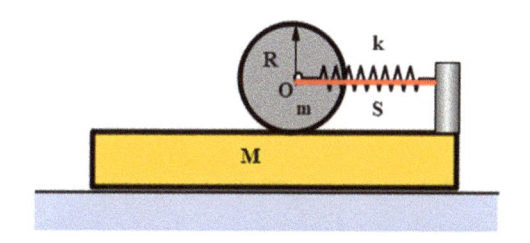

Solution

(a)

Since the net horizontal force acting on the system $(M + m)$ is zero, its horizontal momentum remains conserved. Since both were at rest initially, the initial momentum is zero. Let $u = $ the velocity of m relative to M when the spring is relaxed and v be the recoil velocity of the plank. By conserving the horizontal momentum of the system

$$-Mv + m(u - v) = 0$$

$$\Rightarrow m(u - v) = Mv$$

$$\Rightarrow v = \frac{mu}{M + m}. \qquad (4.44)$$

Conserving the energy we have

$$\Delta K = -\Delta U$$

$$\Rightarrow K_m + K_M = mgh,$$

where

$$K_m = \frac{m}{2}v_m^2 + \frac{mK^2}{2}\omega^2$$

$$= \frac{m}{2}v_m^2 + \frac{mk^2 u^2}{2R^2} (\because \omega = u/R) \text{ and } K_M = \frac{M}{2}v^2.$$

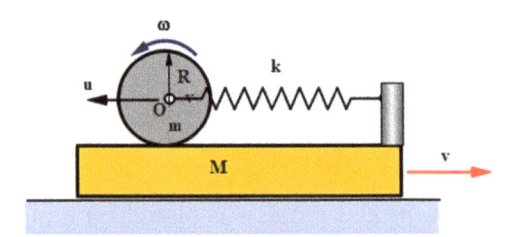

The velocity of m is given as

$$v_m^2 = u^2 + v^2 - 2uv$$

$$\Rightarrow K_m = \frac{1}{2}m(u^2 + v^2 - 2uv) + \frac{mu^2}{2}\left(\frac{K^2}{R^2}\right).$$

Then the change in the kinetic energy of the system $(M + m)$ is

$$\Delta K = K_M + K_m$$

$$= \frac{mv^2}{2} + \frac{1}{2}Mv^2 + \frac{1}{2}mu^2\left(1 + \frac{k^2}{R^2}\right) - muv \qquad (4.45)$$

$$= \frac{1}{2}(M + m)v^2 + \frac{1}{2}mu^2\left(1 + \frac{K^2}{R^2}\right) - muv.$$

Using equations (4.44) and (4.45)

$$\Delta K = \frac{1}{2}(M+m)\left(\frac{mu}{M+m}\right)^2 + \frac{1}{2}mu^2\left(1+\frac{k^2}{R^2}\right) - mu\left(\frac{mu}{M+m}\right)$$

$$= \frac{1}{2}mu^2\left\{1 + \frac{K^2}{R^2} - \frac{m}{M+m}\right\}$$

$$= \frac{1}{2}\left\{\frac{k^2}{R^2} + \frac{M}{M+m}\right\}mu^2$$

$$\Rightarrow \Delta K = \frac{1}{2}\left\{\frac{K^2}{R^2} + \frac{M}{M+m}\right\}mu^2.$$

As the spring becomes relaxed through a distance x, the change in the elastic potential energy of the spring is

$$\Delta U = -\frac{kx^2}{2}.$$

Conserving the energy we have

$$\Delta K = -\Delta U$$

$$\Rightarrow \frac{1}{2}\left\{\frac{K^2}{R^2} + \frac{M}{M+m}\right\}mu^2 = -\left(-\frac{kx^2}{2}\right)$$

$$\Rightarrow u = \sqrt{\frac{k/m}{\frac{K^2}{R^2} + \frac{M}{M+m}}}\, x$$

$$\Rightarrow u = \sqrt{\frac{k/m}{\frac{1}{2} + \frac{M}{M+m}}}\, x.$$

Putting the obtained value of u in equation (4.44) we have

$$v = \left(\frac{m}{M+m}\right)x\sqrt{\frac{k/m}{\frac{1}{2} + \frac{M}{M+m}}}.$$

(b) The annular velocity of the rolling disc is
$$\omega = u/R$$

4-51

$$\Rightarrow \omega = \left(\sqrt{\dfrac{k/m}{\dfrac{K^2}{R^2} + \dfrac{M}{M+m}}} \right)(x/R)$$

$$\Rightarrow \omega = \left(\sqrt{\dfrac{k/m}{\dfrac{K^2}{R^2} + \dfrac{M}{M+m}}} \right)(x/R).$$

Problem 13 A uniform solid cylinder of mass $m_1 = xm$ and radius R is wrapped with a light inextensible string which is connected to a hanging block of mass M after passing over a fixed pulley of mass m and radius r, as shown in the figure. The string does not slip over the cylinder and pulley and the cylinder rolls without sliding on the horizontal surface. At time $t = 0$ the hanging block is released from rest. Find the velocity of the center of mass of the cylinder when the block falls through a distance y. Put $M = 2m$ and $x = 4$.

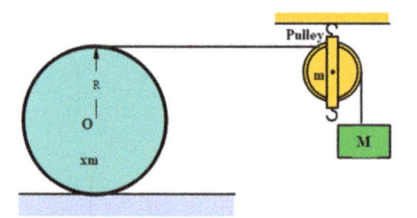

Solution
(a) As the center of mass of the rolling disc moves through a distance y, the string will move through a distance $2y$. As a result the mass M moves down by a distance $2y$. So the velocity of M will be twice that of the center of mass of the cylinder. As the block falls down through a vertical distance $2y$, its change in gravitational potential energy is $-2mgy$. Then the change in gravitational potential energy of the system $(M + m + xm)$ is

$$\Delta U = -2mgy. \qquad (4.46)$$

The change in kinetic energy of the system is

$$\Delta K = K_{\text{cyl}} + K_m + K_M$$
$$= \dfrac{3(xm)v^2}{4} + \dfrac{1}{2}I_m \omega_m^2 + \dfrac{1}{2}m v_M^2,$$

where $v_M = 2v$, $\omega_m = 2v/r$, $I_m = mr^2/2$

$$\Rightarrow \Delta K = 2Mv^2 + \frac{3}{4}(xm)v^2 + 2mv^2. \tag{4.47}$$

The principle of conservation of energy is

$$\Delta U + \Delta K = 0. \tag{4.48}$$

Using equations (4.46), (4.47), and (4.48) we have

$$2Mv^2 + \frac{3}{4}(xm)v^2 + 2mv^2 = 2mgy$$

$$\Rightarrow v^2\left\{2M + \frac{3}{4}(xm) + 2m\right\} = 2mgy$$

$$\Rightarrow v^2\left\{2M + \frac{3}{4}(xm) + 2m\right\} = 2mgy. \tag{4.49}$$

Putting $x = 4$ and $M = 2m$ we have

$$v^2\left\{2(2m) + \frac{3}{4}(4m) + 2m\right\} = 2mgy$$

$$\Rightarrow v^2\{4m + 3m + 2m\} = 2mgy$$

$$\Rightarrow v = \sqrt{2gy}/3.$$

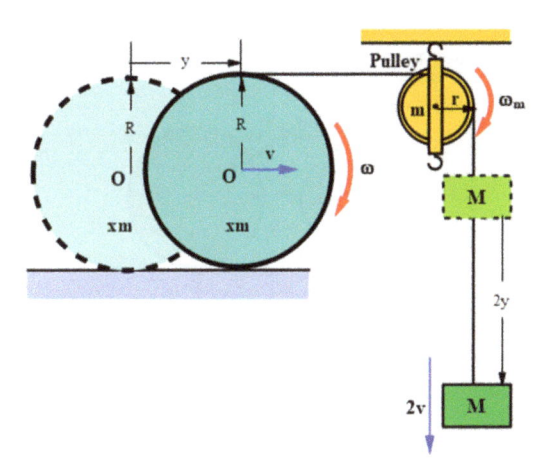

Problem 14 A uniform disc of mass m and radius R is fitted with a light spring of stiffness k at a point P of the ceiling. The disc is given an initial linear and angular velocity such that it rolls without sliding on the horizontal plane. If the disc stops instantaneously at the time of its losing contact with the surface, find the value of linear velocity v.

4-53

Solution

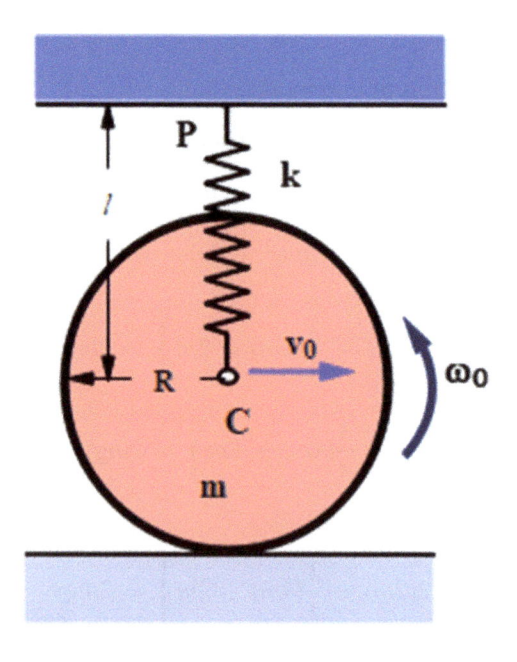

As the disc rolls on the horizontal surface, it moves towards the right, elongating the spring by a distance x, given as

$$x = \frac{l(1 - \cos\theta)}{\cos\theta}. \tag{4.50}$$

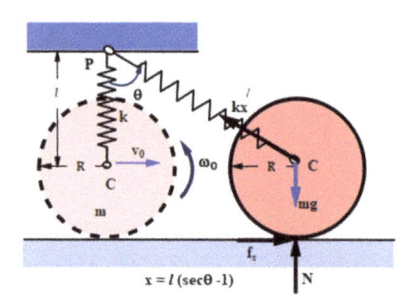

When the rolling disc is lifted off, $N = 0$, so we have

$$kx\cos\theta = mg. \tag{4.51}$$

Using equations (4.50) and (4.51)

$$k\left\{\frac{l(1-\cos\theta)}{\cos\theta}\right\}\cos\theta = mg/l$$

$$\Rightarrow 1 - \cos\theta = \frac{mg}{kl}$$

$$\Rightarrow \cos\theta = 1 - \frac{mg}{kl}. \tag{4.52}$$

Then the elongation of the spring is

$$x = \frac{l(1-\cos\theta)}{\cos\theta} = \frac{l\left(\frac{mg}{kl}\right)}{1-\frac{mg}{kl}} = \frac{mg}{k\left(1-\frac{mg}{kl}\right)}$$

$$\Rightarrow x = \frac{mgl}{kl-mg}. \tag{4.53}$$

In the process of rolling the work done by friction is zero and the other constraint forces such as static friction and the normal reaction do not perform any net work on the spring–disc system. So we can conserve the total mechanical energy of the system.

The change in potential energy is given as

$$\Delta U = \Delta U_{sp} = \frac{kx^2}{2}. \tag{4.54}$$

The change (decrease) in the kinetic energy of the rolling body is given as

$$\Delta K = -\frac{mv_0^2}{2}\left(1+\frac{k'^2}{r^2}\right)$$
$$= -\frac{mv_0^2}{2}\left(1+\frac{1}{2}\right) = -\frac{3mv_0^2}{4}. \tag{4.55}$$

According to the principle of conservation of energy we have

$$\Delta U + \Delta K = 0 \tag{4.56}$$

$$\frac{kx^2}{2} - \frac{3mv_0^2}{4} = 0$$

$$\Rightarrow v_0 = \sqrt{\frac{2k}{3m}}\, x. \tag{4.57}$$

Using equations (4.52) and (4.57)

$$v_0 = \left(\sqrt{\frac{2k}{3m}}\right)\frac{mgl}{kl-mg}$$

$$\Rightarrow v_0 = \frac{l}{\left(\frac{kl}{mg}-1\right)}\sqrt{\frac{2k}{3m}}.$$

Problem 15 A uniform disc of mass M and radius R is fitted with a light spring of natural length l and stiffness k at a point P of the ceiling. The disc is connected to a hanging body of mass m by an inextensible string that passes over the smooth pulley at C. If the body m is released from rest, the disc rolls without sliding on the horizontal plane. Find the speed of the bodies m when the disc is lifted off the ground. Put $k = 5mg/2\ell$

Solution

In the process of rolling the work done by friction is zero and the other constraint forces such as static friction, the normal reaction, and tension do not perform any net work on the system (M + spring + m). So we can conserve the total mechanical energy of the system.

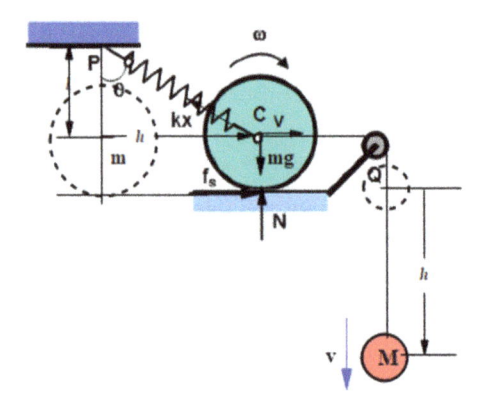

As the rolling body is lifted off when its velocity is v, the change in kinetic energy is

$$\Delta K = +3mv^2/4. \tag{4.58}$$

When the rolling body loses contact with the horizontal surface, we put $N = 0$. Then we can write

$$kx \cos\theta = mg$$

$$\Rightarrow \cos\theta = 1 - \frac{mg}{kl}.$$

Putting $k = \frac{5mg}{2l}$ we have

$$\cos\theta = \frac{3}{5}, \quad \sin\theta = \frac{4}{5}, \quad \text{and} \quad \tan\theta = \frac{4}{3}.$$

4-56

Then, using the conservation of energy, we have
$$\Delta K = -\Delta U$$

$$\text{Or } \frac{3}{4}mv^2 + \frac{1}{2}Mv^2 = \frac{1}{2}Kx^2 + Mgh$$

$$= \frac{1}{2}k\{l(\sec\theta - 1)\}^2 + Mg(l\tan\theta)$$

$$= \frac{1}{2}k\left\{\left(\frac{5}{3} - 1\right)l\right\}^2 + Mgl\left(\frac{4}{3}\right)$$

$$= \frac{2}{9}kl^2 + \frac{4}{3}mgl$$

$$= \frac{2}{9}\left(\frac{5}{2}\frac{mg}{l}\right)l^2 + \frac{4}{3}Mgl$$

$$= \frac{5}{9}mgl + \frac{4}{3}Mgl$$

$$\Rightarrow \frac{3m+2M}{4}v^2 = \frac{(5m+12M)gl}{9}$$

$$\Rightarrow v = \frac{2}{3}\sqrt{\frac{12M+5m}{2M+3m}gl}$$

$$\Rightarrow \omega = v/R = \frac{2}{3R}\sqrt{\frac{12M+5m}{2M+3m}gl}.$$

Rolling bodies on a curved surface

Problem 16 A small uniform solid sphere rolls without sliding along a spherical track from a height $h = R/4$, where R is the radius of curvature of the bottom spherical surface. Find (a) the normal reaction acting on the ball at the top of the spherical surface and (b) the angular position where the sphere will lose contact from the track. Assume that the sphere moves in a vertical plane

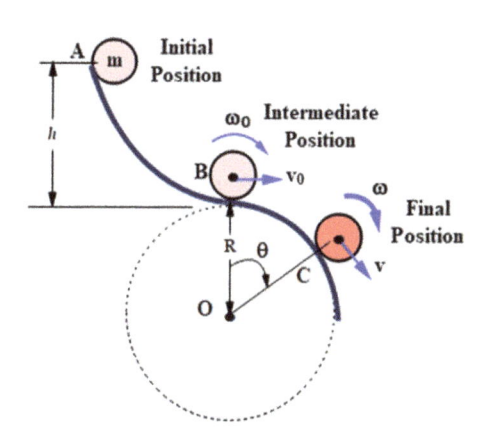

Solution

(a) The body rolls down through a vertical distance of $R/4$ while traveling from A to B. Its gravitational potential energy decreases by

$$\Delta U = -mg\frac{R}{4}.$$

Furthermore, the kinetic energy increases by

$$\Delta K = \frac{7}{10}mv_0^2.$$

Conserving energy for the rolling body,

$$\Delta K = -\Delta U.$$

Using the last three equations we have

$$\frac{7}{10}mv_0^2 = -\left(-mg\frac{R}{4}\right)$$

$$\Rightarrow v_0 = \sqrt{\frac{5}{14}gR}.$$

Then the normal reaction at the top is

$$N = m\left(g - \frac{v_0^2}{R}\right).$$

Putting the obtained value of v_0 we have

$$N = m\left(g - \frac{v_0^2}{R}\right) = m(g - 5g/14) = 9mg/14.$$

(b) Let the sphere further roll from B to C until it leaves the circular track at C. At an angular position θ of the sphere, its center of mass descends by a vertical distance $R(1-\cos\theta)$.

So the change (decrease) in potential energy is

$$\Delta U = -mgR(1 - \cos\theta). \tag{4.59}$$

The kinetic energy of the rod changes (increases) from zero to K_B to K_C. Then the change in kinetic energy is

$$\Delta K = K_C - K_B.$$

Putting $K_B = mgR/4$ and $K_C = (mv^2/2)(1 + 2/5) = 7mv^2/10$ in the last equation we have

4-58

$$\Delta K = \frac{7}{10}mv^2 - mgR/4. \tag{4.60}$$

Conserving the energy we have

$$\Delta U + \Delta K = 0. \tag{4.61}$$

Using the last three equations we have

$$-mgR(1 - \cos\theta) + \left\{\frac{7}{10}mv^2 - mgR/4\right\} = 0$$

$$\Rightarrow \frac{7}{10}mv^2 = mgR(5/4 - \cos\theta)$$

$$\Rightarrow \frac{7}{10}v^2 = gR(5/4 - \cos\theta). \tag{4.62}$$

For losing contact at C, $N = 0$. Then

$$N = m\left(g\cos\theta - \frac{v^2}{R}\right) = 0$$

$$\Rightarrow v^2 = gR\cos\theta. \tag{4.63}$$

Using equations (4.62) and (4.63)

$$\Rightarrow \frac{7}{10}gR\cos\theta = gR(5/4 - \cos\theta)$$

$$\Rightarrow \frac{17}{10}gR\cos\theta = gR(5/4)$$

$$\Rightarrow \cos\theta = 25/34$$

$$\Rightarrow \theta = \cos^{-1}\left(\frac{25}{34}\right).$$

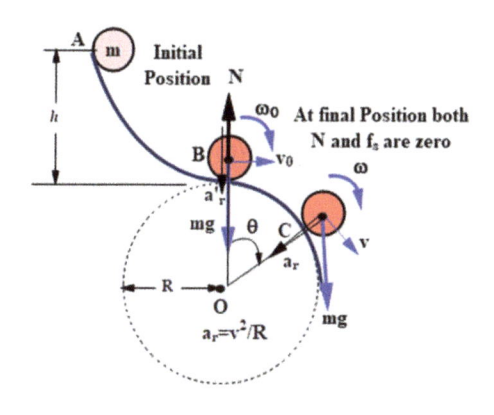

4-59

Problem 17 A uniform cylinder of mass m and radius r and radius of gyration K is released from rest from the top of a wedge of mass M. The sphere rolls without sliding along the circular track of a wedge of radius R. When the cylinder reaches its lowest position, find (a) the relative velocity between the cylinder and wedge, (b) the angular velocity of the cylinder, and (c) the velocities of the cylinder and wedge.

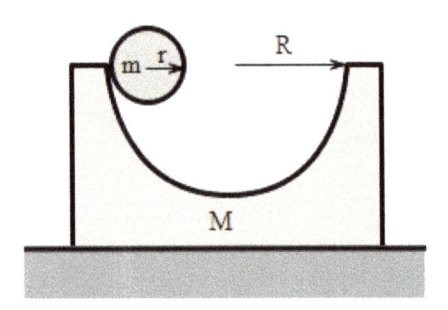

Solution

(a) Let v_1 and v_2 be the velocity of m and M when the cylinder is at its lowest position. Conserving the linear momentum,

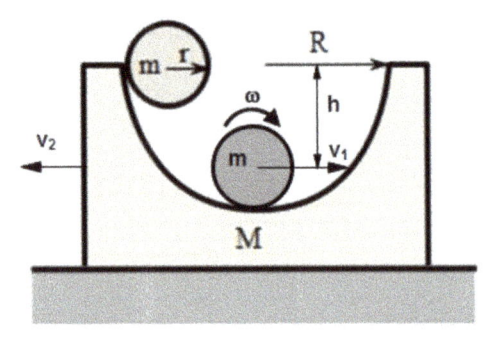

$$P = mv_1 - Mv_2 = 0. \tag{4.64}$$

Conserving the energy we have

$$\frac{1}{2}mv_1^2 + \frac{1}{2}mK^2\omega^2 + \frac{1}{2}Mv_2^2 = mg(R - r). \tag{4.65}$$

Kinematics: We know that

$$r\omega = v_1 + v_2(=v, \text{ say}). \tag{4.66}$$

Now we have three equations and three unknown quantities. Using equations (4.64) and (4.66) we have

4-60

$$v_1 = \frac{Mv}{M+m}, \quad v_2 = \frac{mv}{M+m}. \qquad (4.67)$$

Putting these values of linear velocities from equation (4.67) and angular velocity from equation (4.66) in equation (4.65) and simplifying the factors, we have

$$v = \sqrt{\frac{2g(R-r)}{\frac{M}{M+m} + \frac{K^2}{r^2}}}.$$

(b) Then using equation (4.66) we have

$$\omega = \sqrt{\frac{2g(R-r)}{\frac{Mr^2}{M+m} + K^2}}.$$

(c) The linear velocities of the bodies are

$$v_1 = \frac{Mv}{M+m} = \frac{M}{M+m}\sqrt{\frac{2g(R-r)}{\frac{M}{M+m} + \frac{K^2}{r^2}}}$$

$$v_2 = \frac{mv}{M+m} = \frac{m}{M+m}\sqrt{\frac{2g(R-r)}{\frac{M}{M+m} + \frac{K^2}{r^2}}}.$$

Problem 18 A rough uniform rolling body (a solid sphere, say) of mass m and radius r and radius of gyration K is released from rest from the top of the wedge of mass M. The sphere rolls without sliding along the circular (cylindrical) track of the wedge of radius R from A to B. If the track from B to C is (a) smooth or (b) sufficiently rough, find the maximum height attained by the sphere and wedge.

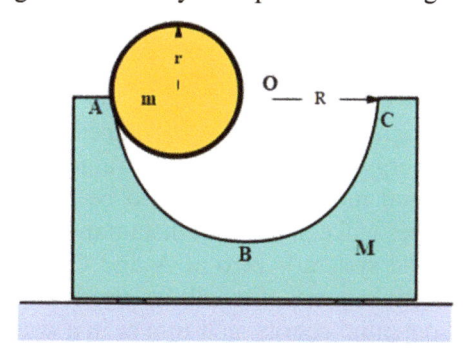

Solution
(a) Let v be the velocity of the rolling body relative to the wedge at the lowest position B. In the last problem, we obtained the expression of v as

$$v^2 = \frac{2g(R-r)}{\frac{M}{M+m} + \frac{k^2}{r^2}}. \tag{4.68}$$

The path from B to C is smooth. So frictional torque about the center of mass of the rolling body is zero. Then the spin angular momentum of the sphere remains conserved. Then the translational kinetic energy of the center of mass of the sphere at the lowest position will be partially converted to the gravitational potential energy at C.

If v is the relative velocity between M and m at C, the translational kinetic energy of the two body system is given as

$$K = \frac{Mmv^2}{2(M+m)}. \tag{4.69}$$

Relative to the frame of the center of mass of the wedge–body ($M + m$) system, the loss of kinetic energy K will be equal to the increase in gravitational potential energy mgh:

$$\frac{Mmv^2}{2(M+m)} = mgh. \tag{4.70}$$

Eliminating v from equation (4.70) by equation (4.68) we have

$$h_{\max} = \frac{2Mg(R-r)}{2(M+m)g\left\{\frac{M}{M+m} + \frac{k^2}{r^2}\right\}}$$

$$= \frac{2Mg(R-r)}{2Mg\left(1 + \frac{k^2(M+m)}{r^2 M}\right)}$$

$$= \frac{g(R-r)}{1 + \frac{2}{5}\frac{(M+m)}{M}} = \frac{5M(R-r)}{7M+2m}.$$

(b) For a sufficiently rough track from B to C rolling takes place, so the kinetic energy of the sphere at C will be zero relative to the wedge. Then conserving the total energy between A and B relative to the wedge, the sphere will go to the same initial height. It can also be obtained by conservation of linear momentum and energy relative to ground frame. Since the initial momentum is zero, the final momentum will also be zero. This will be possible when both the bodies will come to rest instantaneously. This means that the kinetic energy of the system is zero at A and C. Therefore, the change in gravitational potential energy of m will be zero because M does not change its gravitational potential energy as it moves in a straight line horizontally. In other words, m will attain the height $h = R - r$.

Problem 19 A small uniform sphere of mass m and radius r is released from rest from the top of the wedge of mass M which is placed by the side of the vertical wall. The sphere rolls without sliding along the circular track of the wedge of radius R. Find the maximum (a) velocity of the sphere and (b) maximum height attained by the sphere. Assume that the ground is smooth.

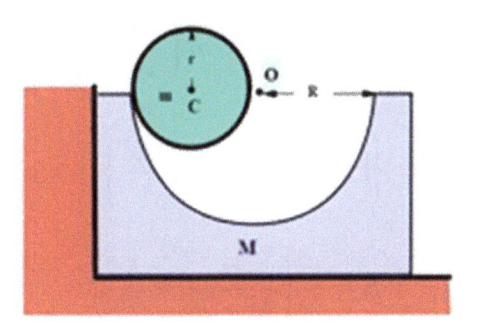

Solution
(a) Until the sphere reaches its lowest position the wedge will remain stationary as the wall prevents it from moving to left. Let v and ω be the linear and angular velocity of the rolling body when it reaches the bottom. Conserving the energy, we have

$$\frac{1}{2}mv^2 + \frac{1}{2}mK^2\omega^2 = mg(R - r). \tag{4.71}$$

Kinematics: We know that, for rolling,

$$r\omega = v. \tag{4.72}$$

Using the equations (4.71) and (4.72) we have

$$\Rightarrow v = \sqrt{\frac{2g(R-r)}{1 + \frac{K^2}{r^2}}}.$$

(b) After the wedge losing contact with the vertical wall, the net horizontal force acting on the system is zero. So the horizontal momentum of the system $(M + m)$ can be conserved. At the maximum height h, say, the relative velocity between the bodies will be zero. Then both the bodies will have the same velocity, v', say. Conserving the momentum between the lowest and highest positions of the sphere, we have

$$mv = (M + m)v'. \tag{4.73}$$

Conserving energy of the system ($M + m$) between the lowest and highest position of the sphere, we have, we have

$$\frac{1}{2}mv^2\left(1 + \frac{K^2}{r^2}\right) = \frac{1}{2}(M + m)v'^2 + mgh. \tag{4.74}$$

Putting v' from equation (4.73) in equation (4.72) we have

$$\frac{1}{2}mv^2\left(1 + \frac{K^2}{r^2}\right) - \frac{1}{2}(M+m)\left(\frac{mv}{M+m}\right)^2 = mgh$$

$$\Rightarrow \frac{1}{2}mv^2\left\{\left(1 - \frac{m}{M+m}\right) + \frac{K^2}{r^2}\right\} = mgh$$

$$\Rightarrow h = \frac{v^2}{2g}\left(\frac{M}{M+m} + \frac{K^2}{r^2}\right).$$

Putting the obtained value of v we have

$$h = \frac{R - r}{\left(1 + \frac{K^2}{r^2}\right)}\left(\frac{M}{M+m} + \frac{K^2}{r^2}\right).$$

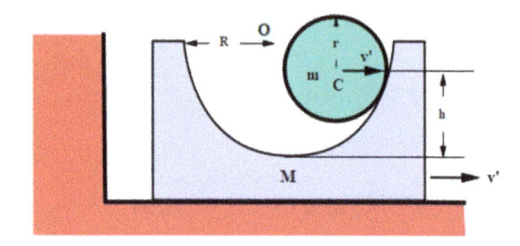

Problem 20 A solid uniform sphere of mass m and radius r is released from the top of a fixed hemisphere of radius R by a gentle push ($v_0 \cong 0$). The coefficient of friction between the sphere and hemisphere is μ. If it rolls without sliding find (a) the angular speed, (b) the normal reaction, (c) the friction acting on the sphere, (d) the coefficient of friction between the sphere and hemispherical surface as a function of the angle θ made by the line joining the centers of the sphere and hemisphere with the vertical, and (e) the position at which the sphere slips and escapes from the hemisphere for (i) sufficient friction and (ii) insufficient friction of $\mu = 1/3$,

$$(\mu \to \infty).$$

Solution

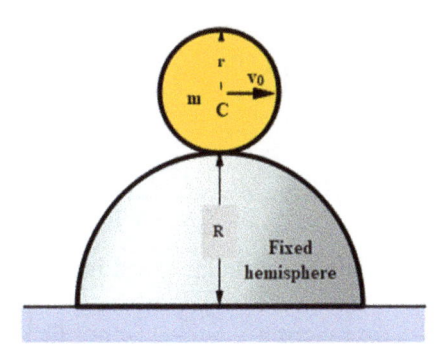

(a) Let v and ω be the linear and angular velocity of the rolling body, respectively, at an angular position θ. The center of mass of the body falls from a height $R + r$ to $(R + r)\cos\theta$. So the decrease in the potential energy of the body is

$$\Delta U = m(R + r)(1 - \cos\theta).$$

The increase in the kinetic energy of the body is

$$\Delta K = \frac{1}{2}mv^2 + \frac{1}{2}mK^2\omega^2.$$

Conserving the energy we have

$$\frac{1}{2}mv^2 + \frac{1}{2}mK^2\omega^2 = mg(R + r)(1 - \cos\theta). \tag{4.75}$$

Kinematics: We know that, for rolling,

$$r\omega = v. \tag{4.76}$$

Using the equations (4.75) and (4.76), and putting $K^2/R^2 = 12/5$, we have

$$\frac{1}{2} \times \frac{7}{5}mr^2\omega^2 = mg(R + r)(1 - \cos\theta)$$

$$\Rightarrow \omega = \sqrt{\frac{10g(R + r)(1 - \cos\theta)}{7r^2}}.$$

(b) The net force along the radial direction is

$$mg\cos\theta - N = ma_r$$

where

$$a_r = \frac{v^2}{(R + r)}$$

$$\Longrightarrow N = m\left\{g\cos\theta - \frac{v^2}{(R+r)}\right\}.$$

Putting $v = r\omega$ we have

$$N = m\left\{g\cos\theta - \frac{r^2\omega^2}{R+r}\right\}. \tag{4.77}$$

Putting the obtained value of angular speed in the equation (4.77) we have

$$N = m\left\{g\cos\theta - \frac{10g}{7}(1-\cos\theta)\right\}$$

$$\Rightarrow N = (17\cos\theta - 10)\frac{mg}{7}.$$

(c) The net torque about the center of mass, that is frictional torque, is given as

$$f_s r = \frac{2}{5}mr^2\alpha \Rightarrow f_s = \frac{2}{5}mr\alpha.$$

The net torque about the point of contact of the sphere, that is gravitational torque, is given as

$$(mg\sin\theta)r = \frac{7}{5}mr^2\alpha \Rightarrow \alpha = \frac{5}{7r}g\sin\theta$$

$$\Rightarrow f_s = \frac{2}{5}mr \cdot \frac{5}{7r}g\sin\theta \Rightarrow f_s = \frac{2}{7}mg\sin\theta.$$

(d) The law of static friction is

$$f_s \leq \mu_s N.$$

(i) Putting the values of f_s and N we have

$$\Rightarrow \frac{2}{7}mg\sin\theta \leq \mu_s(17\cos\theta - 10)\frac{mg}{7}$$

$$\Rightarrow \mu_s \geq \frac{2\sin\theta}{17\cos\theta - 10}.$$

(e) (i) Putting $\mu_s = \infty$ for a sufficiently rough surface we have

$$\theta = \cos^{-1}\frac{10}{17}.$$

This means that the body slips and escapes simultaneously at this position.
(ii) If $\mu_s = \frac{1}{3}$, $17\cos\theta - 10 = 6\sin\theta$

$$\Rightarrow 289\cos^2\theta - 340\cos\theta + 100 = 36 - 36\cos^2\theta$$
$$\Rightarrow 325\cos^2\theta - 340\cos\theta + 64 = 0$$
$$\Rightarrow \cos\theta = \frac{4}{5}.$$

Hence at $\theta = \cos^{-1}\frac{4}{5} = 36°$ and the body slips if $\mu_s = \frac{1}{3}$ but it does not escape at this position.

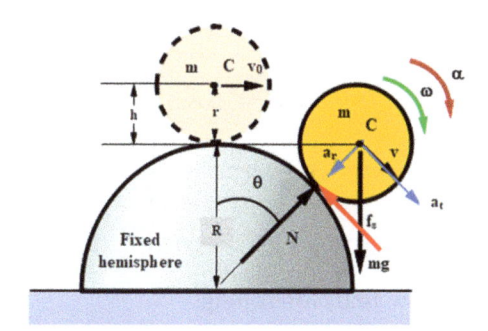

Problem 21 A ball of mass m and radius r rolls without sliding on a rough hemisphere of mass M and radius R which is placed on a smooth horizontal surface, as shown in the figure. (a) Find the velocity of the hemisphere as the function of the angle moved by the line joining the centers of the hemisphere and the ball. (b) At which angular position will the ball lose contact from the hemisphere? (c) Find the velocity of the ball relative to the hemisphere and velocity of the hemisphere at the time of losing contact of the ball. Put $M/m = 29/25$.

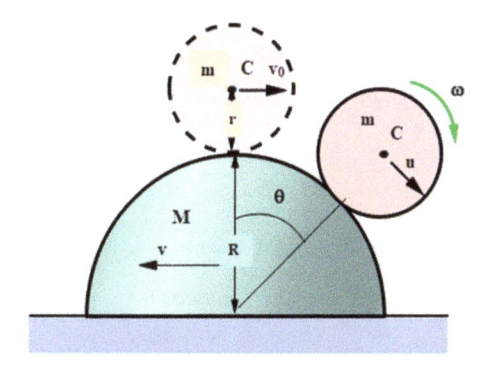

4-67

Solution

(a) It is given that u is the linear velocity of the rolling body relative to the wedge and ω is the angular velocity of the rolling body at an angular position θ. The center of mass of the body drops from a height $R + r$ to $(R + r)\cos\theta$. So the decrease in potential energy of the body is

$$\Delta U = m(R + r)(1 - \cos\theta).$$

The increase in the kinetic energy of the body is

$$\Delta K = \frac{1}{2}mv_m^2 + \frac{1}{2}mr^2\omega^2 + \frac{1}{2}Mv_M^2.$$

Conserving the energy we have

$$-\Delta U = \Delta K$$

$$\Rightarrow \frac{1}{2}mv_m^2 + \frac{1}{4}mr^2\omega^2 + \frac{1}{2}Mv_M^2$$

$$= mg(R + r)(1 - \cos\theta). \tag{4.78}$$

Kinematics: We know that, for rolling,

$$r\omega = u. \tag{4.79}$$

Furthermore, the velocity of the cylinder is

$$v_m^2 = (u^2 + v^2 - 2uv\cos\theta). \tag{4.80}$$

Using the last three equations we have

$$mg(R + r)(1 - \cos\theta) = \frac{1}{2}Mv^2 + \frac{1}{2}m(u^2 + v^2 - 2uv\cos\theta) + \frac{mu^2}{4}$$

$$\Rightarrow \frac{1}{2}(M + m)v^2 + \frac{3}{4}mu^2 - muv\cos\theta = mg(R + r)(1 - \cos\theta). \tag{4.80a}$$

Conserving horizontal momentum,

$$m(u\cos\theta - v) = Mv$$

$$\Rightarrow v = \frac{mu\cos\theta}{M + m}. \tag{4.81}$$

Using (4.80a) and (4.81) we have

$$\Rightarrow \frac{1}{2}(M + m)\left(\frac{mu\cos\theta}{M + m}\right)^2 + \frac{3}{4}mu^2 - mu\left(\frac{mu\cos\theta}{M + m}\right)\cos\theta = mg(R + r)(1 - \cos\theta)$$

$$\Rightarrow \frac{3}{4}mu^2 - \frac{m^2u^2\cos^2\theta}{2(M+m)} = mg(R+r)(1-\cos\theta)$$

$$\Rightarrow \left\{\frac{3(M+m) - 2m\cos^2\theta}{4(M+m)}\right\}mu^2$$

$$= mg(R+r)(1-\cos\theta)$$

$$\Rightarrow u = 2\left\{\sqrt{\frac{(M+m)g(R+r)(1-\cos\theta)}{3M+m(1+2\sin^2\theta)}}\right\}.$$

Putting the value of u in the equation

$$v = \frac{mu\cos\theta}{M+m}$$

we obtain the expression

$$v = 2\sqrt{\frac{m^2(R+r)\cos^2\theta(1-\cos\theta)}{(M+m)\{3M+m(1+2\sin^2\theta)\}}}. \qquad (4.82)$$

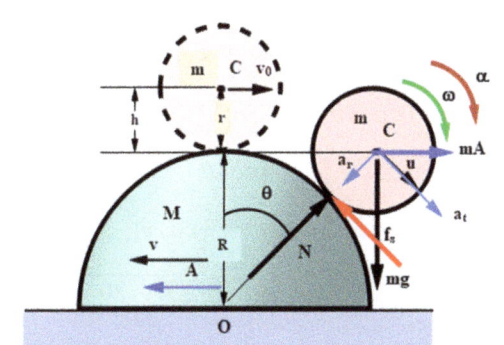

(b) Since $N = 0$, the rolling body loses contact, $f_s = \mu_s N = 0$. Then the wedge will experience zero net force. So at that time the only force acting on the rolling body is gravity. To write the force equation on the rolling body m relative to the accelerating wedge M, we have imposed the pseudo or inertial force mA in addition to all real forces such as gravity, friction etc. The net radially inward force acting on m is

$$mg\cos\theta - N - mA\sin\theta = ma_r \frac{mu^2}{R+r}.$$

If the body m loses contact with the wedge, put $N = 0$ so, the acceleration of the wedge is $A = 0$. Then, we have $mg\cos\theta = \frac{mu^2}{R+r}$

4-69

$$\Rightarrow u = \sqrt{(R+r)g\cos\theta}. \tag{4.83}$$

By using equations (4.82) and (4.83)

$$2\sqrt{\frac{(M+m)g(R+r)(1-\cos\theta)}{3M+m(1+2\sin^2\theta)}} = \sqrt{(R+r)g\cos\theta}$$

$$\Rightarrow \cos\theta = \frac{4(M+m)(1-\cos\theta)}{3M+m(1+2\sin^2\theta)}.$$

Putting $\frac{M}{m} = \frac{29}{25}$ we have

$$\cos\theta = \frac{4\left(\frac{29}{25}+1\right)(1-\cos\theta)}{3\left(\frac{29}{25}\right)+1+2\sin^2\theta}$$

$$\cos\theta = \frac{\frac{216(1-\cos\theta)}{25}}{\frac{87+25+50\sin^2\theta}{25}} = \frac{216(1-\cos\theta)}{112+50\sin^2\theta}$$

$$\Rightarrow 50\sin^2\theta\cos\theta + 112\cos\theta + 216\cos\theta - 210 = 0$$

$$\Rightarrow 50\cos\theta - 50\cos^3\theta + 328\cos\theta - 210 = 0$$

$$\Rightarrow 378\cos\theta - 50\cos^3\theta - 216 = 0$$

$$\Rightarrow 25\cos^3\theta - 184\cos\theta + 108 = 0$$

$$\Rightarrow (5\cos\theta - 3)(5\cos^2\theta + 3\cos\theta - 36) = 0$$

$$\Rightarrow \cos\theta = \frac{3}{5} \text{ or } \cos\theta = \frac{-3\pm\sqrt{9+20\times 36}}{10} > 1$$

$$\Rightarrow \theta = \cos^{-1}\frac{3}{5}.$$

(c) Putting the above value of the angle in equation (4.83) we have

$$u = \frac{\sqrt{15g(R-r)}}{18}, \quad v = \frac{47\sqrt{g(R+r)}}{70}.$$

Problem 22 The cylinder of mass m and radius r is released from rest from the given position. In consequence it rolls without sliding on the circular cut of the rectangular wedge of mass M. The ground is smooth and $R =$ the radius of curvature of the cut.

Find (a) the maximum speed of the wedge, and (b) the maximum normal reaction offered by the ground on the wedge.

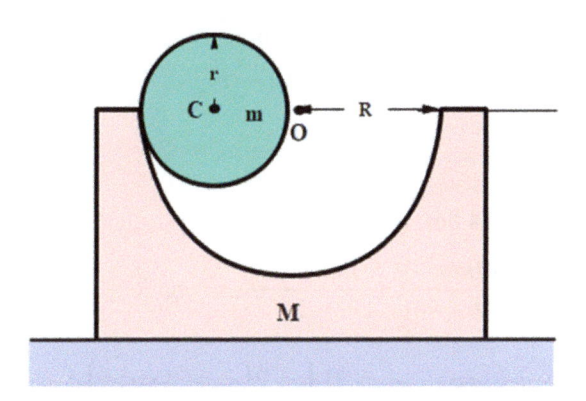

Solution

(a) Let u be the linear velocity of the cylinder relative to the wedge and v be the linear velocity of the wedge when the cylinder reaches its lowest position. Then the angular velocity of the cylinder is

$$v_m = u - v$$

at the bottom. The center of mass of the body falls from a height R to $R - r$. So the decrease in the potential energy of the body is

$$\Delta U = mg(R - r).$$

The increase in the kinetic energy of the body is

$$\Delta K = \frac{1}{2}mv_m^2 + \frac{1}{4}mr^2\omega^2 + \frac{1}{2}Mv_M^2.$$

Conserving the energy we have

$$-\Delta U = \Delta K$$

$$\Rightarrow \frac{1}{2}mv_m^2 + \frac{1}{2}mk^2\omega^2 + \frac{1}{2}Mv_M^2 = mg(R - r). \tag{4.84}$$

We know that

$$v_m = u - v. \tag{4.85}$$

Using equations (4.84) and (4.85) we have

$$\Rightarrow \frac{1}{2}(M + m)v^2 + \frac{1}{2}mk^2\omega^2 + \frac{1}{2}mu^2 - muv$$

$$= mg(R - r). \tag{4.86}$$

Kinematics: We know that, for rolling,

$$r\omega = u. \tag{4.87}$$

Conserving of horizontal momentum,

$$Mv = m(u - v)$$

$$\Rightarrow v = \frac{mu}{M + m}. \tag{4.88}$$

Using equations (4.86), (4.87), and (4.88)

$$\frac{Mmu^2}{2(M+m)} + \frac{mu^2(k/r)^2}{2} = mg(R - r)$$

$$\Rightarrow g(R - r) = \frac{u^2}{2}\left\{\frac{M}{M+m} + (k/r)^2\right\}$$

$$\Rightarrow u = \sqrt{\frac{2g(R - r)}{\left(\frac{M}{M+m} + (k/r)^2\right)}}.$$

Putting the value of u in equation (4.88) the velocity of the wedge is

$$\Rightarrow v = \frac{mu}{M + m}$$

$$\Rightarrow v = \frac{m}{M + m}\sqrt{\frac{2g(R - r)}{\left(\frac{M}{M+m} + (k/r)^2\right)}}.$$

(b) The net upward force acting on m is

$$N - Mg - mg = ma_r = m\left(\frac{u^2}{R - r}\right)$$

$$\Rightarrow N = (M + m)g + m\frac{2g}{\left(\frac{M}{M+m} + \frac{k^2}{r^2}\right)}$$

$$\Rightarrow N = (M + m)g + 2m\frac{(M + m)r^2 g}{Mr^2 + (M + m)k^2}$$

$$\Rightarrow N = (M + m)g\left\{1 + \frac{2}{\frac{M}{m} + \left(\frac{M+m}{m}\right)(k/r)^2}\right\}.$$

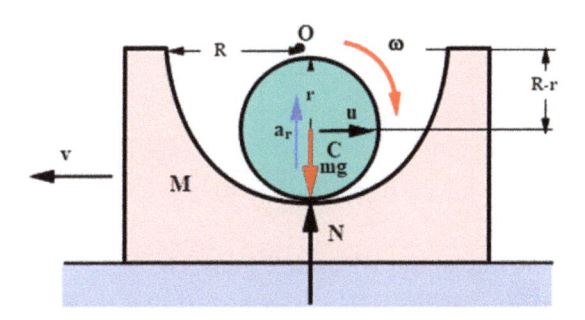

Problem 23 A uniform cylinder of mass m, radius r, and radius of gyration K rolls without sliding from a height H along the curved surface of a wedge of mass M. It leaves the curved track horizontally (due to the smooth edge of the wedge) at a height h and it lands on the ground following a projectile path. If the ground is smooth find (a) the horizontal range, (b) the maximum horizontal range, (c) the kinetic energy of the cylinder at the time of hitting the ground, and (d) the distance between the wedge and cylinder at the time of hitting the ground. Assume that the radius of the cylinder is much less than the height H and $M = m$.

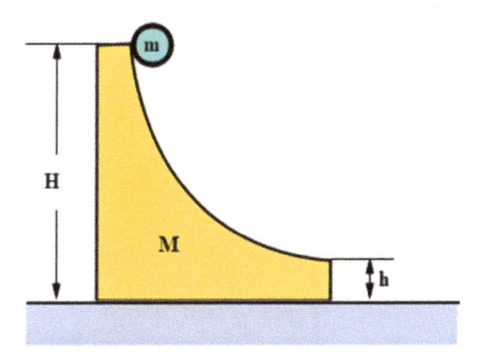

Solution
(a) Referring to last problem, the expression for the velocity of the rolling body relative to the ground is

$$v = \sqrt{\frac{2g(H-h)M}{(M+m)\left\{1 + \frac{K^2}{r^2}\left(1 + \frac{m}{M}\right)\right\}}}.$$

Putting $m = M$ and $K^2/R^2 = 1/2$ we have

$$= \sqrt{\frac{2g(H-h)m}{(m+m)\left\{1 + \frac{1}{2}(1+1)\right\}}}$$

$$= \sqrt{\frac{g(H-h)}{2}}.$$

4-73

Then the horizontal range $= R = vt$

$$= \sqrt{\frac{g(H-h)}{2}} \sqrt{\frac{2h}{g}}$$

$$R = \sqrt{h(H-h)}.$$

(b) The range R is maximum when $d(R^2)/dh = 0$
$$\Rightarrow 2h - H = 0 \Rightarrow h = H/2.$$

Then $R_{\max} = H$.

(c) When $M = m$, $v = u$,

$$\text{then } \omega = \frac{v+u}{r} = \frac{2u}{r}.$$

Then the final kinetic energy is

$$K_f = \frac{1}{2} \times \frac{1}{2} mr^2\omega^2 + \frac{1}{2}mv^2 + mgh$$

$$= \frac{1}{4}m(2v)^2 + \frac{1}{2}mv^2 + mgh$$

$$= \frac{3mv^2}{2} + mgh$$

$$= \frac{3m}{2}\left\{\frac{g(H-h)}{2}\right\} + mgh$$

$$= mg\left[\frac{3(H-h)}{4} + h\right]$$

$$= mg(3H + h), \text{ where } h = \frac{H}{2}$$

$$\Rightarrow K_f = mg\left(3H + \frac{H}{2}\right) = \frac{7}{2}mgH.$$

(d) The velocity of the cylinder relative to the wedge at the time of escaping from the wedge is

$$u_{\text{rel}} = \sqrt{\frac{2g(H-h)}{\left(\frac{M}{M+m}\right) + (K/r)^2}}.$$

Putting $m = M$, $h = H/2$, and $K^2/r^2 = 1/2$ we have

$$= \sqrt{\frac{2g(H - H/2)}{\frac{1}{2} + \frac{1}{2}}}$$

$$= \sqrt{gH}.$$

The time of fall of the cylinder is

$$t = \sqrt{2(H/2)/g} = \sqrt{\frac{H}{g}}.$$

The relative distance is

$$x_{rel} = v_{rel} t = (\sqrt{gH})(\sqrt{H/g}) = H.$$

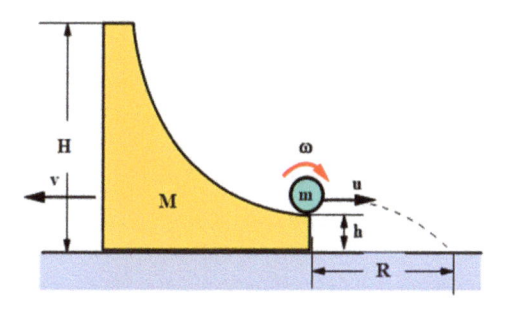

Problem 24 In problem 22, let the wedge enter into a frictional surface as soon as the rolling body reaches the bottom. Find (a) the acceleration of the wedge, the angular acceleration of the cylinder, the static friction in terms of coefficient of kinetic friction between the wedge and ground, and (b) the minimum coefficient friction of static between the rolling body and wedge so that the body will roll on the wedge without sliding at its lowest position. Put $M/m = 2$, the coefficient of kinetic friction $= 133/363$.

Solution
(a) Referring to problem 22, the normal reaction acting on the rolling cylinder at the lowest position is

$$\Rightarrow N' = (M+m)g\left\{1 + \frac{2}{\frac{M}{m} + \left(\frac{M+m}{m}\right)(k/r)^2}\right\}.$$

Putting $K^2/r^2 = 1/2$ and $M/m = 2$ we have

$$\Rightarrow N' = (2m+m)g\left\{1 + \frac{2}{\frac{2}{1} + \left(\frac{2m+m}{m}\right)(1/2)}\right\}$$

$$\Rightarrow N' = 33mg/7.$$

The kinetic friction between the wedge and ground is

$$f_k = \mu_k N' = 33\mu_k mg/7. \tag{4.89}$$

4-75

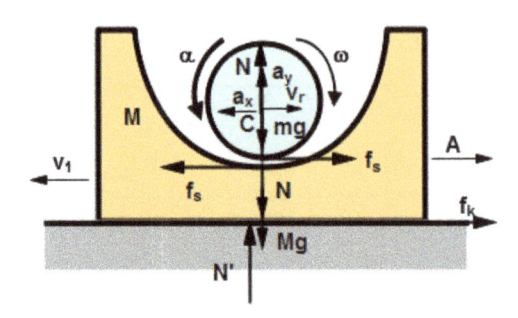

This kinetic friction acts rightward as the wedge moves towards the left as an effect of recoil. Then the static friction will act on the wedge and cylinder towards the left and right, respectively.

The net horizontal force acting on the system $(M + m)$ is

$$F_x(=f_k) = MA + m(A - r\alpha)$$

$$\Rightarrow f_k = (M + m)A - mr\alpha. \tag{4.90}$$

The net torque on the cylinder about its lowest point measured relative to the accelerating wedge is equal to the torque of the pseudo-force mA acting horizontally towards the left on the cylinder, about its lowest point. Then we can write

$$mAr = (3mr^2/2)\alpha$$

$$\Rightarrow A = 3r\alpha/2. \tag{4.91}$$

Using the above equations we have

$$A = 99\mu_k g/49, \ \alpha = 66\mu_k g/49r, \ f_s = 33\mu_k mg/49.$$

(b) The law of static friction is

$$f_s \leqslant \mu_s N. \tag{4.92}$$

The normal reaction between the cylinder and wedge is

$$N = N' - Mg = 33mg/7 - 2mg = 19mg/7. \tag{4.93}$$

Using the equations (4.92) and (4.93) and putting the value of f_s, we have

$$33\mu_k mg/49 \leqslant \mu_s(19mg/7)$$

$$\Rightarrow \mu_s \geqslant (7 \times 33\mu_k)/(19 \times 49)$$

$$\Rightarrow \mu_s \geqslant (7 \times 33(133/363))/(19 \times 49)$$

$$\Rightarrow \mu_s \geqslant 1/11.$$

Problem 25 A ball of mass m, radius r, and radius of gyration K rolls without sliding along the curved track of a fixed wedge from a height h. It leaves the track and launches onto a rough plank of mass M which is fitted with a light spring of stiffness k. If the ground is smooth, find the maximum compression of the spring.

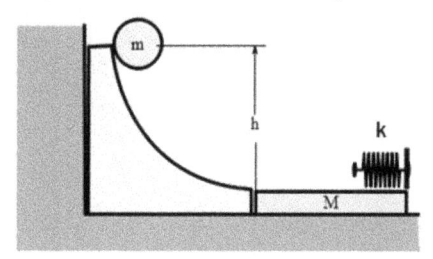

Solution

The ball m rolls down and launches onto the plank with a velocity v_0, say, which is given by conserving the energy of m between its top and lowest positions:

$$\frac{1}{2}mv_0^2(1 + K^2/r^2) = mgh$$

$$\Rightarrow v_0 = \sqrt{2gh/(1 + K^2/r^2)}. \tag{4.94}$$

The velocity of the center of mass of $M + m$ is

$$v_C = \frac{m}{M+m}v_0 = \frac{m}{M+m}\sqrt{2gh/(1 + K^2/r^2)}.$$

As the ball is rolling, its lowest point does not slip on the plank. So neither kinetic friction nor static friction will prevail and the plank will stay at rest just after the ball rolls on the plank without sliding.

As the ball has an initial velocity v_0 and the plank is stationary, it will compress the spring. The compressed spring will retard the motion of the ball by pushing it back and accelerate the plank by pushing it forward. Although there is no relative sliding between the ball and plank, the bodies tend to slide relative to each other due to the effect of the spring force. So static friction comes in to play. However, the net work done by the static friction is zero on the system ($M + m$). Then we can conserve the total mechanical energy of the system to obtain

$$-\Delta U = \Delta K$$

$$\Rightarrow \frac{1}{2}kx^2 = -\frac{1}{2}(M+m)v^2 + \frac{1}{2}mv_0^2(1 + K^2/r^2). \tag{4.95}$$

The bodies move at a common velocity at the maximum compression of the spring which is equal to the velocity of the center of mass given,

$$v_C = \frac{m}{M+m}v_0. \qquad (4.96)$$

Using equations (4.95) and (4.96) we have

$$\Rightarrow \frac{1}{2}kx^2 = \frac{1}{2}mv_0^2 - \frac{1}{2}(M+m)\left(\frac{mv_0}{M+m}\right)^2 + \frac{1}{2}mv_0^2(K^2/r^2)$$

$$\Rightarrow \frac{1}{2}kx^2 = \frac{Mm}{2(M+m)}v_0^2 + \frac{1}{2}mv_0^2(K^2/r^2)$$

$$\Rightarrow \frac{1}{2}kx^2 = \frac{1}{2}mv_0^2\left[\frac{M}{M+m} + \frac{K^2}{r^2}\right]. \qquad (4.97)$$

Using equations (4.95) and (4.97) we have

$$\Rightarrow \frac{1}{2}kx^2 = \frac{1}{2}m\left(\frac{2gh}{1+K^2/r^2}\right)\left(\frac{M}{M+m} + \frac{K^2}{r^2}\right)$$

$$\Rightarrow x = \sqrt{\left(\frac{mgh/k}{1+K^2/r^2}\right)\left(\frac{M}{M+m} + \frac{K^2}{r^2}\right)}.$$

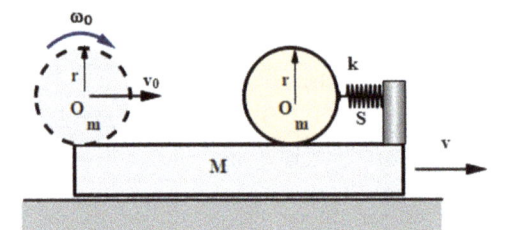

The ball stops translation relative to the plank and stops rotation at the time of maximum compression of the spring

Alternative method

Since the center of mass moves with a constant velocity, the kinetic energy of the center of mass remains conserved. However, the kinetic energy relative to the center of mass frame will decrease to zero at the time of maximum compression of the spring. Then, the loss of kinetic energy will be equal to the gain in elastic potential energy of the spring:

$$\Rightarrow \frac{1}{2}kx^2 = +\frac{1}{2}mv_0^2\left[\frac{M}{M+m} + \frac{K^2}{r^2}\right].$$

Then put the value of v_0 from equation (4.94) in the last expression to get the same result.

Problem 26 A cylinder of mass m and radius r is released from rest from the given position so that it rolls without sliding on a circular track of the a wedge of mass M. If the ground is smooth, find (a) (i) the velocity of the cylinder and the wedge, and (ii) the ratio of rotational and translational kinetic energy of the cylinder when it reaches the bottom of the wedge, and (b) the maximum compression of the spring of stiffness k fitted with the ground. Put $\frac{K^2}{R^2} = \frac{1}{2}$, $r/R = 0.5$, and $m/M = 1$.

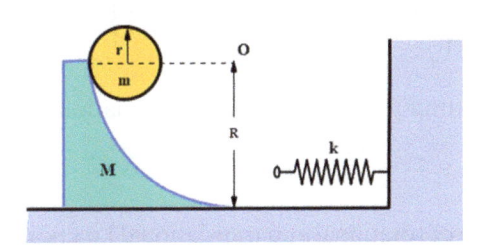

Solution

(a)

(i) Since the net horizontal force acting on the system $(M+m)$, its horizontal momentum remains conserved. Since both were at rest initially, the initial momentum is zero. If m and M have the velocities u (towards right) and v (towards left), respectively, the momentum of the system is

$$-Mv + mu = 0 \Rightarrow Mv = mu. \tag{4.98}$$

Conservation of the energy of the system gives us

$$\frac{Mv^2}{2} + \frac{1}{2}mu^2 + \frac{mK^2\omega^2}{2} = mg(R-r). \tag{4.99}$$

Kinematics: For rolling we write

$$u + v = r\omega. \tag{4.100}$$

Using the equations (4.98) and (4.100)

$$\Rightarrow \left(u + \frac{mu}{M}\right) = R\omega$$

$$\Rightarrow \omega = \left(1 + \frac{m}{M}\right)\frac{u}{r}. \tag{4.101}$$

Putting u from equation (4.98) and ω from equation (4.101) in equation (4.99),

$$\frac{M}{2}\left(\frac{m}{M}u\right)^2 + \frac{1}{2}mu^2 + \frac{mK^2}{2}\left(\frac{u}{r}\right)^2\left(\frac{M+m}{m}\right)^2 = mg(R-r)$$

$$\Rightarrow \frac{u^2}{2}\left[\frac{m^2}{M} + m + \frac{mK^2}{r^2}\left(\frac{M+m}{m}\right)^2\right] = mg(R-r)$$

4-79

$$\Rightarrow \frac{mu^2}{2}\left[\left(1+\frac{m}{M}\right)+\frac{K^2}{r^2}\left(1+\frac{m}{M}\right)^2\right] = mg(R-r)$$

$$\Rightarrow u^2\left(1+\frac{m}{M}\right)\left\{1+\frac{K^2}{r^2}\left(1+\frac{m}{M}\right)\right\} = 2g(R-r)$$

$$\Rightarrow u = \sqrt{\frac{2g(R-r)M}{(M+m)\left\{1+\frac{K^2}{r^2}\left(1+\frac{m}{M}\right)\right\}}} = \sqrt{gR/2}.$$

Then putting the value of u in the equation (4.98) we have

$$v = \frac{m}{M}u \Rightarrow v = u = \sqrt{gR/2}.$$

(ii) The ratio of rotational and translational kinetic energy of the cylinder is

$$K_r/K_t = \frac{mr^2\omega^2}{4} \Big/ \frac{mu^2}{2} = (R\omega/2u)^2. \tag{4.102}$$

From equation (4.101) we have

$$\omega = \left(1+\frac{m}{M}\right)\frac{u}{r} = (1+1)u/(R/2) = 4u/R. \tag{4.103}$$

Using the equations (4.98) and (4.99) we have

$$K_r/K_t = (1/8)(R\omega/u)^2 = 2.$$

(b) Since the ground is smooth, the translational kinetic energy of the cylinder is stored in the form of elastic energy in the spring:

$$\Rightarrow \frac{1}{2}kx^2 = \frac{1}{2}mu^2 \Rightarrow x = (\sqrt{m/k})u.$$

Putting the obtained value of u, we have

$$x = \sqrt{mgR/4k}.$$

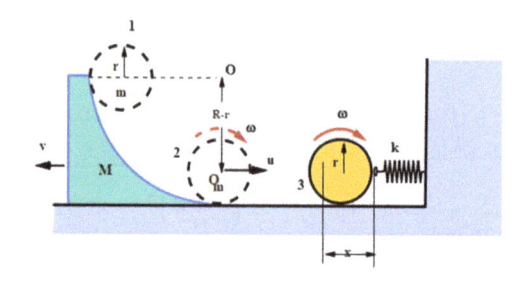

Problem 27 A solid sphere of mass m and radius r is released from rest from a height h from the ground so that it rolls without sliding on the spherical surface of a fixed bowl of radius R. If the maximum normal reaction acting on the sphere is $11mg/6$, find the value of h. Put $r = R/7$.

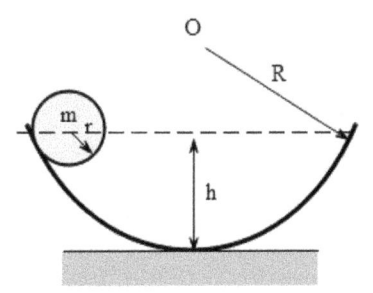

Solution

The normal reaction will be maximum at the lowest position. Let the ball roll down to the lowest position with a velocity, v, say, which is given by conserving the energy between its top and lowest positions:

$$\frac{1}{2}mv^2(1 + K^2/r^2) = mgh$$

$$\Rightarrow v = \sqrt{2gh/(1 + K^2/r^2)}$$

$$= \sqrt{2gh/(1 + 2/5)} = \sqrt{10gh/7}. \qquad (4.104)$$

The normal reaction acting on the sphere is

$$N = m\left(\frac{v^2}{R - r} + g\right).$$

Substituting $N = 11mg/6$ and $r = R/7$ we have

$$11mg/6 = m\left(\frac{v^2}{R - R/7} + g\right)$$

$$\Rightarrow 5g/6 = \frac{7v^2}{6R}$$

$$\Rightarrow v = \sqrt{\frac{5gR}{7}}. \qquad (4.105)$$

Using the equations (4.104) and (4.105) we obtain

$$h = R/2.$$

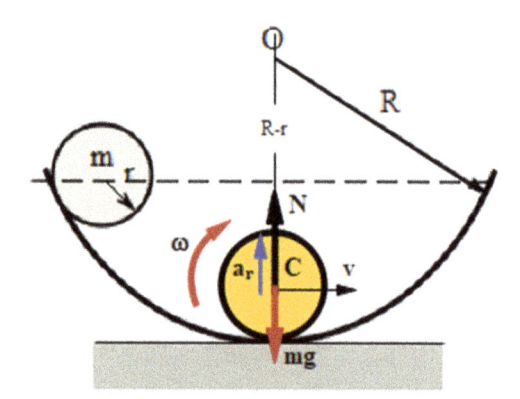

Problem 28 A hollow sphere of radius R and mass m' is rigidly connected to the roof by a rod. A thin uniform spherical shell of mass M, radius r which is completely filed with water of mass m, is released from rest from the given position, as shown in the figure. If the sphere rolls without sliding, find the maximum (a) normal reaction offered by the spherical shell to the hollow sphere and (b) reaction imparted by the hollow sphere to the rod. Put $M = m$, $R = 2r$, and $m' = 2m$.

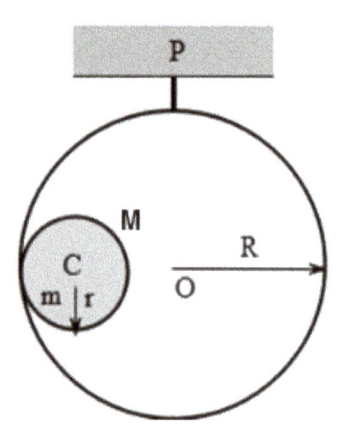

Solution
(a) Let us take the initial position of the center of mass of the ball as the reference level. As the spherical shell rolls down, the water sphere will translate without rotation due to the absence of friction between water and the spherical shell because water is a non-viscous liquid. Then the water sphere has only translational kinetic energy. Conserving the energy of the

4-82

system $(M + m)$, $-(M + m)g(R - r) + \frac{Mv^2}{2}\left(1 + \frac{K^2}{r^2}\right) + \frac{mv^2}{2} = 0$. Putting $k^2/r^2 = 2/3$, $r = R/2$, and $M = m$, we have

$$-(m + m)g(R - R/2) + \frac{mv^2}{2}(1 + 2/3) + \frac{mv^2}{2} = 0$$

$$\Rightarrow v = \sqrt{\frac{3gR}{4}}. \tag{4.106}$$

The normal reaction is

$$N = (M + m)g + \frac{(M + m)v^2}{R - r}$$

$$\Rightarrow N = (m + m)g + \frac{2mv^2}{R - R/2}$$

$$\Rightarrow N = 2m\left(g + \frac{2v^2}{R}\right). \tag{4.107}$$

Using equations (4.106) and (4.107) we have

$$N = 2m\left(g + \frac{2v^2}{R}\right)$$

$$N = 2m(g + 3g/2) = 5mg. \tag{4.108}$$

(b) The vertical force acting on the ceiling at P is

$$F_y = N + m'g. \tag{4.109}$$

Using the equations (4.108) and (4.109) we have

$$\Rightarrow F_y = (5mg) + m'g. \tag{4.110}$$

Putting $m' = 2m$, we have

$$F_y = (5mg) + 2mg = 7mg.$$

4-83

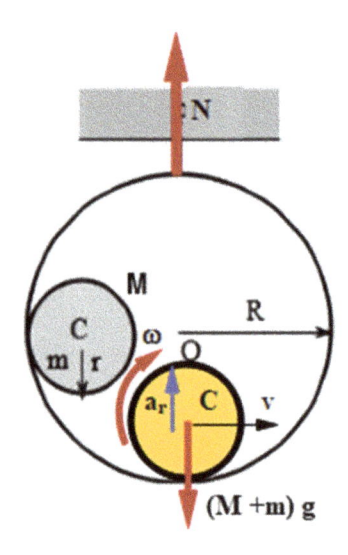

Problem 29 A thin uniform hollow sphere of mass m and radius r is released from a height h so that it rolls down without sliding along a curved track from P to A and completes the vertical circular path AQH. For the solid sphere, find (a) the velocity at H, (b) the normal reaction at H, and (c) the minimum value of h for just completing the circle. Put $r = R/5$, where $R =$ the radius of the curved track.

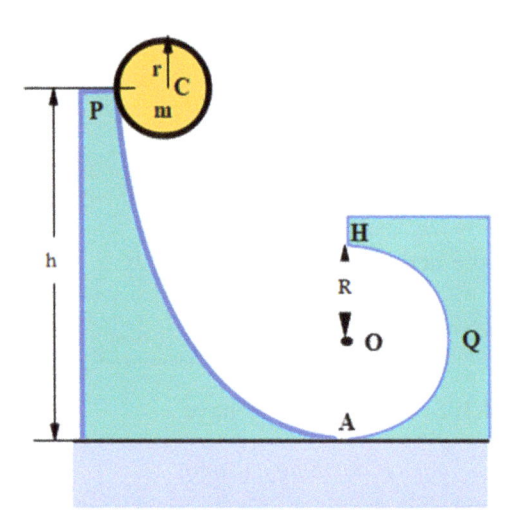

Solution
(a) Let v be the velocity of the rolling sphere at the top of the loop. Then its change in the kinetic energy is

4-84

$$\Delta K = \frac{1}{2}mv^2\left(1 + \frac{K^2}{r^2}\right). \qquad (4.111)$$

The center of mass of the sphere descends through a vertical distance $y = h - 2R + r$. So its change in gravitational potential energy is

$$\Delta U = -mgy = -mg(h - 2R). \qquad (4.112)$$

The equation of energy conservation is

$$\Delta U + \Delta U = 0. \qquad (4.113)$$

Using the last three equations,

$$-mg(h - 2R + r) + \frac{1}{2}mv^2\left(1 + \frac{K^2}{r^2}\right) = 0. \qquad (4.114)$$

The value of h is maximum when v is minimum. This happens when the sphere just loses contact with the track at the highest position. Then the only force acting on it is gravity, which is the net vertical force acting on the sphere at the top, given as $F_y = mg = ma_y$, or $a_y = g$, where $a_y = v^2/(R - r)$. Then we have

$$\frac{mv^2}{(R - r)} = mg$$

$$\Rightarrow v^2 = g(R - r). \qquad (4.115)$$

Using the equations (4.114) and (4.115) we have

$$-mg(h - 2R + r) + \frac{1}{2}mg(R - r)\left(1 + \frac{K^2}{r^2}\right) = 0. \qquad (4.116)$$

Putting $K^2/r^2 = 2/3$ and $r = R/5$ we have

$$\Rightarrow -(h - 2R + R/5) + \frac{1}{2}(R - R/5)\left(1 + \frac{2}{3}\right) = 0$$

$$\Rightarrow h = 37R/15.$$

(b) Conserving the energy between the ground level and top of the track we have

$$\frac{1}{2}mv_0^2\left(1 + \frac{K^2}{r^2}\right) - \frac{1}{2}mv^2\left(1 + \frac{K^2}{r^2}\right)$$

$$= 2mg(R - r). \qquad (4.117)$$

4-85

Using the equations (4.115) and (4.117)

$$\frac{1}{2}mv_0^2\left(1 + \frac{K^2}{r^2}\right)$$

$$= \frac{1}{2}mg(R-r)\left(1 + \frac{K^2}{r^2}\right) + 2mg(R-r)$$

$$= \frac{1}{2}mg(R-r)\left(1 + \frac{K^2}{r^2} + 4\right)$$

$$v_0 = \sqrt{g\left(5 + \frac{K^2}{r^2}\right)\frac{(R-r)}{\left(1 + \frac{K^2}{r^2}\right)}}$$

$$= \sqrt{\frac{g\left(5 + \frac{2}{3}\right)(R - R/5)}{1 + \frac{2}{3}}} = 2\sqrt{17gR}/5.$$

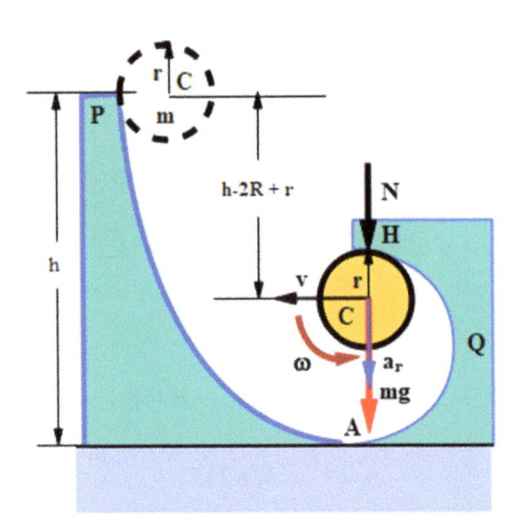

Problem 30 A small uniform solid sphere of mass m, radius of gyration K, and radius r is released from a height H (dotted sphere) so that it rolls down without sliding along a curved track and completes a loop. If the loop offers a downward normal reaction of magnitude $\eta mg(= 2mg)$ at the given position, find the radius of curvature of the track at the given position (bold sphere). Put $h = H/4$

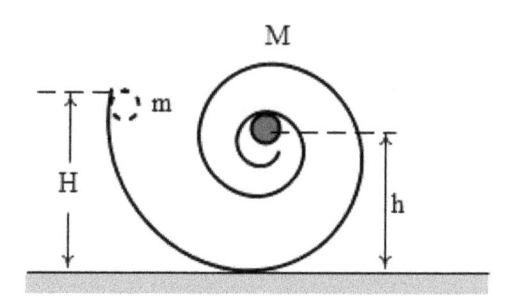

Solution

Let v be the velocity of the rolling sphere at the top of the loop. Then its change in kinetic energy is

$$\Delta K = \frac{1}{2}mv^2\left(1 + \frac{K^2}{r^2}\right). \tag{4.118}$$

The center of mass of the sphere descends through a vertical distance $y = H - h$. So its change in gravitational potential energy is

$$\Delta U = -mgy = -mg(H - h). \tag{4.119}$$

The equation of energy conservation is

$$\Delta U + \Delta U = 0. \tag{4.120}$$

Using the last three equations,

$$-mg(H - h) + \frac{1}{2}mv^2\left(1 + \frac{K^2}{r^2}\right) = 0. \tag{4.121}$$

Let $N =$ the downward pressing force imparted by the track to the ball. Then the net vertical force acting on the sphere at the top, given as $F_y = mg + N = ma_y$, where $a_y = v^2/R$; $R =$ the radius of curvature of the track. Then we have

$$N = \frac{mv^2}{R} - mg. \tag{4.122}$$

Putting $N = \eta mg$ we have

$$\frac{mv^2}{R} = mg(\eta + 1). \tag{4.123}$$

Using the equations (4.121) and (4.123) we have

$$-mg(H - h) + \frac{1}{2}\{mgR(\eta + 1)\}\left(1 + \frac{K^2}{r^2}\right) = 0$$

$$\Rightarrow \{R(\eta + 1)\}(1 + K^2/r^2) = 2(H - h)$$

$$\Rightarrow R = \frac{2(H-h)}{(\eta+1)\}(1+K^2/r^2)}.$$

Putting $K^2/r^2 = 2/5$ and we have

$$\Rightarrow R = \frac{2(H-H/4)}{(2+1)\}(1+2/5)} = 5H/14.$$

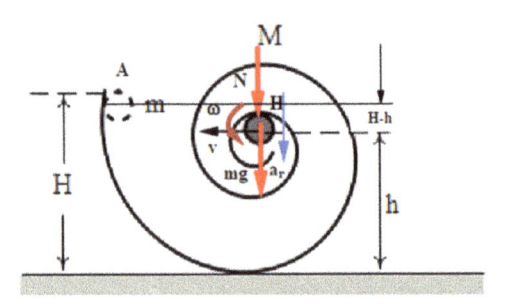

Problem 31 A ring of mass M and radius R hangs by a light string S which is attached to a ceiling. Two identical uniform discs each of mass m and radius r are released from rest from the end of the horizontal diameter of the ring of radius R as shown in the figure. If the discs roll without sliding, find (a) the velocity of the discs, (b) the normal reaction acting on the discs, (c) the friction acting on the discs, (d) the vertical force acting on the ring by the rolling discs, as the function of the angle θ swung by the centers of the discs relative to the center of the ring, and (e) the tension in the string S just before the discs meet. Put $R = 3r$ and $M = 8m$ and assume that there is sufficient friction between each disc and ring so that the disc rolls without slipping.

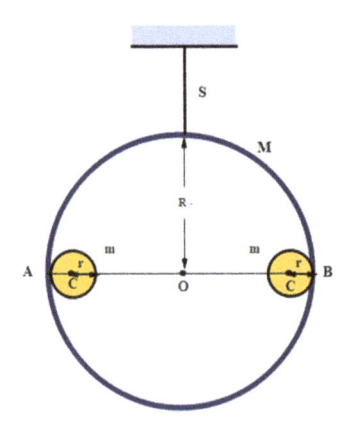

4-88

Solution

(a) Let us take the initial position of the center of mass of the discs as the reference level. As both the discs roll down, at an angular position θ, conserving the energy $-mg(R - r)\sin\theta + \frac{mv^2}{2}(1 + \frac{K^2}{r^2}) = 0$

$$\Rightarrow v = \sqrt{\frac{2g(R - r)\sin\theta}{1 + K^2/r^2}}$$

$$= \sqrt{\frac{2g(R - r)\sin\theta}{1 + 1/2}}$$

$$v = \sqrt{\frac{4g(R - r)\sin\theta}{3}} \text{ (put } r = R/3\text{).} \tag{4.124}$$

(b) The normal reaction is

$$N = mg\sin\theta + \frac{mv^2}{R - r}. \tag{4.125}$$

Using equations (4.124) and (4.125) we have

$$N = \frac{7mg\sin\theta}{3}. \tag{4.126}$$

(c) The expression for static friction is

$$\Rightarrow f_s = \left(\frac{mg\cos\theta}{1 + (r^2/K^2)}\right)$$

$$\Rightarrow f_s = \frac{mg\cos\theta}{1 + 2} = mg\cos\theta/3. \tag{4.127}$$

(d) The vertical force acting on the ring by the rolling discs is

$$F_y' = N\sin\theta + f_s\cos\theta. \tag{4.128}$$

Using the equations (4.126), (4.127), and (4.128) we have

$$F_y' = \{7mg\sin^2\theta\}/3 + \{mg\cos^2\theta\}/3$$

$$\Rightarrow F_y' = (6\sin^2\theta + 1)mg/3. \tag{4.129}$$

(d) From the geometry of the figure the discs will meet at the angle given as

$$\theta = 90° - \sin^{-1}\left\{\frac{r}{R - r}\right\}. \tag{4.130}$$

Putting $r = R/3$ in equation (4.130)

$$\theta = 90° - \sin^{-1}\left\{\frac{R/3}{R - R/3}\right\} = 60°.$$

Putting this angle in the equation (4.129) we have

$$\Rightarrow F_y' = (6\sin^2 60° + 1)mg/3$$
$$= 11mg/6.$$

Adding this value with the weight Mg of the ring, the reaction at the ceiling is

$$F_y = 11mg/6 + Mg.$$

Putting $M = 8m$ we have $F_y = 59mg/6$.

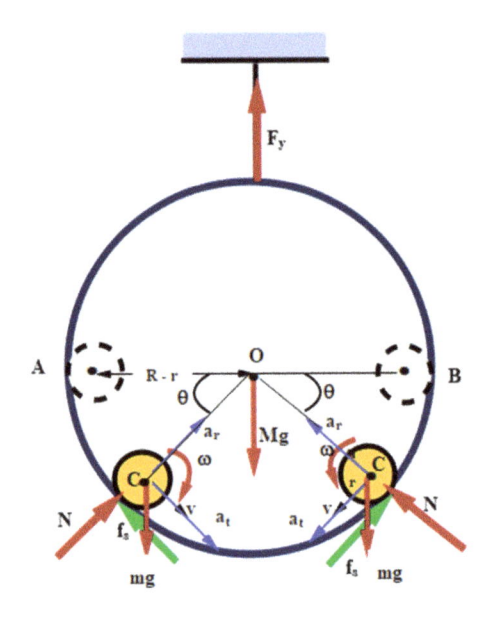

Problem 32 With reference to problem 20, a uniform solid sphere of mass m and radius r is placed at the top of a fixed spherical surface of radius R. The sphere is given a small impulse so that it rolls without sliding on the spherical surface. Find the angle at which the sphere will move with a zero horizontal acceleration or net vertical acceleration. Discuss all possible cases of small and large friction between the spheres.

Solution
Method 1
As both the balls rolls down, at an angular position θ, conserving the energy

$$mg(R+r) = mg(R+r)\cos\theta + \frac{mv^2}{2}\left(1 + \frac{k^2}{r^2}\right)$$

$$\Rightarrow v = \sqrt{\frac{2g(R+r)(1-\cos\theta)}{1+K^2/r^2}}. \tag{4.131}$$

Setting $N = 0$ we have

$$N = 0, \quad mg\cos\theta = \frac{mv^2}{R+r} \tag{4.132}$$

$$\Rightarrow \frac{2g(R+r)(1-\cos\theta)}{1+\frac{k^2}{r^2}} = g(R+r)\cos\theta$$

$$\Rightarrow \cos\theta\left[1 + \frac{2}{1+\frac{k^2}{r^2}}\right] = \frac{2}{1+\frac{k^2}{r^2}}$$

$$\Rightarrow \cos\theta\left[\frac{3+\frac{k^2}{r^2}}{1+\frac{k^2}{r^2}}\right] = \frac{2}{1+\frac{k^2}{r^2}}$$

$$\Rightarrow \theta = \cos^{-1}\left(\frac{2}{3+\frac{k^2}{r^2}}\right).$$

Putting $\frac{k^2}{r^2} = \frac{2}{5}$ we obtain $\theta = \cos^{-1}\frac{10}{17}$.

If the horizontal acceleration will be zero, the net horizontal force acting on the sphere is zero. This means that the net force will be just vertical. This will happen when the body loses contact, $N = 0$, so friction will also vanish. Then the only force acting on the sphere will be the downward gravitational force mg. If the surface is smooth, putting $\frac{k^2}{r^2} = 0$ for $\mu = 0$ (because the ball does not roll due to the absence of friction), we have

$$\theta = \cos^{-1}\frac{2}{3}.$$

Method 2
With reference to problem 20, the static friction is

$$\Rightarrow f_s = \left(\frac{mg\sin\theta}{1+(r^2/K^2)}\right)$$

4-91

$$\Rightarrow f_s = \frac{mg \sin \theta}{1 + 5/2} = 2mg \sin \theta / 7. \qquad (4.133)$$

With reference to problem 20, the normal reaction is

$$mg \cos \theta - N = \frac{mv^2}{R + r}. \qquad (4.134)$$

Putting $K^2/r^2 = 2/5$ in the equation

$$\Rightarrow v = \sqrt{\frac{2g(R + r)(1 - \cos \theta)}{1 + K^2/r^2}}$$

we have the centripetal acceleration

$$\Rightarrow v^2/(R + r) = \frac{10g(1 - \cos \theta)}{7}. \qquad (4.135)$$

Using equations (4.134) and (4.135) we have

$$N = mg(17 \cos \theta - 10)/7. \qquad (4.136)$$

Putting $N = 0$ we have

$$\theta = \cos^{-1} \frac{10}{17}.$$

The horizontal force acting on the sphere is

$$F_x = N \sin \theta - f_s \cos \theta = 0. \qquad (4.137)$$

Using the equations (4.133), (4.136), and (4.137) we have

$$mg(17 \cos \theta - 10)\sin \theta / 7 = 2mg \sin \theta \cos \theta / 7$$

$$\Rightarrow \cos \theta = \frac{2}{3}$$

$$\Rightarrow \theta = \cos^{-1} \frac{2}{3}.$$

This means that at this angle the friction and normal reaction will be given as follows:

$$f_s = 2mg \sin \theta / 7 = 2\sqrt{5}\, mg/21$$

$$N = mg(17 \times 2/3 - 10)/7 = 4mg/21.$$

So at the angle $\theta = \cos^{-1}\frac{2}{3}$, we have shown that their horizontal component will cancel each other. As a result, the net horizontal force acting on the sphere is zero at that position. Surprisingly, this situation will also occur again at the angle $\theta = \cos^{-1}\frac{10}{17}$ when the sphere leaves the circular track ($N = 0$). At that instant the friction will vanish and hence the net horizontal force will be zero. But this situation will occur if the friction between the bodies is sufficient. However, for a finite friction having a coefficient of friction μ between the bodies, the sphere will slip before it loses contact. We can calculate the angle at which the sphere begins to slip by applying the law of static friction, given as

$$f_s \leq \mu_s N. \qquad (4.138)$$

Substituting the values of f_s and N from equations (4.133) and (4.136), respectively, in equation (4.137) we have

$$2mg \sin\theta/7 \leq \mu_s mg(17\cos\theta - 10)/7$$

$$2\sin\theta \leq \mu_s(17\cos\theta - 10). \qquad (4.139)$$

Putting $\mu_s = 1/3$ in equation (4.139), substituting the inequality by an equality for the maximum value of the angle and then simplifying the factors, we have

$$17\cos\theta - 6\sin\theta = 10. \qquad (4.140)$$

By solving the equation (4.140) we have $\theta = 37°$ as obtained in problem 20.

Problem 33 A spherical shell of mass m and radius r has a smooth solid sphere of mass m and radius r, inside it. The sphere is released from the top of a fixed hemisphere of radius R. Assuming that the spherical shell rolls on the hemisphere, find (a) the angle at which it escapes from the hemisphere, (b) the expression for the time taken by the sphere to move from an initial angle θ_0 to an angular position θ, and (c) the coefficient of static friction as the function of angular position θ of the center of mass C of the sphere relative to the center O of the circular track.

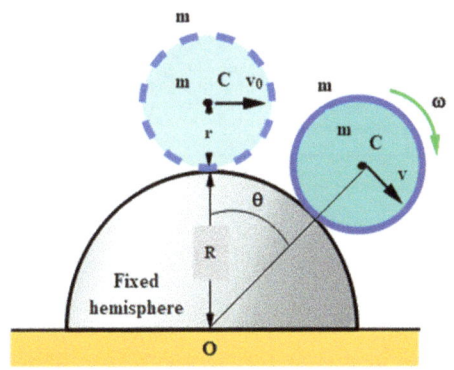

Solution
(a) As the solid sphere is smooth, it does not rotate. It will only translate, possessing translational kinetic energy only. However, the kinetic energy of the thin spherical shell is both translational and rotational. As the center of mass C of the combination falls through a vertical distance

$$h = g(R + r)(1 - \cos\theta).$$

So the potential energy decreases by an amount

$$\Delta U = -(m + m)gy = 2mgy \\ = 2mg(R + r)(1 - \cos\theta). \tag{4.141}$$

The kinetic energy of the system increases by

$$\Delta K = \Delta K_{shell} + \Delta K_{solidsphere} \\ = \frac{1}{2}\left(1 + \frac{2}{3}\right)mv^2 + \frac{1}{2}mv^2 = 4mv^2/3. \tag{4.142}$$

The energy conservation equation is

$$\Delta U + \Delta U = 0. \tag{4.143}$$

Using last three equations we have

$$2mg(R + r)(1 - \cos\theta) = \frac{4}{3}mv^2$$

$$\Rightarrow v^2 = \frac{3}{2}g(R + r)(1 - \cos\theta). \tag{4.144}$$

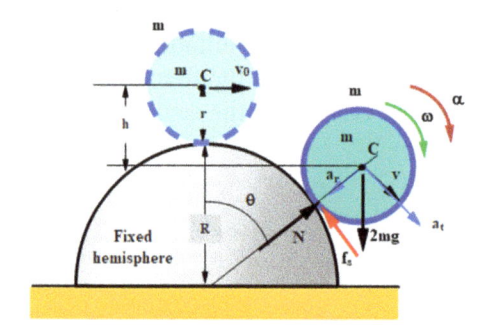

Where the sphere loses contact, $N = 0$,

$$\Rightarrow (2m)g\cos\theta = (2m)v^2/(R + r)$$

$$v^2 = g(R + r)\cos\theta. \tag{4.145}$$

Using equations (4.144) and (4.145) we have

$$\frac{3}{2}g(R+r)(1-\cos\theta) = g(R+r)\cos\theta$$

$$\Rightarrow \frac{3}{2}(1-\cos\theta) = \cos\theta$$

$$\Rightarrow \frac{3}{2} = \frac{5}{2}\cos\theta$$

$$\Rightarrow \theta = \cos^{-1}(3/5) = 53°.$$

(b) Then the angular velocity of the center of mass of the sphere relative to the center of the circular path is

$$\omega_1 = \frac{v}{R+r} = \sqrt{\frac{3g}{2(R+r)}}\sqrt{(1-\cos\theta)}$$

(we have put the value of velocity v from equation (4.144))

$$\Rightarrow \frac{d\theta}{dt} = \sqrt{\frac{3g}{(R+r)}}\sin\frac{\theta}{2}.$$

Separating the variables we have

$$\cosec\frac{\theta}{2}d\theta = \sqrt{\frac{3g}{(R+r)}}dt.$$

The time taken by the sphere to undergo the angular displacement from θ_0 to θ is given by integrating both sides:

$$\left\{\sqrt{\frac{3g}{(R+r)}}\right\}t = \int_{\theta_0}^{\theta}\cosec\frac{\theta}{2}d\theta.$$

The integration is

$$I = \int \cosec\frac{\theta}{2}d\theta = 2\ln\left|\cosec\frac{\theta}{2} - \cot\frac{\theta}{2}\right|$$

$$= 2\ln\left|\frac{1-\cos\frac{\theta}{2}}{\sin\frac{\theta}{2}}\right| = 2\ln\left|\frac{2\sin^2\frac{\theta}{4}}{2\sin\frac{\theta}{4}\cdot\cos\frac{\theta}{4}}\right|$$

$$= 2\ln\left(\tan\frac{\theta}{4}\right).$$

After putting the lower and upper limits we have

$$I = 2\ln\left(\tan\frac{\theta}{4}\right) - 2\ln\left(\tan\frac{\theta_0}{4}\right)$$

4-95

$$I = 2\ln\left\{\left(\tan\frac{\theta}{4}\right)\Big/\left(\tan\frac{\theta_0}{4}\right)\right\}$$

$$\Rightarrow t\left\{\sqrt{\frac{3g}{(R+r)}}\right\} = 2\ln\left\{\frac{\tan\left(\frac{\theta}{4}\right)}{\tan\left(\frac{\theta_0}{4}\right)}\right\}$$

$$\Rightarrow \theta = 4\tan^{-1}\left\{\tan\frac{\theta_0}{4}e^{\left(\sqrt{\frac{3g}{(R+r)}}\right)t/2}\right\}.$$

(c) The normal reaction is

$$N = 2m\left(g\cos\theta - \frac{v^2}{R+r}\right). \tag{4.146}$$

Using equations (4.144) and (4.146)

$$N = 2m[g\cos\theta - 3g(1-\cos\theta)/2]$$

$$\Rightarrow N = mg(5\cos\theta - 3). \tag{4.147}$$

The static friction can be calculated by applying dynamics. Taking the torque about the point of contact of the sphere we have

$$\tau_C = f_s r = mK^2 a/r$$

$$f_s = mK^2 a/r^2. \tag{4.148}$$

Then net force acting on the sphere along the line of motion is

$$2mg\sin\theta - f_s = 2ma. \tag{4.149}$$

Using last two equations we have

$$2mg\sin\theta - mK^2 a/r^2 = 2ma$$

$$\Rightarrow ma(K^2/r^2 + 2) = 2mg\sin\theta$$

$$\Rightarrow a = \frac{2g\sin\theta}{(K^2/r^2) + 2}. \tag{4.150}$$

Using equations (4.148) and (4.150)

$$f_s = m(K^2/r^2)\left[\frac{2g\sin\theta}{(K^2/r^2)+2}\right]$$

$$\Rightarrow f_s = \frac{2mg \sin\theta}{(r^2/K^2) + 2} = \frac{2mg \sin\theta}{(2/3) + 2}$$

$$f_s = \frac{3mg \sin\theta}{4}. \tag{4.151}$$

The law of static friction is

$$f_s \leq \mu_s N. \tag{4.152}$$

Using the equations (4.147), (4.151), and (4.152)

$$\frac{3 \sin\theta}{4} \leq \mu_s \left(5 \cos\theta - 3\right)$$

$$\mu_s \geq \frac{3 \sin\theta}{4(5 \cos\theta - 3)}.$$

Combined motion of a rod in a vertical plane

Problem 34 A uniform rod of mass M and length l is held against the vertical floor when the rod is nearly vertical. After the rod is released from rest, it rotates in the vertical plane. When will the rod make an angle θ with the vertical?

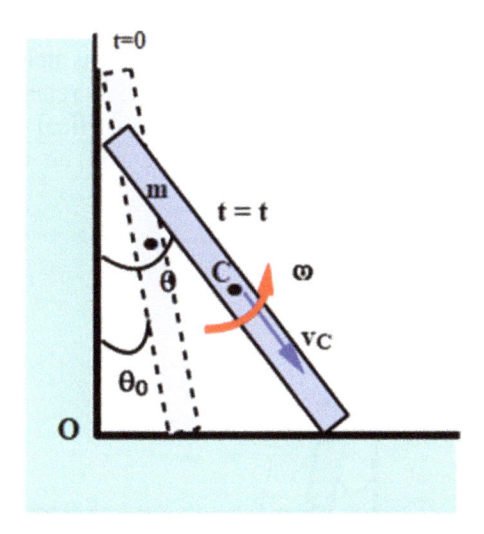

Solution
Following the procedure of problem 35, after conserving energy, we have

$$mg\frac{l}{2}(\cos\theta_0 - \cos\theta) = \frac{1}{6}ml\omega^2$$

$$\Rightarrow \omega = \sqrt{\frac{3g(\cos\theta_0 - \cos\theta)}{l}}.$$

For $\cos\theta_0 \simeq 1 (\because \theta_0 \simeq 0)$

$$\frac{d\theta}{dt} = \sqrt{\frac{3g}{2l} 2\sin^2\frac{\theta}{2}}$$

$$\Rightarrow \left(\sqrt{3\frac{g}{l}}\right)t = \int \frac{d\theta}{\sin\frac{\theta}{2}} = 2\ln\left|\csc\frac{\theta}{2} - \cos\frac{\theta}{2}\right|\Big|_{\theta_0}^{\theta}$$

$$\Rightarrow \sqrt{\frac{3g}{l}} t = 2\ln\left(\frac{1 - \cos\frac{\theta}{2}}{\sin\frac{\theta}{2}}\right) = 2\ln\tan\frac{\theta}{4}\Big|_{\theta_0}^{\theta}$$

$$\Rightarrow \sqrt{\frac{3g}{l}} t = 2\ln\left\{\frac{\tan\frac{\theta}{4}}{\tan\frac{\theta_0}{4}}\right\}$$

$$\Rightarrow t \cong \sqrt{\frac{4l}{3g}} \ln\left\{\frac{\tan\frac{\theta}{4}}{\tan\frac{\theta_0}{4}}\right\}.$$

Problem 35 A uniform rod of mass M and length L is held against the vertical floor when the rod is nearly vertical. After the rod is released from rest from a nearly vertical position ($\varnothing \cong 0$), it rotates in the vertical plane. What will the angular (a) velocity and (b) acceleration of the rod be when the rod rotates through an angle θ?

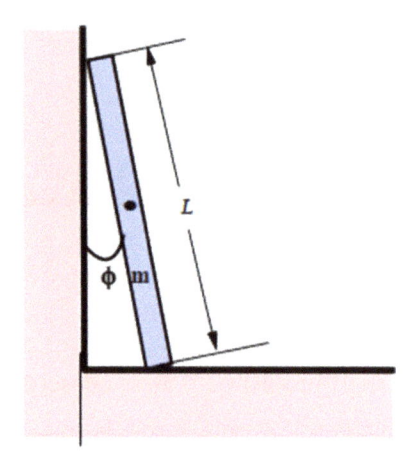

Solution

(a) Let the center of mass of the rod move with a velocity which is the vector sum of its x and y components, given as $\vec{v} = \vec{v}_x + \vec{v}_y$. The coordinates (x, y) of the center of mass (C) are given as

$$x = (L/2)\sin\theta \text{ and } y = (L/2)\cos\theta.$$

By differentiating with time we have

$$v_x = \frac{dx}{dt} = \omega(L/2)\cos\theta$$

and

$$v_y = \frac{dy}{dt} = \omega(L/2)\cos\theta.$$

Then the speed of the center of mass is given as

$$v = \sqrt{v_x^2 + v_y^2} = \omega(L/2).$$

The change (increase) in the kinetic energy of the rod is given as

$$\Delta K = \frac{1}{2}mv_C^2 + \frac{1}{2}I_C\omega^2, \tag{4.153}$$

where $I_C = \frac{1}{12}mL^2$ and $v_C = v = \omega(L/2)$. Putting all these values in equation (4.153) we have

$$\Delta K = \frac{1}{6}m\omega^2 L^2.$$

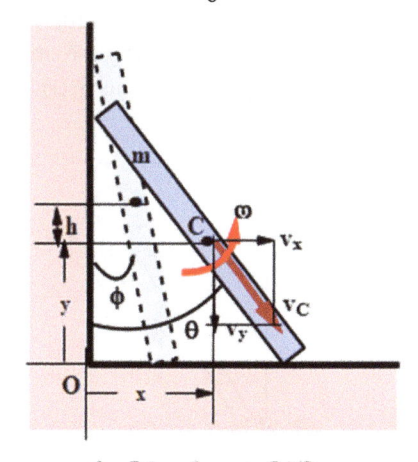

h= L(cosϕ - cosθ)/2

As the center of mass of the rod descends by a vertical distance

$$h = \frac{1}{2}(1 - \cos\theta)L,$$

the change (decrease) in gravitational potential energy of the rod is

$$\Delta U = mgh = mg\frac{L}{2}(1 - \cos\theta). \qquad (4.154)$$

Since the normal reactions at the ends of the rod are perpendicular to the displacements of the corresponding ends, the work done by these forces is zero. Then we can conserve the mechanical energy of the rod. Putting the values of ΔU and ΔK from the last two equations (4.153) and (4.154) in the equation

$$\Delta U + \Delta K = 0$$

and simplifying the factors we have

$$\omega = \sqrt{\frac{3g}{L}(1 - \cos\theta)}.$$

(b) The angular acceleration of the rod is

$$\alpha = \frac{d\omega^2}{2d\theta} = \frac{3g}{2L}\sin\theta.$$

Problem 36 A light rigid rod of mass m and length l is smoothly hinged at the point P of a block of mass M. A particle of mass m is connected with the free end of the rod. The block is kept on a smooth horizontal surface. If the rod is given a gentle push, it rotates in the vertical plane. Find (a) the angular velocity, (b) the angular acceleration of the rod as the function of the angle θ made by the rod with the vertical, and (c) the angular acceleration for the cases (i) $M \ll m$, (ii) $M \gg m$, and (iii) $M = m$, and (d) the locus of the particle m.

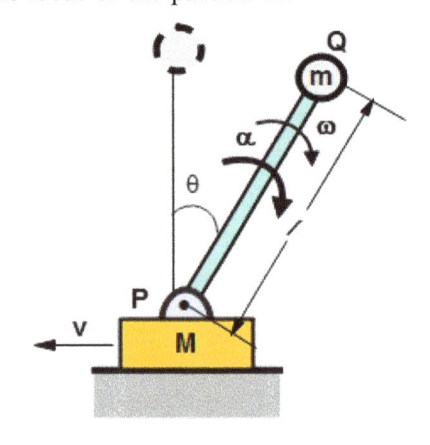

Solution
Method-1
(a) As the rod rotates through an angle θ, the change in gravitational potential energy of the system is

$$\Delta U = -mgl(1 - \cos\theta). \tag{4.155}$$

The kinetic energy of a two-particle system is

$$K = (v_{PC})y - v_C = \frac{Mml^2\omega^2}{2(M+m)} + \frac{1}{2}(M+m)v_C^2. \tag{4.156}$$

Since the block M does not lose contact with the horizontal surface

$$(v_M)_y = \frac{m\omega l}{(M+m)}\sin\theta - v_C = 0$$

$$\Rightarrow v_C = \frac{m\omega l}{M+m}\sin\theta. \tag{4.157}$$

The principle of the conservation of energy is

$$\Delta U + \Delta K = 0. \tag{4.158}$$

Using equations (4.155), (4.156), and (4.157) we have

$$mgl(1-\cos\theta) = \frac{Mml^2\omega^2}{2(M+m)} + \frac{1}{2}(M+m)\left(\frac{m\omega l}{2(M+m)}\sin\theta\right)^2$$

$$\Rightarrow mgl(1-\cos\theta) = \frac{Ml^2\omega^2}{2(M+m)}[M + m\sin^2\theta].$$

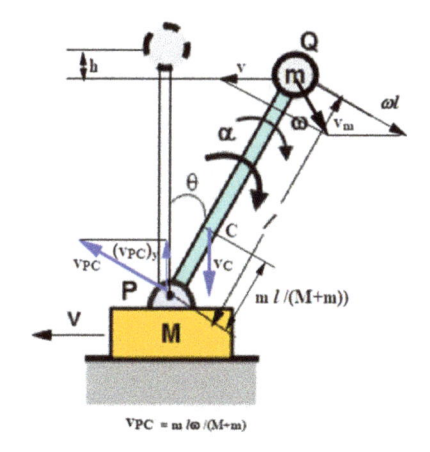

4-101

Method-2

As there is no net horizontal force acting on the rod–block system, the horizontal momentum of the system is conserved. Let v be the recoil velocity of M while the rod rotates in the vertical plane in clockwise sense:

$$P_x = -Mv + m(\omega l \cos\theta - v) = 0$$

$$\Rightarrow v = \frac{m\omega l \cos\theta}{M+m}. \tag{4.159}$$

By energy conservation we have

$$mgl(1 - \cos\theta)$$

$$= \frac{M}{2}v^2 + \frac{1}{2}m(v^2 + \omega^2 l^2 - 2\omega l v \cos\theta)$$

$$\Rightarrow 2mgl(1 - \cos\theta) = (M+m)v^2 + m\omega^2 l^2 - 2m\omega l v \cos\theta$$

$$= (M+m)v^2 + m\omega^2 l^2 - 2m\omega l v \cos\theta. \tag{4.160}$$

Using equations (4.159) and (4.160)

$$+2mgl(1 - \cos\theta) = \frac{m^2\omega^2 l^2 \cos^2\theta}{(M+m)} + m\omega^2 l^2 - 2m\omega l \cos\theta \frac{m\omega l \cos\theta}{M+m}$$

$$= m\omega^2 l^2 - \frac{m^2\omega^2 l^2 \cos^2\theta}{M+m}$$

$$= m\omega^2 l^2 \left[1 - \frac{m\cos^2\theta}{M+m}\right]$$

$$= m\omega^2 l^2 \left\{\frac{M + m\sin^2\theta}{(M+m)}\right\}$$

$$\Rightarrow \omega = \sqrt{\frac{2(M+m)g(1-\cos\theta)}{(M+m\sin^2\theta)l}}.$$

(b) Squaring the angular velocity

$$\omega^2 = \frac{2(M+m)g(1-\cos\theta)}{(M+m\sin^2\theta)l}$$

$$\Rightarrow 2\omega \frac{d\omega}{d\theta} = \frac{2(M+m)g}{l}\left\{\frac{(M+m\sin^2\theta)\sin\theta - (1-\cos\theta)2m\sin\theta\cos\theta}{(M+m\sin^2\theta)^2}\right\}$$

$$\Rightarrow \alpha = \frac{(M+m)g\sin\theta}{l}\left\{\frac{M+m\sin^2\theta-(1-\cos\theta)2m\cos\theta}{(M+m\sin^2\theta)^2}\right\}$$

$$= \frac{(M+m)g\sin\theta}{l}\left\{\frac{M+m(\sin^2\theta-2\cos\theta+2\cos^2\theta)}{(M+m\sin^2\theta)^2}\right\}$$

$$= \frac{(M+m)g\sin\theta}{l(M+m\sin^2\theta)^2}\{M+m(1+\cos^2\theta-2\cos\theta)\}$$

$$\Rightarrow \alpha = \frac{(M+m)g\sin\theta\{M+m(1-\cos\theta)^2\}}{(M+m\sin^2\theta)^2\,l}.$$

(c)

(i) When $M \to \infty$ we have

$$\omega = \sqrt{\frac{2g(1-\cos\theta)}{l}}, \quad \alpha = \frac{g\sin\theta}{l}.$$

(ii) When $M = 0$ we have

$$\alpha = \frac{g\sin\theta}{l}, \quad \omega = \sqrt{\frac{2mg(1-\cos\theta)}{l}}\Big/\sin\theta.$$

(iii) When $M = m$ we have

$$\omega = \sqrt{\frac{2g(1-\cos\theta)}{l}}, \quad \alpha = \frac{2g\sin\theta(2+2\cos^2\theta-2\cos\theta)}{4l(1+\sin^2\theta)^2}$$

$$\Rightarrow \alpha = \frac{g\sin\theta(1+\cos^2\theta-\cos\theta)}{l(1+\sin^2\theta)^2}.$$

(d) Since the net horizontal force acting on the system $(M+m)$ is zero, the center of mass C of the system has zero horizontal acceleration; so, point C moves vertically downward. Let us fix the origin O on the ground that lies on the vertical line along which the center of mass C falls. As the center of mass C does not move horizontally, the coordinates of m are

$$x = \frac{Ml}{M+m}\sin\theta \text{ and } y = l\cos\theta,$$

so, we have

$$\left(\frac{x(M+m)}{Ml}\right)^2 + \left(\frac{y}{l}\right)^2 = l$$

$$\Rightarrow \frac{x^2}{\left(\frac{Ml}{M+m}\right)^2} + \frac{y^2}{l^2} = 1, \text{ where}$$

$a = \frac{Ml}{M+m}$ and $b = l$ to have an ellipse

$$\frac{x^2}{a^2} + \frac{y^2}{b^2} = 1.$$

Problem 37 A uniform rod of mass M and length l is smoothly pivoted with a block of mass m which is free to slide along the smooth horizontal bar. The rod is given an initial angular velocity ω_0 keeping the block stationary. Then the block moves due to the rotating rod. Find the angular velocity of the rod as the function of the angle θ made by the rod with the vertical. Put $M = m$ and $\omega_0 = \sqrt{3g/l}$.

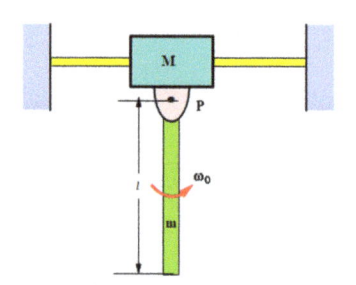

Solution

At any angle of rotation θ of the rod, let the velocity of the block M be v and the angular velocity of the rod be ω. As the net horizontal force acting on the rod–block system is zero, in the horizontal direction, conserving the linear momentum of the system $(M + m)$, we have

$$m\frac{l}{2}\omega_0 = Mv + m\left(v + \frac{\omega l}{2}\cos\theta\right)$$

$$\Rightarrow ml\omega_0 = 2(M + m)v + m\omega l \cos\theta. \tag{4.162}$$

Kinematics: By applying the law of relative velocity and vector addition,

$$v_m^2 = \frac{\omega^2 l^2}{4} + v^2 + \omega l v \cos\theta \tag{4.163}$$

$$\frac{Mv^2}{2} + \frac{m}{2}v_m^2 + \frac{ml^2}{24}\omega^2 = \frac{ml^2\omega_0^2}{6} - mg\frac{l}{2}(1 - \cos\theta)$$

4-104

$$\frac{Mv^2}{2} + \frac{m}{2}v_m^2 + \frac{ml^2}{24}\omega^2 = \frac{ml^2(3g/l)}{6} - mg\frac{l}{2}(1 - \cos\theta)$$

$$\Rightarrow \frac{Mv^2}{2} + \frac{m}{2}v_m^2 + \frac{ml^2}{24}\omega^2 = mg\cos\theta\frac{l}{2}$$

$$\Rightarrow Mv^2 + mv_m^2 + \frac{ml^2}{12}\omega^2 = mgl\cos\theta. \tag{4.164}$$

Using equations (4.163) and (4.164)

$$Mv^2 + m\left\{\frac{\omega l^2}{4} + v^2 + \omega l v\cos\theta\right\} + \frac{\omega l^2}{12}$$

$$= mgl\cos\theta$$

$$(M+m)v^2 + \frac{ml^2}{3}\omega^2 + m\omega l v\cos\theta$$

$$= mgl\cos\theta \tag{4.165}$$

$$\Rightarrow \left\{\frac{ml(\omega_0 - \omega\cos\theta)}{2(M+m)}\right\}^2 (M+m) + \frac{ml^2}{3}\omega^2$$

$$+ m\omega l\cos\theta\frac{ml(\omega_0 - \omega\cos\theta)}{2(M+m)} = mgl\cos\theta$$

$$\Rightarrow \frac{(ml)^2}{4(M+m)}\{\omega_0^2 + \omega^2\cos^2\theta - 2\omega\omega_0\cos\theta$$

$$+ 2\omega\omega_0\cos\theta - 2\omega^2\cos^2\theta\} + \frac{(l\omega)^2}{3} = mgl\cos\theta$$

$$\Rightarrow \frac{ml\{\omega_0^2 - \omega^2\cos^2\theta\}}{4(M+m)} + \frac{l\omega^2}{3} = g\cos\theta$$

$$\Rightarrow \omega^2\left\{\frac{4M + m(1 + 3\sin^2\theta)}{12(M+m)}\right\}$$

$$= g\cos\theta - \frac{ml}{4(M+m)}\omega_0^2$$

$$= g\cos\theta - \frac{ml}{4(m+m)}\left(3g/l\right)$$

4-105

$$= g\cos\theta - \frac{3g}{8}$$

$$\Rightarrow \omega^2 \left\{ \frac{4M + m(1 + 3\sin^2\theta)}{12(M+m)} \right\} = g\cos\theta - \frac{3g}{8}$$

$$\Rightarrow \omega^2 \left\{ \frac{4m + m(1 + 3\sin^2\theta)}{12(m+m)} \right\} = g\cos\theta - \frac{3g}{8}$$

$$\Rightarrow \omega^2 = \frac{3g(8\cos\theta - 3)}{(5 + 3\sin^2\theta)l}$$

$$\Rightarrow \omega = \sqrt{\frac{3g(8\cos\theta - 3)}{(5 + 3\sin^2\theta)l}}.$$

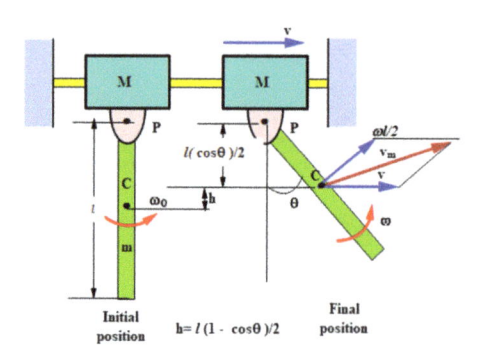

Problem 38 In the previous problem if the rod is made horizontal and then we release the rod–block system from rest, after the rod swings through a right angle it will be vertical. At this position find (a) the angular velocity of the rod and (b) the normal force acting on the block in terms of masses M and m of the block and rod, respectively.

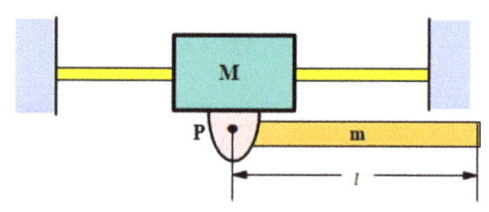

Solution
(a) Since friction is absent, there is no net horizontal force acting on the system $(M + m)$. This means that we can conserve the horizontal momentum of the system. Furthermore, the normal reaction does not perform work. So we can conserve the total mechanical energy of the system.

In order to conserve the horizontal momentum of the system, its center of mass will move vertically. So the block M must slip towards the left along the horizontal bar. As the rod swings through a right angle $\theta = 90°$, the center of mass of the rod falls through a vertical distance $y = (l \sin \theta)/2 = l/2$. Then the decrease in gravitational potential energy of the rod is equal to the increase in the kinetic energy of the rod. According to the principle of conservation of momentum, the horizontal momentum of the system is

$$P_x = -m\{v - (l/2)\omega \sin \theta\} + Mv = 0.$$

Putting the angle $\theta = 90°$ we have

$$v = \frac{ml\omega}{2(M + m)}. \tag{4.166}$$

Conservation of energy: As we know

$$\Delta U + \Delta K = 0$$

$$\Rightarrow mg\frac{l}{2} = \frac{I_C \omega^2}{2} + \frac{1}{2}m\left(v - \frac{l}{2}\omega\right)^2 + \frac{M}{2}v^2$$

$$\Rightarrow mg\frac{l}{2} = \frac{ml^2\omega^2}{2 \times 12} + \frac{M}{2}v^2 + \frac{1}{2}m\left(v - \frac{l}{2}\omega\right)^2$$

$$\Rightarrow mgl = \frac{ml^2\omega^2}{12} + \frac{ml^2\omega^2}{4} + (M + m)v^2 - mvl\omega$$

$$\Rightarrow mgl = \frac{ml^2\omega^2}{3} + (M + m)v^2 - mvl\omega. \tag{4.167}$$

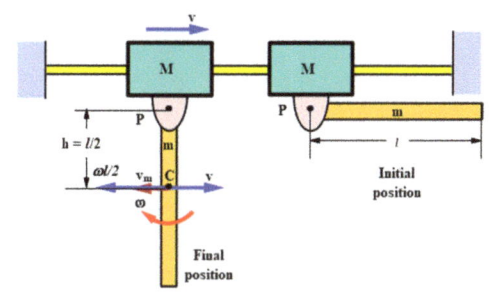

Using equations (4.166) and (4.167) we have

$$mgl = \frac{ml^2\omega^2}{3} + \frac{(M+m)m^2l^2\omega^2}{4(M+m)^2} - ml\omega\frac{ml\omega}{2(M+m)}$$

$$= \frac{ml^2\omega^2}{3} - \frac{m^2l^2\omega^2}{4(M+m)}$$

$$= ml^2\omega^2\left(\frac{1}{3} - \frac{m}{4(M+m)}\right)$$

$$mgl = \frac{(4M+m)ml^2\omega^2}{12(M+m)}$$

$$\Rightarrow \omega = \sqrt{\frac{12(M+m)g}{(4M+m)l}}.$$

(b) $N = Mg + mg + m\frac{l\omega^2}{2}$

$$= (M+m)g + \frac{ml}{2} \times \frac{12g(M+m)}{(4M+m)l}$$

$$= (M+m)g\left\{1 + \frac{6m}{4M+m}\right\}$$

$$= \frac{(7m+4M)(M+m)}{4M+m}g.$$

Problem 39 In problem 37, find (a) the maximum angle swung by the rod for the given value of the initial angular velocity, and (b) the minimum value/s of ω_0 for the rod to cross (a) the horizontal position by swinging and obtuse angle.

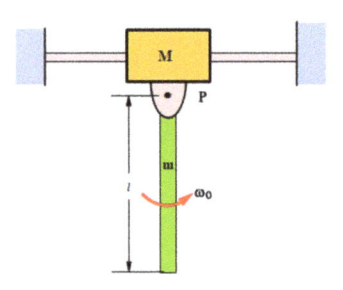

Solution
(a) If the rod will stay at rest momentarily relative to the block at its extreme position, put $\omega = 0$ in the final expression of angular velocity in problem 37 to obtain the maximum angular displacement given as

$$(8\cos\theta - 3) = 0$$

$$\Rightarrow \theta = \cos^{-1}(3/8)$$

(b) Let v = velocity of the block when the rod crosses the horizontal position. Conserving the horizontal momentum of the rod–block system, we have

$$m\frac{l}{2}\omega_0 = (M+m)v$$

$$\Rightarrow v = \frac{ml\omega_0}{2(M+m)}. \qquad (4.168)$$

Kinematics: By applying the law of relative velocity and vector addition, the velocity of the center of mass of the rod is

$$v_m^2 = \frac{\omega^2 l^2}{4} + v^2. \qquad (4.169)$$

Taking the block as the reference level for gravitational potential energy calculation, we can write the principle of conservation of energy as:

a decrease in kinetic energy = an increase in gravitational potential energy

$$\Rightarrow \frac{ml^2\omega_0^2}{6} - \left(\frac{Mv^2}{2} + \frac{ml^2}{24}\omega^2 + \frac{mv_m^2}{2}\right) = mg\frac{l}{2}. \qquad (4.170)$$

Using equations (4.169) and (4.170) we have

$$\frac{Mv^2}{2} + \frac{ml^2}{24}\omega^2 + \frac{m}{2}\left\{\frac{\omega^2 l^2}{4} + v^2\right\} - \frac{ml^2\omega_0^2}{6} = -mg\frac{l}{2}$$

$$\Rightarrow \frac{(M+m)v^2}{2} + \frac{ml^2}{6}(\omega^2 - \omega_0^2) = -mg\frac{l}{2}. \qquad (4.171)$$

Using equations (4.168) and (4.171) we have

$$\frac{(M+m)}{2}\left\{\frac{ml\omega_0}{2(M+m)}\right\}^2 + \frac{ml^2}{3}(\omega^2 - \omega_0^2) = -mgl$$

$$\Rightarrow \frac{m^2 l^2 \omega_0^2}{8(M+m)} + \frac{ml^2}{3}(\omega^2 - \omega_0^2) = -mgl$$

$$\Rightarrow \frac{m^2 l^2 \omega_0^2}{8(M+m)} - \frac{ml^2\omega_0^2}{3} + \frac{ml^2\omega^2}{3} = -mgl$$

$$\Rightarrow ml^2\omega_0^2\left\{\frac{m}{8(M+m)} - \frac{1}{3}\right\} + \frac{ml^2\omega^2}{3} = -mgl$$

4-109

$$\Rightarrow \omega_0^2 \left\{ \frac{m}{4(M+m)} - \frac{1}{3} \right\} + \frac{\omega^2}{3} = -g/l$$

$$\Rightarrow \omega_0^2 \left\{ \frac{3m - 8M - 8m}{24(M+m)} \right\} + \frac{\omega^2}{3} = -g/l$$

$$\Rightarrow -\omega_0^2 \left\{ \frac{5m + 8M}{12(M+m)} \right\} + \frac{\omega^2}{3} = -g/l$$

$$\Rightarrow \omega^2 = \omega_0^2 \left\{ \frac{(5m + 8M)}{8(M+m)} \right\} - 3g/l$$

$$\Rightarrow \omega = \sqrt{\omega_0^2 \left\{ \frac{(5m + 8M)}{8(M+m)} \right\} - 3g/l}.$$

The minimum value of ω_0 for the rod to cross its horizontal position can be given by putting $\omega = 0$. Then we have

$$\omega_0^2 \left\{ \frac{(5m + 8M)}{8(M+m)} \right\} - 3g/l \geq 0$$

$$\Rightarrow \omega_0^2 \left\{ \frac{(5m + 8M)}{8(M+m)} \right\} \geq 3g/l$$

$$\Rightarrow \omega_0 \geq \sqrt{\frac{24g(M+m)}{(8M+5m)l}}.$$

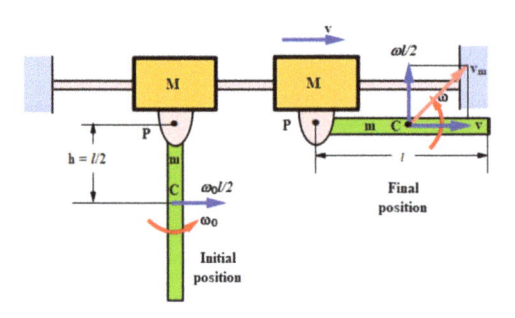

Problem 40 A rod of mass m and length L is kept nearly vertical on a smooth horizontal surface. If it is released from rest, it rotates in a vertical plane. Find (a) the angular velocity, (b) the angular acceleration of the rod, (c) the normal reaction acting on the rod as the function of the angle θ made by the rod with the vertical, and

(d) the equation of trajectory of the top of the rod, assuming the bottom of the rod as the origin.

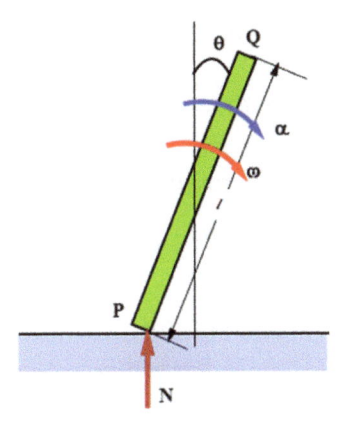

Solution

As there is no net horizontal force acting on the rod, the horizontal momentum of the rod is conserved. This means that the center of mass of the rod will move down vertically. For this to happen, the lowest point of the rod must move backward. Let, at any instant, v be the velocity of the center of mass of the rod while the rod rotates in the vertical plane in a clockwise sense with an angular velocity ω.

(a) As the center of mass descends by a vertical distance of $l/2(1 - \cos\theta)$, the change (decrease) in gravitational potential energy of the rod is

$$\Delta U = -mgl(1 - \cos\theta). \tag{4.172}$$

The kinetic energy of the rod changes (increases) from zero to K. Then the change in kinetic energy is

$$\Delta K = K = \frac{1}{2}\frac{ml^2}{12}\omega^2 + \frac{1}{2}mv^2. \tag{4.173}$$

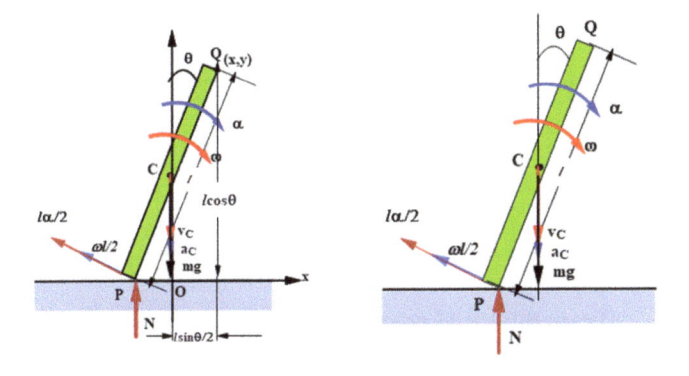

4-111

Conserving the energy we have

$$\Delta U + \Delta K = 0. \tag{4.174}$$

Using last three equations, we have

$$mg(l/2)(1 - \cos\theta) = \frac{1}{2}\frac{ml^2}{12}\omega^2 + \frac{1}{2}mv^2. \tag{4.175}$$

As the lowest point P of the rod does not move vertically,

$$\frac{\omega l \sin\theta}{2} = v. \tag{4.176}$$

Using equations (4.175) and (4.176)

$$mg(l/2)(1 - \cos\theta) = \frac{1}{2}\frac{ml^2}{12}\omega^2 + \frac{1}{2}m\left\{\frac{\omega l \sin\theta}{2}\right\}^2$$

$$\Rightarrow mg(l/2)(1 - \cos\theta) = \frac{ml^2}{8}\omega^2\left\{\frac{1}{3} + \sin^2\theta\right\}$$

$$\Rightarrow \omega^2 = \frac{12g(1 - \cos\theta)}{l(1 + 3\sin^2\theta)}$$

$$\Rightarrow \omega = \sqrt{\frac{12g(1 - \cos\theta)}{l(1 + 3\sin^2\theta)}}. \tag{4.177}$$

(b) Squaring the angular velocity,

$$\omega^2 = \frac{12g(1 - \cos\theta)}{(1 + 3\sin^2\theta)l}.$$

Differentiating both sides with respect to θ we have

$$2\omega\frac{d\omega}{d\theta} = \frac{12g}{l}\left\{\frac{(1 + 3\sin^2\theta)\sin\theta - (1 - \cos\theta)6\sin\theta\cos\theta}{(1 + 3\sin^2\theta)^2}\right\}$$

$$= \frac{12g\sin\theta}{l}\left\{\frac{1 + 3\sin^2\theta - (1 - \cos\theta)6\cos\theta}{(1 + 3\sin^2\theta)^2}\right\}$$

$$= \frac{12g\sin\theta}{l}\left\{\frac{1 + 3\sin^2\theta - 6\cos\theta + 6\cos^2\theta}{(1 + 3\sin^2\theta)^2}\right\}$$

$$= \frac{12g\sin\theta}{l}\left\{\frac{4 + 3\cos^2\theta - 6\cos\theta}{(1 + 3\sin^2\theta)^2}\right\}. \tag{4.178}$$

(c) Since the gravity acts at the center of mass, its torque about the center of mass is zero. The normal reaction will produce a clockwise torque about the center of mass of the rod which causes its rotation. Taking the torque of the normal reaction about the center of mass we have

$$N\left\{\frac{l\sin\theta}{2}\right\} = I_C\alpha = (ml^2/12)\alpha$$

$$N = (ml/6\sin\theta)\alpha. \tag{4.179}$$

Using the equations (4.178) and (4.179)

$$N = \{ml/6\sin\theta\}\frac{6g\sin\theta}{l}\left\{\frac{4-\cos^2\theta-2\cos\theta}{(1+3\sin^2\theta)^2}\right\}$$

$$\Rightarrow N = mg\left\{\frac{4-\cos^2\theta-6\cos\theta}{(1+3\sin^2\theta)^2}\right\}.$$

The right-hand side of the above equation is always greater than zero; so the positive normal reaction physically signifies that the rod will never lose contact with the ground.

Alternative method

The acceleration of the center of mass of the rod is

$$a = dv/dt = \frac{l}{2}(d\omega/dt) + \omega\cos\theta(d\theta/dt)$$

$$= \frac{l}{2}(\alpha\sin\theta + \omega^2\cos\theta). \tag{4.179a}$$

Using the equations (4.177), (4.178), and (4.179) and then putting the above value of acceleration in equation

$$N = m(g-a),$$

and simplify the factors we can get the answer as obtained earlier.

(d) The coordinates of the free end of the rod are

$$x = \frac{l}{2}\sin\theta \text{ and } y = l(\cos\theta).$$

Then

$$\left(\frac{2x}{l}\right)^2 + \left(\frac{y}{l}\right)^2 = 1.$$

It is the equation of an ellipse.

Problem 41 A uniform light rod of length l is smoothly pivoted with a block of mass M which is free to slide along a smooth horizontal bar. A small ball is connected with the free end of the rod. The ball is given an initial horizontal velocity v_0. Find the angular velocity of the rod as the function of the angle θ made by the rod with the vertical. Put $v_0^2 = 6gl$, $l = 2a$ and $M = m$

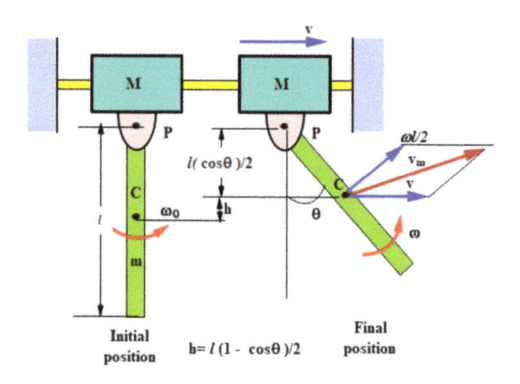

Solution

At any angle of rotation θ of the rod, let the velocity of the block M be v, and the angular velocity of the rod be ω. Then, conserving the linear momentum of the system $(M + m)$, we have

$$m\frac{l}{2}\omega = Mv + m\left(\frac{\omega l}{2}\cos\theta\right)$$

$$\Rightarrow ml\omega_0 = 2(M + m)v + \omega l\cos\theta.\qquad(4.180)$$

Kinematics: By applying the law of relative velocity and vector addition,

$$v_m^2 = \frac{\omega l^2}{4} + v^2 + \omega l v \cos\theta \qquad(4.181)$$

$$\frac{Mv^2}{2} + \frac{m}{2}v_m^2 + \frac{ml^2}{24}\omega^2 = \frac{ml^2\omega^2}{6} - mg\frac{l}{2}(1 - \cos\theta) \qquad(4.182)$$

4-114

$$(M + m)v^2 + \frac{ml^2}{3}\omega^2 + m\omega lv \cos\theta = \frac{ml^2\omega_0^2}{3} - mgl(1 - \cos\theta) \quad (4.183)$$

$$\Rightarrow \left\{\frac{ml(\omega_0 - \omega\cos\theta)}{2(M+m)}\right\}(M+m) + \frac{ml^2}{3}\omega^2$$

$$+ m\omega l\cos\theta\frac{ml(\omega_0 = \omega\cos\theta)}{2(M+m)} + \frac{ml^2\omega^2}{3} = mgl(1 - \cos\theta) + \frac{ml^2\omega_0^2}{6}$$

$$\Rightarrow \frac{ml}{4(M+m)}\left\{\omega_0^2 + \omega^2\cos^2\theta - 2\omega\omega_0\cos\theta + 2\omega\omega_0\cos\theta - 2\omega^2\cos^2\theta\right\}$$

$$+ \frac{l\omega^2}{3} = -g(l - \cos\theta) + \frac{l\omega_0^2}{3}$$

$$\Rightarrow \frac{ml\{\omega_0^2 - \omega^2\cos^2\theta\}}{4(M+m)} + \frac{l\omega^2}{3} = -g(1 - \cos\theta) + \frac{l\omega_0^2}{3}$$

$$\Rightarrow \frac{ml\omega_0^2}{4(M+m)} + l\omega^2\left\{\frac{1}{3} - \frac{m\cos^2\theta}{4(M+m)}\right\} = -g(1 - \cos\theta) + \frac{\omega_0^2 l}{3}$$

$$\Rightarrow \omega^2\left\{\frac{4M + m(1 + 3\sin^2\theta)}{12(M+m)}\right\} = \frac{\omega_0^2 l}{3} = -g(1 - \cos\theta) + \frac{ml\omega_0^2}{4(M+m)}$$

$$\Rightarrow \omega = \sqrt{\frac{12\left\{-g(1 - \cos\theta) + \frac{(4m+m)l\omega_0^2}{12(M+m)}\right\}(M+m)}{\{4M + m(1 + 3\sin^2\theta)\}l}}.$$

Putting $M = m$, $l = 2a$, $\omega_0 = \sqrt{\frac{3g}{a}}$,

4-115

$$\Rightarrow \omega = \sqrt{\dfrac{12\left\{-g(1-\cos\theta)+\dfrac{(4m+m)(2a)\left(\frac{3g}{a}\right)}{12(M+m)}\right\}(M+m)}{\{4M+m(1+3\sin^2\theta)\}l}}$$

$$= \sqrt{\dfrac{12\{-g(1-\cos\theta)+\dfrac{3g}{2\times 4}\}(2)}{(4+1+3\sin^2\theta)l}}$$

$$= \sqrt{\dfrac{12\{-4+4\cos\theta+5\}2}{l(5+3\sin^2\theta)4}}$$

$$= \sqrt{\dfrac{6(1+4\cos\theta)g}{(8-3\cos^2\theta)2a}}$$

$$= \sqrt{\dfrac{3(1+4\cos\theta)g}{(8-3\cos^2\theta)a}}$$

Asymmetric rolling

Problem 42 A bead of mass m is glued at the bottom of a ring of mass M and radius R. If the ring rolls without slipping, and v_0 is the initial velocity of the center of the ring find (a) the velocity of the center of the ring as the function of the angle turned by the the bead relative to the center of the ring (b) the v_0 so that the ring rolls without losing contact with the ground.

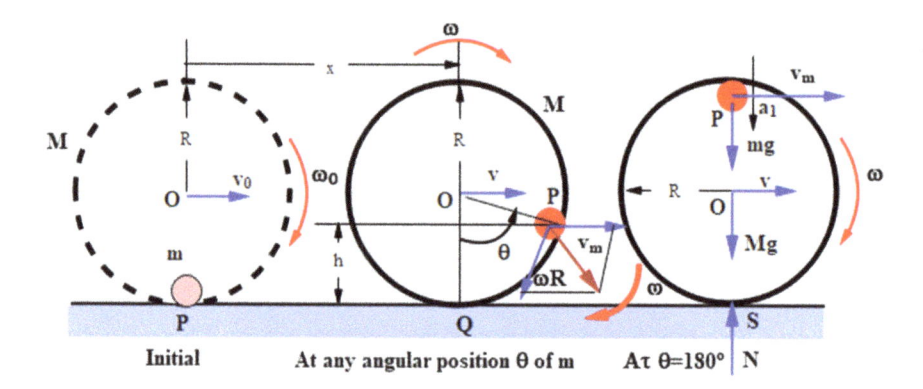

Solution
(a) Since the body is rolling static friction does not perform any work. So we have

$$\Delta U + \Delta K = 0$$

$$\Rightarrow mgR(1 - \cos\theta) + \frac{M}{2}v^2\left(1 + \frac{K^2}{R^2}\right) + \frac{1}{2}mv_m^2 - \frac{1}{2}mv_0^2\left(1 + \frac{K^2}{R^2}\right) = 0$$

$$\Rightarrow \frac{M(v^2 - v_0^2)}{2}\left(1 + \frac{K^2}{R^2}\right) + \frac{1}{2}mv_m^2 + mgR(1 - \cos\theta) = 0, \qquad (4.184)$$

where v_m is the velocity of the particle of mass m, given as

$$v_m^2 = v^2 + v^2 + 2v \cdot v \cos(180 - \theta)$$
$$= 2v^2(1 - \cos\theta). \qquad (4.185)$$

Using the equations (4.184) and (4.185) we have

$$m(v^2 - v_0^2)\left(1 + \frac{k^2}{R^2}\right) + m\{2v^2(1 - \cos\theta)\} + 2mgR(1 - \cos\theta) = 0$$

$$\Rightarrow v^2\left[m\left(1 + \frac{k^2}{R^2}\right) + 2m(1 - \cos\theta)\right] = mv_0^2\left(1 + \frac{k^2}{R^2}\right) - 2mgR(1 - \cos\theta)$$

$$\Rightarrow v = \sqrt{\frac{mv_0^2\left(1 + \frac{k^2}{R^2}\right) - 2mgR(1 - \cos\theta)}{m\left(1 + \frac{k^2}{R^2}\right) + 2m(1 - \cos\theta)}}.$$

Putting $K^2/R^2 = 1$ we have

$$v = \sqrt{\frac{Mv_0^2 - mgR(1 - \cos\theta)}{M + m(1 - \cos\theta)}}. \qquad (4.186)$$

(b) If the hoop does not break off while the bead goes to the top, the normal reaction N must be greater than zero, which is given as

$$(M + m)g - N = ma_1,$$

where

$$a_1 = \frac{v_{rel}^2}{R} = R\omega^2.$$

Putting $N = 0$ in the last equation for break off, we have

$$(M + m)g - N = mR\omega^2,$$

where

$$\omega = \frac{v}{2R}$$

$$\Rightarrow (M+m)g = mR\left(\frac{v}{2R}\right)^2 = \frac{mv^2}{4}$$

$$\Rightarrow v^2 = \frac{4(M+m)}{m}gR. \tag{4.187}$$

Putting this value of v from equation (4.187) in equation (4.186) we have

$$\frac{4(M+m)}{m} = \frac{Mv_0^2 - mgR(1-\cos 180°)}{M + m(1-\cos 180°)}$$

$$\Rightarrow \frac{4(M+m)gR}{m} = \frac{mv_0^2 - 2mgR}{M+2m}$$

$$mv_0^2 = \frac{4(M+2m)(M+mgR)}{m} + 2mgR$$

$$\Rightarrow v_0 = \sqrt{2gR\left(\frac{2M^2 + 4Mm + 5m^2}{Mm}\right)}.$$

Problem 43 A uniform rod of mass m and side $l = 1$ m is in equilibrium on a fixed hemisphere of radius R. If the right-hand edge of the rod gets a downward velocity v as shown in the figure by a gentle tapping, it rolls over the hemisphere without sliding. (a) If the maximum angle of rotation of the rod is $\theta = 36°$ while the rod remains tangential to the hemisphere while rotating in a vertical plane, find the value of v. (b) What is the angular acceleration of the rod as a function of the angle of rotation θ? (c) What is the frequency of oscillation of the rod about the mean position for a small displacement? Put $l = 1$ m, $m = 1$ kg and $g = 10$ m s^{-2}

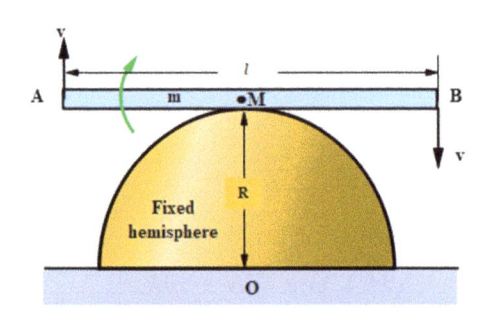

Solution

(a) Let the rod rotate by an angle θ in the process of rolling over the curved surface. At any angle of rotation of the rod, its center of mass moves up by a vertical distance
$$y = R(\cos\theta + \theta\sin\theta).$$
So the change in potential energy is
$$\Delta U = mgy.$$
The kinetic energy of the rod increases from K_0 to K. Then the change in kinetic energy is
$$\Delta K = K - K_0.$$
Conserving the energy, we have
$$\Delta U + \Delta K = 0$$
$$\Rightarrow K - K_0 - mgy = 0, \qquad (4.188)$$

where $K_0 = \frac{1}{2}\frac{ml^2}{12}\omega_0^2$ and $K = \frac{1}{2}\frac{ml^2}{3}\omega^2$ as the rod will rotate a greatest angle and still remain tangential to the curved surface. As the rod will remain at rest instantaneously after rotating a maximum angle, the final kinetic energy is zero.

$$\Rightarrow \frac{1}{2}\frac{ml^2}{12}\omega_0^2 + mgR = mg(R\cos\theta + R\theta\sin\theta)(\because K = 0)$$

$$\Rightarrow \frac{m\omega_0^2 l^2}{24} = mgR(\cos\theta + \theta\sin\theta - 1)$$

$$\Rightarrow \omega_0 = \frac{\sqrt{24gR(\theta\sin\theta + \cos\theta - 1)}}{l}.$$

Putting $\omega_0 = 2v/l$ we have
$$\Rightarrow v = \omega_0 l/2 = \sqrt{6gR(\theta\sin\theta + \cos\theta - 1)},$$
where $\theta = l/2R = 36° = \pi/5$ radians. After calculation, we have $v = 2.9$ rad s^{-1}.

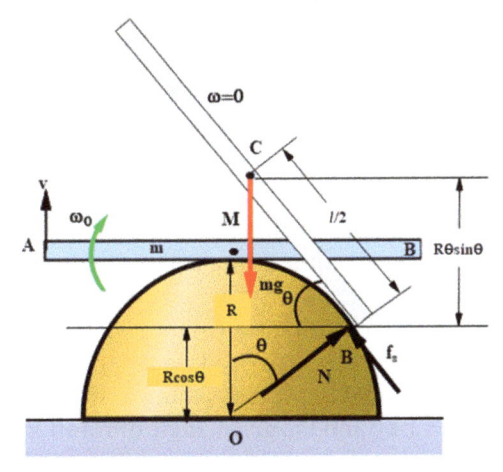

(b) Taking the torque about the point of contact of the rod with the curved surface, we have

$$\tau_P = I_P \alpha$$

$$\Rightarrow mgR\theta \cos\theta = m\left(\frac{l^2}{12} + R^2\theta^2\right)\alpha$$

$$\Rightarrow \vec{\alpha} = \frac{12gR\theta \cos\theta}{l^2 + 12R^2\theta^2}\hat{k}.$$

(c) If $\theta < <, \alpha \cong \frac{12gR}{l^2}\theta.$

According to the theory of oscillation,

$$\alpha = \omega_{osc}^2 \theta$$

$$\Rightarrow \omega_{osc} = \sqrt{\frac{12gR}{l^2}} = \frac{2\sqrt{3gR}}{l}.$$

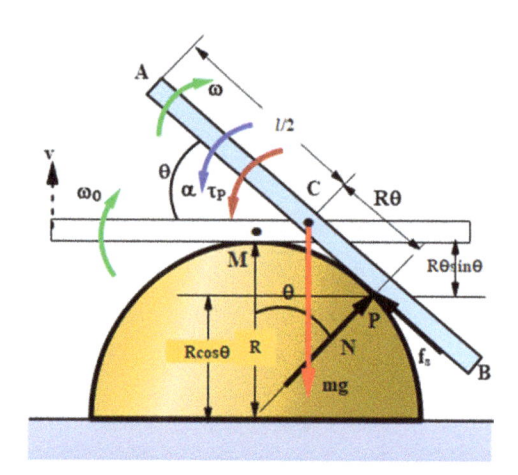

Problem 44 In the last problem, assume that the radius of the curved surface is $R = 3l/2\pi$ and $\omega^2_0 = 72$ (rad s^{-2}). At the maximum angle of inclination with the horizontal of the rod when it still remains tangential to the curved surface, find (a) the angular velocity of the rod, (b) the angular acceleration of the rod, (c) the friction acting on the rod, (d) the normal reaction acting on the rod, and (e) the minimum value of the coefficient of static friction required to prevent sliding of the rod. Put length of the rod $= l = 1$ m, mass of the rod $= 1$ kg and the given angle of rotation is equal to 36°. Take $g = 10$ m s^{-2}.

4-120

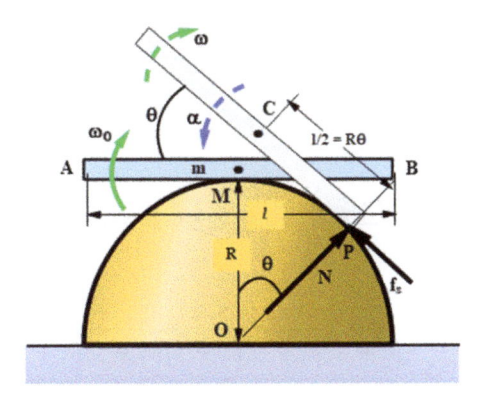

Solution
(a) Let the rod rotate by an angle θ in the process of rolling over the curved surface. At an angle of rotation $\theta = 60°$ of the rod, the height of its center is
$$H = y + h = R\cos\theta + (l/2)\sin\theta.$$

The initial height of the center of mass of the rod is equal to R relative to the point O.

So the change (increase) in potential energy is
$$\Delta U = mg(H - R) = mg(h + y - R)$$
$$= mg\{(l/2)\sin\theta - R(1 - \cos\theta)\}. \tag{4.189}$$

The kinetic energy of the rod changes (decreases) from K_0 to K. Then the change in kinetic energy is
$$\Delta K = K - K_0,$$
where $K_0 = \frac{1}{2}\frac{ml^2}{12}\omega_0^2$ and $K = \frac{1}{2}\frac{ml^2}{3}\omega^2$ as the rod will be at a position corresponding to the maximum angle of rotation:
$$\Delta K = \frac{ml^2}{24}(4\omega^2 - \omega_0^2). \tag{4.190}$$

Conserving the energy we have
$$\Delta U + \Delta K = 0. \tag{4.191}$$

Using the last three equations we have
$$mg\{(l/2)\sin\theta - R(1 - \cos\theta)\}$$
$$+ \frac{ml^2}{24}(4\omega^2 - \omega_0^2) = 0$$

$$\Rightarrow \frac{1}{2}\frac{ml^2}{12}\omega_0^2 + mgR$$

$$= mgR\cos\theta + mg(l/2)\sin\theta + \frac{1}{2}\frac{ml^2}{3}\omega^2$$

$$\Rightarrow \frac{ml^2}{6}\omega^2 = \frac{m\omega_0^2 l^2}{24} - mgR\{\cos\theta + \theta\sin\theta - 1\}$$

$$\Rightarrow \omega^2 = \frac{\omega_0^2}{4} - \frac{6gR(\theta\sin\theta + \cos\theta - 1)}{l^2}$$

$$\Rightarrow \omega = \sqrt{\frac{\omega_0^2}{4} - \frac{6gR(\theta\sin\theta + \cos\theta - 1)}{l^2}},$$

where $\theta = l/2R$. Putting $\omega_0^2 = 72$, $\theta = 36°$ and $R = 0.8$ m and after calculation, we have

$$\Rightarrow \omega = 3.08 \text{ rad s}^{-1},$$

(b) Applying the torque about P we have

$$\tau_P = I_P \alpha$$

$$\Rightarrow mg\frac{l}{2}\cos\theta = \frac{ml^2}{3}\alpha$$

$$\Rightarrow \alpha = \frac{3g}{2l}\cos\theta = \frac{3g}{2l}\cos\frac{l}{2R}\{\because \frac{l}{2} = R\theta\}.$$

4-122

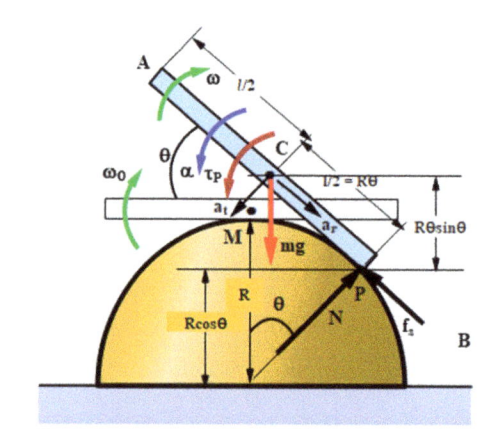

(c) The force equation along the rod is given as

$$mg \sin\theta - f_s = ma_r = m\frac{l}{2}\omega^2$$

$$\Rightarrow f_s = (mg \sin\theta - m\{l/2\}\omega^2) = (1)(10)(3/5) - (1)(1/2)(3)(3) = 1.5 \text{ N}$$

(d) Force equation along the normal to the rod is given as

$$\Rightarrow mg \cos\theta - N = \frac{ml}{2}\alpha$$

$$\Rightarrow (1)(10)(4/5) - N = (1)(1)(12)/2$$

$$\Rightarrow N = 2\text{N}.$$

(e) The law of static friction:

$$\mu_s \leq \frac{(f_s)}{N}$$

$$\Rightarrow \mu_s \leq 1.5/2 = 0.75.$$

So the minimum value of the coefficient of static friction which is designated as the critical value is equal to 0.75 at this position so that the rod will not slip.

Problem 45 A solid rough hemisphere of mass m and radius R is released from the given position. If the ground is (a) sufficiently rough, or (b) smooth find the maximum (i) angular speed and (ii) speed of the center of mass of the hemisphere.

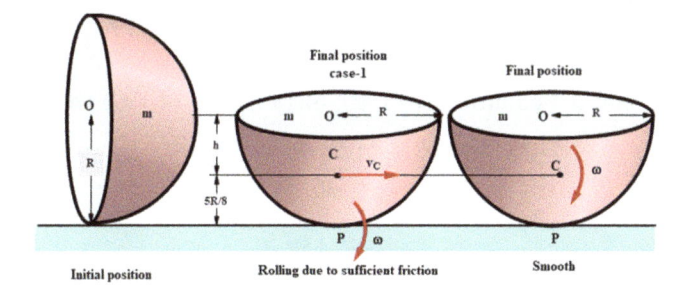

Solution
(a)
    (i) Let the hemisphere roll until its flat surface becomes horizontal. In this process it rotates by an angle $\theta = 90°$ in the process of rolling over the

4-123

flat surface. At an angle of rotation $\theta = 90°$ of the hemisphere, its center of mass descends by a vertical distance $R - 5R/8 = 3R/8$.

So the change (decrease) in potential energy is

$$\Delta U = -3mgR/8. \tag{4.192}$$

The kinetic energy of the rod changes (increases) from zero to K. Then the change in kinetic energy is

$$\Delta K = K = \frac{1}{2} I_P \omega^2. \tag{4.193}$$

By parallel axis theorem we have

$$I_P = I_C + m(CP)^2 \tag{4.194}$$

$$I_O = I_C + m(CO)^2. \tag{4.195}$$

From the equations (4.194) and (4.195) we have

$$I_P = I_O + m(PC^2 - CO^2). \tag{4.196}$$

Putting $I_C = 2mR^2/5$, $PC = 5R/8$, and $OC = 3R/8$ we have

$$I_P = \frac{2mR^2}{5} + m\{(5R/8)^2 - (3R/8)^2\}$$

$$\Rightarrow I_P = \frac{13mR^2}{20}. \tag{4.197}$$

Using equations (4.193) and (4.197)

$$\Delta K = \frac{1}{2} I_P \omega^2 = \frac{13mR^2}{40} \omega^2. \tag{4.198}$$

Conserving the energy, we have

$$\Delta U + \Delta K = 0. \tag{4.199}$$

Using the equations (4.192), (4.193), and (4.199) we have

$$-\frac{3mgR}{8} + \frac{13mR^2}{40} \omega^2 = 0$$

$$\Rightarrow \omega^2 = \frac{15g}{13R}$$

$$\Rightarrow \omega = \sqrt{\frac{15g}{13R}}.$$

(ii) The velocity of the center of mass is
$$v_C = \frac{5R}{8}\omega = \frac{5R}{8}\sqrt{\frac{15g}{13R}} = \frac{5}{8}\sqrt{\frac{15gR}{13}},$$
which is directed towards the right.

(b)
(i) In the absence of friction, the hemisphere cannot roll. However, it rotates anticlockwise about a horizontal axis passing through its center of mass. The center of mass does not move horizontally as there is no horizontal force acting on it. However, due to the effect of gravity, the center of mass moves vertically downwards by a distance of $3R/8$. Let the hemisphere rotate until its flat surface becomes horizontal. In this process it rotates by an angle $\theta = 90°$ on the flat surface. As the center of mass descends by a vertical distance $R - 5R/8 = 3R/8$, as given in equation (4.192), the change (decrease) in potential energy is
$$\Delta U = -3mgR/8. \tag{4.200}$$
The kinetic energy of the hemisphere changes (increases) from zero to K. Then the change in kinetic energy is
$$\Delta K = K = \frac{1}{2}I_C\omega^2. \tag{4.201}$$
By the parallel axis theorem we have
$$I_O = I_C + m(CO)^2$$
$$\Rightarrow I_C = I_O - m(CO)^2.$$
Putting $I_O = 2mR^2/5$ and $OC = 3R/8$ we have
$$I_C = \frac{2mR^2}{5} - m(3R/8)^2 = 83mR^2/320.$$
Then the change in kinetic energy is
$$\Delta K = \frac{1}{2}I_C\omega^2 = \frac{83mR^2}{640}\omega^2. \tag{4.202}$$
Conserving the energy, we have
$$\Delta U + \Delta K = 0. \tag{4.203}$$
Using the equations (4.200), (4.202), and (4.203) we have
$$-\frac{3mgR}{8} + \frac{83mR^2}{640}\omega^2 = 0$$
$$\Rightarrow \omega^2 = \frac{240g}{83R}$$

$$\Rightarrow \omega = \sqrt{\frac{240g}{83R}}.$$

(ii) As there is no net horizontal force acting on the body, its horizontal momentum remains constant. In other words, the horizontal velocity of the center of mass remains constant. Since initially the body was at rest, the initial velocity of the center of mass is zero. Hence, the velocity of the center of mass will be zero for all times. This means that the center of mass just moves up and down by the effect of gravity, whereas the body rotates about its center of mass due to the torque produced by the normal reaction about the center of mass.

Problem 46 A rod of mass m and length R is fitted with the disc of mass M (= m) and radius R, as shown in the figure. The disc is placed on a rough horizontal surface. Let us give a linear velocity to the center of the disc so that it rolls without sliding on the horizontal surface. Find the minimum and maximum initial angular speeds of the disc so that the center of mass of the rod will complete a vertical circle relative to the center of the disc and the disc will roll without losing contact with the ground.

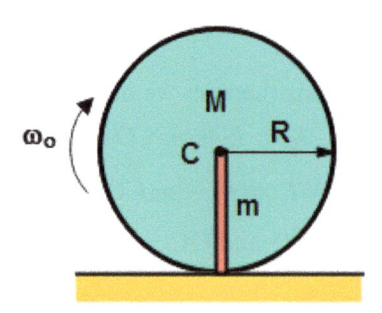

Solution
As the disc rolls on the horizontal surface, the rod also rotates and its center of mass goes up. When the rod becomes vertical, its center of mass rises by a vertical distance R.

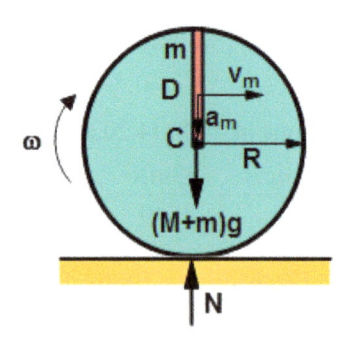

So the change (increase) in potential energy of the system $(M + m)$ is

$$\Delta U = mgR. \qquad (4.204)$$

The kinetic energy of the rod changes (decreases) from K_0 to K. Then the change in kinetic energy is

$$\Delta K = K - K_0. \qquad (4.205)$$

Conserving the energy we have

$$\Delta U + \Delta K = 0. \qquad (4.206)$$

Using the last three equations we have

$$K - K_0 + mgR = 0$$

$$\Rightarrow K_0 - K = mgR. \qquad (4.207)$$

The initial kinetic energy of the system is

$$K_0 = \frac{3}{4}Mv_0^2 + \frac{1}{2} \times \frac{mR^2}{3}\omega_0^2$$

$$= \frac{3}{4}Mv_0^2 + \frac{1}{2} \times \frac{mv_0^2}{3} \quad (\because v_0 = R\omega_0)$$

$$\Rightarrow K_0 = \left\{\frac{3}{4}M + \frac{m}{6}\right\}v_0^2. \qquad (4.208)$$

The final kinetic energy of the system is

$$K = \frac{3}{4}Mv^2 + \frac{1}{2}\frac{mR^2}{12}\omega^2 + \frac{1}{2}mv_m^2.$$

Putting $v_m = 3R\omega/2$ in the last equation we have

$$K = \frac{3}{4}M(R\omega)^2 + \frac{1}{2}\frac{mR^2}{12}\omega^2 + \frac{1}{2}m(3R\omega/2)^2$$

$$= (R\omega)^2\left\{\frac{3}{4}M + \frac{m}{24} + \frac{9m}{8}\right\}$$

$$\Rightarrow K = (R\omega)^2\left\{\frac{3}{4}M + \frac{7m}{6}\right\}. \qquad (4.209)$$

To find the angular velocity ω of the system $(M + m)$ we need to write the force equation in the vertical direction as follows:

$$(M + m)g - N = ma_m + Ma_M. \qquad (4.210)$$

Since the height of the center of mass of the disc remains constant, its vertical acceleration a_M is zero. Then we can write

$$a_M = 0. \tag{4.211}$$

The vertical acceleration a_m of center of mass D of the rod is given as

$$a_D = m\frac{R}{2}\omega^2. \tag{4.212}$$

Using last three equations we have

$$(M + m)g - N = \frac{mR\omega^2}{2}. \tag{4.213}$$

Putting $N = 0$ for losing contact with the horizontal surface we obtain

$$\omega = \sqrt{\frac{2(M + m)g}{mR}}. \tag{4.214}$$

Using the equations (4.207), (4.208), and (4.209) we have

$$\left\{\frac{3}{4}M + \frac{m}{6}\right\}v_0^2 - (R)\{R\omega^2\}\left\{\frac{3}{4}M + \frac{7m}{6}\right\} = mgR. \tag{4.215}$$

Using the equations (4.214) and (4.215) we have

$$\left\{\frac{3}{4}M + \frac{m}{6}\right\}v_0^2 = \left\{\frac{2(M + m)g}{m}\right\}\left\{\frac{3}{4}M + \frac{7m}{6}\right\} + mgR$$

$$\Rightarrow \frac{9M + 2m}{12}v_0^2 = mgR + \frac{9M + 14m}{12} \times \frac{2(M + m)gR}{m}$$

$$\Rightarrow \frac{9M + 2m}{12}v_0^2 = gR\left[\frac{6m^2 + (9M + 14m)(M + m)}{6m}\right]$$

$$\Rightarrow v_0 = v_0|_{max} = \sqrt{\frac{20m^2 + 9M^2 + 23Mm}{(9M + 2m)m}\cdot 2gR}.$$

For the minimum value of v_0, we can put $\omega = 0$ in equation (4.215) (because the rod will finally be at rest at its highest position) to obtain

$$\left\{\frac{3}{4}M + \frac{m}{6}\right\}v_0^2 = mgR$$

$$\Rightarrow v_0|_{min} = \sqrt{\frac{12mgR}{9M + 2m}}.$$

Putting $M = m$ we have

$$v_{0_{max}} = \sqrt{\frac{52gR}{11}}, \quad v_{0_{min}} = \sqrt{\frac{12gR}{11}}$$

$$\Rightarrow \omega_0|_{max} = \sqrt{\frac{52g}{11R}}, \quad \omega_0|_{min} = \sqrt{\frac{12g}{11R}}.$$

Chapter 5

Impulse and momentum of rigid bodies

5.1 The linear momentum of a rigid body

A rigid body is a system of particles rigidly connected with each other. So the linear momentum of a rigid body relative to a fixed point P is given as

$$\vec{P} = M\vec{v}_C,$$

where $\vec{v}_C$ = the velocity of the center of mass of the rigid body.

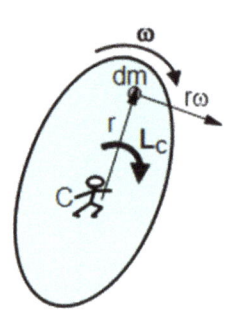

5.2 Angular momentum of a rigid body

5.2.1 Spin angular momentum

The angular momentum of a rigid body about its center of mass C is called spin angular momentum. It is given as

$$\vec{L}_s = I_C \vec{\omega},$$

where $\vec{\omega}$ = the angular velocity of the rigid body and I_C = the moment of inertia of the rigid body about its centroidal axis of rotation. This is also known as *spin* (or internal) angular momentum, denoted by the symbol $\vec{L}_s$.

5.2.2 Orbital angular momentum

The angular momentum of the center of mass C of a rigid body about a fixed point of reference O is called orbital angular momentum. It is given as

$$\vec{L}_{orb} = m\,\vec{r}_C \times \vec{v}_C,$$

where $\vec{r}_C$ = the position and $\vec{v}_C$ = the velocity of the center of mass relative to the point of reference O.

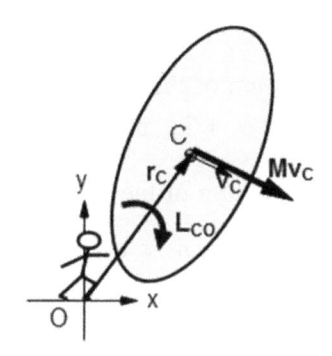

5.2.3 Total angular momentum

The vector sum of the above two angular momenta is called the total angular momentum of the rigid body about the reference point O. It given by the expression
$$\vec{L}_O = I_C\vec{\omega} + m\,\vec{r}_C \times \vec{v}_C.$$

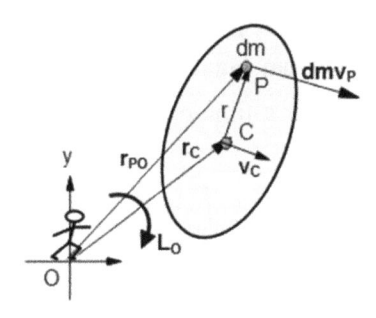

Example 1 The center of a uniform disc of mass $m = 2$ kg and radius $R = 2$ m is located at C (3 m, 4 m). It is spinning with an angular velocity $\omega = 5$ rad s^{-1}, as

5-2

shown in the figure. If its center of mass moves with a velocity $v_C = 4\hat{i} + 3\hat{j}$ m s^{-1} find the total angular momentum about the origin O.

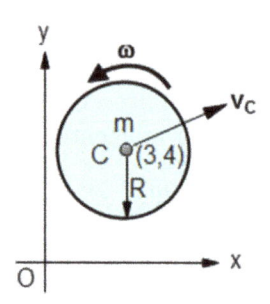

Solution

(a) The spin angular momentum of the disc is $\vec{L}_s = I_C \vec{\omega}$,

where $I_C = \dfrac{mR^2}{2} = \dfrac{(2)(4)}{2} = 4$ kg m^2 and $\vec{\omega} = 5\hat{k}$ rad s^{-1}. This gives $\vec{L}_s = 20\hat{k}$ kg m^2 s^{-1}.

(b) The orbital angular momentum of the disc is

$$\vec{L}_{orb} = m\,\vec{r}_C \times \vec{v}_C \tag{5.1}$$

$$= (3\hat{i} + 4\hat{j}) \times (4\hat{i} + 3\hat{j}) = -14\hat{k} \text{ kg m}^2\text{s}^{-1}.$$

(c) The net angular momentum of the disc about the origin O is given as

$$\vec{L}_O = \vec{L}_s + \vec{L}_{orb}$$

$$= (20\hat{k} - 14\hat{k}) = 6\hat{k} \text{ m}^2 \text{ s}^{-1}.$$

5.3 The impulse–momentum equation

5.3.1 The linear impulse–momentum equation

The linear impulse of a force over a time period t is defined as the time integral of that force given as

$$\vec{I} = \int_0^t \vec{F}\,dt.$$

When many forces act on a rigid body, each force has its own linear impulse. The net linear impulse, that is, the linear impulse of the net force, is numerically equal to the change in the linear momentum of the object. It is expressed as

$$\vec{I}_{net} = \int_0^t \vec{F}_{net}\,dt.$$

This integral form of Newton's second law of translation is known as the linear impulse–momentum equation:
- The net linear impulse of all forces acting on the rigid body is equal to the change in its linear momentum.
- When many forces act on a rigid body, the impulse of a force may not be equal to the change of linear momentum of the body.

5.3.2 The angular impulse–momentum equation

We have learnt that a force can create a torque about a reference point. The net torque about the center of mass, or axis of rotation (permanent or instantaneous), of a rigid body is numerically equal to the rate of change of angular momentum about the same reference point. It is given as

$$\vec{\tau}_{net} = \frac{d\vec{L}_{net}}{dt}.$$

Any combined motion can be split into the combination of a spin (motion relative to the center of mass) and orbital (motion of the center of mass) relative to any reference point, as shown in the following figure.

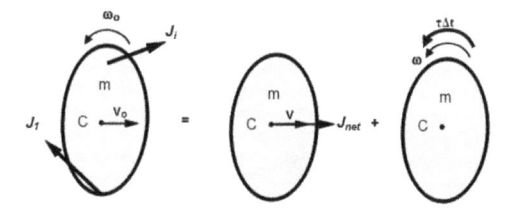

The time integral of the torque is known as the angular impulse of the force about any point of reference, given as

$$\vec{H}_{net} = \int_0^t \vec{\tau}_{net}\, dt.$$

When many forces act on a rigid body, each force generates its own angular impulse. Then the sum of all angular impulses, that is, the net angular impulse, is numerically equal to the change in the angular momentum of an object. It is expressed as

$$\int_0^t \vec{\tau}_{net}\, dt = \Delta \vec{L},$$

where $\int_0^t \vec{\tau}_{net}\, dt = \vec{H}_{net}$ and $\Delta \vec{L} = I(\vec{\omega} - \vec{\omega}_0)$. Then we have

$$\vec{H}_{net} = I(\vec{\omega} - \vec{\omega}_0).$$

This integral form of Newton's second law of rotation is known as the angular impulse–momentum equation:

- The net angular impulse of a force is equal to the change in angular momentum of the rigid body.
- When many forces act on a rigid body, the angular impulse of a force may not be equal to the change of angular momentum of the rigid body.

Example 2 At time $t = 0$, let a sphere of mass m, radius R, and radius of gyration K be released from rest on an inclined plane. After it rolls without sliding for a time t find (a) the friction and (b) the angular impulse relative to (i) the center of mass and (ii) the initial point of contact of the inclined plane with the sphere, (c) (i) the angular velocity of the rigid body and (ii) the velocity of the center of mass of the sphere, by using the impulse–momentum method.

Solution

(a) Let us choose the origin O of the coordinate system as the reference point.

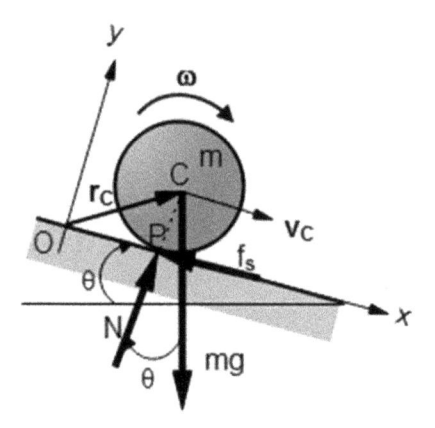

Referring to the free-body diagram the normal reaction is

$$N = mg \cos \theta.$$

The net force acting on the rigid body is equal to

$$F_{net} = mg \sin \theta - f_s.$$

Then the net linear impulse during a time t is given as

$$J = \int_0^t F_{net} \, dt = \int_0^t (mg \sin \theta - f_s) \, dt$$

$$= (mg \sin \theta - f_s) \int_0^t dt$$

$$= (mg \sin \theta - f_s) t.$$

The change in linear moment is equal to mv_C. By using the impulse–momentum equation

$$\vec{I} = \int_0^t \vec{F} dt,$$

we have

$$(mg \sin \theta - f_s)t = mv. \tag{5.2}$$

Since the gravity mg and normal reaction N pass through the center of mass, they cannot produce any torque about the center of mass. So their angular impulse about the center of mass is zero. Then the net angular impulse about the center of mass is given as

$$\vec{J}_{net} = \int_0^t \vec{\tau}_{net} dt,$$

where the net torque about the center of mass is

$$\vec{\tau}_{net} = \vec{f}_s \, R\hat{k}.$$

Then we have

$$\vec{H}_{net} = \int_0^t (-f_s R\hat{k}) dt$$

$$= -f_s R\hat{k} \int_0^t dt = -f_s Rt\hat{k}.$$

The change in angular momentum about the center of mass of the sphere is

$$\Delta \vec{L} = I_C \vec{\omega} = -mK^2 \omega \hat{k}.$$

Now applying the impulse–momentum equation

$$\vec{J}_{net} = \Delta \vec{L},$$

we have

$$f_s Rt = -mK^2 \omega. \tag{5.3}$$

The condition of rolling is

$$v = R\omega. \tag{5.4}$$

Putting the value of v from equation (5.2) and ω from equation (5.3) in equation (5.4) and simplifying the factors, we have

$$f_s = \frac{mg \sin \theta}{1 + \frac{R^2}{K^2}}.$$

(b)

(i) Putting the value of f_s in equation (5.2) we have

$$H_C = f_s Rt = \frac{mgRt \sin\theta}{1 + \frac{R^2}{K^2}}.$$

(ii) As the friction passes through the point of contact P, its angular impulse is zero about P. As $N = mg\cos\theta$, the net angular impulse about a fixed point on the inclined plane is equal to the gravitational impulse during the time t, which is is equal to $mgR \sin\theta \hat{k}$. Then the net angular impulse of the sphere about P is

$$\vec{H}_{net} = mgRt \sin\theta \hat{k}.$$

(c)

(i) The angular momentum of the sphere increases from zero to $\vec{L}_P = I_P \vec{\omega}$. Then the change in angular momentum of the sphere about P is given as

$$\Delta \vec{L}_P = I_P \vec{\omega} = -m(R^2 + K^2)\omega \hat{k}.$$

Applying the impulse–momentum equation $\vec{H}_P = \Delta \vec{L}_P$ about P, we have

$$-mgRt \sin\theta \hat{k} = -m(R^2 + k^2)\omega \hat{k}$$

$$\text{or } \omega = \frac{gRt \sin\theta}{R^2 + K^2}.$$

(ii) The velocity of the center of mass is given as

$$v = R\omega = \frac{gt \sin\theta}{1 + \frac{K^2}{R^2}}.$$

5.4 Conservation of angular momentum

According to the angular impulse–momentum equation the net angular impulse is numerically equal to the change in angular momentum about a given reference point which is given as

$$\vec{H}_{net} = \int_0^t \vec{\tau}_{net} dt = \Delta \vec{L}_P.$$

5-7

When the net torque about the reference point is zero we have $\Delta \vec{L}_P = 0$. This means there is no change in angular momentum about the point of reference P. In other words, the angular momentum remains constant (or conserved) about the reference point. This is called the law of conservation of angular momentum.
- When the net torque acting on a rigid body about a reference point/axis is zero, the angular momentum of the rigid body is conserved about that reference point/axis.

Example 3 A sphere of mass m, radius R, and radius of gyration K is kept on a horizontal surface. The coefficient of kinetic friction between the sphere and surface is μ_k. At time $t = 0$ the sphere is given an initial velocity v_0 where as the initial angular velocity $\omega_0 = 0$. Due to kinetic friction between the sphere and horizontal surface, the sphere will stop sliding and start rolling. (a) Can you conserve angular momentum of the sphere about (i) O and (ii) its center of mass? (b) Using the principle of conservation of angular momentum, find (i) the velocity of the center of mass of the sphere and (ii) the time after when the sphere will begin to roll without sliding.

Solution
(a)
(i) Since N and mg pass through the center of mass of the sphere and $N = mg$, the net torque of these two forces about O is equal to zero. As the sphere slides towards right, the kinetic friction points towards left passing through O. So the frictional torque about O is zero. Since the net torque about O is zero, we can conserve the angular momentum of the sphere about the point O.

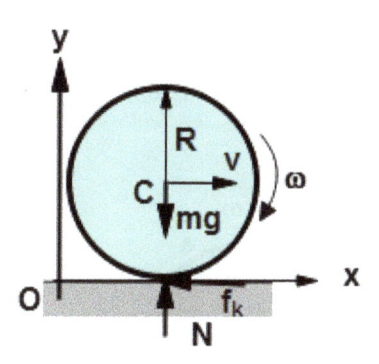

(ii) Since the gravity mg and normal reaction N pass through the center of mass, they cannot produce a torque about the center of mass. Then the net angular impulse about the center of mass is equal to that of the kinetic frictional force, given as

$$\vec{H}_{net} = -\mu_k mg R t \hat{k}.$$

Hence, we cannot conserve the angular momentum about the center of mass.

(b)

(i) Applying the principle of conservation of angular momentum about the point O we have

$$\vec{L}_f = \vec{L}_i,$$

where $\vec{L}_f = mv_0 R\hat{k}$ and

$$\vec{L}_i = mvR\left(1 + \frac{k^2}{R^2}\right)\hat{k}.$$

Then we have

$$v = \frac{v_0}{1 + \frac{k^2}{R^2}}.$$

(ii) The net angular impulse about the center of mass
$= \vec{\tau}_C t = -\mu_k mgRt\hat{k}.$

Putting the obtained value of $\tau_C t$ and $\Delta L_C = mK^2\omega$ in the equation

$$\vec{\tau}_C t = \Delta \vec{L}_C,$$

we have

$$-\mu_k mgRt\hat{k} = mK^2\omega.$$

This gives us

$$t = \omega K^2 / \mu_k gR = R\omega K^2 / \mu_k gR^2.$$

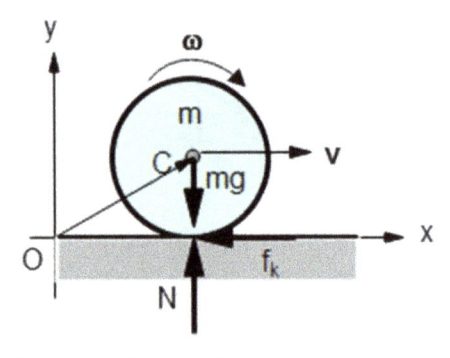

Putting $R\omega = v$ in the last equation we have

$$t = vK^2/\mu_k g\ R^2.$$

Putting the obtained value of v, finally we have

$$t = \frac{v_0}{\left(1 + \frac{R^2}{K^2}\right)\mu_k g}.$$

5.5 The collision of rigid bodies

There are three possible cases of collision with a rigid body. Thr first is the collision of a rigid body with a particle. Second, we will discuss the collision of rigid bodies with a fixed surface. Finally, we will take couple of examples of collisions of a rigid body with another unconstrained (free) rigid body. Let us first talk about the collision of a particle with a rigid body, a rod, say.

5.5.1 The collision of a particle with a rigid body

In the collision of a particle with a rigid body (a rod, say), generally we have two cases. In the first case the rigid body is smoothly pivoted and a particle collides with it. In the second case the rigid body can be set to move freely just after the collision.

Case 1: The rod is pivoted. In the first case the rod gains a momentum just after the particle collides with it. Then the rod collides with the pivot P. Thus there are two collisions taking place almost at the same time because of the extremely small gap between the rod and pivot.

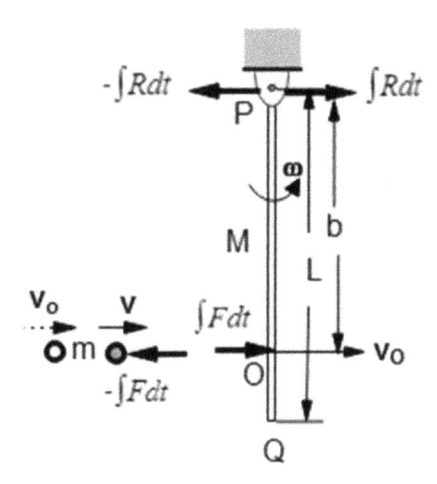

$P \neq C$: During the collision, the action–reaction forces between the particle and rod are internal forces. So they do not produce a net force. But the reaction force (R) offered by the pivot on the rod is external to the rod–particle system. The external impulse ($-\int R dt$) can change the momentum of the system. So, in general, the linear momentum of the rod–particle system is not conserved.

$L = C$: As the reaction force (R) passes through the pivot, its torque about the pivot is zero. Furthermore, the action–reaction pair (F and $-F$) between the rod and particle does not generate a net torque. In other words, the net angular impulse is negligible during a very small time of impact about the pivot. Then we can conserve the angular momentum of the rod–particle system about the pivot just before and after the collision.

NIF: At last we can apply Newton's impact or collision formula at the point of collision. In this case, we have two equations for two unknown quantities, such as the velocity of the particle and angular velocity of the rod just after the collision. Finally, we can solve these equations.

Example 4 A bead of mass m collides with the rod PQ with a velocity v_o at a distance b from the smooth pivot A. The rod has mass $M\,(=m)$, length L and the coefficient of restitution of the collision is e. Then find (a) (i) the angular velocity of the rod and (ii) the velocity of the bead, (b) the value of e for which the bead (i) stops, (ii) moves backward, and (iii) moves forward, just after the collision

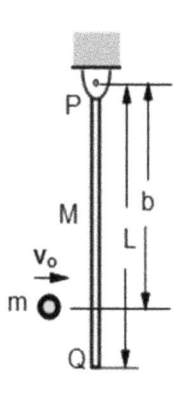

Solution
(a)
(i) $L = C$: According to the above description, we can conserve the angular momentum of the rod–particle system. Let the velocity of the bead be v and the angular velocity of the rod be ω (anticlockwise) just after the collision. Using the law of conservation of angular momentum about P just before and after the collision, we have

$$\vec{L}_f = \vec{L}_i,$$

or

$$mv\,b\hat{k} + \frac{1}{3}ML^2\vec{\omega} = mbv_o\hat{k},$$

or

$$ML\omega = 3M(v_o - v). \qquad (5.5)$$

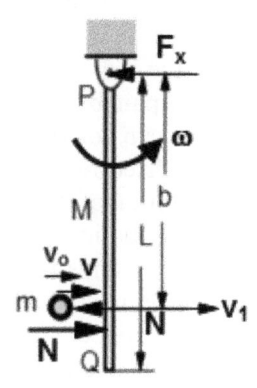

In the free-body diagram we can see that, in general, there is a horizontal reaction force acting at the pivot in response to the impulsive force N acting between the rod and particle during the impact. So the net horizontal force acting F_x on the rod–particle system $(M + m)$ is non-zero. Hence, we cannot conserve the linear momentum of the system.

NIF: At the point of collision we can apply Newton's impact formula

$$-ev_{\text{approach}} = v_{\text{separation}},$$

or

$$-e(v_0 - 0) = (v - v_1),$$

where $v_1 = b\omega$,

$$\text{or } v = b\omega - ev_0. \qquad (5.6)$$

Solving equations (5.5) and (5.6) we have

$$v = \frac{3mb - eML}{ML + 3mb} v_0.$$

(ii) $v = b\omega - ev$

$$\text{or } \frac{3mb - eML}{ML + 3mb} v_0 = b\omega - ev_0,$$

$$\text{or } \omega = \frac{3m(1 + e)v_0}{ML + 3mb}.$$

(b) Putting $v = 0$, $v < 0$, and $v > 0$, we have (i) $e = \frac{3mb}{ML}$, (ii) $e > \frac{3mb}{ML}$, and (iii) $e < \frac{3mb}{ML}$, respectively.

Case 2: The rod is free to move. In this case the pivot is not there. So only one collision takes place between the rod and pivot.

Problems and Solutions in Rotational Mechanics

$P = C$: During the collision, the action–reaction forces between the particle and rod are internal forces. As the net force acting on the rod–particle system is zero, the linear momentum of the rod–particle system is conserved.

$L = C$: The action–reaction pair between the rod and particle does not produce a net torque. Then we can conserve the angular momentum of the rod–particle system about any fixed reference point. But, if we choose the reference point on the line of motion of the center of mass of the rod or rod–particle system or particle, the equation will be more simplified.

NIF: At last we can apply Newton's impact or collision formula at the point of collision. In this case we have three equations for three unknown quantities, such as the angular velocity of the rod, the velocity of the particle, and the center of mass of the rod, just after the collision. Finally, we can solve these equations.

Example 5 A uniform smooth rod of mass $M = 2$ kg and length $l = 1$ m is kept on a horizontal surface. A bead of mass $m = 1$ kg collides with the rod with a velocity u, as shown in the figure. The coefficient of restitution of the collision is $e = 0.5$. Write down all relevant equations.

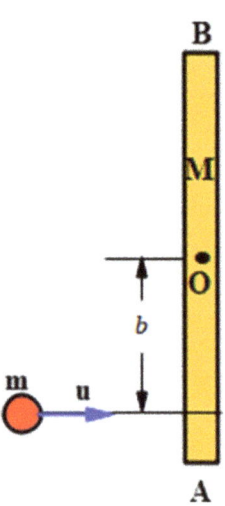

Solution
$L = C$: According to the above description, we can conserve the angular momentum of the rod–particle system. Just after the collision let the velocity of the bead and center of mass of the rod be v (towards the right) and v', respectively, and the angular velocity of the rod be ω (anticlockwise). Let us choose the reference point P on the line of motion of the center of mass of the rod. Conserving the angular momentum of the rod–bead system about O just before and after the collision, we have

$$\vec{L}_f = \vec{L}_i,$$

or $mv\, b\hat{k} + \dfrac{1}{12}ML^2\omega\hat{k} = mub\hat{k}$,

or $ML^2\omega = 12m(u-v)b$ \hfill (5.7)

$P = C$: Conserving the linear momentum of the rod–bead system just before and after the collision, we have

$$\vec{P}_f = \vec{P}_i,$$

or $mv\hat{i} + Mv'\hat{i} = mu\hat{i}$,

or $Mv' = m(u-v)$. \hfill (5.8)

NIF: Let the point of collision of the rod be Q. The velocity of Q just after the collision is given $\vec{v}_Q = \vec{v}_{QC} + \vec{v}_{C}$,
where $\vec{v}_{QC} = b\omega\hat{i}$ and $\vec{v}_C = v'\hat{i}$. Then we have $\vec{v}_Q = (v' + b\omega)\hat{i}$. Applying Newton's impact formula at Q we have

$$-e\vec{v}_{app} = \vec{v}_{sep}.$$

The bead approaches the rod with a velocity $\vec{v}_{app} = u\hat{i}$ and departs with a velocity

$$\vec{v}_{PQ} = \vec{v}_P - \vec{v}_Q = v\hat{i} - (v' + b\omega)\hat{i},$$

or $\vec{v}_{PQ} = \{v - (v' + b\omega)\}\hat{i}$

Putting these values in Newton's collision formula, we have $-eu = v - (v' + b\omega)$,

or $v' = v + eu - b\omega$. \hfill (5.9)

Now we have three equations for three unknown quantities such as v', v, and ω. We can solve for these equations.

5.5.2 Collison of a rigid body with another rigid body

Case 1: The second rigid body is heavy or fixed
$L = C$: Let a rigid body (a rod, say) strike another rigid body (a very heavy object or fixed ground, say). In this case, let us choose the point of impact as the reference point. Since the impulsive force offered by the fixed surface on the rod passes through the point of impact, its torque is zero about the reference point. So about the point of impact we can conserve the angular momentum over the time of impact.

During the impact, other non-impulsive forces (such as gravity) also act on the rigid body. Since the time of impact is extremely small, the impulse of such non-impulsive forces (the product of force and the time of impact) is very small. So we can disregard the impulse of the non-impulsive forces in comparison to that of the impulsive forces.

$P \neq C$ (linear momentum is not conserved): In this case we cannot conserve the linear momentum of the rigid body because a huge force (impulsive force) acts on it during the short time of collision.

NIF (Newton's impact formula): In any collision we can apply Newton's impact formula along the line of impact/collision between the rigid bodies.

Now we have two unknown quantities, namely the velocity v' of the center of mass of the rod and the angular velocity ω of the rigid body. We have two equations, namely the NIF and the conservation of angular momentum about the point of impact. Finally, we can solve two equations for two unknown quantities.

Case 2: The second rigid body is movable
Let rod 1 collide with another free rod 2, at a point (P, say). Since the impact forces pass through the point P, we can conserve the angular momentum of each rod about this reference point. Thus we obtain two equations (one for each body). As the net force acting on the system of two rods is zero, the linear momentum of the system (rod 1 + rod 2) is conserved. This is the third equation we have. We can choose the reference point at the point of impact of the two rods. Thus we can use the principle of conservation of linear and angular momentum of the system.

Problems

Problem 1 An ideal pulley of mass M and radius R is smoothly pivoted at P. An inextensible string connecting two particles of masses m and $2m$ passes over the pulley. The string does not slip on the pulley. If the pulley–particle system is released from rest at time $t = 0$, find (a) the angular impulse acting on the system, and (b) the speed of (i) the particles and (ii) the center of mass of the pulley–particle system as a function of time; assume the pulley as a uniform disc.

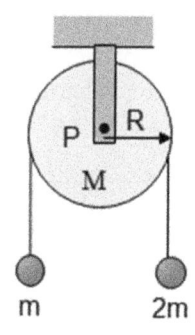

Problems and Solutions in Rotational Mechanics

Solution

(a) The net angular impulse = (gravitational torque) (time) because the net torque produced by the tensions is zero. Then the angular impulse

$$\Rightarrow J = (2m - m)gRt = mgRt \text{ (clockwise)}.$$

(b)

(i) The change in angular momentum of the system is $L = 2\, mvR + mvR + I_0\omega$,

where $v = R\omega$. Then $L = (I_d + 3\, mR^2)\omega$ clockwise.

Equating the net angular impulse with the change in angular momentum, we have

$$\Rightarrow mgRt = (I_d + 3mR^2), \text{ where } I_d = MR^2/2$$

$$\Rightarrow \omega = \frac{2mgt}{(M + 6m)}.$$

The speed of the particles is

$$v = R\omega = \frac{2mgRt}{(M + 6m)}.$$

(ii) The speed of the center of mass of the pulley–particles is

$$v_C = \frac{2mv - mv}{(M + 3m)} = \frac{m}{(M + 3m)}\left\{\frac{2mgRt}{(M + 6m)}\right\} \text{ (vertically downwards)}.$$

Problem 2 A uniform stepped pulley of mass M, radii R and r, and radius of gyration K is smoothly pivoted at O. A light string wrapped on the pulley connects two hanging particles each of mass m. If the string does not slip over the pulley, after releasing the pulley–particle system, using the impulse–momentum equation, find

(a) the angular impulse acting on the pulley–particle system about O, in terms of (a) the relative velocity v between the particles and (b) the angular velocity of the pulley as the function of the time t after releasing the system.

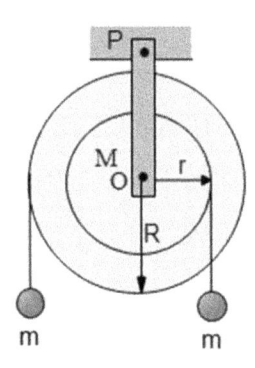

Solution

(a) The velocities of the particles after a time t are $v_1 = R\omega$, $v_2 = r\omega$. It is given that
$$v_1 + v_2 = \omega(R + r) = v.$$
Then
$$\omega = \frac{v}{R + r}.$$
The angular momentum of the system is
$$L_o = (m_1 R^2 + m_2 r^2 + MK^2)\omega$$
$$= \frac{(mR^2 + mr^2 + MK^2)}{R + r}v \text{ (anticlockwise)}.$$
Putting the value of ω
$$\vec{L}_o = (m(R^2 + r^2) + I)\frac{v}{R+r}\hat{k}, \text{ where } I = MK^2.$$

(b) After a time t, the angular impulse is
$$\vec{L} = m(R - r)gt\hat{k}.$$
Applying the angular impulse–momentum equation,
$$(mgR - mgr)t\hat{k} = (mR^2 + mr^2 + MK^2)\omega\hat{k}.$$
Then
$$\vec{\omega} = \frac{mg(R - r)t\hat{k}}{m(R^2 + r^2) + MK^2}.$$

5-17

Problem 3 A uniform disc of mass m and radius R placed on a horizontal surface is given an initial velocity v_0 and initial clockwise spin angular velocity $\omega_0 = 2v_0/R$, as shown in the figure.

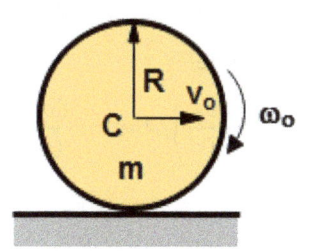

If the coefficient of kinetic friction between the disc and surface is μ, using the impulse–momentum equation, find (a) the final linear and angular velocity of the disc, (b) the time after which the disc will stop sliding, and (c) the distance of relative sliding.

Solution
(a) Let us take a fixed point O on the surface. The torque about O due to the normal reaction and gravity is zero because they are equal and opposite. The torque due to kinetic friction f_k is also zero because it passes through O. Then the net torque and angular impulse about O is zero. So the change in angular momentum of the body about O is zero. In other words, the angular momentum of the body about O remains conserved.

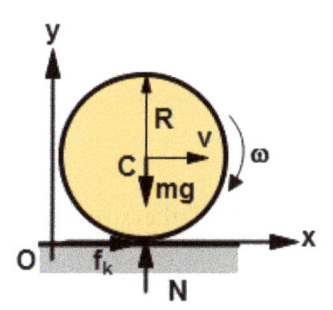

The initial and final angular momenta are $\vec{L_i} = -(mv_0 R + mK^2\omega_0)\hat{k}$ and $\vec{L_f} = m(K^2 + R^2)\vec{\omega_0}$. By conserving the momentum and putting the values of $K^2 = R^2/2$ and $\omega_0 = 2v_0/R$, and simplifying the factors, we have

$$\vec{\omega_f} = -\frac{(v_0 R + K^2\omega_0)}{K^2 + R^2}\hat{k} = -\frac{v_0 R + \frac{R^2}{2}\cdot\frac{2v_0}{R}}{\frac{3R^2}{2}}\hat{k}$$

$$= -\frac{4}{3R}(v_0)\hat{k}.$$

Then the final velocity of the disc is

$$\vec{v_f} = R\omega_f\hat{i} = \left(\frac{v_0 R + K^2\omega_0}{K^2 + R^2}\right)R\hat{i} = \frac{4}{3}v_0\hat{i}.$$

5-18

(b) It will roll after a time

$$t = \frac{\frac{4}{3}v_0 - v_0}{\mu g} = \frac{v_0}{3\mu g}.$$

(c) The distance covered before rolling is

$$s = \left(\frac{16v_0^2}{9}\right)/2\mu g = \frac{8v_0^2}{9\mu g}.$$

Problem 4 A uniform disc of mass m and radius R is placed on a horizontal surface so that its plane remains vertical. The disc is pulled with a constant horizontal force F by a string which does not slip over the disc. If the center of the disc moves by a distance l, assuming rolling without sliding of the disc on the surface, find (a) the velocity of the center of the disc and (b) the angular momentum of the disc, as a function of l relative to the origin O.

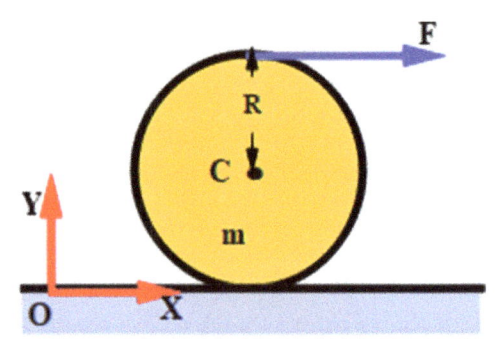

Solution
(a) As the disc rolls, the length of the string unwound will be equal to the distance covered by the center of mass of the disc. So the point of application of the force, that is, the force end of the string moves through a distance $l + l = 2l$. Then the net work done = the work done by the force = $W = F(2l) = 2FL$. The change in kinetic energy of the disc is

$$= \frac{mv_C^2}{2}\left(1 + \frac{1}{2}\right) = \frac{3mv_C^2}{4}.$$

Equating the work with the change in kinetic energy we have

$$v_C = \sqrt{\frac{8Fl}{3m}}$$

(b) The angular momentum of the disc relative to the point O is

$$\vec{L} = -mv_C R\left(1 + \frac{K^2}{R^2}\right)\hat{k}$$

5-19

$$= -mv_C R\left(1 + \frac{1}{2}\right)\hat{k}$$

$$= -\frac{3}{2}mv_C R\hat{k} = -\frac{3}{2}\sqrt{\frac{8Fl}{3m}}R\hat{k}$$

$$= -3R\sqrt{\frac{2Fl}{3m}}\hat{k}.$$

Problem 5 A rod of mass m and length L is smoothly hinged at the point P. A force F continues to act normal to the rod. (i) If the point of application of the force remains constant at a distance b, find the angular velocity of the rod as a function of time. (ii) If the point of application of the force shifts towards the free end of the rod with a constant velocity v relative to the rod, find the angular velocity of the rod when the point of application of the force reaches the end of the rod.

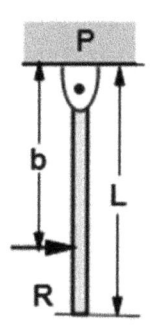

Solution
(i) The angular impulse about the pivot after a time t is Fbt. (b) The change in angular momentum of the rod is $= L = \frac{ml^2}{3}\omega$. Applying the impulse–momentum equation, we have

$$\omega = \frac{3Fbt}{m}.$$

(ii) It is given that that the point of application of force shifts radially away from the axis with a constant velocity v. So at any time t, the radial distance x is given as $\frac{x-b}{v} = t \Rightarrow x = vt + b$.
Applying the impulse–momentum equation we have

5-20

$$\frac{ml^2}{3}\omega = \int F(b+vt)dt$$

$$\frac{ml^2}{3}\omega = F\left(bt + \frac{vt^2}{2}\right),$$

where $t = (l-b)/v$

$$= F\left\{b\left(\frac{l-b}{v}\right) + \frac{v\cdot(l-b)^2}{2v^2}\right\}$$

$$= F\left\{b\frac{(l-b)}{v} + \frac{(l-b)^2}{2v}\right\}$$

$$= F\frac{(l-b)}{v}\left\{b + \frac{l-b}{2}\right\}$$

$$= F\frac{(l^2-b^2)}{2v}$$

$$\Rightarrow \omega = \frac{3F(l^2-b^2)}{2mvl^2}.$$

Problem 6 A string of mass m and length L is wrapped over a pulley of mass m and radius R. The pulley is pivoted smoothly at its center and fixed to the ceiling, as shown in the figure. If we slightly pull the string and let go, find the angular (a) acceleration and (b) velocity of the disc as a function of the length y of the hanging portion of the string. Assume that the string does not slip with the pulley.

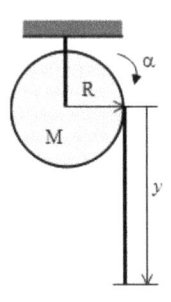

5-21

Solution

(a) The net torque of the hanging portion x of the string about the pivot C is

$$T_{gr} = \int dm'gR = \mu x g R,$$

where $\mu = dm'/dx$.

The change in angular momentum of the disc plus the string about the pivot is

$$\vec{L} = \vec{L}_{hanging} + \vec{L}_{wrapping} + \vec{L}_{disc}$$

$$= m'vR + m''R^2\omega + \frac{MR^2}{2}\omega$$

$$= (m' + m'')R^2\omega + \frac{MR^2}{2}\omega$$

$$\Rightarrow \vec{L} = \left(\frac{2m + M}{2}\right)R^2\omega.$$

Then differentiating with time

$$\frac{d\vec{L}}{dt} = \frac{2m + M}{2}R^2\vec{\alpha}.$$

Using the relation

$$\frac{d\vec{L}}{dt} = \vec{\tau} = I\vec{\alpha},$$

we have

$$|\vec{\alpha}| = \frac{2mgy}{(2m + m)R}.$$

(b) By using energy conservation

$$\Delta U + \Delta K = 0,$$

we have

$$= -\left(\frac{m}{l}y\right)\left(\frac{y}{2}\right)g + \frac{1}{2}\left(\frac{MR^2}{2} + mR^2\right)\omega^2 = 0,$$

or $\omega^2 = \frac{2mgy^2}{(M+2m)R^2}$. Then

$$\omega = \left\{\sqrt{\frac{2mg}{(M+2m)}}\right\}\frac{y}{R}.$$

5-22

Differentiating it with time and simplifying the factors, we can obtain the same answer

$$\alpha = (1/2)d(\omega^2)/dy = \frac{2mgy}{(M+2m)R}.$$

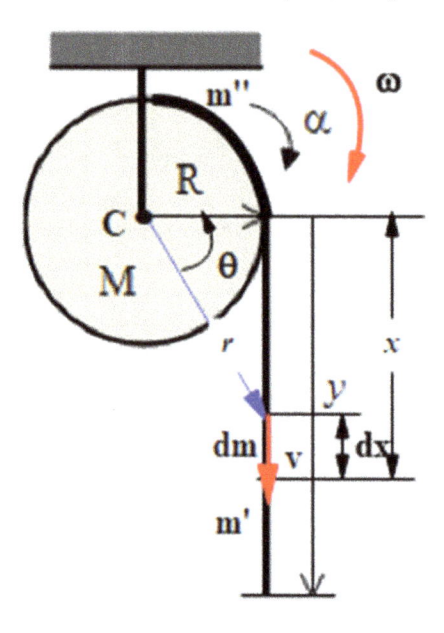

Problem 7 A pulley of mass m, radius R is pivoted smoothly at the top of a vertical light rigid rod of length PM $= H$ which is rigidly fitted with a smooth plank of mass M. A light string wrapped over the pulley is pulled by a horizontal force F. If the string does not slip over the pulley, find (a) the linear and angular velocity of the disc, (b) the kinetic energy as a function of time, and (c) the angular momentum of the of the system $(M+m)$ relative to the origin O, as the function of time for the two positions of the applied force F, given by the bold and dotted arrow.

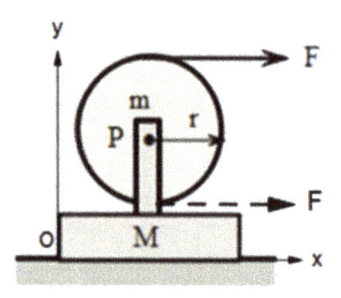

Solution
(a) Using the angular impulse–momentum equation about the pivot P, we have $mK^2\omega = FRt$, which gives

$$\vec{\omega} = -\frac{FR}{mK^2}t\hat{k}.$$

Using the angular impulse–momentum equation about the pivot, we have $Ft = (M + m)v_C$, which gives

$$\vec{v}_C = \frac{Ft}{M+m}\hat{i}.$$

(b) The total kinetic energy $= \frac{m}{2}K^2\omega^2 + \frac{(M+m)}{2}v_C^2$

$$= mk^2\left(\frac{FR}{mk^2}\right)^2 t^2 + \frac{M+m}{2}\left(\frac{Ft}{M+m}\right)^2$$

$$= \frac{F^2t^2R^2}{2mK^2} + \frac{F^2t^2}{2(M+m)}$$

$$K = \frac{F^2t^2}{2m}\left[\frac{R^2}{k^2} + \frac{m}{M+m}\right].$$

(c) The total angular momentum about the origin is and, where $\vec{L_O} = -mv_C H\hat{k}$ and $\vec{L_s} = +mK^2\vec{\omega}$. Now, putting $mv_C = Ft$ and $mK^2\vec{\omega} = Frt\ \hat{k}$, and simplifying the factors we have

$$\vec{L} = -FRt\hat{k}\left\{1 + \frac{mH}{(M+m)R}\right\}.$$

For the force given by the dotted arrow, the spin angular momentum will be positive and the orbital angular momentum will remain clockwise. So the answer will be given as

$$\vec{L} = FRt\hat{k}\left\{1 - \frac{mH}{(M+m)R}\right\}.$$

Problem 8 A drum of mass M and radius R is completely filled with water of mass m. The drum is rolling without sliding with a velocity v on a horizontal surface. The total mass of the two flat surfaces of the drum is m_1 and the mass of the curved surface of the drum is m_2. Find (a) the kinetic energy of the system (drum + water) and (b) the angular momentum of the system about the instantaneous axis of rotation of the drum.

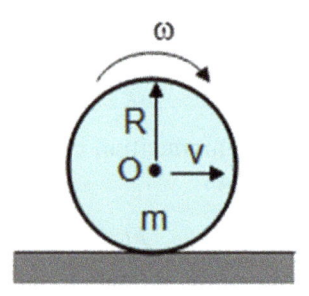

Solution

(a) The instantaneous axis of rotation of the drum is at its lowest point. The drum has two flat faces in the form of discs of mass m_1 and a thin cylinder of mass m_2. The water inside the drum only translates, but cannot rotate due to the absence of friction. Then the kinetic energy of the drum is

$$K = \frac{mv_C^2}{2} + \frac{1}{2}m_1 v_C^2\left(1 + \frac{1}{2}\right) + \frac{1}{2}m_2 v_C^2(1+1) = \frac{v_C^2}{2}\left[m + \frac{3}{2}m_1 + 2m_2\right]$$

$$= (2m + 3m_1 + 4m_2)\frac{v_C^2}{4}, \text{ where } v_C = v.$$

(b) The total angular momentum is

$$\vec{L} = -mv_C R(\hat{k}) + m_1 v_C R\left(1 + \frac{1}{2}\right)(-\hat{k})$$

$$+ m_2 v_C R(1+1)(-\hat{k})$$

$$= -\frac{1}{2}(2m + 3m_1 + 4m_2)v_C R\hat{k}, \text{ where } v_C = v.$$

Problem 9 A body of mass m, radius R, and radius of gyration K has linear velocity v and angular velocity $\omega_0 = \frac{v_0}{R}$, as shown in the figure. Find (a) the angular velocity of the body after a long time and (b) the ratio of the kinetic energy of the body initially and after a long time.

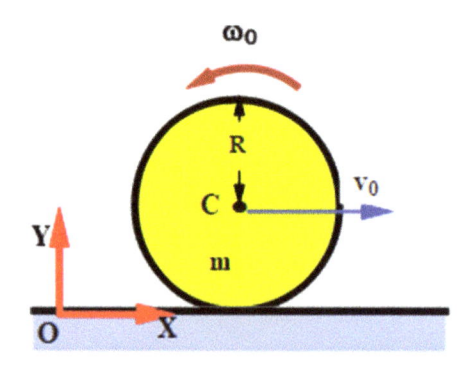

5-25

Solution

(a) The angular momentum of the body remains conserved which is equal to

$$\vec{L} = mK^2\omega_0\hat{k} - mv_0R\hat{k}$$

$$= -mv_0R\left(1 - \frac{K^2}{R^2}\right)\hat{k}$$

$$= mv_0R\left(\frac{K^2}{R^2} - 1\right)\hat{k}.$$

After a long time the body will roll without sliding with a new angular velocity ω, say. So the angular momentum of the body after a long time is $\vec{L} = m(K^2 + R^2)\vec{\omega}$, which can be equated with the initial one and simplified to obtain

$$\omega = \frac{\omega_0(K^2 - R^2)}{(K^2 + R^2)}$$ in a clockwise sense because it is negative.

(b) The initial and final kinetic energy are given as

$$KE_i = K_i = mK^2\omega_0^2 + \frac{1}{2}mv_0^2,$$

where $v_0 = R\omega_0$. Then we have

$$K_i = \frac{mv_0^2}{2}\left(1 + \frac{K^2}{R^2}\right).$$

The final kinetic energy is

$$KE_f = \frac{mv_C^2}{2}\left(1 + \frac{K^2}{R^2}\right).$$

Then their ratio is

$$X = v_0^2/v_C^2.$$

Putting the values of the initial and final velocities v_0 and v_C in the last expression, we have

$$X^2 = \frac{(K^2 + R^2)}{(K^2 - R^2)}.$$

However, the ratio of the magnitude of initial and final angular momenta is one.

Problem 10 A disc of mass m and radius R has an angular velocity ω and a linear velocity $v = R\omega$, as shown in the figure. If the coordinates of the center C of the disc are $3R$ and $2R$, find the total (a) kinetic energy and (b) angular momentum of the disc relative to the origin O.

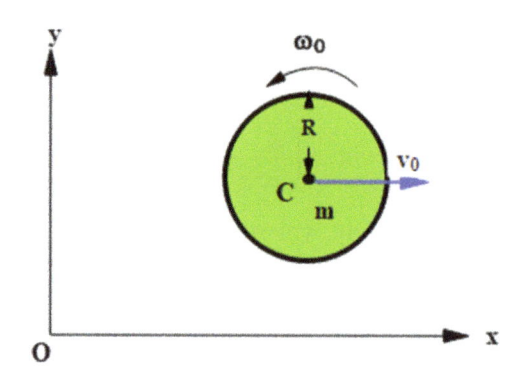

Solution
(a) The kinetic energy is given as
$$K = mK^2\omega_0^2 + \frac{1}{2}mv_0^2,$$
where $v_0 = R\omega_0 = v$. Then we have
$$K = \frac{mv_0^2}{2}\left(1 + \frac{K^2}{R^2}\right) = \frac{1}{2}mv^2\left(1 + \frac{1}{2}\right) = \frac{3}{4}mv^2.$$

(b) The angular momentum about the origin O is $\vec{L} = \left(\frac{mR^2}{2}\right)\left(\frac{v}{R}\hat{k}\right)$
$- mv(2R)\hat{k} = -\frac{3}{2}mvR\hat{k}.$

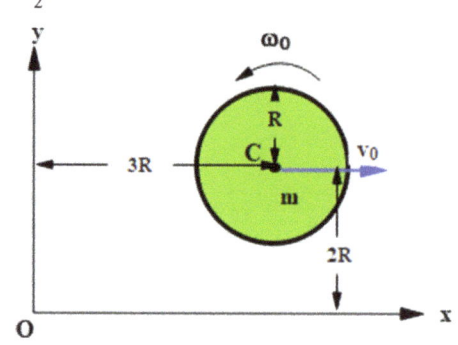

Problem 11 A disc of mass m and radius R is given with an angular velocity ω and a linear velocity v at $t = 0$, as shown in the figure. Just before the disc strikes the ground, it is found that the total angular momentum is zero relative to the point of

5-27

projection O. Then find the frequency f and ω. Put $h = 5$ m, $m = 1$ kg, $R = 0.1$ m, $\theta = \pi/4$ rad, and $v = 10\sqrt{2}$ m s^{-1}.

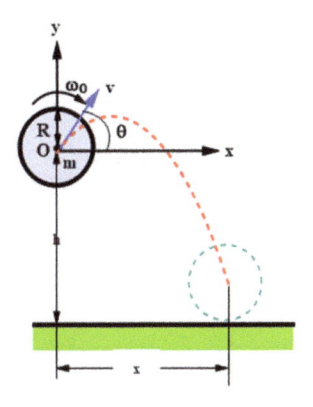

Solution
The disc will hit the ground after a time t given as

$$t = \sqrt{\frac{2}{g}\left(h + \frac{v^2 \sin^2\theta}{2g}\right)} + \frac{v \sin\theta}{g}.$$

Putting $v \sin\theta = 10$ m s^{-1}, $g = 10$, $h = 5$ m, we have
$t = (\sqrt{2} + 1)s$.
The vertical velocity of the disc just before hitting the ground is

$$v_y = \sqrt{v^2 \sin^2\theta + 2gh}$$

$$= \sqrt{100 + 2 \times 10 \times 5} = 10\sqrt{2} \text{ m s}^{-1},$$

whereas the horizontal velocity remains constant as $v \cos\theta = 10$ m s^{-1}. The horizontal displacement $x = v_x t = (v \cos\theta)t = 10(\sqrt{2} + 1)$ m. Then the angular momentum of the center of mass C of the disc just before hitting the ground about the origin or point of projection O is

$$\vec{L}_{orb} = mh'v\cos\theta \hat{k} - mxv_y\hat{k}$$

$$= m\big((h-R)v\cos\theta - xv_y\big)\hat{k}$$

$$= (1)\{10 \times 4.9 - 10(\sqrt{2} + 1)10\sqrt{2}\}\hat{k}$$

$$= -292.42.$$

5-28

The spin angular momentum is

$$L_{spin} = \frac{m}{2}R^2\omega.$$

Since $\vec{L} = \vec{L}_{spin} + \vec{L}_{orb} = 0$ (given)

$$\Rightarrow L_{spin} = L_{orb} = \frac{mR^2\omega}{2} = 292.42$$

$$(1)\left(\frac{1}{10}\right)^2 \frac{2\pi f}{2} = 292.42$$

$$\Rightarrow f = \frac{29242}{\pi} = 9304.27 \text{ per second}$$

and

$$\omega = 58484.27 \text{ rad s}^{-1}.$$

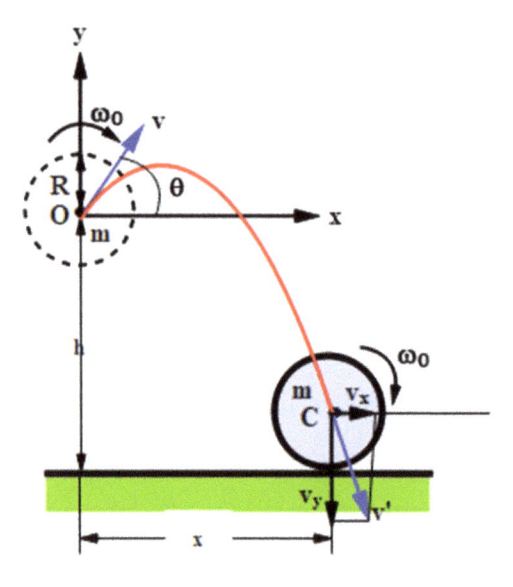

Problem 12 Find the angular momenta of the uniform discs each of mass m and radius R in the following cases. Please note that in the figure:

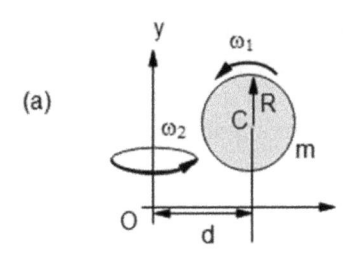

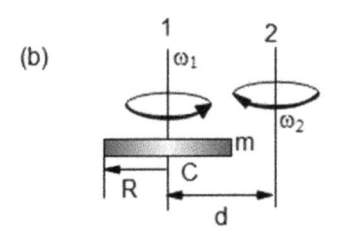

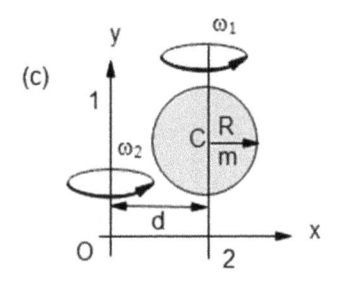

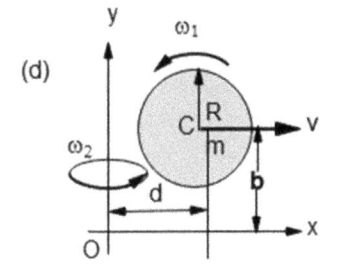

(a) The disc spins about the z-axis and its center of mass C revolves about the y-axis with angular velocities w_1 and w_2, respectively. The z-axis is outward from the page, as shown in the figure.
(b) The disc spins about axis 1 with angular velocity w_1 and the axis 1 revolves about a parallel axis 2 in the opposite sense with an angular velocity w_2.
(c) The disc spins about axis 2 with angular velocity w_1 and axis 2 revolves about a parallel axis 1 in the same sense with an angular velocity w_2.

(d) The disc is spinning about an axis parallel to the z-axis with an angular velocity w_1 and its center of mass moves along the x-axis at the given position with a velocity v. Furthermore, its center of mass is also revolving about the y-axis with an angular velocity w_2.

Solution

(a) The spin angular momentum is $\vec{L}$(spin) $= \frac{mR^2}{2}w_1\hat{k}$ and the orbital angular momentum is $\vec{L}$(orb) $= md^2 w_2 \hat{j}$. Then the total angular momentum is
$\vec{L}$(spin) $+ \vec{L}$(orb) $= \frac{mR^2}{2}w_1\hat{k} + md^2 w_2 \hat{j}$.

(b) Then the total angular momentum is

$$\vec{L}_0 = \frac{mR^2}{2}\vec{w}_{spin} + md^2\vec{w}_{orb}$$

$$\vec{L}_0 = \frac{mR^2}{2}w_1\hat{j} - md^2 w_2 \hat{j}$$

(c) The total angular momentum is

$$\vec{L}_0 = \frac{mR^2}{4}\vec{w}_{spin} + md^2\vec{w}_{orb}$$

$$\vec{L}_0 = \frac{mR^2}{2}w_1\hat{j} + md^2 w_2 \hat{j}$$

$$= m\left(\frac{R^2}{2}w_1 + d^2 w_2\right)\hat{j}.$$

(d) The total angular momentum is

$$\vec{L}_0 = \frac{mR^2}{4}\vec{w}_{spin} + md^2\vec{w}_{orb}$$

$$= \frac{mR^2}{2}w_1(\hat{k}) + \{mvb(-\hat{k}) + md^2 w_2(\hat{j})\}$$

$$= m\left\{\left(\frac{R^2}{2}w_1 - vb\right)\hat{k} + d^2 w_2 \hat{j}\right\}.$$

Problem 13 A uniform rod of mass m and length l is released from the near vertical position such that it undergoes a combined motion in a vertical x–y-plane. When the rod is inclined at an angle θ with a vertical wall, its bottom has a velocity v, as shown in the figure. Find the angular momentum of the rod about (a) the instantaneous axis of rotation and (b) the origin O.

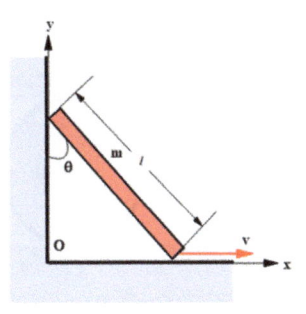

Solution
(a) The the moment of inertia of the rod about the instantaneous axis P is $I_P = m\{(CP)2 + l2/12\}$. We can show that $CP = l/2$ So, $I_P = ml^2/3$. The angular momentum about the instantaneous axis of rotation is

$$\vec{L_P} = I_P \vec{\omega} = \frac{ml^2}{3}\omega\hat{k}$$

$$= \left(\frac{ml^2}{3}\right) \times \left(\frac{v}{l\cos\theta}\right)\hat{k} = \frac{mvl\sec\theta}{3}\hat{k}.$$

(b) The angular momentum about the origin O is

$$\vec{L_O} = \frac{ml^2}{12}\frac{v}{l\cos\theta}\hat{k} - mv_y\frac{l}{2}\sin\theta\hat{k} - mv\frac{1}{2}\sin\theta\hat{k},$$

where $v_x = \frac{\omega l\sin\theta}{2}$ and $v_y = \frac{\omega l\cos\theta}{2}$. Then we have

$$\vec{L_O} = \frac{ml^2}{12}\omega\hat{k} - \hat{k}\left(\frac{ml^2}{4}\omega\sin^2\theta + \frac{ml^2\omega\cos\theta}{4}\right)$$

$$= ml^2\omega\left(\frac{1}{12} - \frac{1}{4}\right)\hat{k} = -\frac{1}{6}ml^2\omega\hat{k}$$

$$= -\frac{1}{6}ml^2\frac{v\hat{k}}{l\cos\theta} = \frac{-mvl\sec\theta}{6}\hat{k}.$$

Problem 14 A uniform cylinder and a solid sphere, each of radius R and masses m and M, respectively, are kept on a rough horizontal plane and are loaded with a plate of mass pm, as shown in the figure. The plate moves with a velocity v and there is no sliding anywhere. If the magnitude of the angular momentum of the system of three bodies about the origin O is equal to $97mvR/20$, find the value of p. Put $M = m/7$.

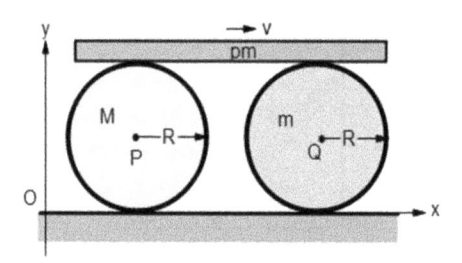

Solution

As there is no slipping between the plate and bodies and the velocity of the plate is equal to v as per the given condition, the centers of the sphere and cylinder move with a velocity $v/2$ and they rotate with an angular velocity $v/2R$. Then the total angular momentum about O is the sum of angular momenta of the plate, cylinder, and solid sphere, respectively, given as

$$L_O = -(pm)v(2R)\hat{k} - \frac{3}{2}mR^2\left(\frac{v}{2R}\right)\hat{k} - \frac{7MR^2}{5}\left(\frac{v}{2R}\right)\hat{k}$$

$$= -mvR\hat{k}\left(\frac{2p}{1} + \frac{3}{4} + \frac{7M/m}{5 \times 2}\right)$$

$$= -mvR\frac{(40p + 15 + 14M/m)}{20}\hat{k}$$

$$= -\frac{mvR}{20}(17 + 40p)\hat{k} \quad \left(\because \frac{M}{m} = \frac{1}{7}\right).$$

Since $|\vec{L_O}| = 97mvR/20$ (given), we have $p = 2$.

Problem 15 The plank A is loaded on a thin hollow sphere and a solid sphere of mass M and m, respectively. These spheres roll on the plank B which is placed on a smooth horizontal surface. There is no relative sliding between the planks A and B and the spheres. If the planks A and B move with velocities v and nv, as shown in the figure, find the angular momentum of the system (planks + spheres) relative to the origin O. (Put $M = 3m$, $x = 2$, $y = 3$, and $n = 3$).

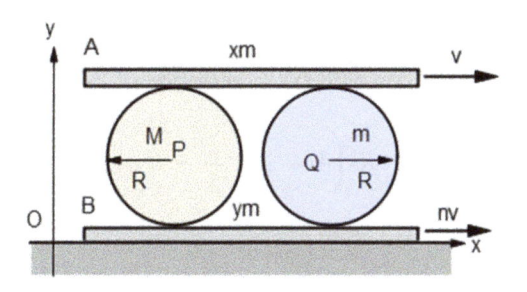

Solution

As there is no slipping between the plates and bodies and the velocity of the plates A and B are equal to v and nv, respectively, as per the given condition, the centers of the sphere and cylinder move with a velocity $v(n + 1)/2$ towards right and the spheres rotate with an angular velocity $v(n - 1)/2R$ in an anticlockwise sense because it is given that n ($= 3$) is greater than one. As the plank B is lying along the x-axis, its angular momentum about the origin will be zero. But the angular momentum of the plate A about O is $-(xm)(v)(2R)\hat{k}$. The spin angular momenta of the hollow and solid spheres are $\frac{2}{3}MR^2\frac{v(n-1)}{2R}\hat{k}$ and $\frac{2}{5}mR^2v\frac{(n-1)}{2R}\hat{k}$, respectively. The angular momenta of the centers of mass of the spheres are $Mv\frac{(n+1)}{2R}(-\hat{k})$ and $m\frac{v(n+1)}{2}R(-\hat{k})$, respectively. Then the total angular momentum about O is the sum of angular momenta of the plate A and spheres, which is equal to the combination of the spin angular momentum of the spheres and the angular momenta of the center of mass of the spheres and the plank (plate) A, given as

$$-(xm)(v)(2R)\hat{k} + \frac{2}{3}MR^2\frac{v(n-1)}{2R}\hat{k}$$

$$+ \frac{2}{5}mR^2v\frac{(n-1)}{2R}\hat{k} + Mv\frac{(n+1)}{2R}(-\hat{k}) + m\frac{v(n+1)}{2}R(-\hat{k}).$$

Putting $M = 3m$, $x = 2$, $n = 3$, we have

$$-(2m)v(2R)\hat{k} + \frac{2}{3}(3m)\frac{v}{2}(3-1)R\hat{k}$$

$$+ \frac{2}{5}mvR\left(\frac{3-1}{2}\hat{k}\right) + 3mv\left(\frac{3+1}{2}\right)R(-\hat{k})$$

$$+ mvR\left(\frac{3+1}{2}\right)(-\hat{k})$$

$$= -4mvR\hat{k} + 2mvR\hat{k} + \frac{2}{5}mvR\hat{k}$$

5-34

$$-6mvR\hat{k} - 2mvR\hat{k}$$

$$= -mvR\hat{k}\left(4 - 2 - \frac{2}{5} + 6 + 2\right)$$

$$= -48mvR/5\ \hat{k}.$$

Problem 16 Find the angular momentum of the system (disc + rod) in each of the following cases. Please note that in figure (i) the given axis of rotation for both bodies is outward from the page (the z-axis) and in figure (ii) the disc is not spinning and the center of the disc revolves around the z-axis with the given angular velocity. Put $M = 3m$ and $L = 2R$, $m = 1$ kg, $R = 0.1$ m, $\omega_1 = \sqrt{g/R}$, and $\omega_2 = 2\sqrt{g/R}$.

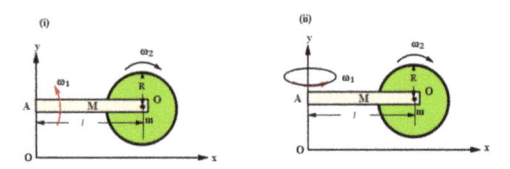

Solution

The angular momentum about O is the sum of the angular momenta of the rod and disc:

$$\vec{L} = \vec{L}_{rod} + \vec{L}_{disc},$$

where $\vec{L}_{rod} = \frac{Ml^2}{3}\vec{\omega}_1$ and $\vec{L}_{disc} = \frac{mR^2}{2}\vec{\omega}_2 + m(\vec{l}\vec{\omega}_1)l$. Then

$$\vec{L} = \frac{Ml^2}{3}\vec{\omega}_1 + \frac{mR^2}{2}\vec{\omega}_2 + m(\vec{l}\vec{\omega}_1)l$$

$$= \left(\frac{Ml^2}{3} + ml^2\right)\vec{\omega}_1 + \frac{mR^2}{2}\vec{\omega}_2$$

$$= \left(\frac{Ml^2}{3} + ml^2\right)\omega_1\hat{k} + \frac{mR^2}{2}(-\omega_2\hat{k})$$

$$= \left\{\frac{(3m)}{3}(2R)^2 + m(2R)^2\right\}\omega_1\hat{k} - \frac{mR^2}{2}\omega_2\hat{k}$$

5-35

$$= 8mR^2\omega_1\hat{k} - \frac{mR^2}{2}\omega_2\hat{k}$$

$$= 8mR^2\left(\sqrt{\frac{g}{R}}\right)\hat{k} - \frac{mR^2}{2}\left(2\sqrt{\frac{g}{R}}\right)\hat{k}$$

$$= mR^2\sqrt{\frac{g}{R}}(8-1)\hat{k} = 7mR\sqrt{gR}\hat{k}$$

$$= 7 \times 1 \times \frac{1}{10} \times \sqrt{10 \times \frac{1}{10}}\hat{k} = 0.7\hat{k} \text{ kg m}^2 \text{ s}^{-2}.$$

For figure (ii) the rod is rotating about the y-axis and the disc spins about the y- and z-axes with a moment of inertia $mR^2/4$ and $mR^2/2$ about its centroidal axes with $\vec{\omega}_1$ and $\vec{\omega}_2$, respectively. Furthermore, the disc has orbital angular momentum as its center of motion revolves about the y-axis with $\vec{\omega}_1$. Then the total angular momentum is given as

$$\vec{L} = \left(\vec{L}_{rod}\right)_{total} + \left(\vec{L}_{disc}\right)_{total}$$

$$= \frac{Ml^2}{3}\vec{\omega}_1 + \vec{L}_{spin_1} + \vec{L}_{spin_2} + \vec{L}_{orb}$$

$$= \frac{Ml^2}{3}\vec{\omega}_1 + \frac{mR^2}{2}\vec{\omega}_2 + \frac{mR^2}{4}\vec{\omega}_1 + ml^2\vec{\omega}_1$$

$$= \vec{\omega}_1\left[\frac{Ml^2}{3} + \frac{mR^2}{4} + ml^2\right] + \vec{\omega}_2\frac{mR^2}{2}$$

$$= \vec{\omega}_1\left[(3m)(2R)^2/3 + \frac{mR^2}{4} + m(2R)^2\right] + \vec{\omega}_2\frac{mR^2}{2}$$

$$= mR^2\left[\left(8 + \frac{1}{4}\right)\vec{\omega}_1 + \vec{\omega}_2/2\right]$$

$$= mR^2[\tfrac{33}{4}\sqrt{\tfrac{g}{R}}\hat{j} - \tfrac{2}{2}\sqrt{\tfrac{g}{R}}\hat{k}]$$

$$= mR\sqrt{gR}\left[\frac{33}{4}\hat{j} - \hat{k}\right]$$

5-36

$$= 1 \times \frac{1}{10}\sqrt{10} \times \frac{1}{10}\left(\frac{33}{4}\hat{j} - \hat{k}\right)$$

$$\Rightarrow \vec{L} = \frac{(33\hat{j} - 4\hat{k})}{40} \text{ kg m}^2\text{s}^{-2}.$$

Problem 17 In a certain binary star system, two stars with identical mass m move in a circular orbit to possess an angular momentum L_1 about the center of mass of the system. In another system three identical stars of the same mass m move around a circular orbit with an angular momentum L_2 about the center of mass of the system. Find the ratio of these angular momenta. Assume the spacing between the stars is equal.

Solution
Method 1

The moment of inertia of a system of two particles of masses m_1 and m_2 separated by a distance l is given as

$$I = \frac{m_1 m_2 l^2}{m_1 + m_2}.$$

Then the angular momentum of this system is

$$L = \frac{m_1 m_2 l^2}{m_1 + m_2}\omega.$$

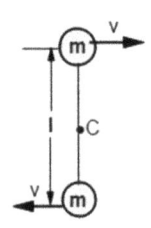

Putting $m_1 = m_2 = m$ and $\omega = 2v/l$, in the last expression, we have

$$L_1 \frac{mml^2}{m+m}\omega = \frac{ml^2}{2} \times \frac{2v}{l}$$

$$L_1 = mvl. \tag{5.10}$$

For a system of three identical particles situated at the vertices of an equilateral triangle of side l, the center of mass will be located at the center C of the circle. So each particle will circulate about C with the radius $r = l/\sqrt{3}$ in a circular orbit.

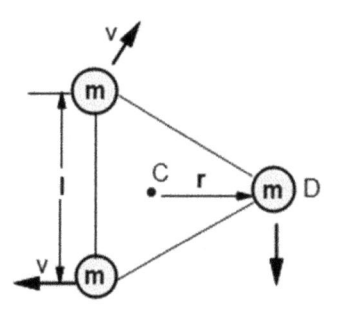

Since $r = \frac{\sqrt{3}l}{2} \times \frac{2}{3} = \frac{l}{\sqrt{3}}$ the angular momentum of the system is

$$L_2 = 3(mvr) = 3mv \cdot \frac{l}{\sqrt{3}}$$

$$= \sqrt{3} mvl. \tag{5.11}$$

Using the equations (5.10) and (5.11)

$$\frac{L_1}{L_2} = \frac{1}{\sqrt{3}}.$$

Method 2

For the two-particle system, the gravity force $\frac{Gm^2}{l^2}$ acting on each particle acts as a centripetal force and provides the centripetal acceleration v_1^2/r. So

$$\frac{Gm^2}{l^2} = \frac{mv_1^2}{r}, \text{ where } r = l/2$$

$$\Rightarrow v_1 = \sqrt{\frac{Gm}{2l}}. \tag{5.12}$$

For the three-particle system the resultant gravity force $(2 \cos 30°)\frac{Gm^2}{l^2}$ acting on each particle acts as a centripetal force providing the centripetal acceleration

$$a_r = v_2^2/r,$$

where $r = l/\sqrt{3}$. So

$$\frac{mV_2^2}{(l/\sqrt{3})} = \left(\frac{Gm^2}{l^2}\right)(2 \cos 30°)$$

$$\Rightarrow V_2 = \sqrt{\frac{Gm}{l}}. \tag{5.13}$$

Then the angular momentum of the two-particle system is

$$L_1 = 2\left(mv_1 \frac{l}{2}\right) = mv_1 l$$

$$L_2 = 3mv_2 l/\sqrt{3} = \sqrt{3}\, mv_2 l.$$

Since the speeds v_1 and v_2 are same we have

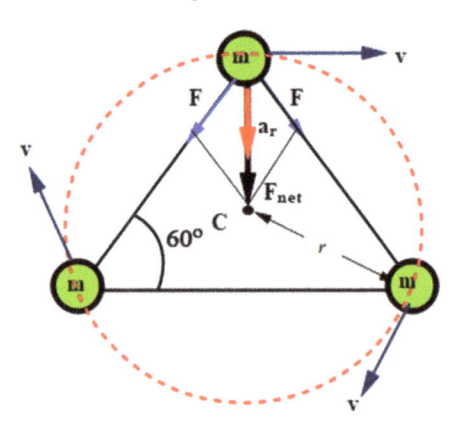

$$\frac{L_2}{L_1} = \sqrt{3}.$$

$F_{net} = 2 F \cos 30° = ma_r = mv^2/r$

Conservation of the angular momentum of rolling bodies

Problem 18 A disc of mass m and radius R is given an initial velocity v_0 and angular velocity $\omega_0 = v_0/R$ as shown in the figure. Due to the friction the disc will stop sliding after some time and thereafter it rolls without sliding. Find (a) the spin angular momentum, (b) the orbital angular momentum, and (c) the total angular momentum of the disc about the points O, P, and Q, and (d) the angular and linear velocity of the disc at the time of its rolling without sliding.

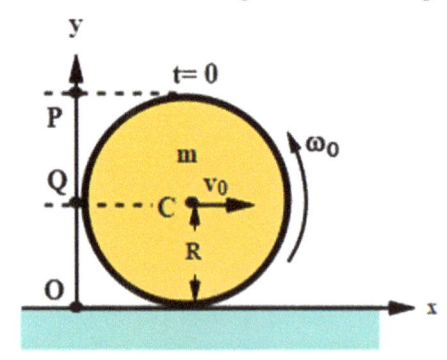

Solution

(a) The spin angular momentum about any point such as O, P, and Q will remain the same, which is given as

$$\vec{L}_{spin} = \frac{mR^2}{2}\omega_0 \hat{k} = mv_0\frac{R}{2}\hat{k}.$$

(b) The orbital angular momenta about O, P, and Q are $mv_0 R(-\hat{k})$, $mv_0 R(+\hat{k})$, and $0\hat{k}$, respectively.

(c) Then, summing the spin and orbital angular momenta, the total angular momentum about O can be given as

$$\vec{L}_O = \frac{mR^2}{2}\omega_0\hat{k} + mv_0 R(-\hat{k}).$$

Putting $\omega_0 = \frac{v_0}{R}$ we have

$$\vec{L}_O = -\left(mv_0 R - \frac{mR^2}{2}\omega_0\right)\hat{k} = -\frac{mv_0 R}{2}\hat{k}.$$

Similarly, the total angular momenta about P and Q are, respectively, given as

$$\vec{L}_P = \frac{mR^2}{2}\omega_0\hat{k} + mv_0 R(\hat{k}) = \frac{3mv_0 R}{2}\hat{k}$$

$$\vec{L}_Q = \frac{mR^2}{2}\omega_0\hat{k} + 0(\hat{k}) = \frac{mv_0 R}{2}\hat{k}.$$

(d) Let the final angular velocity of the body be ω at the time of pure rolling be $\vec{\omega}$. Then its total angular momentum about O is

$$\vec{L}_O = \frac{3mR^2}{2} = \vec{\omega}.$$

We can conserve the total angular momentum about O because the friction acting on the body passes through O and hence it cannot produce any torque about O. Furthermore, the forces N and mg are equal and opposite and collinear, so their net torque about O, P and Q will be zero. However, the frictional torque will not be zero about the other two points P and Q. Hence, we cannot conserve the angular momenta about P and Q. Equating the final angular momenta with that of the initial one about O, we have

$$\frac{3mR^2}{2}\vec{\omega} = -\frac{mv_0 R}{2}$$

$$\Rightarrow \vec{\omega} = -\frac{v_0}{3R}\hat{k}.$$

So the body will roll with a velocity

$$v = \frac{v_0}{3} \text{ towards the right.}$$

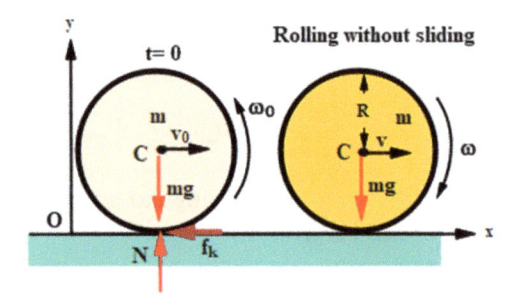

Problem 19 A boy of mass m is standing on the top of a disc of mass M and radius R. If the boy starts walking with a velocity v, assuming (i) zero friction, (ii) sufficient friction, or (iii) finite little friction between the disc and ground, find (a) the linear and angular velocity of the disc just after walking, (b) the work done by the boy in walking, and (c) the work done by friction on the ground until the disc start pure rolling.

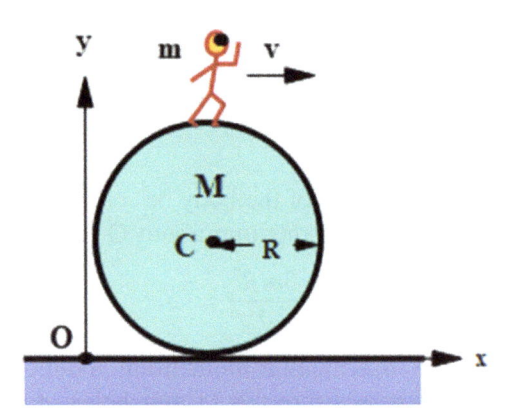

Solution
(a) (i) and (iii) For zero and small friction:
We can conserve the linear momentum exactly for zero friction and approximately for little friction just before and after walking. Let us assume that the center of mass of the disc acquires a velocity $v_M = V$. Conserving linear momentum,

$$P = Mv_M - mv = 0$$

$$\Rightarrow v_M = V = \frac{mv}{M}. \tag{5.14}$$

We can note that the friction acts forward on the boy as he jumps forward and an equal and opposite (backward) friction acts on the disc. The net torque produced by this pair of frictions is zero about any point. So, conserving the angular momentum relative to any point P on the ground, the sum of the angular momenta of both the boy and disc is zero as initially everything was at rest. So just after the jump the total angular momentum about P must be zero. Then we have

$$L_P = 0$$

$$\Rightarrow mv(2R)\,\hat{k} + MVR\hat{k} + \frac{MR^2}{2}\omega\hat{k} = 0$$

$$\Rightarrow 2mvR = MVR + \frac{MR^2}{2}\omega. \tag{5.15}$$

Using the equations (5.14) and (5.15)

$$\Rightarrow mvR = \frac{MR^2}{2}\omega$$

$$\Rightarrow \omega = \frac{2mv}{MR}.$$

For the case (ii) if the surface is so rough that pure rolling starts immediately after the jumping, so the frictional work done will be zero.

Since friction acting on the disc at its bottom passes through the origin P, its torque about the z-axis is zero. If the angular velocity of the disc just after jumping is ω, by conserving the angular momentum relative to P, we have

$$mv(2R) = \frac{3}{2}MR^2\omega$$

$$\text{or } \omega = \frac{4mv}{3MR}.$$

The velocity of the disc just after jumping is

$$V = R\omega = \frac{4mv}{3M}.$$

Even though the surface is a little rough, finally (after a long time) the disc will move with the above velocities in order to obey the law of conservation of angular momentum.

(b) For all cases (i), (ii), and (iii), the work done by the boy in jumping is equal to the change in the kinetic energy of the system just after the jumping. The change in kinetic energy of the system just after jumping is given as

Problems and Solutions in Rotational Mechanics

$$\Delta K_{\text{jump}} = \frac{mv^2}{2} + \frac{MV^2}{2} + \frac{I_C}{2}\omega^2.$$

In the cases (i) and (iii), we have

$$W = \Delta K_{\text{jump}} = \frac{mv^2(M+m)}{2M} + \frac{1}{2} \times \frac{MR^2}{2} \times \frac{4m^2v^2}{M^2R^2}$$

$$= \frac{mv^2}{2}\left[1 + \frac{m}{M} + \frac{2m}{M}\right]$$

$$= \frac{mv^2}{2}\left[1 + \frac{3m}{M}\right].$$

In (ii), putting the values of $V = \dfrac{4mv}{3M}$ and $\omega = \dfrac{4mv}{3MR}$, we have

$$W = \Delta K_{\text{jump}} = \frac{mv^2}{2} + \frac{M(4mv/3M)^2}{2}$$

$$+ \frac{1}{2} \times \frac{MR^2}{2} \times \frac{16m^2v^2}{9M^2R^2}$$

$$= \frac{mv^2}{2}\left[1 + \frac{8m}{3M}\right].$$

(c) For little or finite friction in the case of, there will be work done by friction as the disc slides on the surface for some time until it starts pure rolling. The work done W by friction is numerically equal to the change in the kinetic energy of the system between just after the jumping and the start of pure rolling:

$$W = \Delta K_{\text{sliding}}$$

$$= \frac{1}{2} \times \frac{3}{2}Mv_{\text{f}}^2 - \left(\frac{1}{2}Mv_M^2 + \frac{1}{2} \times \frac{MR^2}{2}\omega^2\right)$$

$$= \frac{M}{2}\left[\frac{3}{2}v_{\text{f}}^2 - v_M^2 - \frac{(R\omega)^2}{2}\right].$$

Putting the obtained values of

$$(v_{\text{f}})_M = \frac{4mv}{3M}, \quad \omega = \frac{2mv}{MR}, \quad \text{and } v_M = \frac{mv}{M},$$

we have

$$W = \frac{M}{2}\left[\frac{3}{2}\left(\frac{4mv}{3M}\right)^2 - \left(\frac{m}{M}v\right)^2 - \frac{1}{2}\left(\frac{2mv}{M}\right)^2\right]$$

$$= \frac{M}{2}\left(\frac{m^2v^2}{M^2}\right)\left(\frac{3}{2} \times \frac{16}{9} - 1 - 2\right)$$

$$= -\frac{m^2v^2}{6M}.$$

However, in the cases (i) and (iii), the work done will be zero because the friction is zero in case (i) and in case of (ii) friction is too large so that it will be a static friction that performs zero work.

Problem 20 An insect of mass m is crawling on a sphere of mass M, radius R, and radius of gyration K with a velocity v_r relative to the sphere. If the ground is (a) smooth or (b) sufficiently rough, find the linear and angular velocity of the sphere.

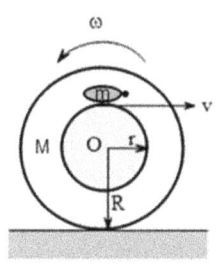

Solution
(a) If the ground is smooth or friction is small, $F = 0$.
So the linear momentum $P = $ constant $= 0$

$$\Rightarrow Mv_C - mv. \tag{5.16}$$

We can also conserve angular momentum about any point, P, say,

$$L_P = \text{constant} = 0$$

$$MK^2\omega = mvr. \tag{5.17}$$

From kinematics we have

$$(v_r - v_C - r\omega) = v. \tag{5.18}$$

Now we have three equations for three unknown quantities such as v, v_r, and ω. From the equations (5.16) and (5.17),

$$v_C = \frac{MK^2\omega}{Mr} = \frac{K^2\omega}{r}. \tag{5.19}$$

From equations (5.16) and (5.18)

$$v_C = \frac{mv}{M} = \frac{m}{M}(v_r - v_C - r\omega)$$

5-44

$$\Rightarrow v_C\left(1 + \frac{m}{M}\right) = \frac{m}{M}(v_r - r\omega). \tag{5.20}$$

Solving (5.19) and (5.20) we have

$$v_C = \frac{mK^2 v_r}{(M+m)K^2 + mr^2}$$

and

$$\omega = \frac{mv_r r}{(M+m)K^2 + mr^2}.$$

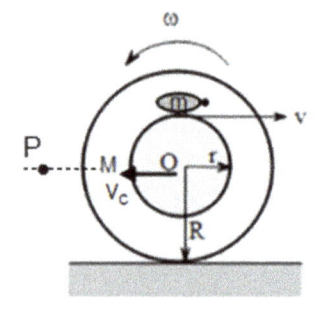

(b) If there is sufficient friction, the body rolls while the insect walks. Let v = the velocity of the insect relative to the ground.

We can conserve angular momentum about any point, P, on the ground. Since $L_P = $ constant $= 0$,

$$M(K^2 + R^2)\omega = mv(R + r). \tag{5.21}$$

From kinematics we have

$$(v_r - v_c - r\omega) = v. \tag{5.22}$$

For rolling we have

$$v_C = R\omega. \tag{5.23}$$

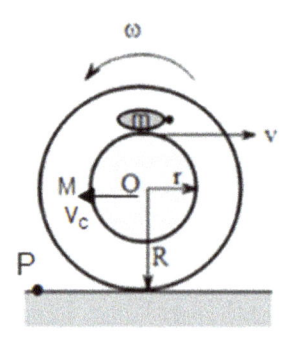

5-45

Since we have three equations for three unknown quantities such as v, v_r, and ω, solving all three equations we have

$$v_C = \frac{mR(R+r)v_r}{M(K^2+R^2)+m(R+r)^2}$$

and

$$\omega = \frac{m(R+r)v_r}{(M+m)K^2+m(R+r)^2}.$$

Problem 21 A sphere of mass m, radius R, and radius of gyration K is kept on a rough horizontal surface. It is given linear and angular velocities v and ω, respectively, so that it returns to its initial position. (a) Find the relation between v_0 and ω_0. (b) What is the final angular velocity in this process? (c) Discuss the nature of motion of the sphere.

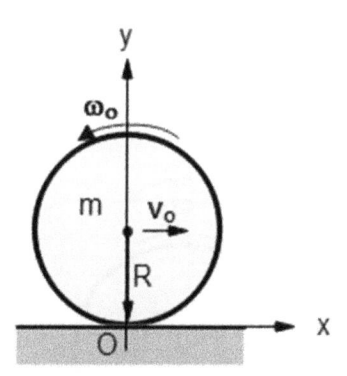

Solution
(a) To conserve the angular momentum about O on the initial angular momentum is

$$\vec{L_O} = mK^2\vec{\omega_0} + mv_0R$$

$$= (mK^2\omega_0 - mv_0R)\hat{k}. \qquad (5.24)$$

It will return if $\vec{L_O}$ is directed outwards

$$\Rightarrow mK^2\omega_0 - mv_0R \geqslant 0 \Rightarrow \omega_0 \geqslant \frac{v_0 R}{K^2}.$$

This means that the initial spin ω_0 must be counter-clockwise which must be greater than $\frac{v_0 R}{K^2}$.

5-46

(b) If the body rolls after some time its angular momentum about O will be
$$\vec{L_f} = -m(K^2 + R^2)\omega \hat{k}.$$
Equating it with the initial angular momentum we have
$$\omega = \frac{v_0 R - K^2 \omega_0}{(K^2 + R^2)}.$$

(c)

(i) The clockwise frictional torque about the center of mass of the body is $\tau_C = \mu mgR$ that generates an angular acceleration
$$\alpha = \tau_C/mK^2 = \frac{\mu mgR}{mK^2}.$$
Since the angular acceleration opposes the initial angular velocity, the body will stop spinning instantaneously after a time
$$t_1 = \omega_0/\alpha = \frac{\omega_0}{\frac{\mu mg}{mK^2}R} = \frac{\omega_0 K^2}{\mu gR}.$$

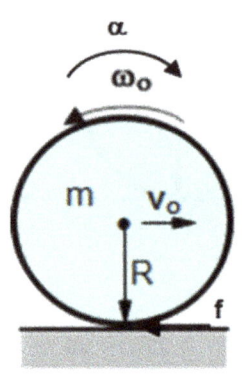

The velocity of the center of mass of the body will be zero at time
$$t_2 = \frac{v_0}{\mu g}.$$
If the body will first stop translation,
$$t_1 \geqslant t_2$$
$$\frac{\omega_0 K^2}{\mu gR} \geqslant \frac{v_0}{\mu g}$$

$$\omega_0 \geqslant \frac{v_0 R}{K^2}.$$

This means that the body still rotates anticlockwise when its center of mass stops instantaneously. As the friction continues to act backward (towards the left), the body will thereafter (after $t = t_2 = \frac{v_0}{\mu g}$) move towards the left retracing its path. While doing so, the body continues to spin anticlockwise until its lowest point stops sliding when the velocity of the center of mass of the body is $v = R\omega$, where

$$\omega = \frac{v_0 R - K^2 \omega_0}{(K^2 + R^2)}.$$

Then, after the time t_3, from the right-hand side extreme position of the body, it will start pure rolling, where t_3 is given by

$$t_3 = \frac{v}{\mu g} = \frac{R\omega}{\mu g} = \frac{R(v_0 R - K^2 \omega_0)}{(K^2 + R^2)\mu g}.$$

So, from starting, the pure rolling begins at time

$$t = t_2 + t_3 = \frac{v_0}{\mu g} + t_3$$

Thereafter the body moves with same velocity

$$v = \frac{R(v_0 R - K^2 \omega_0)}{(K^2 + R^2)}.$$

If the body was given a clockwise initial spin, then it would stop sliding and move with a constant velocity

$$v = \frac{R(v_0 R + K^2 \omega_0)}{(K^2 + R^2)}$$

after a time given as

$$t = (v_0 - v)/\mu g = \frac{v_0}{\mu g} - \frac{R(v_0 R - K^2 \omega_0)}{(K^2 + R^2)\mu g}$$

and continues to move in the same direction.

Problem 22 A sphere of mass m, radius R, and radius of gyration K is kept on a rough horizontal surface. It is given linear and angular velocities as shown in the figure. Find (a) the velocity of the sphere after a long time (b) the time after which it will stop sliding (c) the distance of relative sliding of the sphere.

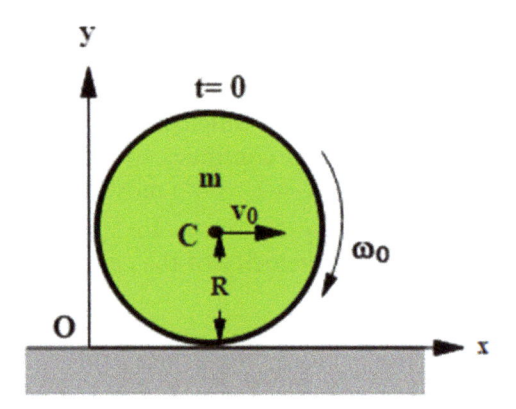

Solution

(a) Conserving the angular momentum about a point O on the ground, the final velocity is given as

$$-\frac{2}{5}mR^2\omega_0\hat{k} - mv_0 R\hat{k} = -\frac{7}{5}mv_f R\hat{k}$$

$$\Rightarrow v_f = \frac{5}{7}\left(\frac{2R\omega_0 + 5v_0}{5}\right)$$

$$= \frac{2R\omega_0 + 5v_0}{7}.$$

(b) The time after the relative sliding will stop is

$$t = \frac{v_0 - v_f}{\mu g} = \frac{v_0 - \frac{2R\omega_0 + 5v_0}{7}}{\mu g}$$

$$= \frac{2v_0 - 2R\omega_0}{7\mu g} = \frac{2(v_0 - R\omega_0)}{7\mu g}.$$

(c) The total distance traversed by the center of mass of the cylinder until it stops sliding is

$$s = \frac{|v_f^2 - v_0^2|}{2\mu g} = \frac{\left|\left(\frac{2R\omega_0 + 5v_0}{7}\right)^2 - v_0^2\right|}{2\mu g}$$

$$= \frac{2(R\omega_0 + 6v_0)(R\omega_0 - v_0)}{49\mu g}.$$

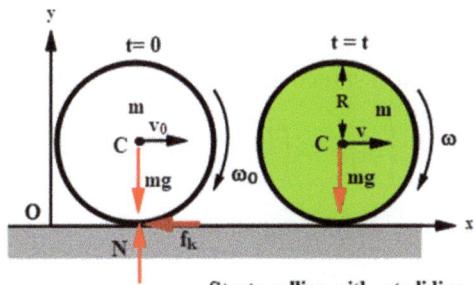

Problems and Solutions in Rotational Mechanics

Problem 23 A uniform solid cylinder of mass m and radius R is given a velocity v_0 and an anticlockwise angular velocity of magnitude $\omega_0 = v_0/R$, as shown in the figure. If the coefficient of friction between the cylinder and ground is μ, find (a) the velocity of its pure rolling, and the time after which the body begins to roll, (b) the total distance of relative sliding, and (c) the mechanical energy loss of the cylinder due to friction.

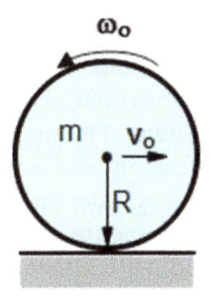

Solution

(a) Referring to problem 21, putting $\frac{K^2}{R^2} = 1/2$ in the obtained expression $\omega_0 \geqslant \frac{v_0 R}{K^2}$, we have

$$\omega_0 \geqslant \frac{2v_0}{R}.$$

Since the given value of $\omega_0 = \frac{v_0}{R}$ is less than $\frac{2v_0}{R}$, the body will not return to its initial position.

Putting $\frac{K^2}{R^2} = 1/2$ and $\omega_0 = \frac{v_0}{R}$ in the expression

$$v = \frac{R(v_0 R - K^2 \omega_0)}{(K^2 + R^2)}$$

obtained in problem 21, we have $v = \frac{v_0}{3}$. This means that the cylinder will roll with a velocity $v = \frac{v_0}{3}$ after a time

5-50

$$t = (v_0 - v)/\mu g = \frac{v_0}{3\mu g}.$$

(b) The cylinder will travel a distance go on many towards right

$$s = \left\{(v_0)^2 - \left(\frac{v_0}{3}\right)^2\right\}/2\mu g = \frac{4v_0^2}{9\mu g}.$$

(c) The loss of kinetic energy during the sliding is

$$|\Delta K| \left\{\frac{1}{2}mv_0^2 + \frac{1}{2}\left(\frac{mR^2}{2}\right)\omega_0^2\right\} - \left\{\frac{1}{2}\left(\frac{3}{2}mR^2\right)\left(\frac{v_0}{3R}\right)^2\right\}$$

$$= \frac{3}{4}mv_0^2 - \frac{mv_0^2}{12}$$

$$= \frac{(9-1)mv_0^2}{12} = \frac{2}{3}mv_0^2.$$

Problem 24 A cylinder rolls on a horizontal surface with a velocity $v_0 = 3$ m s^{-1}. At time $t = 0$, let it be launched onto a horizontal conveyor belt which is moving with a velocity $u = 2v_0$ as shown in the figure. The upper length of the conveyor belt is $AB = L = 5$ m. The coefficient of friction between the cylinder and belt is $\mu = 0.2$. The coefficient of friction between the cylinder and horizontal surface is 0.1. (a) Find the time t_1 at which the cylinder will leave the belt. (b) Prove that the body will roll on the horizontal surface (the ground) with same initial velocity. (c) After what time again will the cylinder begin to roll on the horizontal surface at C after leaving the belt?

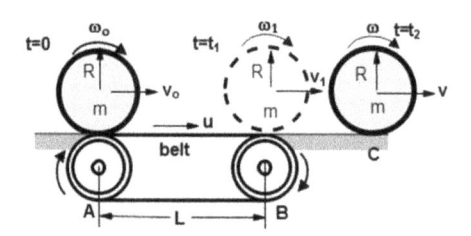

Solution
(a) Since the lowest point of the cylinder has a velocity $v_0 - 2v_0 = -v_0$ (backward) relative to the belt, the friction acting on the cylinder will be forward. So the cylinder moves with a forward acceleration $a = \mu g$. Its velocity at a time t is

$$v = v_0 + \mu g t. \tag{5.25}$$

5-51

As the forward friction produces a clockwise torque of magnitude μmgR about the center of mass of the cylinder, it produces a counterclockwise angular acceleration

$$\alpha = +\frac{\mu mgR}{(mR^2/2)} = +\frac{2\mu g}{R}.$$

Then the angular velocity at time t is

$$\omega = \omega_0 - \frac{2\mu g}{R}t. \tag{5.26}$$

For rolling the condition is

$$v - R\omega = u. \tag{5.27}$$

Using equations (5.25), (5.26), and (5.27)

$$v_0 + \mu gt - R\left(\omega_0 - \frac{2\mu gt}{R}\right) = u$$

$$\Rightarrow v_0 - R\omega_0 + 3\mu gt = u.$$

Putting $v_0 = R\omega_0$ for pure rolling initially,

$$\Rightarrow t = \frac{u}{3\mu g} = \frac{2v_0}{3\mu g} = \frac{2 \times 3}{3(0.2)(10)} = 1s = t_1, \text{ say.}$$

Then the body has the velocity $v = v_0 + \mu gt = v_0 + \frac{u}{3} = v_0 + \frac{2v_0}{3} = \frac{5v_0}{3} = v_1 = 5$ m s^{-1}, towards right at the time of pure rolling, and the angular velocity

$$\omega = \omega_0 - \frac{2\mu gt}{R}$$

$$= \omega_0 - \frac{2u}{3R} = \frac{v_0}{R} - \frac{2(2v_0)}{3R} = -\frac{v_0}{3R}.$$

The negative sign means that the cylinder rotates anticlockwise while moving forward with respect to ground.

The distance covered by the center of mass until it stops sliding is

$$s_1 = \frac{v^2 - v_0^2}{2\mu g} = \left(\frac{25}{9} - 1\right)\frac{v_0^2}{2\mu g}$$

$$\frac{8v_0^2}{9\mu g} = \frac{8 \times (3 \times 3)}{9 \times \frac{1}{5} \times 10} = 4 \text{ m}.$$

The remaining distance is $s_2 = L - s_1 = 5 - 4 = 1$ m which can be traversed in time

$$t_2 = \frac{s_2}{v} = \frac{1}{\frac{5v_0}{3}} = \frac{3}{5 \times 3} = 0.2 \text{ s}$$

Then after a time $= t_1 + t_2 = 1 + 0.2 = 1.2$ s the body leaves the belt.

(b) The body leaves the belt and launches onto the horizontal surface again with a velocity $5v_0/3$ towards the right and an anticlockwise angular velocity $v_0/3R$. Since the angular momentum of the body remains conserved relative to any point of the ground, which is given as

$$\vec{L_f} = -m(K^2/R^2 + 1)v_0 R\hat{k} = -3mv_0 R\hat{k}/2.$$

So, no matter on which moving surface the cylinder is launched, ultimately it will have to roll on the horizontal surface with the same initial velocities to retain the initial angular momentum. We should note that the angular momentum will also remain conserved while the body is moving on the belt, which remains independent of the velocity of the belt.

(c) The time after leaving the belt that the body will roll is given as

$$5v_0/3 - v_0 = \mu g t_3.$$

This gives us

$$t_3 = 2v_0/3\mu g = 2(3)/3(0.1)(10) = 2 \text{ s}.$$

So, from the start, the time after which the body will commence pure rolling is

$$T' = 1.2 + 2 = 3.2 \text{ s}.$$

Problem 25 In the last problem, if a solid sphere is gently placed on the conveyor belt at A, describe its subsequent motion. Put $u = 2v_0 = 7$ m s^{-1}, $\mu = 0.1$ for all contacting surfaces, and $R = 1/4$ m.

Solution

Since the lowest point of the sphere has a velocity $0 - v_0 = -v_0$ (backward) relative to the belt, the friction acting on the sphere will be forward. So the sphere will move with a forward acceleration $a = \mu g$. Its velocity at a time t is

$$v = \mu g t. \tag{5.28}$$

As the forward friction produces an anticlockwise torque of magnitude $\mu m g R$ about the center of mass of the sphere, it produces an clockwise angular acceleration

$$\alpha = \frac{\mu m g R}{(2m R^2/5)} = \frac{5\mu g}{2R}.$$

Then the angular velocity at time t is

$$\omega = \frac{5\mu g}{2R} t. \tag{5.29}$$

For rolling, the condition is

$$v + R\omega = u, \tag{5.30}$$

Using equations (5.28), (5.29), and (5.30)

$$\mu g t + R\left(\frac{5\mu g t}{2R}\right) = u$$

$$\Rightarrow 7\mu g t/2 = u$$

$$\Rightarrow t = \frac{2u}{7\mu g} = \frac{2 \times 7}{7(0.1)(10)} = 2s = t_1 \text{ (let)}.$$

Then the body has the velocity $v = \mu g t = 2$ m s^{-1} towards the right at the time of pure rolling and the angular velocity is

$$\omega = \frac{5\mu g t}{2R} = \frac{5(0.1)(10)(2)}{2(1/4)} = 20 \text{ rad s}^{-1}.$$

This means that the cylinder rotates anticlockwise while moving forward with respect to the ground. However, the angular momentum of the sphere relative to the ground is

$$\vec{L_0} = (mK^2\omega_0 - mv_0 R)\hat{k} = 0\hat{k}.$$

So the sphere will slide after leaving the belt and stop translation and rotation simultaneously after covering a distance $s = \frac{v^2}{2\mu g} = \frac{(2)^2}{2(0.1)(10)} = 2$ m.

Problem 26 A sphere of mass m is kept on the plank of mass M with an initial spin angular velocity ω_0 at time $t = 0$. Due to kinetic friction the sphere rolls without sliding after some time $t = t_1$, say. If the ground is smooth, find (a) the angular velocity of the sphere and (b) the linear velocity of the plank, after the time t_1.

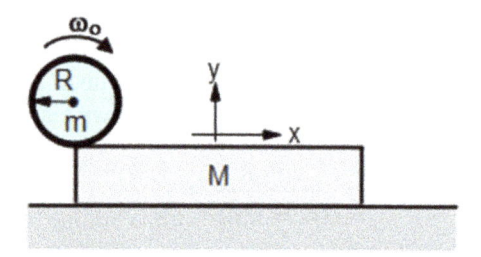

Solution

(a) The friction acting on the sphere is forward and that acting on the plank is backward. So the velocity of the plank will be backward and that of the sphere will be forward (right-ward). As the friction passes through O, its torque is zero. Since the weight mg and normal reaction are equal, opposite, and collinear, their net torque about O is zero. So the angular momentum of the sphere relative to the origin O remains constant. This means

$$L_O = \text{constant},$$

$$\text{or} \quad -mK^2\omega_0 \hat{k} = -mK^2\omega \hat{k} - mRv_C \hat{k}$$

$$\text{or} \quad K^2\omega_0 = K^2\omega + Rv_C. \qquad (5.31)$$

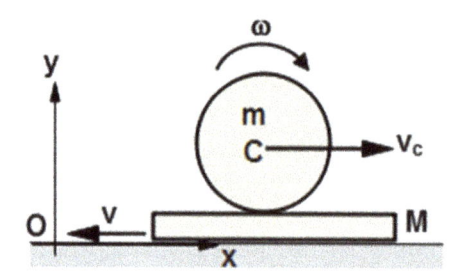

Conservation of the linear momentum of $(M + m)$ gives us

$$mv_C - Mv = 0. \qquad (5.32)$$

For rolling of the sphere on the plank,

$$v_C + v = R\omega. \qquad (5.33)$$

Using equations (5.32) and (5.33) we have

$$M(R\omega - v_C) = mv_C$$

$$\Rightarrow v_C = \frac{MR\omega}{M+m}. \qquad (5.34)$$

Using equations (5.31) and (5.34) we have

$$K^2\omega_0 = K^2\omega + R\left\{\frac{MR\omega}{M+m}\right\}$$

$$\Rightarrow K^2\omega_0 = \omega\left[\frac{(M+m)K^2 + MR^2}{(M+m)}\right]$$

$$\Rightarrow \omega = \frac{\omega_0(M+m)}{m + (1 + R^2/K^2)M} = \frac{\omega_0(M+m)}{m + (1 + 5/2)M}$$

$$= \frac{2\omega_0(M+m)}{7M + 2m}.$$

(b) Since $v = \frac{mR\omega}{M+m}$, putting the value of ω, we have

$$v = \left\{\frac{mR}{M+m}\right\}\left\{\frac{2\omega_0(M+m)}{2m + 7M}\right\} = \frac{2m\omega_0 R}{2m + 7M}$$

Problem 27 Super ball. In the previous problem, let us assume the ball is a *super ball*. There is no energy loss in this case for its elastic collision along the horizontal and vertical directions. If the ball collides with the plank with a vertical downward velocity v_0, assuming that the ground is smooth, find (a) the linear velocity of the center of mass of the sphere, (b) the angular velocity of the sphere, and (c) the linear velocity of the plank just after the collision of the sphere with the plank.

Solution
(a) Conservation of linear momentum of $(M + m)$ gives us $P = C$:

$$mv_C - Mv = 0. \tag{5.35}$$

The angular momentum of the sphere relative to origin $O = L_O = $ constant:

$$\Rightarrow mK^2\omega_0 = mK^2\omega + mvR$$

$$\Rightarrow K^2\omega_0 = K^2\omega + vR. \tag{5.36}$$

Newton's empirical formula:

$$-e\{(-R\omega_0) - 0\} = \{-(R\omega - v_C) - (-v)\}.$$

For the super ball, there is always a perfectly elastic collision; so put $e = 1$ to obtain

$$R\omega_0 = v + v_C - R\omega. \tag{5.37}$$

Now we have three equations for three unknown quantities such as v, v_C, and ω. Let us solve for them.

Putting v_C and ω from equations (5.35) and (5.36), repectively, in equation (5.37) and simplifying the factors,

$$R\omega_0 = v_C + m(v_C/M) - R\{(K^2\omega_0 - vR)/K^2$$

$$\Rightarrow v_C = \frac{2R\omega_0}{1 + \frac{m}{M} + \frac{R^2}{K^2}}$$

$$\Rightarrow v_C = \frac{2R\omega_0}{1 + \frac{m}{M} + \frac{5}{2}} = \frac{2R\omega_0}{\frac{m}{M} + \frac{7}{2}} = \frac{4MR\omega_0}{2m + 7M}.$$

When $M \to \infty$ we have

$$v_C = \frac{2R\omega_0}{1 + \frac{R^2}{K^2}} = \frac{4R\omega_0}{7}.$$

If there is a vertical velocity v_0, say, just before the collision of the ball with the plank, just after the collision the ball will go up vertically with same speed v_0. So the velocity of the center of mass of the ball just after the collision is

$$v'_C = \frac{4MR\omega_0}{2m + 7M}\hat{i} + v_0\hat{j}.$$

(b) From equations (5.35) and (5.37)

$$R\omega_0 = v\left(1 + \frac{m}{M}\right) - R\omega. \qquad (5.38)$$

From equations (5.35) and (5.38)

$$v = \frac{K^2(\omega_0 - \omega)}{R} = \frac{R(\omega + \omega_0)M}{M + m}$$

$$\Rightarrow \omega - \omega_0 = \frac{R^2 M}{K^2(M + m)}(\omega + \omega_0)$$

$$\Rightarrow \omega\left\{\frac{MR^2}{(M+m)K^2} + 1\right\} = \omega_0\left\{1 - \frac{R^2 M}{K^2(M+m)}\right\}$$

$$= \left\{\frac{(M+m)K^2 - MR^2}{(M+m)K^2 + MR^2}\right\}\omega_0$$

$$= \left\{\frac{(M+m)(2R^2/5) - MR^2}{(M+m)(2R^2/5) + MR^2}\right\}\omega_0$$

$$= \left\{\frac{2(M+m) - 5M}{2(M+m) + 5M}\right\}\omega_0$$

$$= \left(\frac{2m - 3M}{2m + 7M}\right)\omega_0.$$

When $M \to \infty$ we have

$$\omega = \left(\frac{K^2 - R^2}{K^2 + R^2}\right)\omega_0 = -3\omega_0/7.$$

The negative sign signifies that the ball will spin opposite to the initial direction just after the impact.

(c) Putting the value of v_C in equation (5.35)

$$v = \frac{mv_C}{M} = \frac{2R\omega_0(m/M)}{1 + \frac{m}{M} + \frac{R^2}{K^2}}$$

$$= \frac{2mR\omega_0}{m + M\left(1 + \frac{R^2}{K^2}\right)}$$

$$= \frac{2mR\omega_0}{m + M\left(1 + \frac{5}{2}\right)}$$

$$= \frac{4mR\omega_0}{2m + 7M}.$$

Note: For an ordinary ball, there is a perfectly inelastic collision ($e = 0$) along the horizontal direction between the ball and plank which appears in the form of rolling after the relative velocity becomes zero due to the effect of kinetic friction. Then

$$R\omega - v_C = v. \tag{5.39}$$

By solving the equations (5.35), (5.36), and (5.37) we will obtain the answers of the last problem, such as

$$\omega = \frac{\omega_0(M+m)}{m + (1 + R^2/K^2)M} = \frac{2\omega_0(M+m)}{2m + 7M}$$

$$v = \left\{\frac{mR}{M+m}\right\}\left\{\frac{2\omega_0(M+m)}{2m + 7M}\right\}$$

$$= \frac{2m\omega_0 R}{2m + 7M}.$$

5-58

Problem 28 A uniform sphere of mass m, radius R, and radius of gyration K is placed just on a plank which is placed on a smooth horizontal surface. The plank is moved with a constant velocity u by an external horizontal force. Find (a) the linear velocity, (b) the time of relative sliding, and (c) the angular velocity of the sphere when it rolls without sliding on the plank, (d) distance covered by the sphere relative to ground till it stops sliding on the plank, (e) the distance of the relative sliding between the sphere and the plank, (f) the heat lost in sliding, (g) the change in kinetic energy of the sphere and (h) the work done by the external agent until the sphere stops sliding on the plank.

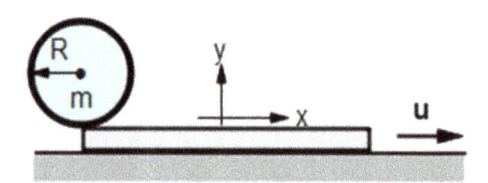

Solution
(a) The angular momentum of the sphere about the origin O is conserved as described in problem 26:

$$L_f = L_i = 0$$

$$\Rightarrow mK^2\omega - mvR = 0$$

$$\Rightarrow \vec{\omega} = \frac{vR}{K^2}\hat{k} \Rightarrow \omega = \frac{vR}{K^2}. \tag{5.40}$$

For rolling of the sphere,

$$v + R\omega = u. \tag{5.41}$$

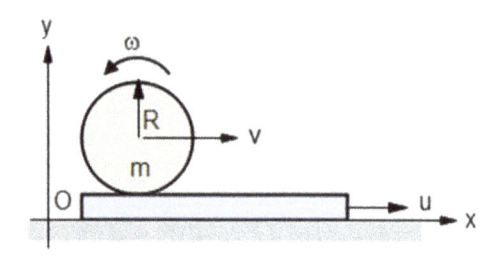

From equations (5.40) and (5.41)

$$v + \left(\frac{vR}{K^2}\right)R = u$$

5-59

$$\Rightarrow v = \frac{u}{1 + \frac{R^2}{K^2}}. \tag{5.42}$$

(b) Since a forward kinetic friction $\mu m g$ acts on the ball, the acceleration of the ball is equal to μg directed along the $+x$-direction. Then the time of relative sliding is

$$t = \frac{v}{\mu g} = \frac{u}{\mu g \left(1 + \frac{R^2}{K^2}\right)}.$$

(c) The final angular velocity of the sphere is

$$\omega = \frac{u - v}{R} = \frac{u}{R}\left(1 - \frac{K^2}{K^2 + R^2}\right)$$

$$\omega = \frac{R^2 u}{R(K^2 + R^2)} = \frac{u}{R\left(1 + \frac{K^2}{R^2}\right)}.$$

(d) The distance covered by the sphere until its relative sliding stops is given as

$$s = \frac{v^2}{2\mu g} = \frac{u^2}{2\mu g \left(1 + \frac{R^2}{K^2}\right)^2}.$$

We can also put the obtained value of time in the equation

$$s = \frac{1}{2}at^2 = \frac{1}{2}\mu g t^2$$

to obtain the same answer.

(e) The distance covered by the sphere relative to the plank till its relative sliding stops is given as

$$s = \frac{(u - v)^2}{2\mu g} = \frac{u^2}{2\mu g \left(1 + \frac{K^2}{R^2}\right)^2}.$$

(f) The heat dissipated = net work done by kinetic friction =
$$Q = W_f = \mu m g s_{rel} = \mu m g \left(\frac{u^2}{2\mu g(1 + R^2/K^2)^2}\right) = \frac{mu^2}{2(1 + R^2/K^2)^2}$$

(g) The change in kinetic energy is

$$\Delta K = \frac{1}{2}mv^2 + \frac{1}{2}mK^2\omega^2 = \frac{1}{2}mv^2\left\{1 + \frac{(K\omega)^2}{v^2}\right\}$$

5-60

Putting the value of $\frac{w}{v} = \frac{R}{K^2}$ from equation (5.40), we have

$$\Delta K = \frac{1}{2}mv^2\left\{1 + K^2\frac{R^2}{K^4}\right\} = \frac{1}{2}mv^2\left\{1 + \frac{R^2}{K^2}\right\}$$

Putting the value of $v = \frac{u}{1 + R^2/K^2}$ from equation (5.42), we have

$$\Delta K = \frac{m}{2}\left(\frac{u}{1 + R^2/K^2}\right)^2\left\{1 + \frac{R^2}{K^2}\right\} = \frac{mu^2}{2(1 + R^2/K^2)}$$

(h) The work done by the external force is

$$W_{ext} = \Delta K + Q = \frac{mu^2}{2(1 + R^2/K^2)} + \frac{mu^2}{2(1 + R^2/K^2)} = \frac{mu^2}{(1 + R^2/K^2)}$$

Alternatively, the work done by the external force against the kinetic friction f to maintain the velocity of the plank constant is given as

$$W_{ext} = fs_1 = \mu mgs_1 = \mu mgut = \mu mgut(\because F = f = \mu mg)$$

Putting the obtained value of $t = \frac{u}{\mu g(1 + R^2/K^2)}$, we have the same expression

$$W_{ext} = \mu mgu \cdot \frac{u}{\mu g(1 + R^2/K^2)} = \frac{mu^2}{(1 + R^2/K^2)}$$

Problem 29 A solid uniform sphere of mass m and radius R is gently placed on a rough plank of mass M and length L with an initial angular velocity ω_0 as shown in the figure of problem 26. If the ground is smooth, (a) find the time and distance of relative sliding and show the results for a very massive plank. (b) What is the minimum value of $L = L_C$ for which the sphere will roll on the plank without sliding? (c) For the cases (i) $L > L_C$ and (ii) $L < L_C$, find the time after which the sphere leaves the plank and the velocity of the sphere at the time of leaving the plank. Assume that the ground is smooth, $\mu =$ the coefficient of friction between the sphere and plank.

Solution

(a) Referring to the solution of problem 26, the velocity of the plank is

$$v = \frac{mR\omega_0}{M + m(1 + R^2/K^2)} = \frac{2mR\omega_0}{2M + 7m}.$$

The kinetic friction between the sphere and plank acts on the plank to right. It increases the velocity of the plank from zero to v during the time t. So, the time after which the sphere stops sliding on the plank is

$$t = \frac{v}{a} = \frac{mR\omega_0}{\frac{\mu mg}{M}\left[m + M\left(1 + \frac{R^2}{K^2}\right)\right]}$$

$$t = \frac{MR\omega_0}{\mu g\left[m + M\left(1 + \frac{R^2}{K^2}\right)\right]}$$

$$= \frac{MR\omega_0}{\mu g\left[m + M\left(1 + \frac{5}{2}\right)\right]}$$

$$t = \frac{2MR\omega_0}{(2m + 7M)\mu g}.$$

The relative acceleration between the sphere and plank is

$$a_{\text{rel}} = \mu g\left(1 + \frac{m}{M}\right).$$

Then the relative distance is

$$L = \frac{1}{2}a_{\text{rel}}t^2$$

$$= \frac{1}{2}\mu g\left(1 + \frac{m}{M}\right)\frac{4M^2R^2\omega_0^2}{(2m + 7M)^2\mu^2 g^2}$$

So we have

$$L = \frac{2M(M + m)R^2\omega_0^2}{(2m + 7M)^2\mu g} = L_C \text{ (critical length)}.$$

When $M \to \infty$ we have

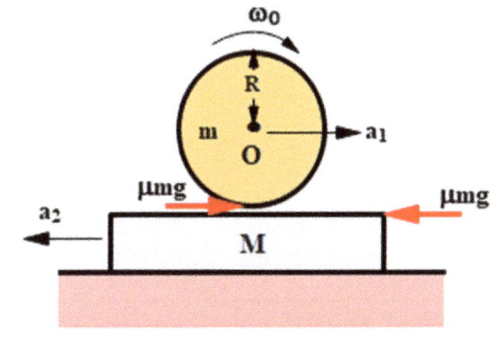

$$L \to \frac{2R^2\omega_0^2}{49\mu g} \text{ and } t \to \frac{2R\omega_0}{7\mu g}.$$

$$a_{\text{rel}} = a_1 + a_2 = \mu g(1 + m/M)$$

(b)

(i) If $L > L_C$, the sphere will leave the plank after a time

$$T = t + \frac{L - L_C}{v_{rel}},$$

where

$$t = \frac{2MR\omega_0}{(2m + 7M)\mu g}$$

and $v_{rel} = R\omega = \frac{2R\omega_0(M+m)}{2m+7M}$ (obtained in problem 26).

The velocity of the sphere at the time of leaving the plank is

$$v_m = \frac{2m\omega_0 R}{2m + 7M} - \frac{2R\omega_0(M+m)}{2m + 7M}$$

$$= \frac{2M\omega_0 R}{2m + 7M}.$$

(ii) If $L < L_C$ the sphere leaves the plank before rolling without sliding after a time

$$t = \sqrt{\frac{2L}{a_{rel}}} = \sqrt{\frac{2L}{\mu g\left(1 + \frac{m}{M}\right)}} = \sqrt{\frac{2ML}{\mu g(M+m)}}$$

and velocity, $v_m = \mu g t = \sqrt{\frac{2\mu MgL}{(M+m)}}$.

Problem 30 A cylinder of mass m, radius R, and radius of gyration K is placed on a trolley car of mass M. The cylinder is given an initial linear and angular velocity v_0 and ω_0, respectively. Assume that the horizontal floor is smooth and μ is the coefficient of friction between the body and surface. Find (a) the angular velocity of the cylinder and linear velocity of the bodies when the cylinder stops sliding on the trolley car (b) fraction of energy lost during the relative sliding between the cylinder and trolley car. Put $M = 2$ m, $u_0 = 2$ m s^{-1}, $v_0 = 1.5$ m s^{-1}, $\omega_0 R = 1$, and $R = 1.5$ m.

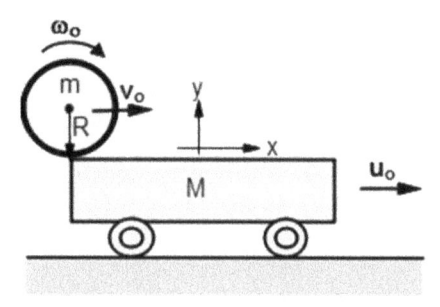

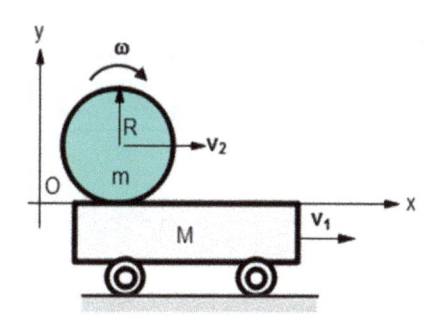

Solution

(a) Conserving angular momentum about O,

$$-(mK^2\omega_0 + mv_0R)\hat{k} = (-mK^2\omega - mv_2R)\hat{k}$$

$$\Rightarrow K^2\omega_0 + v_0R = K^2\omega + v_2R. \qquad (5.43)$$

Conserving linear momentum, $P = C$,

$$Mu_0 + mv_0 = Mv_1 + mv_2. \qquad (5.44)$$

For rolling,

$$R\omega = v_2 - v_1. \qquad (5.45)$$

From equations (5.43) and (5.45)

$$K^2\omega_0 + v_0R = K^2\frac{v_2 - v_1}{R} + v_2R.$$

Multiplying M both sides,

$$M(K^2\omega_0 + v_0R) = \frac{Mv_2}{R}(K^2 + R^2) - \frac{K^2v_1M}{R}. \qquad (5.46)$$

Multiplying $\frac{K^2}{R}$ both sides of equation (5.44) we have

$$\frac{K^2}{R}(Mu_0 + mv_0) = mv_2\frac{K^2}{R} + Mv_1\frac{K^2}{R}. \qquad (5.47)$$

Adding equations (5.46) and (5.47)

$$\frac{v_2}{R}\{mK^2 + M(K^2 + R^2)\}$$

$$= MK^2\omega_0 + \left(MR + \frac{K^2m}{R}\right)v_0 + \frac{MK^2}{R}u_0$$

5-64

$$v_2 = \frac{\left\{MK^2\left(\omega_0 + \frac{u_0}{R}\right) + (MR^2 + mK^2)\frac{v_0}{R}\right\}R}{mK^2 + M(K^2 + R^2)}$$

$$= \frac{MK^2(\omega_0 R + u_0) + (MR^2 + mK^2)v_0}{mK^2 + M(K^2 + R^2)}$$

$$= \frac{M\frac{K^2}{R^2}(\omega_0 R + u_0) + \left(M + m\frac{K^2}{R^2}\right)v_0}{\frac{mK^2}{R^2} + M\left(\frac{K^2}{R^2} + 1\right)}.$$

Putting $\frac{K^2}{R^2} = 1/2$,

$$v_2 = \frac{\frac{M}{2}(\omega_0 R + u_0) + \left(M + \frac{m}{2}\right)v_0}{\frac{m}{2} + \frac{3M}{2}}$$

$$\Rightarrow v_2 = \frac{M(u_0 + \omega_0 R) + (2M + m)v_0}{3M + m}.$$

Putting $M = 2m$, $u_0 = 2$, $\omega_0 R = 1$ m s^{-1}, and

$$v_0 = \frac{1}{5} \text{ m s}^{-1},$$

$$v_2 = \frac{2m(u_0 + \omega_0 R) + 5mv_0}{3 \times 2m + m}$$

$$= \frac{2(u_0 + \omega_0 R) + 5v_0}{7} = \frac{2(2+1) + 5 \times \frac{1}{5}}{7}$$

$$\vec{v}_2 = 1\hat{i} \text{ m s}^{-1}.$$

Using equation (5.43),

$$\omega = \frac{K^2\omega_0 + v_0 R - v_2 R}{K^2}$$

$$\Rightarrow R\omega = R\omega_0 + v_0\frac{R^2}{K^2} - v_2\frac{R^2}{K^2}$$

$$= R\omega_0 + (v_0 - v_2)\frac{R^2}{K^2}$$

$$= 1 + \left(\frac{1}{5} - 1\right)2 = 1 - \frac{8}{5} = -\frac{3}{5}$$

5-65

$$\omega = -\frac{3}{5R} = -\frac{3}{5 \times \frac{1}{5}} = -3 \text{ rad s}^{-1}.$$

This means $\vec{\omega}$ will be opposite to the assumed direction. So $\vec{\omega} = +3\hat{k}$ rad s^{-1}.

Using equation (5.45) $R\omega = v_2 - v_1$

$$\Rightarrow -\frac{3}{5} = 1 - v_1 \Rightarrow v_1 = 1 + \frac{3}{5} = \frac{8}{5} \text{ m s}^{-1}$$

$$\Rightarrow \vec{v_1} = 1.6\hat{i} \text{ m s}^{-1}.$$

(b) The fraction of energy loss

$$\eta = |\Delta E/E| = \frac{\left|\frac{1}{2}m(v_2^2 - v_0^2) + \frac{mR^2}{4}(\omega^2 - \omega_0^2) + \frac{1}{2}M(v_1^2 - u_0^2)\right|}{\frac{1}{2}mv_0^2 + \frac{mR^2\omega_0^2}{4} + \frac{M}{2}u_0^2}$$

$$= 1 - \frac{mv_2^2 + \frac{mR^2}{2}\omega^2 + Mv_1^2}{mv_0^2 + \frac{mR^2}{2}\omega_0^2 + Mu_0^2}$$

$$= 1 - \frac{v_2^2 + \frac{(R\omega)^2}{2} + \frac{M}{m}v_1^2}{v_0^2 + \frac{(R\omega_0)^2}{2} + \frac{M}{m}u_0^2}$$

$$= 1 - \frac{v_2^2 + \frac{(R\omega)^2}{2} + 2v_1^2}{v_0^2 + (R\omega_0)^2 + 2u_0^2}$$

$$= 1 - \frac{1 + \left(\frac{3}{5}\right)^2/2 + 2\left(\frac{8}{5}\right)^2}{\left(\frac{1}{5}\right)^2 + 1 + 2(2)^2} = 1 - \frac{1 + \frac{9}{50} + \frac{128}{25}}{\frac{1}{25} + 1 + 8}$$

$$= 0.303 \text{ (approximately)}.$$

Variable moment of inertia

Problem 31 Loading sand. A thin spherical vessel of mass M half-filled with sand, spins about its vertical axis without friction with an angular velocity ω_0. Sand is poured into it until it is completely filled. If the total mass of the sand is m, find (a)

5-66

the angular velocity after loading, (b) the fraction change in kinetic energy, and (c) the change in total mechanical energy of the system (vessel + sand). Put $\omega_0 = \sqrt{g/R}$ and $M = m$.

Solution
(a) After being completely filled the total sand becomes a solid sphere having a moment of inertia $\frac{2mR^2}{5}$. So after filling, the total moment of inertia will be

$$I = \left(\frac{2MR^2}{3} + \frac{2mR^2}{5}\right).$$

Conserving angular momentum,

$$I_0\omega_0 = I\omega$$

$$\Rightarrow \left(\frac{2MR^2}{3} + \frac{2(m/2)R^2}{5}\right)\omega_0$$

$$= \left\{\frac{2MR^2}{3} + \left(\frac{mR^2}{5}\right)\right\}\omega$$

$$\Rightarrow \frac{10M + 3m}{15}\omega_0 = \frac{(10M + 6m)}{15}\omega$$

$$\Rightarrow \omega = \frac{10M + 3m}{10M + 6m}\omega_0 = 13\omega_0/16.$$

(b) Then

$$\frac{\Delta K}{K} = -\left(1 - \frac{I_0}{I}\right) = -\left(1 - \frac{10M + 3m}{10M + 6m}\right)$$

$$= -\frac{3m}{10M + 6m} = -3/16.$$

(c) The total change in energy is $\Delta E = \Delta K + \Delta U$, where the center of mass of the sand rises from height $\frac{5R}{8}$ to R and the mass of the sand increased from $m/2$ to m. So, we have

$$\Delta U = U_f - U_i = mgR - (m/2)g(5/8) = 11mgR/16$$

and $\Delta K = -\frac{1}{2}I_0\omega_0^2 \frac{3m}{10M+6m}$. Then we have

$$\Delta E = -\frac{1}{2}I_0\omega_0^2 \frac{3m}{10M+6m} + \frac{11mgR}{16}$$

$$= -\frac{1}{2}\frac{(10M+3m)R^2\omega_0^2}{15} \times \frac{3m}{10M+6m} + \frac{11mgR}{16}.$$

Putting $\omega_0 = \sqrt{g/R}$ and $M = m$ we have

$$\Delta E = -\frac{13mgR}{160} + \frac{11mgR}{16} = 97mgR/160.$$

Problem 32 A thin cylindrical drum of mass m, radius R, and radius of gyration K completely filled with sand of mass m spins about its vertical axis without friction with an angular velocity $\omega_0 = 1$ rad s^{-1}. Sand leaks slowly out of it through a hole made at its bottom. When half of the container is emptied, find (a) the angular velocity and (b) the fraction change in mechanical energy, of the system (drum + sand).

Solution
(a) After unloading the sand, the moment of inertia of the sand system changes from $\frac{mR^2}{2}$ to $\frac{(m/2)R^2}{2}$. So the total moment of inertia will change from I_1 to I_2, given as

$$I_i = \left(I_0 + \frac{mR^2}{2}\right)$$

$$I_f = \left(I_0 + \frac{mR^2}{4}\right).$$

Conserving angular momentum,
$$I_i\omega_i = I_f\omega_f = L \text{ (let)}$$

$$\Rightarrow \left(I_0 + \frac{mR^2}{2}\right)\omega_0 = \left(I_0 + \frac{mR^2}{4}\right)\omega,$$

where $I_0 = mK^2$ and $\omega_0 = 1$ rad m s^{-1}

$$\Rightarrow \omega = \frac{2(2I_0 + mR^2)}{4I_0 + mR^2} = \frac{2(2K^2 + R^2)}{4K^2 + R^2}.$$

(b) The change in kinetic energy is

$$\Delta K = \frac{L^2}{2I_f} - \frac{L^2}{2I_i} = -\frac{L^2}{2I_i}\left(1 - \frac{I_i}{I_f}\right).$$

Then the fraction change in kinetic energy is

$$\frac{\Delta K}{K} = -\left(1 - \frac{I_0 + \frac{mR^2}{2}}{I_0 + \frac{mR^2}{4}}\right) = \frac{\frac{mR^2}{2}}{I_0 + \frac{mR^2}{4}} - \frac{mR^2}{I_0 + \frac{mR^2}{4}}$$

$$= \frac{2mR^2}{4I_0 + mR^2} = \frac{2}{\frac{4K^2}{R^2} + 1}.$$

Here the kinetic energy increases due to the increase in the angular velocity.

Problem 33 A cylindrical vessel initially spins with an angular velocity ω_0 about its vertical axis passing through its center of mass. Let us start pouring a viscous oil into the vessel at a constant rate μ, find the angular (a) velocity and (b) acceleration of the system as a function of time t. Assume $I_0 =$ the moment of inertia of the vessel and $R =$ radius of the vessel. Neglect friction between the axis (axle) and vessel.

Solution

(a) The moment of inertia of the rotating oil plus the moment of inertia of the container, after a time t is

$$I = \left(I_0 + \frac{\mu t R^2}{2}\right).$$

Then, conserving the angular momentum,

$$I_0 \omega_0 = I\omega$$

$$I_0 \omega_0 = \left(I_0 + \frac{\mu t R^2}{2}\right)\omega$$

$$\omega = \frac{2I_0 \omega_0}{2I_0 + \mu R^2 t}.$$

(b) Differentiating with respect to time,

$$\alpha = \frac{d\omega}{dt} = -\frac{2I_0 \omega_0 \mu R^2}{(2I_0 + \mu R^2 t)^2}.$$

Problem 34 The cylindrical vessel of radius R completely filled with sand initially spins with an angular velocity ω_0 about a vertical frictionless axle. The initial moment of inertia of the cylinder with sand is I_0. Let us start removing sand out of the vessel at a constant rate μ. Find (a) the angular (i) velocity and (ii) acceleration of the system as a function of time t, and (b) the rate of change of the kinetic energy of the system.

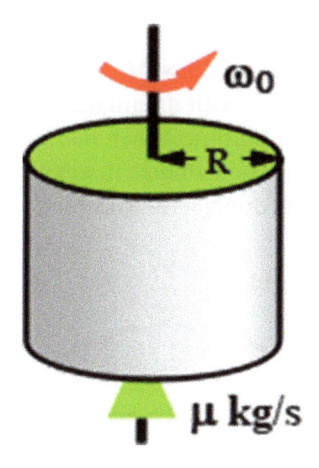

Solution

(a)

(i) The moment of inertia of the rotating oil after a time t is $\frac{\mu t R^2}{2}$. Then, conserving the angular momentum,

$$I_0 \omega_0 = \left(I_0 - \frac{\mu t R^2}{2}\right)\omega \Rightarrow \omega = \frac{2I_0\omega_0}{2I_0 - \mu R^2 t}.$$

(ii) Differentiating with respect to time,

$$\alpha = \frac{d\omega}{dt} = \frac{2I_0\omega_0\mu R^2}{(2I_0 - \mu R^2 t)^2}.$$

(b) Since the kinetic energy is $K = \frac{L^2}{2I}$, differentiating K with time,

$$\frac{dK}{dt} = -\frac{L^2}{2I^2}\frac{dI}{dt} = -\frac{L^2}{2I^2} \times \frac{d(mR^2/2)}{dt}$$

$$= -\frac{L^2 R^2}{4I^2} \times \frac{dm}{dt} = -\frac{L^2}{4I^2}\mu R^2.$$

Putting $I = I_0 - \frac{\mu t R^2}{2}$ and $L = I_0\omega_0$, we have

$$\frac{dK}{dt} = -\frac{\mu I_0^2 R^2 \omega_0^2}{4(I_0 - \mu R^2 t/2)^2}.$$

5-70

Problem 35 A thin spherical drum of mass M completely filled with sand of mass m spins about the smooth vertical axis with an angular velocity ω_0. Sand leaks out of it slowly through a hole made at its bottom. When half of the container is emptied, find (a) the angular velocity, (b) the fraction change in kinetic energy, and (c) the change in mechanical energy of the system (drum + sand). Put $\omega_0 = \sqrt{g/R}$ and $M = m$ in (c).

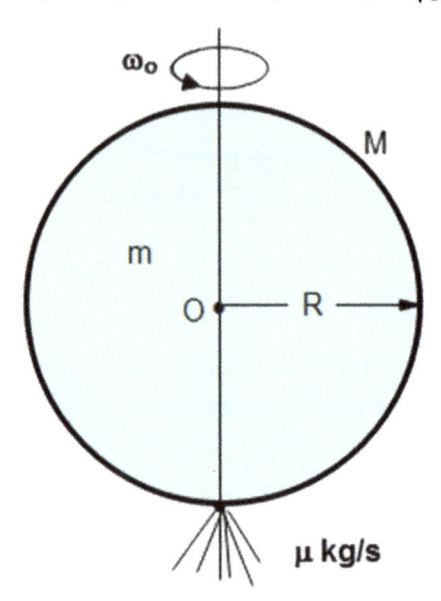

Solution
(a) Conserving the angular momentum,

$$\left(\frac{2}{3}MR^2 + \frac{2}{5}mR^2\right)\omega_0 = \left(\frac{2}{3}MR^2 + \frac{mR^2}{5}\right)\omega$$

$$\omega = \frac{10M + 6m}{10M + 3m}\omega_0.$$

(b) The fraction change in kinetic energy is $\frac{\Delta K}{K} = \frac{K_f - K_i}{K_i} = \frac{K_f}{K_i} - 1 \; (\because K\alpha\frac{1}{I})$

$$= -\left(1 - \frac{I_i}{I_f}\right) = -\left(1 - \frac{10M + 6m}{10M + 3m}\right)$$

$$\frac{\Delta K}{K} = +\frac{3m}{10M + 3m}.$$

(c) The change in total mechanical energy is $\Delta E = \Delta U + \Delta K$.

5-71

As the sand escapes the center of mass, the gravitational potential energy decreases by $\Delta U = mgR - (m/2)g(3\,R/8) = 11mgR/16$. Then

$$\Delta E = -\left(\frac{m}{2}g\frac{11}{8}R\right) + \left(\frac{3m}{10M+3m}\right)K_0$$

$$= -\frac{11mgR}{16} + \frac{3m}{(10M+3m)}\left\{\frac{1}{2}\left(\frac{2MR^2}{3} + \frac{2mR^2}{5}\right)w_0^2\right\}.$$

Putting $w_0^2 = \frac{g}{R}$ and $M = m$

$$\Delta E = -\frac{11mgR}{16} + \frac{3mR^2w_0^2}{30}$$

$$= mgR\left(-\frac{11}{16} + \frac{8}{65}\right)$$

$$= -571mgR/1040.$$

Problem 36 A uniform disc of mass M and radius R is spinning with an angular velocity w_0 about a smooth vertical axis PQ as shown in the figure. An insect of mass m is standing near the axis. The insect is crawling with a relative velocity v_r along the diametrical chute. (a) If a constant torque τ acts on the disc as shown in the figure, find the angular velocity and angular acceleration of the disc, as a function of (i) radial distance and (ii) time. (b) If the disc is made to rotate with a constant angular velocity w, find (i) the torque, (ii) the rate of change of kinetic energy, (iii) the power delivered to the disc by the external agent, (iv) the power loss in the form of heat, and (v) the relation between the power loss and rate of change in kinetic energy.

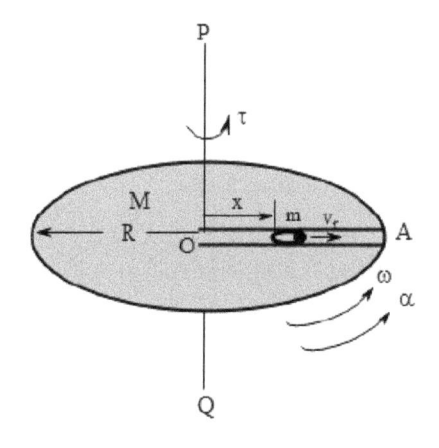

5-72

Solution

(a)

(i) At any radial distance x, the angular momentum is $L = \left(\frac{MR^2}{2} + mx^2\right)\omega$. Then the change in angular momentum is

$$\Delta L = \left(\frac{MR^2}{2} + mx^2\right)\omega - \frac{MR^2}{2}\omega_0.$$

Using the angular impulse–momentum equation,

$$\tau t = \left(\frac{MR^2}{2} + mx^2\right)\omega - \frac{MR^2}{2}\omega_0$$

$$\Rightarrow \omega = \frac{\tau t + \frac{MR^2}{2}\omega_0}{\frac{MR^2}{2} + mx^2} = \frac{2\tau t + MR^2\omega_0}{MR^2 + 2mx^2}$$

$$\Rightarrow \omega = \frac{2Tt + MR^2\omega_0}{MR^2 + 2mv_r^2 t^2}.$$

(ii) Then, differentiating the last expression with time, we have

$$\alpha = \frac{d\omega}{dt} = \frac{(MR^2 + 2mv_r^2 t^2)2\tau - (2\tau t + MR^2\omega_0)4mv_r^2 t}{(MR^2 + 2mv_r^2 t^2)^2},$$

where $v_r t = x$.

(b)

(i) The angular momentum of the system is

$$L = \left(\frac{MR^2}{2} + mx^2\right)\omega.$$

Then, differentiating the last expression with time, we have

$$\frac{dL}{dt} = \omega 2m \cdot x \frac{dx}{dt} \quad (\because \omega = c)$$

$$\Rightarrow \tau = 2m\omega x v_r.$$

(ii) By differentiating the kinetic energy with time we have

$$\frac{dK}{dt} = \frac{\omega^2}{2} \times \frac{dI}{dx}$$

$$= \frac{\omega^2}{2} \times \frac{d}{dx}\left(mx^2 + \frac{MR^2}{2}\right) = m\omega x v_r.$$

(iii) Now, $P_{ext} = \tau\omega = 2m\omega^2 v_r x$.
(iv) Power loss $= P_{ext} - \frac{dK}{dt} = m\omega^2 x v_r$.
(v) So $2\frac{dK}{dt} = P_{ext}$.

Problem 37 A spherical planet of uniform density σ and radius R has a time period T_o, say. In the course of time it is covered by a cosmic dust cloud of thickness $h\,(\ll R)$ and average density ρ. Find the relative change in its time period, that is, $\frac{\Delta T}{T_o}$.

Solution

If the moment of inertia of the layer of dust is I_{layer}, conservation of angular momentum gives, $(I_{earth} + I_{layer})\omega = I_{earth}\omega_0$. Putting $\omega/\omega_o = I_o/I$, $I_{layer} = 2mR^2/3$, and $I_{earth} = 2MR^2/5$, we have

$$\left(\frac{2}{5}MR^2 + \frac{2}{3}mR^2\right)\frac{2\pi}{T} = \left(\frac{2}{5}MR^2\right)\frac{2\pi}{T_o}$$

$$\Rightarrow \frac{T}{T_o} = \left(1 + \frac{5m}{3M}\right) \Rightarrow \frac{\Delta T}{T_o} = \frac{5m}{3M}$$

$$= \frac{5 \times 4\pi R^2 h\rho}{3 \times \frac{4}{3}\pi R^3 \sigma} = \frac{5h\rho}{R\sigma}.$$

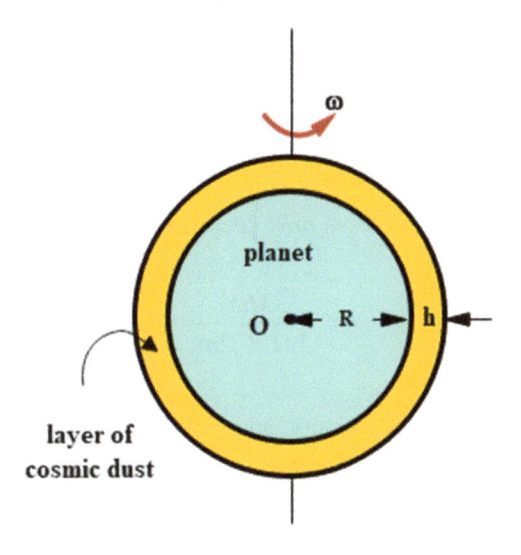

Problem 38 A uniform disc of mass M and radius R is spinning with an angular velocity ω_0 about a smooth vertical axis APB as shown in the figure. A boy of mass m is standing near the axis. (a) If the boy walks with a small relative velocity v_r, say, along the diametrical chute, find (a) angular velocity of the disc when the boy reaches the other end of the chute (b) the fraction of energy loss (c) energy loss of the disc–boy system until the boy reaches the other end of the chute. (d) If friction is absent, find the velocity of the boy relative to the disc when he reaches the other end of the chute assuming that he starts from rest relative to the disc near the axis.

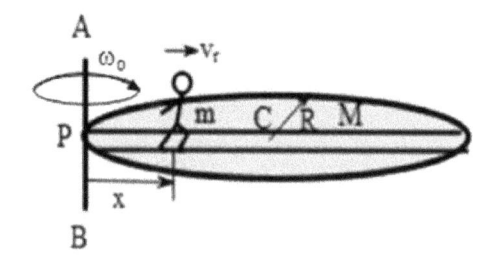

Solution
(a) Conserving the angular momentum, we can write
$$I_f \omega_f = I_i \omega_i,$$
where
$$I_f = \frac{3}{2} MR^2 + mR^2$$
and
$$I_i = \frac{3}{2} MR^2.$$
Then we have
$$\left(\frac{3}{2} MR^2 + mR^2 \right) \omega = \frac{3}{2} MR^2 \omega_0$$
$$\Rightarrow \omega = \frac{3M\omega_0}{3M + 2m}. \tag{5.48}$$

(b) The fractional change in kinetic energy is
$$\eta = -\left(1 - \frac{I_i}{I_f}\right)$$

$$= -\left(1 - \frac{(3/2)MR^2}{3/2MR^2 + mR^2}\right)$$

$$= -\left(1 - \frac{3M}{3M + 2m}\right) = -\frac{2m}{3M + 2m}.$$

(c) The magnitude of change in kinetic energy is

$$|\Delta K| = \left|\frac{L^2}{2I_f} - \frac{L^2}{2I_i}\right|$$

$$= \left|\frac{L^2}{2I_i}\left(\frac{I_i}{I_f} - 1\right)\right|$$

$$= \left\{\frac{1}{2} \times \left(\frac{3}{2}MR^2\right)\omega_0^2\right\}\left(\frac{2m}{3M + 2m}\right)$$

$$= \frac{3Mm\omega_0^2 R^2}{2(3M + 2m)}.$$

(d) If friction is absent, we can conserve the kinetic energy of the system; $K_f = K_i$

$$\Rightarrow \frac{1}{2}m(v_r^2 + R^2\omega^2) + \frac{1}{2}(3MR^2/2)\omega^2$$

$$= \frac{1}{2}(3MR^2/2)\omega_0^2$$

$$\frac{1}{2}mv_r^2 = \frac{1}{2}(3MR^2/2)\omega_0^2 - \frac{1}{2}(mR^2 + \frac{3}{2}MR^2)\omega^2.$$

Putting the value of $\omega = \frac{3M\omega_0}{3M + 2m}$ and simplifying the factors. We have

$$\Rightarrow \frac{1}{2}mv_r^2 = \frac{3Mm\omega_0^2 R^2}{2(3M + 2m)}$$

$$\Rightarrow v_r = \left\{\sqrt{\frac{3M}{3M + 2m}}\right\}R\omega_0.$$

Problem 39 A uniform disc of mass M and radius R is spinning with an angular velocity ω_0 about a smooth vertical axis AMB, as shown in the figure. A boy of mass m is standing near the axis. (a) If the boy walks with a velocity v_r relative to the disc along the diametrical chute from the axis, find (i) the angular velocity and (ii) the acceleration of the disc, as a function of time t. (b) If the boy walks with a velocity v_r relative to the disc along the diametrical chute from the a radial distance $x = R/2$ to the end of the chute, find (i) the change in the kinetic energy of the system (disc plus boy), (ii) the work done by the boy, and (iii) the work done on the boy by the contact forces, until he reaches the other end of the chute. Put $M = 2m$, $b = R/2$. (c) If v_r is constant and the external torque acting on the system $(M + m)$ is zero, find the total number of turns rotated by the disc until the boy reaches the end of the chute.

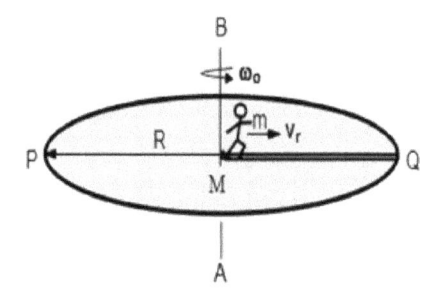

Solution

(a)

(i) Conserving angular momentum at any radial distance x of the boy, for an initial angle velocity ω_0

$$\left(\frac{MR^2}{2} + mx^2\right)\omega = \frac{MR^2}{2}\omega_0 = L$$

$$\Rightarrow \omega = \frac{MR^2\omega_0}{MR^2 + 2mx^2}.$$

(ii) Differentiating with time,

$$\alpha = \frac{d\omega}{dt}$$

$$= -\frac{(MR^2\omega_0)(2m) \times 2x(dx/dt)}{(MR^2 + 2mx^2)^2}$$

$$= \frac{-4MmR^2\omega_0 v_r x}{(MR^2 + 2mx^2)^2} (\because dx/dt = v_r),$$

where $x = v_r t$.

(b)

(i) The change in the kinetic energy of the system is

$$\Delta K = -\left(\frac{I_d\omega_0^2}{2} + \frac{m}{2}v_r^2 + \frac{m}{2}b^2\omega_0^2\right)$$

5-77

$$+\left(\frac{I_d\omega^2}{2} + \frac{m}{2}v_r^2 + \frac{m}{2}R^2\omega^2\right)$$

$$= -\frac{(I_d + mb^2)\omega_0^2}{2} + \frac{1}{2}(I_d + mR^2)\omega^2$$

$$= -\left\{\frac{(I_d + mb^2)\omega_0}{2}\right\}^2 \left[\frac{1}{I_d + mb^2} - \frac{1}{I_d + mR^2}\right]$$

$$= -\frac{(I_d + mb^2)^2\omega_0^2}{2}\left[\frac{m(R^2 - b^2)}{(I_d + mb^2)(I_d + mR^2)}\right]$$

$$= -\left(\frac{I_d + mb^2}{2}\right)\omega_0^2 \frac{m(R^2 - b^2)}{(I_d + mR^2)}.$$

Then we have

$$\frac{\Delta K}{K} = -\frac{m(R^2 - b^2)}{MK^2 + mR^2},$$

where $b = R/2$. Then

$$\Rightarrow \frac{\Delta K}{K} = \frac{m\left(R^2 - \frac{R^2}{4}\right)}{\frac{2mR^2}{2} + mR^2} = \frac{3}{8}.$$

(ii) It is to be noted that we cannot conserve the kinetic energy of the disc–boy system. Since $\Delta K_{system} \neq 0$, the work done by the boy is

$$W_{boy} = \Delta K_{system}$$

$$\frac{m\left(R^2 - \frac{R^2}{4}\right)}{\frac{MR^2}{2} + mK^2}K_0 = \frac{m\frac{3R^2}{4}}{2mR^2}K_0 = \frac{3}{8}K_0.$$

(iii) Work done by contact force acting on the boy

$$= \Delta K_b = -\frac{m}{2}[(v_r^2 + b^2\omega_0^2) - (R^2\omega^2 + v_r^2)]$$

$$= +\frac{m}{2}[R^2\omega^2 - b^2\omega_0^2]. \qquad (5.49)$$

Conserving the angular momentum,

$$\left(\frac{MR^2}{2} + \frac{mR^2}{4}\right)\omega_0 = \left(\frac{MR^2}{2} + mR^2\right)\omega$$

$$\Rightarrow \frac{5mR^2}{4}\omega_0 = 2mR^2\omega \Rightarrow \omega = \frac{5}{8}\omega_0.$$

(iv) Putting this value in the last expression,

$$\Delta K_{boy} = \frac{M}{2}\left[R^2\frac{25}{64}\omega_0^2 - \frac{R^2}{4}\omega_0^2\right]$$

$$= \frac{mR^2\omega_0^2}{8}\left[\frac{25}{16} - 1\right] = \frac{9ml^2\omega_0^2}{128}.$$

Work done on the boy by the contact forces is

$$W = \Delta K_{boy} = \frac{9ml^2\omega_0^2}{128}.$$

(c) When the boy is at the radial distance r after a time t, conserving angular momentum about the axis of rotation,

$$L = I\omega = I_0\omega_0$$

$$\omega = \frac{I_0\omega_0}{I} = \frac{I_0\omega_0}{I_0 + mr^2} = \frac{I_0\omega_0}{I_0 + mv_r^2 t^2}.$$

Integrating both sides,

$$\int_0^\theta d\theta = I_0\omega_0 \int_0^{t=\frac{R}{v_r}} \frac{dt}{I_0 + mv_r^2 t^2}$$

$$\Rightarrow \theta = \frac{I_0\omega_0}{mv_r^2} \int_0^t \frac{dt}{\frac{I_0}{mv_r^2} + t^2}$$

$$= \frac{I_0\omega_0}{mv_r^2} \frac{1}{\sqrt{I_0/mv_r^2}} \tan^{-1}\left(\frac{t}{\sqrt{I_0/mv_r^2}}\right)\Big|_0^{\frac{R}{v_r}}$$

$$= \sqrt{\frac{I_0}{m}}\frac{\omega_0}{v_r}\tan^{-1}\left(\frac{\frac{R}{v_r}\sqrt{m}v_r}{\sqrt{I_0}}\right)$$

$$= \sqrt{\frac{I_0}{m} \frac{\omega_0}{v_r}} \tan^{-1}\left(\frac{\sqrt{m}R}{\sqrt{I_0}}\right)$$

$$N = \frac{\theta}{2\pi} = \frac{1}{2\pi}\left(\sqrt{\frac{M}{2m}}\right)\frac{R\omega_0}{v_r}\tan^{-1}\sqrt{\frac{2m}{M}}.$$

Problem 40 An insect of mass m is resting at the mid-point of a chord AB of a uniform disc of mass M and radius R. The plane of the disc is horizontal and it can rotate freely about a vertical axis passing through its center C. If the insect starts crawling with a velocity v_r relative to the disc, find (a) the work done by the insect, which is numerically equal to the change in kinetic energy of the system $(M + m)$ just after the insect crawls, and (b) the velocity of the insect at B. (Put $M = 2$ m, $h = R/\sqrt{2}$, and $v_r = \sqrt{5/2}$ m s^{-1}.

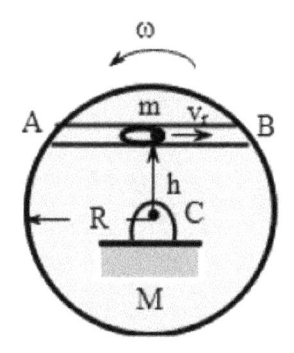

Solution
(a) As everybody was at rest initially, the initial angular momentum is zero. So the angular momentum of the system $(M + m)$ must be zero at any time. This means
$$L_C = 0$$
If we assume that the insect v_r
$$\Rightarrow \frac{MR^2}{2}\omega - m(v_r - h\omega)h = 0$$
$$\Rightarrow \omega = \frac{2mv_r h}{MR^2 + 2mh^2}. \tag{5.50}$$
Then $W = \Delta K$
$$= \frac{1}{2}\frac{MR^2}{2}\omega^2 + \frac{1}{2}v_m^2,$$

where

$$V_m = \frac{MR^2\omega}{2mh}$$

$$= \frac{MR^2}{4}\omega^2 + \frac{m}{2}\left(\frac{MR^2\omega}{2mh}\right)^2$$

$$= \frac{MR^2\omega^2}{4}\left\{1 + \frac{MR^2}{2mh^2}\right\}$$

$$= \frac{MR^2\omega^2}{4}\left(\frac{2mh^2 + mR^2}{2mh^2}\right)$$

$$\Rightarrow W = \frac{MR^2\omega^2}{8mh^2}(2mh^2 + MR^2). \tag{5.51}$$

Using equation (5.50) and (5.51)

$$W = \frac{MR^2(2mh^2 + MR^2)}{8mh^2} \cdot \frac{4m^2 v_r^2 h^2}{(MR^2 + 2mh^2)^2}$$

$$\Rightarrow W = \frac{MmR^2 v_r^2}{2(MR^2 + 2mh^2)}.$$

(b) Putting $h = \frac{R}{2}$ we have

$$W = \frac{MmR^2 v_r^2}{2\left(MR^2 + 2m\frac{R^2}{4}\right)}$$

$$= \frac{Mmv_r^2}{2M + m} = \frac{2}{5}mv_r^2.$$

Conserving the angular momentum of the system at any radial distance r of the insect, we have $L_C = 0$:

$$\frac{M}{2}R^2\omega\hat{k} + m(v_r - \omega r\cos\theta)h\hat{k}M + m(\omega r\sin\theta)x\hat{k} = 0.$$

Putting $r\cos\theta = h$ and $\sin\theta = x$, we have

$$\omega\left\{\frac{MR^2}{2} + m(h^2 + x^2)\right\} = mv_r h$$

$$\Rightarrow \omega = \frac{2mhv_r}{MR^2 + 2(mh^2 + mx^2)}$$

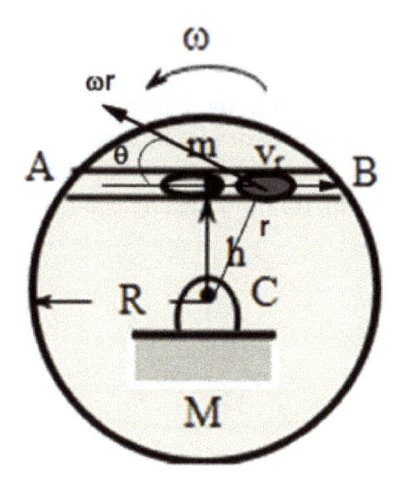

The angular velocity of the disc when the insect travels by a distance x relative to the disc is

$$\omega = \frac{2mv_r h}{MR^2 + 2m(h^2 + x^2)}$$

$$= \frac{2mv_r \frac{R}{\sqrt{2}}}{MR^2 + 2mR^2}$$

$$(\because h^2 + x^2 = R^2, \text{ where } x = \frac{R}{\sqrt{2}})$$

$$= \frac{\sqrt{2}v_r}{4R} = \frac{v_r}{2\sqrt{2}R}.$$

Then the speed of the insect at B is given as

$$v_m = \sqrt{v_r^2 + R^2\omega^2 - 2v_r R\omega \cos 45°}$$

$$= \sqrt{v_r^2 + \frac{v_r^2}{8} - 2v_r\left(\frac{v_r}{2\sqrt{2}}\right)\left(\frac{1}{\sqrt{2}}\right)}$$

$$= v_r\sqrt{\frac{1}{2} + \frac{1}{8}} = v_r\left(\sqrt{5/8}\right).$$

Putting $v_r = (\sqrt{5/2})$ we have

$$v_m = 5/4 = 1.25 \text{ m s}^{-1}.$$

5-82

Problem 41 A disc of mass M, radius R, and moment of inertia I is free to spin about a smooth vertical axis AB, as shown in the figure. An insect of mass m crawls along the perimeter of a circle of diameter a drawn through the center of the disc. (a) Find the angle rotated by the disc when the insect completes a circle. (b) Calculate the angle in (a) if the disc is uniform. (c) Find the angle in (b) by putting $M/m = 2a^2/3R^2$.

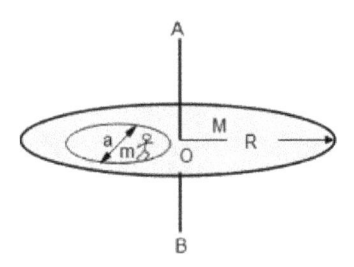

Solution

(a) Let the insect revolve through an angle θ relative to the disc about the axis AB (z-axis) as the plane of the disc is imagined to be lying in the x–y-plane. If the disc rotates through an angle ϕ, the insect will turn through an angle $\beta = \theta - \phi$ relative to the ground frame. So the angular velocities of the disc and insect relative to the ground can be given as following:

$$\omega_{disc} = d\phi/dt$$

$$\omega_{insect} = d\beta/dt = d\theta/dt - d\phi/dt.$$

As there is no external torque acting on the disc–insect system, its angular momentum is conserved. Since the system was stationary, its initial angular momentum is zero. Conserving angular momentum about the axis of rotation we have

$$\vec{L}_{disc} + \vec{L}_{insect} = 0$$

$$\Rightarrow I_{disc}\vec{\omega}_{disc} + I_{insect}\vec{\omega}_{insect} = 0$$

$$\Rightarrow + \frac{MR^2}{2}\frac{d\phi}{dt} - mr^2\frac{d}{dt}(\theta - \phi) = 0$$

$$\Rightarrow \left(\frac{MR^2}{2} + mr^2\right)d\phi = mr^2 d\theta,$$

where $r = a\cos\theta$

$$\Rightarrow \left(\frac{MR^2}{2} + ma^2\cos^2\theta\right)d\phi = ma^2\cos^2\theta\, d\theta.$$

5-83

Separating the variables

$$d\phi = ma^2 \frac{\cos^2\theta d\theta}{\frac{MR^2}{2} + ma^2 \cos^2\theta}$$

$$\Rightarrow d\phi = \frac{ma^2 d\theta}{\sec^2\theta \frac{MR^2}{2} + ma^2}$$

$$= \frac{2ma^2}{MR^2} \cdot \frac{d\theta}{\sec^2\theta + \frac{2MR^2}{MR^2}}$$

$$\Rightarrow d\phi = \frac{\left(\frac{2ma^2}{MR^2} + \sec^2\theta - \sec^2\theta\right)d\theta}{\left(\sec^2\theta + \frac{2ma^2}{MR^2}\right)}$$

$$\Rightarrow d\phi = d\theta - \frac{-\sec^2\theta d\theta}{\left(\sec^2\theta + \frac{2ma^2}{MR^2}\right)}.$$

Integrating both sides,

$$\Rightarrow \phi = \int_0^{\pi/2} d\theta - \int_0^{\pi/2} \frac{\sec^2\theta d\theta}{\left(\frac{2ma^2}{MR^2} + 1\right) + \tan^2\theta}.$$

Let $\tan\theta = x$ and $\sec^2\theta d\theta = dx$

$$= \frac{\pi}{2} - \int \frac{a \, dx}{b + x^2} = \frac{\pi}{2} - \int \frac{a \, dx}{b + x^2}$$

$$= \frac{\pi}{2} - \sqrt{\frac{MR^2}{2ma^2 + MR^2}} \tan^{-1}\frac{\tan\theta}{\sqrt{b}}\Big|_0^{\pi/2}$$

$$= \frac{\pi}{2} - \sqrt{\frac{MR^2}{2ma^2 + MR^2}}(\pi/2 - 0)$$

$$= \frac{\pi}{2}\left(1 - \sqrt{\frac{MR^2}{2ma^2 + MR^2}}\right).$$

(b) Putting $\frac{MR^2}{2} = I$

$$\Rightarrow \theta = \frac{\pi}{2}\left(1 - \sqrt{\frac{I}{ma^2 + I}}\right).$$

(c) Putting $\frac{M}{m} = \frac{2a^2}{3R^2}$ we have

$$\theta = \frac{\pi}{2}\left(1 - \frac{1}{2}\right) = \frac{\pi}{4}.$$

Problem 42 A disc of mass M, radius R, and moment of inertia I is free to spin about a smooth vertical axis AB, as shown in the figure. A boy of mass m walks along the perimeter of the disc. Find the angle rotated by the disc when the boy walks a distance of 3 m along the perimeter of the disc. Put $R = 20$ m and $3M = 8$ m.

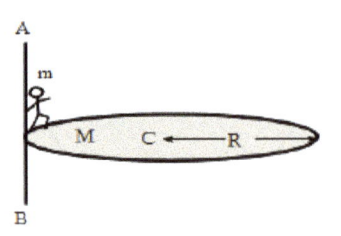

Solution
We have taken the disc in the plane of the paper (x–z plane) and the outward direction is the y-axis, which is the vertical axis of the given diagram in the problem. Let the boy (depicted as a red point) walk with a constant angular velocity relative to the centre C of the disc which is given as

$$\omega_{PC} = \frac{d\phi}{dt}.$$

The angular momentum of the boy is given as

$$L_b = m\left(v_r \sin\frac{\phi}{2} - r\omega\right)r$$

5-85

This is in the clockwise sense in the figure given in this section but in the figure given under the statement it points vertically downward. As both boy and disc were at rest initially, the initial angular momentum is zero. Since no net torque is applied on the disc–boy system, the net angular momentum of the system will be conserved, that is, zero. In order to counteract the angular momentum of the boy about the axis of rotation O, the disc will have to rotate in the opposite (counterclockwise) direction in this figure, but in the figure under the statement it points vertically up. The angular momentum of the disc is ω, its angular momentum about O is

$$L_d = \frac{3MR^2}{2}\omega$$

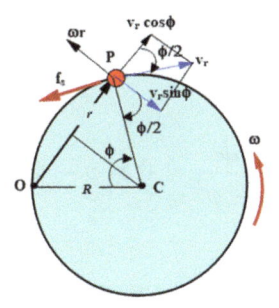

To conserve the angular momentum, equating the angular momenta of the disc and boy about O, we have

$$m\left(v_r \sin\frac{\phi}{2} - r\omega\right)r = \frac{3MR^2}{2}\omega$$

$$\Rightarrow \left(\frac{3MR^2}{2} + mr^2\right)\omega = mv_r r\sin\frac{\phi}{2}$$

$$\Rightarrow \left(\frac{3MR^2}{2} + mr^2\right)\frac{d\theta}{dt} = mv_r r\sin\frac{\phi}{2}$$

$$\Rightarrow \left(\frac{3MR^2}{2m} + r^2\right)\frac{d\theta}{dt} = v_r r\sin\frac{\phi}{2}$$

Putting $r = 2R\sin\frac{\phi}{2}$ and $v_r = R\omega_{PC} = R\frac{d\phi}{dt}$, we have

$$\left\{\frac{3MR^2}{2m} + \left(2R\sin\frac{\phi}{2}\right)^2\right\}\frac{d\theta}{dt} = R\frac{d\phi}{dt}\left(2R\sin\frac{\phi}{2}\right)\sin\frac{\phi}{2}$$

5-86

Separating the variables and simplifying the factors, we have

$$d\theta = \frac{\sin^2\frac{\phi}{2}d\phi}{\left\{\frac{3M}{4m} + 2\sin^2\frac{\phi}{2}\right\}}$$

Integrating both sides, the total angle rotated by the disc is given as

$$\theta = \int_0^\phi \frac{\sin^2\frac{\phi}{2}d\phi}{\left\{\frac{3M}{4m} + 2\sin^2\frac{\phi}{2}\right\}}$$

$$\theta = \frac{1}{2}\int_0^\phi \frac{\sin^2\frac{\phi}{2}}{\frac{3M}{8m} + \sin^2\frac{\phi}{2}} \simeq 0.5 \int_0^\phi \frac{(\phi/2)^2 d\phi}{\frac{3M}{8m} + (\phi/2)^2} \quad (\because \sin\frac{\phi}{2} \simeq \frac{\phi}{2}) \text{ for very small angle.}$$

Putting $3M/8m = 1$, we have

$$\theta = 0.5 \int_0^\phi \frac{\phi^2 d\phi}{4 + \phi^2} = 0.5\left\{\phi - 4\int_0^\phi \frac{d\phi}{4 + \phi^2}\right\}$$

After evaluating the integral,

$$\theta = 0.5\{\phi - 2\tan^{-1}(\phi/2)\}$$

Putting $\varnothing = s/R = 3/20 = 0.15$ rad $= 5.73°$, we have

$$\theta = 0.5\{11.46° - 2\tan^{-1}(0.2/2)\} = 0.02° \text{ (approximately)}.$$

Problem 43 A smooth ball of mass m is given a velocity v_0 from the mid-point of a chute made along a chord at a distance of a from the center of a disc of mass M and radius R. The disc is rotating with a constant angular velocity ω about a smooth vertical axis passing through its center. Find (a) the time after which the ball escapes from the disc, (b) the velocity of the ball at the time of leaving the disc, and (c) the work done by the external agent to rotate the disc with constant angular velocity.

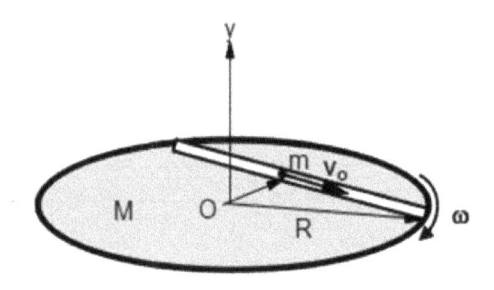

Solution

(a) At a radial distance r, the acceleration of the particle relative to the disc along the chute is

$$a_x = r\omega^2 \cos\theta = \omega^2 x \text{ (because } r\cos\theta = x).$$

Putting $a_x = \frac{v\,dv}{dx}$, cross multiplying, and integrating both sides,

$$\Rightarrow \int_{V_0}^{V} v\,dv = \omega^2 \int_0^x x\,dx$$

$$\Rightarrow v = \sqrt{v_0^2 + \omega^2 x^2}. \tag{5.52}$$

Then putting $v = dx/dt$, separating the variables, and integrating, we have

$$\int_0^{\sqrt{R^2 - a^2}} \frac{dx}{\sqrt{v_0^2 + \omega^2 x^2}} = \int_0^t dt$$

$$\Rightarrow t = \frac{\omega}{v_0} \ln \left| \frac{v_0 + \sqrt{v_0^2 + \omega^2 x^2}}{v_0} \right|$$

$$= \frac{\omega}{v_0} \ln \left| \frac{v_0 + \sqrt{v_0^2 + \omega^2(R^2 - a^2)}}{v_0} \right|.$$

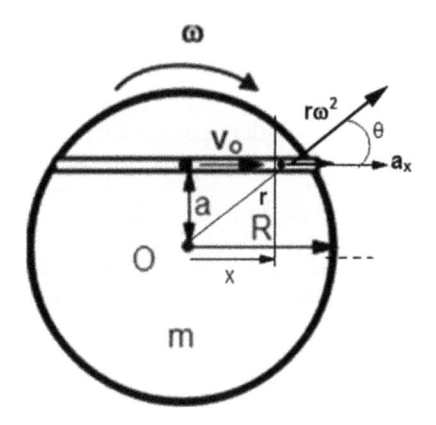

(b) In equation (5.52), putting

$$x^2 = R^2 - a^2,$$

the speed of the ball at the time of leaving the disc is

$$v = \sqrt{v_0^2 + \omega^2(R^2 - a^2)}.$$

(c) Then, the work done by the external agent is

$$W_{ext} = \Delta K = \frac{1}{2}m(v^2 - v_0^2)$$

$$= \frac{1}{2}m\omega^2(R^2 - a^2).$$

Problem 44 A long plank of mass M and length L has a velocity v_0 when it starts rolling on a grid of small cylinders each of mass m and radius R placed side by side. Each cylinder is free to rotate about a fixed horizontal axis. If there are n cylinders in a unit length, find (a) the velocity, (b) the acceleration of the plank as the function of distance x, (c) the displacement x of the plank as the function of time, and (d) the time required by the plank to traverse its entire length into the grid of cylinders. Assume that the mass and radius of each cylinder are m and r, respectively.

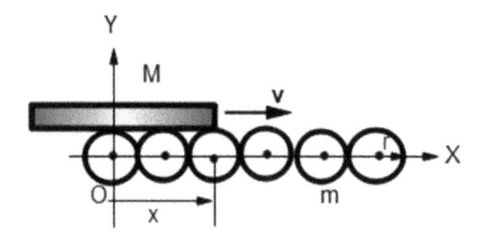

Solution
(a) Since n = the number of cylinders per unit length, the angular momentum of the cylinders touching the plank plus that of the plank is

$$L = (nx)mK^2\omega + Mvr,$$

where K = the moment of inertia of each cylinder.

Then the angular momentum of the plank–cylinder system can be conserved about the origin O. At any displacement x of the plank, conservation of angular momentum gives

$$Mv_0 r = (nx)mK^2\omega + Mvr$$

$$= nmK^2\frac{v}{r}x + Mvr$$

$$\Rightarrow Mv_0 r = \left(\frac{nmK^2}{r^2}x + M\right)vr$$

$$\Rightarrow v = \frac{Mv_0}{\frac{nmK^2}{r^2}x + M}.$$

(b) Then the acceleration of the plank is

$$a = dv/dt = -\frac{(Mv_0)^2 nm\frac{K^2}{r^2}}{\left(nm\frac{K^2}{r^2}x + M\right)^3}.$$

(c) Separating the variables and integrating both sides we have

$$Mv_0 t = \int_0^x \left(nm\frac{K^2}{r^2}x + M\right)dx$$

$$\Rightarrow 0 = \frac{nm}{4}x^2 + Mx - Mv_0 t \left(\because \frac{K^2}{r^2} = 1/2\right)$$

$$\Rightarrow x = \frac{-M \pm \sqrt{M^2 + Mv_0 nmt}}{\frac{nm}{2}}$$

$$\Rightarrow x = 2\left\{\sqrt{1 + \frac{nmv_0 t}{M}} - 1\right\}\frac{M}{nm}.$$

(d) When $x = L$ we have

$$t = \frac{M}{mnv_0}\left\{\left(\frac{Lmn}{2M} + 1\right)^2 - 1\right\}.$$

Problem 45 A disc with a moment of inertia of I_1 is spinning about a smooth vertical axle with an angular velocity w_1. Another disc with a moment of inertia I_2 spinning with an angular velocity w_1 is loaded coaxially onto the first disc. Find (a) the angular velocity of the combination of the discs in a steady state, (b) the angular impulse imparted to the top disc by the bottom one, and (c) the loss of kinetic energy of the combination of the discs.

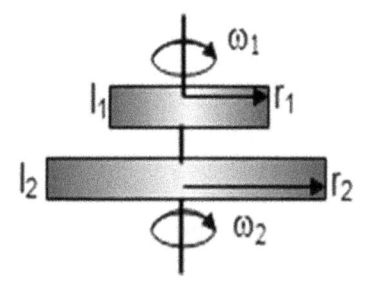

Solution

(a) The total angular momentum remains conserved about the axis of rotation. Since $L_f = L_i$ we have

$$I_1 \vec{\omega}_1 + I_2 \vec{\omega}_2 = I_1 \vec{\omega} + I_2 \vec{\omega}$$

$$\Rightarrow \vec{\omega} = \frac{I_1 \vec{\omega}_1 + I_2 \vec{\omega}_2}{I_1 + I_2}. \tag{5.53}$$

(b) The angular impulse on disc 1 is $\Delta \vec{L}_1 = I_1(\vec{\omega} - \vec{\omega}_1)$

$$= I_1 \left(\frac{I_1 \vec{\omega}_1 + I_2 \vec{\omega}_2}{I_1 + I_2} - \vec{\omega}_1 \right).$$

So the magnitude of angular impulse on each disc is

$$= \frac{I_1 I_2}{I_1 + I_2} |\vec{\omega}_1 - \vec{\omega}_2|.$$

(c) The change in kinetic energy is

$$\Delta K = -\left\{ \left(\frac{1}{2} I_1 \omega_1^2 + \frac{1}{2} I_2 \omega_2^2\right) - \frac{1}{2}(I_1 + I_2)\omega^2 \right\}. \tag{5.54}$$

Using equations (5.53) and (5.54),

$$\Delta K = -\frac{1}{2} \left\{ I_1 \omega_1^2 + I_2 \omega_2^2 - (I_1 + I_2) \frac{(I_1 \vec{\omega}_1 + I_2 \vec{\omega}_2)^2}{(I_1 + I_2)^2} \right\}$$

$$= -\frac{(I_1^2 \omega_1^2 + I_1 I_2 \omega_2^2 + I_1 I_2 \omega_1^2 + I_2^2 \omega_2^2 - I_1^2 \omega_1^2 - I_2^2 \omega_2^2 - 2 I_1 I_2 \vec{\omega}_1 \cdot \vec{\omega}_2)}{2(I_1 + I_2)}$$

$$= -\frac{I_1 I_2}{2(I_1 + I_2)}(\omega_1^2 + \omega_2^2 - 2\vec{\omega}_1 \vec{\omega}_2)$$

$$= -\frac{I_1 I_2}{2(I_1 + I_2)}(\vec{\omega}_1 - \vec{\omega}_2)^2.$$

Problem 46 With reference to the previous problem, (a) if the mass and radii of the discs are m, M, and R, r, respectively, find (i) the energy loss of the combination and (ii) the time after which the relative sliding between the discs will cease. (b) If the moment of inertia and angular velocities of the discs are I, $4I$, and $2\omega_0, \omega_0$, respectively, find the work done by friction on (i) disc 1, (ii) disc 2, and (iii) the total system of two discs.

Solution

(a)

(i) The net work done by friction is
$$W_f = -\Delta K,$$
where
$$\Delta K = \frac{1}{2} \frac{I_1 I_2}{(I_1 + I_2)} |\vec{\omega}_1 - \vec{\omega}_2|^2.$$
So
$$W_f = -\frac{I_1 I_2}{2(I_1 + I_2)} |\vec{\omega}_1 - \vec{\omega}_2|^2,$$
where $I_1 = \frac{mr^2}{2}$ and $I_2 = \frac{MR^2}{2}$
$$\Rightarrow W_f = -\frac{MmR^2r^2 |\vec{\omega}_1 - \vec{\omega}_2|^2}{2(MR^2 + mr^2)}.$$

(ii) To find the time at which relative sliding stops we need to find the frictional torque acting on each disc, which is given as
$$\tau_f = \frac{2\mu mgr}{3} \text{ (as derived earlier in the problem of chapter 3).}$$

If $\omega_1 > \omega_2$ a retarding torque acts on the upper disc. So its angular acceleration is
$$\alpha_1 = \frac{\tau_f}{I_1} = -\frac{2\mu_k mgr}{3\frac{1}{2}mr^2} = -\frac{4\mu_k g}{3r} \text{ (down)}.$$

The torque on the lower disc is accelerating which produces an angular acceleration
$$\alpha_2 = \frac{\tau_f}{I_2} = \frac{4\mu_k mgr}{3MR^2} \text{ (up)}.$$

Applying kinematics on the discs, their angular velocities after a time t are
$$\vec{\omega}'_1 = \vec{\omega}_1 + \vec{\alpha}_1 t \text{ and } \vec{\omega}'_2 = \vec{\omega}_2 + \vec{\alpha}_2 t, \text{ respectively.}$$
At $t = t$ let $\omega'_1 = \omega'_2$ as the relative sliding stops
$$\Rightarrow t(-\vec{\alpha}_1 + \vec{\alpha}_2) = \vec{\omega}_1 - \vec{\omega}_2$$

$$\Rightarrow t = \frac{|\vec{\omega_1} - \vec{\omega_2}|}{|\vec{\alpha_2} - \vec{\alpha_1}|} = \frac{|\omega_1 - \omega_2|}{\alpha_1 + \alpha_2}$$

$$\Rightarrow t = \frac{|\omega_1 - \omega_2|}{\frac{4}{3}\mu_k g R\left(1 + \frac{mr^2}{MR^2}\right)}.$$

Alternative method

(a) Applying the angular impulse–momentum equation,

$$\int_0^t \tau_1 dt = \Delta L_1,$$

where

$$\tau_1 = \tau_2 = \frac{2}{3}\mu_k mgr.$$

So $\frac{2}{3}\mu_k mgrt = I_1\omega_1 - I_1\omega = I_1(\vec{\omega_1} - \vec{\omega})$

$$= I_1\left\{\vec{\omega_1} - \frac{I_1\vec{\omega_1} + I_2\vec{\omega_2}}{(I_1 + I_2)}\right\}$$

$$\Rightarrow \frac{2}{3}\mu_k mgrt = \frac{I_1 I_2 |\vec{\omega_1} - \vec{\omega_2}|}{I_1 + I_2}$$

$$\Rightarrow \frac{2}{3}\mu_k mgrt = \frac{MmR^2r^2 |\vec{\omega_1} - \vec{\omega_2}|}{2(MR^2 + mr^2)}$$

$$\Rightarrow t = \frac{3 |\vec{\omega_1} - \vec{\omega_2}|}{4\mu_k gr\left(1 + \frac{mr^2}{MR^2}\right)}.$$

(b)

(i) If $I_1 = 4I_2 = I$, $\omega_1 = 2\omega_0$, and $\omega_2 = \omega_0$,

$$W_f = -\frac{I_1 I_2 |\vec{\omega_1} - \vec{\omega_2}|^2}{2(I_1 + I_2)} = -\frac{I(4I)|2\omega_0 - \omega_0|^2}{2(I + 4I)}.$$

So $W_f = -\frac{2I\omega_0^2}{5}$. Then

$$W_{f_1} = \Delta K_1 = \frac{1}{2I_1}(\omega^2 - \omega_1^2),$$

where

$$\vec{\omega} = \frac{I_1\vec{\omega_1} + I_2\vec{\omega_2}}{(I_1 + I_2)}$$

$$= \frac{I(2\omega_0) + (4I)\omega_0}{I + 4I} = \frac{6}{5}\omega_0$$

$$\Rightarrow W_{f_1} = -\frac{1}{2}(I)\left(4\omega_0^2 - \frac{36}{25}\omega_0^2\right) = -\frac{32}{25}I\omega_0^2.$$

(ii) $W_{f_2} = \Delta K_2 = \frac{1}{2}I_2\{\omega^2 - \omega_2^2\}$

$$= \frac{1}{2}(4I)\left\{\left(\frac{6}{5}\omega_0\right)^2 - \omega_0^2\right\} = 2I \times \frac{11}{25}\omega_0^2$$

$$\Rightarrow W_{f_2} = \frac{22I\omega_0^2}{25}.$$

(iii) Then

$$W_f = W_{f_1} + W_{f_2}$$

$$= \left(-\frac{32}{25} + \frac{22}{25}\right)I\omega_0^2 = -\frac{2}{5}I\omega_0^2.$$

Problem 47 Two rotating discs of mass M is loaded with another disc of mass m rotating in the opposite direction about same axis. Due to the effect of mutual frictional torque, the discs will stop relative sliding. Eventually they rotate with a common angular velocity w. (a) Find w. (b) After what time will the relative sliding stop between the discs? (c) What fraction of total kinetic energy of the discs is lost due to friction? Assume that R = the radius of each disc.

Solution
(a) We have $\vec{\omega} = \frac{I_1\vec{\omega_1} + I_2\vec{\omega_2}}{(I_1+I_2)}$ (as derived earlier). Putting the given values,

$$\omega = \frac{\left(\frac{MR^2}{2}\omega_0 - \frac{mR^2}{2}\omega_0\right)}{\frac{MR^2}{2} + \frac{mR^2}{2}}$$

$$= \frac{(M - m)}{M + m}\omega_0 \text{ (up)}.$$

(b) The angular impulse–momentum equation,

$$\tau t = \frac{MR^2}{2}\left\{\omega_0 - \frac{\omega_0(M-m)}{M+m}\right\}$$

$$|\tau t| = \frac{MR^2}{2}|\vec{\omega_0}|\frac{2m}{M+m} = \frac{MmR^2\omega_0}{M+m},$$

where the net frictional torque is

$$\tau = \frac{2\mu MgR}{3} \text{ (as derived earlier)}$$

$$t = \frac{3}{2\mu gR}\left\{\frac{Mm}{(M+m)}\right\}R^2\omega_0$$

$$\Rightarrow t = \frac{3MmR\omega_0}{2(M+m)\mu g}.$$

(c) The fraction loss of kinetic energy is

$$-\frac{\Delta K}{K} = \frac{\frac{1}{2}\frac{I_1 I_2}{I_1+I_2}|\vec{\omega_1}-\vec{\omega_2}|}{\frac{1}{2}(I_1+I_2)\omega_0^2}$$

$$= \frac{I_1 I_2}{(I_1+I_2)^2}\cdot\frac{|2\omega_0|^2}{\omega_0^2}$$

$$= \frac{4I_1 I_2}{(I_1+I_2)^2}\omega_0^2 = \frac{4\frac{MR^2}{2}\frac{mR^2}{2}}{\frac{MR^2}{2}+\frac{mR^2}{2}}\omega_0^2$$

$$\Rightarrow -\frac{\Delta K}{K} = \frac{2MmR^2\omega_0^2}{M+m}.$$

Problem 48 A disc with a moment of inertia I is spinning about a smooth vertical axle with an angular velocity $\vec{\omega}_1 = \omega_0 \hat{k}$. There are second and third discs with moments of inertia $2I$ and $3I$ spinning with angular velocities $\vec{\omega}_2 = -\omega_0/2\hat{k}$ and $\vec{\omega}_3 = \frac{\omega_0}{3}\hat{k}$, respectively, are loaded coaxially onto the first disc one by one. Find (a) the common angular velocity of the combination of discs and (b) the angular impulses acting on the discs (c) total work done by friction (d) the fraction of energy lost, after a long time. Assume that there is a negligible friction at the common axis of rotation of the discs.

Solution
(a) The initial angular momentum is
$$\vec{L_i} = I_1\vec{\omega_1} + I_2\vec{\omega_2} + I_3\vec{\omega_3}.$$
The final angular momentum is
$$\vec{L_f} = I_1\vec{\omega} + I_2\vec{\omega} + I_3\vec{\omega} = (I_1 + I_2 + I_3)\vec{\omega}.$$
Conserving the angular momentum $\vec{L_f} = \vec{L_i}$. Putting the values of initial and final angular momenta,
$$(I_1 + I_2 + I_3)\vec{\omega} = I_1\vec{\omega_1} + I_2\vec{\omega_2} + I_3\vec{\omega_3}$$
$$\Rightarrow \vec{\omega} = \frac{I_1\vec{\omega_1} + I_2\vec{\omega_2} + I_3\vec{\omega_3}}{(I_1 + I_2 + I_3)}$$
$$\Rightarrow \vec{\omega} = \frac{(I)\omega_0\hat{k} + (2I)(-\omega_0\hat{k}/2) + (3I)(\omega_0\hat{k}/3)}{(I + 2I + 3I)}$$
$$\Rightarrow \vec{\omega} = \frac{\omega_0\hat{k}}{6}.$$

(b) The impulses on the bodies are given in order as following:
$$I_1(\vec{\omega} - \vec{\omega_1}) = I\left\{\left(\frac{\omega_0}{6}\right) - (\omega_0)\right\}\hat{k} = -\frac{5\omega_0}{6}\hat{k}$$
$$I_2(\vec{\omega} - \vec{\omega_2}) = (2I)\left\{\left(\frac{\omega_0}{6}\right) - \left(-\frac{\omega_0}{2}\right)\right\}\hat{k} = \frac{4\omega_0}{3}\hat{k}$$
$$I_3(\vec{\omega} - \vec{\omega_3}) = (3I)\left\{\left(\frac{\omega_0}{6}\right) - \left(\frac{\omega_0}{3}\right)\right\}\hat{k} = -\frac{I\omega_0}{2}\hat{k}$$
$$I_1(\vec{\omega} - \vec{\omega_1}) = I\left\{\left(\frac{\omega_0}{6}\right) - (\omega_0)\right\}\hat{k} = -\frac{5\omega_0}{6}\hat{k}.$$

The net change in angular momentum $= 0$.

(c) The work done on the first, second, and third discs is given, respectively, as follows:
$$W_1 = \frac{I_1}{2}(\omega^2 - \omega^2)$$
$$= \frac{I}{2}\left\{\left(\frac{\omega_0}{6}\right)^2 - (\omega_0)^2\right\} = -\frac{35I}{72}I\omega_0^2$$

$$W_2 = \frac{I_2}{2}(\omega^2 - \omega'^2)$$

$$= \frac{I_1}{2}(2I)\left\{\left(\frac{\omega_0}{6}\right)^2 - \left(-\frac{\omega_0}{2}\right)^2\right\} = -\frac{2I}{9}\omega_0^2$$

$$W_3 = \frac{I_3}{2}(\omega^2 - \omega'^2)$$

$$= \frac{3I}{2}\left\{\left(\frac{\omega_0}{6}\right)^2 - \left(\frac{\omega_0}{3}\right)^2\right\} = -\frac{I\omega_0^2}{8}.$$

Then summing up all the above three equations of work done, the net work done by friction is $W_f = -\frac{5I\omega_0^2}{6}$.

(d) The total initial kinetic energy is

$$K = \frac{11I\omega_0^2}{12}.$$

Fraction of energy lost

$$-\Delta K/K = \frac{5I\omega_0^2/6}{11I\omega_0^2/12} = \frac{10}{11}.$$

Problem 49 A bead of mass m can slide along the horizontal circular ring of mass M and radius R by a hole made on it. Initially the bead is given a velocity v_0 and the ring is given an initial spin ω_0 as shown in the figure. Due to the friction between the ball and ring, the relative sliding stops after some time and the system $(M + m)$ will move with same angular velocity ω, say. Find the value of (a) the angular velocity of the bodies after a long time, (b) the energy loss of the system, and (c) the angular impulse on each object. Neglect the friction between the vertical axis and the light spokes connecting the ring (not seen in the figure).

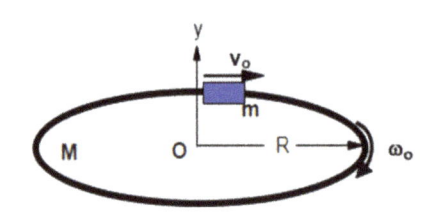

Solution

(a) The common angular velocity ω of M and m can be given by conserving the angular momentum of the system as follows:

$$mv_0 R + MR^2\omega_0 = (MR^2 + mR^2)\omega$$

$$\omega = \frac{mv_0 + MR\omega_0}{(M+m)R}.$$

(b) Then the change in kinetic energy is

$$\Delta K = \frac{1}{2}(M+m)R^2\omega^2 - \left(\frac{1}{2}mv_0^2 + \frac{M}{2}R^2\omega_0^2\right)$$

$$= \frac{1}{2}(M+m)R^2\left(\frac{mv_0 + MR\omega_0}{(M+m)R}\right)^2$$

$$- \left(\frac{1}{2}mv_0^2 + \frac{M}{2}R^2\omega_0^2\right).$$

After simplification we have

$$\Delta K = -\frac{Mm}{2(M+m)}|v_0 - R\omega_0|^2.$$

Alternative method

Let us recast the expression of the change in kinetic energy of a system of two rotating coaxial bodies as derived in problem 45:

$$\Delta K = -\frac{I_1 I_2}{2(I_1 + I_2)}|\vec{\omega}_1 - \vec{\omega}_2|^2$$

$$= -\frac{(mR^2)(MR^2)}{2(mR^2 + MR^2)}|(v_0/R) - \omega_0|^2$$

$$= -\frac{Mm}{2(M+m)}|v_0 - R\omega_0|^2.$$

(c) Angular impulse $J = \frac{I_1 I_2}{I_1 + I_2}|\vec{\omega}_1 - \vec{\omega}_2|$

$$= \frac{MmR^2}{M+m}|v_0/R - \omega_0|$$

$$= \frac{MmR}{M+m}|v_0 - R\omega_0|.$$

Problem 50 A small ball of mass m can slide along a horizontal spoke-less circular ring of mass M and radius R by a hole made on it. Initially the ball is given a velocity v_0 as shown in the figure. Due to the friction between the ball and ring, the relative sliding stops after some time and the system $(M + m)$ will move with same angular velocity w, say. Find (a) the steady state value of w, (b) the speed of the ball as a function of time, (c) the angular velocity of the ring as a function of time, and (d) the time of relative sliding. Neglect the effect of the gravity on the bead.

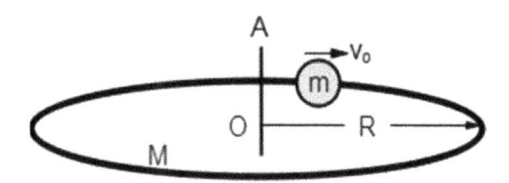

Solution

(a) Conserving the angular momentum after a long time when the system $(M + m)$ rotates with a common angular velocity w_f is given as

$$mv_0 R = (mR^2 + MR^2)w_f$$

$$\Rightarrow w_f = \frac{mv_0}{(M + m)R}. \tag{5.55}$$

(b) If $v =$ the velocity of the ball relative to the ground, the normal reaction between the ball and ring is $N = \frac{mv^2}{r}$. Then the kinetic friction acting on the ball is

$$f_B = \mu_k N = \mu_k \frac{mv^2}{R}.$$

This decelerates the ball providing a tangential acceleration

$$a_t = \frac{f_k}{m} = -\mu_k \frac{v^2}{R}.$$

Then the rate of change of speed of the ball is

$$\frac{dv}{dt} = -\mu_k \frac{v^2}{R}.$$

Separating the variables and integrating,

$$R \int_{v_0}^{v} \frac{dv}{v^2} = -\mu_k \int_0^t dt$$

5-99

$$\Rightarrow \left(\frac{1}{v} - \frac{1}{v_0}\right) = \mu_k t/R$$

$$\Rightarrow \frac{1}{v} = \frac{1}{v_0} + \mu_k \frac{t}{R}$$

$$\Rightarrow \frac{1}{v} = \frac{1 + \mu_k v_0 \frac{t}{R}}{v_0}$$

$$\Rightarrow v_f = \frac{Rv_0}{R + \mu_k v_0 t} \tag{5.56}$$

(c) The torque produced by the friction on the ring is

$$\tau_M = m\mu_k \frac{v^2}{R} \times R = m\mu_k v^2,$$

which gives an angular acceleration

$$\alpha_M = \tau_M/I = \frac{mv^2 \mu_k}{MR^2}$$

$$\Rightarrow \frac{d\omega}{dt} = \frac{\mu_k m}{MR^2}\left(\frac{Rv_0}{R + \mu_k v_0 t}\right)^2$$

$$\Rightarrow \int_0^\omega d\omega = \frac{m\mu_k R}{(MR^2)} \int_0^t \frac{dt}{(R + \mu_k v_0 t)^2}$$

$$\Rightarrow \omega = \frac{m\mu_k R^2 v_0^2}{MR^2 v_0 \mu_k}\left(\frac{1}{R} - \frac{1}{\mu_k v_0 t}\right)$$

$$\Rightarrow \omega = \frac{mv_0}{MR}\left(1 - \frac{R}{R + \mu_k v_0 t}\right). \tag{5.57}$$

(d) Since the final angular velocity of the ring and ball will be

$$\omega_f = \frac{mv_0}{(M + m)R}.$$

By using equations (5.55) and (5.57),

$$\Rightarrow \omega = \frac{mv_0}{(M + m)R} = \frac{mv_0}{MR}\left(1 - \frac{R}{R + \mu_k v_0 t}\right)$$

$$\Rightarrow \frac{R}{R + \mu_k v_0 t} = \frac{m}{M + m}$$

$$\Rightarrow \frac{R + \mu_k v_0 t}{R} = \frac{M + m}{m}$$

$$\Rightarrow \frac{\mu_k v_0 t}{R} = \frac{M}{m} \Rightarrow t = \frac{RM}{m \mu_k v_0}.$$

Alternative method

From equation (5.55) and (5.56),

$$\frac{m v_0}{M + m} = \frac{R v_0}{R + \mu_k v_0 t}$$

$$\Rightarrow R + \mu_k v_0 t = \frac{M + m}{m} R$$

$$\Rightarrow \mu_k v_0 t = \frac{M}{m} R \Rightarrow t = \frac{MR}{\mu_k m v_0}.$$

Problem 51 A man of mass m stands on a stationary disc of mass M and radius R at a radial distance r of the disc. The man starts walking in a circle of radius r with a velocity v_r relative to the disc. Find (a) the angular velocity of the disc, (b) the work done by friction on (i) the man, (ii) the disc, (c) the work done by the man, (d) the static friction acting between the man and disc, (e) the coefficient of static friction between the man and disc, and (f) if the disc stops after the man starts walking, find the initial angular velocity of disc. Assume sufficient friction between the man and disc to avoid slipping and $I = $ the moment of inertia of the disc.

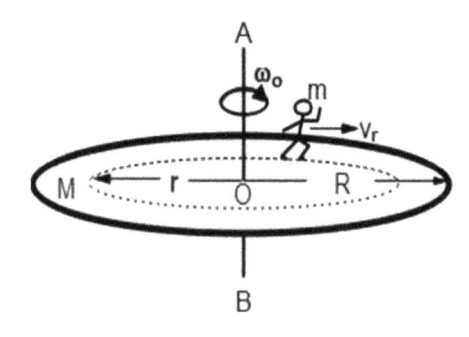

5-101

Solution

(a) The frictional torque will change the angular momentum of both the disc and man. However, the total angular momentum will be equal to the zero because initially the bodies were at rest. So $\vec{L_d} + \vec{L_m} = 0$.

After the man starts walking, the velocity of the man relative to the disc will be given as

$$v_m = (v_r - r\omega).$$

Then the angular momentum of the man is $L_m = mv_m r = m(v_r - r\omega)r$, which is vertically downwards. The disc will rotate in the opposite sense whose angular velocity is vertically up. The angular momentum of the disc is

$$L_d = I_d \omega = I\omega$$

Conserving the angular momentum of the disc–man system,

$$I\omega\hat{j} - m(v_r - r\omega)r\hat{j} = 0$$

$$\Rightarrow \omega = \frac{mv_r r}{I + mr^2}.$$

(b)

(i) Putting the velocity of the man (relative to ground)

$$v_m = \frac{I\omega}{mr},$$

in the expression of the work–energy theorem, the work done by friction on the man is

$$W_1 = \frac{1}{2}mv_m^2 = \frac{1}{2}m\left(\frac{I\omega}{mr}\right)^2 = \frac{I^2\omega^2}{2mr^2}.$$

Putting the obtained value of ω, we have

$$W_1 \frac{I^2}{2mr^2} \frac{m^2 v_r^2 r^2}{(I + mr^2)^2}$$

$$\Rightarrow W_1 = \frac{mv_r^2}{2\left(1 + \frac{mr^2}{I}\right)^2}.$$

Putting the obtained value of $I = MR^2/2$,

$$W_1 = \frac{mv_r^2}{2} \frac{1}{(1 + 2mr^2/MR^2)^2}.$$

(ii) The work done by the friction on the disc in increasing its kinetic energy is

$$W_2 = \frac{I\omega^2}{2} = \frac{I}{2}\left(\frac{mv_r r}{I + mr^2}\right)^2$$

$$= \frac{mv_r^2 r^2}{2I} \cdot \frac{1}{\left(1 + \frac{mr^2}{I}\right)^2}$$

$$\Rightarrow W_2 = \frac{mv_r^2 r^2}{MR^2}\left(\frac{1}{1 + 2mr^2/MR^2}\right)^2 \text{ obtained after putting } I = \frac{mR^2}{2}.$$

(c) The total work done by the man is

$$W_m = \Delta K = \frac{1}{2}I\omega^2 + \frac{1}{2}mv_m^2$$

$$= \frac{I}{2}\omega^2 + \frac{1}{2}m\left(\frac{I\omega}{mr}\right)^2 = \frac{I}{2}\omega^2\left\{1 + \frac{I}{mr^2}\right\}$$

$$= \frac{I}{2}\left(\frac{mv_r r}{I + mr^2}\right)^2\left(\frac{I + mr^2}{mr^2}\right)$$

$$= \frac{I}{2(I + mr^2)}(mv_r^2)$$

$$\Rightarrow W_m = \frac{mv_r^2}{2\left(1 + \frac{mr^2}{I}\right)}.$$

(d) Here we should remember that the net work done by friction (in the case of walking, static friction) is zero. So, ultimately, the man (internal elastic forces etc) is responsible for a change in the kinetic energy of the system. The static friction acting between the man and disc is

$$f_s = mv_m^2/r,$$

where

$$v_m = \frac{I\omega}{mr}$$

$$\Rightarrow f_s = m\frac{I^2\omega^2}{m^2 r^3},$$

where

$$\omega = \frac{mv_r r}{I + mr^2}$$

$$\Rightarrow f_s = \frac{mI^2}{m^2r^3}\left\{\frac{m^2v_r^2 r^2}{(I+mr^2)^2}\right\}$$

$$= \frac{mv_r^2}{r\left(1+\frac{mr^2}{I}\right)^2}.$$

(e) Since $f_s \leq \mu_s mg$

$$\Rightarrow \mu_s \geq \frac{f_s}{mg}$$

$$\Rightarrow \mu_s \geq \frac{v_r^2}{\left\{gr\left(1+\frac{mr^2}{I}\right)^2\right\}}.$$

(f) If the disc has an initial angular velocity of w_0 and the disc will come to rest just after walking, then putting

$$\omega = \omega_0 - \frac{mv_r r}{I+mr^2} = 0.$$

We obtain the value of initial angular velocity

$$\omega_0 = \frac{mv_r r}{I+mr^2}.$$

Problem 52 A heavy disc of mass m and radius R rotates about the vertical axis with an angular velocity w. A coin of mass m and radius r is gently placed on the disc at a distance d from the axis of rotation of the disc. If the coefficient of friction between the coin and disc is μ, find (a) the final angular velocity of the coin and (b) the torque required to maintain the constant angular velocity of the disc.

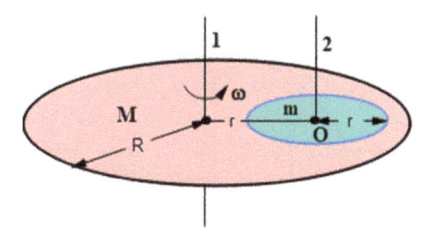

Solution
We have taken the disc in the plane of the paper (x–z plane; x-axis points to right and z-axis points down) and the outward direction is the y-axis which is the vertical axis of the given diagram in the statement of the problem. Initially the angular

velocity of the disc is zero as it is just lowered on to the turntable which is made to rotate with a constant angular velocity w. Here, we should note that the axis of rotation of the disc is made fixed by a smooth vertical axle. We mention that the disc is named as '1' and the turntable is named as '2'. At time $t = 0$ (just after placing the disc on the turntable), let us take a point 1 on the disc and another point 2 on the turntable in contact with the point 1. Since the disc is stationary (not rotating) at $t = 0$, the velocity of 1 relative to 2 is equal to velocity of 1 relative to the centre C of the turntable. Since the point C is at rest, $v_{12} = v_{PC} = v_P$ at P and $v_{12} = v_{QC} = v_Q$ at Q. This means that at P and Q the kinetic friction df and df' must act in the direction of v_P and v_Q respectively, as shown in the figure. Then, it is easy to see that these two elementary frictional forces df and df' acting at P and Q of the disc, respectively, will produce a counterclockwise torque. Similarly, considering the torques of such elementary friction pairs about O, we can understand that the disc will begin to rotate under the action of a counterclockwise torque, in the direction of rotation of the turntable.

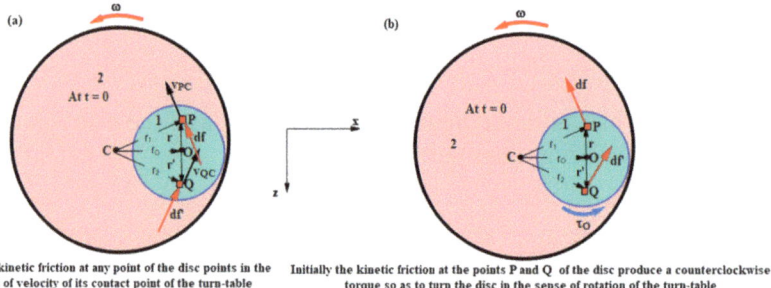

Initially the kinetic friction at any point of the disc points in the direction of velocity of its contact point of the turn-table

Initially the kinetic friction at the points P and Q of the disc produce a counterclockwise torque so as to turn the disc in the sense of rotation of the turn-table

After the disc attains an angular velocity w', say, the velocity of turntable relative to the disc at any point P, say, is given as

$$\vec{v}_{21} = \vec{v}_2 - \vec{v}_1 = \vec{\omega} x \vec{r}_1 - \vec{\omega}' x \vec{r}.$$
$$= \vec{\omega} x (\vec{r}_{CO} + \vec{r}) - \vec{\omega}' x \vec{r} = \vec{\omega} x \vec{r}_{CO} + \vec{\omega} x \vec{r} - \vec{\omega}' x \vec{r}$$
$$= \vec{\omega} x \vec{r}_{CO} + \vec{\omega} x \vec{r} - \vec{\omega}' x \vec{r} = \vec{\omega} x \vec{r}_{CO} + (\vec{\omega} - \vec{\omega}') x \vec{r}$$

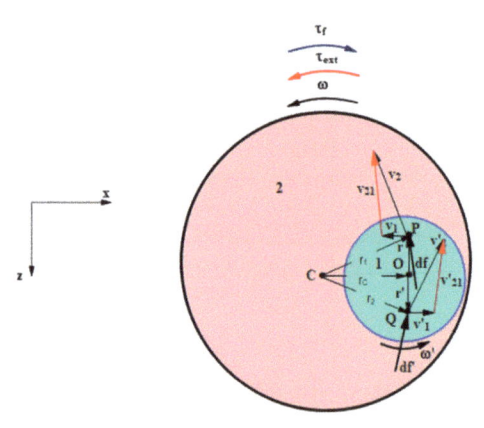

Before reaching the steady state, the kinetic friction df at P and and df' at Q will produce a gradually reducing counterclockwise torque about O such that the angular velocity increses from zero to ω'

As the disc is continually acted upon by a torque in the direction of the rotation of the turntable, the angular velocity for the disc will go on increasing till it will be equal to that of the turntable ($\vec{\omega}' = \vec{\omega}$). Then the second term of the last equation, that is, $(\vec{\omega} - \vec{\omega}') \times \vec{r} = 0$. Then we have

$$\vec{v}_{12} = \vec{\omega} \times \vec{r}_{CO} = \omega d \hat{k}$$

This tells us that at the steady state ($\vec{\omega}' = \vec{\omega}$), the relative velocity between the turntable and the disc will be same (in both magnitude and direction) at all contact points, as shown in the figure.

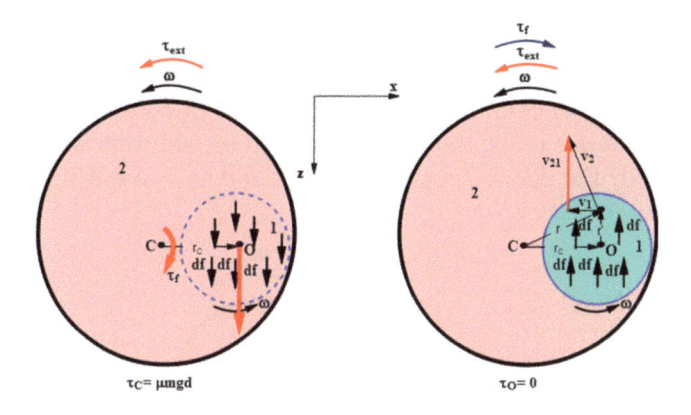

The velocity of the turntable relative to the disc is given as

$$\vec{v}_{21} = -\vec{v}_{12} = -\omega d \hat{k}$$

As the kinetic friction opposes the relative sliding, the elementary frictions df at each point of the disc must oppose $\vec{v}_{12}$ or favour $\vec{v}_{21}$. Then, the all elementary friction 'df' will point upward in the direction of $\vec{v}_{12}$. As all the elementary frictions are parallel, their net torque about the center C of the disc will be zero. So, thereafter the disc will move with the terminal angular velocity which is equal to the angular velocity ω of the turntable. In this steady state, the net friction f acting at the center of mass O of the disc in the upward direction and an equal and opposite (downward) friction acts on the turntable at the contact point of O at a distance of $r_O = d$ which is given as

$$\vec{f} = \int d\vec{f} = -\mu m g \hat{k}$$

This friction produces a clockwise torque against the applied or external torque. As the turntable is made to rotate with constant angular velocity by the external agent, its angular acceleration is zero; then the net torque acting on the turntable is zero. So, the external or applied torque on the turntable is equal and opposite to the frictional torque which is given as

$$\vec{\tau}_{Ext} = - \vec{\tau}_f = \mu mg \hat{j}$$

So, the external torque is a counterclockwise torque when we view on the paper or computer screen. But it is directed vertically up when we refer to the figure given in the statement of the problem. In other words, when we view the rotating turntable from above, the external torque will be counterclockwise.

Problem 53 A uniform disc of mass m and radius r is smoothly pivoted at its center O by a uniform rod of mass m and length R. The rod and disc can rotate about a smooth vertical axes passing through the points A and O, respectively, while they lie in the horizontal plane (the plane of the page). The initial angular velocities of the rod and disc are given as w_1 and w_2, respectively. If the disc slows down to half of its initial angular speed due to the friction at O, at that instant what will be the angular velocity of the rod? Put $R/r = 2$, $w_1 = 3$ rad s^{-1}, and $w_2 = 4$ rad s^{-1}.

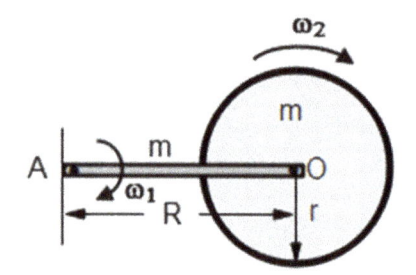

Since the angular velocity of the rod decreases to $\frac{w_1}{2}$, let the angular velocity of the disc be $\vec{\omega}' = - w'\hat{k}$ (clockwise, say), so that the total angular momentum remains conserved. Then, the final angular momentum is

$$\vec{L}_f = \vec{L}_{rod} + \vec{L}_{disc} = - \frac{mR^2 w'}{3}\hat{k} + \frac{mr^2(w_2/2)}{2}(-\hat{k}) - mR(Rw')\hat{k}$$

$$\Rightarrow \vec{L}_f = = - \left(\frac{mR^2 w'}{3} + \frac{mr^2 w_2}{4} + mR^2 w' \right)\hat{k}$$

5-107

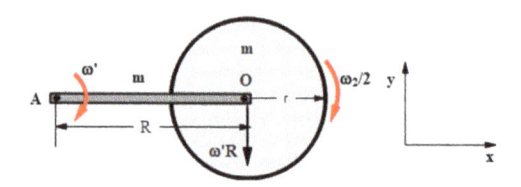

Equating the initial and final angular momentum about A, we have,

$$\Rightarrow -\left(\frac{mR^2\omega'}{3} + \frac{mr^2\omega_2}{4} + mR^2\omega'\right)\hat{k} = -\left(\frac{mR^2\omega_1}{3} + \frac{mr^2\omega_2}{2} + mR^2\omega_1\right)\hat{k}$$

$$\Rightarrow \left(\frac{4R^2\omega'}{3}\right) = \left(\frac{4R^2\omega_1}{3} + \frac{r^2\omega_2}{4}\right)$$

$$\Rightarrow \omega' = \frac{16\omega_1 + 3(r/R)^2\omega_2}{16}$$

$$\Rightarrow \omega' = \frac{16(3) + 3(1/2)^2(4)}{16} = \frac{51}{16} \text{ rad s}^{-1}$$

Positive sign of the result shows that the rod will rotate in the assumed clockwise direction.

Problem 54 In the previous problem if the rod has an anticlockwise spin ω_1 so that the net angular momentum is zero about the axis of rotation of the rod, (a) find the angular velocity ω_2 of the disc. (b) If the disc slows down to half of its initial angular speed due to the friction at O, at that instant what will be the angular velocity of the rod?

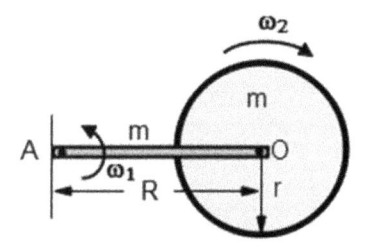

Solution

(a) For the total angular momentum to be zero, $\frac{mR^2}{3}\vec{\omega_1} + \frac{mr^2}{2}\vec{\omega_2} + mR^2\vec{\omega_1} = 0$

$$\Rightarrow 4\vec{\omega_1}\frac{R^2}{3} + \frac{r^2}{2}\vec{\omega_2} = 0$$

$$\Rightarrow 4\frac{\omega_1 R^2}{3}\hat{k} - \frac{r^2}{2}\omega_2\hat{k} = 0$$

$$\Rightarrow \frac{\omega_1}{\omega_2} = \frac{3}{8}\left(\frac{r}{R}\right)^2 = \frac{3}{8}\left(\frac{r}{2r}\right)^2 = \frac{3}{32}$$

$\Rightarrow \omega_2 = \frac{32}{3}\omega_1 = \frac{32}{3} \times 3 = 32$ rad s^{-1} in a clockwise sense.

(b) For the new angular velocities, $\vec{\omega'}_1$, $\vec{\omega'}_2$, the total angular momentum is

$$\frac{mR^2}{3}(\vec{\omega'}_1) + \frac{mr^2}{2}\vec{\omega'}_2 + mR^2\vec{\omega'}_1 = 0$$

5-108

$$\Rightarrow \frac{\omega'_1}{\omega'_2} = \frac{3}{32}.$$

Putting $\omega'_2 = \frac{\omega_2}{2} = 32/2 = 16$ we have

$$\Rightarrow \omega'_1 = \frac{3}{2} \text{ rad s}^{-1}$$

Problem 55 A uniform disc of mass M and radius R is pivoted at its center O by a uniform rod of mass m and length R as shown in the figure. The rod and disc can rotate about a smooth vertical axis passing through the point O while they lie in the horizontal plane (the plane of the page). The magnitude of the initial angular velocities of the rod and disc is given as ω_0. If there is friction between the disc and rod, find (a) the common angular velocity in a steady state and (b) common angular velocity if $m/M = 2, 6, 1$ and 18, (c) the fraction loss of kinetic energy of the system (disc + rod) putting the four given values of m/M, (d) torque acting on the disc, (e) time after which the relative sliding between the disc and rod will stop.

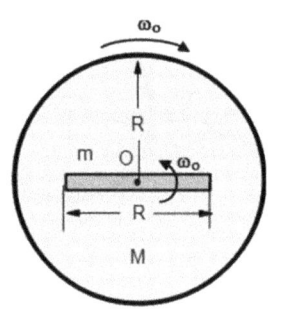

Solution
(a) Let the common angular velocity be ω. By conserving angular momentum,

$$\vec{L_f} = \vec{L_i}$$

$$\Rightarrow \left(\frac{MR^2}{2} + \frac{mR^2}{12}\right)\vec{\omega} = \left(\frac{mR^2}{12}\omega_0 - \frac{MR^2}{2}\omega_0\right)\hat{k}$$

$$\Rightarrow \frac{R^2}{2}\left(M + \frac{m}{6}\right)\vec{\omega} = \frac{R^2}{2}\omega_0\left(\frac{m}{6} - M\right)$$

$$\Rightarrow \vec{\omega} = \frac{\omega_0 \hat{k}(m - 6M)}{(m + 6M)} = \left\{\frac{\frac{m}{M} - 6}{\frac{m}{M} + 6}\right\}\omega_0 \hat{k}.$$

(b) If $\frac{m}{M} = 2$,

$$\vec{\omega} = \frac{2-6}{2+6}\omega_0\hat{k} = -\frac{4}{8}\omega_0\hat{k} = -\frac{\omega_0}{2}\hat{k}.$$

If $\frac{m}{M} = 6$, $\vec{\omega} = \frac{6-6}{6+6}\omega_0\hat{k} = 0\hat{k}.$

If $\frac{m}{M} = 18$, $\vec{\omega} = \frac{18-6}{18+6}\omega_0\hat{k} = \frac{12}{24}\omega_0\hat{k} = \frac{\omega_0}{2}\hat{k}.$

If $\frac{m}{M} = 1$, $\vec{\omega} = \frac{1-6}{1+6}\omega_0\hat{k} = -\frac{5}{7}\omega_0\hat{k}.$

(c) $\Delta K = \frac{I_1 I_2}{2(I_1 + I_2)} |\vec{\omega}_1 - \vec{\omega}_2|^2$

$$\Delta K = -\frac{I_1 I_2}{2(I_1 + I_2)}|\omega_0 + \omega_0|^2 = -\frac{2I_1 I_2}{(I_1 + I_2)}|\omega_0|^2$$

and

$$K = \frac{I_1 + I_2}{2}|\omega_0|^2.$$

Then the fraction loss of energy is

$$\eta = -\frac{\Delta K}{K} = \frac{2I_1 I_2}{(I_1 + I_2)}|\omega_0|^2 \bigg/ \frac{(I_1 + I_2)}{2}|\omega_0|^2 = \frac{4I_1 I_2}{(I_1 + I_2)^2}$$

$$= \frac{4\left(\frac{MR^2}{2}\right)\left(\frac{mR^2}{12}\right)}{\left(\frac{MR^2}{2} + \frac{mR^2}{12}\right)^2} = \frac{24Mm}{(6M + m)^2}.$$

Then $\frac{\Delta K}{K} = \frac{24Mm}{(6M+m)^2} = \frac{24m/M}{(6+m/M)^2}.$

Putting $\frac{m}{M} = 2, 6, 18, 1$, the fraction loss of energy $= \eta = -\frac{\Delta K}{K} = 3/4, 1,$ 3/4, and 24/49, respectively.

(d) The torque exerted by the friction df acting on the element dm of the rod, about O, is

$$d\tau_f = x df,$$

where $df = \mu(dm)g$ and $dm = (m/R)dx.$

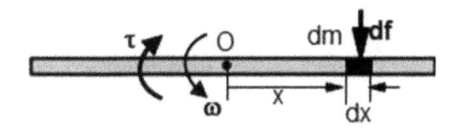

5-110

Then, integrating, the frictional total torque is

$$T_f = 2\int_0^{R/2} x\mu g \frac{m}{R}\, dx = \frac{2\mu g m}{R}dx\cdot\frac{x^2}{2}\Big|_0^{R/2}$$

$$= \frac{m\mu g}{R} \times \frac{R^2}{4} = \frac{\mu m g R}{4}.$$

(e) Then the angular acceleration of the rod is $\alpha = \dfrac{T_f}{I_{rod}} = \dfrac{\mu m g R/4}{(mR^2/12)} = \dfrac{3\mu g}{R}$.
The time of relative sliding of the rod is

$$\vec{w}_f = \vec{w}_i + \vec{\alpha} t.$$

By substituting the obtained values for angular velocity and acceleration of the rod,

$$\frac{w_0(m - 6M)}{(m + 6M)}\hat{k} = w_0\hat{k} - \frac{3\mu g}{R}\hat{k}t.$$

The direction of angular acceleration is opposite to the initial angular velocity of the rod. Therefore, it is taken as negative (clockwise):

$$\Rightarrow t = \frac{4MRw_0}{(m + 6M)\mu g}.$$

Problem 56 A uniform disc of mass m_1 and radius R is pivoted smoothly with another uniform disc of mass m and length r at a distance of a from the axis of rotation A of the bigger disc. The discs can rotate about smooth vertical axes A and B passing through its center of mass while both discs lie in the horizontal plane, as shown in the figure. The initial angular velocities of the disc are given as w_1 and w_2, respectively. Due to the friction between the discs they rotate with a common angular velocity w. Assume that the rod is smoothly pivoted with the fixed axis and the small disc rotate about the axis B which is fixed relative to the bigger disc. Find (a) the final (steady state) angular velocities of the bodies and (b) the fraction loss of the kinetic energy of the system. Put $w_1 = 4$ rad s^{-1} and $w_2 = 2$ rad s^{-1}, $R = 1$ m, $a = OC = 0.5$ m, $r = 0.25$ m, $m_1 = 10$ kg, and $m_2 = 32$ kg.

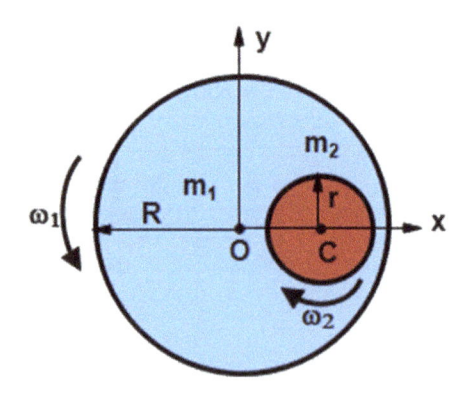

5-111

Solution

(a) The initial angular momentum is

$$\vec{L_i} = \left\{\frac{m_1 R^2}{2}\omega_1 \hat{k}\right\} + \frac{m_2 r^2}{2}\omega_2(-\hat{k}) + m_2 a v_c \hat{k}$$

$$= \left\{\frac{m_1 R^2}{2}\omega_1 - \frac{m_2 r^2 \omega_2}{2} + m_2 a(\omega_1 a)\right\}\hat{k}$$

$$= \left(\frac{m_1 R^2}{2}\omega_1 - \frac{m_2 r^2 \omega_2}{2} + m_2 a^2 \omega_1\right)\hat{k}$$

$$= \left(\frac{10 \times 1 \times 4}{2} - \frac{32}{2} \times \frac{1}{16} \times 2 + 32 \times \frac{1}{4} \times 4\right)\hat{k}$$

$$= 50\hat{k} \text{ kg m}^2 \text{ s}^{-1}.$$

After the relative sliding stops the moment of inertia of the system about the axis of rotation of the bigger disc is

$$I_0 = \left(\frac{m_1 R^2}{2}\right) + \left(\frac{m_2 r^2}{2} + m_2 a^2\right)$$

$$= \frac{10 \times 1}{2} + 32\left(\frac{1}{16 \times 2} + \frac{1}{4}\right)$$

$$= 5 + 9 = 14 \text{ kg m}^2.$$

Then $\vec{\omega} = \frac{\vec{L_i}}{I_0} = \frac{50}{14}\hat{k} = \frac{25}{7}\hat{k}$ rad s^{-1}.

(b) The change in kinetic energy is

$$|\Delta K| = |K_i - K_f|,$$

where

$$K_i = \frac{1}{2}\left\{\frac{m_1 R^2}{2}\omega_1^2 + \frac{m_2 r^2}{2}\omega_2^2 + m_2 v_C^2\right\}$$

$$= \frac{1}{2}\left\{\frac{m_1 R^2 \omega_1^2}{2} + \frac{m_2 r^2 \omega_2^2}{2} + m_2 a^2 \omega_1^2\right\}$$

$$= \frac{1}{2}\left\{\left(\frac{m_1 R^2}{2} + m_2 a^2\right)\omega_1^2 + \frac{m_2 r^2 \omega_2^2}{2}\right\} = \frac{1}{2}\left[\left\{\frac{10 \times 1}{2} + 32\left(\frac{1}{4}\right)\right\}(16) + \left\{32\left(\frac{1}{16}\right)(1/2)(4)\right\}\right]$$

$$= 106 \text{ J}.$$

The final kinetic energy is

$$K_f = \frac{L^2}{2I_f} = \frac{(50)^2}{2(14)} = 625/7 \text{ J}.$$

The loss of kinetic energy is

$$K_i - K_f = 106 - 625/7 = (117/7) \text{ J}.$$

Then the fraction loss of kinetic energy is

$$(K_i - K_f)/K_i = (117/7)/106 = 117/742.$$

Problem 57 A disc with a groove PQ can rotate freely in a horizontal plane about the vertical axis AB. A boy of mass m begins to walk with a constant velocity v_r relative to the groove from the position P, as shown in the figure, find the angle rotated by the disc when the boy reaches the other end Q of the groove.

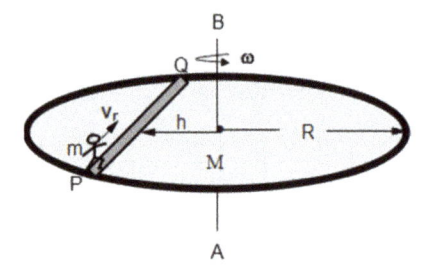

Solution
When the boy walks through a distance x relative to the disc, the disc will rotate in the opposite sense to the sense of turning of the boy (depicted as a point) relative to the axis so as to conserve the angular momentum of the disc–boy system. The components of velocity of the boy along and perpendicular to the groove are $(v_r - \omega r \cos\theta)$ and $(\omega r \sin\theta)$, respectively. Considering the angular momenta of these two components, the total angular momentum of the system is equal to the initial angular momentum, that is, zero. So, $\vec{L} = -m(v_r - \omega r \cos\theta)h\hat{K} + (m\omega r \sin\theta)x\hat{K} + Mk^2\omega\hat{K} = 0.$

5-113

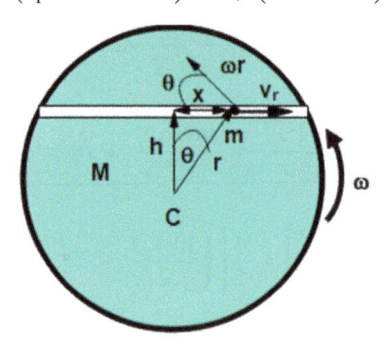

Putting $r\cos\theta = h$ and $\sin\theta = x$

$$\Rightarrow \omega = \frac{mv_r h}{m(h^2 + x^2) + MK^2}$$

$$\Rightarrow \frac{d\theta}{dt} = \frac{mv_r h}{mh^2 + MK^2 + mv_r^2 t^2}$$

$$\Rightarrow d\theta = \frac{mv_r h\, dt}{mh^2 + MK^2 + mv_r^2 t^2}$$

$$\Rightarrow \int_0^\theta d\theta = \int_0^t \frac{mv_r h\, dt}{mh^2 + MK^2 + mv_r^2 t^2}$$

$$\Rightarrow \theta = mv_r h\left(\frac{1}{ab}\right)\tan^{-1}\frac{bt}{a},$$

where $mh^2 + MK^2 = a^2$ and $mv_r^2 = b^2$.
Then we have

$$\theta = \sqrt{\frac{mh^2}{mh^2 + MK^2}}\,\tan^{-1}\sqrt{\frac{m(R^2 - h^2)}{mh^2 + MK^2}}.$$

Putting $K^2 = \frac{R^2}{2}$, $h^2 = \frac{R^2}{2}$, and $M = m$, we have

$$\theta = \sqrt{\frac{m\frac{R^2}{2}}{m\frac{R^2}{2} + m\frac{R^2}{2}}}\,\tan^{-1}\sqrt{\frac{m\left(R^2 - \frac{R^2}{2}\right)}{m\frac{R^2}{2} + \frac{mR^2}{2}}} \Rightarrow \theta = \frac{1}{\sqrt{2}}\tan^{-1}\frac{1}{\sqrt{2}}.$$

Conservation of energy and angular momentum

Problem 58 A disc of mass M having a diametrical groove can rotate freely in a horizontal plane about a vertical axis passing through its center. Two small spheres each of mass m are released from rest from the center of the disc. Let us give an initial angular velocity ω_0 to the disc. Find (a) the angular velocity ω of the disc, (b) the velocity of the spheres relative to (i) the ground and (ii) the disc when the balls reach the end of the groove, and (c) the net work done by the internal forces acting on the balls till they reach the ends of the groove. Put $M = 2m$.

Solution

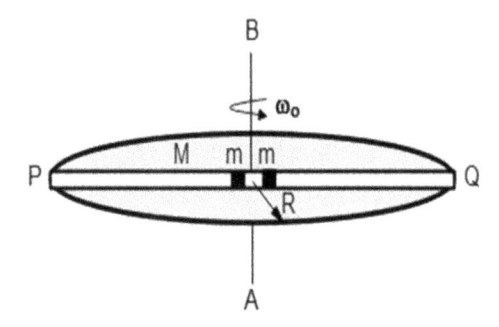

(a) Let the disc have angular velocity ω and the velocity of the balls relative to ground be v when the balls reach the end of the groove.

Conserving the angular momentum of the disc–ball system, we have

$$\omega\left(2mR^2 + \frac{MR^2}{2}\right) = \frac{MR^2}{2}\omega_0$$

$$\Rightarrow \omega = \left(\frac{M}{M + 4m}\right)\omega_0. \tag{5.58}$$

(b)

(i) Conserving kinetic energy,

$$2 \times \frac{m}{2}v^2 + \frac{1}{2}\frac{MR^2}{2}\omega^2 = \frac{M}{4}R^2\omega_0^2$$

$$\Rightarrow mv^2 = \frac{MR^2}{4}(\omega_0^2 - \omega^2). \tag{5.59}$$

Using equations (5.58) and (5.59) we have

$$mv^2 = \frac{MR^2}{4}\omega_0^2\left[1 - \left(\frac{M}{M+4m}\right)^2\right]$$

$$mv^2 = \frac{2(M + 2m)4m}{(M + 4m)^2}\left(\frac{MR^2\omega_0^2}{4}\right)$$

$$\Rightarrow v = \left\{\frac{\sqrt{2(M + 2m)M}}{M + 4m}\right\}R\omega_0.$$

(ii) If $M = 2m$,

$$v = \frac{\sqrt{2 \times 4m \times 2m}}{6m}R\omega_0$$

5-115

$$= \frac{4m}{6m} R\omega_0 = \frac{2}{3} R\omega_0.$$

If v_r is the velocity of the balls relative to the disc, conserving kinetic energy, we have

$$\frac{L^2}{2I} + 2 \times \frac{mv_r^2}{2} = \frac{L^2}{2I_0}.$$

Then

$$2 \times \frac{mv_r^2}{2} = \frac{L^2}{2I_0}\left(1 - \frac{I_0}{I}\right)$$

$$= \frac{MR^2}{2}\omega_0^2 \left(1 - \frac{\frac{MR^2}{2}}{\frac{MR^2}{2} + 2mR^2}\right)$$

$$mv_r^2 = \frac{MR^2\omega_0^2}{2}\left\{\frac{2 \times 2mR^2}{(M + 4m)R^2}\right\}$$

$$\Rightarrow v_r = \left\{\sqrt{\frac{2M}{M+4m}}\right\} R\omega_0 = \frac{R\omega_0}{\sqrt{3}}.$$

Alternative method
The velocity of the balls relative to the disc is given as

$$v_r^2 + R^2\omega^2 = v^2.$$

Putting $v = \frac{2}{3} R\omega_0$ and $\omega = \frac{1}{3}\omega_0$ in the last expression, we can obtain the same answer.

(c) The net work done by the internal forces acting on the balls until they reach the ends of the groove is

$$W_{\text{int}} = 2(mv^2/2) = mv^2 = m\left(\frac{2R\omega_0}{3}\right)^2$$

$$= \frac{4mR^2\omega_0^2}{9}.$$

Problem 59 A ring of mass M and radius R has two light spokes OP and OQ along which two identical smooth beads of mass m can slide. The ring is free to rotate in a horizontal plane about the vertical axis AB. If we give an initial spin ω_0, find the velocity of each bead when they reach the perimeter of the ring relative to (a) the ring and (b) ground. Assume that initially the beads are near the axis and $M = m$.

5-116

Solution

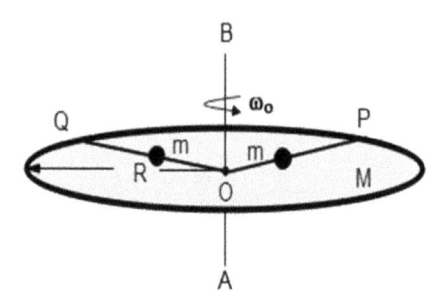

(a) Let the disc have angular velocity ω and the velocity of the balls relative to ground be v when the balls reach the end of the groove.

Conserving angular momentum,

$$L = (2mR^2 + MR^2)\omega = MR^2\omega_0$$

$$\Rightarrow \omega = \frac{M\omega_0}{M + 2m} = \frac{\omega_0}{3}.$$

If v_r is the velocity of the balls relative to the disc, conserving kinetic energy, we have

$$\frac{L^2}{2I} + 2 \times \frac{mv_r^2}{2} = \frac{L^2}{2I_0}$$

$$\Rightarrow 2 \times \frac{mv_r^2}{2} = \frac{L^2}{2I_0}\left(1 - \frac{I_0}{I}\right)$$

$$\Rightarrow mv_r^2 = \frac{MR^2}{2}\omega_0^2\left(1 - \frac{MR^2}{(2m + M)R^2}\right)$$

$$\Rightarrow mv_r^2 = \frac{MR^2}{2}\omega_0^2\left(\frac{2m}{2m + M}\right)$$

$$\Rightarrow v_r = \frac{R\omega_0}{\sqrt{2\frac{m}{M} + 1}} = \frac{R\omega_0}{\sqrt{2 + 1}} = \frac{R\omega_0}{\sqrt{3}}.$$

(b) $v_m = \sqrt{v_r^2 + R^2\omega^2} = \sqrt{\frac{R^2\omega^2}{3} + \frac{R^2\omega_0^2}{9}}$

$$= \frac{2}{3}R\omega_0.$$

Problem 60 A smooth bead of mass m slides along the rod of mass M and length L which is freely rotating about the vertical axis. If the initial spin of the rod is ω_0, find (a) the kinetic energy of the bead when it is at the radial distance r and (b) at the time of escaping from the rod, (c) the differential equation for the motion of the bead relative to the rod, and (d) the results in (a), (b), and (c) when the rod is too massive compared to the slender. Assume that initially the bead is situated near the axis.

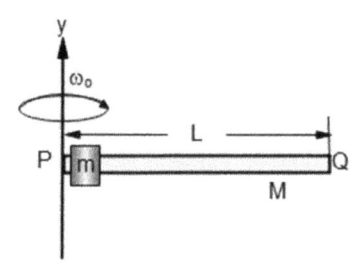

Solution
(a) As there is no external torque acting on the system (rod + bead), conserving angular momentum of the system when the bead is at a radial distance r, we have

$$L = \left(mr^2 + \frac{ML^2}{3}\right)\omega = \frac{ML^2}{3}\omega_0$$

$$\Rightarrow \omega = \frac{ML^2}{3mr^2 + ML^2}\omega_0. \tag{5.60}$$

As the kinetic energy of the system is conserved, the change in kinetic energy of the bead is given as

$$\Delta K_{bead} = |\Delta K_{rod}| = \frac{I_{rod}(\omega_0^2 - \omega^2)}{2}$$

$$= \frac{ML^2}{6}\omega_0^2\left\{1 - \left(\frac{\omega}{\omega_0}\right)^2\right\}. \tag{5.61}$$

Using equations (5.60) and (5.61)

$$\Delta K_{slender} = \frac{ML^2\omega_0^2}{6}\left[1 - \left(\frac{ML^2}{3mr^2 + ML^2}\right)^2\right].$$

5-118

(b) Putting $r = L$ we have

$$K = \frac{ML^2\omega_0^2}{6}\left[1 - \left(\frac{M}{3m+M}\right)^2\right]$$

$$= \frac{ML^2\omega_0^2}{6}\left[1 - \left(\frac{1}{\frac{3m}{M}+1}\right)^2\right].$$

(c) The differential equation for the bead is given as

$$\frac{d^2r}{dt^2} - r\omega^2 = 0,$$

where the angular velocity of the rod is given as

$$\omega = \frac{ML^2\omega_0}{3mr^2 + ML^2}.$$

Then the differential equation of the bead is given as

$$\frac{d^2r}{dt^2} = \frac{M^2L^4\omega_0^2 r}{(3mr^2 + ML^2)^2} = 0$$

$$\Rightarrow \frac{d^2r}{dt^2} = \frac{M^2L^4\omega_0^2 r}{(3mr^2 + ML^2)^2}.$$

(d)
  (i) If the rod is too massive, its angular velocity remains practically constant; $\omega \cong \omega_0$.
  (ii) Hence, the decrease in its kinetic energy is very small; in this sense, the kinetic energy of the bead can be imagined as a practically constant quantity. However, the sum of change in kinetic energy of the bead and rod is always zero. If we rotate the rod with a constant angular velocity by applying an external torque, it will do some work on the rod–particle (bead) system. As a result, the kinetic energy of the bead will increase. Therefore, we can neither apply the conservation of angular momentum nor the conservation of the kinetic energy of the system.

  In other words, if the kinetic energy of the rod which is given as

$$K = \frac{ML^2\omega_0^2}{6}$$

will remain constant, the kinetic energy of the bead will increase through centrifugal action. The radial velocity of the bead v_r can be calculated as follows:

$$\int v_r dv_r = \omega_0^2 \int r dr$$

$$\Rightarrow v_r^2 = \omega_0^2(r^2 - r_0^2),$$

where r_0 is very small.

(iii) Then the differential equation is

$$\frac{d^2 r}{dt^2} = \omega_0^2 r.$$

Problem 61 A bead of mass m is released from rest from $r = l/2$ in a tube of mass M and length l which is rotating about the vertical axis. If the the tube (a) is given an initial angular velocity ω_0 and released, and (b) maintains a constant angular velocity ω_0, find the velocity of the bead relative to ground at the time of escaping from the tube.

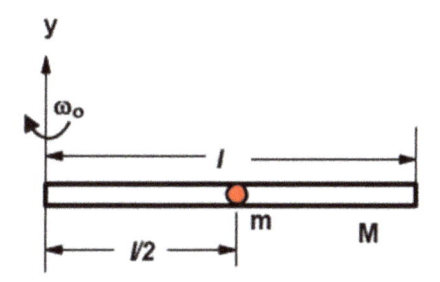

Solution

(a) In this case there is no external torque acting on the tube–bead system ($M + m$). So the energy of the system remains constant.

Conserving kinetic energy of the system at the time of escaping the tube we have

$$\frac{m}{2}\left(\frac{l}{2}\omega_0\right)^2 + \frac{Ml^2}{6}\omega_0^2 = \frac{Ml^2}{6}\omega^2 + \frac{1}{2}mv^2,$$

where $v = $ the speed of the bead at the time of escaping the rod,

$$\Rightarrow \frac{mv^2}{2} = \frac{Ml^2}{6}(\omega_0^2 - \omega^2) + \frac{ml^2\omega_0^2}{8}$$

$$= \frac{Ml^2}{6}\omega_0^2\left\{1 - \left(\frac{\omega}{\omega_0}\right)^2\right\} + \frac{ml\omega_0^2}{8}. \qquad (5.62)$$

5-120

Conservation of angular momentum yields

$$w_0\left(\frac{Ml^2}{3} + \frac{ml^2}{4}\right) = w\left(\frac{ml^2}{3} + ml^2\right) \Rightarrow w = \frac{w_0(4M + 3m)}{4(M + 3m)}. \tag{5.63}$$

Using equations (5.62) and (5.63)

$$\tfrac{1}{2}mv^2 = \frac{Ml^2w_0^2}{6}\left\{1 - \left[\frac{4M + 3m}{4(M + 3m)}\right]^2\right\} + \frac{mlw_0^2}{8}$$

$$\Rightarrow \tfrac{1}{2}mv^2 = \frac{Ml^2w_0^2}{6}\frac{(9m)(8M + 15m)}{\{4(M + 3m)\}^2} + \frac{ml^2w_0^2}{8}$$

$$\Rightarrow v = \sqrt{l^2w_0^2\left[\frac{3M(8M + 15m)}{16(M + 3m)^2} + \frac{1}{4}\right]}.$$

(b) When $M \to \infty$, the angular velocity remains constant; so, $v^2 = l^2w_0^2\left[\frac{24}{16} + \frac{1}{4}\right]v = \sqrt{7}w_0l/2$.

Problem 62 A tube of mass M and length l rotates about the vertical axis with an initial angular velocity w. A small ball of mass m (= 1 kg) is released from rest from the axis of rotation as shown in the figure so that it collides with the spring fitted with the tube. Find the value of w if the maximum compression of the spring is equal to $l/4$. Assume $k =$ the stiffness of the spring $= 36$ N m^{-1} and $M = 3$ m.

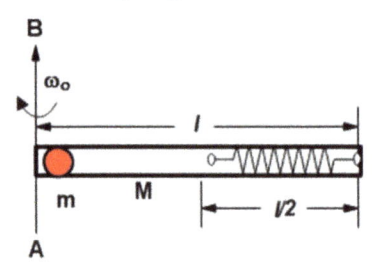

Solution
At the time of maximum compression x of the spring, let us assume that the angular velocity of the tube is w. Then by conserving the angular momentum of the system $(M + m)$ we have

$$L = w\left\{m\left(l - \tfrac{l}{2} + x\right)^2 + \frac{Ml^2}{3}\right\} = \frac{Ml^2}{3}w_0 \Rightarrow w = \frac{Ml^2w_0}{3m\left(l - \tfrac{l}{2} + x\right)^2 + Ml^2}.$$

Conserving the total energy (kinetic plus spring potential energy)

$$\frac{L^2}{2I_0} = \frac{L^2}{2I} + \frac{k}{2}x^2$$

5-121

$$\Rightarrow \frac{L^2}{2I_0}\left(1 - \frac{I_0}{I}\right) = \frac{k}{2}x^2.$$

Putting the values of the ratio of the initial and final moments of inertia of the system, that is, I_0/I, in the last equation,

$$\Rightarrow \frac{k}{2}x^2 = \frac{1}{2}\frac{Ml^2}{3}\omega_0^2\left\{1 - \frac{Ml^2}{3m\left(l - \frac{l}{2} + x\right)^2 + Ml^2}\right\}.$$

Putting $m = 1$ kg, $K = 36$ N m^{-1} and $x = \frac{l}{4}$ in the last expression and simplifying the factors, we have $\omega_0 = 2.5$ rad s^{-1}.

Problem 63 A smooth tube of mass M and length L is free to rotate about the vertical axis. It carries a block of mass m which starts from $b = l/2$ being attached to a light spring of stiffness k, as shown in the figure. If the rod (a) is given an initial angular velocity ω_0 and released, and (b) maintains a constant angular velocity ω_0, find the velocity of the bead relative to ground at the time of escaping from the rod. Put $\omega_0 = \sqrt{k/m}$ and $M = m$. Assume that initially the spring is relaxed

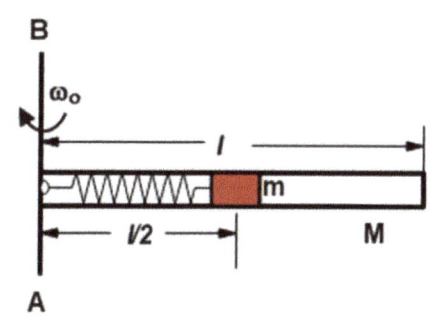

Solution

(a) When the block reaches the other end of the tube, the elongation of the spring $=x = l/2$. Let us assume that the angular velocity of the tube is ω at that instant. Then by conserving the angular momentum of the system $(M + m)$ we have

$$\left(\frac{ml^2}{4} + \frac{Ml^2}{3}\right)\omega_0 = \left(\frac{\frac{Ml^2}{3} + ml^2}{3}\right)\omega.$$

If $M = m$

$$\Rightarrow \left(\frac{ml^2}{4} + \frac{ml^2}{3}\right)\omega_0 = \left(\frac{ml^2}{3} + ml^2\right)\omega$$

5-122

$$\Rightarrow \frac{7\omega_0}{12} = \frac{4}{3}\omega \Rightarrow \omega = \frac{7}{16}\omega_0.$$

Conserving the energy of the system $(M+m)$,

$$\frac{k}{2}(\frac{l}{2})^2 + \frac{m}{2}v_r^2 = \Delta K$$

$$= \frac{L^2}{2I_0} - \frac{L^2}{2I} = \frac{L^2}{2I_0}(1 - \frac{I_0}{I}) = \frac{1}{2}(\frac{7}{12}ml^2)\omega_0^2(1 - \frac{7}{16})$$

$$\Rightarrow \frac{kl^2}{8} + \frac{mv_r^2}{2} = \frac{1}{2} \times \frac{7}{12} \times \frac{9}{16}ml^2\omega_0^2 \Rightarrow \frac{mv_r^2}{2} = \frac{l^2}{8}(\frac{63m\omega_0^2}{48} - k)$$

$$\Rightarrow v_r = \frac{l}{2}\sqrt{\frac{63}{48}\omega_0^2 - \frac{k}{m}}.$$

Putting $\omega_0 = \sqrt{\frac{k}{m}}$ we have

$$v_r = \frac{l}{2}(\frac{15}{18})\omega_0 = \frac{\sqrt{5}\omega_0 l}{8}.$$

(b) Imposing the centrifugal force $m(l/2 + x)\omega_0^2$ and subtracting the spring force kx, the net radially outward force $= F_{\text{radial}} = m(\frac{l}{2} + x)\omega_0^2 - kx = \frac{mv_r dv_r}{2mdx}$

Separating the variables and integrating both sides, we have

$$\frac{v_r^2}{2} = \omega_0^2(\frac{l}{2}x + \frac{x^2}{2}) - \frac{kx^2}{2m}.$$

Putting $x = \frac{l}{2}$ we have

$$\frac{v_r^2}{2} = \omega_0^2(\frac{l^2}{4} + \frac{l^2}{8}) - \frac{kl^2}{8m}$$

$$\Rightarrow \frac{v_r^2}{2} = \frac{\omega_0^2 l^2}{4}(1 + \frac{1}{2}) - \frac{kl^2}{8m} \Rightarrow \frac{v_r^2}{2} = \frac{3\omega_0^2 l^2}{8} - \frac{kl^2}{8m}$$

$$\Rightarrow \frac{v_r^2}{2} = \frac{l^2}{4}(\frac{3\omega_0^2}{2} - \frac{k}{2m}).$$

Putting $\frac{k}{m} = \omega_0^2$ we have

$$v_r^2 = \frac{l^2}{2}(\frac{3\omega_0^2}{2} - \frac{\omega_0^2}{2}) = \frac{l^2}{2}\omega_0^2$$

$$\Rightarrow v_r = \frac{\omega_0 l}{\sqrt{2}}$$

Problem 64 In the previous question let us assume that the initial length of the spring ($= b$) is very small compared to its maximum elongation. Let us release the tube after giving an initial angular velocity ω_0. Find the maximum extension of the spring. Assume that the tube is long enough such that the block cannot reach the free end of the tube.

Solution

Let x = maximum elongation of the spring. Equating the angular momentum of the system at the the time of maximum extension of the spring with the initial angular momentum,

$$\left(mx^2 + \frac{ml^2}{3}\right)\omega = \frac{Ml^2}{3}\omega_0$$

$$\Rightarrow \omega = \frac{Ml^2\omega_0}{3mx^2 + Ml^2} \tag{5.64}$$

The change in kinetic energy of the system (M + spring + m) = $\Delta K = \left(\dfrac{L^2}{2I_0}\right)\left(\dfrac{I_0}{I} - 1\right)$

$$= \frac{1}{2} \times \frac{Ml^2}{3}\omega_0^2\left(\frac{\frac{Ml^2}{3}}{\frac{Ml^2}{3} + mx^2} - 1\right)$$

$$\Rightarrow \frac{k}{2}x^2 = \frac{Ml^2\omega_0^2}{6}\left(\frac{3mx^2}{Ml^2 + 3mx^2}\right)$$

$$\Rightarrow k = \frac{Mml^2\omega_0^2}{Ml^2 + 3mx^2} = \frac{m\omega_0^2}{1 + \frac{3mx^2}{Ml^2}}$$

$$\Rightarrow x = l\sqrt{\frac{M}{3m}\left(\frac{m\omega_0^2}{k} - 1\right)}.$$

Problem 65 A smooth tube of mass M and length l is free to rotate about the vertical axis. It carries a block of mass m at a radial distance r. If the the rod is given a horizontal impulse J at its free end as shown in the figure, find (a) the angular velocity ω of the rod (i) just after giving the impulse and (ii) when the ball reaches the other end of the tube, and (b) the velocity of the ball relative to the tube at the time of touching the other end of the rod.

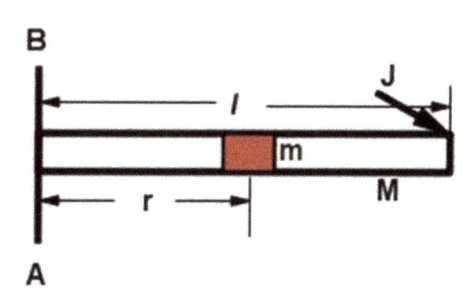

Solution
(a)
(i) The angular impulse of J is

$$Jl = \left(mr^2 + \frac{Ml^2}{3}\right)\omega_0 = \left(m + \frac{M}{3}\right)l^2\omega$$

$$\Rightarrow \omega_0 = \frac{3Jl}{3mr^2 + Ml^2}.$$

(ii) The final angular velocity of the tube is

$$\omega = \frac{3J}{(3m + M)l}.$$

(b) Let the block move with a velocity v_r relative to the tube. Conservation of energy yields,

$$\frac{L^2}{2I_f} + \frac{m}{2}v_r^2 = \frac{L^2}{2I_i}$$

$$\Rightarrow \frac{mv_r^2}{2} = \frac{L^2}{2I_i}\left(1 - \frac{I_i}{I_f}\right)$$

$$\Rightarrow mv_r^2 = \frac{J^2l^2}{I_i}\left(\frac{I_f - I_i}{I_f}\right)$$

$$= \frac{J^2l^2}{I_i} \cdot \frac{\left\{\frac{Ml^2}{3} + ml^2 - \frac{Ml^2}{3} - mr^2\right\}}{\left(\frac{M}{3} + m\right)l^2}$$

$$\Rightarrow mv_r^2 = \frac{3J^2m(l^2 - r^2)}{(Ml^2 + 3ml^2)(M + 3m)}$$

$$\Rightarrow v_r = J\sqrt{\frac{3(l^2 - r^2)}{(Ml^2 + 3ml^2)(M + 3m)}}.$$

Problem 66 In problem 60, if the initial angular velocity of the rod is ω_0 and the coefficient of friction between the ball and bottom surface of the tube is (a) finite so that the ball slides (even though it rotates ignore its rotational kinetic energy) and (b) very large so that the ball rolls without sliding (here take the rotational kinetic energy into consideration) between the tube and the spheres, find the velocity of the spheres relative to ground at the time of escaping from the disc. Assume that the particle was initially at a small distance from the axis of rotation and μ is the coefficient of kinetic friction.

Solution
(a) Equating the angular momentum of the system at the the time of escaping of the ball from the tube with the initial angular momentum, we have

$$\left(ml^2 + \frac{Ml^2}{3}\right)\omega = \frac{Ml^2}{3}\omega_0$$

$$\Rightarrow \omega = \frac{M\omega_0}{M + 3m}. \tag{5.65}$$

The change in kinetic energy of the sphere and ball are ΔK_{sphere} and ΔK_{ball}, respectively. Applying the work–energy theorem, we have

$$W_{\text{friction}} = \Delta K = \Delta K_{\text{tube}} + \Delta K_{\text{ball}} \tag{5.66}$$

$$\Rightarrow W_{\text{friction}} = -\frac{1}{2}\frac{Ml^2}{3}(\omega_0^2 - \omega^2) + \Delta K_{\text{ball}}$$

$$\Rightarrow -\mu mgl = -\frac{Ml^2}{6}\omega_0^2\left\{1 - \left(\frac{\omega}{\omega_0}\right)^2\right\} + K_{\text{ball}}$$

$$= -\frac{Ml^2\omega_0^2}{6}\left[1 - \left(\frac{M}{M + 3m}\right)^2\right] + K_{\text{ball}}. \tag{5.67}$$

Using the equations (5.65) and (5.67) we have

$$\Rightarrow -\mu mgl = -\frac{Ml^2\omega_0^2(2M + 3m)(3m)}{6(M + 3m)^2} + K_{\text{ball}}$$

$$\Rightarrow K_{\text{ball}} = \frac{Mml^2\omega_0^2(2M + 3m)}{2(M + 3m)^2} - \mu mgL$$

$$\Rightarrow v_{\text{ball}} = \sqrt{2\left\{\frac{Ml^2\omega_0^2(2M + 3m)}{2(M + 3m)^2}\right\} - 2\mu gL}$$

$$= \sqrt{\frac{Ml^2\omega_0^2(2M + 3m)}{(M + 3m)^2} - 2\mu gL}.$$

(b) If the ball rolls without sliding we can conserve the kinetic energy of the system. Putting $W_{\text{friction}} = 0$ in equation (5.66) we have

$$K_{\text{ball}} = \frac{Mml^2\omega_0^2(2M + 3m)}{2(M + 3m)^2}. \tag{5.68}$$

The kinetic energy of the rolling ball is

$$K_{\text{ball}} = \frac{m}{2}v^2\left(1 + \frac{2}{5}\right) = \frac{7mv^2}{10}. \tag{5.69}$$

Using equations (5.68) and (5.69) we have

$$\Rightarrow \frac{7mv^2}{10} = \frac{Mml^2\omega_0^2(2M + 3m)}{2(M + 3m)^2}$$

$$\Rightarrow v = \left\{\frac{\sqrt{\frac{5}{7}M(2M + 3m)}}{(M + 3m)}\right\} l\omega_0.$$

Problem 67 A smooth tube of mass M and length L having two small ball of mass m near the vertical axis is given an initial angular velocity ω_0. Find the velocity of the spheres relative to ground at the time of escaping from the tube. Put $M = 2m$.

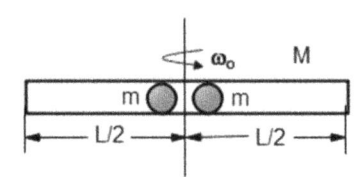

Solution
Equating the angular momentum of the system at the the time of escaping of the balls from the tube with the initial angular momentum, we have

$$\left(2 \times \frac{ml^2}{4} + \frac{Ml^2}{12}\right)\omega = \frac{Ml^2}{12}\omega_0$$

$$\Rightarrow \omega = \frac{M\omega_0}{M + 6m}. \tag{5.70}$$

The change in kinetic energy of the sphere and ball are ΔK_{sphere} and ΔK_{ball}, respectively. Applying the work–energy theorem, we have

$$0 = \Delta K = \Delta K_{\text{tube}} + \Delta K_{\text{ball}} \tag{5.71}$$

$$\Rightarrow 0 = -\frac{1}{2}\frac{Ml^2}{12}(w_0^2 - w^2) + \Delta K_{ball}$$

$$\Rightarrow 0 = -\frac{Ml^2}{24}w_0^2\left\{1 - \left(\frac{w}{w_0}\right)^2\right\} + K_{ball}. \tag{5.72}$$

Using equations (5.70) and (5.72) we have

$$-\frac{Ml^2w_0^2}{24}\left[1 - \left(\frac{M}{M+6m}\right)^2\right] + K_{ball} = 0 \tag{5.73}$$

$$\Rightarrow K_{ball} = \frac{Mml^2w_0^2(2M+3m)}{2(M+6m)^2}$$

$$\Rightarrow v_{ball} = \sqrt{\left\{\frac{Ml^2w_0^2(2M+3m)}{2(M+6m)^2}\right\}}$$

$$= \frac{lw_0\sqrt{M(2M+3m)}}{2(M+6m)}$$

$$= \frac{lw_0\sqrt{m(2\times 2m + 3m)}}{2(2m+6m)} \left(\because M = 2m\right)$$

$$= \frac{\sqrt{7}}{16}lw_0.$$

Problem 68 A uniform ring of mass M and radius R is spinning with an angular velocity w_0 about a smooth vertical axis APB. Two smooth beads each of mass m are released from the diametrical opposite positions as shown in the figure. Find the relative velocity between the beads just before they meet.

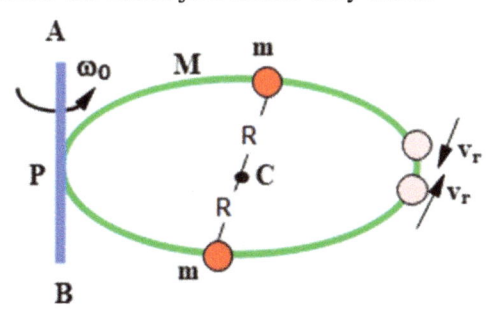

Solution

Let the velocity of each bead relative to the ring just before meeting be v_r. Then the final angular momentum due to the beads is

$$L_{\text{beads}} = m(v_r + 2R\omega)(2R) - m(v_r - 2R\omega)(2R)$$

$$= 8mR^2\omega \text{ (up)}.$$

The final angular momentum of the ring is

$$L_{\text{ring}} = 2MR^2\omega \text{ (up)}.$$

Adding these two angular momenta, the net final angular momentum is

$$L_f = 2(M + 4m)R^2\omega.$$

The initial angular momentum of the ring–bead system is

$$L_i = 2MR^2\omega_0 + 2mr^2\omega_0 \text{ (up)}.$$

Putting $r = \sqrt{2}R$ we have

$$L_i = 2MR^2\omega_0 + 4mR^2\omega_0 = 2(M + 2m)R^2\omega_0.$$

Equating the final angular momentum with the initial angular momentum, we have

$$2(M + 2m)R^2\omega_0 = 2(M + 4m)R^2\omega$$

$$\Rightarrow \omega = \frac{(M + 2m)}{M + 4m}\omega_0. \tag{5.74}$$

If v_r = velocity of each bead relative to the ring just before colliding, the velocities of beads 1 and 2 relative to ground are $(v_r + 2R\omega)$ and $(v_r - 2R\omega)$, respectively. The total kinetic energy of the beads is

$$K = \tfrac{1}{2}m(v_r - 2R\omega)^2 + \tfrac{1}{2}m(v_r + 2R\omega)^2$$

$$= \tfrac{1}{2}m\left\{(v_r - 2R\omega)^2 + (v_r + 2R\omega)^2\right\}$$

$$= \tfrac{1}{2}m\left\{2(v_r^2 + 4R^2\omega^2)\right\} = m(v_r^2 + 4R^2\omega^2)$$

Conserving the kinetic energy of the system,

$$m(v_r^2 + 4R^2\omega^2) + \tfrac{1}{2}(2MR^2)\omega^2 = \tfrac{1}{2}(2MR^2 + 2mr^2)\omega_0^2$$

Putting $r = \sqrt{2}R$, we have

$$mv_r^2 + (M + 4m)R^2\omega^2 = (M + 2m)R^2\omega_0^2$$

$$\Rightarrow mv_r^2 = (M + 2m)R^2\omega_0^2\left(1 - \frac{(M + 4m)}{(M + 2m)}\frac{\omega^2}{\omega_0^2}\right) \tag{5.75}$$

Using equations (5.74) and (5.75) we have

$$\Rightarrow mv_r^2 = (M+2m)R^2\omega_0^2\left(1 - \frac{(M+4m)}{(M+2m)}\frac{(M+2m)^2}{(M+4m)^2}\right)$$

$$\Rightarrow mv_r^2 = (M+2m)R^2\omega_0^2\left(1 - \frac{(M+2m)}{(M+4m)}\right)$$

$$\Rightarrow v_r^2 = (M+2m)R^2\omega_0^2\left(\frac{2}{M+4m}\right)$$

$$\Rightarrow v_r = R\omega_0\sqrt{\frac{2(M+2m)}{M+4m}}$$

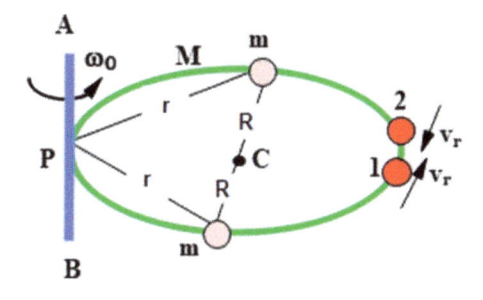

Problem 69 A rigid vertical circular loop of mass M and radius R can rotate freely about its vertical diameter. An initial angular speed ω_0 is given to the loop when two beads each of mass m are situated at the bottom of the ring as shown in the figure.
(a) If the beads will describe a maximum acute angle θ, find the value of ω_0.
(b) Describe the subsequent motion of the beads.

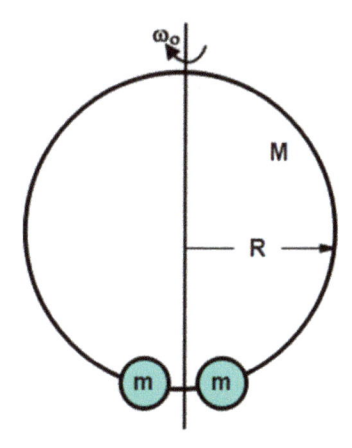

Solution
(a) At the angular position θ of the beads, the distance of the beads from the axis of rotation is $r = R\sin\theta$. As the beads do not slide relative to the ring,

5-130

their moment of inertia relative to the axis of rotation can be given as $2mr^2$. Hence the total moment of inertia of the system is

$$I = (1/2)MR^2 + 2mr^2.$$

If the angular velocity of the system $(M + 2m)$ is ω at the highest position of the beds, conserving the angular momentum about the axis of rotation, we have

$$\left(2mr^2 + \frac{MR^2}{2}\right)\omega = \frac{MR^2}{2}\omega_0 = L$$

$$\Rightarrow \omega = \frac{MR^2\omega_0}{4mr^2 + MR^2},$$

where $r = R\sin\theta$

$$\Rightarrow \omega = \frac{MR^2\omega_0}{4mR^2\sin^2\theta + MR^2}$$

$$\Rightarrow \omega = \frac{M\omega_0}{4m\sin^2\theta + M}. \tag{5.76}$$

Conserving energy, we have

$$\Delta K + \Delta U = 0$$

$$\Rightarrow \frac{1}{2}\frac{MR^2}{2}\omega_0^2 = \frac{1}{2} \times \left(\frac{MR^2}{2} + 2mR^2\sin^2\theta\right)\omega^2 + 2mgh.$$

Putting $h = R(1 - \cos\theta)$ and simplifying the factors, we have

$$MR\omega_0^2 = (M + 4m\sin^2\theta)R\omega^2 + 8mg(1 - \cos\theta). \tag{5.77}$$

Using equations (5.76) and (5.77) we have

$$MR\omega_0^2 = (M + 4m\sin^2\theta)R\left(\frac{M\omega_0}{4m\sin^2\theta + M}\right)^2$$

$$+ 8mg(1 - \cos\theta)$$

$$\Rightarrow MR\omega_0^2\left(1 - \frac{M}{4m\sin^2\theta + M}\right) = 8mg(1 - \cos\theta)$$

$$\Rightarrow MR\omega_0^2\left(\frac{\sin^2\theta}{4m\sin^2\theta + M}\right) = 2g(1 - \cos\theta)$$

$$\Rightarrow \omega_0 = \sqrt{\frac{2g(4m\sin^2\theta + M)(1 - \cos\theta)}{MR\sin^2\theta}}.$$

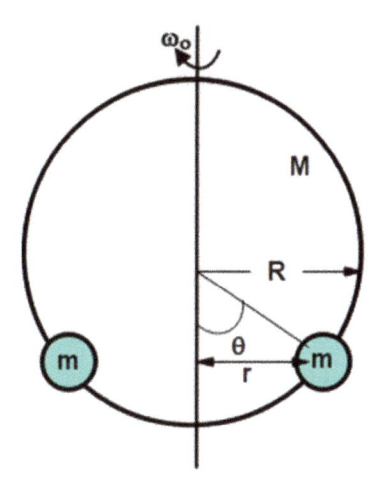

Alternative method

The rise in gravitational potential energy is equal to the decrease in the kinetic energy of rotation of the system,

$$2mgh = \frac{L_0^2}{2I_0} - \frac{L^2}{2I} = \frac{L^2}{2I_0} - \frac{L^2}{2I}$$

$$= \frac{MR\omega_0^2}{4}\left(1 - \frac{\frac{MR^2}{2}}{4mr^2 + MR^2}\right)$$

$$\Rightarrow 2mgh = \frac{MR^2\omega_0^2}{4}\left(\frac{4mr^2}{4mr^2 + MR^2}\right)$$

$$\Rightarrow gR(1 - \cos\theta) = \frac{MR^2\omega_0^2}{2}\left(\frac{\sin^2\theta}{4m\sin^2\theta + M}\right)$$

$$\Rightarrow \omega_0 = \sqrt{\frac{2g(4m\sin^2\theta + M)(1 - \cos\theta)}{MR\sin^2\theta}}.$$

(b) The beads will oscillate between the given maximum angular position and the lowest (initial) position.

Problem 70 In the last problem, if when the angular velocity of the loop is maintained constant, let $\omega_0 = \omega = $ constant. (a) If the maximum angle swung by the beads is θ, find the work done by the external agent until the beads reach their highest positions. (b) Discuss the nature of motion of the beads for different values of the angular speeds of the rotating loop. Assume $\omega_0 = \sqrt{g/R}$ and $\omega_0 = \sqrt{2g/R}$.

Solution

(a) If the initial angular velocity of the ring is kept constant, we have to apply an external torque. So we can neither conserve the angular momentum nor mechanical energy. However, we can apply Newton's second law and the work–energy theorem to solve this problem as follows.

As the beads rise by a height h from the lowest point of the loop, the rise in gravitational energy is

$$\Delta U = 2mgh = 2mgR(1 - \cos\theta). \tag{5.78}$$

The change in kinetic energy of the system is

$$\Delta K = \frac{1}{2} \times \left(\frac{MR^2}{2} + 2mR^2 \sin^2\theta\right)\omega^2 - \frac{1}{2}\frac{MR^2}{2}\omega^2$$

$$= \frac{1}{2} \times (2mR^2\sin^2\theta)\omega^2 = mR^2\omega^2\sin^2\theta. \tag{5.79}$$

According to the work–energy theorem, the work done by the external agent is

$$W_{\text{ext}} = \Delta K + \Delta U. \tag{5.80}$$

Using the last three equations, we have

$$\Rightarrow W_{\text{ext}} = mR^2\omega^2\sin^2\theta + 2mgR(1 - \cos\theta). \tag{5.81}$$

Let us now find the value of ω.

At any angular position θ, the forces acting on each bead are weight mg, centrifugal force $mr\omega^2$ (observed from the rotating frame), and the normal reaction N offered by the loop. Resolving the forces along the direction of relative velocity v_r tangent to the rotating circular loop, the net tangential force acting on each bead measured from the rotating loop, is

$$F_t = mr\omega^2 \cos\theta - mg\sin\theta = \frac{mv_r \, dv_r}{R d\theta}$$

Putting $r = R\sin\theta$ we have

$$mR\omega^2 \sin\theta \cos\theta - mg\sin\theta = \frac{mv_r \, dv_r}{R d\theta}$$

$$\Rightarrow R\omega^2 \sin\theta \cos\theta - g\sin\theta = \frac{v_r \, dv_r}{R d\theta}. \tag{5.82}$$

Separating the variables and integrating both sides we have

$$\int_0^\theta (R\omega^2 \sin\theta \cos\theta - g\sin\theta) R d\theta = \int_0^0 v_r \, dv_r$$

$$\Rightarrow \int_0^\theta \left(\frac{1}{2}R\omega^2 \sin 2\theta - g\sin\theta\right) d\theta = 0.$$

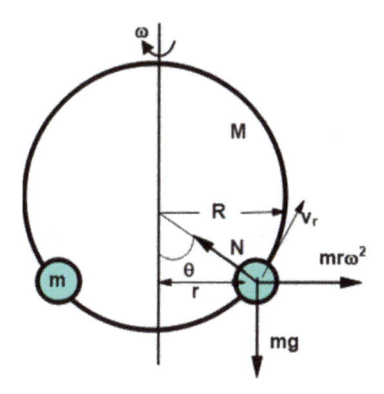

Since the beads will stop relative to the loop at the height position, we have put the upper limit of $v_r = 0$. Needless to mention, the initial value of $v_r = 0$ as the beads were at rest at the bottom of the loop initially.
After evaluating the integral we have

$$\frac{1}{4}R\omega^2(1 - \cos 2\theta) = g(1 - \cos \theta)$$

$$\Rightarrow \frac{1}{2}R\omega^2\sin^2\theta = g(1 - \cos \theta)$$

$$\Rightarrow \omega = \sqrt{2g(1 - \cos \theta)/R}\,\mathrm{cosec}\,\theta. \tag{5.83}$$

Using equations (5.81) and (5.83) we have

$$W_{ext} = mR^2\frac{2g(1 - \cos \theta)}{R\sin^2\theta}\sin^2\theta + 2mgR(1 - \cos \theta) = 2mgR(1 - \cos \theta) + 2mgR(1 - \cos \theta)$$

$$= 4mgR(1 - \cos \theta).$$

(b) In equation (5.82), putting the tangential acceleration of the beads relative to the loop as zero, for stable equilibrium of the beads we have

$$R\omega^2 \sin \theta \cos \theta - g \sin \theta = \frac{v_r dv_r}{Rd\theta} = 0$$

$$\Rightarrow (R\omega^2 \cos \theta - g)\sin \theta = 0.$$

This gives the following two solutions.
$\sin \theta = 0$, we have the lowest position as one solution. On the other hand, we have the stable equilibrium, given as

$$(R\omega^2 \cos \theta - g) = 0$$

5-134

$$\cos\theta = g/R\omega^2.$$

If the bead moves up
$\cos\theta = g/R\omega^2 \leqslant 1$. This gives us the condition

$$\omega \geqslant \sqrt{g/R}.$$

Hence, for any value of angular speed less than $\sqrt{g/R}$, the beads will not move up relative to the loop. If $\omega = \sqrt{2g/R}$ we have the equilibrium angle given as

$$\cos\theta = g/R\omega^2 = 1/2$$

$$\Rightarrow \theta = 60°.$$

Putting $\omega = \sqrt{2g/R}$ in equation (5.83) we have

$$\omega = \sqrt{2g(1 - \cos\theta)/R}\,\mathrm{cosec}\,\theta = \sqrt{2g/R}$$

$$\Rightarrow (1 - \cos\theta)\cos\theta = 0$$

$$\Rightarrow \theta = 0° \text{ and } 90°.$$

This means that the beads will oscillate between the above two extreme positions (the lowest point and the horizontal diametrical point of the rotating loop) via a mean position at $\theta = 60°$.

Problem 71 A rigid vertical circular loop of mass M and radius R can rotate freely about its vertical diameter. An initial angular speed $\omega_0 = (g/R)^{1/2}$ is given to the loop when two beads each of mass m are situated at the top of the ring as shown in the figure. Find the velocity of the beads relative to (i) ground and (ii) the loop, after they describe a right angle relative to the center of the loop. Put $M = m$.

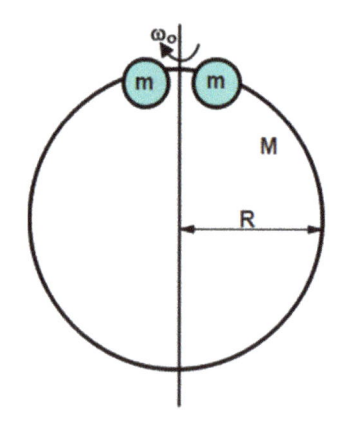

Solution

(i) When the beads describe a right angle relative to the center of the ring, the moment of inertia of each bead is equal to mR^2 and moment of inertia of the ring about the vertical axis is $MR^2/2$. By conserving the angular momentum of the system $(M + 2m)$, we have

$$\frac{MR^2}{2}\omega_0 = \left(\frac{MR^2}{2} + 2mR^2\right)\omega$$

$$\omega = \frac{M\omega_0}{M + 4m}. \tag{5.84}$$

As the beads come down, the total gravitational potential energy decreases by $mgR = mgR = 2mgR$. Conserving energy, we have

$$\frac{1}{2}\left(\frac{MR^2}{2}\right)\omega_0^2 = \frac{1}{2}\left(\frac{MR^2}{2}\right)\omega^2 + K_m - 2mgR, \text{ where } k_m = \text{total kinetic energy of the beads}$$

$$\Rightarrow K_m = \frac{M}{4}R^2(\omega_0^2 - \omega^2) + 2mgR$$

$$= \frac{M}{4}R^2\omega_0^2\left[1 - \left(\frac{\omega}{\omega_0}\right)^2\right] + 2mgR. \tag{5.85}$$

Using equations (5.84) and (5.85)

$$K_m = \frac{MR^2\omega_0}{4}\left[1 - \left(\frac{M}{M+4m}\right)^2\right] + 2mgR$$

$$= \frac{MR^2\omega_0^2}{4} \times \frac{4m \times 2(M+2m)}{(M+4m)^2} + 2mgR$$

$$= \frac{2Mm(M+2m)R^2\omega_0^2}{(M+4m)^2} + 2mgR.$$

Putting $\omega_0 = \sqrt{\frac{g}{R}}$ and $M = m$ we have

$$\frac{2 \times 1}{2}mv_m^2 = \left(\frac{6}{25} + 2\right)mgR$$

$$v = \sqrt{25\,gR} = \frac{\sqrt{56gR}}{5}$$

(ii) Let v_r be the velocity of each bead relative to the ring when they revolve by 900 relative to the center C of the ring. By applying the vector addition, the velocity v of each bead can be given as

$$v_r^2 + w^2 R^2 = v^2$$

$$\Rightarrow v_r^2 + \frac{w_0^2 R^2}{25} = \frac{56}{25} gR$$

$$\Rightarrow v_r^2 = \frac{56-1}{25} gR$$

$$\Rightarrow v_r = (11/5)^{1/2} gR.$$

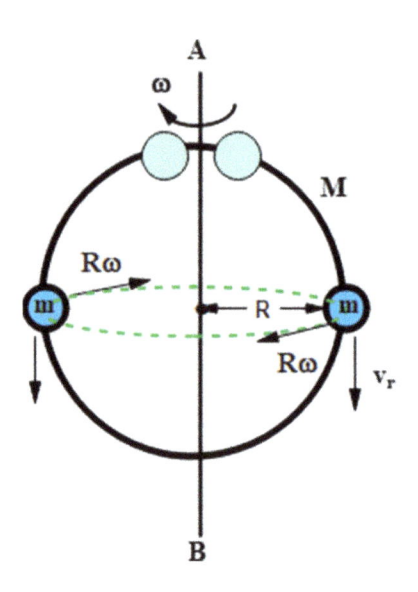

Problem 72 A uniform hollow cone of mass M, base radius R, and height H has a smooth groove along its slant surface as shown in the figure. The cone is freely rotating about its vertical centroidal axis with an angular velocity $w = (g/R)^{1/2}$. A smooth bead of mass m is released from rest from the top of the cone. The bead slides down along the groove made on the cone. Find the speed the bead reaches at the bottom of the cone. Put $H = 2R$.

Solution

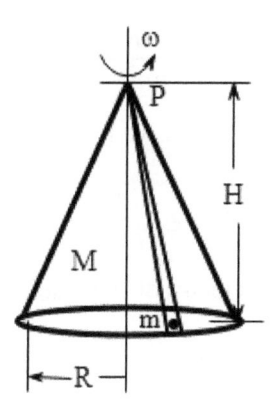

The moment of inertia of the hollow cone is $I = MR^2/2$ about the given axis. If ω' is the angular velocity of the cone when the bead reaches the bottom of the cone, conserving the angular momentum of the cone–bead system, we have

$$\frac{MR^2}{2}\omega = \left(\frac{MR^2}{2} + mR^2\right)\omega'$$

$$\omega' = \frac{M\omega}{M+2m}. \tag{5.86}$$

Let the bead have kinetic energy K_m at the bottom of the cone. As it descends through a vertical distance H, the change in gravitational potential energy is equal to $-mgh$. Conserving the energy of the system,

$$\frac{1}{2}\left(\frac{MR^2}{2}\right)\omega^2 = \frac{1}{2}\left(\frac{MR^2}{2}\right)\omega'^2 + K_m - mgH$$

$$\Rightarrow K_m = \frac{M}{4}R^2(\omega^2 - \omega'^2) + mgH$$

$$\Rightarrow K_m = \frac{M}{4}R^2\omega^2\left[1 - \left(\frac{\omega'}{\omega}\right)^2\right] + mgH. \tag{5.87}$$

Using equations (5.86) and (5.87) we have

$$K_m = \frac{mR^2\omega}{4}\left[1 - \left(\frac{M}{M+2m}\right)^2\right] + mgH$$

$$K_m = \frac{MR^2\omega^2}{4} \times \frac{4m(M+m)}{(M+2m)^2} + mgH$$

$$= \frac{Mm(M+m)R^2\omega^2}{(M+2m)^2} + mgH.$$

Putting $\omega = \sqrt{\frac{g}{R}}$, $M = m$, and $H = 2R$ we have $\frac{1}{2}mv_m^2 = (\frac{2}{9} + 2)mgR$

$$\Rightarrow v_m = \sqrt{\frac{40gR}{9}}.$$

The kinematical relation of the velocity of the bead is

$$v_r^2 + \omega'^2 R^2 = v_m^2. \tag{5.88}$$

Putting $\omega' = \omega/3$ (obtained by putting $M = m$ in equation (5.86) and the obtained value of v_m in the equation (5.88) we have

$$\Rightarrow v_r^2 + \frac{\omega^2 R^2}{9} = \frac{40}{9}gR$$

$$\Rightarrow v_r^2 = \frac{40-1}{9}gR$$

$$\Rightarrow v_r = \sqrt{39gR}/3.$$

Problem 73 A solid sphere of mass $m_1 = M$, radius R, and charge Q and a thin spherical shell of mass $m_2 = m$, radius R, and charge $-q$ are placed at a distance $r = 4R$. The spheres are made of dialectric material and the mass and charge distribution are uniform. If the system of two spheres is released from rest, they roll without sliding on the insulated rough horizontal surface. Find (a) their angular velocities and (b) the relative velocity between the centers of the spheres, just before they meet. Put $M/m = 5/7$, $Q = 7$ micro-coulombs, $q = 10$ micro-coulombs, $M = 1$ kg, and $R = 0.25$ m.

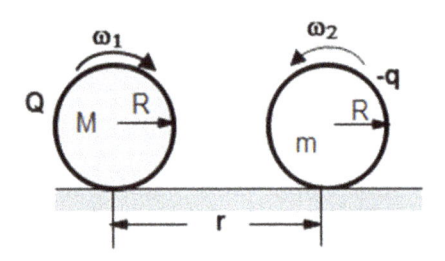

Solution
(a) The rolling without sliding requires static friction to prevail. The total work done by static friction is zero. So we can conserve the energy of the system

of two spheres. This means that the kinetic energy of the spheres increases at the expense of the electrostatic potential energy by electrostatic attraction. Symbolically,

$$|\Delta K| = |\Delta U|. \tag{5.89}$$

About any point on the ground we can conserve the angular momentum of the system of two spheres. Since the spheres were at rest initially, at any instant, the net angular momentum contributed by the rolling spheres will be zero. This means that the angular momenta of the spheres are equal and opposite, which can be given as

$$I_2 \omega_2 = I_1 \omega_1 = L, \tag{5.90}$$

where I_1 and I_2 are the moments of inertia of the spheres about their lowest points, given as

$$I_1 = \frac{7}{5} m_1 R^2 \text{ and } I_2 = \frac{5}{3} m_2 R^2.$$

Using equations (5.89) and (5.90) we have

$$\Rightarrow \frac{L^2}{2I_1} + \frac{L^2}{2I_2} = |\Delta U|$$

$$\Rightarrow \frac{L^2}{2I_1}\left(1 + \frac{I_1}{I_2}\right) = |\Delta U|$$

$$\Rightarrow \frac{1}{2} I_1 \omega_1^2 \left(1 + \frac{I_1}{I_2}\right) = \frac{Qq}{4\pi\varepsilon_0}\left(\frac{1}{r_f} - \frac{1}{r_i}\right),$$

where $1/r_i = 1/(4R)$ as the initial separation is equal to $4R$ and $1/r_f = 1/(2R)$ as they begin to touch just before meeting. Furthermore, $\frac{I_1}{I_2} = \frac{\frac{7}{5} m_1 R^2}{\frac{5}{3} m_2 R^2} = \frac{21}{25}\frac{m_1}{m_2} = \frac{21}{25} \times \frac{5}{7} = \frac{3}{5}$

$$\Rightarrow \frac{1}{2} I_1 \omega_1^2 \left(1 + \frac{3}{5}\right) = \frac{KQq}{4R}\frac{1}{r_f}$$

$$\Rightarrow \omega_1^2 \frac{4I_1}{5} = \frac{KQq}{4R}$$

$$\Rightarrow \omega_1 = \sqrt{\frac{5KQq}{2 \times 8RI_1}} = \sqrt{\frac{5KQq}{16R\left(\frac{7}{5}m_1 R^2\right)}}$$

$$= \sqrt{\frac{5 \times 9 \times 10^9 \times 7 \times 10^{-6} \times 10 \times 10^{-6}}{16 \times \frac{1}{4} \times \frac{7}{5} \times 1 \times \left(\frac{1}{4}\right)^2}}$$

$$= \sqrt{\frac{5 \times 9 \times 10^{-3} \times 5 \times 16 \times 10}{2 \times 2}}$$

$$= \sqrt{\frac{9 \times 10^{-3} \times 100 \times 2 \times 10}{2}} = 3 \text{ rad s}^{-1}.$$

From equation (5.90),

$$w_2 = \frac{I_1 w_1}{I_2} = \frac{3}{5} \times 3 = \frac{9}{5} = 1.8 \text{ rad s}^{-1}.$$

(b) The relative velocity between the centers of the spheres is

$$v_{rel} = v_1 + v_2$$

$$= R(w_1 + w_2) = \frac{1}{4}(3 + 1.8)$$

$$= \frac{4.8}{4} = 1.2 \text{ m s}^{-1}.$$

Problem 74 At time $t = 0$, a disc of mass M and radius R has an initial angular velocity w_0 while a constant torque τ about the vertical axis is acting as shown in the figure. A smooth block of mass m is moved with a velocity v_r relative to the disc. Find the angular velocity and angular acceleration of the disc at a radial position x and time t.

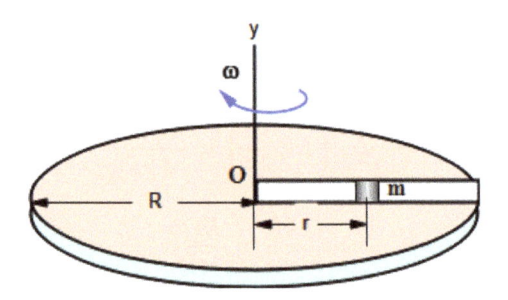

Solution
Method 1
At any radial distance x of the block, the angular momentum of the system $(M + m)$ is

$$L = (I + mx^2)w.$$

5-141

The initial angular momentum of the system is
$$L_0 = I\omega_0.$$
The change in angular momentum of the system is
$$\Delta L = L - L_0 = (I + mx^2)\omega - I\omega_0.$$
The angular impulse–momentum equation is
$$\tau t = \Delta L = (I + mx^2)\omega - I\omega_0$$
$$\Rightarrow \omega = \frac{\tau t + I\omega_0}{I + mx^2},$$
where $x = v_r t$ for the block.

Then, the angular acceleration of the disc is
$$\alpha = \frac{d\omega}{dt}$$
where we can put the obtained value of angular velocity and differentiate with respect to time to obtain the angular acceleration

Method 2
According to Newton's second law of rotation for a system of particles,
$$\tau = \frac{dL}{dt}$$
$$= \frac{d}{dt}(I + mx^2)\omega$$
$$\Rightarrow \tau = (I + mx^2)\frac{d\omega}{dt} + 2\omega m x v_r$$
$$\Rightarrow \alpha = \frac{\tau - 2\omega m x v_r}{I + mx^2}, \text{ where}$$
$$x = v_r t \text{ and } \omega = \frac{\tau t + I\omega_0}{I + mv_r^2 t^2}.$$

Note: It should be noted that, for a non-rigid system (a block slides on a disc), the torque is given as
$$\frac{dL}{dt} = (I + mx^2)\alpha + 2\omega m x v_r$$
which is not equal to $I_p \alpha$. So, for a non-rigid system, we cannot write
$$\tau = I\alpha.$$

Problem 75 A disc of mass M and radius R is rigidly welded with a uniform rod of mass m and length R, as shown in the figure. If the disc rolls with an angular velocity ω_0 as shown in the figure, at the given position find (a) the minimum value of ω_0 so that the disc breaks off at the given position and (b) the angular momentum of the combined body (disc + rod) about O at the given position and (c) minimum magnitude of the initial angular velocity so that the disc will go on rolling without bouncing (or breaking off). Put $M = 2$ m, $m = 3$ kg, and $R = 0.8$ m.

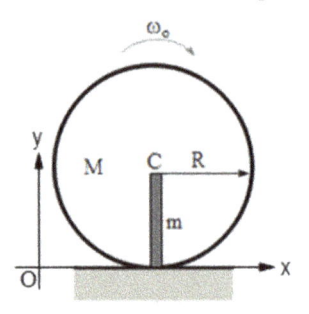

Solution

(a) Referring to the free-body diagram at the highest position of the mid-position of the center of mass of the system, the net vertical force acting on the system $(M + m)$ is

$$N - Mg - mg = M(a_1) + m(-a_2).$$

Putting $a_1 = 0$ for the disc, $N = 0$ (for disc to lose contact with the ground) and for the rod $a_2 = a_m = \frac{R}{2}\omega^2$ (vertically downward), in the last equation, and simplifying the factors, we have

$$\omega = \sqrt{\frac{2(M+m)g}{mR}}. \tag{5.91}$$

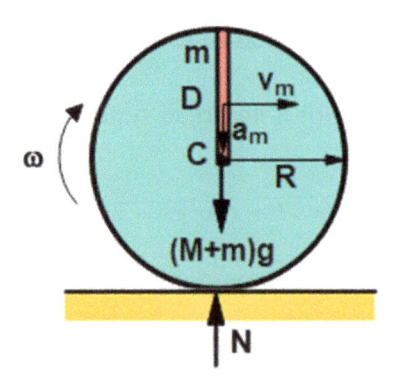

5-143

(b) The angular momentum of the system when the center of mass of the system is at its highest position is
$$\vec{L}_f = \vec{L}_{\text{disc}} + \vec{L}_{\text{rod}},$$
where
$$\vec{L}_{\text{disc}} = -\frac{3MR^2}{2}\omega\hat{k}$$

$$\vec{L}_{\text{rod}} = -mv_C\left(\frac{3}{2}R\right)\hat{k} - \frac{mR^2}{12}\omega\hat{k}$$

$$= -m\left(\frac{3}{2}R\omega\right)\left(\frac{3}{2}R\right)\hat{k} - \frac{mR^2}{12}\omega\hat{k}$$

$$= -\frac{9mR^2}{4}\omega\hat{k} - \frac{mR^2}{12}\omega\hat{k} = -\frac{7mR^2}{3}\omega\hat{k}.$$

Adding these two angular momenta we have
$$\vec{L}_f = -\left(\frac{3MR^2\omega}{2} + \frac{7mR^2}{3}\omega\right)\hat{k}$$

$$= -\frac{(14m + 9M)}{6}R^2\omega\hat{k}. \tag{5.92}$$

Using equations (5.91) and (5.92) we have
$$\vec{L}_f = -\frac{(14m + 9M)}{6}R^2\sqrt{\frac{2(M+m)g}{mR}}\hat{k}.$$

Putting $M = 2m$ we have
$$\vec{L}_0 = -\frac{14m + 18m}{6}R^2\sqrt{\frac{2(2m+m)g}{3mR}}\hat{k}$$

$$= -\frac{32}{6}mR^2 \times \sqrt{\frac{2g}{R}}\hat{k}$$

$$= -\frac{16}{3}mR\sqrt{2gR}\hat{k}$$

$$= -\frac{16}{3} \times 3 \times \frac{4}{5}\sqrt{2 \times 10 \times \frac{4}{5}}\hat{k}$$

5-144

$$= -\frac{16 \times 4}{5} \times 4\hat{k}$$

$$= -\frac{256}{5}\hat{k} = -51.2\hat{k} \text{ kg m}^2 \text{ s}^{-1}.$$

(c) Conserving the total energy, we have

$$K_i + U_i = K_f + U_f$$

$$\Rightarrow \frac{3}{4}MR^2\omega_0^2 + \frac{MR^2}{24}\omega_0^2 + \frac{1}{2}mv_0^2 + mg\frac{R}{2}$$

$$= \frac{3}{4}MR^2\omega_m^2 + \frac{MR^2}{24}\omega_m^2 + \frac{1}{2}mv^2 + mg\frac{3R}{2}.$$

Putting $v_0 = $ the initial velocity of the center of mass of the rod $= R\omega_0/2$ and $v = $ the final velocity of the center of mass of the rod $= 3R\omega_m/2$, we have

$$\Rightarrow \frac{3}{4}MR^2\omega_0^2 + \frac{mR^2}{24}\omega_0^2 + \frac{1}{2}m\left(\frac{R}{2}\omega_0\right)^2 - mgR$$

$$= \frac{3}{4}MR^2\omega_m^2 + \frac{MR^2}{24}\omega_m^2 + \frac{m}{2}\left(\frac{3}{2}R\omega_m\right)^2$$

$$\Rightarrow \left\{\frac{3}{4}M + \left(\frac{m}{24} + \frac{m}{8}\right)\right\}R^2\omega_0^2 - mgR$$

$$= \left\{\frac{3}{4}M + \left(\frac{m}{24} + \frac{9}{8}m\right)\right\}R^2\omega_m^2$$

$$\Rightarrow \frac{(9M + 2m)}{12}\omega_0^2$$

$$= \left(\frac{14m + 9M}{12}\right)\omega_m^2 + \frac{mg}{R}. \tag{5.93}$$

Using equations (5.91) and (5.93) we have

$$\frac{9M + 2m}{12}\omega_0^2 = \frac{14m + 9M}{12} \times \frac{2(M + m)g}{mR} + \frac{mg}{R}.$$

Putting $M = 2m$ we have

$$\frac{5m}{3}\omega_0^2 = \frac{32m}{12} \times \frac{2 \times 3mg}{mR} + \frac{mg}{R}$$

$$= \frac{48g}{3R} + \frac{g}{R}$$

$$\Rightarrow \frac{5}{3}\omega_0^2 = \frac{51g}{3R}$$

$$\Rightarrow \omega_0 = \sqrt{\frac{51g}{5R}} = \sqrt{\frac{51 \times 10}{5 \times \frac{4}{5}}}$$

$$= \sqrt{\frac{255}{2}} \text{ rad s}^{-1}.$$

Problem 76 A disc of mass m and radius R moves by the effect of gravity and the tensions of the strings which are wrapped over the disc. If the disc rolls over the strings without slipping, find (a) the velocity of the center of mass of the disc, and (b) the angular momentum of the disc relative to the instantaneous axis of rotation at the given position in terms of the given angular velocity and angular position of the strings (c) magnitude of angular velocity and the velocity of center of mass of the disc if the disc falls through a vertical distance h to attain the given angular position.

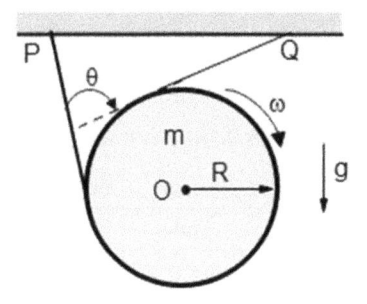

Solution
Method 1
(a) Referring to the diagram, we can equate the velocity of A and B along the respective string to zero because the strings are connected to the fixed points Q and P of the roof, respectively. For this purpose, let us assume that the horizontal and vertical components of the velocity of the center of mass of the disc are v_x and v_y, respectively. The velocities of A and B relative to O have magnitude ωR pointing towards Q and P, respectively, along the strings, as shown in the figure.

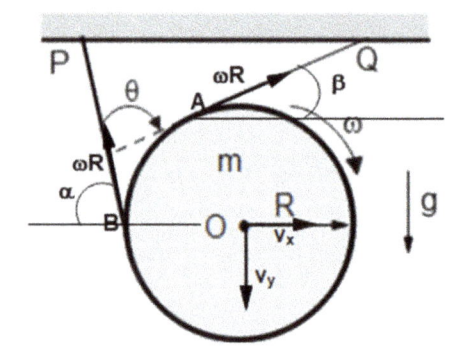

The tangential component of velocity of A is
$$(v_A)_t = v_Q = 0$$
$$\Rightarrow v_x \cos\beta + \omega R - v_y \sin\beta = 0. \tag{5.94}$$

The tangential component of the velocity of B is
$$(v_B)_t = v_P = 0$$
$$\Rightarrow \omega R - v_x \cos\alpha - v_y \sin\alpha = 0. \tag{5.95}$$

Multiplying equations (5.94) and (5.95) by $\cos\alpha$ and $\cos\beta$, respectively, we have
$$\cos\alpha(-v_x \cos B + v_y \sin B = \omega R) \tag{5.96}$$

$$\cos\beta(v_x \cos\alpha + v_y \sin\alpha = \omega R). \tag{5.97}$$

Adding equations (5.96) and (5.97) we have
$$v_y(\sin\beta \cos\alpha + \cos\beta \sin\alpha) = \omega R(\cos\alpha + \cos\beta)$$
$$\Rightarrow v_y = \frac{\omega R(\cos\alpha + \cos\beta)}{\sin(\alpha + \beta)}.$$

From equation (5.94)
$$v_x = \frac{v_y \sin\beta - \omega R}{\cos\beta}$$
$$= \frac{\omega R}{\cos\beta}\left\{\frac{(\cos\alpha + \cos\beta)}{\sin(\alpha + \beta)} - 1\right\}$$
$$= \omega R\left\{\frac{\cos\alpha \sin\beta + \cos\beta \sin\beta - \sin\alpha \cos\beta - \cos\alpha \sin\beta}{\cos\beta \sin(\alpha + \beta)}\right\}$$
$$v_x = \frac{\sin\beta - \sin\alpha}{\sin(\alpha + \beta)}\omega R.$$

5-147

Then the velocity of the center of mass O of the disc is

$$v = \sqrt{v_x^2 + v_y^2}$$

$$v = \sqrt{\frac{(\sin\beta - \sin\alpha)^2 + (\cos\alpha + \cos\beta)^2}{\sin(\alpha+\beta)}} \, \omega R$$

$$= \frac{\sqrt{2 + 2\cos(\alpha+\beta)}}{\sin(\alpha+\beta)} \omega R$$

$$= \frac{2\cos\left(\frac{\alpha+\beta}{2}\right)}{2\sin\left(\frac{\alpha+\beta}{2}\right)\cos\left(\frac{\alpha+\beta}{2}\right)} \omega R$$

$$= \frac{\omega R}{\sin\left(\frac{\alpha+\beta}{2}\right)} = \frac{\omega R}{\sin\left(\frac{\pi}{2} - \frac{\theta}{2}\right)}.$$

We can show that

$$\left(\frac{\alpha+\beta}{2}\right) = \left(\frac{180-\theta}{2}\right).$$

So finally we have

$$v = \omega R \sec\frac{\theta}{2}$$

Method 2

The velocity of the points A and B along AQ and BP is zero. Then these points move perpendicular to their respective strings, shown in the figure as v_A and v_B, respectively. If we draw the perpendiculars at A and B onto their velocity vectors, they will meet at the point C. So the point C is the instantaneous axis of rotation of the disc at the given position. Since the point C is at rest at the given position, the velocity of the center of mass O of the disc must be perpendicular to the line CO, which is given as

$$v_C = (CO)\omega.$$

We can easily show that

$$CO = R \csc\left(\frac{\alpha+\beta}{2}\right).$$

Then we have

$$v = v_O = \left\{R \csc\left(\frac{\alpha+\beta}{2}\right)\right\}\omega.$$

Putting $(\frac{\alpha+\beta}{2}) = (\frac{180-\theta}{2})$ we have

$$v = \omega R \csc\left(\frac{\pi}{2} - \frac{\theta}{2}\right)$$

$$= \omega R \sec\frac{\theta}{2}.$$

(b) The angular momentum of the disc is $\vec{L_C} = -m(OC)v_0\hat{k} + I_0\omega(-\hat{k})$

$$= -\left\{mR\sec\frac{\theta}{2}v_C + mK^2\omega\right\}\hat{k}$$

$$= -\left\{mR\sec\frac{\theta}{2}\left(\omega R\sec\frac{\theta}{2}\right) + mK^2\omega\right\}\hat{k}$$

$$= -mR^2\omega\hat{k}\left(\sec^2\frac{\theta}{2} + \frac{K^2}{R^2}\right).$$

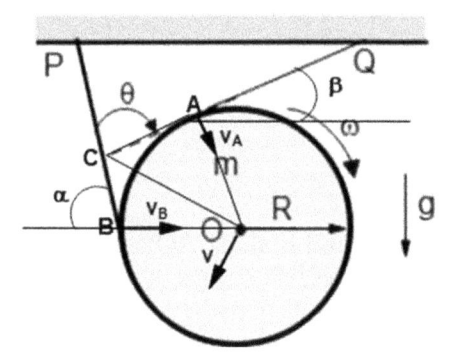

(c) Conserving the energy, we have $\frac{m}{2}v_0^2 + \frac{m}{2}K^2\omega^2 = mgh$.

Putting $v_O = 2R(\sec\frac{\theta}{2})\omega$ we have $\frac{m\omega^2}{2}(R^2\sec^2\frac{\theta}{2} + K^2) = mgh$,

$$\omega = \sqrt{\frac{2gh}{\sec^2\frac{\theta}{2} + \frac{K^2}{R^2}}}.$$

Then the velocity of O is

$$v_C = 2R\omega \sec\frac{\theta}{2}$$

5-149

$$=2R\sec\frac{\theta}{2}\sqrt{\frac{2gh}{\sec^2\frac{\theta}{2}+\frac{K^2}{R^2}}}.$$

Problem 77 A man of mass M stands at the axis of the turn-table, with his arms stretched horizontally holding small dumb-bells of mass m in each hand. The initial kinetic energy of rotation of the system is E. Let the man bring his arms close to his body and slowly lower the dumb-bells until his hands remain vertically down. In the process of dropping his hands, the moment of inertia of the man changes from I_0 to I and the center of mass of the boy drops by a height h. Neglecting the friction at the axis of rotation of the turn-table, find the work done by the man in the process of dropping his hands.

Solution
Here the system is the man plus turn-table.
The initial angular momentum (before dropping the hands) of the system is

$$L_0 = I_0\omega_0.$$

The final angular momentum (after dropping the hands) of the system is

$$L_0 = I\omega.$$

Conserving the angular momentum,

$$I\omega = I_0\omega_0 \text{ (let)}.$$

The change in kinetic energy of the system is

$$\Delta K = \frac{L^2}{2I} - \frac{L^2}{2I_0} = \frac{L^2}{2I_0}\left(\frac{I_0}{I} - 1\right).$$

As the center of mass of the system falls through a height h, the change in the gravitational potential energy of the system is

$$\Delta U = -(M + 2m)gh.$$

Then the work done by the man is

$$W_{man} = \Delta U + \Delta K = -(M + 2m)gh + \frac{L^2}{2I_0}\left(\frac{I_0}{I} - 1\right).$$

Putting $L^2/2I_0 = E$ we have

$$W_{man} = \Delta U + \Delta K = -(M + 2m)gh + E\left(\frac{I_0}{I} - 1\right).$$

Problem 78 Six balls each having mass m hang from a ceiling by six identical light inextensible strings each of length l. The first ball is pulled to a height h and released. It collides elastically with the second ball, the second ball with the third ball, and so on. As a result, the sixth ball will be shifted. The distance between two adjacent strings is equal to x. Find (a) the angular impulse and (b) the torque exerted by the additional downward tension in the string connected to the last (left-hand) ball just after the fifth collision, relative to the point of suspension of the first ball.

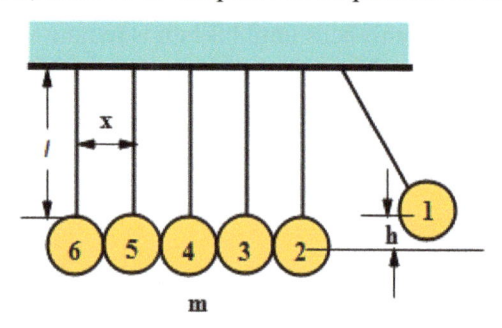

Solution

(a) The momentum delivered by the first ball is

$$\Delta P = mv = m\sqrt{2gh},$$

which is directed towards the left.
Then the angular impulse about the point of suspension is
$$\overrightarrow{\Delta L} = -mvl\hat{k} = -ml\sqrt{2gh}\,\hat{k}.$$

(b) The extra torque about the point of suspension is

$$\tau = (\Delta T)(5x),$$

where the extra tension just after the collision is

$$\Delta T = \frac{mv^2}{l} = \frac{m(2gh)}{l}$$

$$\Rightarrow \vec{\tau} = 10\frac{mgh}{l}\hat{k}.$$

Problem 79 A horizontal impulse I acts on a rough sphere of mass m, radius R, and radius of gyration K, which is placed on a horizontal surface. If the surface is smooth, find (a) the inner velocity of O and the angular velocity of the body, (b) the angular momentum of the body relative to a point fixed with the ground, (c) the kinetic energy, (d) the instantaneous axis of rotation, and (e) the condition for rolling just after the application of the impulse.

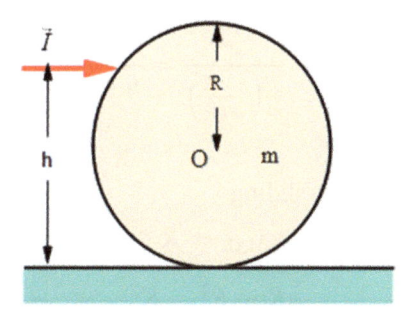

Solution

(a) The center of mass of the body moves with

$$v_C = \frac{I}{m}.$$

The angular impulse about the center of mass of the body is equal to $I(h - R)$, which changes the angular momentum of the body about its center of mass by $mK^2\omega$.

Applying the impulse–momentum equation we have

$$mK^2 \omega = I(h - R).$$

Then the angular velocity of the body is

$$\omega = \frac{I(h - R)}{mK^2}.$$

(b) The angular momentum of the body about a fixed point on the ground is

$$\vec{L} = -Ih\hat{k}.$$

(c) The kinetic energy just after striking the sphere is

$$K = \frac{m}{2}v_C^2 + \frac{m}{2}K^2\omega^2.$$

Putting the values of v_C and ω we have

$$K = \frac{m}{2}\left(\frac{I}{m}\right)^2 + \frac{m}{2}K^2\frac{I^2(h-r)^2}{m^2K^4}$$

$$= \frac{I^2}{2m}\left\{1 + \frac{(h-R)^2}{K^2}\right\}.$$

(d) The position of the instantaneous axis of rotation is

$$x_{IAR} = \frac{v_C}{\omega}.$$

Putting in the values of v_C and ω we have

$$x_{IAR} = \frac{\frac{I}{m}}{\frac{I}{m}\left(\frac{h-R}{mK^2}\right)} = \frac{K^2}{h-R}.$$

(e) If the body rolls without sliding

$$x_{IAR} = R$$

$$hR - R^2 = K^2$$

$$\Rightarrow h = \frac{R^2 + K^2}{R}.$$

Problem 80 A ball of radius R and radius of gyration K is affected at a height h by an impulse $I = F\delta t$. Find (a) the angular momentum of the body relative to a point fixed with the ground on the right-hand side of the body, (b) the linear and angular velocity just after the cuing, (c) the linear and angular velocity at the time of rolling due to friction, and (d) the value of h so that the body rolls just after the cuing. Discuss the cases when (i) $h > h_C$ and (ii) $h < h_C$.

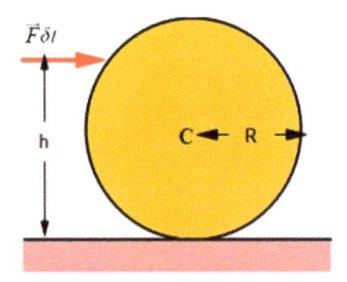

Solution

(a) Relative to a fixed point on the ground

$$\vec{L} = -Ih\hat{k} = \text{constant}.$$

(b) Referring to the last problem, just after applying the impulse, the linear and angular velocities are

$$v_C = \frac{I}{m} \text{ (towards the right)}$$

and

$$\omega = \frac{I(h-R)}{mK^2} \text{ (clockwise)}.$$

(c) If it rolls due to friction, the angular momentum of the body is $L = Ih = m(K^2 + R^2)\omega$ = constant relative to the ground,

$$\Rightarrow \omega = \frac{Ih}{m(K^2 + R^2)}.$$

The velocity of the center of mass of the rolling body will be

$$v_C = R\omega = \frac{IhR}{m(K^2 + R^2)}.$$

(d) For rolling $v_P = 0$; so we have

$$v_C = R\omega.$$

Putting the values of v_C and ω we have

$$\Rightarrow I/m = R\left\{\frac{I(h-R)}{mK^2}\right\}$$

$$\Rightarrow h = \frac{K^2}{R} + R.$$

1. If $h > \frac{K^2}{R} + R$, friction acts forward.
2. If $h = \frac{K^2}{R} + R$, friction $= 0$.
3. If $h < \frac{K^2}{R} + R$, friction acts backward.

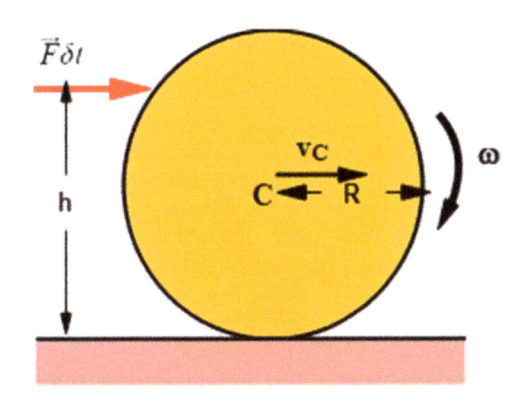

Inelastic impact by a connecting string

Problem 81 A bead of mass m is connected to a disc by an inextensible thread which is wrapped over the disc. The disc is pivoted smoothly at its center O, as shown in the figure. If the bead is given a velocity v_0 while the string is little bit slack, find (a) the angular velocity of the disc just after the impact, (b) the (i) fraction change in kinetic energy and (ii) angular impulse during the impact, (c) the tension of the string just after the impact, and (d) change in kinetic energy of the system during the impact. Assume that the friction is present between the ground and bodies (bead and disc) having a coefficient of friction μ.

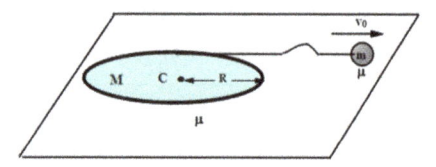

Solution

(a) We can approximately conserve the angular momentum of the disc–particle system during the impact because the angular impulse due to friction is negligible during the small time of impact. If $v =$ the velocity of the particle and $\omega =$ the angular velocity of the disc just after the string is taut, conserving angular momentum about O, we have

$$mvR + \frac{MR^2}{2}\omega = mv_0 R,$$

where $v = R\omega$

$$\Rightarrow \left(m + \frac{M}{2}\right)R^2\omega = mv_0 R$$

$$\Rightarrow \omega = \frac{2mv_0}{(2m+M)}.$$

(b)

(i) The fraction change in kinetic energy is

$$\frac{\Delta K}{K} = \frac{I_2}{I_1 + I_2}.$$

Putting $I_1 = mR^2$ and $I_2 = \frac{MR^2}{2}$ we have

$$\frac{\Delta K}{K} = -\frac{M}{M+2m}.$$

(ii) The angular impulse is $J =$ the change in angular momentum

$$\Rightarrow J = \frac{M}{2}R^2\omega = \left(\frac{M}{2}R^2\right)\left\{\frac{2mv_0}{(M+2m)R}\right\}$$

$$\Rightarrow J = \frac{Mmv_0 R}{(M+2m)}.$$

5-155

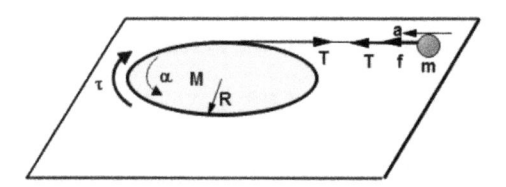

(c) The frictional torque acting on the thin ring of mass dm and of thickness dr and radius r is $d\tau_f = \mu dmgr$.

Integrating, the total frictional torque acting on the disc is

$$\tau_f = \int \mu dmgr = \int_0^R \mu gr 2\pi r dr \frac{M}{\pi R^2} = \frac{2\mu MgR}{3}.$$

Then the angular acceleration of the disc is

$$\alpha = \frac{-TR + \tau_f}{\frac{MR^2}{2}} = \frac{2(\tau_f - TR)}{MR^2}. \tag{5.98}$$

Applying the force equation on the particle

$$\mu mg + T = ma, \tag{5.99}$$

where

$$a = R\alpha. \tag{5.100}$$

Using the equations (5.98), (5.99), and (5.100)

$$\mu g + \frac{T}{m} = R\frac{2(\tau_f - TR)}{MR}$$

$$\Rightarrow T\left(\frac{1}{m} + \frac{2}{M}\right) = \frac{2\tau_f}{MR} - \mu g,$$

where

$$\tau_f = \frac{2\mu MgR}{3}$$

$$\Rightarrow T\left(\frac{M + 2m}{Mm}\right) = \frac{2 \times 2\mu MgR}{3MR} - \mu g$$

$$\Rightarrow T\left(\frac{M + 2m}{Mm}\right) = \frac{\mu g}{3}$$

$$\Rightarrow T = \frac{\mu Mmg}{3(M + 2m)}.$$

(d) The change in kinetic energy of the system (disc + particle) is

$$\Delta K = \frac{1}{2}mv^2 + \frac{1}{2}\left(\frac{MR^2}{2}\right)\omega^2 - \frac{1}{2}mv_0^2$$

$$= \left(\frac{MR^2}{2} + \frac{MR^2}{4}\right)\frac{(2mv_0)^2}{(2m+M)^2 R^2} - \frac{1}{2}mv_0^2$$

$$= \frac{1}{2}\frac{(2m+M)R^2}{2} \times \frac{4m^2 v_0^2}{(2m+M)^2 R^2} - \frac{1}{2}mv_0^2$$

$$= \frac{m}{2}v_0^2\left(\frac{2m - 2m - M}{M + 2m}\right)^2$$

$$= \frac{-Mmv_0^2}{2(2m+M)}.$$

Problem 82 A disc is smoothly pivoted at O and placed on a horizontal surface. The string PQ is slightly loosened and a bead of mass m is given a velocity v. Find (a) the (i) angular velocity of the disc and (ii) the linear velocity of the particle just after the string is taut, (b) the tension in the string, (c) the linear acceleration of m, (d) the angular acceleration of the disc (pulley) just after the string is taut, and (e) the fraction of energy loss during the impact. Assume that no friction exists between all contacting surfaces and the string does not slip over the disc.

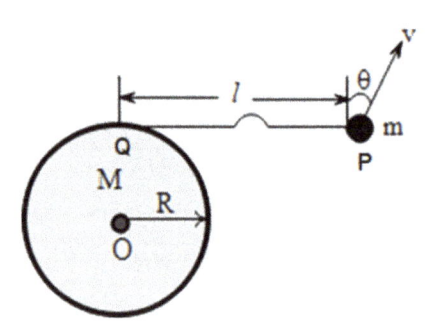

Solution
(a)
    (i) Since the net torque acting on the system $(M+m)$ about O is zero, its angular momentum about O must remain conserved. Conserving the angular momentum about O,

$$\left(\vec{L_0}\right)_i = \left(\vec{L_0}\right)_f. \tag{5.101}$$

The initial angular momentum about O is

$$\left(\vec{L_0}\right)_i = mvl\cos\theta\hat{k} - mRv\sin\theta\hat{k}. \tag{5.102}$$

The angular momentum about O just after the string is taut is

$$(\vec{L}_O)_f = mvl\cos\theta \hat{k} - mv'R\sin\theta\hat{k} - \frac{MR^2}{2}\omega\hat{k}, \qquad (5.103)$$

where $v' = R\omega$

Using last three equations and cancelling the term $mvl\cos\theta\hat{k}$ from both sides, we have

$$\left(\frac{MR^2}{2} + mR^2\right)\omega = mRv\sin\theta$$

$$\Rightarrow \omega = \frac{2mv\sin\theta}{(M+2m)R}. \qquad (5.104)$$

(ii) Since the impact does not take place perpendicular to the string, the perpendicular component of velocity, that is, $v\cos\theta$, remains constant during the impact. Then the velocity of the particle is

$$v_m = \sqrt{R^2\omega^2 + v^2\cos^2\theta}. \qquad (5.105)$$

Using equations (5.104) and (5.105) we have

$$\Rightarrow v_m = \sqrt{\left[\frac{2mv\sin\theta}{M+2m}\right]^2 + v^2\cos^2\theta}$$

$$= \left\{\sqrt{\frac{4m^2}{(M+2m)^2}\sin^2\theta + \cos^2\theta}\right\}v.$$

(b) The tension pulls the particle towards the left with an acceleration a_P, say. Then

$$T = ma_P,$$

where a_P = acceleration of P along the string, which is given as

$$\vec{a}_P = \vec{a}_{PQ} + \vec{a}_Q = \{-l\omega^2 + R\alpha\}\hat{i}.$$

Then the net force acting on the bead m is

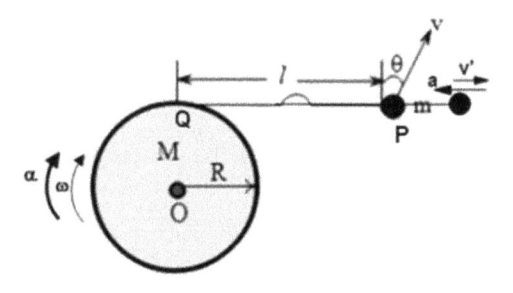

$$-T = ma_P = -m(l\omega^2 - R\alpha). \qquad (5.106)$$

The torque of tension T about O is

$$TR = \frac{MR^2}{2}\alpha$$

$$\Rightarrow T = \frac{MR\alpha}{2}. \tag{5.107}$$

Using the equations (5.106) and (5.107)

$$T = m\left[l\omega^2 - \frac{2T}{M}\right]$$

$$\Rightarrow T\left[1 + \frac{2m}{M}\right] = ml\omega^2$$

$$\Rightarrow T = \frac{ml\omega^2}{1 + \frac{2m}{M}}. \tag{5.108}$$

Using equations (5.104) and (5.108) the tension in the string is

$$T = \frac{Mml\omega^2}{M + 2m} = \frac{Mlm}{M + 2m}\left\{\frac{2mv\sin\theta}{(M+2m)R}\right\}^2$$

$$= \frac{Mml(4m^2v^2\sin^2\theta)}{(M+2m)(M+2m)^2 R^2}$$

$$\Rightarrow T = \frac{4Mm^2v^2 l\sin^2\theta}{(M+2m)^3 R^2}.$$

(c) From equation (5.106) the acceleration of the particle is given as

$$a_m = \frac{T}{m} = \frac{4Mm^2v^2 l\sin^2\theta}{(M+2m)^3 R^2}.$$

(d) From equation (5.107) the angular acceleration of the disc is

$$\alpha = \frac{2T}{MR} = \frac{8Mm^2v^2 l\sin^2\theta}{\{(M+2m)^3 R^2\}MR}$$

$$= \frac{8m^2v^2 l\sin^2\theta}{(M+2m)^3 R^3}.$$

(e) The change in the kinetic energy of the system is

$$\Rightarrow \Delta K = \frac{1}{2}\left(\frac{M+2m}{2}\right)R^2\omega^2 - \frac{mv^2}{2}\sin^2\theta. \tag{5.109}$$

Using equations (5.104) and (5.109) we have

$$\Delta K = \frac{1}{2}\left(\frac{M+2m}{2}\right)R^2 \frac{4m^2v^2 \sin^2\theta}{(M+2m)^2 R^2} - \frac{mv^2}{2}\sin^2\theta$$

$$= -\sin^2\theta \frac{mv^2}{2}\left[\frac{2m}{M+2m} - 1\right]$$

$$= -\frac{Mmv^2 \sin^2\theta}{2(M+2m)}.$$

Then the fraction loss of kinetic energy is

$$\eta = \left|\frac{\Delta K}{K}\right| = \frac{\frac{Mmv^2 \sin^2\theta}{2(M+2m)}}{\frac{1}{2}mv^2}$$

$$\Rightarrow \left|\frac{\Delta K}{K}\right| = \frac{M \sin^2\theta}{M+2m}.$$

Problem 83 A disc is smoothly pivoted at O on a horizontal surface. A string PQ of length $2l$ is slackened and a bead of mass m is given a velocity v. Just after the string becomes taut, find (a) the angular velocity of the disc, (b) the tension in the string, (c) the fraction of energy loss during the impact. Assume that all contacting surfaces are smooth and the string does not slip over the disc.

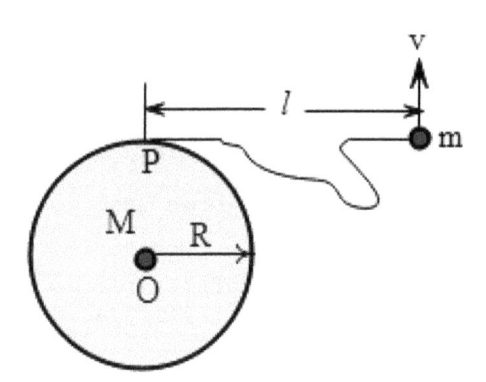

Solution
(a) Just before the string is taut, the angle θ made by the string with the line of motion of the bead is equal to $30°$. The impulse $T\delta t$ developed in the string during the collision changes the momentum of both the bead and disc. The angular impulse, that is, $R(T\delta t)\sin\theta$ acting on the disc about the pivot O

5-160

changes the angular momentum of the disc from zero to $I\omega$. The impulse –momentum equation for the disc is

$$R(T\delta t)\sin 30° = \frac{MR^2}{2}\omega$$

$$\Rightarrow R(T\delta t) = MR^2\omega. \tag{5.110}$$

The linear impulse $(T\delta t)$ acting on the particle m changes its linear momentum from $mv\cos\theta$ to mv_1 (say) along the string. Then, the impulse –momentum equation for the bead is

$$-T\delta t = m(v_1 - v\cos\theta)$$

$$\Rightarrow -T\delta t = m(v_1 - v\cos 30°)$$

$$-T\delta t = m\left(v_1 - \frac{\sqrt{3}}{2}v\right). \tag{5.111}$$

Since the impact takes place along the string, the component velocity of the bead perpendicular to the string remains conserved. Then we can write

$$v' = v\sin\theta = v\sin 30° = v/2. \tag{5.112}$$

Equating the components of velocities of the ends of the string along its length,

$$v_1 = \omega R \sin\theta = \omega R \sin 30° = \omega R/2. \tag{5.113}$$

Adding equations (5.110) and (5.111) we have

$$MR\omega + m\left(v_1 - \frac{\sqrt{3}}{2}v\right) = 0. \tag{5.114}$$

Using equations (5.113) and (5.114) we have

$$MR\omega + m\left(\frac{R\omega}{2} - \frac{\sqrt{3}}{2}v\right) = 0$$

$$\Rightarrow \omega = \frac{\sqrt{3}\,mv}{(2M+m)R}. \tag{5.115}$$

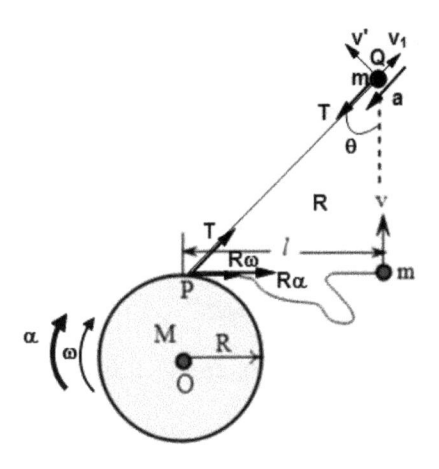

(b) The tension in the string is

$$\vec{T} = m(\vec{a}'_{QP} + \vec{a}'_P), \qquad (5.116)$$

where the acceleration of Q relative to P is given as

$$a'_{QP} = \frac{v_{QP}^2}{l} = \frac{(\omega R \cos\theta + v')^2}{l}$$

$$= \frac{(\omega R \cos\theta + v \sin\theta)^2}{l} \left(\because v' = v \sin\theta \right).$$

Putting $\theta = 30°$ we have

$$a'_{QP} = \frac{1}{l}\left(\frac{v}{2} + \frac{\sqrt{3}\omega R}{2}\right)^2$$

$$= \frac{1}{4l}(v + \sqrt{3}\omega R)^2. \qquad (5.117)$$

The radial acceleration of the particle P is

$$a'_P = R\omega^2 \cos\theta - R\alpha \sin\theta.$$

Putting $\theta = 30°$ we have

$$a'_P = \frac{R}{2}(\sqrt{3}\omega^2 - \alpha). \qquad (5.118)$$

Using the equations (5.116), (5.117), and (5.118) we have

$$T = m\left\{\frac{1}{4l}(v + \sqrt{3}R\omega)^2 + \frac{R}{2}(\sqrt{3}\omega^2 - \alpha)\right\}. \qquad (5.119)$$

Just after the string is taut, the tension in the string pulls the revolving bead towards the center of its revolution. The torque produced by tension about O is

$$(T \sin \theta)R = \frac{MR^2}{2}\alpha$$

$$(T \sin 30°)R = \frac{MR^2}{2}\alpha$$

$$\Rightarrow T = MR\alpha. \tag{5.120}$$

Using equations (5.114), (5.119), and (5.120) we have

$$T = m\left[\frac{1}{4l}\left\{v + \sqrt{3}R\frac{\sqrt{3}mv}{(2M+m)R}\right\}^2 + \frac{R}{2}\left\{\sqrt{3}\frac{3m^2v^2}{(2M+m)^2R^2} - \frac{T}{MR}\right\}\right]$$

$$T + \frac{Tm}{2M} = m\left[\frac{1}{4l}\left\{v + \sqrt{3}R\frac{\sqrt{3}mv}{(2M+m)R}\right\}^2 + \frac{R}{2}\left\{\sqrt{3}\frac{3m^2v^2}{(2M+m)^2R^2}\right\}\right]$$

$$\Rightarrow T = \frac{Mm^2v^2}{(m+2M)^3}\left\{\frac{2(M+2m)^2}{l} + \frac{3\sqrt{3}m^2}{R}\right\}.$$

(c) The change in kinetic energy is

$$\Delta K = \frac{m}{2}\left(v_1^2 + v'^2\right) + \frac{MR^2}{4}\omega^2 - \frac{1}{2}mv^2$$

$$= \frac{m}{2}\left\{(v\sin\theta)^2 + (\omega R \sin\theta)^2\right\} + \frac{MR^2}{4}\omega^2 - \frac{1}{2}mv^2.$$

Putting $\theta = 30°$ we have

$$\Delta K = \frac{\omega^2 R^2}{4}\left[\frac{m}{2} + M\right] - \frac{3}{8}mv^2. \tag{5.121}$$

Using equations (5.115) and (5.121) we have

$$\Delta K = \frac{3m^2v^2}{(2M+m)^2R^2}\cdot\frac{R^2}{4}\left(\frac{m+2M}{2}\right) - \frac{3}{8}mv^2$$

$$\Rightarrow \frac{3mv^2}{8}\left\{\frac{m}{2M+m} - 1\right\} = -\frac{3mv^2}{8}\left(\frac{2M}{2M+m}\right)$$

5-163

$$\Rightarrow \quad -\frac{\Delta K}{K} = \frac{3}{4}\frac{2M}{2M+m} = \frac{3M}{2(2M+m)}.$$

Problem 84 A block of mass m is connected to a pulley of mass M, moment of inertia I, and radius R by an inextensible string which is wrapped over the pulley. The pulley is smoothly hinged at its center. If the block is released from rest from the given position, an impact takes place between the pulley and ball by the impulsive tension of the string during the impact. Find (a) the angular velocity of the pulley just after the impact, (b) the change in energy during the impact, and (c) the fraction of energy lost during the impact. Assume that the string does not slip over the pulley and the pulley is a uniform disc. Put $M = 2m$.

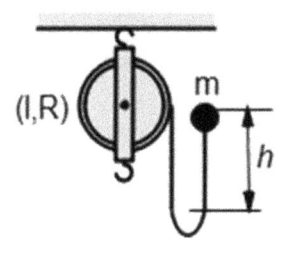

Solution
(a) Since the impulsive tension is an internal force in the pulley–ball system, it cannot change the angular momentum of the system due to zero net torque acting about the pivot. So, conserving the angular momentum of the system during the impact, we have

$$L = mv_0 R = I\omega + mvR, \qquad (5.122)$$

where $\omega =$ the angular velocity of the pulley and $v =$ the velocity of the ball, just after the impact.

Since the string is inextensible,

$$v = R\omega. \qquad (5.123)$$

Using equations (5.122) and (5.123) we have

$$L = mv_0 R = (I + mR^2)\omega. \qquad (5.124)$$

As the string will be taut after the ball falls through a distance $2h$, the velocity of the ball just before the impact is

$$v_0 = \sqrt{2g(2h)} = 2\sqrt{gh}. \qquad (5.125)$$

Using equations (5.124) and (5.125) we have

$$\omega = \frac{mv_0 R}{I + mR^2} = \frac{2mR\sqrt{gh}}{I + mR^2}. \tag{5.126}$$

Putting $I = MR^2/2 = (2m)R^2/2 = mR^2$. We have

$$\omega = \frac{2mR\sqrt{gh}}{mR^2 + mR^2} = \frac{\sqrt{gh}}{R}.$$

(b) The change in kinetic energy during the impact is

$$\Delta K = \frac{L^2}{2I_{\text{net}}} - \frac{1}{2}mv_0^2. \tag{5.127}$$

Using equations (5.126) and (5.127) we have

$$\Delta K = \frac{m^2 v_0^2 R^2}{2(I + mR^2)} - \frac{1}{2}mv_0^2$$

$$= -\frac{mv_0^2}{2}\left(\frac{I}{I + mR^2}\right) = -\frac{m4gh}{2}\frac{I}{I + mR^2}$$

$$= -\frac{2Imgh}{(I + mR^2)}.$$

Putting $I = mR^2$ we have

$$\Delta K = -mgh.$$

(c) Then the fraction of energy lost is

$$\eta = -\frac{\Delta K}{K} = \frac{I}{I + mR^2}.$$

Putting $I = mR^2$ we have

$$\eta = \frac{mR^2}{mR^2 + mR^2} = 1/2.$$

Problem 85 Two small balls each of mass m are connected to a pulley of moment of inertia I and radius R, as shown in the figure. Another small ball of mass m is released from rest from a height h as shown in the figure. The coefficient of restitution of the collision is e. Find (a) the angular velocity of the pulley just after the collision, (b) the value of e if the freely falling ball stops just after the collision, and (c) the velocity of the balls just after ball C sticks to ball C just after the collision. Assuming the the string does not slip over the pulley during the impact and the collision between B and C is nearly head-on.

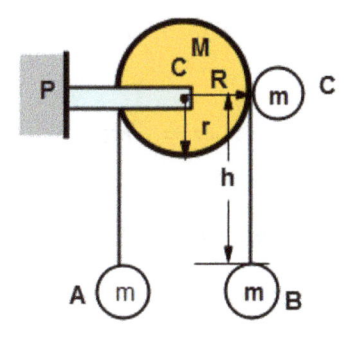

Solution

(a) The freely falling ball C collides with the ball B with a velocity
$$v_0 = \sqrt{2gh}.$$

Let the balls B and C move down just after the collision with velocities v and v', respectively.

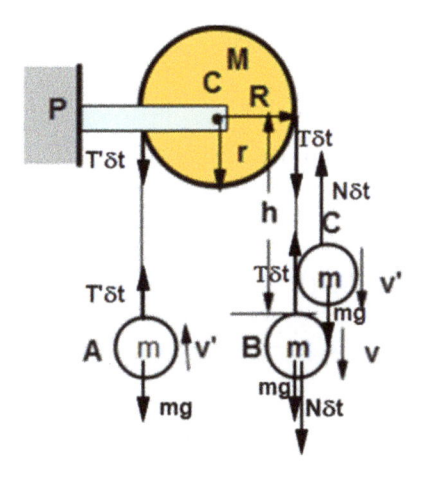

Applying Newton's impact formula we have
$$-e(-v_0) = (-v') - (-v)$$
$$\Rightarrow ev_0 = v - v'. \tag{5.128}$$

Let us consider the impulses $T'\delta t$ and $T\delta t$ imparted by the string on the bodies A and B, respectively, to write the following impulse–momentum equations.

For pulley M:
$$(T - T)'\delta tR = I\omega. \tag{5.129}$$

For A:
$$(-mg + T')\delta t = mv. \tag{5.130}$$

For B:
$$(-mg + T)\delta t = -mv. \tag{5.131}$$

For C:
$$(N - mg)\delta t = m\{(-v') - (-v_0)\}$$
$$\Rightarrow (N - mg)\delta t = m(v_0 - v'). \tag{5.132}$$

Adding equations (5.131) and (5.132) we have
$$(-2mg + T)\delta t = -mv + m(v_0 - v'). \tag{5.133}$$

Subtracting equation (5.130) from equation (5.133) we have
$$(-mg + T - T')\delta t = -2mv + m(v_0 - v') \tag{5.134}$$
$$\Rightarrow (T - T')\delta t \approx -2mv + m(v_0 - v')(\because mg\delta t \cong 0).$$

From equations (5.129) and (5.134)
$$\frac{I\omega}{R} = -2mv + m(v_0 - v'). \tag{5.135}$$

As the string does not slip over the pulley,
$$\omega = \frac{v}{R}. \tag{5.136}$$

Using equations (5.128), (5.135), and (5.136) we have
$$v = \frac{(1+e)v_0}{3 + \frac{I}{MR^2}} \tag{5.137}$$

$$\Rightarrow \omega = \frac{v}{R} = \frac{(1+e)v_0/R}{(3 + I/MR^2)}. \tag{5.138}$$

(b) Using equations (5.128) and (5.137) we have
$$v' = v_0 \left\{ \frac{1 - \left(2 + \frac{I}{MR^2}\right)e}{3 + \frac{I}{MR^2}} \right\}. \tag{5.139}$$

If the body stops just after the collision, put $v' = 0$ in equation (5.139) to obtain
$$e = \frac{1}{4 + \frac{I}{mR^2}}.$$

5-167

(c) If the body sticks after the collision, put $e = 0$ in equation (5.139) to obtain

$$v = v' = \frac{v_0}{3 + \frac{I}{mR^2}}.$$

Problem 86 A small ball of mass m is connected to a disc of mass M and radius R by a light inextensible string of length l_0. The disc is pivoted smoothly at C. If the ball is given a velocity v_0 perpendicular to the string, the disc will rotate and the ball will move such that the string will never slacken. Eventually, the string gets wrapped over the disc and the ball will hit the disc. Just before the ball hits the disc find (a) the angular velocity of the disc, (b) the velocity of the ball, and (c) the velocity of hitting if $l_0 = R$. Neglect gravity and friction.

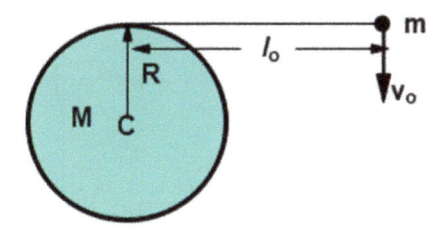

Solution
(a) The ball moves in a curve such that it has one component of its velocity that is tangential and the other component of its velocity will be radial just before it hits the disc, as shown in the figure. As the string is inextensible, the work done by tension is zero. In other words, the kinetic energy of the ball–disc system remains constant. Let ω be the angular velocity of the disc due to the torque produced by the tension and $v =$ the radial velocity of the ball just before the ball hits the disc. Conserving the kinetic energy we have

$$\frac{MR^2}{4}\omega^2 + \frac{m}{2}(R^2\omega^2 + v^2) = \frac{1}{2}mv_0^2$$

$$\Rightarrow \frac{1}{2}\left(\frac{MR^2}{2} + mR^2\right)\omega^2 + \frac{mv^2}{2} = mv_0^2$$

$$\Rightarrow \left(\frac{M + 2m}{2}\right)R^2\omega^2 + mv^2 = mv_0^2. \tag{5.140}$$

Conserving the angular momentum of the system $(M + m)$ about C, we have

$$mv_0 l_0 = mR\omega \cdot R + \frac{MR^2}{2}\omega = \left(\frac{MR^2}{2} + mR^2\right)\omega$$

$$\Rightarrow \left(\frac{M+2m}{2}\right)R^2\omega = mv_0 l_0 \qquad (5.141)$$

$$\Rightarrow \omega = \frac{2mv_0 l_0}{(M+2m)R^2}.$$

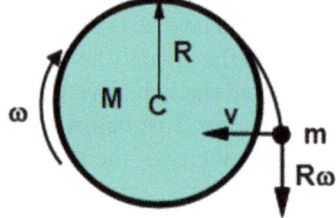

(b) Using equations (5.140) and (5.141)

$$\left(\frac{M+2m}{2}\right)\frac{R^2 m^2 v_0^2 l_0^2}{\left(\frac{M+2m}{2}\right)^2 R^4} + mv^2 = mv_0^2$$

$$\Rightarrow v_0^2 - \frac{2mv_0^2 l_0^2}{(M+2m)R^2} = v^2$$

$$\Rightarrow v = v_0\sqrt{1 - \frac{2ml_0^2}{R^2(M+2m)}}.$$

(c) If $l_0 = R$, $v = \sqrt{\frac{M}{M+2m}}$.

Problem 87 A disc of mass m is kept on a horizontal plane being smoothly pivoted at its center C, as shown in the figure. A ball of mass m is connected to the disc by an inextensible light string which is wrapped over the disc. A bead of mass m collides with ball. Find the tension in the string, (b) angular acceleration of the disc, (c) acceleration of the bead just after the collision. Assume the coefficient of restitution of the collision as $e = 0.5$ and the pulley is smoothly pivoted at its center.

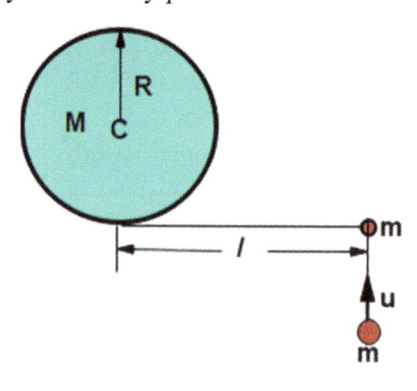

Solution

(a) Just after the bead sticks to the ball they will move as a combined unit with a velocity $v = mu/(m+m) = u/2$. So the string will be taut with a tension T, say, which generates a torque TR producing an angular acceleration α about the center C of the disc:

$$\tau_C = I_C \alpha$$

$$\Rightarrow TR = \frac{MR^2}{2}\alpha$$

$$\Rightarrow T = \frac{MR}{2}\alpha. \qquad (5.142)$$

Applying Newton's second law on the combined mass $2m$,

$$\vec{T} = 2m\vec{a_P} = 2m(\vec{a_{PQ_r}} + \vec{a_{Q_r}})$$

$$\Rightarrow -T = 2m(-a_{PQ_r} + a_{Q_r})$$

$$\Rightarrow T = 2m(a_{PQ_r} - a_{Q_r})$$

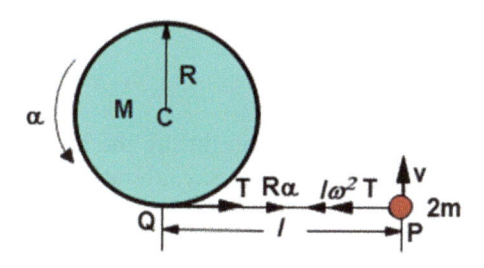

$$\Rightarrow T = 2m(l\omega^2 - R\alpha). \qquad (5.143)$$

The angular velocity ω of the point P relative to Q is

$$\omega = \frac{u}{2l}. \qquad (5.144)$$

Using equations (5.143) and (5.144) we have

$$\Rightarrow T = 2m\left\{\frac{u^2}{4l} - R\alpha\right\}.$$

Using equations (5.142) and (5.143) we have

$$T = 2m\left\{\frac{u^2}{4l} - \frac{2T}{M}\right\}$$

5-170

$$\Rightarrow T\left\{1 + \frac{4m}{M}\right\} = \frac{mu^2}{2l}$$

$$\Rightarrow T = \frac{Mmu^2}{2(M+4m)l}.$$

(b) Then the angular acceleration is

$$\alpha = \frac{2T}{MR} = \frac{mu^2}{(M+4m)lR}.$$

(c) The acceleration of the beads is

$$a_m = \frac{T}{m} = \frac{Mu^2}{2(M+4m)l}.$$

Problem 88 A rod PQ of mass M and length l is pivoted smoothly at its mid-point C. It is connected with a particle of mass m by an inextensible light string tied at the end Q of the rod. The string is slightly slack and the rod is given an angular velocity ω_0, as shown in the figure. Find (a) the angular velocity of the rod, (b) the velocity of the particle just after the string gets taut, (c) the impulse acting on the particle, (d) the energy transmitted to the bead, and (e) the fraction of energy lost during the collision. Assume that the rod and particle move on a smooth horizontal surface.

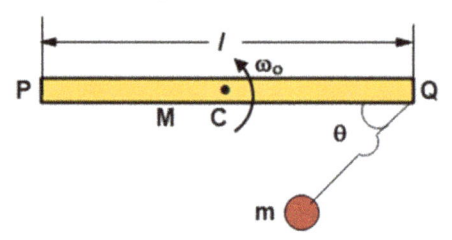

Solution

(a) The angular impulse of the tension during the impact changes the angular momentum of the rod from ω_0 to ω, say, about the pivot. Then we can write the impulse–momentum equations as

$$-T\delta t \sin\theta \frac{l}{2} = \frac{Ml^2}{12}(\omega - \omega_0). \tag{5.145}$$

The impulse of the tension changes the velocity of the ball from zero to v, say. Then its impulse–momentum will be

$$T\delta t = mv. \tag{5.146}$$

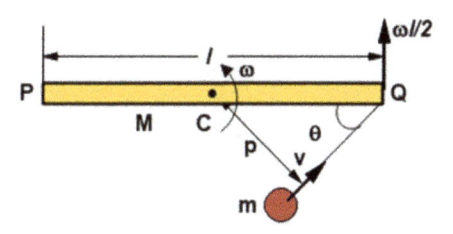

Since the string is inextensible, the velocities of the bead m and end Q of the rod along the string must be equal,

$$\omega \frac{l}{2} \sin \theta = v. \tag{5.147}$$

From equations (5.145) and (5.146)

$$-mv \sin \theta \frac{l}{2} = \frac{Ml^2}{12}(\omega - \omega_0). \tag{5.148}$$

From equations (5.147) and (5.148)

$$\frac{Ml^2}{12}\omega + mv \sin \theta \frac{l}{2} = \frac{Ml^2}{12}\omega_0$$

$$\Rightarrow \omega \left[\frac{Ml^2}{12} + \frac{ml^2}{4} \sin^2 \theta \right] = \frac{Ml^2}{12}\omega_0$$

$$\Rightarrow \omega = \frac{M\omega_0}{(M + 3m \sin^2 \theta)}.$$

(b) Then $v = \omega \frac{l}{2} \sin \theta = \dfrac{M\omega_0 l \sin \theta}{2(M + 3m\sin^2\theta)}.$

(c) The impulse is

$$I = mv = \frac{Mm\omega_0 l \sin \theta}{2(M + 3m\sin^2\theta)}.$$

(d) The energy transmitted to the bead is

$$K = \frac{1}{2}mv^2 = \frac{m}{2}\left(\frac{M\omega_0 l \sin \theta}{M + 3m\sin^2\theta}\right)^2.$$

(e) $\eta = \frac{\Delta K}{K} = -1(1 - \frac{K_f}{K_i})$

$$K_f = \frac{1}{2}mv^2 + \frac{Ml^2}{24}\omega^2,$$

5-172

where

$$v = \frac{\omega l}{2}\sin\theta$$

$$\text{Then, } K_f = \frac{1}{2}m\frac{\omega^2 l}{4}\sin^2\theta + \frac{Ml^2\omega^2}{24}$$

$$= \frac{\omega^2 l^2}{8}\left(m\sin^2\theta + \frac{M}{3}\right)$$

$$= \frac{M^2\omega_0 l^2\sin^2\theta}{24(M+3m\sin^2\theta)}.$$

Now we can find the fraction loss of energy, which can also be found in another method, as given below

Alternative method

During the interaction between the rod and the bead, the tension in the string is an internal force in the rod–bead system. Hence, the net torque produced by the tension is zero about the pivot. So the angular momentum of the system is conserved about C:

$$L = \frac{Ml^2}{12}\omega_0 = \frac{Ml^2}{12}\omega + mvp,$$

where $p = \frac{l}{2}\sin\theta$

$$\Rightarrow \frac{Ml^2}{12}\omega_0 = \frac{Ml^2}{12}\omega + mv\frac{l}{2}\sin\theta,$$

where $v = \frac{\omega l}{2}\sin\theta$

$$\Rightarrow \left(\frac{Ml^2}{12}\right)\omega_0 = \frac{Ml^2}{12} + \frac{ml^2}{4}\sin^2\theta\,\omega$$

$$\Rightarrow \omega = \frac{M\omega_0}{M+3m\sin^2\theta}.$$

The linear velocity of the bead is

$$v = \frac{\omega l}{2}\sin\theta = \frac{M\omega_0 l\sin\theta}{2(M+3m\sin^2\theta)}.$$

Then the impulse is

$$I = mv = \frac{Mm\omega_0 l\sin\theta}{2(M+3m\sin^2\theta)}.$$

The change in the kinetic energy of the system is

$$\Delta K = \frac{L^2}{2I_0}\left(1 - \frac{I_0}{I}\right)$$

$$\Rightarrow \left|\frac{\Delta K}{K_0}\right| = 1 - \frac{I_0}{I} = \left|\frac{I - I_0}{I}\right|$$

$$= \frac{\frac{Ml^2}{4}\sin^2\theta}{\frac{Ml^2}{12} + \frac{Ml^2}{4}\sin^2\theta}$$

$$= \frac{3m\sin^2\theta}{M + 3\sin^2\theta}.$$

Problem 89 A rod PQ of mass M and length l is pivoted smoothly at C. It is connected to a particle of mass m by an inextensible string AQ of length $l/2$ tied at the end Q. The string is slightly slack and the rod is given an angular velocity ω_0 and simultaneously the particle is given a linear velocity v_0, as shown in the figure. Find (a) the angular velocity of the rod, (b) the impulse, velocity of the particle and velocity of the center of mass of the rod of the particle, and (c) the tension in the string (d) the acceleration of the particle, (e) the acceleration of the point P of the rod, just after the string becomes taut. Assume that the bodies move in a smooth horizontal surface.

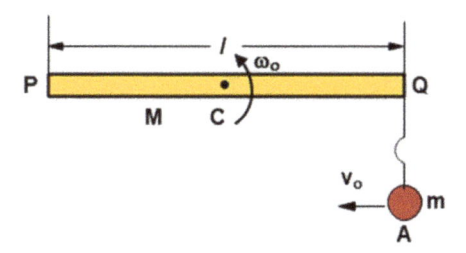

Solution
(a) Method 1
Since the tension in the string is an internal force in the rod–particle system, the net torque produced by the tension is zero about the pivot. So the angular momentum of the system $(M + m)$ is conserved about the pivot C just before and after the impact. Since the velocity v_0 is perpendicular to the tension in the string, the linear momentum of the particle remains constant during the impact. So, we can write:

$$\frac{Ml^2}{12}\omega - mv_0l + \frac{mvl}{2} = \frac{Ml^2}{12}\omega_0 - mv_0l$$

5-174

$$\Rightarrow \frac{Ml}{6}(\omega_0 - \omega) = mv. \tag{5.149}$$

Conserving the linear momentum of the system, we have

$$Mv_C = mv. \tag{5.150}$$

If the string will be taut just after the collision, equating the velocities of A and Q along the string AQ, we have

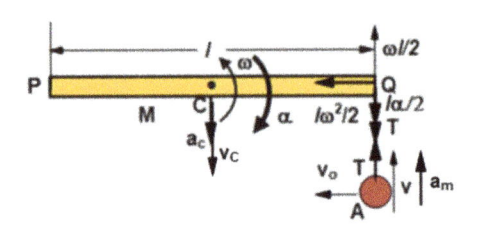

$$v_A = v_Q = v_{QC} - v_C$$

$$\Rightarrow v = l\omega/2 - v_C. \tag{5.151}$$

Using equations (5.150) and (5.151) we have

$$\left(1 + \frac{m}{M}\right)v = \frac{l\omega}{2}. \tag{5.152}$$

Using equations (5.149) and (5.152) we have

$$\frac{Ml}{6}(\omega_0 - \omega) = \frac{Mml\omega}{2(M+m)}$$

$$\Rightarrow \left\{\frac{m}{(M+m)} + \frac{1}{3}\right\}\omega = \frac{\omega_0}{3}$$

$$\Rightarrow \omega = \frac{(M+m)\omega_0}{(M+4m)}. \tag{5.152a}$$

(b) During the impact, let us consider the impulses $T\delta t$ imparted by the string on the particle A and the rod to write the following impulse–momentum equations.
For disc M:

$$-T\delta t l/2 = (Ml^2/12)(\omega - \omega_0) \tag{5.153}$$

$$T\delta t = Mv_C. \tag{5.154}$$

For *m*:

$$T\delta t = mv. \tag{5.155}$$

Putting ω from equation (5.153), v_C from equation (5.154), and v from equation (5.155) in equation (5.151), we have

$$\frac{T\delta t}{m} = \left(\omega_0 - \frac{6T\delta t}{Ml}\right)\frac{l}{2} - \frac{T\delta t}{M}$$

$$\Rightarrow T\delta t = \frac{Mml\omega_0}{2(M+4m)}. \tag{5.156}$$

Then the velocity of the particle *m* is

$$v = T\delta t/m = \frac{Ml\omega_0}{2(M+4m)}. \tag{5.157}$$

Then the velocity of the center of the rod is

$$v_C = T\delta t/M = \frac{ml\omega_0}{2(M+4m)}. \tag{5.158}$$

Putting the impulse $T\delta t$ from equation (5.158) in equation (5.152) we can also obtain same value of ω.

(c) Just after the impact, when the string is taut, let T be the tension in the string generated by the motion of the particle A relative to the end Q of the rod. Applying Newton's second law of translation and rotation, let us write the following equations.

For rod *M*:

The torque about the point C rotates the rod with a clockwise angular acceleration α, given as

$$Tl/2 = (Ml^2/12)\alpha$$

$$T = Ml\alpha/6. \tag{5.159}$$

As the string pulls the rod down, the acceleration of the center of mass C of the rod points vertically down, given as

$$T = Ma_C. \tag{5.160}$$

For *m*:

As the string pulls the particle up, the acceleration of the particle A points vertically up, given as

$$T = ma_m. \tag{5.161}$$

As the string pulls the particle A up, its upward acceleration is given by writing the accelerations along the vertical direction:

$$\vec{a}_A(=\vec{a}_m) = \vec{a}_{AQ} + \vec{a}_{QC} + \vec{a}_C$$

$$\Rightarrow a_m \hat{j} = \frac{v_0^2}{l}\hat{j} - \left(a_C + \frac{l\alpha}{2}\right)\hat{j}$$

$$\Rightarrow a_m = \frac{v_0^2}{l} - \left(a_C + \frac{l\alpha}{2}\right). \tag{5.162}$$

Putting $l\alpha/2$ from equation (5.159), a_C from equation (5.160), and a_m from equation (5.161) in equation (5.162) we have

$$\frac{T}{m} = \frac{v_0^2}{l} - \left(\frac{3T}{M} + \frac{T}{M}\right)$$

$$\Rightarrow T\left(\frac{4m}{M} + 1\right) = \frac{mv_0^2}{l}$$

$$\Rightarrow T = \frac{Mmv_0^2}{(4m+M)l}. \tag{5.163}$$

(d) Then the acceleration of the particle is

$$a_A = T/m = \frac{Mv_0^2}{(4m+M)l} \text{ (up)}.$$

(e) Then the acceleration of the point P is

$$\vec{a}_P = \vec{a}_{PC} + \vec{a}_C$$

$$= \frac{l\omega^2}{2}\hat{i} + \frac{l\alpha}{2}\hat{j} - \frac{T}{M}\hat{j} \tag{5.164}$$

$$= \frac{l\omega^2}{2}\hat{i} + \left(\frac{l\alpha}{2} - \frac{T}{M}\right)\hat{j}.$$

Putting $l\alpha/2$ from equation (5.159), ω^2 from equation (5.152a), and T from equation (5.163) in equation (5.164) we have

$$\vec{a}_P = \frac{l}{2}\left(\frac{M\omega_0}{M+3m}\right)^2 \hat{i} + \frac{2mv_0^2}{(4m+M)l}\hat{j}.$$

The collision of particles with a rod

Problem 90 A bead of mass m collides with a ball of mass m which is connected to another ball of mass M by a rigid light rod of length l. If the bead sticks to the ball just after the collision, find (a) the angular velocity of the rod, (b) the energy loss during the collision, and (c) the fraction of energy lost during the collision.

Solution

(a) As the colliding red ball sticks to the yellow ball just after the collision, they will move as a combined unit of mass $2m$. Let the rod have an angular velocity ω and the colliding balls move as a combined unit with velocities v just after the impact, as shown in the figure.

The angular momentum of the system $(M + 2m)$ about the center of mass is conserved:

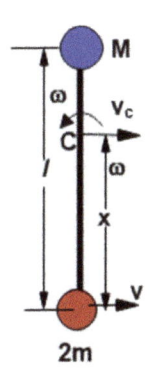

$$L_C = mux = I_C\omega. \tag{5.165}$$

The distance of the center of mass from $2m$ is

$$x = \frac{Ml}{M + 2m}. \tag{5.166}$$

The moment of inertia about C is

$$I_C = \frac{M(2m)l^2}{2m + M}. \tag{5.167}$$

Using all three equations we have

$$\Rightarrow mu\left(\frac{Ml}{M+2m}\right) = \frac{M(2m)l^2}{2m+M}\omega$$

5-178

$$\Rightarrow \omega = \frac{u}{2l}. \tag{5.168}$$

Conserving the linear momentum of the system,

$$(M + 2m)v_C = mu.$$

Then the velocity of the center of mass of the system is

$$v_C = \frac{mu}{M + 2m}. \tag{5.169}$$

(b) The change in kinetic energy is

$$K_f - K_i = K_{rot} + K_{trans} - K_{initial}$$

$$= \frac{1}{2}\left(\frac{2Mml^2}{M + 2m}\right)\omega^2 + \frac{1}{2}(M + 2m)v_C^2 - \frac{m}{2}u^2. \tag{5.170}$$

Using equations (5.167), (5.168), and (5.170) we have

$$\Delta K = \frac{1}{2}\frac{2Mml^2}{M + 2m}\left(\frac{u^2}{4l^2}\right) + \frac{1}{2}(M + 2m)\frac{m^2u^2}{(M + 2m)^2} - \frac{m}{2}u^2$$

$$= \frac{mu^2}{2}\left(\frac{M}{2(M + 2m)}\right) + \frac{mu^2}{2}\left(\frac{m}{M + 2m} - 1\right)$$

$$= \frac{mu^2}{2}\left[\frac{M + 2m - 2M - 4m}{2(M + 2m)}\right] = -\frac{mu^2}{4}.$$

(c) Then the fraction of energy lost is

$$\eta = -\frac{\Delta K}{K} = \frac{1}{2}.$$

Alternative method: Since the collision takes place between two particles each of mass m,

$$\Delta K = \frac{1}{2}\frac{Mm}{M + m}u^2 = \frac{mu^2}{4}.$$

Hence $\frac{\Delta K}{K} = 50\%$.

Problem 91 In the last problem, let the coefficient of restitution of the collision be e. Just after the collision (a) write the equations of the bodies, (b) the angular speed of the rod just after the collision, (c) the tension in the rod, (d) the acceleration of the connected bodies M and m, and (e) the energy loss during the collision.

Solution

(a) Let the rod have an angular velocity ω and the colliding balls have velocities v_1 and v_2 just after the impact, as shown in the figure. Conserving the angular momentum about any fixed point P that lies on the dotted line passing through the line of motion of the center of mass C of the system $(M + m)$, we can write

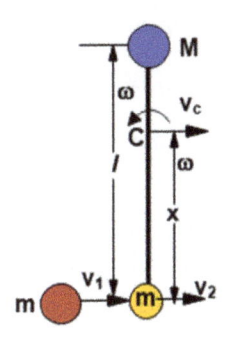

$$mux = \frac{Mml^2\omega}{M + m} + mv_1 x$$

$$\Rightarrow m(u - v_1)x = \frac{Mml^2\omega}{M + m}.$$

Putting $x = Ml/(M + m)$ we have

$$(u - v_1)\frac{Ml}{M + m} = \frac{Ml^2\omega}{M + m}$$

$$\Rightarrow u - v_1 = \omega l. \qquad (5.171)$$

Applying the Newton's impact (or collision) formula (NIF) for collision,

$$-eu = v_1 - v_2. \qquad (5.172)$$

Kinematics:

$$v_2 = v_C + \omega x. \qquad (5.173)$$

Conserving the linear momentum of the system $(M + 2m)$

$$mu = mv_1 + (M + m)v_C. \qquad (5.174)$$

(b) Using equations (5.173) and (5.174)

$$mu = m_1 v_1 + (M + m)(v_2 - \omega x). \qquad (5.175)$$

Using equations (5.172) and (5.175)

$$mu = mv_1 + (M + m)(v_1 + eu - \omega x). \qquad (5.176)$$

Using equations (5.171) and (5.176)
$$mu = (2m + M)(u - \omega l) + (M + m)eu - (M + m)\omega x.$$
By putting $(M + m)\omega x = Ml\omega$ we have
$$\omega l(M + m) = (2m + M)u - mu + (M + m)eu$$
$$= (M + m)u + (M + m)eu$$
$$\Rightarrow 2\omega l = u(1 + e)$$
$$\Rightarrow \omega = \frac{u(1 + e)}{2l}. \tag{5.177}$$

N.B. you can directly get this answer by writing the equation of conservation of the linear momentum of the two colliding particles each of mass m to have $v_1 + v_2 = v_0$ and then writing NIF: $v_1 - v_2 = -ev_0$ and finally you can solve these two equations.

(c) The tension in the rod can be given as
$$T = mx\omega^2 = m\frac{Ml}{M + m}\omega^2. \tag{5.178}$$

Using equations (5.176) and (5.177) we have
$$T = \frac{Mml}{M + m}\left\{\frac{u(1 + e)}{2l}\right\}^2$$
$$= \frac{Mmu^2(1 + e)^2}{4(M + m)l}.$$

(d) The acceleration of the particle of mass m and M can be given as
$$a_m = \frac{T}{m} = \frac{Mu^2(1 + e)^2}{4(M + m)l}$$
$$a_M = \frac{T}{M} = \frac{Mu^2(1 + e)^2}{4(M + m)l}.$$

(e) The expression for the loss of kinetic energy is
$$\Delta K = -\frac{1}{2}\frac{Mm(1 - e^2)u^2}{M + m}.$$

Then the fraction of energy lost is
$$\eta = -\frac{\Delta K}{K} = \left\{\frac{1}{2}\frac{Mm(1 - e^2)u^2}{M + m}\right\}\frac{1}{\left\{\frac{1}{2}mu^2\right\}}$$

$$= \frac{(1-e^2)M}{M+m}$$

$$\Rightarrow \frac{\Delta K}{K} = \frac{(1-e^2)}{2}.$$

Problem 92 Two particles of mass m and $2m$ are attached rigidly at the ends of a light rigid rod of length l. The rod–particle system is placed on a smooth horizontal surface. A linear impulse $J = F\delta t$ acts at the mid-point of the rod, as shown in the figure. Find (a) the velocity of the center of mass of the rod, (b) the angular velocity of the rod, and (c) the kinetic energy of the system just after the application of the impulse.

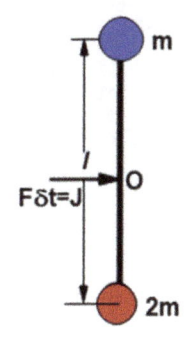

Solution
(a) The position of the center of mass C of the system (rod + particles) from the particle $2m$ is

$$x' = \frac{ml}{2m+m} = \frac{l}{3}.$$

Then we have

$$x_C = \frac{l}{2} - x' = \frac{l}{2} - \frac{l}{3} = \frac{l}{6}.$$

The applied impulse is

$$J = Mv_C,$$

where v_C = the velocity of C

$$\Rightarrow \vec{v}_C = \frac{J}{m+2m}\hat{i} = \frac{J}{3m}\hat{i}.$$

(b) Let the rod have an angular velocity ω due to the angular impulse of J about C, which is given as

$$Jx = J\left(\frac{l}{6}\right) = \frac{(2m)(m)}{2m+m}l^2\omega$$

$$\Rightarrow \frac{Jl}{6} = \frac{2}{3}ml^2\omega$$

$$\Rightarrow \omega = \frac{J}{4ml}.$$

(c) The kinetic energy of the system is

$$K = K_{\text{trans}} + K_{\text{rot}}$$
$$= \frac{1}{2}(m + 2m)v_C^2 + \frac{1}{2}I_C\omega^2$$
$$= \frac{1}{2}(m + 2m)v_C^2 + \frac{1}{2}\frac{I_C^2\omega^2}{I_C}$$
$$= \frac{1}{2}(3m)v_C^2 + \frac{J^2x^2}{2I_C}$$
$$= \frac{J^2}{2(3m)} + \frac{J^2\frac{l^2}{36}}{\frac{2m.m}{2m+m}l^2}$$
$$= \frac{J^2}{6m} + \frac{J^2\frac{l^2}{36}}{2 \times \frac{2}{3}ml^2} = \frac{J^2}{6m}\left(1 + \frac{1}{8}\right)$$
$$= \frac{9J^2}{6 \times 8m} = \frac{3J^2}{16m}.$$

Problem 93 A rod PQ of mass M and length l is smoothly pivoted at P. A putty of mass m collides at the lowest point Q of the rod with a horizontal velocity v_0 and sticks to the rod after the collision. Find (a) the angular velocity of the rod just after the collision, (b) the fraction of energy lost during the collision, (c) the maximum angle (assumed to be acute) that the rod swings after the collision, (d) the reaction at the pivot offered by the rod just after the collision and (e) the value of v_0 if the rod swings up to the horizontal. (Put $M = m$).

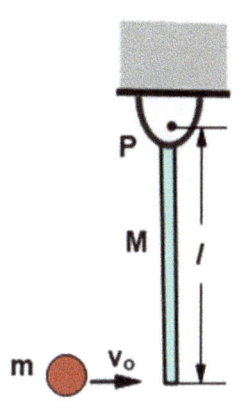

Solution
(a) Let the rod rotate with an angular velocity ω just after the collision.

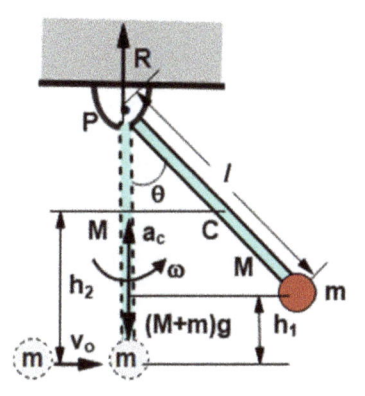

Conserving angular momentum about the pivot O,
$$L = mv_0 l = I\omega.$$

As the putty sticks to the rod just after the collision, the moment of inertia of the system $(M + m)$ after the collision is
$$I = \left(\frac{Ml^2}{3} + ml^2\right)$$

$$\Rightarrow \omega = \frac{3mv_0}{(M+3m)l} = \frac{3v_0}{4l} (\because M = m). \tag{5.179}$$

5-184

(b) The fraction change in kinetic energy is

$$\frac{\Delta K}{K} = \frac{\frac{L^2}{2I} - \frac{L^2}{2I_0}}{\frac{L^2}{2I_0}} = \left(\frac{I_0}{I} - 1\right)$$

$$= \frac{ml^2}{ml^2 + \frac{Ml^2}{3}} - 1$$

$$= \frac{3m}{3m + M} - 1$$

$$= -\frac{M}{M + 3m} = -\frac{1}{4}.$$

(c) After the collision we can apply the energy conservation given by

$$\Delta U + \Delta K = 0. \tag{5.180}$$

As the putty rises through a height h_1 and the center of mass of the rod rises through a height h_2, the change in gravitational potential energy is

$$\Delta U = mgh_1 + Mgh_2$$
$$= mgl(1 - \cos\theta) + \frac{Mgl}{2}(1 - \cos\theta) \tag{5.181}$$
$$= (1 - \cos\theta)\left(\frac{M + 2m}{2}\right)gl.$$

The kinetic energy of the system $(M + m)$ decreases to zero at the angle θ. Then we can write

$$\Delta K = -\frac{1}{2}I_p\omega^2 = -\frac{1}{2}\left(m + \frac{M}{3}\right)l^2\omega^2. \tag{5.182}$$

Using equations (5.180), (5.181), and (5.182)

$$\frac{1}{2}\left(m + \frac{M}{3}\right)l^2\omega^2 = (1 - \cos\theta)\left(\frac{M + 2m}{2}\right)gl. \tag{5.183}$$

Using equations (5.179) and (5.183)

$$(1 - \cos\theta)\left(\frac{M + 2m}{2}\right)gl = \frac{(3m + M)l^2}{6} \cdot \frac{9m^2v_0^2}{(M + 3m)^2l^2}$$

$$\Rightarrow (1 - \cos\theta)\left(\frac{M + 2m}{2}\right)gl = \frac{3m^2v_0^2}{2(M + 3m)}$$

5-185

$$\Rightarrow 2\sin^2\frac{\theta}{2} = \frac{3m^2v_0^2}{(M+2m)(M+3m)gl}$$

$$\Rightarrow 2\sin^2\frac{\theta}{2} = \frac{3m^2v_0^2}{(M+2m)(M+3m)gl}$$

$$\Rightarrow \theta = 2\sin^{-1}\left[\left\{\sqrt{\frac{3}{2(M+2m)(M+3m)gl}}\right\}mv_0\right]$$

$$= 2\sin^{-1}\sqrt{\frac{3m^2v_0^2}{2\times 3m\times 4m\times gl}} \quad (\because M=m)$$

$$\Rightarrow \theta = 2\sin^{-1}\frac{v_0}{2\sqrt{2gl}}. \tag{5.184}$$

(d) Referring to the free-body diagram, the reaction force R is given as
$$R - (M+m)g = (M+m)a_C = Ma_1 + ma_2$$

$$\Rightarrow R - (M+m)g = M\frac{l}{2}\omega^2 + ml\omega^2$$

$$\Rightarrow R = (M+2m)\frac{l\omega^2}{2} + (M+m)g. \tag{5.185}$$

Using equations (5.179) and (5.185)
$$R = (M+2m)\frac{l}{2}\frac{9m^2v_0^2}{(M+3m)^2 l} + (M+m)g$$

$$= (M+m)g\left[1 + \frac{9m^2v_0^2(M+2m)}{2g(M+3m)^2 l(M+m)}\right].$$

If $M = m$ we have
$$R = 2mg\left[I + \frac{9m^2v_0^2(3m)}{2g(16m^2)l(2m)}\right]$$

$$= 2mg\left[1 + \frac{27v_0^2}{64gl}\right].$$

(e) If the rod rotates by a right angle, put $\theta = 90°$ in equation (5.184) to obtain
$$\frac{\pi}{2} = 2\sin^{-1}\frac{v_0}{2\sqrt{2gl}}$$

5-186

$$\Rightarrow \frac{1}{\sqrt{2}} = \frac{v_0}{2\sqrt{2gl}}$$

$$\Rightarrow v_0 = 2\sqrt{gl}.$$

Problem 94 A bead of mass m collides with the uniform rod of mass M and length l, as shown in the figure. If the bead sticks to the rod just after the collision, find (a) the angular speed of the rod just after the collision and (b) the energy loss during the collision.

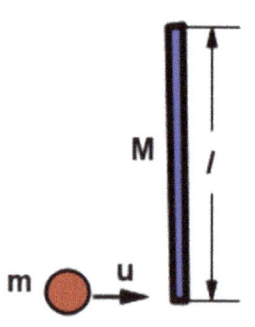

Solution

(a) Let the rod have an angular velocity w just after the impact. Conserving the angular momentum about the point P lying on the line of motion of the center of mass C of the rod–particle system, we have

$$mux = I_C w. \qquad (5.186)$$

The distance of the center of mass of the system from the bottom of the rod is

$$x = \frac{Ml}{2(M+m)}. \qquad (5.187)$$

The moment of inertia of the system about the center of mass C is

$$I_C = \frac{Ml^2}{12} + \frac{Mml^2}{4(M+m)}$$

$$I_C = \frac{Ml^2}{12} + \frac{Mml^2}{4(M+m)}$$

$$= \frac{Ml^2}{4}\left(\frac{1}{3} + \frac{m}{(M+m)}\right)$$

$$\Rightarrow I_C = \frac{Ml^2}{12}\left(\frac{M+4m}{M+m}\right). \tag{5.188}$$

Using equations (5.186), (5.187), and (5.188) we have

$$\frac{muMl}{2(M+m)} = \frac{Ml^2}{12}\left(\frac{M+4m}{M+m}\right)\omega$$

$$\Rightarrow \omega = \frac{6mu}{(M+4m)l}. \tag{5.189}$$

Conserving the linear momentum,

$$(M+m)v_C = mu$$

$$\Rightarrow v_C = \frac{mu}{M+m}. \tag{5.190}$$

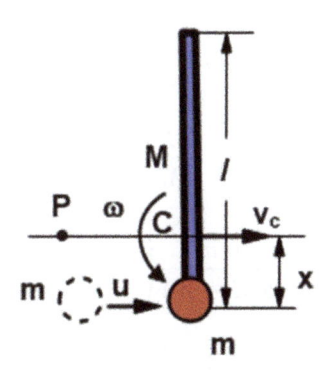

(b) The change in kinetic energy during the collision is

$$\Delta K = \frac{1}{2}(M+m)v_C^2 + \frac{1}{2}I_C\omega^2 - \frac{1}{2}mu^2. \tag{5.191}$$

Using equations (5.189), (5.190), and (5.191) we have

$$\Delta K = \frac{1}{2}(M+m)\left(\frac{mu}{M+m}\right)^2 +$$

$$\frac{1}{2}\left\{\frac{1}{12}\left(\frac{M+4m}{M+m}\right)Ml^2\right\}\left\{\frac{6mu}{(M+4m)l}\right\}^2 - \frac{1}{2}mu^2$$

$$= -\frac{Mmu^2}{2(M+m)} + \frac{3Mm^2u^2}{2(M+m)(M+4m)}$$

$$= \frac{Mmu^2}{2(M+m)}\left\{-1 + \frac{3m}{(M+4m)}\right\}$$

$$= -\frac{Mmu^2}{2(M+4m)}.$$

Alternative method

$$\Delta K = -\frac{Mmu^2}{2(M+m)} + \frac{1}{2I_C}(I_C\omega)^2$$

$$= -\frac{Mmu^2}{2(M+m)} + \frac{1}{2}\frac{(mux)^2}{I_C}. \tag{5.192}$$

Using equations (5.187), (5.189), and (5.192) we have

$$\Delta K = -\frac{Mmu^2}{2(M+m)} + \left(\frac{Ml^2}{24}\right)\left(\frac{M+4m}{M+m}\right)\left\{\frac{36m^2u^2}{(M+4m)^2l^2}\right\}$$

$$= -\frac{Mmu^2}{2(M+m)} + \frac{M(M+4m)}{2(M+m)} \times \frac{3m^2u^2}{(M+4m)^2}$$

$$\Rightarrow \frac{\Delta K}{K} = \frac{Mmu^2}{2(M+m)}\left\{-1 + \frac{3m}{M+4m}\right\}$$

$$= \frac{Mmu^2}{2(M+m)}\left[\frac{-M-4m+3m}{M+4m}\right]$$

$$\Rightarrow \Delta K = -\frac{Mmu^2}{2(M+4m)}$$

$$\Rightarrow \frac{\Delta K}{K} = -\frac{M}{M+4m}.$$

Problem 95 A small ball of mass m collides with a uniform rod of mass M and length l, as shown in the figure. The coefficient of restitution of the collision is e. Find (a) the angular velocity of the rod just after the collision and (b) the condition for which the beads stops and moves forward and backward.

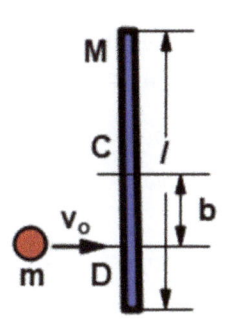

Solution

(a) Let the rod have an angular velocity w and the ball attains a velocity v_1 and the center of mass C of the rod gets a velocity v_2 just after the impact.

As the net torque is zero on the rod–ball system $(M + m)$, we can conserve the angular momentum of the system about any point lying on the line of motion of the center of mass of the rod. Then we can write

$$mv_0 b = mbv_1 + \frac{Ml^2}{12}w$$

$$\Rightarrow m(v_0 - v_1)b = \frac{Ml^2}{12}w$$

$$\Rightarrow m(v_0 - v_1) = \frac{Ml^2 w}{12b}. \tag{5.193}$$

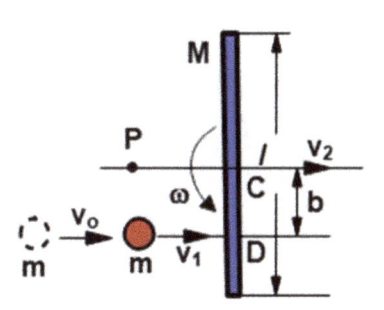

NIF for the collision between the point D of the rod and ball:

$$-e(v_0) = v_1 - v_D. \tag{5.194}$$

Kinematics:

$$v_D = v_2 + bw. \tag{5.195}$$

From equations (5.194) and (5.195)

$$-e(v_0) = v_1 - (v_2 + bw). \tag{5.196}$$

Conserving the linear momentum of the rod–particle system, we have
$$mv_0 = Mv_2 + mv_1. \qquad (5.197)$$

From equations (5.193) and (5.196)
$$-ev_0 = v_1 - b\omega - \{m(v_0 - v_1)/M\}$$

$$\Rightarrow \left(\frac{m}{M} - e\right)v_0 = v_1\left(1 + \frac{m}{M}\right) - b\omega. \qquad (5.198)$$

From equations (5.193) and (5.198)
$$\left(\frac{m}{M} - e\right)v_0 = \left(v_0 - \frac{Ml\omega^2}{12bm}\right)\left(1 + \frac{m}{M}\right) - b\omega$$

$$\omega\left[\frac{Ml^2}{12mb}\left(1 + \frac{m}{M}\right) + b\right] = \left\{\left(1 + \frac{m}{M}\right) - \frac{m}{M} + e\right\}v_0$$

$$\Rightarrow \omega b = \frac{(1+e)v_0}{\left\{\frac{Ml^2(M+m)}{12Mmb^2} + 1\right\}}$$

$$= \frac{(1+e)v_0}{\frac{(M+m)l^2}{12mb^2} + 1}$$

$$\Rightarrow \omega = \frac{(1+e)\left(\frac{v_0}{b}\right)}{\frac{(M+m)l^2}{12mb^2} + 1}.$$

(b) If the ball stops just after the collision, put $v_1 = 0$ in equation (5.193) to obtain
$$\frac{Ml^2}{12m}\omega = mv_0.$$

Putting the obtained value of ω we have
$$\frac{Ml^2}{12b}\left\{\frac{(1+e)\frac{v_0}{b}}{\frac{(M+m)l^2}{12mb^2} + 1}\right\} = \frac{m}{v_0}$$

$$\Rightarrow \frac{Ml^2(1+e)}{12b^2\frac{(M+m)l^2 + 12mb^2}{12mb^2}} = m$$

$$\Rightarrow \frac{Mml^2(1+e)}{(M+m)l^2 + 12mb^2} = m$$

5-191

$$\Rightarrow Ml^2 + Ml^2 e = Ml^2 + ml^2 + 12mb^2$$

$$\Rightarrow Ml^2 e = m(l^2 + 12b^2)$$

$$\Rightarrow e = \frac{m}{M}\left(1 + \frac{12b^2}{l^2}\right)$$

If $e < \frac{m}{M}\left(1 + \frac{12b^2}{l^2}\right)$, $v_1 \rightarrow$ (rightward)

If $e = \frac{m}{M}\left(1 + \frac{12b^2}{l^2}\right)$, $v_1 = 0$ (stops)

If $e > \frac{m}{M}\left(1 + \frac{12b^2}{l^2}\right)$, $v_1 \leftarrow$ (leftward).

Problem 96 In the previous problem let $b = y = 1/3$, $M = m$, $e = 0$ (inelastic collision), and the radius of gyration of the rod be K. Find (a) the angular velocity of the rod just after the collision and (b) the fraction loss in the kinetic energy in the collision.

Solution

(a) The general expression for ω is

$$mv_0\left(\frac{yM}{M+m}\right) = \left(MK^2 + \frac{Mmy^2}{M+m}\right)\omega,$$

where $y = $ the distance of the point of collision from the mid-point of the rod:

$$\Rightarrow \omega = \frac{mv_0 y}{(M+m)K^2 + my^2} = \frac{mv_0 \frac{l}{3}}{(m+m)\frac{l^2}{12} + m\left(\frac{l}{3}\right)^2} = \frac{6v_0}{5l} (\because M = m).$$

(b) As we derived in problem 136,

$$\left|\frac{\Delta K}{K}\right| = \frac{MK^2}{(M+m)K^2 + my^2}$$

$$= \frac{\frac{ml^2}{12}}{(m+m)\frac{l^2}{12} + m\left(\frac{l}{3}\right)^2} = \frac{\frac{1}{12}}{\frac{1}{6} + \frac{1}{9}}$$

$$= \frac{1}{12} \times \frac{9 \times 6}{9+6} = \frac{9}{15 \times 2} = \frac{3}{10} = 30\%.$$

Alternative method:
Due to the absence of any net external force acting on the system, conserving the linear momentum, we have

$$v_C = \frac{mv_0}{M+m} = \frac{mv_0}{m+m} = \frac{v_0}{2}.$$

The kinetic energy of the center of mass is

$$K_C = \frac{1}{2}(M+m)v_C^2 = \frac{1}{2}(m+m)\left(\frac{v_0}{2}\right)^2 = \frac{mv_0^2}{4}.$$

The kinetic energy of the center of mass remains conserved because the momentum and velocity of the center of mass remains constant.

Just after the impact, the kinetic energy of the system relative to the center of mass is

$$K'_{C_f} = \frac{I_C}{2}\omega^2 = \frac{1}{2}\left(\frac{Mmy^2}{M+m} + MK^2\right)\omega^2$$

$$= \frac{1}{2}\left\{\frac{m}{2}\left(\frac{L}{3}\right)^2 + \frac{ml^2}{12}\right\}\left(\frac{v_0}{5L}\right)^2 = \frac{mv_0^2}{10}.$$

Just before the impact, the kinetic energy of the system relative to the center of mass is

$$K'_{C_i} = \frac{1}{2}\frac{Mmv_0^2}{M+m} = \frac{mv_0^2}{4}(\because M = m).$$

Then $\Delta K = \Delta K' = K'_{C_f} - K'_{C_i}$

$$= \frac{mv_0^2}{10} - \frac{mv_0^2}{4} = -\frac{6mv_0^2}{40} = -\frac{3mv_0^2}{20}.$$

Then the fraction loss in kinetic energy in the collision is

$$\left|\frac{\Delta K}{K_i}\right| = \frac{\frac{3mv_0^2}{20}}{\frac{mv_0^2}{2}} = \frac{3}{10} = 30\%.$$

Problem 97 A bead of mass m collides with a rod of mass M and length l, at a distance $b = l/3$ from the mid-point C of the rod, as shown in the figure. If the bead comes to rest just after the collision, find (a) the coefficient of restitution of collision between the rod and the bead, (b) the energy lost and (c) the fraction of energy lost during the collision. Put $m/M = 3/10$.

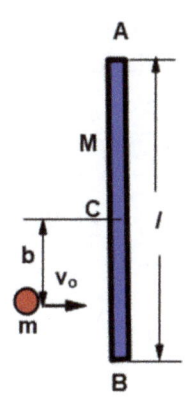

Solution
(a) With reference to problem 95,

$$v_1 = v_0 - \frac{Ml^2}{12mb}\omega = 0$$

$$\Rightarrow \omega = \frac{12mbv_0}{Ml^2}. \tag{5.199}$$

Again, we have obtained in problem 95,

$$\omega = \frac{\frac{(1+e)v_0}{b} \times 12mb^2}{(M+m)l^2 + 12mb^2}$$

$$\Rightarrow \omega = \frac{12(1+e)mv_0 b}{(M+m)l^2 + 12mb^2}. \tag{5.200}$$

Using equations (5.199) and (5.200)

$$\frac{12(1+e)mv_0 b}{(M+m)l^2 + 12mb^2} = \frac{12mbv_0}{Ml^2}$$

$$\Rightarrow \frac{(1+e)}{(M+m)l^2 + 12mb^2} = \frac{1}{Ml^2}$$

$$\Rightarrow (1+e)Ml^2 = (M+m)l^2 + 12mb^2$$

$$\Rightarrow e = \frac{m}{M}\left(1 + \frac{12b^2}{l^2}\right)$$

$$= \frac{3}{10}\left(1 + \frac{12l^2/9}{l^2}\right) = 0.7.$$

5-194

(b) The energy lost in the collision is
The loss of kinetic energy during the collision is

$$\Delta K = \frac{L^2}{2I_C} + \frac{mv_C^2}{2} - \frac{mv_0^2}{2}$$

$$\Delta K = \frac{(mv_0 b)^2}{2(Ml^2/12)} + \frac{m(mv_0/M)^2}{2} - \frac{mv_0^2}{2}$$

$$= \left\{ \left(\frac{6mb^2}{Ml^2} \right) + \left(\frac{m}{M} \right)^2 - 1 \right\} \frac{mv_0^2}{2} \qquad (5.201)$$

$$= \left\{ \left(\frac{6 \times 3l^2/9}{10l^2} \right) + \left(\frac{3}{10} \right)^2 - 1 \right\} \frac{mv_0^2}{2}$$

$$= \left\{ \left(\frac{1}{5} \right) + \left(\frac{3}{10} \right)^2 - 1 \right\} \frac{mv_0^2}{2}$$

$$= -\frac{71 mv_0^2}{200}.$$

(c) The fraction of energy lost in the collision is

$$\eta = -\Delta K/K = \frac{71 mv_0^2}{200} \bigg/ \frac{mv_0^2}{2}$$

$$= 0.71.$$

Problem 98 A small ball of mass m collides perpendicular to the rod of mass $M\ (=m)$ and length l, as shown in the figure. If the ball sticks to the rod find (a) the angular and linear velocity of the rod just after the collision, and (b) the energy loss in the collision.

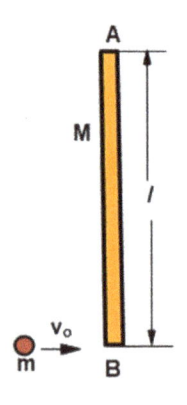

5-195

Solution

(a) Referring to problem 95 we have the following expressions:

$$\omega = \frac{12(1+e)mv_0 b}{(M+m)l^2 + 12mb^2}$$

$$v = \frac{m(l^2 + 12b^2) - Mel^2}{(M+m)l^2 + 12mb^2} v_0.$$

Putting $e = 0$, $b = l/2$, and $M = m$ we have the following results:

$$\omega = \frac{12mv_0 l/2}{(M+m)l^2 + (1/4)12ml^2} = \frac{6v_0}{5l}$$

$$v = \frac{m(4l^2) - (0)}{2ml^2 + 12(1/4)ml^2} v_0 = \frac{4}{5} v_0.$$

(b) Referring to problem 104 and derivation in problem 136, the fraction loss of kinetic energy for a perpendicular hit an angle of $90°$ is

$$\frac{-\Delta K}{K} = \frac{MK^2}{(M+m)K^2 + mb^2}$$

$$= \frac{K^2}{2K^2 + (1/4)l^2} (\because M = m, b = (l/2))$$

$$\left\{ \because MK^2 = \frac{Ml^2}{12} \Rightarrow K^2 = \frac{l^2}{12} \right\}.$$

Then finally we have

$$\frac{-\Delta K}{K} = \frac{K^2}{2K^2 + (1/4)l^2} = \frac{\frac{l^2}{12}}{2 \times \frac{l^2}{12} + (1/4)l^2}$$

$$= \frac{1}{5}.$$

Problem 99 A rod AC of mass m and length $2l$ is welded to a light rod of length $BC = l$ to form a composite rod AB. A particle of mass m collides with the end B of the composite rod, as shown in the figure. If the particle sticks to the rod just after the collision, find (a) the angular speed and linear velocity of the center of mass of the rod just after the collision, (b) the change in kinetic energy of the rod–particle system, and (c) the fraction of energy lost during the collision.

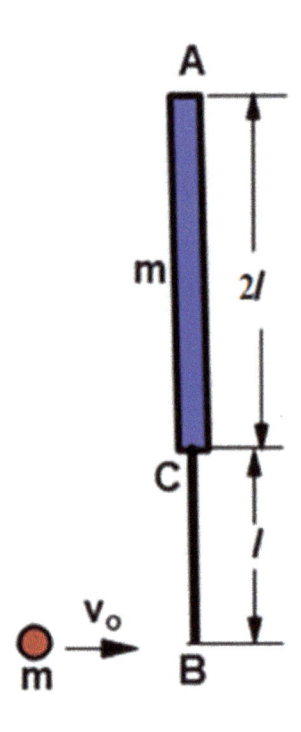

Solution
(a) The center of mass of the rod–particle system will lie at the point C of the composite rod after the particle m sticks to rod at point B just after the collision. Let the rod rotate about the center of mass C with an angular velocity w, the point C moves with a velocity v_C as shown in the figure.
Conserving the angular momentum about P we can write

$$mv_0 l = I_C w,$$

where the moment of inertia of the system about C is

$$I_C = ml^2 + \frac{m(2l)^2}{12} + ml^2 = \frac{7}{3}ml^2.$$

Then we have

$$mv_0 l = \frac{7}{3}ml^2 w$$

$$\Rightarrow w = \frac{3v_0}{7l}. \tag{5.202}$$

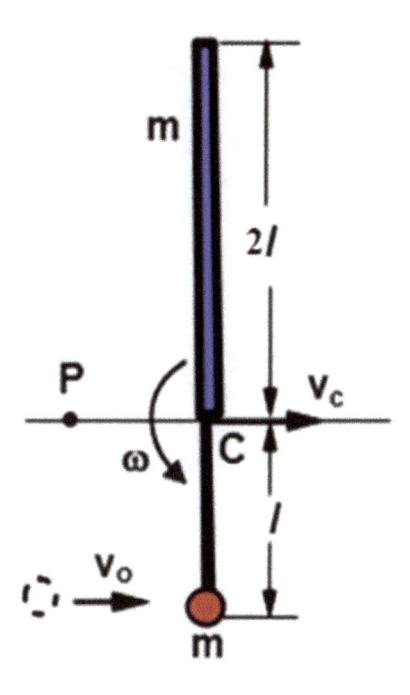

The conservation of linear momentum of the system yields

$$mv_0 = (m + m)v_C$$

$$\Rightarrow v_C = v_0/2. \tag{5.203}$$

(b) The change in kinetic energy in the collision is

$$\Delta K = \frac{1}{2}(m + m)v_C^2 + \frac{1}{2}I_C\omega^2 - \frac{1}{2}mv_0^2. \tag{5.204}$$

Using equations (5.202), (5.203), and (5.204) we have

$$\Delta K = \frac{2m(v_0/2)^2}{2} + \frac{1}{2}\left(\frac{7ml^2}{3}\right)\left(\frac{9v_0^2}{49l^2}\right) - \frac{1}{2}mv_0^2$$

$$= \frac{mv_0^2}{2}\left(-\frac{1}{2} + \frac{3}{7}\right) = -\left(\frac{mv_0^2}{28}\right).$$

(c) Then the fraction of energy lost is

$$\eta = -\frac{\Delta K}{K} = \frac{1}{14}.$$

5-198

Problem 100 A uniform wooden bar of mass M and length l is smoothly pivoted at its mid-point. A mud pellet of mass m strikes the end of the bar with a velocity v_0 and sticks to the rod, as shown in the figure. (a) Find (i) the speed of the rod just after the collision, (ii) the energy loss, and (iii) the fraction of energy lost during the collision. (b) If the rod can rotate in the vertical plane through a right angle $\theta = 90°$, find the minimum value of v. (c) If the reaction at the pivot will be equal to zero when the pellet reaches the top, find v.

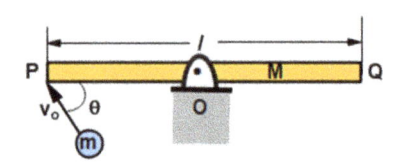

Solution
(a)
(i) Let the rod have an angular velocity ω_0 just after the impact. Conserving the angular momentum about the pivot O we have

$$L_0 = mv\frac{l}{2}\sin\theta = \left(\frac{ml^2}{4} + \frac{Ml^2}{12}\right)\omega_0$$

$$\Longrightarrow \omega_0 = \frac{6mv\sin\theta}{(3m+M)l}. \tag{5.205}$$

(ii) The change in kinetic energy in the collision is

$$\Delta K = \frac{L^2}{2I} - \frac{1}{2}mv^2$$

$$= \frac{\left(mv\frac{l}{2}\sin\theta\right)^2}{2\left(\frac{ml^2}{4} + \frac{Ml^2}{12}\right)} - \frac{mv^2}{2}$$

$$= -\frac{mv^2}{2}\left[1 - \frac{\sin^2\theta}{4} \times \frac{12m}{(M+3m)}\right]$$

$$= -\frac{M + 3m\cos^2\theta}{M + 3m}\left(\frac{mv^2}{2}\right).$$

(iii) Then the fraction of energy lost is

$$\eta = \frac{|\Delta K|}{K} = \frac{M + 3m\cos^2\theta}{M + 3m}.$$

(b) As the pellet just goes to the top, final the angular velocity of the rod will be zero. Then, conserving energy, we have

$$\frac{1}{2}I\omega_0^2 = mg\frac{l}{2}$$

$$\Rightarrow \omega_0 = \sqrt{\frac{mgl}{I}}, \qquad (5.206)$$

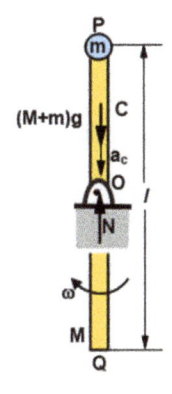

where the moment of inertia about O is

$$I = I_O = \left(\frac{3m + M}{12}\right)l^2.$$

From equations (5.205) and (5.206) we have

$$\Rightarrow \frac{6mv \sin \theta}{(3m + M)l} = \sqrt{\frac{mgl}{\left(\frac{3m+M}{12}\right)l^2}}$$

So, $v = \dfrac{1}{6 \sin \theta}\sqrt{\dfrac{12gl(3m + M)}{m}}$

$$\Rightarrow v = \frac{\sqrt{\frac{(M+3m)}{3m}gl}}{\sin \theta}.$$

(c) When the pellet reaches the top, the reaction at the pivot is

$$(M + m)g - N = (M + m)a_C$$

$$= M(0) + m\frac{l\omega^2}{2}. \qquad (5.207)$$

5-200

Putting $N = 0$ as the reaction at the pivot will be zero in equation (5.207), we have

$$(M + m)g = m\frac{l\omega^2}{2}. \tag{5.208}$$

Conserving the energy of the system between the top and bottom positions of the pellet just after the impact, we have

$$\frac{1}{2}I(\omega_0^2 - \omega^2) = mg\frac{l}{2}$$

$$\Rightarrow \omega_0^2 = \omega^2 + \frac{mgl}{I}. \tag{5.209}$$

Using equations (5.208) and (5.209) we have

$$-\left\{\frac{2(M + m)g}{ml}\right\} = -\omega_0^2 + \frac{mgl}{I}. \tag{5.210}$$

Using equations (5.205) and (5.210) we have

$$-\left\{\frac{2(M + m)g}{ml}\right\} = -\left\{\frac{6mv \sin \theta}{(3m + M)l}\right\}^2 + \frac{mgl}{I}$$

$$\Rightarrow \left\{\frac{6mv \sin \theta}{(3m + M)l}\right\}^2 = \left\{\frac{2(M + m)g}{ml}\right\} + \frac{mgl}{I}$$

$$\Rightarrow v = \frac{(3m + M)l}{6m \sin \theta}\sqrt{\left\{\frac{2(M + m)g}{ml}\right\} + \frac{mgl}{I}},$$

where the moment of inertia about O is

$$I = \left(\frac{3m + M}{12}\right)l^2$$

$$\Rightarrow v = \frac{(3m + M)l}{6m \sin \theta}\sqrt{\left\{\frac{2(M + m)g}{ml}\right\} + \frac{mgl}{\left(\frac{3m + M}{12}\right)l^2}}$$

$$\Rightarrow v = \frac{(3m + M)}{6m \sin \theta}\sqrt{\left\{\frac{2(M + m)}{m} + \frac{12m}{(3m + M)}\right\}gl}$$

$$\Rightarrow v = \frac{(3m + M)}{6m \sin \theta}\sqrt{\left\{\frac{(M + m)}{m} + \frac{6m}{(3m + M)}\right\}2gl}.$$

5-201

Problem 101 A ball of mass m collides with a rod AB at a distance b from the smooth pivot A. The rod has mass M (= m) and length l, and the coefficient of restitution of the collision is e. Then find (a) the angular velocity of the rod just after the collision, (b) the condition for which the ball moves forward, backward, and stops just after the collision, (c) the height to which the center of mass rises, and (d) the value of b for which the linear momentum of the rod−ball system remains conserved during the collision without any transverse reaction at the pivot.

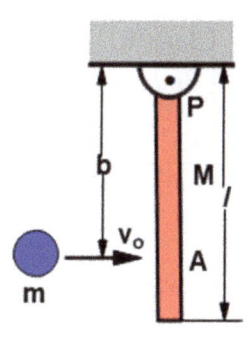

Solution

(a) Let the rod have an angular velocity w and the ball attains a velocity v just after the impact.

Conserving angular momentum about the pivot O we have

$$mv_0 b = mvb + \frac{Ml^2}{3}w. \tag{5.211}$$

The NIF for the collision is

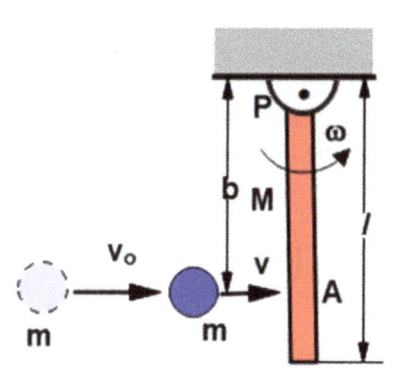

$$-ev_0 = v - bw. \tag{5.212}$$

Using equations (5.211) and (5.212) we have $mv_0 b = mb(bw - ev_0) + \frac{Ml^2}{3}w$

$$\Rightarrow mv_0 b(1+e) = \left(mb^2 + \frac{Ml^2}{3}\right)w$$

5-202

$$\Rightarrow \omega = \frac{3mv_0b(1+e)}{3mb^2 + Ml^2}. \tag{5.213}$$

(b) Using equations (5.212) and (5.213) we have

$$v = b\omega - ev_0$$

$$\Rightarrow v = b\left\{\frac{3mv_0b(1+e)}{3mb^2 + Ml^2}\right\} - ev_0$$

$$\Rightarrow v = \frac{3mb^2 - eMl^2}{3mb^2 + Ml^2}v_0. \tag{5.214}$$

1. If $e < \dfrac{3mb^2}{Ml^2}$, $v \to$ (forward).
2. If $e = \dfrac{3mb^2}{Ml^2}$, $v = 0$ (stops).
3. If $e > \dfrac{3mb^2}{Ml^2}$, $v \leftarrow$ (backward).

(c) Putting $e = 0$ and conserving energy, we have that the loss in kinetic energy is equal to the gain in gravitational potential energy as the center of mass rises to a height h_C. Then we can write

$$\frac{I_0}{2}\omega^2 = M_{\text{total}}gh_C$$

$$\Rightarrow h_C = \frac{L_P^2}{2I(m_{\text{total}}g)},$$

where $L_P = mv_0b$, $m_{\text{total}} = M + m$, and $I = \dfrac{Ml^2}{3} + mb^2$.
Then the height attained by the center of mass of the rod–particle system is

$$h_C = \frac{m^2v_0^2b^2}{2\left(\frac{Ml^2}{3} + mb^2\right)(M+m)g}$$

$$= \frac{3m^2v_0^2b^2}{2(Ml^2 + 3mb^2)(M+m)g}.$$

(d) To conserve linear momentum, the reaction force acting on the rod must be zero at the pivot. This is possible when there will not be any collision between the pivot and the rod. Then the velocity of the rod at the pivot must be zero (without any pivot). So we can write

$$\vec{v}_P = \vec{v}_C + \vec{v}_{PC} = \left(v_C - \frac{l}{2}\omega\right)\hat{i} = 0$$

$$\Rightarrow v_C = \frac{l}{2}\omega. \tag{5.215}$$

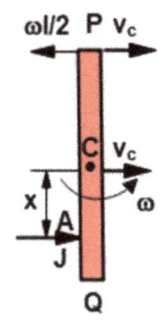

If the horizontal reaction force is zero at the pivot P during the impact, we can think that the rod is free (neglecting the gravity for the sake of simplicity as the gravity is not an impulsive force). For the free rod, the velocity of its center of mass due to the impulse J (say) imparted by the ball is

$$v_C = J/m. \tag{5.216}$$

The angular velocity of the rod is given as

$$\omega = \frac{Jx}{I_C} = \frac{Jx}{(Ml^2/12)} = \frac{12Jx}{Ml^2}. \tag{5.217}$$

Putting the obtained values of v_C from equation (5.216) and ω from equation (5.217) in equation (5.215), we have

$$\Rightarrow \frac{J}{M} = \frac{l}{2}\frac{12Jx}{Ml^2}$$

$$\Rightarrow x = \frac{l}{6}.$$

Then distance of the point of collision from the pivot P is

$$b = \frac{l}{2} + x = \frac{l}{2} + \frac{l}{6} = \frac{2l}{3}.$$

Problem 102 A uniform rod of mass M and length l pivoted smoothly at P is given an angular velocity ω such that it collides with the ball of mass m at the bottom of the rod. The coefficient of restitution of the collision is e. Find (a) the angular velocity of the rod, (b) velocity of the ball just after the collision.

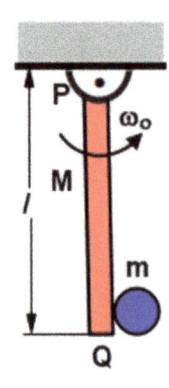

Solution

(a) Let the rod have an angular velocity w and the ball attains a velocity v just after the impact.

The NIF (Newton's impact formula) for the collision is

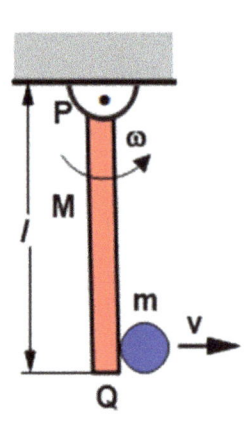

$$-elw_0 = lw - v. \tag{5.218}$$

Angular momentum conservation about the pivot O gives

$$\frac{Ml^2}{3}w_0 = \frac{Ml^2}{3}w + mvl. \tag{5.219}$$

Using equations (5.218) and (5.219)

$$\frac{Ml^2}{3}w_0 = \frac{Ml^2}{3}w + ml^2(w + ew_0)$$

$$\Rightarrow w\left(\frac{M}{3} + m\right)l^2 = w_0\left(\frac{M}{3} - em\right)l^2$$

$$\Rightarrow \omega = \frac{M - 3em}{M + 3m}\omega_0.$$

(b) The velocity of the particle is

$$v = (\omega + e\omega_0)l$$

$$= \left(\frac{M - 3em}{M + 3m}\omega_0 + e\omega_0\right)l$$

$$\Rightarrow v = \frac{l\omega_0 M(1 + e)}{M + 3m}.$$

Problem 103 A rod of mass M and length l is pivoted smoothly at its top so that it is free to rotate in its vertical plane. When the rod is released from rest, after the center of mass of the rod falls through a vertical distance h, the rod collides with a ball of mass m. If the ball sticks to the rod just after the collision, find (a) the maximum height raised by the center of mass of the rod–ball system after the collision, (b) the angular velocity of the rod just after the collision, (c) the energy lost in the collision. Assume $\theta =$ the initial angle made by the rod with the vertical.

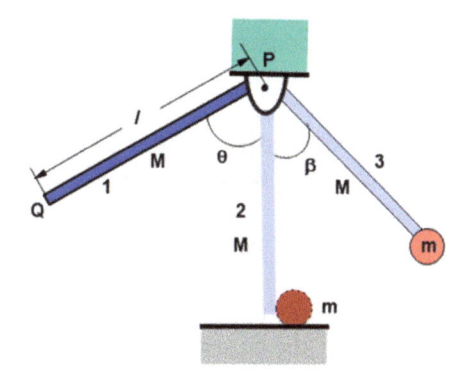

Solution
(a) Just before the impact the rod was not attached to the ball. So the moment of inertia of the system before the collision is given as

$$I_0 = \frac{Ml^2}{3}. \tag{5.220}$$

After the impact the ball sticks to the rod. So the moment of inertia of the system after the impact is given as

$$I = \frac{(M + 3m)l^2}{3}. \tag{5.221}$$

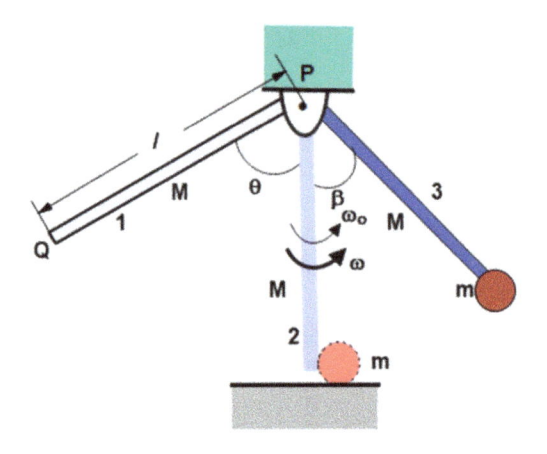

From equations (5.220) and (5.221) we have

$$\Rightarrow I_0/I = \frac{M}{M + 3m}. \tag{5.222}$$

Let the rod have an angular velocity ω_0 and ω just before and after the collision, respectively.

The angular momentum conservation about the pivot O is

$$I_0 \omega_0 = I \omega. \tag{5.223}$$

Conserving the energy between the initial position 1 of the rod and its position 2 just before colliding with the ball, we have

$$\frac{I_0}{2} \omega_0^2 = Mgh. \tag{5.224}$$

Conserving the energy between the right-hand extreme position 3 of the rod and its position 2 just after the collision with the ball, we have

$$\frac{I}{2} \omega^2 = (M + m)gh'. \tag{5.225}$$

From equations (5.224) and (5.225) we have

$$\frac{I\omega^2}{I_0 \omega_0^2} = \frac{h'}{h} \left(\frac{M + m}{M} \right)$$

$$\Rightarrow h' = \left(\frac{I}{I_0} \right) \left(\frac{\omega^2}{\omega_0^2} \right) \left(\frac{M}{M + m} \right) h. \tag{5.226}$$

From equations (5.223) and (5.226) we have

$$h' = \left(\frac{I}{I_0}\right)\left(\frac{I_0}{I}\right)^2\left(\frac{M}{M+m}\right)h$$

$$\Rightarrow h' = \left(\frac{I_0}{I}\right)\left(\frac{M}{M+m}\right)h. \qquad (5.227)$$

From equations (5.222) and (5.227) we have

$$h' = \left(\frac{M}{M+3m}\right)\left(\frac{M}{M+m}\right)h$$

$$\Rightarrow h' = \frac{M^2 h}{(M+3m)(M+m)}.$$

(b) Let the rod have an angular velocity ω and the ball attains a velocity v just after the impact.

The NIF (Newton's impact/collision formula) for the collision is

$$-e l \omega_0 = l\omega - v.$$

Putting $e = 0$ for inelastic collision we have

$$v = l\omega. \qquad (5.228)$$

The angular momentum conservation about the pivot O is

$$\frac{Ml^2}{3}\omega_0 = \frac{Ml^2}{3}\omega + mvl. \qquad (5.229)$$

Using the equations (5.228) and (5.229)

$$\frac{Ml^2}{3}\omega_0 = \frac{Ml^2}{3}\omega + ml^2\omega$$

$$\Rightarrow \omega = \frac{M}{M+3m}\omega_0. \qquad (5.230)$$

(c) The loss of kinetic energy during the collision is

$$\Delta K = \frac{1}{2}I_0\omega_0^2 - \frac{1}{2}I\omega^2$$

$$= \frac{1}{2}I_0\omega_0^2\left(1 - \frac{I\omega^2}{I_0\omega_0^2}\right)$$

$$= \frac{1}{2}I_0\omega_0^2\left(1 - \frac{I^2\omega^2 I_0}{I_0^2\omega_0^2 I}\right)$$

$$\Delta K = \frac{1}{2}I_0\omega_0^2\left(1 - \frac{I_0}{I}\right)\left(\because \frac{I\omega}{I_0\omega_0} = 1\right). \tag{5.231}$$

From equations (5.222) and (5.231) we have

$$\Delta K = \frac{1}{2}I_0\omega_0^2\left(1 - \frac{M}{M + 3m}\right)$$

$$\Delta K = \frac{1}{2}I_0\omega_0^2\left(-\frac{3m}{M + 3m}\right). \tag{5.232}$$

From equations (5.224) and (5.232) we have

$$\Delta K = (Mgh)\left(-\frac{3m}{M + 3m}\right)$$

$$\Rightarrow \Delta K = -\frac{3Mmgh}{M + 3m}.$$

Alternative method:
The loss of kinetic energy during the collision is

$$\Delta K = \frac{L^2}{2I_0}\left(1 - \frac{I_0}{I}\right).$$

Putting I/I_0 from equation (5.222) we have

$$\Delta K = \frac{L^2}{2I_0}\left(1 - \frac{M}{M + 3m}\right)$$

$$\Delta K = \frac{L^2}{2I_0}\left(-\frac{3m}{M + 3m}\right).$$

Putting in $L^2/2I_0 = I_0\omega_0^2/2 = Mgh$ from equation (5.224), we have the same answer, given as

$$\Delta K = -\frac{3Mmgh}{M + 3m}.$$

Problem 104 A ball of mass m collides with a smooth rod AB of mass M and length l, with a velocity v_0, at an angle $\theta = 37°$, as shown in the figure. If the ball sticks to the rod just after the collision, find (a) the angular velocity of the rod and (b) the fraction loss in energy in the collision. Put $b = l/3$ and $M = m$.

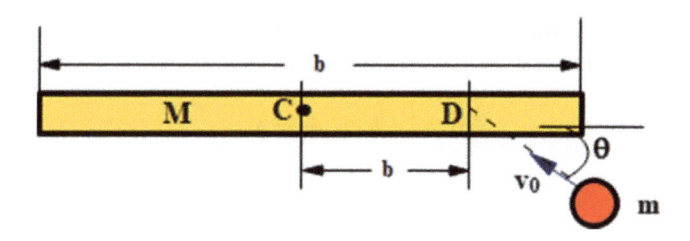

Solution

(a) Let the rod have an angular velocity ω just after the collision. By conserving angular momentum about the center of mass O:

$$L = C$$

$$L_O = mu\left(\frac{\sin\theta Mb}{M+m}\right) = \left(MK^2 + \frac{Mmb^2}{m+M}\right)\omega$$

$$\Rightarrow \omega = \frac{mub\sin\theta}{(M+m)K^2 + mb^2}.$$

Putting $M = m$ we have

$$\omega = \frac{mv_0\frac{l}{3}\frac{3}{5}}{2m\frac{l^2}{12} + \frac{ml^2}{9}} = \frac{\frac{1}{5}v_0}{\frac{1}{6}+\frac{1}{9}l}$$

$$= \frac{9 \times 6}{5 \times 15}\frac{v_0}{l} = 0.72\frac{v_0}{l}.$$

Conserving the linear momentum of the system $(M+m)$ we have

$$(M+m)v_C = mv_0$$

$$\Rightarrow v_C = \frac{mv_0}{M+m} = \frac{v_0}{2}(\because M = m).$$

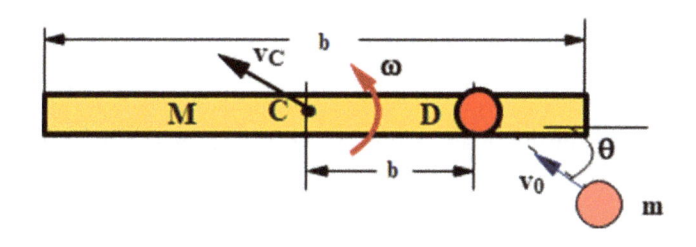

(b) The change in kinetic energy is equal to the decrease in translational kinetic energy plus increase in rotational kinetic energy. It is given as

$$\Delta K = \Delta K_{\text{tran}} + \Delta K_{\text{rot}}$$

$$= \frac{-Mmv_0^2}{2(M+m)} + \frac{L^2}{2I_C}$$

$$= \frac{Mmv_0^2}{2(M+m)} + \frac{m^2 v_0^2 \sin^2 \theta M^2 b^2}{2M\left\{K^2 + \frac{mb^2}{M+m}\right\}(M+m)^2}$$

$$= \frac{-Mmv_0^2}{2(M+m)} + \frac{Mm^2 v_0^2 b \sin^2 \theta}{2\{(M+m)K^2 + mb^2\}(M+m)}$$

$$= \frac{-Mmv_0^2}{2(M+m)}\left[1 - \frac{mb^2 \sin^2 \theta}{(M+m)K^2 + mb^2}\right].$$

Then the fraction of energy lost is

$$\eta = \frac{|\Delta K|}{K} = \frac{M}{M+m}\left[1 - \frac{mb^2 \sin^2 \theta}{(M+m)K^2 + mb^2}\right].$$

Now, after evaluation, we have

$$\frac{\Delta K}{K} = \frac{1}{2}\left[1 - \frac{m\frac{9}{25} \times \frac{l^2}{9}}{2m \times \frac{l^2}{12} + \frac{ml^2}{9}}\right]$$

$$= \frac{1}{2}\left[1 - \frac{\frac{1}{25}}{\frac{1}{6}+\frac{1}{9}}\right] = \frac{1}{2}\left[1 - \frac{1}{25} \times \frac{54}{15}\right]$$

$$= \frac{107}{125}.$$

Problem 105 A ball of mass m collides with a velocity v at one end of a uniform rod of mass M and length l. A block of mass m is attached rigidly at the other end of the rod. If the ball sticks to the rod after the collision, find (a) the velocity of the center of mass of the system $(M + 2m)$, (b) the angular velocity of the rod just after the collision, (c) velocities of the block A and ball B just after the collision, and (b) the fraction of energy loss in the collision. Put $M = 2m$.

Solution

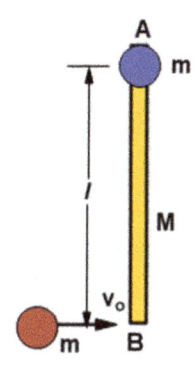

(a) The center of mass of the system $(M + 2m)$ is the mid-point C of the rod.
Conservation of the linear momentum of the rod–particle system:

$$P = mv_1 + mv_2 + Mv_C = mv_0$$

$$= \left(v_C - \frac{l\omega}{2}\right) + m\left(v_C + \frac{l\omega}{2}\right) + Mv_C = mv_0$$

$$\Rightarrow (M + 2m)v_C = mv_0$$

$$\Rightarrow v_C = \frac{mv_0}{M + 2m} = \frac{mv_0}{2m + 2m} = \frac{v_0}{4}(\because M = 2m).$$

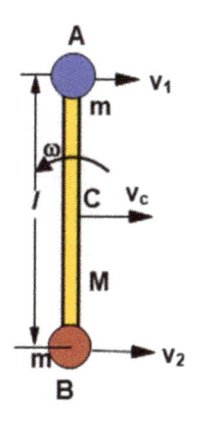

(b) Conservation of the angular momentum of the rod–particle system about C:

$$\left\{2\left(\frac{ml^2}{4}\right) + \frac{Ml^2}{12}\right\}\omega = mv_0\frac{l}{2}$$

$$\Rightarrow \left(\frac{ml^2}{2} + \frac{Ml^2}{12}\right)\omega = mv_0\frac{l}{2}$$

$$\Rightarrow \omega = \frac{6mv_0}{(M + 6m)l} = \frac{3v_0}{4l}(\because M = 2m).$$

(c) The velocity of A is

$$v_1 = v_C - \frac{l\omega}{2} = \frac{v_0}{2} - \frac{1}{2}\left(\frac{3v_0}{4l}\right) = \frac{v_0}{8} \text{ forward.}$$

The velocity of B is

$$v_1 = v_C + \frac{l\omega}{2} = \frac{v_0}{2} + \frac{1}{2}\left(\frac{3v_0}{4l}\right) = \frac{7v_0}{8} \text{ forward.}$$

(d) The kinetic energy of the system just after the collision is

$$K = \frac{1}{2}(M + 2m)v_C^2 + \frac{1}{2} \times \left(\frac{M}{12}l^2 + 2x\frac{m}{4}l^2\right)\omega^2$$

$$= \frac{1}{2}(2m + 2m)\left(\frac{v_0}{4}\right)^2 + \frac{1}{2} \times \left(\frac{2m}{12}l^2 + \frac{m}{2}l^2\right)\left(\frac{3v_0}{4l}\right)^2$$

$$= \frac{mv_0^2}{8} + \frac{3mv_0^2}{16} = \frac{5}{16}mv_0^2.$$

Then the change in kinetic energy is

$$\Delta K = K - K_0 = \frac{5}{16}mv_0^2 - \frac{1}{2}mv_0^2 = -\frac{3}{16}mv_0^2$$

So, the fraction of energy loss $= -\Delta K/K_0 = \frac{3}{16}mv_0^2/(mv_0^2/2) = 3/8.$

Problem 106 A uniform rod of mass M and length $l = 4$ m moves with an angular velocity $\omega_0 = 1$ rad s^{-1} collides elastically with a particle of mass m which moves with a velocity $u = 1$ m s^{-1}. If the velocity of the rod just before the collision is $v = 1$ m s^{-1}, find the velocities of the bodies just after the collision. Assume that all contacting surfaces are smooth with elastic collision and $b = 1$ m and $M = m = 1$ kg.

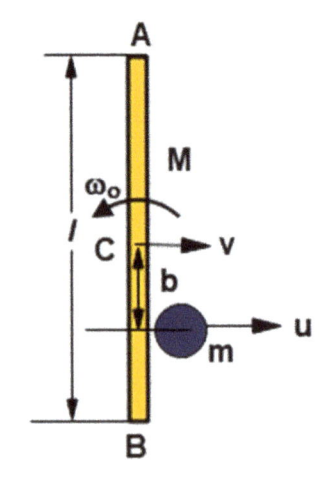

Solution
Referring to the FBD (free-body diagram) we can write the following equations.

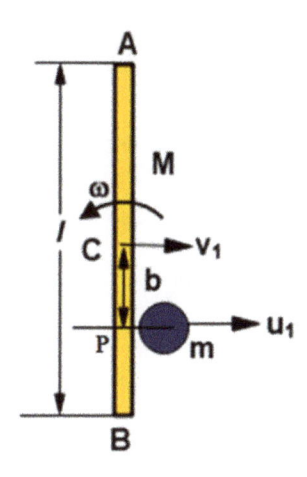

The NIF for the collision:
Just after the collision let the velocity of the particle be u_1 and the velocity of the center of mass of the rod be v_1 and the angular velocity of the rod be w. Then we have

$$-e(u_P - u) = v_P - u_1, \tag{5.233}$$

where u_P and v_P are the velocities of the colliding point P of the rod just before and after the collision, given as follows:

$$u_P = v + bw_0 \tag{5.234}$$

$$v_P = v_1 + bw. \tag{5.235}$$

Using the equations (5.233), (5.234), and (5.235) we have

$$-e(bw_0 + v - u) = bw + v_1 - u_1$$

$$\Rightarrow -(1)(1 \times 1 + 1 - 1) = w + v_1 - u_1$$

$$\Rightarrow u_1 - v_1 - w = 1. \tag{5.236}$$

Conserving angular momentum about C relative to ground,

$$\frac{Ml^2}{12} w_0 + mub = \frac{Ml^2}{12} w + mu_1 b$$

$$\Rightarrow \frac{16}{12} \times 1 + 1 \times 1 = \frac{16}{12} w + u_1 \tag{5.237}$$

$$\Rightarrow 4 + 3 = 4w + 3u_1$$

$$\Rightarrow 3u_1 + 4w = 7. \tag{5.238}$$

Momentum conservation:

$$Mv + mu = Mv_1 + mu_1$$

$$\Rightarrow mv + mu = mv_1 + mu_1$$

$$\Rightarrow 1 + 1 = u_1 + v_1$$

$$\Rightarrow u_1 + v_1 = 2. \tag{5.239}$$

Adding equations (5.236) and (5.239)

$$(2u_1 - w) = 3 \tag{5.240}$$

Multiplying the last equation by 4 on both sides, we have $8u_1 - 4w = 12.$ (5.241)

Solving the last two equations,

$$11w = 5$$

$$\Rightarrow w = \frac{5}{11} \text{ rad s}^{-1}$$

$$2u_1 = 3 + \frac{5}{11} = 3 + \frac{5}{11} = \frac{38}{11}$$

$$\Rightarrow u_1 = \frac{19}{11} \text{ m s}^{-1}$$

$$v_1 = 2 - u_1 = 2 - \frac{19}{11} = +\frac{3}{11} \text{ m s}^{-1}.$$

Problem 107 A light rod of length l is pivoted smoothly at P. A uniform plate of mass M and length b is connected to the rod by a light rod. A wet mud pellet of mass m collides with the plate with a velocity v_0 and sticks to the plate just after the collision. Find the angular velocity of the center of the plate relative to P just after the collision, if the rod is (a) rigidly and (b) smoothly connected with the plate at its center of mass. Assume that the string swings through an acute angle. Put $l = 2b$ and $M = 3m$.

Solution

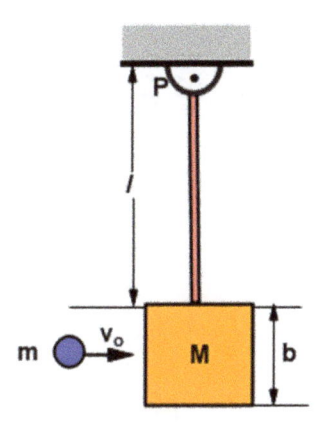

(a) If the plate is rigidly connected with the rod, the moment of inertia of the system $(M + m)$ is

$$I = I_{\text{plate}} + I_{\text{bullet}}$$

$$= \left\{ I_C + M\left(l + \frac{b}{2}\right)^2 \right\} + mr^2,$$

where the distance $r \; (= PQ)$ of the particle

$$r^2 = \left(l + \frac{b}{2}\right)^2 + \left(\frac{b}{2}\right)^2$$

$$= \left\{ M\left(\frac{b^2 + b^2}{12}\right) + M\left(l + \frac{b}{2}\right)^2 \right\} + m\left\{ \left(l + \frac{b}{2}\right)^2 + \left(\frac{b}{2}\right)^2 \right\}$$

$$= \frac{Mb^2}{6} + (M + m)\left(2b + \frac{b}{2}\right)^2 + \frac{mb^2}{4} (\because l = 2b).$$

Putting $M = 3m$ we have

$$I = \frac{3mb^2}{6} + 4m \times \frac{25b^2}{4} + \frac{mb^2}{4} = \frac{103}{4}mb^2.$$

Then, conserving angular momentum,

$$\frac{103}{4}mb^2\omega = m\left(l + \frac{b}{2}\right)v_0$$

$$= m\left(2b + \frac{b}{2}\right)v_0 = \frac{5mbv_0}{2}$$

$$\Rightarrow \omega = \frac{10v_0}{103b}.$$

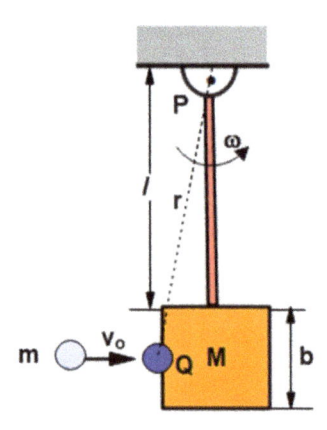

(b) If the plate is pivoted smoothly with the rod, the plate will not rotate, so it will act as a particle. Thus the moment of inertia of the system is

$$I = (M+m)\left(2b + \frac{b}{2}\right)^2 = (3m+m) \times \frac{25b^2}{4}$$

$$= 25mb^2.$$

Then, conserving the angular momentum about the pivot, we have

$$mv_0\left(2b + \frac{b}{2}\right) = I\omega = 25\,mb^2\omega$$

$$\Rightarrow m\left(\frac{5b}{2}\right)v_0 = 25mb^2\omega$$

$$\Rightarrow \omega = \frac{v_0}{10b}.$$

Problem 108 In the last problem if the coefficient of restitution is e, just after the collision, find the angular velocity of the plate if it is connected rigidly to the rod. Put $M = 3m$.

Solution
The moment of inertia of the plate about P is

$$I_P = \left\{I_c + M\left(l + \frac{b}{2}\right)^2\right\},$$

where the distance r (= PQ) of the particle

$$I_C = \frac{M(b^2+b^2)}{12} = \frac{Mb^2}{6}$$

$$\Rightarrow I_P = \left\{\frac{Mb^2}{6} + M\left(2b + \frac{b}{2}\right)^2\right\} (\because l = 2b)$$

$$\Rightarrow I_P = \frac{77Mb^2}{12}. \tag{5.242}$$

Applying Newton's impact formula we have

$$-ev_0 = \left(v - \frac{5\omega b}{2}\right)$$

$$\Rightarrow 5\omega b - 2v = 2ev_0. \tag{5.243}$$

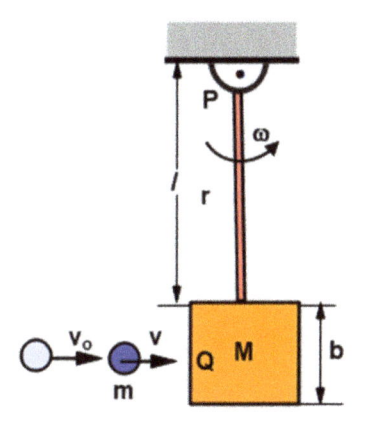

Conserving angular momentum about the pivot P,

$$I_P\omega + mv\frac{5b}{2} = mv_0\frac{5b}{2}. \tag{5.244}$$

Using equations (5.242) and (5.244) we have

$$\frac{77Mb^2}{12}\omega + mv\frac{5b}{2} = mv_0\frac{5b}{2}$$

$$\Rightarrow 77M\omega b + 30mv = 30mv_0$$

$$\Rightarrow 77 \times 3m\omega b + 30mv = 30mv_0 \; (\because M = 3m)$$

$$231\omega b = 30(v_0 - v). \tag{5.245}$$

Also we have

$$5\omega b = 2(ev_0 + v). \tag{5.246}$$

From equations (5.245) and (5.246)

$$\frac{77\omega b}{10} + \frac{5\omega b}{2} = (e+1)v_0$$

$$\Rightarrow \frac{\omega b(77+25)}{10} = (e+1)v_0$$

$$\Rightarrow \omega = \frac{5v_0(e+1)}{51b}.$$

Problem 109 A rod of mass m and length l is kept horizontal in a vertical plane. The end P of the rod is connected to a fixed point O by a light inextensible string OP. An impulse J is given at the end Q of the rod at an acute angle θ to the rod. Find (a) the angular velocity of the rod, (b) the kinetic energy of the rod, (c) the velocity of P and Q, and (d) the acceleration of P just after giving the impulse.

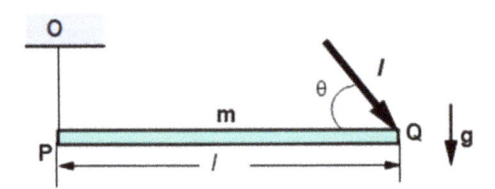

Solution
(a) Let us assume that the string is absent. If the rod has an angular velocity ω and its center of mass has linear velocity $v_y = I\sin\theta/m$.
Then the velocity of the end P of the rod is

$$v_P = \frac{\omega l}{2} - \frac{I\sin\theta}{m}. \quad (5.247)$$

Conserving angular momentum about the point C of the rod,

$$I_y \frac{l}{2} = I(\sin\theta)\frac{l}{2} = \frac{ml^2}{12}\omega$$

$$\Rightarrow \omega = \frac{6I\sin\theta}{ml}. \quad (5.248)$$

Using equations (5.247) and (5.248) we have

$$v_P = \frac{3I\sin\theta}{m} - \frac{I\sin\theta}{m} = \frac{2I\sin\theta}{m} \text{ (up)}.$$

5-219

As the point P of the rod moves up, the string will be slackened. Then the rod will be acted upon by the impulse only, there will be zero impulse due to the string. So the rod will rotate with the angular velocity

$$\omega = \frac{6I \sin \theta}{ml}.$$

in a clockwise sense as shown in the figure.

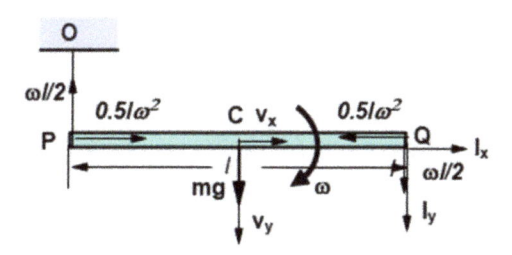

(b) The kinetic energy of the rod just after applying the impulse is

$$K = \frac{1}{2} I \omega^2 + \frac{1}{2} m v_C^2. \tag{5.249}$$

Using equations (5.247), (5.248), and (5.249) we have

$$K = \frac{1}{2} \frac{ml^2}{12} \left(\frac{36 I^2 \sin^2 \theta}{m^2 l^2} \right) + \frac{1}{2} \frac{I^2}{m}$$

$$\Rightarrow K = \frac{I^2}{2m}(1 + 3 \sin^2 \theta).$$

(c) Just after the application of the impulse, the string becomes slackened and gravity produces a zero torque on the rod about its center of mass because it acts at the center of mass of the rod, so

$$\tau_C = 0 \Rightarrow \alpha = 0.$$

Then the velocity of P is

$$\vec{v}_P = \left(\frac{\omega l}{2} - v_y \right) \hat{j} + v_x \hat{i}$$

$$= \left(\frac{3I \sin \theta}{m} - \frac{I \sin \theta}{m} \right) \hat{j} + \frac{I \cos \theta}{m} \hat{i}$$

$$= \frac{I}{m}(\cos \theta \hat{i} + 2 \sin \theta \hat{j})$$

$$\Rightarrow v_P = \frac{I}{m}\sqrt{1 + 3\sin^2\theta}.$$

The center of mass of the rod moves with velocity

$$\vec{v}_C = \vec{v}_x + \vec{v}_y = \frac{\vec{I}}{m}.$$

The velocity of the end Q is

$$v_Q = v_x\hat{i} + \left(v_y + \frac{l}{2}\omega\right)\hat{j}$$

$$= \frac{I\cos\theta}{m}\hat{i} + \left(\frac{I\sin\theta}{m} + \frac{3I\sin\theta}{m}\right)\hat{j}$$

$$\Rightarrow v_Q = \frac{I}{m\sqrt{\cos^2\theta + 16\sin^2\theta}}$$

$$= \frac{I}{m}\sqrt{1 + 15\sin^2\theta}.$$

(d) The acceleration of P relative to C is

$$a_{PC} = \frac{l}{2}\omega^2 = \frac{l}{2}\frac{36I^2\sin^2\theta}{m^2l^2}$$

$$= \frac{18I^2\sin^2\theta}{m^2l},$$

which is directed towards the right.

As the center of mass has a downward acceleration g, the acceleration of P relative to the ground is

$$\vec{a}_P = \vec{a}_{PC} + \vec{a}_C = \frac{18I^2\sin^2\theta}{m^2l}\hat{i} - g\hat{j}.$$

Similarly, the acceleration of Q relative to the ground is

$$\vec{a}_Q = \vec{a}_{QC} + \vec{a}_C = \frac{18I^2\sin^2\theta}{m^2l}\hat{i} - g\hat{j}.$$

Problem 110 In the previous problem, if the impulse is reversed, find (a) the angular velocity of the rod, (b) the velocity of P and Q, and (c) the kinetic energy of the rod just after giving the impulse.

5-221

Solution

(a) In this case the string will be taut. Then the impulsive tension in the string generates an angular impulse about the center of mass C of the rod.

The net linear impulse acting on the rod along the x- and y-directions are given as follows:

$$I \sin \theta + T\delta t = mv_y \tag{5.250}$$

$$I \cos \theta = mv_{qx}. \tag{5.251}$$

The net angular impulse about the center of mass of the rod is

$$(I \sin \theta - T\delta t)\frac{l}{2} = \frac{ml^2\omega}{12}$$

$$\Rightarrow I \sin \theta - T\delta t = \frac{ml\omega}{6}. \tag{5.252}$$

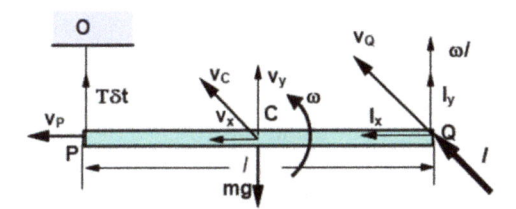

If the string is taut, the point P does not move vertically; $(v_P)_y = 0$

$$v_y = \frac{l}{2}\omega. \tag{5.253}$$

Using equations (5.250), (5.252), and (5.253) we have

$$\Rightarrow (I \sin \theta + T\delta t)/m = 3(I \sin \theta - T\delta t)/m$$

$$\Rightarrow T\delta t = \frac{1}{2} I \sin \theta. \tag{5.254}$$

The position impulse of the tension physically signifies that the string will be taut.

Using equations (5.252) and (5.254) we have

$$I \sin \theta - \frac{1}{2} I \sin \theta = \frac{ml\omega}{6}$$

$$\Rightarrow \omega = \frac{3I \sin \theta}{ml}.$$

Alternative method
Let us assume that the string becomes slackened. So the applied angular impulse about the center of mass of the rod rotates it with an assumed angular velocity ω', given as

$$I \sin \theta \frac{l}{2} = \frac{ml^2}{12}\omega'$$

$$\Rightarrow \omega' = \frac{6I \sin \theta}{ml}.$$

As we know,

$$(v_P)_y = -\omega'\frac{l}{2} + v_y,$$

where v_y is the vertical velocity of C is

$$v_y = \frac{(I \sin \theta)}{m}.$$

Putting ω' and v_C we have

$$\Rightarrow v_P = -\frac{3I \sin \theta}{ml} + \frac{I \sin \theta}{ml} = -\frac{2I \sin \theta}{ml}.$$

The point P should move down in the absence of the string, but it cannot do so due to the presence of the string. So the string does not become slackened.

Then the angular impulse about the point P of the rod is equal to angular momentum of the rod about the point P:

$$\Rightarrow I_P = \frac{ml^2}{3}\omega = (I \sin \theta)l$$

$$\Rightarrow \omega = \frac{3I \sin \theta}{ml}.$$

(b) Then the center of mass of the rod moves with a velocity

$$v_C = \sqrt{v_{C_x}^2 + v_{C_y}^2} = \sqrt{v_x^2 + v_y^2}.$$

Now we know that

$$v_{C_y} = v_y = \frac{\omega l}{2} = \frac{3I \sin \theta}{2m}.$$

From equation (5.251) we have

$$v_x = \frac{I \cos \theta}{m}.$$

5-223

Hence, the velocity of the center of mass C of the rod is

$$v_C = \sqrt{v_x^2 + v_y^2} = \sqrt{\left(\frac{I\cos\theta}{m}\right)^2 + \left(\frac{3I\sin\theta}{2m}\right)^2}$$

$$= \frac{I}{2m}\sqrt{4\cos^2\theta + 9\sin^2\theta}$$

$$= \frac{I}{2m}\sqrt{4 + 5\sin^2\theta}.$$

The velocity of the point P of the rod is

$$v_P = \sqrt{v_{P_x}^2 + v_{P_y}^2}$$

$$= \sqrt{\left(\frac{I\cos\theta}{m}\right)^2 + (0)^2}$$

$$= \frac{I\cos\theta}{m} \text{ (towards the left).}$$

The velocity of the point Q of the rod is

$$v_Q = \sqrt{v_{QP}^2 + v_P^2}$$

$$= \sqrt{(\omega l)^2 + \left(\frac{I\cos\theta}{m}\right)^2}$$

$$= \sqrt{\left(\frac{3I\sin\theta}{ml}l\right)^2 + \left(\frac{I\cos\theta}{m}\right)^2}$$

$$= \frac{I}{m}\sqrt{1 + 8\sin^2\theta}.$$

(c) The kinetic energy of the rod is

$$K = \frac{1}{2}I_C\omega^2 + \frac{1}{2}mv_C^2$$

$$= \frac{1}{2}\frac{ml^2}{12}\left(\frac{9I^2\sin^2\theta}{m^2l^2}\right) + \frac{1}{2}m\left(v_{C_x}^2 + v_{C_y}^2\right)$$

$$\left(\because v_{C_y} = \frac{\omega l}{2} = \frac{3I\sin\theta}{2m}\right)$$

$$\Rightarrow K = \frac{3\,I^2 \sin^2 \theta}{8\,m} + \frac{1}{2}m\left\{\left(\frac{I^2 \cos^2 \theta}{m^2}\right) + \frac{9I^2 \sin^2 \theta}{4m^2}\right\}$$

$$= \frac{I^2}{8m}[3\sin^2\theta + 4\cos^2\theta + 9\sin^2\theta]$$

$$= (2\sin^2\theta + 1)\frac{I^2}{2m}.$$

Problem 111 A uniform rod of mass M and length $L = 2a$ is placed on a smooth horizontal surface. It is given a spin angular velocity w and simultaneously collided by two particles A and B of masses m and $2m$ with velocities u and v, respectively, as shown in the figure. If the particles stick to the rod just after the collision, find the angular speed of the rod and linear velocity of the center of mass of the rod just after the collision. Put $u = 3$ m s^{-1}, $v = 2$ m s^{-1}, $M = 3m$, $a = 1m$, $w_0 = 2$ rad s^{-1}, and $\theta = 30°$.

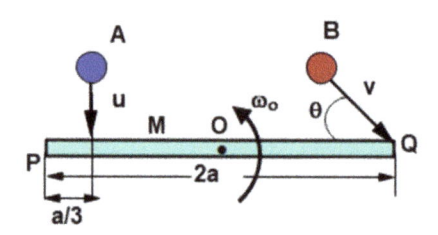

Solution
Let the rod have an angular velocity w and its center of mass O has linear velocity v_1 just after the collision.
Conserving the angular momentum about O we have

$$\vec{L}_O = C.$$

The initial angular momenta are given as

$$\vec{L}_i = \left\{mu\frac{2a}{3} + 3m\frac{(2a)^2}{12}w_0 - 2m(v\sin\theta)a\right\}\hat{k}$$

$$= ma\left(\frac{2u}{3} + aw_0 - 2v\sin\theta\right)\hat{k}$$

$$= ma\left(\frac{2}{3} \times 3 + 1 \times 2 - 2 \times 2 \times \frac{1}{2}\right)\hat{k}$$

Then, we have

$$\vec{L}_i = 2ma\hat{k}.$$

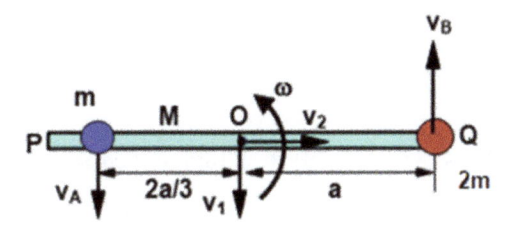

The final angular momentum of the system (rod + particles) about O is given as

$$\vec{L}_f = \left\{ m(v_A)\frac{2a}{3} + \left(\frac{Ma^2}{3}\right)\omega + 2m(v_B)a \right\}\hat{k}$$

$$\vec{L}_f = \left\{ m\left(v_1 + \frac{2a\omega}{3}\right)\frac{2a}{3} + \left(\frac{Ma^2}{3}\right)\omega + 2m(a\omega - v_1)a \right\}\hat{k}$$

$$= ma\left[\left(\frac{2}{3}v_1 - 2v_1\right) + \omega\left(\frac{4a}{9} + 2a + a\right)\right]\hat{k}$$

$$= ma\left[-\frac{4v_1}{3} + \frac{31}{9}\omega a\right]\hat{k}$$

$$= ma\frac{-12v_1 + 31\omega a}{9}\hat{k}.$$

Then we have

$$\frac{-12v_1 + 31\omega a}{9}ma = 2ma$$

$$\Rightarrow -12v_1 + 31\omega a = 18. \tag{5.255}$$

Conserving the vertical linear momentum,

$$mu + 2mv\sin\theta$$

$$= m\left(v_1 + \frac{2a\omega}{3}\right) + 3mv_1 + 2m(-a\omega + v_1)$$

$$\Rightarrow m \times 3 + 2m \times 2 \times \frac{1}{2}$$

$$= 6mv_1 + ma\omega\left(\frac{2}{3} - 2\right)$$

$$\Rightarrow 5 = 6v_1 - \frac{4}{3}\omega a$$

5-226

$$18v_1 - 4wa = 15. \tag{5.256}$$

Multiply the equation (5.255) by 3 and equation (5.256) by 2 and then add them to obtain

$$3(-12v_1 + 31wa = 18) + 2(18v_1 - 4wa = 15) = 0$$

$$\Rightarrow (93 - 8)wa = 30 + 54$$

$$\Rightarrow w = \frac{84}{85a} = \frac{84}{85} \text{ rad s}^{-1}.$$

From equation (5.256),

$$v_1 = \frac{15 + 4w_1 a}{18}$$

$$= \frac{15 + 4(84/85)}{18} = \frac{1611}{1530}.$$

Problem 112 A uniform rod of mass M and length $l = 2a = 2$ m is placed on a smooth horizontal surface. It is given a spin angular velocity $w = 2$ rad s^{-1} and simultaneously collided by two stationary beads A and B of masses $2m$ and m, respectively, as shown in the figure. If the particles stick to the rod just after the collision, find the angular speed of the rod just after the collision. Put $M = 3m$.

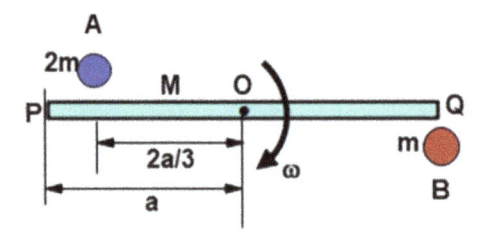

Solution
Let the rod have an angular velocity w_1 and its center of mass O have linear velocity v_1 just after the collision.
Conserving the linear momentum, $\vec{P} = \vec{C}$

$$\Rightarrow 2m(-v_A) + Mv_B + m(v_1) = 0$$

$$\Rightarrow 2m\left(v_1 - \frac{2aw_1}{3}\right) + Mv_1 + m(v_1 + aw_1) = 0$$

$$\Rightarrow v_1(3m + M) + w_1 a(m - 4M) = 0$$

$$\Rightarrow v_1(3m + 3m) + w_1 a\left(-\frac{m}{3}\right)$$

$$\Rightarrow 6v_1 m = w_1 \frac{ma}{3}$$

$$\Rightarrow 18v_1 = w_1 a$$

$$\Rightarrow 18v_1 = w_1(1) = w_1. \tag{5.257}$$

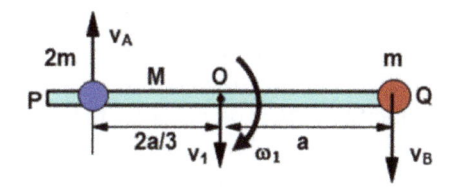

Conserving angular momentum about O,
$\vec{L_O} = C$. The initial angular momentum is

$$\vec{L_i} = -\frac{Ml^2}{12} w \hat{k}.$$

The final angular momentum is

$$\vec{L_f} = -2m\left(\frac{2aw_1}{3} - v_1\right)\frac{2a}{3}\hat{k}$$

$$-m(v_1 + w_1 a)a\hat{k} - \frac{Ml^2}{12} w_1 \hat{k}.$$

Conserving the angular momentum about O and simplifying the factors, we have

$$\Rightarrow a^2 w = \frac{4ma}{3}\left(\frac{2aw_1}{3} - v_1\right)$$

$$+ ma(v_1 + w_1 a) + 3m\frac{a^2}{3}w_1$$

$$\Rightarrow aw = \frac{8}{9}aw_1 + w_1 a + aw_1 - \frac{4}{3}v_1 + v_1$$

$$\Rightarrow aw = \left(\frac{8}{9} + 2\right)aw_1 - \frac{v_1}{3}$$

$$\Rightarrow 26aw_1 - 3v_1 = 9aw$$

Or, $26w_1 - 3v_1 = 9(1)(2) = 18.$ \hfill (5.258)

Using equations (5.257) and (5.258)

$$26w_1 - \frac{3w_1}{18} = 108$$

$$\Rightarrow w_1(156 - 1) = 108$$

$$\Rightarrow w_1 = \frac{18}{155}.$$

Problem 113 A smooth uniform rod of mass $M = 8m$ and length $L = 6a$ is kept on a smooth horizontal surface. Two particles of masses m and $2m$ strike the rod simultaneously and stick to the rod after the collision, as shown in the figure. Find (a) the linear and angular velocities of the rod just after the collision and (b) the ratio of translational kinetic energy of the rod and rotational kinetic energy of the rod–particle system just after the collision. Put $v_0 = 11$ m s^{-1}, $a = 1/5$ m, and $M = 8m$.

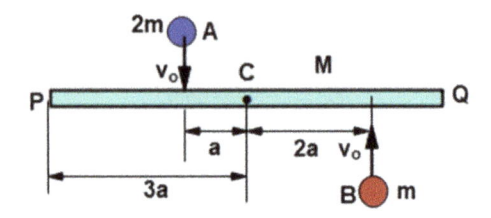

Solution
(a) Let the rod has an angular velocity w and its center of mass C has linear velocity v just after the collision. Conserving the linear momentum,

$$2m(v + aw) + 8mv + m(v - 2aw)$$

$$= 2mv_0 - mv_0$$

$$\Rightarrow 11v = v_0 \Rightarrow v = \frac{v_0}{11} = \frac{11}{11} = 1 \text{ m s}^{-1}$$

Conserving the angular momentum of the rod–particles system about the center of mass C of the rod, we have

$$2m(v + aw)a + (8m)\left(\frac{36a^2}{12}\right)w$$

$$- m(v - 2aw)(2a) = 2mv_0 a + mv_0(2a)$$

$$\Rightarrow 2ma(v + aw) + 24ma^2w - 2ma(v - 2aw)$$

$$= 4mv_0 a$$

5-229

$$\Rightarrow v + a\omega + 12a\omega - v + 2a\omega = 2v_0$$

$$\Rightarrow 15a\omega = 2v_0 \Rightarrow \omega = \frac{2v_0}{15a}$$

$$= \frac{2 \times 11}{15 \times \frac{1}{5}} = \frac{22}{3} \text{ rad s}^{-1}.$$

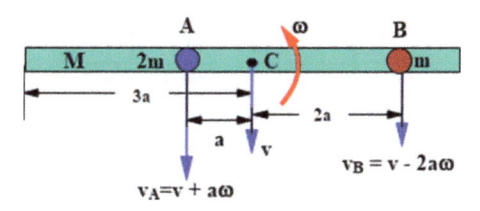

(b) Then the translational kinetic energy of the rod is

$$K_t = \frac{1}{2}(8m)v^2 = \frac{8mv^2}{2}.$$

The rotational kinetic energy of the rod–particle system is

$$K_{rot} = \frac{1}{2}\{(Ml^2/12) + 2ma^2 + m(2a)^2\}\omega^2$$

$$= \frac{1}{2}\{(8m)(6a)^2/12 + 2ma^2 + 4ma^2\}\omega^2$$

$$= \frac{1}{2}\{(8m)(6a)^2/12 + 2ma^2 + 4ma^2\}\omega^2$$

$$\Rightarrow K_{rot} = 15ma^2\omega^2.$$

Then the ratio of translational and rotational energy is

$$\frac{K_t}{K_{rot}} = (8mv^2/2)/(15ma^2\omega^2)$$

$$= \frac{8}{30}(v/a\omega)^2 = \frac{8}{30}(1)/\left\{\left(\frac{15}{22}\right)^2\right\}$$

$$= \frac{15}{121}.$$

Problem 114 A uniform rod PQ of mass m and length $l = 2a$ is kept on a smooth horizontal plane. A horizontal impulse $I = 2mu$ is given at the end Q at an angle $\theta = 30°$ to the rod. At the same time a bead of mass m collides inelastically at A with a velocity u perpendicular to the rod at a distance $a/2$ from the center C of the rod, as shown in the figure. Find (a) the velocity of the center of mass of the rod–particle system, (b) the velocity of center of mass of the rod and the angular velocity of the rod just after the impact. Put $M = 3m$.

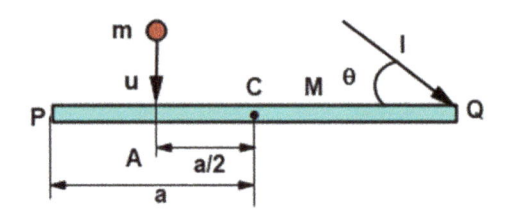

Solution
(a) If v and v' be the velocities of the center of mass of the rod–particle system, conserving the linear momentum in the x- and y-directions, we have

$$v_y = \frac{I \sin\theta + mu}{M + m} = \frac{(2mu)(0.5) + mu}{3m + m} = \frac{u}{2}$$

$$v_x = v' = \frac{I \cos\theta}{M + m} = \frac{(2mu)\cos 30°}{3m + m} = \frac{\sqrt{3}\,u}{4}.$$

(b) Let the rod have an angular velocity w and its center of mass C have linear velocity v just after the collision. Writing the linear impulse–momentum equations,

$$I \sin\theta + mu = Mv + m(aw/2 + v)$$

$$\Rightarrow I \sin\theta + mu = (M + m)v + maw/2$$

$$\Rightarrow (2mu)\sin 30° + mu = (3m + m)v + maw/2$$

$$\Rightarrow 8v + aw = 4u. \tag{5.259}$$

The initial angular momentum about C is

$$(L_0)_i = mu\frac{a}{2} - (I \sin\theta)a.$$

The final angular momentum about C is

$$(L_C)_f = m\left(v + \frac{aw}{2}\right)\frac{a}{2} + \frac{Ma^2}{3}w.$$

5-231

Conserving the angular momentum,

$$mv\frac{a}{2} + \left(\frac{Ma^2}{3} + \frac{ma^2}{4}\right)\omega = mu\frac{a}{2} - 2mua(1/2)$$

$$\Rightarrow mv\frac{a}{2} + \left(\frac{3ma^2}{3} + \frac{ma^2}{4}\right)\omega = -mu\frac{a}{2}$$

$$\Rightarrow 2v + 5a\omega = -2u. \tag{5.260}$$

Solving equations (5.259) and (5.260) we have

$$v = 8u/19 \text{ and } \omega = -12u/19a.$$

Since we obtained a negative result, the angular velocity will be directed opposite to the assumed anticlockwise sense. Hence the rod will rotate in clockwise sense.

Problem 115 A smooth uniform wooden bar of mass $M = 3$ kg and length $L = 2a$ is kept on a smooth horizontal surface. Two particles A and B of masses $m_1 = 1$ kg and $m_2 = 2$ kg strike the rod simultaneously with velocities $u = 2$ m s^{-1} and $v = 1$ m s^{-1}, as shown in the figure. If the particles stick to the rod after the collision find (a) the linear and angular velocities of the rod just after the collision and (b) the fraction of energy loss during the collision. Put $b = 1/2$ m and $a = 1$ m.

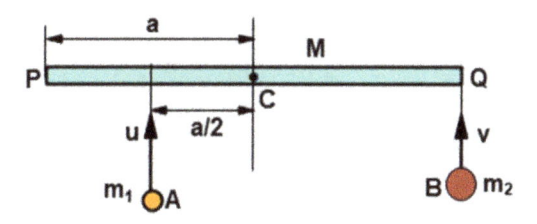

Solution
(a) Let the rod have an angular velocity ω and its center of mass C have linear velocity v_1 just after the collision. Conservation of linear momentum $\overrightarrow{P} = C$

$$m_1 v_A + M v_1 + m_2 v_B = m_1 u + m_2 v,$$

where $v_A = v_1 - \omega/2$ and $v_B = v_1 - \omega a$

5-232

$$v_1(m_1 + m_2 + M) + \omega(-m_1 a/2 + m_2 a)$$
$$= m_1 u + m_2 v$$
$$+v_1(1 + 2 + 3) + \omega\{-1 \times (1/2) + 2 \times 1\} = 1 \times 2 + 2 \times 1 = 4$$
$$\Rightarrow +6v_1 + (3/2)\omega = 4$$
$$\Rightarrow 3\omega + 12v_1 = 8. \tag{5.261}$$

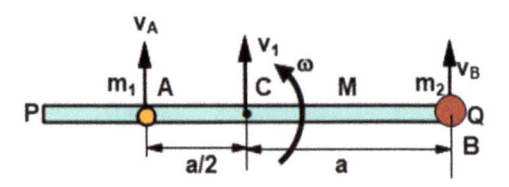

Conserving the angular momentum about C we have
$$-m_1 v_A b + M v_1 + m_2 v_B a + (Ma^2/3)\omega$$
$$= -m_1 u b + m_2 v a, \quad \text{where } b = a/2$$
$$\Rightarrow -m_1(v_1 - b\omega)b + m_2(v_1 + a\omega)a + \frac{Ma^2}{3}\omega$$
$$= -m_1 u b + m_2 v a$$
$$\Rightarrow v_1(-m_1 b + m_2 a) + \omega\left(m_1 b^2 + m_2 a^2 + \frac{Ma^2}{3}\right)$$
$$= -m_1 u b + m_2 v b$$
$$\Rightarrow v_1\left(-1 \times \frac{1}{2} + 2 \times 1\right) + \omega\left(1 \times \frac{1}{4} + 2 \times 1 + 3\frac{1^2}{3}\right)$$
$$= -1 \times 2 \times \frac{1}{2} + 2 \times 1 \times 1$$
$$\Rightarrow \frac{3}{2}v_1 + \omega\left(\frac{13}{4}\right) = 1$$
$$\Rightarrow 6v_1 + 13\omega = 4. \tag{5.262}$$

From equations (5.261) and (5.262)

$$\Rightarrow \frac{8 - 3\omega}{12} = \frac{4 - 13\omega}{6}$$

$$\Rightarrow 8 - 3\omega = 8 - 26\omega$$

$$\Rightarrow \omega = 0.$$

This means that the rod will not rotate just after the collision because the initial angular impulse acting on the rod is zero about the center of mass of the rod–particles system but not necessarily about the center of mass of the rod.

Then

$$v_1 = \frac{2}{3} \text{ m s}^{-1}.$$

(b) The change in the kinetic energy of the system is

$$\Delta K = \frac{1}{2} m_1 (v_1 - b\omega)^2 + \frac{1}{2} m_2 (v_1 + a\omega)^2$$

$$- \frac{1}{2} m_1 u^2 - \frac{1}{2} m_2 v^2$$

$$= \frac{1}{2}(m_1 + m_2) v_1^2 - \frac{1}{2} m_1 u^2 - \frac{1}{2} m_2 v^2 \quad (\because \omega = 0)$$

$$= \frac{1}{2}(1 + 2)\frac{4}{9} - \frac{1}{2} \times 1 \times 4 - \frac{1}{2} \times 2 \times 1$$

$$= 2/3 - 3 = -(7/3) \text{ J}.$$

Then the fraction of loss in energy is

$$\Rightarrow \eta = -\frac{\Delta K}{K_i} = \frac{7/3}{3} = 7/9.$$

Problem 116 A horizontal rod of mass M and length $l = 3a$ is smoothly pivoted at P. It is in equilibrium being loaded with a block of mass m. A ball of mass $2m$ collides with the bar after falling from a height H. If the ball sticks to the rod just after the collision and the block leaves the rod just after the collision, find the maximum height attained by the block. Put $H = 11$ m.

Solution

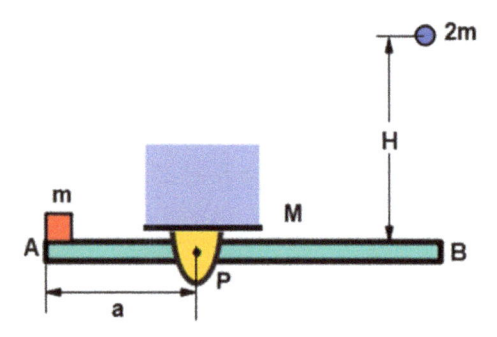

For equilibrium, the net torque about the pivot is zero. Then we can equate the gravitational torques of the rod and block about P we obtain

$$mga = Mg\frac{a}{2}.$$

Then

$$M = 2m. \tag{5.263}$$

Let the ball collide at B with a velocity v_0 after falling through a distance H. So we have

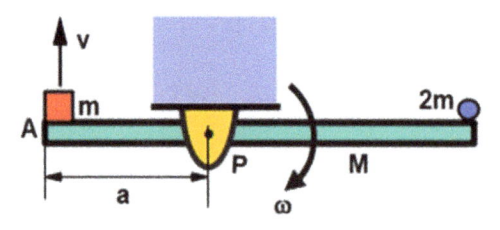

$$v_0^2 = 2gH. \tag{5.264}$$

Let the rod have an angular velocity w and its center of mass C have linear velocity v just after the collision. Conserving the angular momentum about P we have

$$L_P = \text{constant}$$

$$mv_0(2a) = w\{I_A + I_B + I_{\text{rod}}\}$$

$$\Rightarrow mv_0(2a) = w\left\{ma^2 + (2m)(2a)^2 + \frac{M(3a)^2}{12} + \frac{Ma^2}{4}\right\}$$

$$= w\left\{9ma^2 + \left(\frac{3}{2} + \frac{1}{2}\right)ma^2\right\}(\because M = 2m)$$

5-235

$$\Rightarrow mv_0 2a = 11ma^2\omega$$

$$\Rightarrow a\omega = v = \frac{2v_0}{11}. \tag{5.265}$$

This means that the block will move vertically up with velocity v just after losing contact with the rod just after the collision. If it goes to a height h we have

$$v^2 = 2gh. \tag{5.266}$$

Using equations (5.264), (5.265), and (5.266) we have

$$h = \frac{4}{121}H = \frac{4}{121} \times 11 = \frac{4}{11} \text{ m}.$$

Problem 117 A horizontal bar AB of mass M is smoothly pivoted at its mid-point P. A block of mass $3m$ is placed at the end A of the bar. The bar is in equilibrium by being connected to a string S, as shown in the figure. (a) Find the tension in the string. (b) After falling from a height $H = 5$ m, a ball of mass m collides with the bar at a distance b from the pivot. If the ball sticks to the rod just after the collision and the block leaves the rod just after the collision, find the maximum height attained by the block. Put $M = 3m$, $b = 3a/4$, and $c = 2a/3$.

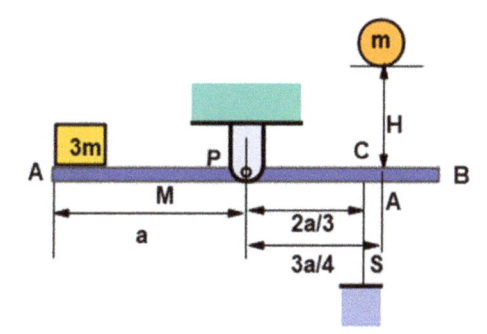

Solution
(a) Equating the torques about O,

$$T(2a/3) = (3mga)$$

$$\Rightarrow T = \frac{3mg \times a}{2a/3} = \frac{9mg}{2}.$$

(b) Let the rod have an angular velocity ω just after the collision.
Conserving the angular momentum about the pivot,

$$mv_0(3a/4) = I_p\omega$$

5-236

$$\Rightarrow mv_0(3a/4) = \left\{\frac{Ma^2}{3} + (3m)a^2 + m\left(\frac{3a}{4}\right)^2\right\}\omega$$

$$= \left(ma^2 + 3ma^2 + \frac{9ma^2}{16}\right)\omega \quad (\because M = 3m)$$

$$\Rightarrow mv_0(3a/4) = 73\frac{ma^2}{16}\omega$$

$$\Rightarrow \omega = \frac{12v_0}{73a}.$$

The block will leave the bar with a velocity given as

$$v = a\omega = \frac{12v_0}{73},$$

where the velocity

$$v_0 = \sqrt{2 \times g \times H} = \sqrt{2 \times 10 \times 5} = 10 \text{ m s}^{-1}$$

$$\Rightarrow v = \frac{12v_0}{73} = \frac{12 \times 10}{73} = \frac{120}{73}.$$

Then the maximum height attained by the block is

$$h_{max} = \frac{v^2}{2g} = \frac{120 \times 120}{2 \times 10 \times 73 \times 73} = \frac{720}{5329} \text{ m} = 0.135 \text{ m (approximately)}.$$

Problem 118 A light rigid horizontal bar AB is pivoted smoothly at P. A block of mass M is placed at the end A of the bar, while the bar is in equilibrium with the help of an edge. A ball of mass m collides inelastically with the bar after falling from a height $H = 5$ m. Find the value of x for which (a) the angular velocity of the rod and (b) the velocity of the block will be maximum just after the impact, and (c) the maximum height attained by the block.

Solution

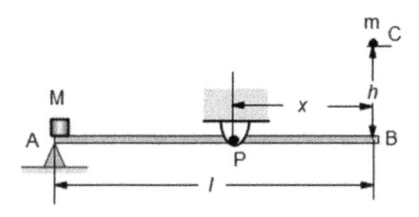

(a) Let the rod has an angular velocity ω just after the collision. The block will leave the bar with a velocity given as

$$v = (l - x)\omega. \tag{5.267}$$

Conserving the angular momentum about the pivot P we have

$$mv_0 x = I_p \omega$$

$$\Rightarrow mv_0 x = \{mx^2 + M(l - x)^2\}\omega$$

$$\Rightarrow \omega = \frac{mv_0 x}{mx^2 + M(l - x)^2}. \tag{5.268}$$

If the angular velocity is maximum put

$$\frac{d\omega}{dx} = 0. \tag{5.269}$$

Using the equations (5.268) and (5.269) we have

$$\frac{d}{dx}\left\{\frac{mv_0 x}{mx^2 + M(l - x)^2}\right\} = 0$$

$$\Rightarrow \frac{d}{dx}\left\{\frac{x}{M(l - x)^2 + mx^2}\right\} = 0$$

$$\Rightarrow M(l - x)^2 + mx^2 = x\{2mx - 2(M)(l - x)\}$$

$$\Rightarrow Ml^2 + Mx^2 - 2Mlx + mx^2$$

$$= 2mx^2 + 2Mx^2 - 2Mlx$$

$$\Rightarrow Mx^2 + mx^2 = Ml^2$$

$$\Rightarrow x = \sqrt{\frac{M}{M + m}}\, l$$

$$= \sqrt{\frac{1}{1+1}} l = \frac{l}{\sqrt{2}} (\because M = m).$$

(b) If the height is maximum then the velocity (v) and momentum (mv) must be maximum. The velocity of the block just after the impact is given as

$$v = (l-x)\omega = (l-x)\frac{mv_0 x}{M(l-x)^2 + mx^2}$$

$$= v = mv_0 \frac{x(l-x)}{M(l-x)^2 + mx^2}. \tag{5.270}$$

If v is maximum, put $dv/dx = 0$. Then we have

$$\{M(l-x)^2 + mx^2\}\{l-2x\} = \{x(l-x)\}\{2mx - 2M(l-x)\}$$

$$\Rightarrow \{(l-x)^2 + x^2\}(l-2x)$$

$$= x(l-x)\{2x - 2l + 2x\}$$

$$\Rightarrow (l^2 + 2x^2 - 2lx)(l-2x) = 2x(l-x)(2x-l)$$

$$\Rightarrow (l - 2x(\{l^2 + 2x^2 - 2lx - (2xl - x^2)\})$$

$$\Rightarrow 3x^2 - 4lx + l^2 = 0$$

$$\Rightarrow x = \frac{4l \pm \sqrt{16l^2 - 4 \times 3l^2}}{6} = \frac{4l \pm 2l}{6}$$

$$\Rightarrow x = l \text{ or } x = \frac{l}{3}.$$

(c) Putting $x = \frac{l}{3}$ in equation (5.270), we have

$$v_{max} = v_0 \frac{\frac{l}{3} \times \frac{2l}{3}}{\frac{l^2}{9} + \frac{4l^2}{9}} = \frac{v_0 \left(\frac{2l^2}{9}\right)}{\frac{5l^2}{9}}$$

$$\Rightarrow v_{max} = \frac{2v_0}{5}.$$

Then the maximum height attained by the block is

$$h_{max} = (v_{max})^2 / 2g = \frac{4}{25}\left(\frac{v_0^2}{2g}\right) = \frac{4}{25}h.$$

5-239

Problem 119 A light rigid rod AB, pivoted smoothly at P, is placed on a smooth horizontal surface. A ball of mass m_1 collides with the bar at the end B with a velocity v_0. A block of mass m_2 is placed at the end A of the bar which is connected to a horizontal light spring of stiffness k fitted with a rigid surface at C. If the coefficient of restitution of collision is e, the block m_2 leaves the rod and moves towards left compressing the spring. Find the maximum compression of the spring. Put $x = 1/3$, $e = 0.5$, $m_1 = m$, $e = 1/2$ and $m_2 = 2m$.

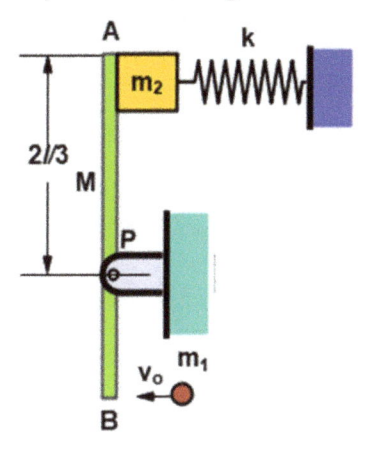

Solution
Let the rod have an angular velocity w and the ball has a velocity v towards right, just after the elastic collision.
The NIF (Newton's impact/collision formula) for the collision is

$$-ev_0 = -v - xw$$

$$\Rightarrow ev_0 = v + xw. \tag{5.271}$$

Conserving the angular momentum about the pivot P

$$-m_1 v_0 x = m_1 v x - m_2 v'(l - x),$$

where $v' = (l - x)w$

$$\Rightarrow m_1 v_0 x = -m_1 v x + m_2 (l - x)^2 w. \tag{5.272}$$

Using equations (5.271) and (5.272)

$$ev_0 - xw = \frac{m_2(l - x)^2 w - m_1 v_0 x}{m_1 x}$$

$$\Rightarrow m_1 v_0 x (1 + e) = \{m_1 x^2 + m_2(l - x)^2\} w$$

$$\Rightarrow w = \frac{m_1 v_0 x (1 + e)}{m_1 x^2 + m_2(l - x)^2}.$$

5-240

If $x = \frac{l}{3}$, $m_1 = m$, $m_2 = 2m$

$$\omega = \frac{mv_0 \frac{l}{3}\left(1 + \frac{1}{2}\right)}{m\frac{l^2}{9} + 2m\frac{4l^2}{9}} = \frac{v_0/2}{l} = \frac{v_0}{2l}.$$

Then the velocity of the block just after the collision is

$$v' = \left(\frac{2l}{3}\right)\omega = \frac{v_0}{2l} \times \frac{2l}{3} = \frac{v_0}{3}.$$

Hence the compression of the spring is given as

$$\frac{k}{2}x^2 = \frac{1}{2}(2m)v'^2$$

$$\Rightarrow x = (\sqrt{2m/k})v'$$

$$= (\sqrt{2m/k})(v_0/3) \quad (\because v' = v_0/3).$$

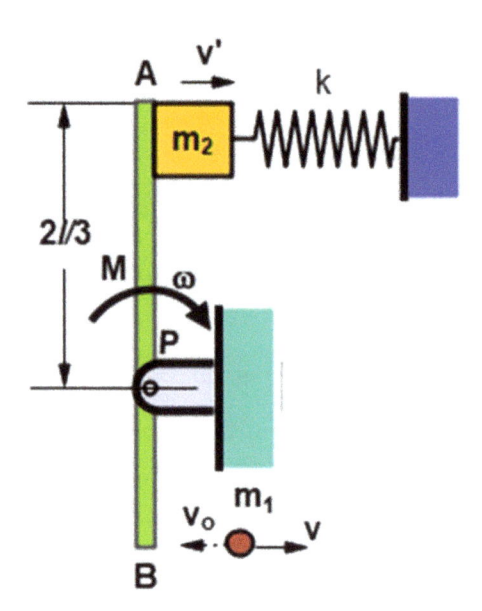

Problem 120 A uniform horizontal rod PQ of mass m and length l is smoothly pivoted at P. If it is released from rest when it was horizontal (position 1 as shown in the figure), it swings in a vertical plane and collides with a smooth ball of mass m placed on a horizontal surface. After the collision, if the rod swings through an angle of 60° and the block just completes the vertical circle following the smooth semi-circular path, (a) assuming $l = 2R$ find the value(s) of M/m and (b) the value of l/R.

5-241

Solution

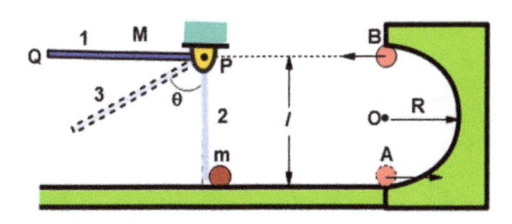

(a) Let the rod have an angular velocity ω and the ball has a velocity v towards the right just after the collision.

The NIF for the collision is

$$-e(0 - l\omega_0) = v - (-l\omega)$$

$$\Rightarrow \omega_0 = \frac{l\omega + v}{el}. \tag{5.273}$$

Conserving the angular momentum about the pivot,

$$\frac{Ml^2}{3}\omega_0 = -\frac{Ml^2}{3}\omega + mvl. \tag{5.274}$$

Using equation (5.273) and (5.274)

$$\left(\frac{Ml^2}{3}\omega_0\right)/el\omega_0 = \left\{\left(mvl - \frac{Ml^2\omega}{3}\right)/(v + l\omega)\right\}$$

$$\Rightarrow \frac{M}{3e} = \frac{mv - \frac{M}{3}l\omega}{v + l\omega}$$

$$\Rightarrow e = \frac{(v + l\omega)M}{3mv - Ml\omega}. \tag{5.275}$$

We know that for completing a loop from A to B, the minimum velocity required at A is

$$v = \sqrt{5gR}. \tag{5.276}$$

Furthermore, for the rod to swing through an angle θ, conserving the energy of the rod between positions 2 and 3

$$\frac{Ml^2}{6}\omega^2 = Mg(l/2)(1 - \cos\theta)$$

$$\Rightarrow \omega l = \sqrt{3gl(1 - \cos\theta)}$$

5-242

$$\Rightarrow l\omega = \sqrt{3gl(1 - \cos 60°)} = \sqrt{3gl/2}. \qquad (5.277)$$

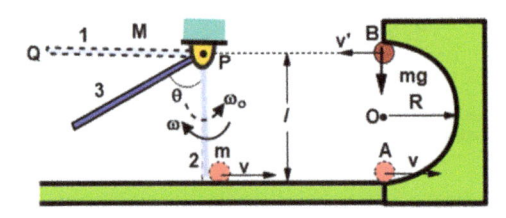

Putting v from equation (5.276) and ω from equation (5.277) in equation (5.275) we have

$$e = \frac{M\left(\sqrt{5gR} + \sqrt{\frac{3gl}{2}}\right)}{3m\sqrt{5gR} - M\sqrt{\frac{3gl}{2}}}.$$

Putting $l = 2R$ and $e \leqslant 1$ we have

$$e = \frac{M(\sqrt{5} + \sqrt{3})}{3\sqrt{5}m - \sqrt{3}M} \leqslant 1$$

$$\Rightarrow M/m \leqslant \frac{3\sqrt{5}}{\sqrt{5} + 2\sqrt{3}}.$$

(b) Solving equations (5.273) and (5.274)

$$\omega = \frac{\omega_0(3em - M)}{M + 3m}. \qquad (5.278)$$

The velocity of the block just after the impact is

$$v = l(\omega + e\omega_0)$$

$$= l\left[\frac{\omega_0(3em - M)}{M + 3m} + e\omega_0\right]$$

$$= l\omega_0\left[\frac{3em - M + eM + 3em}{M + 3m}\right]$$

$$= l\omega_0\left\{\frac{M(1 + e)}{M + 3m}\right\}. \qquad (5.279)$$

Using equations (5.276) and (5.279) we have

$$v = \sqrt{5gR} = \frac{l\omega_0 M(1 + e)}{M + 3m}$$

$$\Rightarrow \omega_0 = \frac{M+3m}{Ml(1+e)}\sqrt{5gR}. \tag{5.280}$$

Conservation of the energy between position 1 and 2 of the rod during its downward journey gives

$$\frac{Ml^2}{6}\omega_0^2 = Mg\frac{l}{2}$$

$$\Rightarrow \omega_0 = \sqrt{\frac{3g}{l}}. \tag{5.281}$$

Using the equations (5.280) and (5.281)

$$\sqrt{\frac{3g}{l}} = \frac{M+3m}{Ml(1+e)}\sqrt{5gR}$$

$$\Rightarrow \frac{l}{R} = \frac{5(M+3m)^2}{3(1+e)^2 M^2}.$$

Problem 121 A uniform rod of mass m and length l is connected rigidly to a square plate of mass M hanging from a smooth pivot. If a bead of mass m strikes the hanging plate and sticks to it, find (a) the velocity of the plate just after the collision and (b) the maximum height attained by the center of mass of the system (rod + plate + particle) after the collision. Put $b = l$.

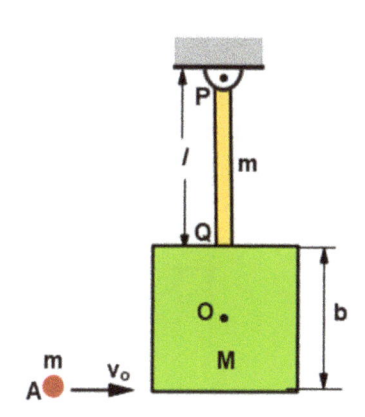

Solution
(a) Let the rod have an angular velocity ω just after the collision. The moment of inertia of the rod–particle system just after the impact is

$$I_P = m\left\{\frac{b^2}{4} + (l+b)^2\right\} + \frac{ml^2}{3} + M\left(\frac{b^2}{6} + \left(l+\frac{b}{2}\right)^2\right)$$

$$= m\left(\frac{b^2}{4} + 4b^2\right) + \frac{mb^2}{3} + M\left(\frac{b^2}{6} + \frac{9b^2}{4}\right)$$

$$= \frac{51mb^2 + 4mb^2 + 29mb^2}{12} = \frac{84mb^2}{12} = 7mb^2.$$

Conserving the angular momentum about P we have

$$mv_0(l + b) = 7mb^2\omega$$

$$\omega = 2v_0/7b.$$

(b) Conservation of energy:
Let the center of mass rise through a height h_C:

$$\frac{1}{2}I_P\omega^2 = m_{total}gh_C$$

$$\Rightarrow \frac{I_P^2\omega^2}{2I_P} = m_{total}gh_C$$

$$\Rightarrow \frac{(2mv_0b)^2}{2 \times 7mb^2} = (m + m + m)gh_C$$

$$\Rightarrow h_C = 2v_0^2/21g.$$

Problem 122 In the previous problem, if the rod is substituted by a light string, find (a) the angular velocity of the rod and (b) the linear velocity of the point O and the bead just after the collision.

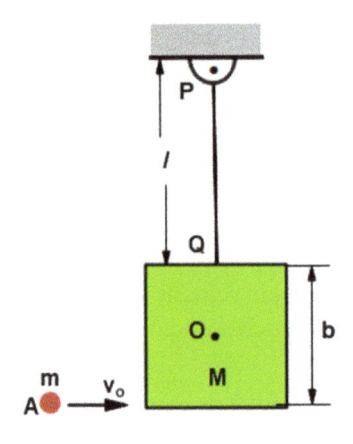

Solution

(a) Let the plate have an angular velocity ω just after the collision and the velocity of center of mass of the system $(M + m)$ move with a velocity v_C. Let I_C be the moment of inertia of the rod–particle system just after the impact about its center of mass C. Conserving the angular momentum during the impact about any point lying on the line of motion of center of mass C we can write

$$mv_0 y_C \hat{k} = I_C \omega \hat{k}. \tag{5.282}$$

The moment of inertia of the plate–bead system about its center of mass C we have

$$I_C = \frac{M}{12}(b^2 + b^2) + \frac{Mm(b/\sqrt{2})^2}{M + m}$$

$$= \frac{Mb^2}{6} + \frac{Mmb^2}{2(M + m)}$$

$$= \frac{Mb^2}{2}\left\{\frac{1}{3} + \frac{m}{(M + m)}\right\}. \tag{5.283}$$

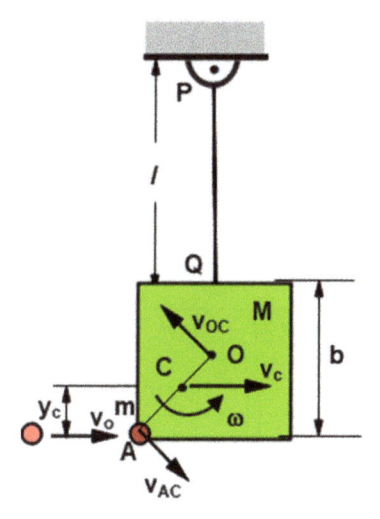

The value of y_C is given as

$$y_C = \frac{bmM \cos 45°}{\sqrt{2}(M + m)} = \frac{Mb}{2(M + m)}. \tag{5.284}$$

Using equations (5.282), (5.283), and (5.284) we have

$$\frac{Mmbv_0}{2(M+m)} = \frac{Mb^2}{2}\left(\frac{1}{3} + \frac{m}{M+m}\right)\omega$$

$$\Rightarrow \frac{mv_0}{(M+m)} = \frac{b}{3}\left\{\frac{M+4m}{M+m}\right\}\omega$$

$$\Rightarrow \omega = \frac{3mv_0}{(M+4m)b}. \tag{5.285}$$

(b) By linear momentum conservation,

$$v_C = \frac{mv_0}{M+m}. \tag{5.286}$$

The velocity of the point O is

$$\vec{v}_O = \vec{v}_{OC} + \vec{v}_C. \tag{5.287}$$

The velocity of O relative to the center of mass C of the system $(M+m)$ is

$$\vec{v}_{OC} = \omega(OC)\frac{1}{\sqrt{2}}(-\hat{i} + \hat{j}). \tag{5.288}$$

Using equations (5.286), (5.287), and (5.288) we have

$$\vec{v}_O = \left(v_C - \omega(OC)\frac{1}{\sqrt{2}}\right)\hat{i} + \omega(OC)\frac{1}{\sqrt{2}}\hat{j}. \tag{5.289}$$

Using equations (5.285), (5.286), and (5.289) we have

$$\vec{v}_O = \left(\frac{3mv_0}{(M+4m)} - \frac{3mv_0}{(M+4m)b}\frac{OC}{\sqrt{2}}\right)\hat{i}$$

$$+ \frac{3mv_0}{(M+4m)b}\frac{OC}{\sqrt{2}}\hat{j}$$

$$= \frac{3mv_0}{(M+4m)b}\left\{\left(1 - \frac{OC}{b\sqrt{2}}\right)\hat{i} + \frac{OC}{b\sqrt{2}}\hat{j}\right\}.$$

Putting $OC = \dfrac{mb}{\sqrt{2}(M+m)}$ in the last equation we have

$$\vec{v}_O = \frac{3mv_0}{(M+4m)b}\left\{\left(1 - \frac{m}{2(M+m)}\right)\hat{i} + \frac{m}{2(M+m)}\hat{j}\right\}$$

$$= \frac{3mv_0}{(M+4m)b}\left\{\left(\frac{2M+m}{2(M+m)}\right)\hat{i} + \frac{m}{2(M+m)}\hat{j}\right\}$$

$$= \frac{3mv_0}{2(M+4m)(M+m)b}\{(2M+m)\hat{i} + m\hat{j}\}.$$

The velocity of the point of the bead A is

$$\vec{v_A} = \vec{v_{AC}} + \vec{v_C}. \tag{5.290}$$

The velocity of A relative to the center of mass C of the system $(M+m)$ is

$$\vec{v_{AC}} = \omega(AC)\frac{1}{\sqrt{2}}(\hat{i} - \hat{j}). \tag{5.291}$$

Using equations (5.286), (5.290), and (5.291) we have

$$\vec{v_A} = \left(v_C + \omega(AC)\frac{1}{\sqrt{2}}\right)\hat{i} - \omega(AC)\frac{1}{\sqrt{2}}\hat{j}. \tag{5.292}$$

Using equations (5.285), (5.286), and (5.291) we have

$$\vec{v_A} = \left(\frac{3mv_0}{(M+4m)} + \frac{3mv_0}{(M+4m)b}\frac{AC}{\sqrt{2}}\right)\hat{i}$$

$$+ \frac{3mv_0}{(M+4m)b}\frac{OC}{\sqrt{2}}\hat{j}$$

$$= \frac{3mv_0}{(M+4m)b}\left\{\left(1 - \frac{OC}{b\sqrt{2}}\right)\hat{i} + \frac{OC}{b\sqrt{2}}\hat{j}\right\}.$$

Putting $AC = \frac{Mb}{\sqrt{2}(M+m)}$ in the last equation, we have

$$\vec{v_A} = \frac{3mv_0}{(M+4m)b}\left\{\left(1 + \frac{M}{2(M+m)}\right)\hat{i} - \frac{M}{2(M+m)}\hat{j}\right\}$$

$$= \frac{3mv_0}{(M+4m)b}\left\{\left(\frac{3M+2m}{2(M+m)}\right)\hat{i} - \frac{M}{2(M+m)}\hat{j}\right\}$$

$$= \frac{3mv_0}{2(M+4m)(M+m)b}\{(3M+m)\hat{i} - 2m\hat{j}\}.$$

Problem 123 A thin rod of mass m and length l is pivoted smoothly at P. A rectangular plank of mass $M (= 2m/3)$ and length l is welded to a thin rod to form a composite rod. A jet of water of density ρ strikes the rectangular rod elastically with a velocity v such that the composite rod remains in equilibrium. If the effective area on which the water collides is equal to A, find the magnitude of v. Put $\theta = 60°$.

5-248

Solution

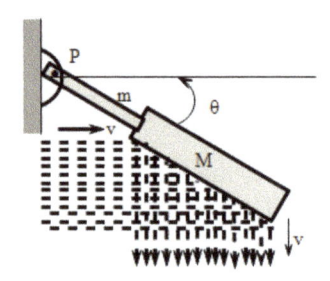

Consider a thin horizontal strip of the plate on which water is imparting its momentum on an elementary strip of area dA of the plate. During an elementary time δt, the momentum delivered by an elementary segment of water of mass dm having velocity v on the plate is

$$\delta P = 2\delta m v \sin \theta.$$

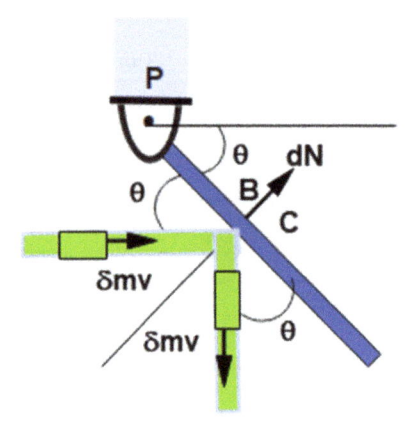

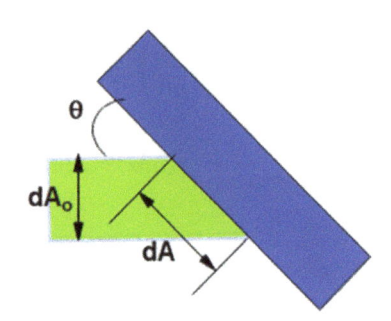

Then the force acting on an elementary strip of the plate is

$$dN = \frac{2\delta mv \sin \theta}{\delta t} = 2\left(\frac{\delta m}{\delta t}\right)v \sin \theta$$

$$= 2(\rho \delta A_0 v)v \sin \theta$$

$$\Rightarrow dN = 2\delta A_0 \rho v^2 \sin \theta$$

$$= 2(\delta A \sin \theta)\rho v \sin \theta (\because \delta A_0 = \delta A \sin \theta).$$

Then, integrating, the force acting on the plate due to the liquid is

$$N = 2A\rho v^2 \sin^2 \theta.$$

This force produces an counter-clockwise torque about P so as to balance the clockwise gravitational torque. So the composite rod remains in equilibrium. The center of force of the impact force of water is at a distance of $3l/2$ from the pivot. Equating the gravitational torque with the torque of the impact force of water on the plate about P we have

$$N\frac{3l}{2} = mg\frac{l}{2}\cos \theta + Mg\frac{3l}{2}\cos \theta$$

$$\Rightarrow 2A\rho v^2 \sin^2 \theta \frac{3l}{2} = \left(mg + \frac{3Mg}{2}\right)l \cos \theta$$

$$\Rightarrow 3A\rho v^2 \sin^2 \theta = (2m + 3M)\frac{g}{2} \cos \theta$$

$$\Rightarrow 3A\rho v^2 \sin^2 60° = (2m + 3M)\frac{g}{2} \cos 60°$$

$$\Rightarrow 9A\rho v^2 = g(3M + 2m)$$

$$\Rightarrow v = \frac{1}{3}\sqrt{\frac{g(3M + 2m)}{A\rho}}$$

$$= \frac{2}{3}\sqrt{\frac{mg}{A\rho}} (\because 3M = 2m).$$

Problem 124 A rigid horizontal rod AB of length l and mass M is pivoted smoothly at its mid-point P. A ball of mass m_1 is released from a height h such that it collides at the end B of the rod and sticks to the rod just after the collision. As a result, the other ball of mass m_2 attached at the end A of the rod will go to its highest position when the rod will be vertical. Find (a) the value of h/l and (b) the fraction of energy lost during the collision. Put $M = 3$ kg, $m_1 = 1$ kg, and $m_2 = 2$ kg.

Solution

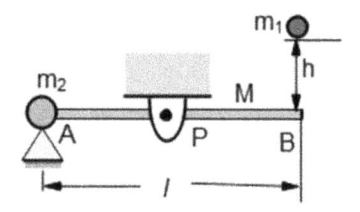

(a) Let the rod rotate with an angular velocity ω just after the collision. Conserving the angular momentum about the pivot,

$$m_1 v_0 \frac{l}{2} = I_P \omega. \qquad (5.293)$$

As the body A moves up its gravitational potential energy rises by $m_2 gl/2$ and that of B decreases by $m_1 gl/2$. So the change in gravitational potential energy is

$$\Delta U = (m_2 - m_1) g \frac{l}{2}.$$

As the rod will come to rest when A will reach the vertical position, the kinetic energy decreases to zero. So the change in kinetic energy is

$$\Delta K = -\frac{1}{2} I_P \omega^2.$$

Conserving energy gives us

$$\Delta K + \Delta U = 0$$

$$\Rightarrow -\frac{1}{2} I_P \omega^2 + (m_2 - m_1) g \frac{l}{2} = 0$$

$$\Rightarrow \frac{(I_P \omega)^2}{I_P} = (m_2 - m_1) gl. \qquad (5.294)$$

Using equations (5.293) and (5.294)

$$\frac{\left(m_1 v_0 \frac{l}{2}\right)^2}{I_P} = (m_2 - m_1) gl$$

$$\Rightarrow \frac{m_1^2 v_0^2 l}{4 I_P} = (m_2 - m_1) gl$$

$$\Rightarrow \frac{m_1^2 (2gh) l^2}{4 \left\{ \frac{Ml^2}{12} + (m_1 + m_2) \frac{l^2}{4} \right\}} = (m_2 - m_1) gl$$

$$\Rightarrow \frac{2m_1^2 h}{\frac{M}{3} + (m_1 + m_2)} = (m_2 - m_1)gl$$

$$\Rightarrow \frac{h}{l} = \frac{(m_2 - m_1)\left(\frac{M}{3} + m_1 + m_2\right)}{2m_1^2}.$$

After evaluation we have

$$\frac{h}{l} = \frac{(2-1)\left(\frac{3}{3} + 1 + 2\right)}{2 \times (1)^2} = 2.$$

(b) The fraction of energy lost is

$$\eta = -\frac{\frac{L^2}{2I_f} - \frac{L^2}{2I_i}}{\frac{L^2}{2I_i}} = -\left(\frac{I_i}{I_f} - 1\right)$$

$$= 1 - \frac{m_1 \frac{l^2}{4}}{(m_1 + m_2)\frac{l^2}{4} + \frac{Ml^2}{12}}$$

$$= 1 - \frac{m_1}{m_1 + m_2 + \frac{M}{3}}$$

$$= 1 - \frac{1}{1 + 2 + \frac{3}{3}} = 1 - \frac{1}{4} = \frac{3}{4}.$$

Problem 125 A vertical rod of mass m is pivoted smoothly with a block of mass M which is free to slide along a smooth horizontal bar by a hole made in it. If a small ball of mass m collides at the bottom of the rod and sticks to it, find the angular velocity of the rod and the velocity of the block just after the collision. Put $M = m$.

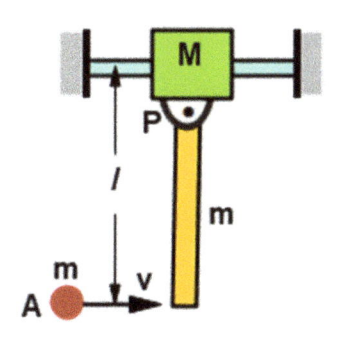

Solution

Let the rod rotate with an angular velocity ω, the block M move with a velocity v_1, the center of mass C of the rod have a velocity v_C, and the ball move with a velocity v_2 just after the collision.

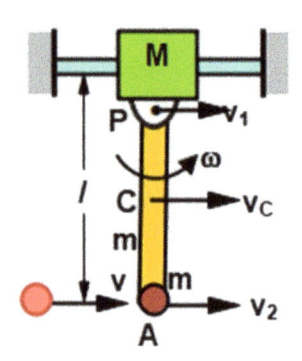

Momentum conservation of the system $(M + 2m)$ gives

$$mv_C + Mv_1 + mv_2 = mv,$$

where

$$v_C = v_1 + \frac{l}{2}\omega, \text{ and}$$

$$v_2 = v_1 + l\omega$$

$$\Rightarrow m\left(v_1 + \frac{l}{2}\omega\right) + Mv_1 + m(v_1 + l\omega) = mv$$

$$\Rightarrow v_1(M + 2m) + 3m\frac{l}{2}\omega = mv$$

$$\Rightarrow v_1(m + 2m) + 3m\frac{l}{2}\omega = mv$$

$$\Rightarrow 6v_1 + 3l\omega = 2v. \tag{5.295}$$

Conservation the of angular momentum about P, $L_P = C$, gives

$$\Rightarrow mvl = m(v_1 + l\omega)l + \frac{ml^2}{12}\omega + m\left(v_1 + \frac{\omega}{2}\right)\frac{l}{2}\omega$$

$$\Rightarrow mvl = \frac{3}{2}mv_1l + ml^2\omega\left(\frac{12}{12} + \frac{1}{12} + \frac{1 \times 3}{4 \times 3}\right)$$

5-253

$$\Rightarrow \frac{3}{2}v_1 + \frac{16}{12}l\omega = v$$

$$\Rightarrow 9v_1 + 8l\omega = 6v. \quad (5.296)$$

Solving equations (5.295) and (5.296) we have $v_1 = -2v/21$ and $\omega = (6/7)v/l$. The negative sign of v_1 means that its direction will be opposite to that of the assumed one.

Problem 126 Two identical rods each of mass m and length l are welded to form a composite (L-shaped) rod which is kept on a smooth horizontal surface. A ball of mass m collides with the composite rod inelastically with a velocity v_0, as shown in the figure. Find (a) the angular velocity of the composite rod, (b) the velocity of the particle m relative to the ground, (c) the acceleration of the particle m, (d) the force acting on the ball just after the collision, and (e) the fraction of energy lost during the collision.

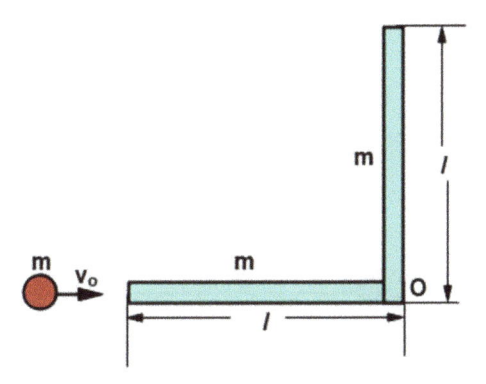

Solution
(a) The coordinates of the center of mass of the system (rod + particle) can be given as

$$x_C = \frac{\left(ml + m\frac{l}{2}\right)}{3m} = \frac{l}{2}$$

$$y_C = \frac{\frac{ml}{2}}{3m} = \frac{l}{6}.$$

The moment of inertia of the system about its center of mass is

$$I_C = I_1 + I_2 + I_3$$

5-254

$$= mr_1^2 + m\left(\frac{l^2}{12} + r_2^2\right) + m\left(\frac{l^2}{12} + r_3^2\right)$$

$$= m\left[r_1^2 + r_2^2 + r_3^2 + \frac{l^2}{6}\right]$$

$$= m\left[\left(\frac{l^2}{4} + \frac{l^2}{36}\right) + \frac{l^2}{36} + \left(\frac{l^2}{9} + \frac{l^2}{4}\right) + \frac{l^2}{6}\right]$$

$$= ml^2\left[\frac{1}{2} + \frac{1}{18} + \frac{1}{9} + \frac{1}{6}\right]$$

$$= ml^2\left[\frac{9+1+2+3}{18}\right] = \frac{5}{6}ml^2.$$

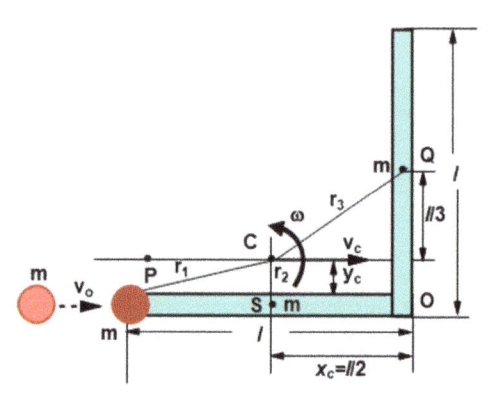

Let the rod rotate with an angular velocity ω and the center of mass C of the rod −ball system have a velocity v_C and the ball move with a velocity v_1 and acceleration a_1, just after the collision.

Conserving the angular momentum about the center of mass C we have

$$\vec{L}_C = \vec{C}$$

$$+mv_0 r_2 \hat{k} = I_C \vec{\omega}$$

$$\Rightarrow \vec{\omega} = \frac{mv_0 r_2}{I_C}\hat{k} = \frac{mv_0(l/6)}{(5ml^2/6)}\hat{k} = v_0/5l.$$

(b) Conserving the linear momentum of the system $(M + 2m)$

$$\vec{v}_C = \frac{mv_0}{m+2m}\hat{i} = \frac{v_0}{3}\hat{i}.$$

5-255

We know that the velocity of the ball (particle 1) relative to C is
$$\vec{v}_{1C} = r_1\omega \cos\theta \hat{i} + r_1\omega \sin\theta(-\hat{j})$$
$$= \frac{l}{6}\omega \hat{i} + \frac{l}{2}\omega(-\hat{j})$$
$$= \frac{l}{6}\omega \hat{i} + \frac{l}{2}\omega(-\hat{j}) = \frac{l}{2}\omega\left(\frac{\hat{i}}{3} - \hat{j}\right).$$

The velocity of particle 1 (the ball) relative to ground is
$$\vec{v}_1 = \vec{v}_{1C} + \vec{v}_C = \frac{l}{2}\frac{v_0}{(5l)}\left(\frac{\hat{i}}{3} - \hat{j}\right) + \frac{v_0}{3}\hat{i}$$
$$= \frac{v_0}{30}(11\hat{i} - 3\hat{j}).$$

(c) The acceleration of the particle is
$$|\vec{a}_P| = r_1\omega^2 = \left(\sqrt{\frac{5}{18}}l\right)(v_0/5l)^2 = (\sqrt{5/2})\frac{v_0^2}{75l}.$$

(d) The force at the point is
$$R = ma_C = \left(\sqrt{5/2}\right)\frac{mv_0^2}{75l}.$$

(e) The change in kinetic energy is
$$\Delta K = \frac{1}{2}(3m)v_C^2 + \frac{1}{2}\left(\frac{5}{6}ml^2\right)\left(\frac{v_0}{5l}\right)^2 - \frac{1}{2}mv_0^2$$
$$= mv^2\left(\frac{1}{6} + \frac{1}{60} - \frac{1}{2}\right) = -\frac{19}{60}mv_0^2 \text{ obtained after putting } v_C = v_0/3.$$

The fraction of energy lost is
$$\eta = \left|\frac{-\Delta K}{K}\right| = \frac{19}{30}.$$

Problem 127 Two identical rods each of mass m and length l are welded to form a composite (T-shaped) rod which is kept on a smooth horizontal surface. A ball of mass m collides with the composite rod with a velocity v_0, as shown in the figure. If e is the coefficient of restitution of the collision, find (a) the velocity of the ball, (b) the velocity of the center of mass of the T-shaped rod, and (c) the angular velocity of the rod just after the collision.

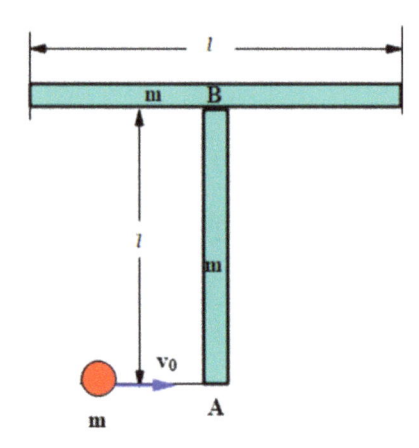

Solution

(a) The center of mass C of the composite rod is located at a distance of $l/4$ from the mid-point of the vertical rod, as shown in the following figure. The moment of inertia of the rod–particle system is

$$I_C = I_1 + I_2$$

(where I_1 and I_2 are the moment of inertia of the composite rod and particle about C, respectively)

$$= \frac{ml^2}{12} + m\left(\frac{l}{4}\right)^2 + \frac{ml^2}{12} + m\left(\frac{l}{4}\right)^2$$

$$= \frac{2ml^2}{4}\left(\frac{1}{3} + \frac{1}{4}\right) = \frac{2 \times 7}{48}ml^2 = \frac{7}{24}ml^2.$$

5-257

Conserving the angular momentum about the point P,
$$L_P = C$$
$$\Rightarrow mv_0 \frac{3}{4}l = mv\frac{3}{4}l + I_C \omega$$
$$\Rightarrow \frac{3}{4}mv_0 l = m \times \frac{3}{4}lv + \frac{7ml^2}{24}\omega$$
$$\Rightarrow 3(v_0 - v) = \frac{7}{6}l\omega. \tag{5.297}$$

Conservation of the linear momentum $P = C$ gives
$$mv_0 = (mv) + (2m)v_C$$
$$\Rightarrow v_C = \frac{v_0 - v}{2}. \tag{5.298}$$

From equations (5.297) and (5.298)
$$3(2v_C) = \frac{7l\omega}{6}$$
$$\Rightarrow v_C = \frac{7l\omega}{36}. \tag{5.299}$$

The NIF for the collision is
$$-e(u_A - u_B) = v_A - v_B$$
$$-e[v_0 - 0] = v - \left(\frac{3l\omega}{4} + v_C\right)$$
$$v = -ev_0 + \left(\frac{3l\omega}{4} + v_C\right). \tag{5.300}$$

Using equations (5.299) and (5.300) we have
$$v = -ev_0 + \frac{3l\omega}{4} + \frac{7l\omega}{36}$$
$$v = -ev_0 + \frac{l\omega}{4}\left(3 + \frac{7}{9}\right) = -ev_0 + \frac{34l\omega}{36}$$
$$\Rightarrow v = -ev_0 + \frac{17l\omega}{18}. \tag{5.301}$$

Using equations (5.297) and (5.302)

$$v = -ev_0 + \frac{17}{18}\left\{\frac{18}{7}(v_0 - v)\right\}$$

$$\Rightarrow -\frac{17}{7}v_0 + ev_0 = -v - \frac{17}{7}v = -\frac{24}{7}v$$

$$v_0(7e - 17) = -24v$$

$$\Rightarrow v = \frac{v_0(17 - 7e)}{24}.$$

1. If $e = 0$, $v = \frac{17}{24}v_0$.
2. If $e = 1$, $v = \frac{5}{12}v_0$.

(b) Then using equation (5.298)

$$v_C = \frac{v_0 - v}{2} = \frac{24v_0 - v_0(17 - 7e)}{2 \times 24}$$

$$= \frac{7v_0(1 + e)}{48}.$$

(c) $\omega = \frac{36v_C}{7l} = \frac{7v_0(1 + e)}{48l} \times \frac{36}{7}$

$$\vec{\omega} = \frac{3v_0(1 + e)}{4l}\hat{k}.$$

If we put $e = 0$, angular velocity of the rod is equal to $3v_0/4l$. However, if we put $e = 0$, the linear velocity of the center of mass C will be equal to $7v_0/48l$ which may not be same when you do the problem from the basics by assuming that the particle will stick to the end **B** of the rod. This is because the system will change from the rod to rod–particle. So, the center of mass of the system will change from C to G.

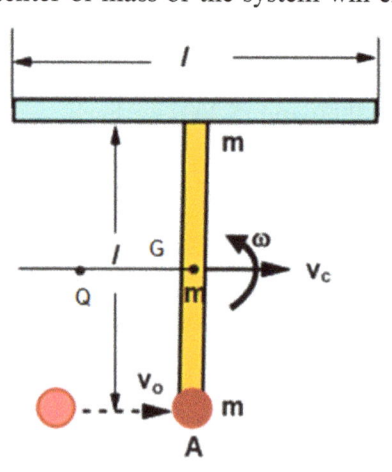

5-259

Problem 128 Two identical rods each of mass m and length l are welded to form a composite (L-shaped) rod which is kept on a smooth horizontal surface. The composite rod is pivoted smoothly at the point P. A ball of mass m collides with the composite rod and sticks to the rod. Find (a) the angular velocity of the rod (b) the fraction of kinetic energy lost in the collision (c) the reaction at the pivot just after the collision.

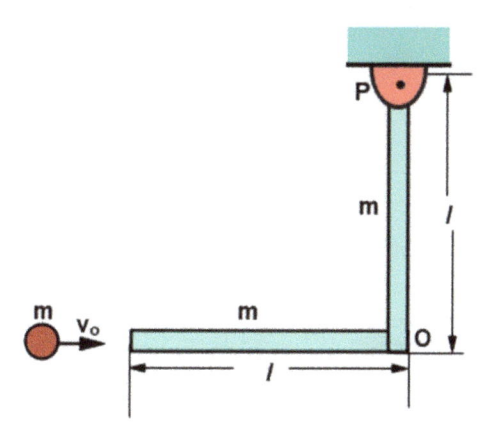

Solution

(a) Let the rod have angular velocity ω just after the collision. The coordinates of the center of mass of the rod–particle system relative to the pivot P are

$$x_C = \frac{m(-l/2) + m(-l)}{3m} = -l/2$$

$$y_C = \frac{m(-l/2) + m(-l) + m(-l)}{3m} = -5l/6.$$

The moment of inertia about the pivot is

$$I_P = \frac{ml^2}{3} + \left\{\frac{ml^2}{12} + r_1^2\right\} + mr_2^2$$

$$= \frac{ml^2}{3} + m\left\{\frac{l^2}{12} + \left(l^2 + \frac{l^2}{4}\right)\right\} + m(l^2 + l^2)$$

$$= ml^2\left(\frac{1}{3} + \frac{1}{12} + \frac{5}{4} + 2\right)$$

$$= ml^2\left(\frac{4 + 1 + 15 + 24}{12}\right) = \frac{44}{12}ml^2 = \frac{11}{3}ml^2.$$

5-260

Conserving the angular momentum about the pivot P,

$$L_P = \frac{11}{3}ml^2\omega = mv_0 l$$

$$\Rightarrow \omega = \frac{3v_0}{11l}.$$

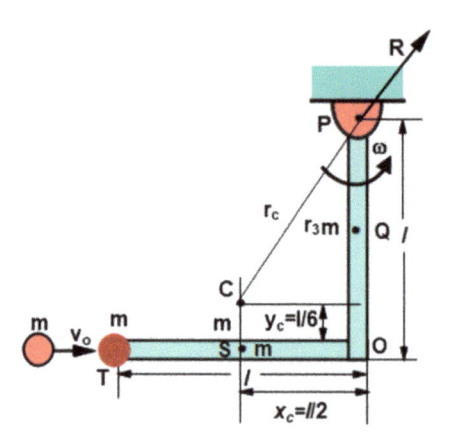

(b) Then the fraction change in kinetic energy is

$$\frac{\Delta K}{K} = \frac{\frac{L_P^2}{2I_P} - \frac{1}{2}mv_0^2}{\frac{1}{2}mv_0^2}$$

$$= \frac{1}{\left(\frac{1}{2}mv_0^2\right)}\left(\frac{m^2v_0^2 l^2}{2 \times \frac{11}{3}ml^2} - \frac{1}{2}mv_0^2\right)$$

$$= \left(\frac{3}{11} - 1\right) = -\frac{8}{11}.$$

(c) The reaction force at the pivot is

$$R = m_{\text{total}} r_C \omega^2 = (3m)\left\{\sqrt{\left(\frac{5l}{6}\right)^2 + \left(\frac{l}{2}\right)^2}\right\}\omega^2$$

$$= 3m \times \frac{l}{6}(\sqrt{34})\omega^2 = \sqrt{34}\frac{l\omega^2}{2}$$

5-261

$$\Rightarrow R = m\sqrt{\frac{34}{4}\left\{1\frac{9v_0^2}{121l^2}\right\}}$$

$$= \frac{9\sqrt{34}}{242}\frac{mv_0^2}{l}.$$

The x-component of R is

$$R_x = R\sin\theta$$

$$= \frac{9\sqrt{34}}{242} \times \frac{\frac{l}{2}}{\frac{\sqrt{34}}{6}}\frac{mv_0^2}{l}$$

$$\Rightarrow R_x = \frac{27mv_0^2}{242l}.$$

Similarly, the y-component of R is

$$R_y = R\cos\theta = \frac{9\sqrt{34}}{242} \times \frac{\frac{5l}{6}}{\frac{\sqrt{34}}{6}}\frac{mv_0^2}{l}.$$

Then we can write

$$\vec{R} = \frac{9mv_0^2}{242l}(3\hat{i} + 5\hat{j}).$$

Problem 129 A rigid horizontal bar of mass M and length l is smoothly pivoted at its mid-point P. An insect of mass m falls from a height h while the bar is horizontal. The insect collides inelastically with the rod at its right-hand end and sticks to the rod just after the collision. If the horizontal and vertical reaction at the pivot just after the collision of the insect with the rod are equal in magnitude, find h/l. Put $M/m = 12$.

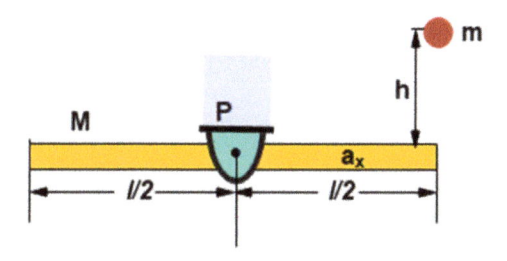

Solution

Let the rod have angular velocity ω just after the insect collides inelastically at C, as shown by the red dot. The components of the reaction force are given as follows:

$$R_x = mx\omega^2 \tag{5.302}$$

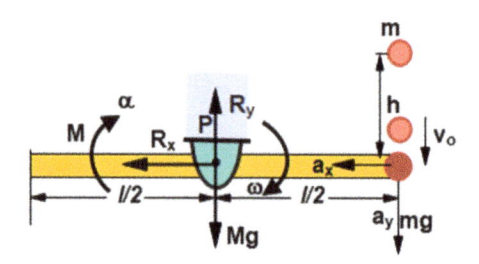

$$R_y = (N+m)g - mx\alpha, \text{ where } x = l/2. \tag{5.303}$$

As per the given condition,

$$R_x = R_y$$

$$\Rightarrow ma_x = (M+m)g - ma_y$$

$$\Rightarrow m(l/2)\omega^2 = (M+m)g - m(l/2)\alpha$$

$$\Rightarrow m(l/2)\omega^2 = (12m+m)g - m(l/2)\alpha$$

$$\Rightarrow \omega^2 = 26g/l - \alpha. \tag{5.304}$$

Just after the collision the insect its weight generates a torque $mg/2$ about the pivot P; so, the angular acceleration of the rod is

$$\alpha = T_P/I_P = \frac{mg(l/2)}{\frac{(12m)l^2}{12} + m(l/2)^2}$$

$$\Rightarrow \alpha = \frac{g(l/2)}{(l^2 + l^2/4)} = 2g/5l. \tag{5.305}$$

Just before the collision at a distance $x = l/2$ from the pivot, the angular momentum of the system is

$$L_i = mv_0 l/2.$$

Just after the collision, the angular momentum of the system is

$$L_f = \left(\frac{Ml^2}{12} + m(l/2)^2\right)\omega.$$

Conserving the angular momenta,
$$L_f = L_i$$
$$\Rightarrow \left(\frac{Ml^2}{12} + m(l/2)^2\right)\omega = mv_0(l/2)$$
$$\Rightarrow m(l^2 + l^2/4)\omega = mv_0(l/2) (\because M = 12m)$$
$$\Rightarrow \omega = \frac{2v_0}{5l}. \qquad (5.306)$$

Using equations (5.304), (5.305), and (5.306)
$$\left(\frac{2v_0}{5l}\right)^2 = 26g/l - (2g/5l)$$

$$\frac{4v_0^2}{25l^2} = 26g/l - (2g/5l).$$

Putting $v_0 = \sqrt{2gh}$ we have
$$\frac{4(2gh)}{25l^2} = 26g/l - (2g/5l)$$

$$\Rightarrow \frac{4h}{25} = 13l - (l/5) = 64l/5$$

$$\Rightarrow \frac{h}{l} = 80.$$

Problem 130 A rigid horizontal bar of mass M and length l is smoothly pivoted at its mid-point P. An insect of mass m falls from a height h while the bar is horizontal. The insect collides inelastically with the rod at a distance $b = l/4$ from the pivot. If the insect crawls away from the pivot just after the collision so that the rod moves with a constant angular velocity and if the rod rotates through a right angle when the insect reaches the end of the rod, find h/l. Put $M/m = 12$.

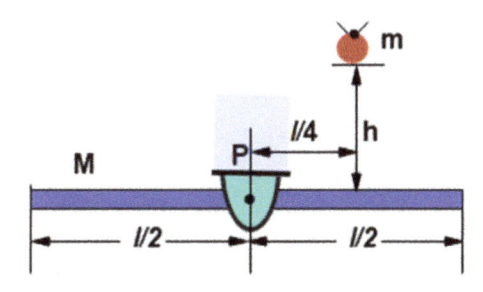

Solution

At any distance x from the pivot, the angular momentum of the system is

$$L = \left(\frac{Ml^2}{12} + mx^2\right)\omega.$$

The rate of change of the angular momentum of the system is

$$\frac{dL}{dt} = \omega \frac{d}{dt}\left(\frac{Ml^2}{12} + mx^2\right).$$

If the angular velocity is maintained as constant we have

$$\frac{dL}{dt} = \omega \frac{d}{dt}\left(\frac{Ml^2}{12} + mx^2\right)$$

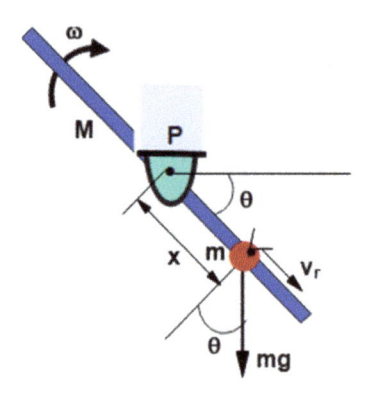

$$\Rightarrow \frac{dL}{dt} = 2m\omega x \frac{dx}{dt}. \tag{5.307}$$

The net torque acting on the system about P is responsible for changing its angular momentum. Newton's second law of rotation for a non-rigid system is

$$\tau_P = mgx \cos\theta = \frac{dL}{dt}. \tag{5.308}$$

Using equations (5.307) and (5.308) we have

$$mgx \cos\omega t = 2\omega m x \frac{dx}{dt}$$

$$\Rightarrow \int_0^t \cos\omega t\, dt = \frac{2\omega}{g} \int_{b=\frac{l}{4}}^x dx$$

5-265

$$\Rightarrow \frac{1}{\omega}\sin\omega t\bigg|_0^t = \frac{2\omega}{g}\left(-\frac{l}{4}+x\right)$$

$$\Rightarrow \sin\omega t = \frac{2\omega^2}{g}\left(x-\frac{l}{4}\right)$$

$$\Rightarrow x = \frac{g}{2\omega^2}\sin\omega t + \frac{l}{4}. \tag{5.309}$$

Since the insect reaches the end of the rod when it rotates through a right angle, if we put $x = \frac{l}{2}$ and $\omega t = \frac{\pi}{2}$ in the last equation we have

$$\frac{l}{2} = \frac{g}{2\omega^2}\sin\frac{\pi}{2} + \frac{l}{4}.$$

Then $\frac{l}{4} = \frac{g}{2\omega^2}$

$$\Rightarrow \omega = \sqrt{\frac{2g}{l}}. \tag{5.310}$$

The angular momentum just before the collision is

$$L_i = \frac{mv_0 l}{4}.$$

The angular momentum just after the collision is

$$L_f = \left(\frac{Ml^2}{12} + \frac{ml^2}{16}\right)\omega$$

$$\Rightarrow L_f = \left(\frac{12ml^2}{12} + \frac{ml^2}{16}\right)\omega$$

$$\Rightarrow L_f = \frac{17ml^2}{16}\omega.$$

Conserving the angular momenta of the system $(M + m)$ just before and after the collision about P we have

$$\frac{17ml^2}{16}\omega = \frac{mv_0 l}{4}$$

$$\Rightarrow \omega = \frac{4v_0}{17l}. \tag{5.311}$$

Using equations (5.310) and (5.311) we have

$$\frac{4v_0}{17l} = \sqrt{\frac{2g}{l}}$$

$$\Rightarrow v_0 = \frac{17}{4}\sqrt{2gl}. \tag{5.312}$$

As the insect falls from a height h just before the collision,

$$\Rightarrow v_0 = \sqrt{2gh}. \tag{5.313}$$

From equations (5.312) and (5.313) we have

$$\Rightarrow \sqrt{2gh} = \frac{17}{4}\sqrt{2gl}$$

$$\Rightarrow h = 289l/16$$

$$\Rightarrow \frac{h}{l} = 289/16.$$

Problem 131 A rod of mass M and length $l\,(=2a)$ connected to two particles each of mass m, is hinged smoothly at P such that it can rotate in the vertical plane. A ball of mass m collides with the rod at C at a distance $a/2$ from the pivot, after falling from a height h. The coefficient of restitution of the collision is $e = 0.5$. If the ball rod rotates through an angle of $90°$ when the ball reaches its highest point after the collision, find the value of h/l. Put $M = 3m$.

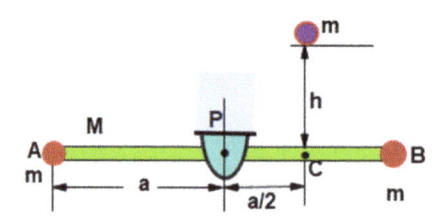

Solution
Let the rod have an angular velocity w and the ball move up with a velocity v just after the collision. Just before the collision let the velocity of the ball be v_0.
The NIF (Newton's impact/collision formula) for the collision is

$$\left(v + \frac{aw}{2}\right) = ev_0 = \frac{v_0}{2}$$

$$\Rightarrow 2v + aw = 2ev_0 = v_0. \tag{5.314}$$

Conserving the angular momentum about the pivot,

$$-\left(2ma^2 + \frac{Ma^2}{3}\right)w + mva = -mv_0\frac{a}{2}$$

5-267

$$\Rightarrow 2(mva - 3ma^2w) = -mv_0 a$$

$$\Rightarrow 2v - 6aw = -v_0. \tag{5.315}$$

Using equations (5.314) and (5.315)

$$\Rightarrow 2v - 6aw = -\left(\frac{2v + aw}{2e}\right)$$

$$\Rightarrow 2v\left(1 + \frac{1}{2e}\right) = \left(6 - \frac{1}{2e}\right)aw$$

$$\Rightarrow 2v(1 + 1) = (6 - 1)aw$$

$$\Rightarrow 4v = 5aw.$$

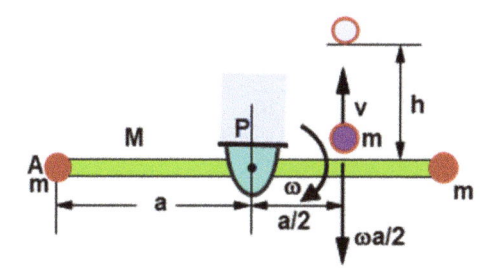

Putting the above value in equation (5.314),

$$2v + \frac{4v}{5} = 2ev_0 = v_0$$

$$\Rightarrow v = \frac{5v_0}{14}. \tag{5.316}$$

Then

$$\omega = \frac{4v}{5a} = \frac{4}{5a} \times \frac{5v_0}{14}$$

or

$$\omega = \frac{2v_0}{7a}. \tag{5.317}$$

The time for the motion of the ball is v/g and the time of rotation of the angle of $\pi/2$ radians is equal to $\pi/2\omega$, equating these two values of time we have

$$\frac{v}{g} = \frac{\pi}{2\omega}. \tag{5.318}$$

5-268

Using equations (5.316), (5.317) and (5.318), we have

$$\frac{1}{g}\left(\frac{5v_0}{14}\right) = \frac{\pi}{2\left(\frac{2v_0}{7a}\right)}$$

$$\Rightarrow v_0^2 = \frac{49\pi g a}{10}. \qquad (5.319)$$

Putting $v_0^2 = 2gh$ we have

$$\Rightarrow 2gh = \frac{49\pi g a}{10}$$

$$\Rightarrow \frac{h}{a} = \frac{49\pi}{20}$$

$$\Rightarrow \frac{h}{l} = \frac{h}{2a}$$

$$= \frac{49\pi}{2 \times 20} = \frac{49\pi}{40}.$$

Problem 132 A rod PQ of mass m and length l is pivoted smoothly at a point P of the ceiling. Another rod AB of mass M and length l is welded with the rod PQ at its mid-point. A solid sphere of mass m and radius r and a block of mass $2m$ is connected rigidly to the rod AB, as shown in the figure. The system (rods + sphere + block) is in equilibrium by a string NR tied to the rod PQ at its mid-point R. (a) Find the tension in the string. (b) Let a particle of mass m hit the block with a velocity u with a coefficient of restitution e of the collision. If we assume that the string was at the verge of breaking due to its maximum tension, it breaks just after the collision. Then, find the angular velocity of the rods just after the collision. (Put $13M = 4m$, $u = 13.81$ m s^{-1}, $e = 1/2$, $r = l/4$, and $l = 1$m).

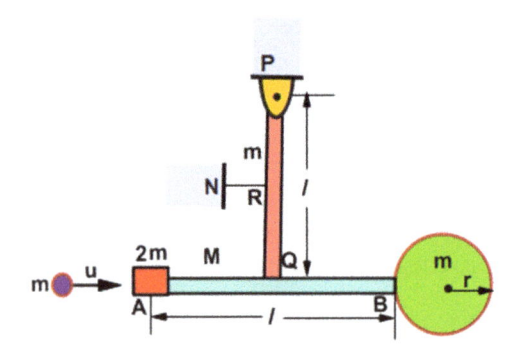

5-269

Solution

Let the rod have an angular velocity ω and the ball move towards the right with a velocity u_1 just after the collision.

Conserving the angular momentum about the pivot just before and just after the collision,

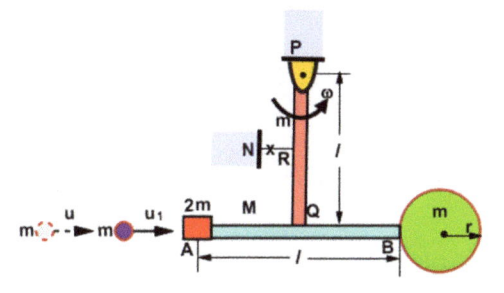

$$I\omega + mu_1 l = mul. \tag{5.320}$$

The NIF for the collision is

$$-eu = (u_1 - \omega l) \Rightarrow \omega l - u_1 = eu. \tag{5.321}$$

From equations (5.320) and (5.321)

$$\frac{mul - I\omega}{ml} = \omega l - eu$$

$$\Rightarrow \omega l \left(\frac{I}{ml^2} + 1\right) = u(1 + e)$$

$$\Rightarrow \omega = \frac{\frac{u}{l}(1 + e)ml^2}{I + ml^2}. \tag{5.322}$$

The moment of inertia of the vertical rod is

$$I_1 = \frac{ml^2}{3}.$$

The moment of inertia of the horizontal rod is

$$I_2 = \frac{Ml^2}{12} + Ml^2 = \frac{13Ml^2}{12}.$$

The moment of inertia of the sphere is

$$I_3 = \frac{2}{5}mr^2 + m\left\{l^2 + \left(\frac{l}{2} + r\right)^2\right\}$$

$$= \frac{2}{5}m\left(\frac{l^2}{16}\right) + m\left(\frac{9l^2}{16} + l^2\right) (\because r = l/4)$$

$$= \frac{127}{80}ml^2.$$

The moment of inertia of the particle (small block) is

$$I_4 = 2m\left\{\left(\frac{l}{2}\right) + l^2\right\} = \frac{5ml^2}{2}.$$

Then the total moment of inertia is

$$I = \frac{4 \times ml^2}{3 \times 4} + \frac{13M}{12}l^2 + \frac{127}{80}ml^2 + \frac{5}{2}ml^2$$

$$= \frac{8ml^2}{12} + \frac{127}{80}ml^2 + \frac{5}{2}ml^2 (\because 13M = 4m)$$

$$= ml^2\left(\frac{2}{3} + \frac{127}{80} + \frac{5}{2}\right) = \frac{1141}{240}ml^2. \tag{5.323}$$

Using equations (5.322) and (5.323) the angular velocity of the rod is

$$\omega = \frac{(240)(1+e)u}{1381l} = \frac{\left(1+\frac{1}{2}\right)(13.81)(240)}{1381 \times 1} = 3.6 \text{ rad s}^{-1}$$

obtained after putting $u = 13.81$ m s^{-1}

$$l = 1\,m \text{ and } e = \frac{1}{2}$$

Problem 133 A uniform rod of mass M and length l is smoothly pivoted at P. If the rod is released from rest, it rotates in the vertical plane and collides with a block of mass m which is attached to a spring of stiffness k. The coefficient of restitution of the collision is e. Find the maximum compression of the spring if (a) the maximum height attained by the center of mass of the rod is h after the collision and (b) the rod is released from rest when it makes an angle θ_0 with the vertical.

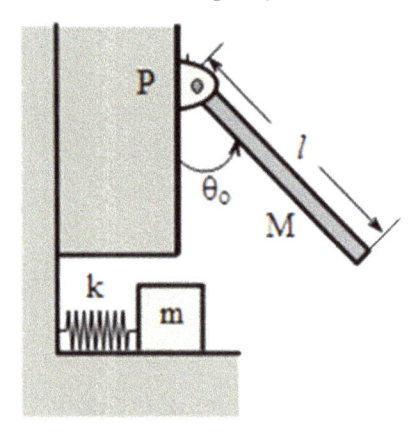

Solution

(a) Let the angular velocity of the rod just before and after the collision be ω_0 (clockwise) and ω (anticlockwise), respectively, and the ball move towards left with a velocity v just after the collision.

Applying the NIF (Newton's impact/collision formula) for the collision we have

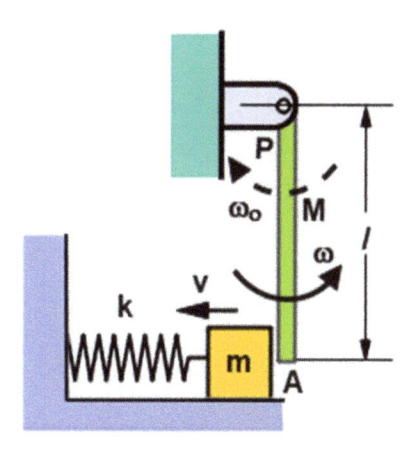

$$(v + l\omega) = el\omega_0. \tag{5.324}$$

Angular momentum conservation about the pivot gives

$$I(\omega + \omega_0) = mvl$$

$$\frac{Ml^2}{3}(\omega + \omega_0) = mvl$$

$$Ml(\omega + \omega_0) = 3mv. \tag{5.325}$$

Using equations (5.324) and (5.325)

$$\frac{3mu}{Ml} - \omega = \frac{v + l\omega}{el}$$

$$\frac{v}{l}\left(\frac{3m}{M} + \frac{1}{e}\right) = \omega\left(1 + \frac{1}{e}\right)$$

$$v\frac{3me + M}{Me} = \frac{(e+1)\omega l}{e}$$

$$\Rightarrow v(M + 3me) = M(e+1)\omega l. \tag{5.326}$$

5-272

Energy conservation:

$$\frac{1}{2}mv^2 = \frac{k}{2}x^2 \tag{5.327}$$

$$\frac{1}{2}\frac{Ml^2}{3}\omega^2 = Mgh. \tag{5.328}$$

Using equations (5.327) and (5.328)

$$\frac{\frac{1}{2}mv^2}{\frac{1}{2}\frac{Ml^2}{3}\omega^2} = \left(\frac{k}{2}x^2\right)/Mgh$$

$$\Rightarrow \frac{kx^2}{2gh} = \frac{3m}{M}\left(\frac{v}{l\omega}\right)^2$$

$$\Rightarrow x = \left(\frac{v}{l\omega}\right)\sqrt{\frac{6mgh}{Mk}}. \tag{5.329}$$

Using equations (5.326) and (5.329)

$$x = \frac{M(e+1)}{M+3me}\sqrt{\frac{2mgh}{Mk}}.$$

(b) Putting the value of $l\omega$ from equation (5.326) in equation (5.324) we have

$$v + l\omega = el\omega_0$$

$$\Rightarrow v + \frac{v(M+3me)}{M(e+1)} = el\omega_0$$

$$\Rightarrow v\{M(e+1) + M + 3me\} = Me(e+1)l\omega_0$$

$$\Rightarrow v = \frac{Me(e+1)l\omega_0}{M(e+2)+3me}. \tag{5.330}$$

Conserving energy for the spring–block system we have

$$\frac{1}{2}kx^2 = \frac{1}{2}mv^2$$

$$\Rightarrow x = \left(\sqrt{\frac{m}{k}}\right)v. \tag{5.331}$$

Using equations (5.330) and (5.331)

$$x = \frac{Me(e+1)l\omega_0}{M(e+2)+3me}\sqrt{\frac{m}{k}}. \tag{5.332}$$

Conserving energy for the rod between its initial position and the position just before the collision we have

$$\frac{Ml^2\omega_0^2}{2 \times 3} = Mg\frac{l}{2}(1 - \cos\theta_0)$$

$$\Rightarrow \omega_0 = \sqrt{\frac{3g(1 - \cos\theta_0)}{l}}. \tag{5.333}$$

Using equations (5.332) and (5.333)

$$x = \frac{Me(e+1)}{M(e+2) + 3me}\sqrt{\frac{3glm}{k}(1 - \cos\theta_0)}.$$

Problem 134 A uniform rod of mass M and length $L = 2l$ is pivoted smoothly at its mid-point P. A wet mud pellet of mass m falls from a height h and collides with the rod at C and sticks to the rod just after the collision. Find the maximum reaction force at the pivot. Put $M/m = 3/4$ and $h/l = 1/4$.

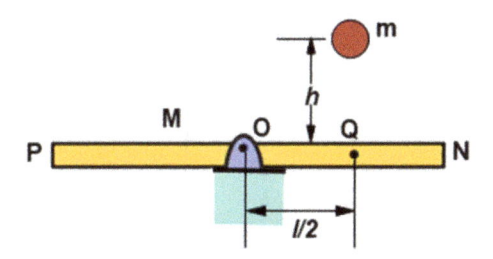

Solution
Let the angular velocity of the rod just after the collision be ω_0. The moment of inertia of the rod–particle system about the pivot is

$$I = \frac{Ml^2}{3} + \frac{ml^2}{4} = \frac{ml^2}{2} (\because M/3 = m/4)$$

$$\Rightarrow \frac{I}{ml^2} = \frac{1}{2}. \tag{5.334}$$

Let the angular velocity of the rod be ω (anticlockwise) when the mud pellet reaches its lowest position. Conserving the energy of the rod–pellet system between the horizontal and vertical positions, we have

$$\frac{I}{2}(\omega^2 - \omega_0^2) = mgl/2$$

$$\Rightarrow \omega^2 = \frac{mgl}{I} + \omega_0^2. \tag{5.335}$$

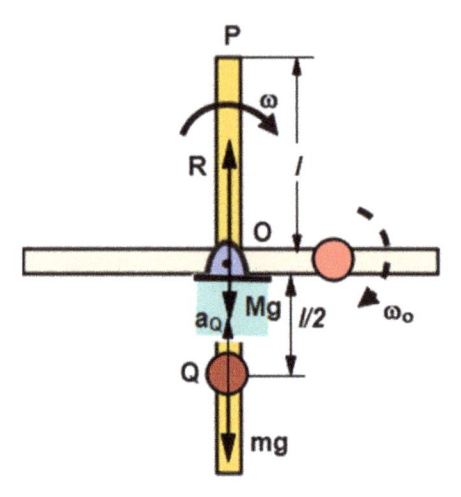

Let the pellet hit the rod with a velocity v. Conserving angular momentum about the pivot just before and after the collision,

$$I\omega_0 = mv\frac{l}{2}$$

$$\Rightarrow \omega_0 = \frac{\frac{mvl}{2}}{I} = \frac{mvl}{2\left(\frac{ml^2}{2}\right)} = \frac{v}{l}$$

$$\Rightarrow \omega_0^2 = \frac{v^2}{l^2} = \frac{2gh}{l^2}(\because v^2 = 2gh). \tag{5.336}$$

The force equation for the system $(M + m)$ is

$$R = m\frac{l}{2}\omega^2 + (M + m)g$$

$$= m\frac{l}{2}\omega^2 + \left(\frac{3}{4} + 1\right)mg(\because M = 3m/4)$$

$$\Rightarrow R = \frac{ml\omega^2}{2} + \frac{7}{4}mg. \tag{5.337}$$

Recasting equation (5.335) we have

$$\omega^2 = \frac{mgl}{I} + \omega_0^2 = \frac{mgl}{\frac{ml^2}{2}} + \omega_0^2$$

$$\Rightarrow \omega^2 = \omega_0^2 + 2\frac{g}{l}. \tag{5.338}$$

ÀUsing equation (5.336) and equation (5.338)

$$w^2 = \frac{2gh}{l^2} + \frac{2g}{l} = \frac{2g}{l}\left(1 + \frac{h}{l}\right)$$

$$\Rightarrow w^2 = \frac{2g}{l}\left(1 + \frac{h}{l}\right). \tag{5.339}$$

Using equations (5.337) and (5.339)

$$\Rightarrow R = \frac{2mlg}{2l}\left(1 + \frac{h}{l}\right) + \frac{7}{4}mg$$

$$= mg\left(1 + \frac{h}{l}\right) + \frac{7}{4}mg$$

$$= mg\left(1 + \frac{7}{4} + \frac{h}{l}\right)$$

$$= mg\left(\frac{11}{4} + \frac{h}{l}\right)$$

$$= \left(\frac{11}{4} + \frac{1}{4}\right)mg$$

$$= 3mg.$$

Problem 135 A composite rod PQ is comprised of two identical rods PC and CQ each of mass m and length l joined end to end at C by a weak welding. The composite rod is pivoted smoothly at P. If the rod is released from rest when it is horizontal at position 1, it snaps at the mid-point C at its vertical position 2. In consequence, the rod QC becomes detached from the rod PQ. Describe the motion of the rods after the snapping.

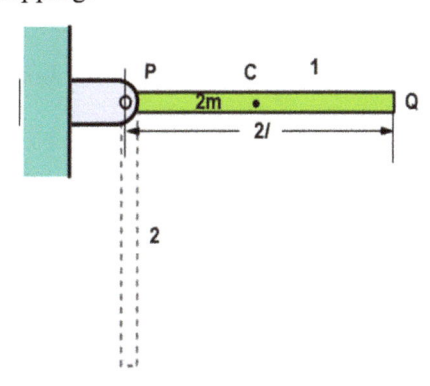

Solution

Let the center of mass C of the lower piece CQ of the rod PQ move with a velocity v and the upper piece PC and lower piece CQ have angular velocities ω_1 and ω_2, respectively, just after the snapping of the rod at C.

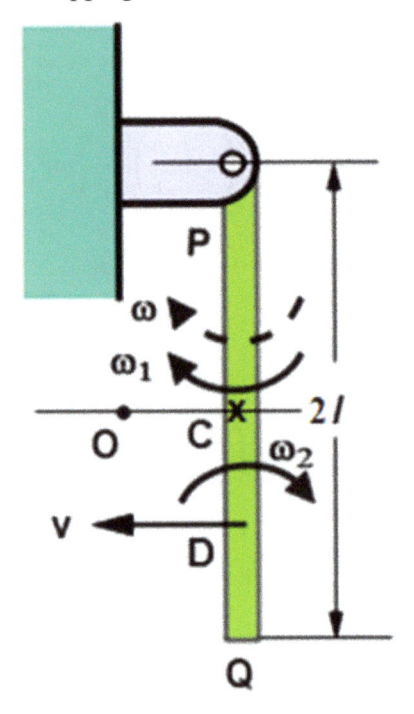

Conserving the angular momentum of both rods about P we have

$$\frac{2m(2l)^2}{3}\omega = \frac{ml^2\omega_1}{3} + \frac{ml^2\omega_2}{12} + \frac{3mvl}{2}$$

$$\Rightarrow \frac{8}{3}l\omega = \frac{l}{3}\omega_1 + \frac{\omega_2 l}{12} + \frac{3v}{2}$$

$$\Rightarrow 32l\omega = (4\omega_1 + \omega_2)l + 18v. \quad (5.340)$$

Conserving the angular momentum of the bottom rod about O we have

$$\frac{7ml^2}{3}\omega = \frac{ml^2}{12}\omega_2 + \frac{mvl}{2}$$

$$\Rightarrow \frac{7}{3}l\omega = \frac{\omega_2}{12}l + \frac{v}{2}$$

$$\Rightarrow 28l\omega = \omega_2 + 6v$$

$$\Rightarrow 6v = (28\omega - \omega_2)l. \quad (5.341)$$

Kinematics: The velocity of top end of QC and bottom end of PC just before and after the snapping remains the same. So we can write

$$v - \frac{l}{2}\omega_2 = l\omega_1$$

$$2v = (2\omega_1 + \omega_2)l. \qquad (5.342)$$

From equations (5.340) and (5.342)

$$\frac{32\omega - (4\omega_1 + \omega_2)}{18} = \frac{2\omega_1 + \omega_2}{2}$$

$$\Rightarrow 22\omega_1 + 10\omega_2 = 32\omega$$

$$\Rightarrow 11\omega_1 + 5\omega_2 = 16\omega. \qquad (5.343)$$

From equations (5.341) and (5.342)

$$\frac{28\omega - \omega_2}{6} = \frac{2\omega_1 + \omega_2}{2}$$

$$\Rightarrow 6\omega_1 + 4\omega_2 = 28\omega$$

$$\Rightarrow (3\omega_1 + 2\omega_2 = 14\omega). \qquad (5.344)$$

Solving equations (5.343) and (5.344) we have

$$\omega_1 = -\frac{38\omega}{7} \text{ and } \omega_2 = \frac{106\omega}{7}.$$

The negative sign signifies that the upper part of the rod will rotate with an anticlockwise angular velocity. The lower part will rotate with a clockwise angular velocity. Furthermore, the center of mass of the lower part will move horizontally towards the left with a velocity v which can be calculated as follows.

Putting the value of $\omega_2 = \frac{106\omega}{7}$ in equation (5.341) we have

$$6v = \left(28\omega - \frac{106\omega}{7}\right)l$$

$$= \left(\frac{196\omega - 106}{7}\right)\omega l = \frac{90\omega l}{7}$$

$$\Rightarrow v = \frac{15\omega l}{7}.$$

The value of ω can be determined by conserving energy between the positions 1 and 2 as follows. The principle of energy conservation is given as

$$\Delta K + \Delta U = 0$$

$$\Rightarrow \frac{I\omega^2}{2} - (2m)g(2l/2)$$

5-278

$$\Rightarrow w^2 = \frac{4mgl}{I}. \tag{5.345}$$

Putting $I = (2m)(2l)^2/3 = 8ml^2/3$ in equation (5.345) we have

$$\Rightarrow w^2 = \frac{4mgl}{8ml^2/3} = \frac{3g}{2l}$$

$$w = \sqrt{\frac{3g}{2l}}.$$

Putting this value of w in the following obtained values such as

$$w_1 = -\frac{38w}{7} \text{ and } w_2 = \frac{106w}{7} \text{ and } v = \frac{15wl}{7},$$

we obtain their final values.

Problem 136 A ball of mass m collides with the smooth uniform rod AB of mass M and radius of gyration K, with a velocity v_0 perpendicular to the rod at a distance of b from the mid-point O of the rod, as shown in the figure. If the ball sticks to the rod just after the collision, find the (a) angular velocity of the rod just after the collision (b) change in energy in the collision (c) fraction loss of energy in the collision.

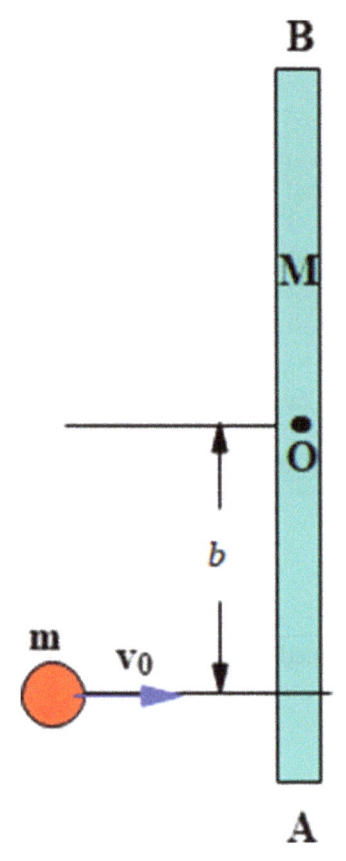

Solution

(a) The centre of mass C of the system $(M + m)$ is located at a distance

$$x = \frac{Mb}{M + m} \tag{5.346}$$

from the point of impact D.

The moment of inertia of the system about C is

$$I_C = mx^2 + M(b - x)^2 + MK^2 \tag{5.347}$$

Using equations (5.346) and (5.347),

$$I_C = m\left(\frac{Mb}{M + m}\right)^2 + M\left(\frac{mb}{M + m}\right)^2 + MK^2 = \frac{Mmb^2}{M + m} + MK^2 \tag{5.348}$$

The angular momentum of the system just before the collision about P is

$$\vec{L}_{i_i} = mv_0 x \hat{k} \tag{5.349}$$

Using equations (5.346) and (5.348),

$$\vec{L}_i = \frac{Mmb}{M + m} v_0 \hat{k} \tag{5.350}$$

The angular momentum of the system just after the collision about P is

$$\vec{L}_f = I_C \omega \hat{k} \tag{5.351}$$

Using equations (5.348) and (5.351),

$$\vec{L}_f = \left(\frac{Mmb^2}{M + m} + MK^2\right) \omega \hat{k} \tag{5.352}$$

Conservation of angular momentum about the center of mass C of the system $(M + m)$,

$$\vec{L}_f = \vec{L}_i \tag{5.353}$$

Using equations (5.350), (5.352) and (5.353),

$$\left(\frac{Mmb^2}{M + m} + MK^2\right) \omega \hat{k} = \frac{Mmb}{M + m} v_0 \hat{k}$$

$$\Rightarrow \omega = \frac{mb}{(M + m)K^2 + mb^2} v_0$$

(b) Conserving the linear momentum, we have
$$(M+m)v_C = mv_0$$
$$\Rightarrow v_C = mv_0/(M+m) \tag{5.354}$$

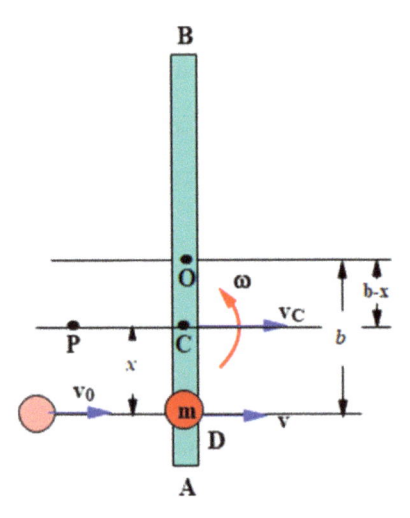

The final kinetic energy (just after the collision) of the system $(M+m)$ is

$$K_f = K_{\text{Trans}} + K_{\text{Rot}} = \frac{(M+m)}{2}v_C^2 + \frac{I_C}{2}\omega^2 = \frac{(M+m)}{2}v_C^2 + \frac{(I_C\omega)^2}{2I_C} \tag{5.355}$$

$$\Rightarrow K_f = \frac{(M+m)}{2}v_C^2 + \frac{(L_f)^2}{2I_C} = \frac{(M+m)}{2}v_C^2 + \frac{(L_i)^2}{2I_C} \quad (\because L_f = L_i)$$

The initial kinetic energy (just before the collision) of the system $(M+m)$ is

$$K_i = \frac{m}{2}v_0^2 \tag{5.356}$$

Using equations (5.355) and (5.356), the change in kinetic energy is

$$\Delta K = K_f - K_i = \frac{(M+m)}{2}v_C^2 + \frac{(L_i)^2}{2I_C} - \frac{m}{2}v_0^2 \tag{5.356}$$

Using the equations 5.354 and 5.356,

$$\Delta K = \frac{(M+m)}{2}\left(\frac{mv_0}{M+m}\right)^2 + \frac{(L_i)^2}{2I_C} - \frac{m}{2}v_0^2$$

$$\Rightarrow \Delta K = -\frac{Mmv_0^2}{2(M+m)} + \frac{(L_i)^2}{2I_C} \tag{5.356}$$

Using equations (5.348), (5.350) and (5.356),

$$\Rightarrow \Delta K = -\frac{Mmv_0^2}{2(M+m)} + \frac{\left(\frac{Mmbv_0}{(M+m)}\right)^2}{2\left(\frac{Mmb^2}{M+m} + MK^2\right)}$$

$$\Rightarrow \Delta K = -\frac{Mmv_0^2}{2(M+m)} + \frac{Mm^2b^2v_0^2}{2\{mb^2 + (M+m)K^2\}(M+m)}$$

$$\Rightarrow \Delta K = -\frac{Mmv_0^2}{2(M+m)}\left[1 - \frac{mb^2}{\{mb^2 + (M+m)K^2\}}\right]$$

$$\Rightarrow \Delta K = -\frac{Mmv_0^2 K^2}{2\{mb^2 + (M+m)K^2\}}$$

(c) The fraction of energy lost is

$$\Rightarrow \eta = -\Delta K/K = \frac{Mmv_0^2 K^2}{\{mb^2 + (M+m)K^2\}2(mv_0^2/2)}$$

$$\Rightarrow \eta = \frac{MK^2}{mb^2 + (M+m)K^2}.$$